Applied Probability and Statistics (Continued)

BARTHOLOMEW · Stochastic Models for Social Processes, *Second Edition*
BENNETT and FRANKLIN · Statistical Analysis in Chemistry and the Chemical Industry
BHAT · Elements of Applied Stochastic Processes
BOX and DRAPER · Evolutionary Operation: A Statistical Method for Process Improvement
BROWNLEE · Statistical Theory and Methodology in Science and Engineering, *Second Edition*
BURY · Statistical Models in Applied Science
CHERNOFF and MOSES · Elementary Decision Theory
CHOW · Analysis and Control of Dynamic Economic Systems
CLELLAND, deCANI, BROWN, BURSK, and MURRAY · Basic Statistics with Business Applications, *Second Edition*
COCHRAN · Sampling Techniques, *Second Edition*
COCHRAN and COX · Experimental Designs, *Second Edition*
COX · Planning of Experiments
COX and MILLER · The Theory of Stochastic Processes, *Second Edition*
DANIEL and WOOD · Fitting Equations to Data
DAVID · Order Statistics
DEMING · Sample Design in Business Research
DODGE and ROMIG · Sampling Inspection Tables, *Second Edition*
DRAPER and SMITH · Applied Regression Analysis
DUNN and CLARK · Applied Statistics: Analysis of Variance and Regression
ELANDT-JOHNSON · Probability Models and Statistical Methods in Genetics
FLEISS · Statistical Methods for Rates and Proportions
GOLDBERGER · Econometric Theory
GROSS and CLARK · Survival Distributions
GROSS and HARRIS · Fundamentals of Queueing Theory
GUTTMAN, WILKS and HUNTER · Introductory Engineering Statistics, *Second Edition*
HAHN and SHAPIRO · Statistical Models in Engineering
HALD · Statistical Tables and Formulas
HALD · Statistical Theory with Engineering Applications
HARTIGAN · Clustering Algorithms
HOEL · Elementary Statistics, *Third Edition*
HOLLANDER and WOLFE · Nonparametric Statistical Methods
HUANG · Regression and Econometric Methods
JAGERS · Branching Processes with Biological Applications
JOHNSON and KOTZ · Distributions in Statistics
Discrete Distributions
Continuous Univariate Distributions-1
Continuous Univariate Distributions-2
Continuous Multivariate Distributions
JOHNSON and LEONE · Statistics and Experimental Design: In Engineering and the Physical Sciences, Volumes I and II
LANCASTER · The Chi Squared Distribution

continued on back

Linear Models

S. R. SEARLE

Professor of Biological Statistics
Biometrics Unit
N.Y. State College of Agriculture
Cornell University, Ithaca, N.Y.

John Wiley & Sons, Inc.

New York · London · Sydney · Toronto

Library of Congress Catalog Card Number: 70-138919

ISBN 0 471 76950 9

Printed in the United States of America

10 9 8 7 6 5 4

Preface

This book describes general procedures of estimation and hypothesis testing for linear statistical models and shows their application for unbalanced data (i.e., unequal-subclass-numbers data) to certain specific models that often arise in research and survey work. In addition, three chapters are devoted to methods and results for estimating variance components, particularly from unbalanced data. Balanced data of the kind usually arising from designed experiments are treated very briefly, as just special cases of unbalanced data. Emphasis on unbalanced data is the backbone of the book, designed to assist those whose data cannot satisfy the strictures of carefully managed and well-designed experiments.

The title may suggest that this is an all-embracing treatment of linear models. This is not the case, for there is no detailed discussion of designed experiments. Moreover, the title is not *An Introduction to . . .*, because the book provides more than an introduction; nor is it *. . . with Applications*, because, although concerned with applications of general linear model theory to specific models, few applications in the form of real-life data are used. Similarly, *. . . for Unbalanced Data* has also been excluded from the title because the book is not devoted exclusively to such data. Consequently the title *Linear Models* remains, and I believe it has brevity to recommend it.

My main objective is to describe linear model techniques for analyzing unbalanced data. In this sense the book is self-contained, based on prerequisites of a semester of matrix algebra and a year of statistical methods. The matrix algebra required is supplemented in Chapter 1, which deals with generalized inverse matrices and allied topics. The reader who wishes to pursue the mathematics in detail throughout the book should also have some knowledge of statistical theory. The requirements in this regard are supplemented by a summary review of distributions in Chapter 2, extending to sections on the distribution of quadratic and bilinear forms and the singular multinormal distribution. There is no attempt to make this introductory material complete. It serves to provide the reader with foundations for developing results for the general linear model, and much of the detail of this and other chapters can be omitted by the reader whose training in mathematical statistics is sparse. However, he must know Theorems 1 through 3 of Chapter 2, for they are used extensively in succeeding chapters.

Chapter 3 deals with full-rank models. It begins with a simple explanation of regression (based on an example) and proceeds to multiple regression, giving a unified treatment for testing a general linear hypothesis. After dealing with various aspects of this hypothesis and special cases of it, the chapter ends with sections on reduced models and other related topics. Chapter 4 introduces models not of full rank by discussing regression on dummy (0, 1) variables and showing its equivalence to linear models. The results are well known to most statisticians, but not to many users of regression, especially those who are familiar with regression more in the form of computer output than as a statistical procedure. The chapter ends with a numerical example illustrating both the possibility of having many solutions to normal equations and the idea of estimable and non-estimable functions.

Chapter 5 deals with the non-full-rank model, utilizing generalized inverse matrices and giving a unified procedure for testing any testable linear hypothesis. Chapters 6 through 8 deal with specific cases of this model, giving many details for the analysis of unbalanced data. Within these chapters there is detailed discussion of certain topics that other books tend to ignore: restrictions on models and constraints on solutions (Sections 5.6 and 5.7); singular covariance matrices of the error terms (Section 5.8); orthogonal contrasts with unbalanced data (Section 5.5g); the hypotheses tested by F-statistics in the analysis of variance of unbalanced data (Sections 6.4f, 7.1g, and 7.2f); analysis of covariance for unbalanced data (Section 8.2); and approximate analyses for data that are only slightly unbalanced (Section 8.3). On these and other topics, I have tried to coordinate some ideas and make them readily accessible to students, rather than continuing to leave the literature relatively devoid of these topics or, at best, containing only scattered references to them. Statisticians concerned with analyzing unbalanced data on the basis of linear models have talked about the difficulties involved for many years but, probably because the problems are not easily resolved, little has been put in print about them. The time has arrived, I feel, for trying to fill this void. Readers may not always agree with what is said, indeed I may want to alter some things myself in due time but, meanwhile, if this book sets readers to thinking and writing further about these matters, I will feel justified. For example, there may be criticism of the discussion of F-statistics in parts of Chapters 6 through 8, where these statistics are used, not so much to test hypotheses of interest (as described in Chapter 5), but to specify what hypotheses are being tested by those F-statistics available in analysis of variance tables for unbalanced data. I believe it is important to understand what these hypotheses are, because they are not obvious analogs of the corresponding balanced data hypotheses and, in many cases, are relatively useless.

The many numerical illustrations and exercises in Chapters 3 through 8 use hypothetical data, designed with easy arithmetic in mind. This is because I agree with C. C. Li (1964) who points out that we do not learn to solve quadratic equations by working with something like

$$683125x^2 + 1268.4071x - 213.69825 = 0$$

just because it occurs in real life. Learning to first solve $x^2 + 3x + 2 = 0$ is far more instructive. Whereas real-life examples are certainly motivating, they usually involve arithmetic that becomes as cumbersome and as difficult to follow as is the algebra it is meant to illustrate. Furthermore, if one is going to use real-life examples, they must come from a variety of sources in order to appeal to a wide audience, but the changing from one example to another as succeeding points of analysis are developed and illustrated brings an inevitable loss of continuity. No apology is made, therefore, for the artificiality of the numerical examples used, nor for repeated use of the same example in many places. The attributes of continuity and of relatively easy arithmetic more than compensate for the lack of reality by assuring that examples achieve their purpose of illustrating the algebra.

Chapters 9 through 11 deal with variance components. The first part of Chapter 9 describes random models, distinguishing them from fixed models by a series of examples and using the concepts, rather than the details, of the examples to make the distinction. The second part of the chapter is the only occasion where balanced data are discussed in depth: not for specific models (designs) but in terms of procedures applicable to balanced data generally. Chapter 10 presents methods currently available for estimating variance components from unbalanced data, their properties, procedures, and difficulties. Parts of these two chapters draw heavily on Searle (1971). Finally, Chapter 11 catalogs results derived by applying to specific models some of the methods described in Chapter 10, gathering together the cumbersome algebraic expressions for variance component estimators and their variances in the 1-way, 2-way nested, and 2-way crossed classifications (random and mixed models), and others. Currently these results are scattered throughout the literature. The algebraic expressions are themselves so lengthy that there would be little advantage in giving numerical illustrations. Instead, extra space has been taken to typeset the algebraic expressions in as readable a manner as possible.

All chapters except the last have exercises, most of which are designed to encourage the student to reread the text and to practice and become thoroughly familiar with the techniques described. Statisticians, in their consulting capacity, are much like lawyers. They do not need to remember every technique exactly, but must know where to locate it when needed and be able to understand it once found. This is particularly so with the techniques of unbalanced data analysis, and so the exercises are directed towards impressing on the reader the methods and logic of establishing the techniques rather than the details of the results themselves. These can always be found when needed.

No computer programs are given. This would be an enormous task, with no certainty that such programs would be optimal when written and even less chance by the time they were published. While the need for good programs is obvious, I think that a statistics book is not the place yet for such programs. Computer programs printed in books take on the aura of quality and authority, which, even if valid initially, soon becomes outmoded in today's fast-moving computer world.

The chapters are long, but self-contained and liberally sign-posted with sections, subsections, and sub-subsections—all with titles (see Contents).

My sincere thanks go to many people for helping with the book: the Institute of Statistics at Texas A. and M. University which provided me with facilities during a sabbatical leave (1968–1969) to do most of the initial writing; R. G. Cornell, N. R. Draper, and J. S. Hunter, the reviewers of the first draft who made many helpful suggestions; and my colleagues at Cornell who encouraged me to keep going. I also thank D. F. Cox, C. H. Goldsmith, A. Hedayat, R. R. Hocking, J. W. Rudan, D. L. Solomon, N. S. Urquhart, and D. L. Weeks for reading parts of the manuscript and suggesting valuable improvements. To John W. Rudan goes particular gratitude for generous help with proof reading. Grateful thanks also go to secretarial help at both Texas A. and M. and Cornell Universities, who eased the burden enormously.

S. R. SEARLE

Ithaca, New York
October, 1970

Linear Models

A WILEY PUBLICATION IN MATHEMATICAL STATISTICS

List of Chapters

Contents

2. Distributions and Quadratic Forms (*Continued*)

Linear Models

CHAPTER 1

GENERALIZED INVERSE MATRICES

1. INTRODUCTION

The application of generalized inverse matrices to linear statistical models is of relatively recent occurrence. As a mathematical tool such matrices aid in understanding certain aspects of the analysis procedures associated with linear models, especially the analysis of unbalanced data, a topic to which considerable attention is given in this book. An appropriate starting point is therefore a summary of the features of generalized inverse matrices that are important to linear models. Other ancillary results in matrix algebra are also discussed.

a. Definition and existence

A generalized inverse of a matrix $\mathbf{A}$ is defined, in this book, as any matrix $\mathbf{G}$ that satisfies the equation

$$\mathbf{AGA} = \mathbf{A}. \tag{1}$$

The name "generalized inverse" for matrices $\mathbf{G}$ defined by (1) is unfortunately not universally accepted, although it is used quite widely. Names such as "conditional inverse", "pseudo inverse" and "*g*-inverse" are also to be found in the literature, sometimes for matrices defined as is $\mathbf{G}$ of (1) and sometimes for matrices defined as variants of $\mathbf{G}$. However, throughout this book the name "generalized inverse" of $\mathbf{A}$ is used exclusively for any matrix $\mathbf{G}$ satisfying (1).

Notice that (1) does not define $\mathbf{G}$ as "the" generalized inverse of $\mathbf{A}$ but as "a" generalized inverse. This is because $\mathbf{G}$, for a given matrix $\mathbf{A}$, is not unique. As shown below, there is an infinite number of matrices $\mathbf{G}$ that satisfy (1) and so we refer to the whole class of them as generalized inverses of $\mathbf{A}$.

One way of illustrating the existence of $\mathbf{G}$ and its non-uniqueness starts with the equivalent diagonal form of $\mathbf{A}$. If $\mathbf{A}$ has order $p \times q$ the reduction to this diagonal form can be written as

$$\mathbf{P}_{p\times p}\mathbf{A}_{p\times q}\mathbf{Q}_{q\times q} = \mathbf{\Delta}_{p\times q} \equiv \begin{bmatrix} \mathbf{D}_{r\times r} & \mathbf{0}_{r\times(q-r)} \\ \mathbf{0}_{(p-r)\times r} & \mathbf{0}_{(p-r)\times(q-r)} \end{bmatrix}$$

or, more simply, as

$$\mathbf{PAQ} = \mathbf{\Delta} = \begin{bmatrix} \mathbf{D}_r & \mathbf{0} \\ \mathbf{0} & \mathbf{0} \end{bmatrix}.$$

As usual, **P** and **Q** are products of elementary operators [see, for example, Searle (1966), Sec. 5.7], r is the rank of **A** and $\mathbf{D}_r$ is a diagonal matrix of order r. In general, if $d_1, d_2, \ldots, d_r$ are the diagonal elements of any diagonal matrix **D** we will use the notation $\mathbf{D}\{d_i\}$ for $\mathbf{D}_r$; i.e.,

$$\mathbf{D}_r \equiv \begin{bmatrix} d_1 & 0 & \cdots & 0 \\ 0 & d_2 & \cdots & 0 \\ & & \ddots & \vdots \\ 0 & \cdots & 0 & d_r \end{bmatrix} \equiv \operatorname{diag}\{d_i\} \equiv \mathbf{D}\{d_i\} \qquad \text{for} \quad i = 1, \ldots, r. \tag{2}$$

Furthermore, as in $\mathbf{\Delta}$, null matrices will be represented by the symbol **0**, with order being determined by context on each occasion.

Derivation of **G** comes easily from $\mathbf{\Delta}$. Analogous to $\mathbf{\Delta}$ we define $\mathbf{\Delta}^-$ (to be read as "Δ minus") as

$$\mathbf{\Delta}^- = \begin{bmatrix} \mathbf{D}_r^{-1} & \mathbf{0} \\ \mathbf{0} & \mathbf{0} \end{bmatrix}.$$

Then, as shown below,

$$\mathbf{G} = \mathbf{Q\Delta^-P} \tag{3}$$

satisfies (1). Hence **G** is a generalized inverse of **A**. Clearly **G** as given by (3) is not unique, for neither **P** nor **Q** by their definition is unique; neither is $\mathbf{\Delta}$ *nor* $\mathbf{\Delta}^-$, and therefore $\mathbf{G} = \mathbf{Q\Delta^-P}$ is not unique.

Before showing that **G** does satisfy (1), note from the definitions of $\mathbf{\Delta}$ and $\mathbf{\Delta}^-$ given above that

$$\mathbf{\Delta\Delta^-\Delta} = \mathbf{\Delta}. \tag{4}$$

Hence, by the definition implied in (1), we can say that $\mathbf{\Delta}^-$ is a generalized inverse of $\mathbf{\Delta}$, an unimportant result in itself but one which leads to **G** satisfying (1). To show this we use $\mathbf{\Delta}$ to write

$$\mathbf{A} = \mathbf{P}^{-1}\mathbf{\Delta Q}^{-1}, \tag{5}$$

the inverses $\mathbf{P}^{-1}$ and $\mathbf{Q}^{-1}$ existing because $\mathbf{P}$ and $\mathbf{Q}$ are products of elementary operators and hence non-singular. Then (3), (4) and (5) give

$$\mathbf{AGA} = \mathbf{P}^{-1}\mathbf{\Delta}\mathbf{Q}^{-1}\mathbf{Q}\mathbf{\Delta}^{-}\mathbf{P}\mathbf{P}^{-1}\mathbf{\Delta}\mathbf{Q}^{-1} = \mathbf{P}^{-1}\mathbf{\Delta}\mathbf{\Delta}^{-}\mathbf{\Delta}\mathbf{Q}^{-1} = \mathbf{P}^{-1}\mathbf{\Delta}\mathbf{Q}^{-1} = \mathbf{A};$$

i.e., (1) is satisfied. Hence $\mathbf{G}$ is a generalized inverse of $\mathbf{A}$.

Example. For

$$\mathbf{A} = \begin{bmatrix} 4 & 1 & 2 \\ 1 & 1 & 5 \\ 3 & 1 & 3 \end{bmatrix}$$

a diagonal form is obtained using

$$\mathbf{P} = \begin{bmatrix} 0 & 1 & 0 \\ 1 & -4 & 0 \\ -\frac{2}{3} & -\frac{1}{3} & 1 \end{bmatrix} \quad \text{and} \quad \mathbf{Q} = \begin{bmatrix} 1 & -1 & 1 \\ 0 & 1 & -6 \\ 0 & 0 & 1 \end{bmatrix},$$

so that

$$\mathbf{PAQ} = \mathbf{\Delta} = \begin{bmatrix} 1 & 0 & 0 \\ 0 & -3 & 0 \\ 0 & 0 & 0 \end{bmatrix} \quad \text{and} \quad \mathbf{\Delta}^{-} = \begin{bmatrix} 1 & 0 & 0 \\ 0 & -\frac{1}{3} & 0 \\ 0 & 0 & 0 \end{bmatrix}.$$

Hence

$$\mathbf{G} = \mathbf{Q}\mathbf{\Delta}^{-}\mathbf{P} = \tfrac{1}{3}\begin{bmatrix} 1 & -1 & 0 \\ -1 & 4 & 0 \\ 0 & 0 & 0 \end{bmatrix}.$$

The reader should verify that $\mathbf{AGA} = \mathbf{A}$.

It is to be emphasized that generalized inverses exist for rectangular matrices as well as for square ones. This is evident from the formulation of $\mathbf{\Delta}_{p\times q}$. However, for $\mathbf{A}$ of order $p \times q$, we define $\mathbf{\Delta}^{-}$ as having order $q \times p$, the null matrices therein being of appropriate order to make this so. As a result $\mathbf{G}$ has order $q \times p$.

Example. Consider

$$\mathbf{B} = \begin{bmatrix} 4 & 1 & 2 & 0 \\ 1 & 1 & 5 & 15 \\ 3 & 1 & 3 & 5 \end{bmatrix},$$

the same as $\mathbf{A}$ in the previous example except for an additional column.

With **P** as given earlier and **Q** now taken as

$$\mathbf{Q} = \begin{bmatrix} 1 & -1 & 1 & 5 \\ 0 & 1 & -6 & -20 \\ 0 & 0 & 1 & 0 \\ 0 & 0 & 0 & 1 \end{bmatrix} \quad \text{and} \quad \mathbf{PBQ} = \mathbf{\Delta} = \begin{bmatrix} 1 & 0 & 0 & 0 \\ 0 & -3 & 0 & 0 \\ 0 & 0 & 0 & 0 \end{bmatrix},$$

$\mathbf{\Delta}^-$ is then taken as

$$\mathbf{\Delta}^- = \begin{bmatrix} 1 & 0 & 0 \\ 0 & -\frac{1}{3} & 0 \\ 0 & 0 & 0 \\ 0 & 0 & 0 \end{bmatrix}, \quad \text{so that} \quad \mathbf{G} = \mathbf{Q\Delta^-P} = \begin{bmatrix} \frac{1}{3} & -\frac{1}{3} & 0 \\ -\frac{1}{3} & \frac{4}{3} & 0 \\ 0 & 0 & 0 \\ 0 & 0 & 0 \end{bmatrix}.$$

b. An algorithm

Another way of computing **G** is based on knowing the rank of **A**. Suppose it is r and that **A** can be partitioned in such a way that its leading $r \times r$ minor is non-singular, i.e.,

$$\mathbf{A}_{p\times q} = \begin{bmatrix} \mathbf{A}_{11} & \mathbf{A}_{12} \\ \mathbf{A}_{21} & \mathbf{A}_{22} \end{bmatrix}$$

where $\mathbf{A}_{11}$ is $r \times r$ of rank r. Then a generalized inverse of **A** is

$$\mathbf{G}_{q\times p} = \begin{bmatrix} \mathbf{A}_{11}^{-1} & \mathbf{0} \\ \mathbf{0} & \mathbf{0} \end{bmatrix},$$

where the null matrices are of appropriate order to make **G** be $q \times p$. To see that **G** is a generalized inverse of **A**, note that

$$\mathbf{AGA} = \begin{bmatrix} \mathbf{A}_{11} & \mathbf{A}_{12} \\ \mathbf{A}_{21} & \mathbf{A}_{21}\mathbf{A}_{11}^{-1}\mathbf{A}_{12} \end{bmatrix}.$$

Now, by the way in which **A** has been partitioned, $[\mathbf{A}_{21} \quad \mathbf{A}_{22}] = \mathbf{K}[\mathbf{A}_{11} \quad \mathbf{A}_{12}]$ for some matrix **K**. Therefore $\mathbf{K} = \mathbf{A}_{21}\mathbf{A}_{11}^{-1}$ and so $\mathbf{A}_{22} = \mathbf{KA}_{12} = \mathbf{A}_{21}\mathbf{A}_{11}^{-1}\mathbf{A}_{12}$. Hence $\mathbf{AGA} = \mathbf{A}$.

Example. A generalized inverse of

$$\mathbf{A} = \begin{bmatrix} 1 & 2 & 5 & 2 \\ 3 & 7 & 12 & 4 \\ 0 & 1 & -3 & -2 \end{bmatrix}, \quad \text{having rank 2,} \quad \text{is } \mathbf{G} = \begin{bmatrix} 7 & -2 & 0 \\ -3 & 1 & 0 \\ 0 & 0 & 0 \\ 0 & 0 & 0 \end{bmatrix}.$$

There is no need for the non-singular minor of order r to be in the leading position. Suppose it is not. Let $\mathbf{R}$ and $\mathbf{S}$ represent the elementary row and column operations respectively to bring it to the leading position. Then $\mathbf{R}$ and $\mathbf{S}$ are products of elementary operators with

$$\mathbf{RAS} = \mathbf{B} = \begin{bmatrix} \mathbf{B}_{11} & \mathbf{B}_{12} \\ \mathbf{B}_{21} & \mathbf{B}_{22} \end{bmatrix}$$

where $\mathbf{B}_{11}$ is non-singular of order r. Then

$$\mathbf{F} = \begin{bmatrix} \mathbf{B}_{11}^{-1} & \mathbf{0} \\ \mathbf{0} & \mathbf{0} \end{bmatrix}$$

is a generalized inverse of $\mathbf{B}$ and $\mathbf{G}_{q\times p} = \mathbf{SFR}$ is a generalized inverse of $\mathbf{A}$. Now $\mathbf{R}$ and $\mathbf{S}$ are products of elementary operators that interchange rows (or columns); i.e., $\mathbf{R}$ and $\mathbf{S}$ are products of matrices that are identity matrices with rows (or columns) interchanged. Therefore $\mathbf{R}$ and $\mathbf{S}$ are identity matrices with rows (or columns) in a different sequence from that found in $\mathbf{I}$. Such matrices are known as *permutation matrices* and are orthogonal; i.e.,

$$\mathbf{R} = \mathbf{I} \text{ with its rows in a different sequence}$$
$$\equiv \text{permutation matrix}$$

and $$\mathbf{R}'\mathbf{R} = \mathbf{I}. \tag{6}$$

The same is true for $\mathbf{S}$, and so from $\mathbf{RAS} = \mathbf{B}$ we have

$$\mathbf{A} = \mathbf{R}'\mathbf{BS}' = \mathbf{R}'\begin{bmatrix} \mathbf{B}_{11} & \mathbf{B}_{12} \\ \mathbf{B}_{21} & \mathbf{B}_{22} \end{bmatrix}\mathbf{S}'.$$

Clearly, so far as $\mathbf{B}_{11}$ is concerned, this product represents the operations of returning the elements of $\mathbf{B}_{11}$ to their original positions in $\mathbf{A}$. Now consider $\mathbf{G}$: we have

$$\mathbf{G} = \mathbf{SFR} = (\mathbf{R}'\mathbf{F}'\mathbf{S}')' = \left\{\mathbf{R}'\begin{bmatrix} (\mathbf{B}_{11}^{-1})' & \mathbf{0} \\ \mathbf{0} & \mathbf{0} \end{bmatrix}\mathbf{S}'\right\}'.$$

In this, analogous to the form of $\mathbf{A} = \mathbf{R}'\mathbf{BS}'$, the product involving $\mathbf{R}'$ and $\mathbf{S}'$ in $\mathbf{G}'$ represents putting the elements of $(\mathbf{B}_{11}^{-1})'$ into the corresponding positions (of $\mathbf{G}'$) that the elements of $\mathbf{B}_{11}$ occupied in $\mathbf{A}$. Hence an algorithm for finding a generalized inverse of $\mathbf{A}$ by this method is as follows.

(i) In $\mathbf{A}$, of rank r, find any non-singular minor of order r. Call it $\mathbf{M}$ (using the symbol $\mathbf{M}$ in place of $\mathbf{B}_{11}$).

(ii) Invert $\mathbf{M}$ and transpose the inverse: $(\mathbf{M}^{-1})'$.

(iii) In $\mathbf{A}$ replace each element of $\mathbf{M}$ by the corresponding element of $(\mathbf{M}^{-1})'$; i.e., if $a_{ij} = m_{st}$, the (s, t)th element of $\mathbf{M}$, then replace a_{ij} by $m^{t,s}$, the (t, s)th element of $\mathbf{M}^{-1}$, equivalent to the (s, t)th element of the transpose of $\mathbf{M}^{-1}$.

(iv) Replace all other elements of $\mathbf{A}$ by zero.

(v) Transpose the resulting matrix.

(vi) The result is $\mathbf{G}$, a generalized inverse of $\mathbf{A}$.

Note that this procedure is *not* equivalent, in (iii), to replacing elements of $\mathbf{M}$ in $\mathbf{A}$ by the elements of $\mathbf{M}^{-1}$ (and others by zero) and then in (v) transposing. It is if $\mathbf{M}$ is symmetric. Nor is it equivalent to replacing, in (iii), elements of $\mathbf{M}$ in $\mathbf{A}$ by elements of $\mathbf{M}^{-1}$ (and others by zero) and then in (v) not transposing (see Exercise 5). In general, the algorithm must be carried out exactly as described.

One case where it can be simplified is when $\mathbf{A}$ is symmetric. Then any principal minor of $\mathbf{A}$ is symmetric and the transposing in both (iii) and (v) can be ignored. The algorithm can then become as follows.

(i) In $\mathbf{A}$, of rank r and symmetric, find any non-singular *principal* minor of order r. Call it $\mathbf{M}$.

(ii) Invert $\mathbf{M}$.

(iii) In $\mathbf{A}$ replace each element of $\mathbf{M}$ by the corresponding element of $\mathbf{M}^{-1}$.

(iv) Replace all other elements of $\mathbf{A}$ by zero.

(v) The result is $\mathbf{G}$, a generalized inverse of $\mathbf{A}$.

However, when $\mathbf{A}$ is symmetric and a non-symmetric non-principal minor is used for $\mathbf{M}$, then the general algorithm must be used.

Example. The matrix

$$\mathbf{A}_1 = \begin{bmatrix} 4 & 1 & 2 & 0 \\ 1 & 1 & 5 & 15 \\ 3 & 1 & 3 & 5 \end{bmatrix}$$

has the following matrices, among others, as generalized inverses:

$$\begin{bmatrix} 0 & 0 & 0 \\ 0 & -\frac{3}{2} & \frac{5}{2} \\ 0 & \frac{1}{2} & -\frac{1}{2} \\ 0 & 0 & 0 \end{bmatrix}, \quad \begin{bmatrix} 0 & 0 & 0 \\ 0 & -\frac{5}{10} & \frac{15}{10} \\ 0 & 0 & 0 \\ 0 & \frac{1}{10} & -\frac{1}{10} \end{bmatrix} \quad \text{and} \quad \begin{bmatrix} \frac{5}{20} & 0 & 0 \\ 0 & 0 & 0 \\ 0 & 0 & 0 \\ -\frac{3}{20} & 0 & \frac{4}{20} \end{bmatrix},$$

derived from inverting the 2×2 minors

$$\begin{bmatrix} 1 & 5 \\ 1 & 3 \end{bmatrix}, \quad \begin{bmatrix} 1 & 15 \\ 1 & 5 \end{bmatrix} \quad \text{and} \quad \begin{bmatrix} 4 & 0 \\ 3 & 5 \end{bmatrix} \quad \text{respectively.}$$

Similarly,

$$\mathbf{A}_2 = \begin{bmatrix} 2 & 2 & 6 \\ 2 & 3 & 8 \\ 6 & 8 & 22 \end{bmatrix} \quad \text{has} \quad \begin{bmatrix} 0 & 0 & 0 \\ 0 & 11 & -4 \\ 0 & -4 & \frac{3}{2} \end{bmatrix}$$

as a generalized inverse.

These derivations of a matrix **G** that satisfies (1) are by no means the only ways in which such a matrix can be computed. For matrices of small order they can be satisfactory, but for those of large order other methods may be preferred. Some of these are discussed subsequently. Most methods involve, of course, the same kind of numerical problems as are incurred in calculating the regular inverse $\mathbf{A}^{-1}$ of a non-singular matrix **A**. Despite this, the generalized inverse has importance because of its general application to non-square matrices and to square, singular matrices. In the special case that **A** is non-singular $\mathbf{G} = \mathbf{A}^{-1}$, as one would expect, and in this case **G** is unique.

The fact that **A** has a generalized inverse even when it is singular or rectangular has particular application in the problem of solving equations, e.g., of solving $\mathbf{Ax} = \mathbf{y}$ for **x** when **A** is singular or rectangular. In situations of this nature the use of a generalized inverse **G** leads, as we shall see, very directly to a solution. And this is of great importance in the study of linear models, wherein such equations arise quite frequently. For example, when a model can be written as $\mathbf{y} = \mathbf{Xb} + \mathbf{e}$, the least squares procedure for estimating **b** often leads to equations $\mathbf{X'X\hat{b}} = \mathbf{X'y}$ where the matrix $\mathbf{X'X}$ is singular. Hence the solution cannot be written as $(\mathbf{X'X})^{-1}\mathbf{X'y}$; but using a generalized inverse of $\mathbf{X'X}$ a solution can be obtained directly and its properties studied.

Since the use of generalized inverse matrices in solving linear equations is the application of prime interest so far as linear models are concerned, the procedures involved are now outlined. Following this, some general properties of generalized inverses are discussed.

2. SOLVING LINEAR EQUATIONS

a. Consistent equations

A convenient starting point from which to develop the solution of linear equations using a generalized inverse is the definition of consistent equations.

Definition. The linear equations $\mathbf{Ax} = \mathbf{y}$ are defined as being consistent if any linear relationships existing among the rows of **A** also exist among the corresponding elements of **y**.

As a simple example, the equations

$$\begin{bmatrix} 1 & 2 \\ 3 & 6 \end{bmatrix}\begin{bmatrix} x_1 \\ x_2 \end{bmatrix} = \begin{bmatrix} 7 \\ 21 \end{bmatrix}$$

are consistent: in the matrix on the left the second row is thrice the first, and this is also true of the elements on the right. But the equations

$$\begin{bmatrix} 1 & 2 \\ 3 & 6 \end{bmatrix}\begin{bmatrix} x_1 \\ x_2 \end{bmatrix} = \begin{bmatrix} 7 \\ 24 \end{bmatrix}$$

are not consistent. Further evidence of this is seen by writing them in full:

$$x_1 + 2x_2 = 7 \quad \text{and} \quad 3x_1 + 6x_2 = 24.$$

As a consequence of the first, $3x_1 + 6x_2 = 21$, which cannot be true if the second is to hold. The equations are therefore said to be inconsistent.

The formal definition of consistent equations does not demand that linear relationships must exist among the rows of **A**, but if they do then the definition does require that the same relationships also exist among the corresponding elements of **y** for the equations to be consistent. For example, when $\mathbf{A}^{-1}$ exists, the equations $\mathbf{Ax} = \mathbf{y}$ are always consistent, for there are no linear relationships among the rows of **A** and therefore none that the elements of **y** must satisfy.

The importance of the concept of consistency lies in the following theorem: linear equations can be solved if, and only if, they are consistent. Proof can be established from the above definition of consistent equations [see, for example, Searle (1966), Sec. 6.2, or Searle and Hausman (1970), Sec. 7.2]. Since it is only consistent equations that can be solved, discussion of a procedure for solving linear equations is hereafter confined to equations that are consistent. The procedure is described in a series of theorems.

b. Obtaining solutions

The link between a generalized inverse of the matrix **A** and consistent equations $\mathbf{Ax} = \mathbf{y}$ is set out in the following theorem adapted from Rao (1962).

Theorem 1. Consistent equations $\mathbf{Ax} = \mathbf{y}$ have a solution $\mathbf{x} = \mathbf{Gy}$ if and only if $\mathbf{AGA} = \mathbf{A}$.

Proof. If the equations $\mathbf{Ax} = \mathbf{y}$ are consistent and have $\mathbf{x} = \mathbf{Gy}$ as a solution, write $\mathbf{a}_j$ for the jth column of **A** and consider the equations $\mathbf{Ax} = \mathbf{a}_j$. They have a solution: the null vector with its jth element set equal to unity. Therefore the equations $\mathbf{Ax} = \mathbf{a}_j$ are consistent. Furthermore, since consistent equations $\mathbf{Ax} = \mathbf{y}$ have a solution $\mathbf{x} = \mathbf{Gy}$, it follows that consistent equations $\mathbf{Ax} = \mathbf{a}_j$ have a solution $\mathbf{x} = \mathbf{Ga}_j$. Therefore $\mathbf{AGa}_j = \mathbf{a}_j$; and this is true for all values of j, i.e., for all columns of **A**. Hence $\mathbf{AGA} = \mathbf{A}$.

Conversely, if $\mathbf{AGA} = \mathbf{A}$, then $\mathbf{AGAx} = \mathbf{Ax}$, and when $\mathbf{Ax} = \mathbf{y}$ this gives $\mathbf{AGy} = \mathbf{y}$, i.e., $\mathbf{A(Gy)} = \mathbf{y}$. Hence $\mathbf{x} = \mathbf{Gy}$ is a solution of $\mathbf{Ax} = \mathbf{y}$, and the theorem is proved.

Theorem 1 indicates how a solution to consistent equations may be obtained: find any matrix **G** satisfying $\mathbf{AGA} = \mathbf{A}$, i.e., find **G** as any generalized inverse of **A**, and then **Gy** is a solution. However, as Theorem 2 shows, **Gy** is not the only solution. There are, indeed, many solutions whenever **A** is anything other than a square, non-singular matrix.

Theorem 2. If **A** has q columns and if **G** is a generalized inverse of **A**, then the consistent equations $\mathbf{Ax} = \mathbf{y}$ have solution

$$\tilde{\mathbf{x}} = \mathbf{Gy} + (\mathbf{GA} - \mathbf{I})\mathbf{z}, \tag{7}$$

where **z** is any arbitrary vector of order q.

Proof.
$$\begin{aligned} \mathbf{A}\tilde{\mathbf{x}} &= \mathbf{AGy} + (\mathbf{AGA} - \mathbf{A})\mathbf{z} \\ &= \mathbf{AGy}, \text{ because } \mathbf{AGA} = \mathbf{A}, \\ &= \mathbf{y}, \text{ by Theorem 1;} \end{aligned}$$

i.e., $\tilde{\mathbf{x}}$ satisfies $\mathbf{Ax} = \mathbf{y}$ and hence is a solution. The notation $\tilde{\mathbf{x}}$ emphasizes that $\tilde{\mathbf{x}}$ is a solution, distinguishing it from the general vector of unknowns **x**.

Note that the solution $\tilde{\mathbf{x}}$ involves an element of arbitrariness because **z** is an arbitrary vector: **z** can have any value at all and $\tilde{\mathbf{x}}$ will still be a solution to $\mathbf{Ax} = \mathbf{y}$. No matter what value is given to **z**, the expression for $\tilde{\mathbf{x}}$ given in (7) satisfies $\mathbf{Ax} = \mathbf{y}$. Furthermore, this will be so for whatever generalized inverse of **A** is used for **G**.

Example. Consider the equations $\mathbf{Ax} = \mathbf{y}$ as

$$\begin{bmatrix} 5 & 3 & 1 & -4 \\ 8 & 5 & 2 & 3 \\ 21 & 13 & 5 & 2 \\ 3 & 2 & 1 & 7 \end{bmatrix} \begin{bmatrix} x_1 \\ x_2 \\ x_3 \\ x_4 \end{bmatrix} = \begin{bmatrix} 6 \\ 8 \\ 22 \\ 2 \end{bmatrix}, \tag{8}$$

so defining **A**, **x** and **y**. It will be found that

$$\mathbf{G} = \begin{bmatrix} 5 & -3 & 0 & 0 \\ -8 & 5 & 0 & 0 \\ 0 & 0 & 0 & 0 \\ 0 & 0 & 0 & 0 \end{bmatrix}$$

is a generalized inverse of $\mathbf{A}$ satisfying $\mathbf{AGA} = \mathbf{A}$, and the solution (7) is

$$\tilde{\mathbf{x}} = \mathbf{Gy} + (\mathbf{GA} - \mathbf{I})\mathbf{z}$$

$$= \begin{bmatrix} 6 \\ -8 \\ 0 \\ 0 \end{bmatrix} + \left\{ \begin{bmatrix} 1 & 0 & -1 & -29 \\ 0 & 1 & 2 & 47 \\ 0 & 0 & 0 & 0 \\ 0 & 0 & 0 & 0 \end{bmatrix} - \mathbf{I} \right\} \begin{bmatrix} z_1 \\ z_2 \\ z_3 \\ z_4 \end{bmatrix}$$

$$= \begin{bmatrix} 6 - z_3 - 29z_4 \\ -8 + 2z_3 + 47z_4 \\ -z_3 \\ -z_4 \end{bmatrix} \tag{9}$$

where z_3 and z_4 are arbitrary. This means that (9) is a solution to (8) no matter what values are given to z_3 and z_4. For example, putting $z_3 = 0 = z_4$ gives

$$\tilde{\mathbf{x}}_1' = [6 \quad -8 \quad 0 \quad 0] \tag{10}$$

and putting $z_3 = -1$ and $z_4 = 2$ gives

$$\tilde{\mathbf{x}}_2' = [-51 \quad 84 \quad 1 \quad -2]. \tag{11}$$

It will be found that both $\tilde{\mathbf{x}}_1$ and $\tilde{\mathbf{x}}_2$ satisfy (8). That (9) does satisfy (8) for all values of z_3 and z_4 can be seen by substitution. For example, the left-hand side of the first equation is then

$$\begin{aligned} &5(6 - z_3 - 29z_4) + 3(-8 + 2z_3 + 47z_4) + (-z_3) - 4(-z_4) \\ &= 30 - 24 + z_3(-5 + 6 - 1) + z_4(-145 + 141 + 4) \\ &= 6 \end{aligned}$$

as it should be.

The $\mathbf{G}$ used earlier is not the only generalized inverse of the matrix $\mathbf{A}$ in (8). Another is

$$\dot{\mathbf{G}} = \begin{bmatrix} 0 & 0 & 0 & 0 \\ 0 & -5 & 2 & 0 \\ 0 & 13 & -5 & 0 \\ 0 & 0 & 0 & 0 \end{bmatrix}$$

for which (7) becomes

$$\dot{\mathbf{x}} = \dot{\mathbf{G}}\mathbf{y} + (\dot{\mathbf{G}}\mathbf{A} - \mathbf{I})\dot{\mathbf{z}}$$

$$= \begin{bmatrix} 0 \\ 4 \\ -6 \\ 0 \end{bmatrix} + \left\{ \begin{bmatrix} 0 & 0 & 0 & 0 \\ 2 & 1 & 0 & -11 \\ -1 & 0 & 1 & 29 \\ 0 & 0 & 0 & 0 \end{bmatrix} - \mathbf{I} \right\} \begin{bmatrix} \dot{z}_1 \\ \dot{z}_2 \\ \dot{z}_3 \\ \dot{z}_4 \end{bmatrix}$$

$$= \begin{bmatrix} -\dot{z}_1 \\ 4 + 2\dot{z}_1 - 11\dot{z}_4 \\ -6 - \dot{z}_1 + 29\dot{z}_4 \\ -\dot{z}_4 \end{bmatrix} \tag{12}$$

for arbitrary values $\dot{z}_1$ and $\dot{z}_4$. This too, it will be found, satisfies (8).

c. Properties of solutions

One might now ask about the relationship, if any, between the two solutions (9) and (12) found by using the two generalized inverses $\mathbf{G}$ and $\dot{\mathbf{G}}$. Both satisfy (8) for an infinite number of sets of values of z_3, z_4 and $\dot{z}_1$, $\dot{z}_4$. The basic question is: Do the two solutions generate, through allocating different sets of values to the arbitrary values z_3 and z_4 in $\tilde{\mathbf{x}}$ and $\dot{z}_1$ and $\dot{z}_4$ in $\dot{\mathbf{x}}$, the same series of vectors that satisfy $\mathbf{Ax} = \mathbf{y}$? The answer is "yes". This is so because, on putting $\dot{z}_1 = -6 + z_3 + 29z_4$ and $\dot{z}_4 = z_4$, the solution in (12) becomes identical to that in (9). Hence (9) and (12) both generate the same sets of solutions to (8)

The relationship between solutions using $\mathbf{G}$ and those using $\dot{\mathbf{G}}$ is that, on putting

$$\mathbf{z} = (\mathbf{G} - \dot{\mathbf{G}})\mathbf{y} + (\mathbf{I} - \dot{\mathbf{G}}\mathbf{A})\dot{\mathbf{z}},$$

$\tilde{\mathbf{x}}$ reduces to $\dot{\mathbf{x}}$.

A stronger result, which concerns generation of all solutions from $\tilde{\mathbf{x}}$, is contained in the following theorem.

Theorem 3. For the consistent equations $\mathbf{Ax} = \mathbf{y}$ all solutions are, for any specific $\mathbf{G}$, generated by $\tilde{\mathbf{x}} = \mathbf{Gy} + (\mathbf{GA} - \mathbf{I})\mathbf{z}$, for arbitrary $\mathbf{z}$.

Proof. Let $\mathbf{x}^*$ be any solution to $\mathbf{Ax} = \mathbf{y}$. Choose $\mathbf{z} = (\mathbf{GA} - \mathbf{I})\mathbf{x}^*$ and it will be found that $\tilde{\mathbf{x}}$ reduces to $\mathbf{x}^*$. Thus, by appropriate choice of $\mathbf{z}$, any solution to $\mathbf{Ax} = \mathbf{y}$ can be put in the form of $\tilde{\mathbf{x}}$.

The importance of this theorem is that one need derive only one generalized inverse of $\mathbf{A}$ in order to be able to generate all solutions to $\mathbf{Ax} = \mathbf{y}$. There are no solutions other than those that can be generated from $\tilde{\mathbf{x}}$.

Having established a method for solving linear equations and shown that they can have an infinite number of solutions, we ask two questions: What relationships exist among the solutions and to what extent are the solutions linearly independent (LIN)? Since each solution is a vector of order q there can, of course, be no more than q LIN solutions. In fact there are fewer, as Theorem 4 shows. But first, a lemma.

Lemma 1. Let $\mathbf{H} = \mathbf{GA}$ where the rank of $\mathbf{A}$, denoted by $r(\mathbf{A})$, is r, i.e., $r(\mathbf{A}) = r$; and $\mathbf{A}$ has q columns. Then $\mathbf{H}$ is idempotent with rank r and $r(\mathbf{I} - \mathbf{H}) = q - r$.

Proof. $\mathbf{H}^2 = \mathbf{GAGA} = \mathbf{GA} = \mathbf{H}$, showing that $\mathbf{H}$ is idempotent. Furthermore, by the rule for the rank of a product matrix, $r(\mathbf{H}) = r(\mathbf{GA}) \leq r(\mathbf{A})$. Similarly, because $\mathbf{AH} = \mathbf{AGA} = \mathbf{A}$, we have $r(\mathbf{H}) \geq r(\mathbf{A})$. Therefore $r(\mathbf{H}) = r(\mathbf{A}) = r$. And since $\mathbf{H}$ is idempotent so is $\mathbf{I} - \mathbf{H}$, of order q, so that $r(\mathbf{I} - \mathbf{H}) = \text{tr}(\mathbf{I} - \mathbf{H}) = q - \text{tr}(\mathbf{H}) = q - r(\mathbf{H}) = q - r$.

Theorem 4. When $\mathbf{A}$ is a matrix of q columns and rank r, and when $\mathbf{y}$ is a non-null vector, the number of LIN solutions to the consistent equations $\mathbf{Ax} = \mathbf{y}$ is $q - r + 1$.

Proof. Writing $\mathbf{H} = \mathbf{GA}$, the solutions to $\mathbf{Ax} = \mathbf{y}$ are, from Theorem 2,

$$\tilde{\mathbf{x}} = \mathbf{Gy} + (\mathbf{H} - \mathbf{I})\mathbf{z}.$$

Now because $r(\mathbf{H} - \mathbf{I}) = q - r$, there are only $(q - r)$ arbitrary elements in $(\mathbf{H} - \mathbf{I})\mathbf{z}$ for arbitrary $\mathbf{z}$; the other r elements are linear combinations of those $q - r$. Therefore there are only $(q - r)$ LIN vectors $(\mathbf{H} - \mathbf{I})\mathbf{z}$ and using them in $\tilde{\mathbf{x}}$ gives $(q - r)$ LIN solutions. For $i = 1, 2, \ldots, q - r$ let $\tilde{\mathbf{x}}_i = \mathbf{Gy} + (\mathbf{H} - \mathbf{I})\mathbf{z}_i$ be these solutions. $\tilde{\mathbf{x}} = \mathbf{Gy}$ is also a solution. Assume it is linearly dependent on the $\tilde{\mathbf{x}}_i$, so that for scalars λ_i, $i = 1, 2, \ldots, q - r$, not all of which are zero,

$$\mathbf{Gy} = \sum \lambda_i \tilde{\mathbf{x}}_i = \sum \lambda_i [\mathbf{Gy} + (\mathbf{H} - \mathbf{I})\mathbf{z}_i]. \tag{13}$$

Then

$$\mathbf{Gy} = \mathbf{Gy} \sum \lambda_i + \sum \lambda_i [(\mathbf{H} - \mathbf{I})\mathbf{z}_i]. \tag{14}$$

Now the left-hand side of (14) contains no $\mathbf{z}$'s. Therefore, on the right-hand side the second term must be null. But since the $(\mathbf{H} - \mathbf{I})\mathbf{z}_i$ are LIN this can be true only if every λ_i is zero. This means (13) is no longer true for some λ_i non-zero. Therefore $\mathbf{Gy}$ is independent of the $\tilde{\mathbf{x}}_i$; hence $\mathbf{Gy}$ and $\tilde{\mathbf{x}}_i$ for $i = 1, 2, \ldots, q - r$ form a set of $(q - r + 1)$ LIN solutions. When $q = r$ there is but one solution, corresponding to the existence of $\mathbf{A}^{-1}$, and that solution is $\mathbf{x} = \mathbf{A}^{-1}\mathbf{y}$.

This theorem means that $\tilde{\mathbf{x}} = \mathbf{Gy}$ and $\tilde{\mathbf{x}} = \mathbf{Gy} + (\mathbf{H} - \mathbf{I})\mathbf{z}$ for $(q - r)$ LIN vectors $\mathbf{z}$ are LIN solutions of $\mathbf{Ax} = \mathbf{y}$. All other solutions will be linear combinations of those forming a set of LIN solutions. A means of constructing solutions as linear combinations of other solutions is contained in the following theorem.

Theorem 5. If $\tilde{\mathbf{x}}_1, \tilde{\mathbf{x}}_2, \ldots, \tilde{\mathbf{x}}_s$ are any s solutions of consistent equations $\mathbf{Ax} = \mathbf{y}$ for which $\mathbf{y} \neq \mathbf{0}$, then any linear combination of these solutions $\mathbf{x}^* = \sum \lambda_i \tilde{\mathbf{x}}_i$ is also a solution of the equations if, and only if, $\sum \lambda_i = 1$, the summation being for $i = 1, 2, \ldots, s$.

Proof. Because

$$\mathbf{x}^* = \sum \lambda_i \tilde{\mathbf{x}}_i,$$
$$\mathbf{Ax}^* = \mathbf{A} \sum \lambda_i \tilde{\mathbf{x}}_i = \sum \lambda_i \mathbf{A}\tilde{\mathbf{x}}_i.$$

And because $\tilde{\mathbf{x}}_i$ is a solution, $\mathbf{A}\tilde{\mathbf{x}}_i = \mathbf{y}$ for all i, so giving

$$\mathbf{Ax}^* = \sum \lambda_i \mathbf{y} = \mathbf{y}(\sum \lambda_i). \tag{15}$$

Now if $\mathbf{x}^*$ is a solution of $\mathbf{Ax} = \mathbf{y}$ then $\mathbf{Ax}^* = \mathbf{y}$, and by comparison with (15) this means, $\mathbf{y}$ being non-null, that $\sum \lambda_i = 1$. Conversely, if $\sum \lambda_i = 1$, equation (15) implies that $\mathbf{Ax}^* = \mathbf{y}$, namely that $\mathbf{x}^*$ is a solution. So the theorem is proved.

Notice that Theorem 5 is in terms of any s solutions. Hence for any number of solutions, whether LIN or not, any linear combination of them is itself a solution provided the coefficients in that combination sum to unity.

Corollary. When $\mathbf{y} = \mathbf{0}$, $\mathbf{Gy}$ is null and there are only $q - r$ non-null LIN solutions to $\mathbf{Ax} = \mathbf{0}$; also, $\sum \lambda_i \tilde{\mathbf{x}}_i$ is a solution of $\mathbf{Ax} = \mathbf{0}$ for any values of the λ_i's.

Example (*continued*). It can be shown that the value of $r(\mathbf{A}) = r$ for $\mathbf{A}$ defined in (8) is $r = 2$. Therefore there are $q - r + 1 = 4 - 2 + 1 = 3$ LIN solutions to (8). Two are shown in (10) and (11), with (10) being the solution $\mathbf{Gy}$ when the value $\mathbf{z} = \mathbf{0}$ is used. Another solution, putting $\mathbf{z}' = [0 \quad 0 \quad 0 \quad 1]$ in (9), is

$$\tilde{\mathbf{x}}_3' = [-23 \quad 39 \quad 0 \quad -1].$$

Thus $\tilde{\mathbf{x}}_1$, $\tilde{\mathbf{x}}_2$ and $\tilde{\mathbf{x}}_3$ are LIN solutions and any other solution will be a linear combination of these three. For example, with $\mathbf{z}' = [0 \quad 0 \quad -1 \quad 0]$ the solution (9) becomes

$$\tilde{\mathbf{x}}_4' = [7 \quad -10 \quad 1 \quad 0]$$

and it can be seen that

$$\tilde{\mathbf{x}}_4 = 2\tilde{\mathbf{x}}_1 + \tilde{\mathbf{x}}_2 - 2\tilde{\mathbf{x}}_3,$$

the coefficients on the right-hand side, 2, 1 and -2, summing to unity in accord with Theorem 5.

A final theorem relates to an invariance property of the elements of a solution. It is important in the study of linear models because of its relationship with what is known as estimability, discussed in Chapter 5. Without worrying about details of estimability here, we give the theorem and refer to it later as needed. The theorem is due to Rao (1962) and it concerns linear combinations of the elements of a solution vector: certain combinations are invariant to whatever solution is used.

Theorem 6. The value of $\mathbf{k}'\tilde{\mathbf{x}}$ is invariant to whatever solution of $\mathbf{Ax} = \mathbf{y}$ is used for $\tilde{\mathbf{x}}$ if and only if $\mathbf{k}'\mathbf{H} = \mathbf{k}'$ (where $\mathbf{H} = \mathbf{GA}$ and $\mathbf{AGA} = \mathbf{A}$).

Proof. For a solution $\tilde{\mathbf{x}}$ given by Theorem 2

$$\mathbf{k}'\tilde{\mathbf{x}} = \mathbf{k}'\mathbf{Gy} + \mathbf{k}'(\mathbf{H} - \mathbf{I})\mathbf{z}.$$

This is independent of the arbitrary $\mathbf{z}$ if $\mathbf{k}'\mathbf{H} = \mathbf{k}'$; and since any solution can be put in the form $\tilde{\mathbf{x}}$ by appropriate choice of $\mathbf{z}$, the value of $\mathbf{k}'\tilde{\mathbf{x}}$ for any $\tilde{\mathbf{x}}$ is $\mathbf{k}'\mathbf{Gy}$ provided that $\mathbf{k}'\mathbf{H} = \mathbf{k}'$.

It may not be entirely clear that when $\mathbf{k}'\mathbf{H} = \mathbf{k}'$ the value of $\mathbf{k}'\tilde{\mathbf{x}} = \mathbf{k}'\mathbf{Gy}$ is invariant to whichever of the many generalized inverses is used for the matrix $\mathbf{G}$. We therefore clarify this point. First, by Theorem 4 there are $(q - r + 1)$ LIN solutions of the form $\tilde{\mathbf{x}} = \mathbf{Gy} + (\mathbf{H} - \mathbf{I})\mathbf{z}$. Let these solutions be $\tilde{\mathbf{x}}_i$ for $i = 1, 2, \ldots, q - r + 1$. Suppose for some other generalized inverse, $\mathbf{G}^*$ say, we have a solution

$$\mathbf{x}^* = \mathbf{G}^*\mathbf{y} + (\mathbf{H}^* - \mathbf{I})\mathbf{z}^*.$$

Then, since the $\tilde{\mathbf{x}}_i$'s are a LIN set of $(q - r + 1)$ solutions, $\mathbf{x}^*$ must be a linear combination of them; that is, there is a set of scalars λ_i, for $i = 1, 2, \ldots, q - r + 1$, such that

$$\mathbf{x}^* = \sum_{i=1}^{q-r+1} \lambda_i\tilde{\mathbf{x}}_i$$

where not all the λ_i's are zero and for which, by Theorem 5, $\sum \lambda_i = 1$.

Proving the sufficiency part of the theorem demands showing that $\mathbf{k}'\tilde{\mathbf{x}}$ is the same for all solutions $\tilde{\mathbf{x}}$ when $\mathbf{k}'\mathbf{H} = \mathbf{k}'$. Note that when $\mathbf{k}'\mathbf{H} = \mathbf{k}'$,

$$\mathbf{k}'\tilde{\mathbf{x}} = \mathbf{k}'\mathbf{H}\tilde{\mathbf{x}} = \mathbf{k}'\mathbf{HGy} + \mathbf{k}'(\mathbf{H}^2 - \mathbf{H})\mathbf{z} = \mathbf{k}'\mathbf{HGy} = \mathbf{k}'\mathbf{Gy}.$$

Therefore $\mathbf{k}'\tilde{\mathbf{x}}_i = \mathbf{k}'\mathbf{Gy}$ for all i, and

$$\mathbf{k}'\mathbf{x}^* = \mathbf{k}'\sum \lambda_i\tilde{\mathbf{x}}_i = \sum \lambda_i\mathbf{k}'\tilde{\mathbf{x}}_i = \sum \lambda_i\mathbf{k}'\mathbf{Gy} = \mathbf{k}'\mathbf{Gy}(\sum \lambda_i) = \mathbf{k}'\mathbf{Gy} = \mathbf{k}'\tilde{\mathbf{x}}_i;$$

i.e., for any solution at all, $\mathbf{k}'\tilde{\mathbf{x}} = \mathbf{k}'\mathbf{Gy}$ if $\mathbf{k}'\mathbf{H} = \mathbf{k}'$. To prove the necessity part of the theorem choose $\mathbf{z}^* = \mathbf{0}$ in $\mathbf{x}^*$. Then

$$\begin{aligned}\mathbf{k}'\mathbf{x}^* = \mathbf{k}'\mathbf{Gy} &= \mathbf{k}'\sum \lambda_i\tilde{\mathbf{x}}_i = \mathbf{k}'\sum \lambda_i[\mathbf{Gy} + (\mathbf{H} - \mathbf{I})\mathbf{z}_i] \\ &= \mathbf{k}'\mathbf{Gy}(\sum \lambda_i) + \mathbf{k}'\sum \lambda_i(\mathbf{H} - \mathbf{I})\mathbf{z}_i \\ &= \mathbf{k}'\mathbf{Gy} + \mathbf{k}'\sum \lambda_i(\mathbf{H} - \mathbf{I})\mathbf{z}_i.\end{aligned}$$

Hence $\mathbf{k}' \sum \lambda_i(\mathbf{H} - \mathbf{I})\mathbf{z}_i = 0$. But the λ_i are not all zero and the $(\mathbf{H} - \mathbf{I})\mathbf{z}_i$ are LIN. Therefore this last equation can be true only if $\mathbf{k}'(\mathbf{H} - \mathbf{I}) = 0$, i.e., $\mathbf{k}'\mathbf{H} = \mathbf{k}'$. Hence $\mathbf{k}'\mathbf{x}^*$ for *any* solution $\mathbf{x}^*$ equals $\mathbf{k}'\mathbf{G}\mathbf{y}$ if and only if $\mathbf{k}'\mathbf{H} = \mathbf{k}'$. This proves the theorem conclusively.

Example. In deriving (9),

$$\mathbf{H} = \mathbf{G}\mathbf{A} = \begin{bmatrix} 1 & 0 & -1 & -29 \\ 0 & 1 & 2 & 47 \\ 0 & 0 & 0 & 0 \\ 0 & 0 & 0 & 0 \end{bmatrix}$$

and for

$$\mathbf{k}' = [3 \quad 2 \quad 1 \quad 7] \tag{16}$$

it will be found that $\mathbf{k}'\mathbf{H} = \mathbf{k}'$. Therefore $\mathbf{k}'\tilde{\mathbf{x}}$ is invariant to whatever solution is used for $\tilde{\mathbf{x}}$. Thus from (10) and (11)

$$\mathbf{k}'\tilde{\mathbf{x}}_1 = 3(6) + 2(-8) + 1(0) + 7(0) = 2$$

and

$$\mathbf{k}'\tilde{\mathbf{x}}_2 = 3(-51) + 2(84) + 1(1) + 7(-2) = 2,$$

and in general, from (9),

$$\begin{aligned} \mathbf{k}'\tilde{\mathbf{x}} &= 3(6 - z_3 - 29z_4) + 2(-8 + 2z_3 + 47z_4) + 1(-z_3) + 7(-z_4) \\ &= 18 - 16 + z_3(-3 + 4 - 1) + z_4(-87 + 94 - 7) \\ &= 2. \end{aligned}$$

So too does $\mathbf{k}'\dot{\mathbf{x}}$ have the same value for, from (12),

$$\begin{aligned} \mathbf{k}'\dot{\mathbf{x}} &= 3(-\dot{z}_1) + 2(4 + 2\dot{z}_1 - 11\dot{z}_4) + 1(-6 - \dot{z}_1 + 29\dot{z}_4) + 7(-\dot{z}_4) \\ &= 8 - 6 + \dot{z}_1(-3 + 4 - 1) + \dot{z}_4(-22 + 29 - 7) \\ &= 2. \end{aligned}$$

There are, of course, many values of $\mathbf{k}'$ that satisfy $\mathbf{k}'\mathbf{H} = \mathbf{k}'$. For each of these, $\mathbf{k}'\tilde{\mathbf{x}}$ is invariant to the choice of $\tilde{\mathbf{x}}$; i.e., for two such vectors $\mathbf{k}_1'$ and $\mathbf{k}_2'$ say, $\mathbf{k}_1'\tilde{\mathbf{x}}$ and $\mathbf{k}_2'\tilde{\mathbf{x}}$ are different but each has a value that is the same for all values of $\tilde{\mathbf{x}}$. Thus in the example $\mathbf{k}_1' = \mathbf{k}_1'\mathbf{H}$, where

$$\mathbf{k}_1' = [1 \quad 2 \quad 3 \quad 65]$$

is different from (16); and

$$\mathbf{k}_1'\tilde{\mathbf{x}}_1 = 1(6) + 2(-8) + 3(0) + 65(0) = -10$$

is different from $\mathbf{k}'\tilde{\mathbf{x}}$ for $\mathbf{k}'$ of (16). But $\mathbf{k}_1'\tilde{\mathbf{x}} = -10$ for every $\tilde{\mathbf{x}}$.

The invariance of $\mathbf{k}'\tilde{\mathbf{x}}$ to $\tilde{\mathbf{x}}$ holds for any $\mathbf{k}'$ for which $\mathbf{k}'\mathbf{H} = \mathbf{k}'$, as shown in Theorem 6. Two corollaries of the theorem follow.

Corollary 1. $\mathbf{k}'\tilde{\mathbf{x}}$ is invariant to $\tilde{\mathbf{x}}$ for $\mathbf{k}'$ of the form $\mathbf{k}' = \mathbf{w}'\mathbf{H}$, for arbitrary $\mathbf{w}'$. (Idempotency of $\mathbf{H}$ ensures that $\mathbf{k}' = \mathbf{w}'\mathbf{H}$ satisfies $\mathbf{k}'\mathbf{H} = \mathbf{k}'$.)

Corollary 2. There are only r LIN vectors $\mathbf{k}'$ for which $\mathbf{k}'\tilde{\mathbf{x}}$ is invariant to $\tilde{\mathbf{x}}$. [Because $r(\mathbf{H}) = r$ there are in $\mathbf{k}' = \mathbf{w}'\mathbf{H}$ of order q exactly $q - r$ elements that are linear combinations of the other r. Therefore for arbitrary vectors $\mathbf{w}'$ there are only r LIN vectors $\mathbf{k}' = \mathbf{w}'\mathbf{H}$.] We return to this point when discussing estimable functions in Chapter 5.

The concept of a generalized inverse has now been defined and its use in solving linear equations explained. We next briefly discuss the generalized inverse itself, its various definitions and some of its properties. Extensive review of generalized inverses and their applications is to be found in Boullion and Odell (1968) and the approximately 350 references listed there.

3. THE PENROSE INVERSE

Penrose (1955), in extending the work of Moore (1920), shows that for any matrix $\mathbf{A}$ there is a unique matrix $\mathbf{K}$ which satisfies the following four conditions:

$$
\begin{aligned}
\mathbf{AKA} &= \mathbf{A} && \text{(i)} \\
\mathbf{KAK} &= \mathbf{K} && \text{(ii)} \\
(\mathbf{KA})' &= \mathbf{KA} && \text{(iii)} \\
(\mathbf{AK})' &= \mathbf{AK} && \text{(iv)}.
\end{aligned}
\qquad (17)
$$

We refer to these as Penrose's conditions and to $\mathbf{K}$ as the (unique) Penrose inverse; more correctly it is the Moore-Penrose inverse. Penrose's proof of the existence of $\mathbf{K}$ satisfying these conditions is lengthy but instructive. It rests upon two lemmas relating to matrices having real (but not complex) numbers as elements, lemmas that are used repeatedly in what follows.

Lemma 2. $\mathbf{X}'\mathbf{X} = \mathbf{0}$ implies $\mathbf{X} = \mathbf{0}$.

Lemma 3. $\mathbf{PX}'\mathbf{X} = \mathbf{QX}'\mathbf{X}$ implies $\mathbf{PX}' = \mathbf{QX}'$.

The first of these is true because $\mathbf{X}'\mathbf{X} = \mathbf{0}$ implies that sums of squares of elements of each row are zero and hence the elements themselves are zero. Lemma 3 is proved by applying Lemma 2 to

$$(\mathbf{PX}'\mathbf{X} - \mathbf{QX}'\mathbf{X})(\mathbf{P} - \mathbf{Q})' = (\mathbf{PX}' - \mathbf{QX}')(\mathbf{PX}' - \mathbf{QX}')' = \mathbf{0}.$$

Proof of the existence and uniqueness of $\mathbf{K}$ starts by noting that (i) and (iii) imply $\mathbf{AA}'\mathbf{K}' = \mathbf{A}$. Conversely, if $\mathbf{AA}'\mathbf{K}' = \mathbf{A}$ then $\mathbf{KA}(\mathbf{KA})' = \mathbf{KA}$, showing that $\mathbf{KA}$ is symmetric, namely that (iii) is true; and using this in

$\mathbf{AA'K'} = \mathbf{A}$ leads to (i). Thus (i) and (iii) are true if and only if $\mathbf{AA'K'} = \mathbf{A}$, equivalent to

$$\mathbf{KAA'} = \mathbf{A'}. \tag{18}$$

Similarly, (ii) and (iv) are true if and only if

$$\mathbf{KK'A'} = \mathbf{K}. \tag{19}$$

Hence any $\mathbf{K}$ satisfying (18) and (19) also satisfies the Penrose conditions.

Before showing how $\mathbf{K}$ can be derived we show that it is unique. For if it is not, assume that some other matrix $\mathbf{M}$ satisfies the Penrose conditions. Then from conditions (i) and (iv) in terms of $\mathbf{M}$ we would have

$$\mathbf{A'AM} = \mathbf{A'} \tag{20}$$

and (ii) and (iii) would lead to

$$\mathbf{A'M'M} = \mathbf{M}. \tag{21}$$

Therefore, on substituting (20) into (19) and using (19) again we have

$$\mathbf{K} = \mathbf{KK'A'} = \mathbf{KK'A'AM} = \mathbf{KAM};$$

and on substituting (21) into this and using (18) and (21) we get

$$\mathbf{K} = \mathbf{KAM} = \mathbf{KAA'M'M} = \mathbf{A'M'M} = \mathbf{M}.$$

Therefore $\mathbf{K}$ satisfying (18) and (19) is unique and satisfies the Penrose conditions; we derive its form by assuming that

$$\mathbf{K} = \mathbf{TA'} \tag{22}$$

for some matrix $\mathbf{T}$. Then (18) is satisfied if

$$\mathbf{TA'AA'} = \mathbf{A'}; \tag{23}$$

and since satisfaction of (18) also implies (i) we have $\mathbf{AKA} = \mathbf{A}$, i.e.,

$$\mathbf{A'K'A'} = \mathbf{A'}.$$

Therefore $$\mathbf{TA'K'A'} = \mathbf{TA'},$$

which is $$\mathbf{KK'A'} = \mathbf{K},$$

namely (19). Thus we have proved that if $\mathbf{K} = \mathbf{TA'}$ as in (22), with $\mathbf{T}$ being any matrix satisfying (23), then $\mathbf{K}$ satisfies (18) and (19) and hence the Penrose conditions.

There remains the derivation of a suitable $\mathbf{T}$. This is done as follows. Consider $\mathbf{A'A}$: it is square and so are its powers. And for some integer t there will be, as a consequence of the Cayley-Hamilton theorem [see, e.g., Searle (1966), Sec. 7.5e], a series of scalars $\lambda_1, \lambda_2, \ldots, \lambda_t$, not all zero, such that

$$\lambda_1\mathbf{A'A} + \lambda_2(\mathbf{A'A})^2 + \cdots + \lambda_t(\mathbf{A'A})^t = \mathbf{0}.$$

If λ_r is the first λ in this identity that is non-zero, then $\mathbf{T}$ is defined as

$$\mathbf{T} = (-1/\lambda_r)[\lambda_{r+1}\mathbf{I} + \lambda_{r+2}(\mathbf{A'A}) + \cdots + \lambda_t(\mathbf{A'A})^{t-r-1}]. \tag{24}$$

To show that this satisfies (23) note that, by direct multiplication,

$$\begin{aligned}\mathbf{T}(\mathbf{A'A})^{r+1} &= (-1/\lambda_r)[\lambda_{r+1}(\mathbf{A'A})^{r+1} + \lambda_{r+2}(\mathbf{A'A})^{r+2} + \cdots + \lambda_t(\mathbf{A'A})^t]\\ &= (-1/\lambda_r)[-\lambda_1\mathbf{A'A} - \lambda_2(\mathbf{A'A})^2 - \cdots - \lambda_r(\mathbf{A'A})^r].\end{aligned}$$

Since, by definition, λ_r is the first non-zero λ in the series $\lambda_1, \lambda_2, \ldots$, the above reduces to

$$\mathbf{T}(\mathbf{A'A})^{r+1} = (\mathbf{A'A})^r, \tag{25}$$

and repeated use of Lemma 3 reduces this to (23). Thus $\mathbf{K} = \mathbf{TA'}$ with $\mathbf{T}$ as defined in (24) satisfies (23) and hence is the unique generalized inverse satisfying all four of the Penrose conditions.

Example. For

$$\mathbf{A} = \begin{bmatrix} 1 & 0 & 2 \\ 0 & -1 & 1 \\ -1 & 0 & -2 \\ 1 & 2 & 0 \end{bmatrix} \quad \text{we have} \quad \mathbf{A'A} = \begin{bmatrix} 3 & 2 & 4 \\ 2 & 5 & -1 \\ 4 & -1 & 9 \end{bmatrix}.$$

Then, by the Cayley-Hamilton theorem,

$$66(\mathbf{A'A}) - 17(\mathbf{A'A})^2 + (\mathbf{A'A})^3 = \mathbf{0}$$

and so $\mathbf{T}$ is taken as

$$\mathbf{T} = (-1/66)(-17\mathbf{I} + \mathbf{A'A}) = (1/66)\begin{bmatrix} 14 & -2 & -4 \\ -2 & 12 & 1 \\ -4 & 1 & 8 \end{bmatrix}$$

and

$$\mathbf{K} = \mathbf{TA'} = (1/66)\begin{bmatrix} 6 & -2 & -6 & 10 \\ 0 & -11 & 0 & 22 \\ 12 & 7 & -12 & -2 \end{bmatrix}$$

is the Penrose inverse of $\mathbf{A}$ satisfying (17).

An alternative procedure for deriving $\mathbf{K}$ has been suggested by Graybill et al. (1966). Their method is to find $\mathbf{X}$ and $\mathbf{Y}$ such that

$$\mathbf{AA'X} = \mathbf{A} \quad \text{and} \quad \mathbf{A'AY} = \mathbf{A'} \tag{26}$$

and then

$$\mathbf{K} = \mathbf{XAY}. \tag{27}$$

Proof that **K** satisfies the four Penrose conditions depends upon using (26) and Lemma 3 to show that **AXA** = **A** = **AYA**.

4. OTHER DEFINITIONS

It is clear that the Penrose inverse **K** is not easy to compute, especially when **A** has many columns, because then the application of the Cayley-Hamilton theorem to **A'A** for obtaining **T** will be tedious. However, as has already been shown, only the first Penrose condition needs to be satisfied in order to have a matrix useful for solving linear equations. And in pursuing the topic of linear models it is found that this is the only condition really needed. It is for this reason that a generalized inverse of **A** has been defined as any matrix **G** that satisfies **AGA** = **A**, a definition that is retained throughout this book. Nevertheless, a variety of names are to be found in the literature, both for **G** and for other matrices satisfying fewer than all four of the Penrose conditions. A set of descriptive names is given in Table 1.1.

TABLE 1.1. SUGGESTED NAMES FOR MATRICES SATISFYING SOME OR ALL OF THE PENROSE CONDITIONS

Conditions Satisfied (Eq. 17)	Name of Matrix	Symbol
i	Generalized inverse	$\mathbf{A}^{(g)}$
i and ii	Reflexive generalized inverse	$\mathbf{A}^{(r)}$
i, ii and iii	Normalized generalized inverse	$\mathbf{A}^{(n)}$
i, ii, iii and iv	Penrose inverse	$\mathbf{A}^{(p)}$

In the notation of Table 1.1 $\mathbf{A}^{(g)} = \mathbf{G}$, the generalized inverse already defined and discussed, and $\mathbf{A}^{(p)} = \mathbf{K}$, the Penrose inverse. This has also been called the pseudo inverse and the *p*-inverse by various authors. The suggested definition of a normalized generalized inverse in Table 1.1 is not universally accepted. As given there, it is used by Urquhart (1968), whereas Goldman and Zelen (1964) call it a "weak" generalized inverse. An example of such a matrix is a left inverse **L** such that **LA** = **I**. The description "normalized" has also been used by Rohde (1966) for a matrix satisfying conditions (i), (ii) and (iv). An example of this kind of matrix is the right inverse **R** for which **AR** = **I**.

Using the symbols of Table 1.1 it can be seen that

$$\mathbf{A}^{(g)} \supset \mathbf{A}^{(r)} \supset \mathbf{A}^{(n)} \supset \mathbf{A}^{(p)},$$

namely that the set of matrices $\mathbf{A}^{(g)}$ includes all those that are reflexive, $\mathbf{A}^{(r)}$, which in turn includes all the normalized generalized inverses $\mathbf{A}^{(n)}$, which includes the unique $\mathbf{A}^{(p)} = \mathbf{K}$. Relationships between the four can be established as follows:

$$\begin{aligned} \mathbf{A}^{(r)} &= \mathbf{A}^{(g)}\mathbf{A}\mathbf{A}^{(g)} \\ \mathbf{A}^{(n)} &= \mathbf{A}'(\mathbf{A}\mathbf{A}')^{(g)} \\ \mathbf{A}^{(p)} &= \mathbf{A}'(\mathbf{A}\mathbf{A}')^{(g)}\mathbf{A}(\mathbf{A}'\mathbf{A})^{(g)}\mathbf{A}'. \end{aligned} \tag{28}$$

That these expressions satisfy the appropriate conditions can be proved by repeated use of Lemma 3 of Sec. 3 or by Theorem 7 of Sec. 5, which follows.

5. SYMMETRIC MATRICES

The study of linear models frequently leads to equations of the form $\mathbf{X}'\mathbf{X}\hat{\mathbf{b}} = \mathbf{X}'\mathbf{y}$ that have to be solved for $\hat{\mathbf{b}}$. Special attention is therefore given to properties of a generalized inverse of the symmetric matrix $\mathbf{X}'\mathbf{X}$.

a. Properties of a generalized inverse

Four useful properties of a generalized inverse of $\mathbf{X}'\mathbf{X}$ are contained in the following theorem.

Theorem 7. When $\mathbf{G}$ is a generalized inverse of $\mathbf{X}'\mathbf{X}$, then

(i) $\mathbf{G}'$ is also a generalized inverse of $\mathbf{X}'\mathbf{X}$;
(ii) $\mathbf{X}\mathbf{G}\mathbf{X}'\mathbf{X} = \mathbf{X}$; i.e., $\mathbf{G}\mathbf{X}'$ is a generalized inverse of $\mathbf{X}$;
(iii) $\mathbf{X}\mathbf{G}\mathbf{X}'$ is invariant to $\mathbf{G}$;
(iv) $\mathbf{X}\mathbf{G}\mathbf{X}'$ is symmetric, whether $\mathbf{G}$ is or not.

Proof. By definition, $\mathbf{G}$ satisfies

$$\mathbf{X}'\mathbf{X}\mathbf{G}\mathbf{X}'\mathbf{X} = \mathbf{X}'\mathbf{X}. \tag{29}$$

Transposing gives $\mathbf{X}'\mathbf{X}\mathbf{G}'\mathbf{X}'\mathbf{X} = \mathbf{X}'\mathbf{X}$, and so (i) is established; and applying Lemma 3 yields (ii). To substantiate (iii) suppose that $\mathbf{F}$ is some other generalized inverse, different from $\mathbf{G}$. Then (ii) gives $\mathbf{X}\mathbf{G}\mathbf{X}'\mathbf{X} = \mathbf{X}\mathbf{F}\mathbf{X}'\mathbf{X}$ and the use of Lemma 3 then yields $\mathbf{X}\mathbf{G}\mathbf{X}' = \mathbf{X}\mathbf{F}\mathbf{X}'$; i.e., $\mathbf{X}\mathbf{G}\mathbf{X}'$ is the same for all generalized inverses of $\mathbf{X}'\mathbf{X}$. Finally, to prove (iv) consider $\mathbf{S}$ as a symmetric generalized inverse of $\mathbf{X}'\mathbf{X}$. Then $\mathbf{X}\mathbf{S}\mathbf{X}'$ is symmetric. But $\mathbf{X}\mathbf{S}\mathbf{X}' = \mathbf{X}\mathbf{G}\mathbf{X}'$ and therefore $\mathbf{X}\mathbf{G}\mathbf{X}'$ is symmetric. Hence the theorem is proved.

Corollary. Applying part (i) of the theorem to its other parts shows that

$$\mathbf{X}\mathbf{G}'\mathbf{X}'\mathbf{X} = \mathbf{X}, \qquad \mathbf{X}'\mathbf{X}\mathbf{G}\mathbf{X}' = \mathbf{X}' \qquad \text{and } \mathbf{X}'\mathbf{X}\mathbf{G}'\mathbf{X}' = \mathbf{X}';$$

$$\mathbf{X}\mathbf{G}'\mathbf{X}' = \mathbf{X}\mathbf{G}\mathbf{X}'; \text{ and } \mathbf{X}\mathbf{G}'\mathbf{X}' \text{ is symmetric.}$$

It is to be emphasized that not all generalized inverses of a symmetric matrix are symmetric. This is evident from the general algorithm given at the end of Sec. 1. For example, applying that algorithm to

$$\mathbf{A}_2 = \begin{bmatrix} 2 & 2 & 6 \\ 2 & 3 & 8 \\ 6 & 8 & 22 \end{bmatrix} \quad \text{gives} \quad \begin{bmatrix} 2 & -1\frac{1}{2} & 0 \\ 0 & 0 & 0 \\ -\frac{1}{2} & \frac{1}{2} & 0 \end{bmatrix}$$

as a non-symmetric generalized inverse of the symmetric matrix $\mathbf{A}_2$. However, Theorem 7 and its corollary very largely enable us to avoid difficulties that this lack of symmetry of generalized inverses of $\mathbf{X'X}$ might otherwise appear to involve. For example, if $\mathbf{G}$ is a generalized inverse of $\mathbf{X'X}$ and $\mathbf{P}$ is some other matrix,

$$(\mathbf{PXGX'})' = \mathbf{XG'X'P'} = \mathbf{XGX'P'}$$

not because $\mathbf{G}$ is symmetric (which, in general, it is not) but because $\mathbf{XG'X'} = \mathbf{XGX'}$. An example of *this*, for $\mathbf{A}_2$ shown above, is

$$\mathbf{A}_2 = \begin{bmatrix} 2 & 2 & 6 \\ 2 & 3 & 8 \\ 6 & 8 & 22 \end{bmatrix} = \begin{bmatrix} 1 & 1 & 0 \\ 1 & 1 & 1 \\ 3 & 3 & 2 \end{bmatrix} \begin{bmatrix} 1 & 1 & 3 \\ 1 & 1 & 3 \\ 0 & 1 & 2 \end{bmatrix} = \mathbf{X'X}.$$

Then $\mathbf{XGX'} =$

$$\begin{bmatrix} 1 & 1 & 3 \\ 1 & 1 & 3 \\ 0 & 1 & 2 \end{bmatrix} \begin{bmatrix} 2 & -1\frac{1}{2} & 0 \\ 0 & 0 & 0 \\ -\frac{1}{2} & \frac{1}{2} & 0 \end{bmatrix} \begin{bmatrix} 1 & 1 & 0 \\ 1 & 1 & 1 \\ 3 & 3 & 2 \end{bmatrix} = \frac{1}{2} \begin{bmatrix} 1 & 1 & 0 \\ 1 & 1 & 0 \\ 0 & 0 & 2 \end{bmatrix} = \mathbf{XG'X'}$$

$$= \begin{bmatrix} 1 & 1 & 3 \\ 1 & 1 & 3 \\ 0 & 1 & 2 \end{bmatrix} \begin{bmatrix} 2 & 0 & -\frac{1}{2} \\ -1\frac{1}{2} & 0 & \frac{1}{2} \\ 0 & 0 & 0 \end{bmatrix} \begin{bmatrix} 1 & 1 & 0 \\ 1 & 1 & 1 \\ 3 & 3 & 2 \end{bmatrix}.$$

b. Two methods of derivation

In addition to the methods given in Sec. 1, two methods discussed by John (1964) are sometimes pertinent to linear models. They depend on the regular inverse of a non-singular matrix:

$$\mathbf{S}^{-1} = \begin{bmatrix} \mathbf{X'X} & \mathbf{H'} \\ \mathbf{H} & \mathbf{0} \end{bmatrix}^{-1} = \begin{bmatrix} \mathbf{B}_{11} & \mathbf{B}_{12} \\ \mathbf{B}_{21} & \mathbf{B}_{22} = \mathbf{0} \end{bmatrix}. \qquad (30)$$

$\mathbf{H}$ used here, in keeping with John's notation, is *not* the matrix $\mathbf{GA}$ used earlier. Where $\mathbf{X'X}$ has order p and rank $p - m$ $(m > 0)$, the matrix $\mathbf{H}$ is any matrix of order $m \times p$ that is of full row rank with its rows also LIN of those of

$\mathbf{X'X}$. [The existence of such a matrix is assured by considering m vectors of order p that are LIN of any set of $p - m$ LIN rows of $\mathbf{X'X}$. Furthermore, if these rows constitute $\mathbf{H}$ in such a way that the m LIN rows of $\mathbf{H}$ correspond in $\mathbf{S}$ to the m rows of $\mathbf{X'X}$ that are linear combinations of that set of $p - m$ rows, then $\mathbf{S}^{-1}$ of (30) exists.] With (30) existing the two matrices

$$\mathbf{B}_{11} \text{ and } (\mathbf{X'X} + \mathbf{H'H})^{-1} \text{ are generalized inverses of } \mathbf{X'X}. \tag{31}$$

Three useful lemmas help in proving these results.

Lemma 4. The matrix $\mathbf{T} = [\mathbf{I}_r \quad \mathbf{U}]$ has rank r for any matrix $\mathbf{U}$, of r rows.

Proof. Elementary operations carried out on $\mathbf{T}$ to find its rank will operate on $\mathbf{I}_r$, none of whose rows (or columns) can be made null by such operations. Therefore $r(\mathbf{T}) \nless r$ and so $r(\mathbf{T}) = r$.

Lemma 5. If $\mathbf{X}_{N\times p}$ has rank $p - m$ for $m > 0$, then there exists a matrix $\mathbf{D}_{p\times m}$ such that $\mathbf{XD} = \mathbf{0}$ and $r(\mathbf{D}) = m$.

Proof. Let $\mathbf{X} = [\mathbf{X}_1 \quad \mathbf{X}_2]$ where $\mathbf{X}_1$ is $N \times (p - m)$ of full column rank. Then the columns of $\mathbf{X}_2$ are linear combinations of those of $\mathbf{X}_1$ and so, for some matrix $\mathbf{C}$, of order $(p - m) \times m$, the sub-matrices of $\mathbf{X}$ satisfy $\mathbf{X}_2 = \mathbf{X}_1\mathbf{C}$. Letting $\mathbf{D}' = [-\mathbf{C}' \quad \mathbf{I}_m]$, which by Lemma 4 has rank m, we then have $\mathbf{XD} = \mathbf{0}$ and $r(\mathbf{D}) = m$ and the lemma is proved: a matrix $\mathbf{D}$ exists.

Lemma 6. For $\mathbf{X}$ and $\mathbf{D}$ of Lemma 5 and $\mathbf{H}$ of order $m \times p$ with full row rank, $\mathbf{HD}$ has full rank if and only if the rows of $\mathbf{H}$ are LIN of those of $\mathbf{X}$.

Proof. (i) Given $r(\mathbf{HD}) = m$, assume that the rows of $\mathbf{H}$ depend on those of $\mathbf{X}$. Then $\mathbf{H} = \mathbf{KX}$ for some $\mathbf{K}$, and $\mathbf{HD} = \mathbf{KXD} = \mathbf{0}$. This cannot be so, because $r(\mathbf{HD}) = m$. Therefore the assumption is false and so the rows of $\mathbf{H}$ are LIN of those of $\mathbf{X}$.

(ii) Given that the rows of $\mathbf{H}$ are LIN of those of $\mathbf{X}$, the matrix $\begin{bmatrix}\mathbf{X}\\ \mathbf{H}\end{bmatrix}$, of order $(N + m) \times p$, has full column rank. Therefore it has a left inverse, $[\mathbf{U} \quad \mathbf{V}]$ say [Searle (1966), Sec. 5.13], and so

$$\mathbf{UX} + \mathbf{VH} = \mathbf{I}, \qquad \text{i.e.,} \qquad \mathbf{UXD} + \mathbf{VHD} = \mathbf{D}; \qquad \text{or} \qquad \mathbf{VHD} = \mathbf{D},$$

using Lemma 5. But $r(\mathbf{D}_{p\times m}) = m$ and therefore $\mathbf{D}$ has a left inverse, $\mathbf{E}$ say, and so $\mathbf{EVHD} = \mathbf{ED} = \mathbf{I}_m$. Therefore $r(\mathbf{HD}) \geq m$ and so, because $\mathbf{HD}$ is $m \times m$, $r(\mathbf{HD}) = m$, and the lemma is proved.

Proof of (31) can now be established. First, it is necessary to show that in (30) $\mathbf{B}_{22} = \mathbf{0}$. In multiplying the two sides of (30) we get an identity matrix:

$$\mathbf{X'XB}_{11} + \mathbf{H'B}_{21} = \mathbf{I} \quad \text{and} \quad \mathbf{X'XB}_{12} + \mathbf{H'B}_{22} = \mathbf{0}, \tag{32}$$

$$\mathbf{HB}_{11} = \mathbf{0} \quad \text{and} \quad \mathbf{HB}_{12} = \mathbf{I}. \tag{33}$$

Pre-multiplying (32) by $\mathbf{D}'$ and using Lemmas 5 and 6 leads to

$$\mathbf{B}_{21} = (\mathbf{D}'\mathbf{H}')^{-1}\mathbf{D}' \qquad \text{and} \qquad \mathbf{B}_{22} = \mathbf{0}. \tag{34}$$

Then, from (32) and (34)

$$\mathbf{X}'\mathbf{X}\mathbf{B}_{11} = \mathbf{I} - \mathbf{H}'(\mathbf{D}'\mathbf{H}')^{-1}\mathbf{D}' \tag{35}$$

and post-multiplying this by $\mathbf{X}'\mathbf{X}$ shows, from Lemma 5, that $\mathbf{B}_{11}$ is a generalized inverse of $\mathbf{X}'\mathbf{X}$. Furthermore, making use of (33), (35) and Lemmas 5 and 6 gives

$$(\mathbf{X}'\mathbf{X} + \mathbf{H}'\mathbf{H})[\mathbf{B}_{11} + \mathbf{D}(\mathbf{D}'\mathbf{H}'\mathbf{H}\mathbf{D})^{-1}\mathbf{D}'] = \mathbf{I}.$$

Pre- and post-multiplying $(\mathbf{X}'\mathbf{X} + \mathbf{H}'\mathbf{H})^{-1}$ obtained from this by $\mathbf{X}'\mathbf{X}$ then shows that $(\mathbf{X}'\mathbf{X} + \mathbf{H}'\mathbf{H})^{-1}$ is a generalized inverse of $\mathbf{X}'\mathbf{X}$.

It can also be shown that $\mathbf{B}_{11}$ satisfies the second of Penrose's conditions, (ii) in (17), but $(\mathbf{X}'\mathbf{X} + \mathbf{H}'\mathbf{H})^{-1}$ does not; and neither generalized inverse in (31) satisfies condition (iii) or (iv).

John (1964) refers to Graybill (1961, p. 292) and to Kempthorne (1952, p. 79) in discussing $\mathbf{B}_{11}$ and to Plackett (1960, p. 41) and Scheffé (1959, p. 19) in discussing $(\mathbf{X}'\mathbf{X} + \mathbf{H}'\mathbf{H})^{-1}$, in terms of defining generalized inverses of $\mathbf{X}'\mathbf{X}$ as being matrices $\mathbf{G}$ for which $\mathbf{b} = \mathbf{G}\mathbf{X}'\mathbf{y}$ is a solution of $\mathbf{X}'\mathbf{X}\mathbf{b} = \mathbf{X}'\mathbf{y}$. By Theorem 1 they then satisfy condition (i), as has just been shown. Rayner and Pringle (1967) also discuss these results, indicating that $\mathbf{D}$ of the preceding discussion may be taken as $(\mathbf{X}'\mathbf{X} + \mathbf{H}'\mathbf{H})^{-1}\mathbf{H}'$. This, as Chipman (1964) shows (see Exercise 13), means that $\mathbf{H}\mathbf{D} = \mathbf{I}$ and so (35) becomes

$$\mathbf{X}'\mathbf{X}\mathbf{B}_{11} = \mathbf{I} - \mathbf{H}'\mathbf{H}(\mathbf{X}'\mathbf{X} + \mathbf{H}'\mathbf{H})^{-1}, \tag{36}$$

a simplified form of Rayner and Pringle's equation (7). The relationship between the two generalized inverses of $\mathbf{X}'\mathbf{X}$ shown in (31) is therefore that indicated in (36). Note also that Lemma 6 is equivalent to Theorem 3 of Scheffé (1959, p. 17).

6. ARBITRARINESS IN A GENERALIZED INVERSE

The existence of many generalized inverse matrices $\mathbf{G}$ that satisfy $\mathbf{AGA} = \mathbf{A}$ has been emphasized. We here examine the nature of the arbitrariness in such generalized inverses, as discussed by Urquhart (1969a). Some lemmas concerning rank are given first.

Lemma 7. A matrix of full row rank r can be written as a product of matrices, one being of the form $[\mathbf{I}_r \quad \mathbf{S}]$ for some matrix $\mathbf{S}$, of r rows.

Proof. Suppose $\mathbf{B}_{r\times q}$ has full row rank r and contains an $r \times r$ non-singular minor, $\mathbf{M}$ say. Then, for some matrix $\mathbf{L}$ and some permutation matrix **Q** [see (6)], we have $\mathbf{BQ} = [\mathbf{M} \quad \mathbf{L}]$, so that

$$\mathbf{B} = \mathbf{M}[\mathbf{I} \quad \mathbf{M}^{-1}\mathbf{L}]\mathbf{Q}^{-1} = \mathbf{M}[\mathbf{I} \quad \mathbf{S}]\mathbf{Q}^{-1}, \qquad \text{for} \qquad \mathbf{S} = \mathbf{M}^{-1}\mathbf{L}.$$

Lemma 8. $\mathbf{I} + \mathbf{KK}'$ has full rank for any non-null matrix **K**.

Proof. Assume that $\mathbf{I} + \mathbf{KK}'$ does not have full rank. Then its columns are not LIN and there exists a non-null vector **u** such that

$$(\mathbf{I} + \mathbf{KK}')\mathbf{u} = \mathbf{0}, \qquad \text{so that} \qquad \mathbf{u}'(\mathbf{I} + \mathbf{KK}')\mathbf{u} = \mathbf{u}'\mathbf{u} + \mathbf{u}'\mathbf{K}(\mathbf{u}'\mathbf{K})' = 0.$$

But $\mathbf{u}'\mathbf{u}$ and $\mathbf{u}'\mathbf{K}(\mathbf{u}'\mathbf{K})'$ are both sums of squares of real numbers. Hence their sum is zero only if their elements are, i.e., only if $\mathbf{u} = \mathbf{0}$. This contradicts the assumption. Therefore $\mathbf{I} + \mathbf{KK}'$ has full rank.

Lemma 9. When **B** has full row rank $\mathbf{BB}'$ is non-singular.

Proof. As in Lemma 7, write $\mathbf{B} = \mathbf{M}[\mathbf{I} \quad \mathbf{S}]\mathbf{Q}^{-1}$ where $\mathbf{M}^{-1}$ exists. Then, because **Q** is a permutation matrix and thus orthogonal, $\mathbf{BB}' = \mathbf{M}(\mathbf{I} + \mathbf{SS}')\mathbf{M}'$ which, by Lemma 8 and because $\mathbf{M}^{-1}$ exists, is non-singular.

Corollary. When **B** has full column rank $\mathbf{B}'\mathbf{B}$ is non-singular.

Consider now a matrix $\mathbf{A}_{p\times q}$ of rank r, less than both p and q. **A** contains at least one non-singular minor of order r, which we will assume is the leading minor. There is no loss of generality in this assumption because if it is not true, the algorithm of Sec. 1b will always yield a generalized inverse of **A** from a generalized inverse of $\mathbf{B} = \mathbf{RAS}$ for permutation matrices **R** and **S**, where **B** has its leading $r \times r$ minor non-singular. Discussion of generalized inverses of **A** is therefore confined to **A** having its leading $r \times r$ minor non-singular. Accordingly, **A** is partitioned as

$$\mathbf{A} = \begin{bmatrix} (\mathbf{A}_{11})_{r\times r} & (\mathbf{A}_{12})_{r\times(q-r)} \\ (\mathbf{A}_{21})_{(p-r)\times r} & (\mathbf{A}_{22})_{(p-r)\times(q-r)} \end{bmatrix}. \tag{37}$$

Then, with $\mathbf{A}_{11}^{-1}$ existing, **A** can be written as

$$\mathbf{A} = \begin{bmatrix} \mathbf{I} \\ \mathbf{A}_{21}\mathbf{A}_{11}^{-1} \end{bmatrix} \mathbf{A}_{11}[\mathbf{I} \quad \mathbf{A}_{11}^{-1}\mathbf{A}_{12}] \tag{38}$$

$$= \mathbf{LA}_{11}\mathbf{M}, \qquad \text{with} \quad \mathbf{L} = \begin{bmatrix} \mathbf{I} \\ \mathbf{A}_{21}\mathbf{A}_{11}^{-1} \end{bmatrix} \text{ and } \mathbf{M} = [\mathbf{I} \quad \mathbf{A}_{11}^{-1}\mathbf{A}_{12}]. \tag{39}$$

Since, from Lemma 4, **L** has full column rank and **M** has full row rank, Lemma 9 shows that

$$(\mathbf{L}'\mathbf{L})^{-1} \text{ and } (\mathbf{MM}')^{-1} \text{ exist.} \tag{40}$$

The arbitrariness in a generalized inverse of **A** is investigated by means of this partitioning. Thus, on substituting (39) into **AGA** = **A** we get

$$\mathbf{LA}_{11}\mathbf{MGLA}_{11}\mathbf{M} = \mathbf{LA}_{11}\mathbf{M}.$$

Pre-multiplication by $\mathbf{A}_{11}^{-1}(\mathbf{L}'\mathbf{L})^{-1}\mathbf{L}'$ and post-multiplication by $\mathbf{M}'(\mathbf{MM}')^{-1}\mathbf{A}_{11}^{-1}$ then gives

$$\mathbf{MGL} = \mathbf{A}_{11}^{-1}. \tag{41}$$

Whatever the generalized inverse is, suppose it is partitioned as

$$\mathbf{G} = \begin{bmatrix} (\mathbf{G}_{11})_{r\times r} & (\mathbf{G}_{12})_{r\times(p-r)} \\ (\mathbf{G}_{21})_{(q-r)\times r} & (\mathbf{G}_{22})_{(q-r)\times(p-r)} \end{bmatrix} \tag{42}$$

of order $q \times p$, conformable for multiplication with **A**. Then substituting (42) and (39) into (41) gives

$$\mathbf{G}_{11} + \mathbf{A}_{11}^{-1}\mathbf{A}_{12}\mathbf{G}_{21} + \mathbf{G}_{12}\mathbf{A}_{21}\mathbf{A}_{11}^{-1} + \mathbf{A}_{11}^{-1}\mathbf{A}_{12}\mathbf{G}_{22}\mathbf{A}_{21}\mathbf{A}_{11}^{-1} = \mathbf{A}_{11}^{-1}. \tag{43}$$

This is true whatever the generalized inverse may be. Therefore, for any matrices $\mathbf{G}_{11}$, $\mathbf{G}_{12}$, $\mathbf{G}_{21}$ and $\mathbf{G}_{22}$ that satisfy (43), **G** as given in (42) will be a generalized inverse of **A**. Therefore, on substituting from (43) for $\mathbf{G}_{11}$, we have

$$\mathbf{G} = \begin{bmatrix} \mathbf{A}_{11}^{-1} - \mathbf{A}_{11}^{-1}\mathbf{A}_{12}\mathbf{G}_{21} - \mathbf{G}_{12}\mathbf{A}_{21}\mathbf{A}_{11}^{-1} - \mathbf{A}_{11}^{-1}\mathbf{A}_{12}\mathbf{G}_{22}\mathbf{A}_{21}\mathbf{A}_{11}^{-1} & \mathbf{G}_{12} \\ \mathbf{G}_{21} & \mathbf{G}_{22} \end{bmatrix} \tag{44}$$

as a generalized inverse of **A** for any matrices $\mathbf{G}_{12}$, $\mathbf{G}_{21}$ and $\mathbf{G}_{22}$ of appropriate order. Thus is the arbitrariness of a generalized inverse characterized.

Certain consequences of (44) can be noted. One is that by making $\mathbf{G}_{12}$, $\mathbf{G}_{21}$ and $\mathbf{G}_{22}$ null, $\mathbf{G} = \begin{bmatrix} \mathbf{A}_{11}^{-1} & \mathbf{0} \\ \mathbf{0} & \mathbf{0} \end{bmatrix}$, a form discussed earlier. Another is that when **A** is symmetric **G** is not necessarily symmetric. Only when $\mathbf{G}_{12} = \mathbf{G}_{21}'$ and $\mathbf{G}_{22}$ is symmetric will **G** be symmetric. And when $p \geq q$, **G** can have full row rank q even if $r < q$; for example, if $\mathbf{G}_{12} = -\mathbf{A}_{11}^{-1}\mathbf{A}_{12}\mathbf{G}_{22}$, $\mathbf{G}_{21} = \mathbf{0}$ and $\mathbf{G}_{22}$ has full row rank, then **G** also has full row rank; in general, then, the rank of **G** can exceed that of **A**. In particular, this means that singular matrices can have non-singular generalized inverses.

The arbitrariness evident in (44) prompts investigating the relationship of one generalized inverse to another. It is simple: if $\mathbf{G}_1$ is a generalized inverse of **A** then so is

$$\mathbf{G} = \mathbf{G}_1\mathbf{A}\mathbf{G}_1 + (\mathbf{I} - \mathbf{G}_1\mathbf{A})\mathbf{X} + \mathbf{Y}(\mathbf{I} - \mathbf{A}\mathbf{G}_1) \tag{45}$$

for any **X** and **Y**. Pre- and post-multiplication of (45) by **A** shows that this is so.

The importance of (45) is that it provides a method for generating all generalized inverses of $\mathbf{A}$. They can all be put in the form of (45). To see this we need only show for some other generalized inverse, $\mathbf{G}_2$ say, different from $\mathbf{G}_1$, that there exist values of $\mathbf{X}$ and $\mathbf{Y}$ giving $\mathbf{G} = \mathbf{G}_2$. Putting $\mathbf{X} = \mathbf{G}_2$ and $\mathbf{Y} = \mathbf{G}_1\mathbf{A}\mathbf{G}_2$ achieves this.

The form of $\mathbf{G}$ in (45) is entirely compatible with the partitioned form given in (44). For, if we take $\mathbf{G}_1 = \begin{bmatrix} \mathbf{A}_{11}^{-1} & \mathbf{0} \\ \mathbf{0} & \mathbf{0} \end{bmatrix}$ and partition $\mathbf{X}$ and $\mathbf{Y}$ in the same manner as $\mathbf{G}$, then (45) becomes

$$\mathbf{G} = \begin{bmatrix} \mathbf{A}_{11}^{-1} - \mathbf{A}_{11}^{-1}\mathbf{A}_{12}\mathbf{X}_{21} - \mathbf{Y}_{12}\mathbf{A}_{21}\mathbf{A}_{11}^{-1} & -\mathbf{A}_{11}^{-1}\mathbf{A}_{12}\mathbf{X}_{22} + \mathbf{Y}_{12} \\ \mathbf{X}_{21} - \mathbf{Y}_{22}\mathbf{A}_{21}\mathbf{A}_{11}^{-1} & \mathbf{X}_{22} + \mathbf{Y}_{22} \end{bmatrix}. \tag{46}$$

This characterizes the arbitrariness even more specifically than does (44). Thus for the four sub-matrices of $\mathbf{G}$ shown in (42) we have:

Sub-matrix	Source of Arbitrariness
$\mathbf{G}_{11}$	$\mathbf{X}_{21}$ and $\mathbf{Y}_{12}$
$\mathbf{G}_{12}$	$\mathbf{X}_{22}$ and $\mathbf{Y}_{12}$
$\mathbf{G}_{21}$	$\mathbf{X}_{21}$ and $\mathbf{Y}_{22}$
$\mathbf{G}_{22}$	$\mathbf{X}_{22}$ and $\mathbf{Y}_{22}$

This means that, in the partitioning of

$$\mathbf{X} = \begin{bmatrix} \mathbf{X}_{11} & \mathbf{X}_{12} \\ \mathbf{X}_{21} & \mathbf{X}_{22} \end{bmatrix} \quad \text{and} \quad \mathbf{Y} = \begin{bmatrix} \mathbf{Y}_{11} & \mathbf{Y}_{12} \\ \mathbf{Y}_{21} & \mathbf{Y}_{22} \end{bmatrix}$$

implicit in (45), the first set of rows in the partitioning of $\mathbf{X}$ does not enter into $\mathbf{G}$, and neither does the first set of columns of $\mathbf{Y}$.

It has been shown earlier (Theorem 3) that all solutions to $\mathbf{Ax} = \mathbf{y}$ can be generated from $\tilde{\mathbf{x}} = \mathbf{Gy} + (\mathbf{GA} - \mathbf{I})\mathbf{z}$, where $\mathbf{z}$ is the infinite set of arbitrary vectors of order q. We now show that all solutions can also be generated from $\tilde{\mathbf{x}} = \mathbf{Gy}$, where $\mathbf{G}$ is the infinite set of generalized inverses indicated in (45). First, a lemma.

Lemma 10. If $\mathbf{z}_{q\times 1}$ is arbitrary and $\mathbf{y}_{p\times 1}$ is known and non-null, there exists an arbitrary matrix $\mathbf{X}$ such that $\mathbf{z} = \mathbf{Xy}$.

Proof. Since $\mathbf{y} \neq \mathbf{0}$ one element, y_k say, will be non-zero. Writing $\mathbf{z} = \{z_i\}$ and $\mathbf{X} = \{x_{ij}\}$ for $i = 1, \ldots, q$ and $j = 1, \ldots, p$, let $x_{ij} = z_i/y_k$ for $j = k$ and $x_{ij} = 0$ otherwise. Then $\mathbf{Xy} = \mathbf{z}$ and $\mathbf{X}$ is arbitrary.

Now we have the theorem on generating solutions.

Theorem 8. For all possible generalized inverses $\mathbf{G}$ of $\mathbf{A}$, $\tilde{\mathbf{x}} = \mathbf{Gy}$ generates all solutions to the consistent equations $\mathbf{Ax} = \mathbf{y}$.

Proof. For the generalized inverse $\mathbf{G}_1$, solutions to $\mathbf{Ax} = \mathbf{y}$ are $\tilde{\mathbf{x}} = \mathbf{G}_1\mathbf{y} + (\mathbf{G}_1\mathbf{A} - \mathbf{I})\mathbf{z}$, where $\mathbf{z}$ is arbitrary. Let $\mathbf{z} = -\mathbf{Xy}$ for some arbitrary $\mathbf{X}$. Then

$$\begin{aligned}\tilde{\mathbf{x}} &= \mathbf{G}_1\mathbf{y} - (\mathbf{G}_1\mathbf{A} - \mathbf{I})\mathbf{Xy} \\ &= \mathbf{G}_1\mathbf{y} - \mathbf{G}_1\mathbf{AG}_1\mathbf{y} + \mathbf{G}_1\mathbf{AG}_1\mathbf{y} + (\mathbf{I} - \mathbf{G}_1\mathbf{A})\mathbf{Xy} \\ &= [\mathbf{G}_1\mathbf{AG}_1 + (\mathbf{I} - \mathbf{G}_1\mathbf{A})\mathbf{X} + \mathbf{G}_1(\mathbf{I} - \mathbf{AG}_1)]\mathbf{y} \\ &= \mathbf{Gy}\end{aligned}$$

where $\mathbf{G}$ is exactly the form given in (45) using $\mathbf{G}_1$ for $\mathbf{Y}$.

7. OTHER RESULTS

Procedures for inverting partitioned matrices are well known [e.g., Searle (1966), Sec. 8.7]. In particular, the inverse of the partitioned full rank symmetric matrix

$$\mathbf{M} = \begin{bmatrix}\mathbf{X}' \\ \mathbf{Z}'\end{bmatrix}[\mathbf{X} \quad \mathbf{Z}] = \begin{bmatrix}\mathbf{X}'\mathbf{X} & \mathbf{X}'\mathbf{Z} \\ \mathbf{Z}'\mathbf{X} & \mathbf{Z}'\mathbf{Z}\end{bmatrix} \equiv \begin{bmatrix}\mathbf{A} & \mathbf{B} \\ \mathbf{B}' & \mathbf{D}\end{bmatrix}, \text{ say,} \tag{47}$$

can, for

$$\mathbf{W} = (\mathbf{D} - \mathbf{B}'\mathbf{A}^{-1}\mathbf{B})^{-1} = [\mathbf{Z}'\mathbf{Z} - \mathbf{Z}'\mathbf{X}(\mathbf{X}'\mathbf{X})^{-1}\mathbf{X}'\mathbf{Z}]^{-1},$$

be written as

$$\begin{aligned}\mathbf{M}^{-1} &= \begin{bmatrix}\mathbf{A}^{-1} + \mathbf{A}^{-1}\mathbf{BWB}'\mathbf{A}^{-1} & -\mathbf{A}^{-1}\mathbf{BW} \\ -\mathbf{WB}'\mathbf{A}^{-1} & \mathbf{W}\end{bmatrix} \\ &= \begin{bmatrix}\mathbf{A}^{-1} & \mathbf{0} \\ \mathbf{0} & \mathbf{0}\end{bmatrix} + \begin{bmatrix}-\mathbf{A}^{-1}\mathbf{B} \\ \mathbf{I}\end{bmatrix}\mathbf{W}[-\mathbf{B}'\mathbf{A}^{-1} \quad \mathbf{I}].\end{aligned} \tag{48}$$

The analogy of (48) for generalized inverses, when $\mathbf{M}$ is symmetric but singular, has been derived by Rohde (1965). On defining $\mathbf{A}^-$ and $\mathbf{Q}^-$ as generalized inverses of $\mathbf{A}$ and $\mathbf{Q}$ respectively, where $\mathbf{Q} = \mathbf{D} - \mathbf{B}'\mathbf{A}^-\mathbf{B}$, then a generalized inverse of $\mathbf{M}$ is

$$\begin{aligned}\mathbf{M}^- &= \begin{bmatrix}\mathbf{A}^- + \mathbf{A}^-\mathbf{BQ}^-\mathbf{B}'\mathbf{A}^- & -\mathbf{A}^-\mathbf{BQ}^- \\ -\mathbf{Q}^-\mathbf{B}'\mathbf{A}^- & \mathbf{Q}^-\end{bmatrix} \\ &= \begin{bmatrix}\mathbf{A}^- & \mathbf{0} \\ \mathbf{0} & \mathbf{0}\end{bmatrix} + \begin{bmatrix}-\mathbf{A}^-\mathbf{B} \\ \mathbf{I}\end{bmatrix}\mathbf{Q}^-[-\mathbf{B}'\mathbf{A}^- \quad \mathbf{I}].\end{aligned} \tag{49}$$

It is to be emphasized that the generalized inverses referred to here are just as have been defined throughout, namely satisfying only the first of Penrose's four conditions. (In showing that $\mathbf{MM}^-\mathbf{M} = \mathbf{M}$, considerable use is made of Theorem 7.)

The regular inverse of the product **AB** is $\mathbf{B}^{-1}\mathbf{A}^{-1}$ when **A** and **B** are non-singular. But there is no analogous result for generalized inverses. When one matrix is non-singular, **B** say, we have $\mathbf{B}^{-1}\mathbf{A}^{-}$ as a generalized inverse of **AB**, as indicated by Rohde (1964). Greville (1966) considers the situation for unique generalized inverses $\mathbf{A}^{(p)}$ and $\mathbf{B}^{(p)}$ and gives five separate conditions under which $(\mathbf{AB})^{(p)} = \mathbf{B}^{(p)}\mathbf{A}^{(p)}$; but one would hope for conditions less complex than Greville's in the case of generalized inverses $\mathbf{A}^{-}$ and $\mathbf{B}^{-}$ satisfying just the first of the Penrose conditions. What can be shown is that $\mathbf{B}^{-}\mathbf{A}^{-}$ is a generalized inverse of **AB** if and only if $\mathbf{A}^{-}\mathbf{ABB}^{-}$ is idempotent. Also, if the product **AB** is itself idempotent then it has **AB**, $\mathbf{AA}^{-}$, $\mathbf{B}^{-}\mathbf{B}$ and $\mathbf{B}^{-}\mathbf{BAA}^{-}$ as generalized inverses. Other problems of possible interest are the generalized inverse of $\mathbf{A}^{k}$ in terms of that of **A**, for integer k, and the generalized inverse of $\mathbf{XX}'$ in terms of that of $\mathbf{X}'\mathbf{X}$.

8. EXERCISES

1. Reduce the matrices

$$\mathbf{A} = \begin{bmatrix} 2 & 3 & 1 & -1 \\ 5 & 8 & 0 & 1 \\ 1 & 2 & -2 & 3 \end{bmatrix} \quad \text{and} \quad \mathbf{B} = \begin{bmatrix} 1 & 2 & 3 & -1 \\ 4 & 5 & 6 & 2 \\ 7 & 8 & 10 & 7 \\ 2 & 1 & 1 & 6 \end{bmatrix}$$

to diagonal form and find a generalized inverse of each.

2. Find generalized inverses of **A** and **B** in Exercise 1 by inverting non-singular minors.

3. Find a generalized inverse of each of the following matrices:
   (*a*) **PAQ** when **P** and **Q** are nonsingular.
   (*b*) **GA** when **G** is a generalized inverse of **A**.
   (*c*) $k\mathbf{A}$, when k is a scalar.
   (*d*) **ABA**, when **ABA** is idempotent.
   (*e*) **J**, when **J** is square, with every element unity.

4. What kinds of matrices
   (*a*) are their own generalized inverses?
   (*b*) have their transposes as a generalized inverse?
   (*c*) have an identity matrix for a generalized inverse?
   (*d*) have every matrix of order $p \times q$ for a generalized inverse?
   (*e*) have only non-singular generalized inverses?

5. Section 1b contains an algorithm for deriving a generalized inverse of any matrix. Prove that in general neither, nor both, of the matrix transpositions in

steps (iii) and (v) of the algorithm can be omitted, but that when $\mathbf{M} = \mathbf{M}'$ the transposition in (iii) can be. Illustrate these results using $\mathbf{A}_1$ and $\mathbf{A}_2$ given in Sec. 1b. Also, find a generalized inverse of $\mathbf{A}_2$ using a non-symmetric non-principal minor for $\mathbf{M}$.

6. Explain why equations (*a*) $\mathbf{Ax} = \mathbf{0}$ and (*b*) $\mathbf{X}'\mathbf{Xb} = \mathbf{X}'\mathbf{y}$ are always consistent.

7. For $\mathbf{A}$ and $\mathbf{B}$ of Exercise 1 find general solutions to

$$\mathbf{Ax} = \begin{bmatrix} -1 \\ -13 \\ -11 \end{bmatrix} \quad \text{and to} \quad \mathbf{Bx} = \begin{bmatrix} 14 \\ 23 \\ 32 \\ -5 \end{bmatrix}.$$

8. If $\mathbf{z} = (\mathbf{G} - \mathbf{F})\mathbf{y} + (\mathbf{I} - \mathbf{FA})\mathbf{w}$, where $\mathbf{G}$ and $\mathbf{F}$ are generalized inverses of $\mathbf{A}$, show that the solution $\tilde{\mathbf{x}} = \mathbf{Gy} + (\mathbf{GA} - \mathbf{I})\mathbf{z}$ to $\mathbf{Ax} = \mathbf{y}$ reduces to $\tilde{\mathbf{x}} = \mathbf{Fy} + (\mathbf{FA} - \mathbf{I})\mathbf{w}$.

9. If $\mathbf{Ax} = \mathbf{y}$ are consistent equations and $\mathbf{F}$ and $\mathbf{G}$ are both generalized inverses of $\mathbf{A}$ find, in its simplest form, a solution for $\mathbf{w}$ to the equations

$$(\mathbf{I} - \mathbf{GA})\mathbf{w} = (\mathbf{F} - \mathbf{G})\mathbf{y} + (\mathbf{FA} - \mathbf{I})\mathbf{z}.$$

10. If $\mathbf{A}$ has full column rank, show that its generalized inverses are also left inverses satisfying the first three Penrose conditions.

11. Prove that equation (25) reduces to equation (23).

12. Find the Penrose inverse of $\begin{bmatrix} 1 & 0 & 2 \\ 2 & -1 & 5 \\ 0 & 1 & -1 \\ 1 & 3 & -1 \end{bmatrix}$.

13. Suppose $\mathbf{X}$ has order $n \times p$ and rank $p - m$ $(m > 0)$. If $\mathbf{H}$, of order $m \times p$ and full row rank, has its rows LIN of those of $\mathbf{X}$, show for

$$\mathbf{D} = (\mathbf{X}'\mathbf{X} + \mathbf{H}'\mathbf{H})^{-1}\mathbf{H}'$$

that $\mathbf{XD} = \mathbf{0}$ and $\mathbf{HD} = \mathbf{I}$. [*Hint:* Partition $\mathbf{X}' = [\mathbf{X}_1' \;\; \mathbf{X}_2']$ so that $\mathbf{X}_1$ has full row rank $p - m$, and then use the inverse of $[\mathbf{X}_1' \;\; \mathbf{H}']$. See Chipman (1964). Equation (36) is a consequence.]

14. By direct multiplication show that $\mathbf{AGA} = \mathbf{A}$ for $\mathbf{A}$ and $\mathbf{G}$ given in (37) and (44) respectively.

15. Develop a non-singular generalized inverse of a singular matrix, proving that it is both non-singular and a generalized inverse.

16. Show that the rank of a generalized inverse of $\mathbf{A}$ does not necessarily have the same rank as $\mathbf{A}$ and that it is the same if and only if it is a reflexive inverse [Rohde (1966)].

17. Show that $\mathbf{M}^-$ of equation (49) is a generalized inverse of $\mathbf{M}$ in (47).

18. Prove that $\mathbf{B}^-\mathbf{A}^-$ is a generalized inverse of $\mathbf{AB}$ if and only if $\mathbf{A}^-\mathbf{ABB}^-$ is idempotent.

19. Why is $\mathbf{X}^-\mathbf{y}$ a solution for $\mathbf{b}$ to $\mathbf{X}'\mathbf{Xb} = \mathbf{X}'\mathbf{y}$?

20. If $\mathbf{P}_{m\times q}$ and $\mathbf{D}_{m\times m}$ have rank m, show that $\mathbf{D}^{-1} = \mathbf{P}(\mathbf{P}'\mathbf{DP})^-\mathbf{P}'$.

21. If $\mathbf{G}$ is a generalized inverse of $\mathbf{A}_{p\times q}$ show that $\mathbf{G} + \mathbf{Z} - \mathbf{GAZAG}$ generates
 (i) all generalized inverses of $\mathbf{A}$, and
 (ii) all solutions to consistent equations $\mathbf{Ax} = \mathbf{y}$ as $\mathbf{Z}$ ranges over all matrices of order $q \times p$ [Urquhart (1969)].

22. When $\mathbf{PAQ} = \begin{bmatrix} \mathbf{D} & \mathbf{0} \\ \mathbf{0} & \mathbf{0} \end{bmatrix}$ show that $\mathbf{G} = \mathbf{Q}\begin{bmatrix} \mathbf{D}^{-1} & \mathbf{X} \\ \mathbf{Y} & \mathbf{Z} \end{bmatrix}\mathbf{P}$ is a generalized inverse of $\mathbf{A}$. Under what conditions does $\mathbf{GAG} = \mathbf{G}$? Use $\mathbf{G}$ to answer Exercise 15.

23. Using $\mathbf{AGA} = \mathbf{A}$ find a generalized inverse of $\mathbf{AB}$ when $\mathbf{B}$ is orthogonal and of $\mathbf{LA}$ when $\mathbf{L}$ is non-singular.

24. What is the Penrose inverse of a symmetric idempotent matrix?

25. Use the idempotency of $\mathbf{H} = \mathbf{GA}$ to prove Corollary 2 of Theorem 6.

CHAPTER 2

DISTRIBUTIONS AND QUADRATIC FORMS

1. INTRODUCTION

Analysis of variance techniques involve partitioning a total sum of squares into component sums of squares whose ratios (under appropriate distributional conditions) lead to F-statistics suitable for testing certain hypotheses. When discussing linear models generally, especially where unbalanced data (data having unequal subclass numbers) are concerned, it is convenient to think of sums of squares involved in this process as quadratic forms in the observations. In this context very general theorems can be established, of which familiar analyses of variance and associated F-tests are then just special cases. An introductory outline[1] of the general procedure is easily described.

Suppose $\mathbf{y}_{n\times 1}$ is a vector of n observations. Then $\mathbf{y}'\mathbf{y} = \sum_{i=1}^{n} y_i^2$ is the total sum of squares of the observations which gets partitioned into component sums of squares in an analysis of variance. Let $\mathbf{P}$ be an orthogonal matrix

$$\mathbf{PP}' = \mathbf{P}'\mathbf{P} = \mathbf{I}, \tag{1}$$

and partition $\mathbf{P}$ row-wise into k sub-matrices $\mathbf{P}_i$, of order $n_i \times n$, for $i = 1, 2, \ldots, k$, with $\sum_{i=1}^{k} n_i = n$; i.e.,

$$\mathbf{P} = \begin{bmatrix} \mathbf{P}_1 \\ \mathbf{P}_2 \\ \cdot \\ \cdot \\ \cdot \\ \mathbf{P}_k \end{bmatrix} \quad \text{and} \quad \mathbf{P}' = [\mathbf{P}_1' \;\; \mathbf{P}_2' \;\; \cdots \;\; \mathbf{P}_k']. \tag{2}$$

[1] Kindly brought to my notice by D. L. Weeks.

Then

$$\mathbf{y}'\mathbf{y} = \mathbf{y}'\mathbf{I}\mathbf{y} = \mathbf{y}'\mathbf{P}'\mathbf{P}\mathbf{y} = \sum_{i=1}^{k} \mathbf{y}'\mathbf{P}_i'\mathbf{P}_i\mathbf{y}. \tag{3}$$

In this way $\mathbf{y}'\mathbf{y}$ is partitioned into k sums of squares

$$\mathbf{y}'\mathbf{P}_i'\mathbf{P}_i\mathbf{y} = \mathbf{z}_i'\mathbf{z}_i = \sum_{j=1}^{n_i} z_{ij}^2 \qquad \text{for} \quad i = 1, \ldots, k$$

where

$$\mathbf{z}_i = \mathbf{P}_i\mathbf{y} = \{z_{ij}\} \qquad \text{for} \quad j = 1, 2, \ldots, n_i.$$

Each of these sums of squares corresponds to the lines in an analysis of variance (with, as we shall see, degrees of freedom equal to the rank of $\mathbf{P}_i$), having $\mathbf{y}'\mathbf{y}$ as the total sum of squares. The general nature of results to be developed in this chapter can be demonstrated for the k terms $\mathbf{y}'\mathbf{P}_i'\mathbf{P}_i\mathbf{y}$ of (3). First, for example, in Corollary 2.1 of Theorem 2 we show that if the elements of the $\mathbf{y}$ vector are normally and independently distributed with zero mean and variance σ^2, then $\mathbf{y}'\mathbf{A}\mathbf{y}/\sigma^2$, where $\mathbf{A}$ has rank r, has a χ^2-distribution with r degrees of freedom if and only if $\mathbf{A}$ is idempotent. This is just the property that the matrix $\mathbf{P}_i'\mathbf{P}_i$ has in (3): $\mathbf{P}_i'\mathbf{P}_i\mathbf{P}_i'\mathbf{P}_i = \mathbf{P}_i'(\mathbf{P}_i\mathbf{P}_i')\mathbf{P}_i = \mathbf{P}_i'\mathbf{I}\mathbf{P}_i = \mathbf{P}_i'\mathbf{P}_i$ because $\mathbf{P}'\mathbf{P} = \mathbf{I}$ as in (1). Thus each $\mathbf{P}_i'\mathbf{P}_i$ in (3) is idempotent and therefore each term $\mathbf{y}'\mathbf{P}_i'\mathbf{P}_i\mathbf{y}/\sigma^2$ in (3) has a χ^2-distribution. Second, in Theorem 4 we prove that when the elements of $\mathbf{y}$ are normally distributed as just described, then $\mathbf{y}'\mathbf{A}\mathbf{y}$ and $\mathbf{y}'\mathbf{B}\mathbf{y}$ are independent if and only if $\mathbf{AB} = \mathbf{0}$. This too is true for the terms in (3) for, with $i \neq j$, $\mathbf{P}_i\mathbf{P}_j' = \mathbf{0}$ from (1) and (2) and so

$$\mathbf{P}_i'\mathbf{P}_i\mathbf{P}_j'\mathbf{P}_j = \mathbf{0}.$$

Hence the terms in (3) are independent; and since they all have χ^2-distributions their ratios, suitably modified by degrees of freedom, can be F-distributions. In this way tests of hypotheses are established.

Example. Corresponding to a vector of 4 observations consider

$$\mathbf{P} = \left[\begin{array}{cccc} 1/\sqrt{4} & 1/\sqrt{4} & 1/\sqrt{4} & 1/\sqrt{4} \\ \hdashline 1/\sqrt{2} & -1/\sqrt{2} & 0 & 0 \\ 1/\sqrt{6} & 1/\sqrt{6} & -2/\sqrt{6} & 0 \\ 1/\sqrt{12} & 1/\sqrt{12} & 1/\sqrt{12} & -3/\sqrt{12} \end{array}\right] = \left[\begin{array}{c} \mathbf{P}_1 \\ \hdashline \mathbf{P}_2 \end{array}\right] \tag{4}$$

partitioned as shown. Then it is clear that $\mathbf{P}$ is orthogonal and that

$$\mathbf{P}_1\mathbf{y} = (1/\sqrt{4})\sum_{i=1}^{4} y_i = \sqrt{4}\,\bar{y}.$$

Hence in terms of (3)

$$q_1 \equiv \mathbf{y}'\mathbf{P}_1'\mathbf{P}_1\mathbf{y} = 4\bar{y}^2,$$

and it will be found that

$$q_2 \equiv \mathbf{y}'\mathbf{P}_2'\mathbf{P}_2\mathbf{y} = \sum_{i=1}^{4} y_i^2 - 4\bar{y}^2 = \sum_{i=1}^{4} (y_i - \bar{y})^2.$$

Therefore, when the elements of $\mathbf{y}$ are normally and independently distributed with zero mean and unit variance, q_1 and q_2 each have χ^2-distributions, as is well known. Furthermore, from the orthogonality of $\mathbf{P}$ it is obvious that $\mathbf{P}_1'\mathbf{P}_2 = \mathbf{0}$ and so q_1 and q_2 are also distributed independently. In this way

$$F = \frac{4\bar{y}^2/1}{\left(\sum_{i=1}^{4} y_i^2 - 4\bar{y}^2\right)\Big/ 3}$$

provides an F-test for the hypothesis that the mean of the y-variable is zero.

The matrix $\mathbf{P}$ in (4) is a fourth-order Helmert matrix. Its general characteristics are as follows. Writing

$$\mathbf{H}_{n\times n} = \begin{bmatrix} \mathbf{h}' \\ \mathbf{H}_0 \end{bmatrix} \begin{matrix} 1 \times n \\ (n-1) \times n \end{matrix}$$

for a Helmert matrix of order n,

$$\mathbf{h}' = \text{first row of } \mathbf{H}_{n\times n} = (1/\sqrt{n})\mathbf{1}_n'$$

where

$$\mathbf{1}_n' = [1 \quad 1 \quad \cdots \quad 1],$$

a vector of n 1's, and

$$\mathbf{H}_0 = \text{last } (n-1) \text{ rows of } \mathbf{H}_{n\times n}$$

with $\mathbf{H}_0$ having its rth row as

$$\left[\frac{1}{\sqrt{r(r+1)}}\mathbf{1}_r' \qquad \frac{-r}{\sqrt{r(r+1)}} \qquad \mathbf{0}_{(n-r-1)\times 1}\right] \qquad \text{for } r = 1, 2, \ldots, n-1.$$

It is clear that $\mathbf{H}_{n\times n}$ is orthogonal, that $\mathbf{y}'\mathbf{h}\mathbf{h}'\mathbf{y} = n\bar{y}^2$ and, by induction, it is readily shown that $\mathbf{y}'\mathbf{H}_0'\mathbf{H}_0\mathbf{y} = \sum_{i=1}^{n} y_i^2 - n\bar{y}^2$. Further properties of Helmert matrices are to be found in Lancaster (1965).

2. SYMMETRIC MATRICES

An expression of the form $\mathbf{x}'\mathbf{A}\mathbf{y}$ is called a *bilinear form*. It is a homogeneous second-degree function of the first degree in each of the x's and y's. For example,

$$\mathbf{x}'\mathbf{A}\mathbf{y} = [x_1 \quad x_2]\begin{bmatrix} 4 & 8 \\ -2 & 7 \end{bmatrix}\begin{bmatrix} y_1 \\ y_2 \end{bmatrix}$$

$$= 4x_1y_1 + 8x_1y_2 - 2x_2y_1 + 7x_2y_2.$$

When $\mathbf{x}$ is used in place of $\mathbf{y}$ the expression becomes $\mathbf{x}'\mathbf{A}\mathbf{x}$; it is then called a *quadratic form* and is a quadratic function of the x's:

$$\begin{aligned}\mathbf{x}'\mathbf{A}\mathbf{x} &= [x_1 \quad x_2]\begin{bmatrix} 4 & 8 \\ -2 & 7 \end{bmatrix}\begin{bmatrix} x_1 \\ x_2 \end{bmatrix} \\ &= 4x_1^2 + (8 - 2)x_1x_2 + 7x_2^2 \\ &= 4x_1^2 + (3 + 3)x_1x_2 + 7x_2^2 \\ &= [x_1 \quad x_2]\begin{bmatrix} 4 & 3 \\ 3 & 7 \end{bmatrix}\begin{bmatrix} x_1 \\ x_2 \end{bmatrix}.\end{aligned}$$

In this way any quadratic form $\mathbf{x}'\mathbf{A}\mathbf{x}$ can be written as $\mathbf{x}'\mathbf{A}\mathbf{x} = \mathbf{x}'\mathbf{B}\mathbf{x}$ where $\mathbf{B} = \frac{1}{2}(\mathbf{A} + \mathbf{A}')$ is symmetric. Furthermore, whereas any quadratic form can be written as $\mathbf{x}'\mathbf{A}\mathbf{x}$ for an infinite number of matrices, each can be written in only one way as $\mathbf{x}'\mathbf{B}\mathbf{x}$ for $\mathbf{B}$ symmetric. For example,

$$4x_1^2 + 6x_1x_2 + 7x_2^2 = [x_1 \quad x_2]\begin{bmatrix} 4 & 3 + a \\ 3 - a & 7 \end{bmatrix}\begin{bmatrix} x_1 \\ x_2 \end{bmatrix}$$

for any value of a, but only when $a = 0$ is the matrix involved symmetric. This means that for any particular quadratic form there is only one, unique matrix such that the quadratic form can be written as $\mathbf{x}'\mathbf{A}\mathbf{x}$ with $\mathbf{A}$ being symmetric. Because of the uniqueness of this symmetric matrix all further discussion of quadratic forms $\mathbf{x}'\mathbf{A}\mathbf{x}$ is confined to the case of $\mathbf{A}$ being symmetric.

3. POSITIVE DEFINITENESS

A property of some quadratic forms used repeatedly in what follows is that of positive definiteness. A quadratic form $\mathbf{x}'\mathbf{A}\mathbf{x}$ is said to be *positive definite* if it is positive for all values of $\mathbf{x}$ except $\mathbf{x} = \mathbf{0}$; i.e., if

$$\mathbf{x}'\mathbf{A}\mathbf{x} > 0 \qquad \text{for all } \mathbf{x}, \text{ except } \mathbf{x} = \mathbf{0},$$

then $\mathbf{x}'\mathbf{A}\mathbf{x}$ is positive definite. And the corresponding (symmetric) matrix is also described as positive definite.

Example.

$$\begin{aligned}\mathbf{x}'\mathbf{A}\mathbf{x} &= [x_1 \quad x_2 \quad x_3]\begin{bmatrix} 3 & 5 & 1 \\ 5 & 13 & 0 \\ 1 & 0 & 1 \end{bmatrix}\begin{bmatrix} x_1 \\ x_2 \\ x_3 \end{bmatrix} \\ &= 3x_1^2 + 13x_2^2 + x_3^2 + 10x_1x_2 + 2x_1x_3\end{aligned}$$

can be rearranged as

$$\mathbf{x}'\mathbf{A}\mathbf{x} = (x_1 + 2x_2)^2 + (x_1 + 3x_2)^2 + (x_1 + x_3)^2$$

which is positive for any (real) values of the x's except $x_1 = 0 = x_2 = x_3$, i.e., except for $\mathbf{x} = \mathbf{0}$ (in which case $\mathbf{x}'\mathbf{A}\mathbf{x}$ is always zero). Hence $\mathbf{x}'\mathbf{A}\mathbf{x}$ is positive definite (abbreviated p.d.).

A slight relaxation of the above definition concerns $\mathbf{x}'\mathbf{A}\mathbf{x}$ when its value is either positive *or* zero for all $\mathbf{x} \neq \mathbf{0}$. We define an $\mathbf{x}'\mathbf{A}\mathbf{x}$ of this nature as being *positive semi-definite* (abbreviated p.s.d.) when

$$\mathbf{x}'\mathbf{A}\mathbf{x} \geq 0 \qquad \text{for all } \mathbf{x} \neq \mathbf{0}, \text{ with } \mathbf{x}'\mathbf{A}\mathbf{x} = 0 \text{ for at least one } \mathbf{x} \neq \mathbf{0}.$$

Under these conditions $\mathbf{x}'\mathbf{A}\mathbf{x}$ is a p.s.d. quadratic form and the corresponding symmetric matrix $\mathbf{A}$ is a p.s.d. matrix. This definition is widely accepted [e.g., Graybill (1961, p. 3) and Rao (1965, p. 31)], although not universally so. For example, a definition used by Scheffé (1959, p. 398) is that $\mathbf{A}$ is a p.s.d. matrix when $\mathbf{x}'\mathbf{A}\mathbf{x} \geq 0$ for all $\mathbf{x} \neq \mathbf{0}$ without demanding that $\mathbf{x}'\mathbf{A}\mathbf{x} = 0$ for at least one non-null $\mathbf{x}$. Hence this definition includes matrices that we have defined as either p.d. or p.s.d. We will call such matrices *non-negative definite* (n.n.d.) in keeping, for example, with Rao (1965, p. 31). Thus n.n.d. matrices are either p.d. or p.s.d.

Example.

$$\mathbf{x}'\mathbf{A}\mathbf{x} = [x_1 \quad x_2 \quad x_3]\begin{bmatrix} 37 & -2 & -24 \\ -2 & 13 & -3 \\ -24 & -3 & 17 \end{bmatrix}\begin{bmatrix} x_1 \\ x_2 \\ x_3 \end{bmatrix}$$

$$= (6x_1 - 4x_3)^2 + (x_1 - 2x_2)^2 + (3x_2 - x_3)^2$$

is zero when $\mathbf{x}' = [2, 1, 3]$. Hence $\mathbf{x}'\mathbf{A}\mathbf{x}$ and $\mathbf{A}$ are positive semi-definite. On the other hand,

$$\mathbf{y}'\mathbf{y} = \mathbf{y}'\mathbf{I}\mathbf{y} = \sum_{i=1}^{n} y_i^2$$

is positive definite because it is zero only when $\mathbf{y} = \mathbf{0}$. But

$$\mathbf{y}'\mathbf{y} - n\bar{y}^2 = \mathbf{y}'(\mathbf{I} - n^{-1}\mathbf{J}_n)\mathbf{y} = \sum_{i=1}^{n} y_i^2 - n\bar{y}^2,$$

where $\mathbf{J}_n$ is square a matrix of order n with every element equal to one, is a positive semi-definite quadratic form because it is zero not only if $\mathbf{y} = \mathbf{0}$ but also if every element of $\mathbf{y}$ is the same, i.e., if $\mathbf{y} = \alpha\mathbf{1}$ for any α.

Lemmas concerning positive (semi-)definite [abbreviated p.(s.)d.] matrices that we will subsequently utilize are as follows.

Lemma 1. The symmetric matrix $\mathbf{A}$ is positive definite if and only if all its principal leading minors have positive determinants.

Proof. Proofs of this lemma are available in many texts. Most are lengthy and contribute little to the mainstream of our work here, and so are omitted. An elegant inductive proof by Seelye (1958) is commended to the interested reader.

Corollary. Positive definite matrices are non-singular. (They are one of their own principal leading minors and therefore have non-zero, indeed positive, determinants.) Note that the converse of this corollary is not true: non-singular matrices are not, in general, positive definite.

Lemma 2. For $\mathbf{P}$ non-singular, $\mathbf{P}'\mathbf{A}\mathbf{P}$ is or is not positive (semi-)definite according as $\mathbf{A}$ is or is not p.(s.)d.

Proof. Let $\mathbf{y} = \mathbf{P}^{-1}\mathbf{x}$ and consider $\mathbf{x}'\mathbf{A}\mathbf{x} = \mathbf{y}'\mathbf{P}'\mathbf{A}\mathbf{P}\mathbf{y}$. When $\mathbf{x} = \mathbf{0}$, $\mathbf{y} = \mathbf{0}$ and $\mathbf{x}'\mathbf{A}\mathbf{x} = \mathbf{y}'\mathbf{P}'\mathbf{A}\mathbf{P}\mathbf{y} = 0$. And for $\mathbf{x} \neq \mathbf{0}$, $\mathbf{y} \neq \mathbf{0}$ and $\mathbf{y}'\mathbf{P}'\mathbf{A}\mathbf{P}\mathbf{y} \geq 0$ according as $\mathbf{x}'\mathbf{A}\mathbf{x} \geq 0$. Hence $\mathbf{P}'\mathbf{A}\mathbf{P}$ is p.(s.)d. according as $\mathbf{A}$ is p.(s.)d.

Lemma 3. Latent roots of a positive (semi-)definite matrix are all positive (non-negative, i.e., zero or positive).

Proof. Suppose λ and $\mathbf{u} \neq \mathbf{0}$ are a latent root and vector respectively of $\mathbf{A}$ with $\mathbf{A}\mathbf{u} = \lambda\mathbf{u}$. Then consider $\mathbf{u}'\mathbf{A}\mathbf{u} = \mathbf{u}'\lambda\mathbf{u} = \lambda\mathbf{u}'\mathbf{u}$ for $\mathbf{u} \neq \mathbf{0}$. When $\mathbf{A}$ is p.d., $\mathbf{u}'\mathbf{A}\mathbf{u} > 0$ and so $\lambda\mathbf{u}'\mathbf{u} > 0$, i.e., $\lambda > 0$; hence all latent roots of a p.d. matrix are positive. When $\mathbf{A}$ is p.s.d., $\mathbf{u}'\mathbf{A}\mathbf{u} \geq 0$ with $\mathbf{u}'\mathbf{A}\mathbf{u} = 0$ for at least one $\mathbf{u} \neq \mathbf{0}$, i.e., $\lambda = 0$ for at least one $\mathbf{u} \neq \mathbf{0}$; hence all latent roots of a p.s.d. matrix are zero or positive. This proves the lemma.

Corollary. Positive semi-definite matrices are singular. (They have one or more zero latent roots and therefore a zero determinant.) The converse is not true: singular matrices are not, in general, positive semi-definite.

Lemma 4. A symmetric matrix is positive definite if and only if it can be written as $\mathbf{P}'\mathbf{P}$ for a non-singular $\mathbf{P}$.

Proof. If $\mathbf{A} = \mathbf{P}'\mathbf{P}$ for $\mathbf{P}$ non-singular, then $\mathbf{A}$ is symmetric and $\mathbf{x}'\mathbf{A}\mathbf{x} = \mathbf{x}'\mathbf{P}'\mathbf{P}\mathbf{x}$, which is the sum of squares of the elements of $\mathbf{P}\mathbf{x}$. Hence $\mathbf{x}'\mathbf{A}\mathbf{x} > 0$ for all $\mathbf{P}\mathbf{x} \neq \mathbf{0}$ and $\mathbf{x}'\mathbf{A}\mathbf{x} = 0$ for all $\mathbf{P}\mathbf{x} = \mathbf{0}$. But $\mathbf{P}\mathbf{x} = \mathbf{0}$ only when $\mathbf{x} = \mathbf{0}$, because $\mathbf{P}^{-1}$ exists. Hence $\mathbf{x}'\mathbf{A}\mathbf{x} > 0$ for all $\mathbf{x} \neq \mathbf{0}$ and $\mathbf{x}'\mathbf{A}\mathbf{x} = 0$ only for $\mathbf{x} = \mathbf{0}$. Therefore $\mathbf{A}$ is p.d.

The necessary condition is established by noting that for $\mathbf{A}$ being symmetric there exists a matrix $\mathbf{Q}$ such that $\mathbf{Q}\mathbf{A}\mathbf{Q}'$ is a diagonal matrix with only 0's and 1's in its diagonal. But if $\mathbf{A}$ is positive definite it has full rank. Therefore $\mathbf{Q}\mathbf{A}\mathbf{Q}' = \mathbf{I}$ and so, because $\mathbf{Q}$ is non-singular, $\mathbf{A} = \mathbf{Q}^{-1}\mathbf{Q}^{-1\prime}$ which is of the form $\mathbf{P}'\mathbf{P}$.

Lemma 5. $\mathbf{A}'\mathbf{A}$ is positive definite when $\mathbf{A}$ has full column rank and it is positive semi-definite otherwise.

Proof. Consider $\mathbf{x}'\mathbf{A}'\mathbf{A}\mathbf{x}$, equal to the sum of squares of elements of $\mathbf{Ax}$. When $\mathbf{A}$ has full column rank $\mathbf{Ax} = \mathbf{0}$ only when $\mathbf{x} = \mathbf{0}$, and so $\mathbf{x}'\mathbf{A}'\mathbf{A}\mathbf{x} > 0$ for all $\mathbf{x} \neq \mathbf{0}$; i.e., $\mathbf{A}'\mathbf{A}$ is p.d. And when $\mathbf{A}$ has less than full column rank $\mathbf{Ax} = \mathbf{0}$ for some $\mathbf{x} \neq \mathbf{0}$, for which $\mathbf{x}'\mathbf{A}'\mathbf{A}\mathbf{x}$ will also be zero, and $\mathbf{A}'\mathbf{A}$ is then p.(s.)d.

Corollary. $\mathbf{AA}'$ is positive definite when $\mathbf{A}$ has full row rank and it is positive semi-definite otherwise.

Lemma 6. A sum of positive (semi-)definite matrices is positive (semi-) definite.

Proof. Consideration of $\mathbf{x}'\mathbf{A}\mathbf{x} = \mathbf{x}'(\sum_i \mathbf{A}_i)\mathbf{x}$ makes this clear.

Lemma 7. A symmetric matrix $\mathbf{A}$, of order n and rank r, can be written as $\mathbf{LL}'$ where $\mathbf{L}$ is $n \times r$ of rank r; i.e., $\mathbf{L}$ has full column rank.

Proof.

$$\mathbf{PAP}' = \begin{bmatrix} \mathbf{D}_r^2 & \mathbf{0} \\ \mathbf{0} & \mathbf{0} \end{bmatrix} = \begin{bmatrix} \mathbf{D}_r \\ \mathbf{0} \end{bmatrix} [\mathbf{D}_r \quad \mathbf{0}]$$

for some orthogonal $\mathbf{P}$, where $\mathbf{D}_r^2$ is diagonal of order r. Hence

$$\mathbf{A} = \mathbf{P}' \begin{bmatrix} \mathbf{D}_r \\ \mathbf{0} \end{bmatrix} [\mathbf{D}_r \quad \mathbf{0}]\mathbf{P} = \mathbf{LL}'$$

where $\mathbf{L}' = [\mathbf{D}_r \quad \mathbf{0}]\mathbf{P}$ of order $r \times n$ and full row rank; i.e., $\mathbf{L}$ is $n \times r$ of full column rank. Note also that although $\mathbf{LL}' = \mathbf{A}$, $\mathbf{L}'\mathbf{L} = \mathbf{D}_r^2$. Also, $\mathbf{L}'$ is real only when $\mathbf{A}$ is n.n.d., for only then are the non-zero elements of $\mathbf{D}_r^2$ positive.

Lemma 8. A symmetric matrix having latent roots equal to 0 and 1 is idempotent.

Proof. A symmetrix matrix $\mathbf{X}$ can always be expressed in canonical form under orthogonal similarity $\mathbf{U}'\mathbf{X}\mathbf{U} = \mathbf{D}$ where $\mathbf{D}$ is diagonal, with diagonal elements being the latent roots of $\mathbf{X}$ [see, e.g., Searle (1966), Sec. 7.8]. When these roots are 0 or 1

$$\mathbf{U}'\mathbf{X}\mathbf{U} = \begin{bmatrix} \mathbf{I} & \mathbf{0} \\ \mathbf{0} & \mathbf{0} \end{bmatrix}$$

from which it is trivial to show that $\mathbf{X}^2 = \mathbf{X}$.

Lemma 9. If $\mathbf{A}$ and $\mathbf{V}$ are symmetric and $\mathbf{V}$ is positive definite, then $\mathbf{AV}$ having latent roots 0 and 1 implies that $\mathbf{AV}$ is idempotent.

Proof. $|\mathbf{AV} - \lambda\mathbf{I}| = 0$ has roots 0 and 1. By Lemma 4, $\mathbf{V} = \mathbf{P}'\mathbf{P}$ for some non-singular matrix $\mathbf{P}$. ($|\mathbf{A}|$ is the determinant of $\mathbf{A}$.) Therefore

$$|\mathbf{P}||\mathbf{AV} - \lambda\mathbf{I}||\mathbf{P}^{-1}| = 0 \qquad \text{has roots 0 and 1;}$$

i.e.,

$$|\mathbf{PAP}' - \lambda\mathbf{I}| = 0 \qquad \text{has roots 0 and 1.}$$

Thus $\mathbf{PAP}'$ has latent roots 0 and 1. But $\mathbf{PAP}'$ is symmetric (because $\mathbf{A}$ is). Therefore by Lemma 8, $\mathbf{PAP}'$ is idempotent, i.e., $\mathbf{PAP}'\mathbf{PAP}' = \mathbf{PAP}'$. Hence, because $\mathbf{P}$ is non-singular, $\mathbf{AP}'\mathbf{PAP}'\mathbf{P} = \mathbf{APP}'$; i.e., $\mathbf{AVAV} = \mathbf{AV}$, showing the idempotency of $\mathbf{AV}$.

4. DISTRIBUTIONS

For the sake of reference and establishing notation, certain salient features of commonly used statistical distributions are now summarized. No attempt is made at completeness or full rigor. Any number of texts [e.g., Graybill (1961), Wilks (1962), Mood and Graybill (1963) and Hogg and Craig (1965)] give the pertinent details with which, it is assumed, the reader will be familiar. What follows will serve only to remind him of these things.

a. Multivariate density functions

In considering n random variables $X_1, X_2, \ldots, X_n$, for which $x_1, x_2, \ldots, x_n$ represents a set of realized values we write the cumulative density function as

$$Pr(X_1 \leq x_1, X_2 \leq x_2, \ldots, X_n \leq x_n) = F(x_1, x_2, \ldots, x_n). \tag{5}$$

Then the density function is

$$f(x_1, x_2, \ldots, x_n) = \frac{\partial^n}{\partial x_1\, \partial x_2 \ldots \partial x_n} F(x_1, x_2, \ldots, x_n). \tag{6}$$

Conditions which $f(x_1, x_2, \ldots, x_n)$ must satisfy are

$$f(x_1, x_2, \ldots, x_n) \geq 0 \quad \text{for} \quad -\infty < x_i < \infty \quad \text{for all } i$$

and

$$\int_{-\infty}^{\infty} \cdots \int_{-\infty}^{\infty} f(x_1, x_2, \ldots, x_n)\, dx_1\, dx_2 \ldots dx_n = 1.$$

The marginal density function for what might be called the "last $n - k$ x's" is $f(x_1, x_2, \ldots, x_n)$ after integrating out the first k x's, i.e., the marginal of $x_{k+1}, \ldots, x_n$ is

$$g(x_{k+1}, \ldots, x_n) = \int_{-\infty}^{\infty} \cdots \int_{-\infty}^{\infty} f(x_1, \ldots, x_k, x_{k+1}, \ldots, x_n)\, dx_1 \ldots dx_k. \tag{7}$$

The conditional distribution, for the "first k x's" given the "last $n - k$ x's" is the ratio of $f(x_1, x_2, \ldots, x_n)$ to the marginal for the "last $n - k$ x's"; i.e.,

$$f(x_1, \ldots, x_k \mid x_{k+1}, \ldots, x_n) = \frac{\text{density function of all } n\ x\text{'s}}{\text{marginal density of ``last } n - k\ x\text{'s''}}$$

$$= \frac{f(x_1, x_2, \ldots, x_n)}{g(x_{k+1}, \ldots, x_n)}. \tag{8}$$

Use of the words "first" and "last" in these descriptions implies no rigid sequencing of the variables; they are merely convenient aids to identification.

b. Moments

The kth moment about zero of the ith variable is $E(x_i^k)$, the expected value of the kth power of x_i:

$$\mu_{x_i}^{(k)} = E(x_i^k) = \int_{-\infty}^{\infty} x_i^k g(x_i)\, dx_i$$

and on substituting from (7) for $g(x_i)$ this gives

$$\mu_{x_i}^{(k)} = \int_{-\infty}^{\infty} \cdots \int_{-\infty}^{\infty} x_i^k f(x_1, x_2, \ldots, x_n)\, dx_1\, dx_2 \ldots dx_n. \tag{9}$$

In particular, when $k = 1$, the superscript (k) is usually omitted and μ_i is written for $\mu_i^{(1)}$.

The covariance between the ith and jth variables for $i \neq j$ is

$$\begin{aligned}\sigma_{ij} &= E(x_i - \mu_i)(x_j - \mu_j)\\ &= \int_{-\infty}^{\infty}\int_{-\infty}^{\infty} (x_i - \mu_i)(x_j - \mu_j) g(x_i, x_j)\, dx_i\, dx_j\\ &= \int_{-\infty}^{\infty} \cdots \int_{-\infty}^{\infty} (x_i - \mu_i)(x_j - \mu_j) f(x_1, x_2, \ldots, x_n)\, dx_1 \ldots dx_n,\end{aligned} \tag{10}$$

and similarly the variance of the ith variable is

$$\begin{aligned}\sigma_{ii} \equiv \sigma_i^2 &= E(x_i - \mu_i)^2\\ &= \int_{-\infty}^{\infty} (x_i - \mu_i)^2 g(x_i)\, dx_i\\ &= \int_{-\infty}^{\infty} \cdots \int_{-\infty}^{\infty} (x_i - \mu_i)^2 f(x_1, x_2, \ldots, x_n)\, dx_1 \ldots dx_n.\end{aligned} \tag{11}$$

Variances of and covariances between the variables in the vector

$$\mathbf{x}' = [x_1 \quad x_2 \quad \ldots \quad x_n]$$

are given in (10) and (11). Arraying these variances and covariances as the elements of a matrix gives the *variance-covariance matrix of the x's* as

$$\text{var}(\mathbf{x}) = \mathbf{V} = \{\sigma_{ij}\} \qquad \text{for} \quad i, j = 1, 2, \ldots, n.$$

Diagonal elements of $\mathbf{V}$ are variances and off-diagonal elements are covariances.

Notation. The variance of a scalar random variable x will be written as $v(x)$, whereas the variance-covariance matrix of a vector of random variables $\mathbf{x}$ will be denoted by var($\mathbf{x}$).

The vector of means corresponding to $\mathbf{x}'$ is

$$E(\mathbf{x}') = \boldsymbol{\mu}' = [\mu_1 \quad \mu_2 \quad \cdots \quad \mu_n]$$

and so, by the definition of variance and covariance,

$$\text{var}(\mathbf{x}) = E[(\mathbf{x} - \boldsymbol{\mu})(\mathbf{x} - \boldsymbol{\mu})'] = \mathbf{V}. \tag{12}$$

Furthermore, since the correlation between the ith and jth variables is $\sigma_{ij}/\sigma_i\sigma_j$, the matrix of correlations is

$$\mathbf{R} = \left\{\frac{\sigma_{ij}}{\sigma_i\sigma_j}\right\} = \mathbf{D}\{1/\sigma_i\}\mathbf{V}\mathbf{D}\{1/\sigma_i\} \qquad \text{for} \quad i, j = 1, \ldots, n \tag{13}$$

where, using (2) of Sec. 1.1, the $\mathbf{D}$'s are diagonal matrices with elements $1/\sigma_i$ for $i = 1, 2, \ldots, n$. Clearly the diagonal elements of $\mathbf{R}$ are all unity, and $\mathbf{R}$ is symmetric. It is known as the *correlation matrix*.

The matrix $\mathbf{V}$ is non-negative definite. To see that this is so, consider $\mathbf{t}'\mathbf{V}\mathbf{t}$ for some non-null vector $\mathbf{t}$. Then $\mathbf{t}'\mathbf{V}\mathbf{t} = \sum_i\sum_j t_i t_j \sigma_{ij} = v(\sum_i t_i x_i) = v(\mathbf{t}'\mathbf{x})$ which, by the definition of a variance, is positive, unless $\mathbf{t}'\mathbf{x}$ is identically zero in which case $v(\mathbf{t}'\mathbf{x}) = \mathbf{t}'\mathbf{V}\mathbf{t} = 0$. Hence $\mathbf{V}$ is a n.n.d. matrix. $\mathbf{R}$ is also n.n.d., because in (13) all the σ's are positive.

c. Linear transformations

When the variables $\mathbf{x}$ are transformed to variables $\mathbf{y}$ by the linear transformation $\mathbf{y} = \mathbf{T}\mathbf{x}$, moments of $\mathbf{y}$ are easily derived; for example,

$$\mu_y = \mathbf{T}\mu_x \qquad \text{and} \qquad \text{var}(\mathbf{y}) = \mathbf{T}\mathbf{V}\mathbf{T}'. \tag{14}$$

When making a transformation of this nature that involves a non-singular $\mathbf{T}$, an integral involving the differentials $dx_1, dx_2, \ldots, dx_n$ is transformed by substituting for the x's in terms of the y's and by replacing the differentials by $\|\mathscr{J}\|\, dy_1 dy_2 \ldots dy_n$, where $\|\mathscr{J}\|$ is the Jacobian of the x's with respect to the y's. The Jacobian matrix is defined as $\mathscr{J} = \{\partial x_i/\partial y_j\}$ for $i, j = 1, 2, \ldots, n$ and $\|\mathscr{J}\|$ is the absolute value of the determinant $|\mathscr{J}|$. Because $\mathbf{x} = \mathbf{T}^{-1}\mathbf{y}$, this means $\mathscr{J} = \mathbf{T}^{-1\prime}$ and so $\|\mathscr{J}\| = 1/\|\mathbf{T}\|$. Hence when the transformation from

$\mathbf{x}$ to $\mathbf{y}$ is $\mathbf{y} = \mathbf{T}\mathbf{x}$ the product of differentials

$$dx_1 \ldots dx_n \text{ is replaced by } (dy_1 \ldots dy_n)/\|\mathbf{T}\|. \tag{15}$$

This is the procedure, for example, in deriving the density function of $\mathbf{y} = \mathbf{T}\mathbf{x}$ from that of $\mathbf{x}$. First, substitute from $\mathbf{x} = \mathbf{T}^{-1}\mathbf{y}$ for each x_i in $f(x_1, x_2, \ldots, x_n)$. Suppose the resulting function of the y's is written as $f(\mathbf{T}^{-1}\mathbf{y})$. Then, because

$$\int_{-\infty}^{\infty} \cdots \int_{-\infty}^{\infty} f(x_1, x_2, \ldots, x_n)\, dx_1 \ldots dx_n = 1$$

the transformation gives

$$\int_{-\infty}^{\infty} \cdots \int_{-\infty}^{\infty} f(\mathbf{T}^{-1}\mathbf{y})(1/\|\mathbf{T}\|)\, dy_1 \ldots dy_n = 1.$$

But now suppose $h(y_1, y_2, \ldots, y_n)$ is the density function of the y's. Then

$$\int_{-\infty}^{\infty} \cdots \int_{-\infty}^{\infty} h(y_1, y_2, \ldots, y_n)\, dy_1 \ldots dy_n = 1.$$

By comparison we therefore find

$$h(y_1, y_2, \ldots, y_n) = \frac{f(\mathbf{T}^{-1}\mathbf{y})}{\|\mathbf{T}\|} \tag{16}$$

Example. If

$$y_1 = 3x_1 - 2x_2$$
$$y_2 = 5x_1 - 4x_2$$

is the transformation $\mathbf{y} = \mathbf{T}\mathbf{x}$, then $\|\mathbf{T}\| = 2$; and

$$h(y_1, y_2) = \tfrac{1}{2} f[x_1 = 2y_1 - y_2,\ x_2 = \tfrac{1}{2}(5y_1 - 3y_2)].$$

d. Moment generating functions

Moments, and relationships between distributions, are often derived by means of moment generating functions. In the univariate case the moment generating function (abbreviated m.g.f.) of the random variable x, written as a function of t, is, on omitting due attention to the definition of t [see, e.g., Mood and Graybill (1963, p. 114)],

$$\begin{aligned} M_x(t) &= E(e^{tx}) \\ &= \int_{-\infty}^{\infty} e^{tx} f(x)\, dx \\ &= \int_{-\infty}^{\infty} (1 + tx + t^2x^2/2 + t^3x^3/3! + \ldots) f(x)\, dx \\ &= [1 + t\mu_x^{(1)} + (t^2/2)\mu_x^{(2)} + (t^3/3!)\mu_x^{(3)} \ldots]. \end{aligned} \tag{17}$$

Hence
$$\mu_x^{(k)} = \left.\frac{\partial^k M_x(t)}{\partial t^k}\right|_{t=0} \tag{18}$$

i.e., the kth moment of x is the kth partial differential of the m.g.f. with respect to t, evaluated at the point $t = 0$. Likewise, for some function of x, $h(x)$ say, the m.g.f. of $h(x)$ is

$$M_{h(x)}(t) = E(e^{th(x)}) = \int_{-\infty}^{\infty} e^{th(x)} f(x)\, dx \tag{19}$$

and the kth moment about zero of the function is

$$\mu_{h(x)}^{(k)} = \left.\frac{\partial^k M_{h(x)}(t)}{\partial t^k}\right|_{t=0}. \tag{20}$$

In multivariate situations similar results hold. The m.g.f. of the joint distribution of n variables utilizes a vector of parameters $\mathbf{t}' = [t_1 \quad t_2 \quad \ldots \quad t_n]$:

$$\begin{aligned} M_{\mathbf{x}}(\mathbf{t}) &= E(e^{t_1x_1+t_2x_2+\ldots+t_nx_n}) \\ &= Ee^{\mathbf{t}'\mathbf{x}} \\ &= \int_{-\infty}^{\infty} \cdots \int_{-\infty}^{\infty} e^{\mathbf{t}'\mathbf{x}} f(x_1, x_2, \ldots, x_n)\, dx_1 \ldots dx_n. \end{aligned} \tag{21}$$

And the m.g.f. of a scalar function of the elements of $\mathbf{x}$, the quadratic $\mathbf{x}'\mathbf{A}\mathbf{x}$ say, is

$$\begin{aligned} M_{\mathbf{x}'\mathbf{A}\mathbf{x}}(t) &= E(e^{t\mathbf{x}'\mathbf{A}\mathbf{x}}) \\ &= \int_{-\infty}^{\infty} \cdots \int_{-\infty}^{\infty} e^{t\mathbf{x}'\mathbf{A}\mathbf{x}} f(x_1, x_2, \ldots, x_n)\, dx_1 \ldots dx_n. \end{aligned} \tag{22}$$

As well as yielding the moments of a distribution the m.g.f. also has other important uses, two of which shall be invoked repeatedly. First, if two random variables have the same m.g.f. they have the same density function. This is done under wide regularity conditions whose details are omitted here [see Mood and Graybill (1963), for example]. Second, two random variables are independent if their joint m.g.f. factorizes into the product of their two separate m.g.f.'s. This means that if

$$M_{(x_1,x_2)}(t_1, t_2) = M_{x_1}(t_1) M_{x_2}(t_2)$$

then x_1 and x_2 are independent.

Although not used in this book, the reader will elsewhere encounter characteristic functions. They are derived formally by using (it) in place of t in the m.g.f.'s, where $i = \sqrt{-1}$.

e. Univariate normal

When the random variable X has a normal distribution with mean μ and variance σ^2, we will write "x is $N(\mu, \sigma^2)$", or $x \sim N(\mu, \sigma^2)$. The density function of x is then

$$f(x) = \frac{1}{\sigma\sqrt{2\pi}} e^{-\frac{1}{2}(x-\mu)^2/\sigma^2}, \qquad \text{for} \quad -\infty < x < \infty,$$

wherein application of (9) and (11) will show that $E(x) = \mu$ and $E(x - \mu)^2 = \sigma^2$. And, in accord with (17), the m.g.f. of x is

$$\begin{aligned} M_x(t) &= (1/\sigma\sqrt{2\pi})\int_{-\infty}^{\infty} \exp[tx - \tfrac{1}{2}(x - \mu)^2/\sigma^2]\, dx \\ &= (1/\sigma\sqrt{2\pi})\int_{-\infty}^{\infty} \exp -\tfrac{1}{2}\{[x - (\mu + t\sigma^2)]^2 - (2\mu t\sigma^2 + t^2\sigma^4)\}/\sigma^2\, dx \\ &= e^{\mu t + \frac{1}{2}t^2\sigma^2}(1/\sigma\sqrt{2\pi})\int_{-\infty}^{\infty} \exp[-(x - \mu - t\sigma^2)^2/2\sigma^2]\, dx \\ &= e^{\mu t + \frac{1}{2}t^2\sigma^2}. \end{aligned}$$

From (18) it is then easily established that $\mu_x^{(1)} = \mu$, and $\mu_x^{(2)} = \sigma^2 + \mu^2$, so that $E(x - \mu)^2 = \mu_x^{(2)} - \mu^2 = \sigma^2$.

f. Multivariate normal

(*i*) *Density function.* When the random variables in $\mathbf{x}' = [x_1 \quad x_2 \quad \dots \quad x_n]$ have a multivariate normal distribution with vector of means $\boldsymbol{\mu}$ and variance-covariance matrix $\mathbf{V}$, we write "$\mathbf{x}$ is $N(\boldsymbol{\mu}, \mathbf{V})$" or "$\mathbf{x} \sim N(\boldsymbol{\mu}, \mathbf{V})$". When $E(x_i) = \mu$ for all i then $\boldsymbol{\mu} = \mu\mathbf{1}$; and if the x_i's are mutually independent, all with the same variance σ^2, then $\mathbf{V} = \sigma^2\mathbf{I}$ and we write "$\mathbf{x}$ is $N(\mu\mathbf{1}, \sigma^2\mathbf{I})$". This is equivalent to the more usual notation $NID(\mu, \sigma^2)$, but by retaining the matrix notation of $N(\mu\mathbf{1}, \sigma^2\mathbf{I})$ we emphasize that this is just a special case of the general multivariate normal $N(\boldsymbol{\mu}, \mathbf{V})$,

At this stage we confine ourselves to the case when $\mathbf{V}$ is positive definite. The multivariate normal density function is then

$$f(x_1, x_2, \dots, x_n) = \frac{e^{-\frac{1}{2}(\mathbf{x}-\boldsymbol{\mu})'\mathbf{V}^{-1}(\mathbf{x}-\boldsymbol{\mu})}}{(2\pi)^{\frac{1}{2}n}\, |\mathbf{V}|^{\frac{1}{2}}} \tag{23}$$

(*ii*) *Aitken's integral.* A result in integral calculus that is particularly applicable to any discussion of the multivariate normal distribution is Aitken's integral. It is as follows. For $\mathbf{A}$ being a positive definite symmetric matrix of order n

$$\int_{-\infty}^{\infty} \cdots \int_{-\infty}^{\infty} e^{-\frac{1}{2}\mathbf{x}'\mathbf{A}\mathbf{x}}\, dx_1 \dots dx_n = (2\pi)^{\frac{1}{2}n}\, |\mathbf{A}|^{-\frac{1}{2}}. \tag{24}$$

To establish this result, note that because $\mathbf{A}$ is positive definite there exists a non-singular matrix $\mathbf{P}$ such that $\mathbf{P}'\mathbf{A}\mathbf{P} = \mathbf{I}_n$. Hence $|\mathbf{P}'\mathbf{A}\mathbf{P}| = |\mathbf{P}|^2|\mathbf{A}| = 1$ and so $|\mathbf{P}| = |\mathbf{A}|^{-\frac{1}{2}}$; and letting $\mathbf{x} = \mathbf{P}\mathbf{y}$ gives $\mathbf{x}'\mathbf{A}\mathbf{x} = \mathbf{y}'\mathbf{P}'\mathbf{A}\mathbf{P}\mathbf{y} = \mathbf{y}'\mathbf{y}$ and so, from (15),

$$\int_{-\infty}^{\infty}\cdots\int_{-\infty}^{\infty} e^{-\frac{1}{2}\mathbf{x}'\mathbf{A}\mathbf{x}}\, dx_1 \ldots dx_n = \int_{-\infty}^{\infty}\cdots\int_{-\infty}^{\infty} e^{-\frac{1}{2}\mathbf{y}'\mathbf{y}}\, dy_1 \ldots dy_n/\|\mathbf{P}^{-1}\|$$

$$= |\mathbf{P}|\int_{-\infty}^{\infty}\cdots\int_{-\infty}^{\infty} \exp\left(-\frac{1}{2}\sum_{i=1}^{n} y_i^2\right) dy_1 \ldots dy_n$$

$$= |\mathbf{A}|^{-\frac{1}{2}}\prod_{i=1}^{n}\left\{\int_{-\infty}^{\infty} e^{-\frac{1}{2}y_i^{\,2}}\, dy_i\right\}$$

$$= (2\pi)^{\frac{1}{2}n}\, |\mathbf{A}|^{-\frac{1}{2}}.$$

Direct application of this result to (23) shows that

$$\int_{-\infty}^{\infty}\cdots\int_{-\infty}^{\infty} f(x_1, x_2, \ldots, x_n)\, dx_1 \ldots dx_n = (2\pi)^{\frac{1}{2}n}\, |\mathbf{V}^{-1}|^{-\frac{1}{2}}/(\sqrt{2\pi})^n\, |\mathbf{V}|^{\frac{1}{2}} = 1,$$

as one would expect.

(*iii*) *Moment generating function*. As in (21) the m.g.f. for the multivariate normal distribution is

$$M_{\mathbf{x}}(\mathbf{t}) = (2\pi)^{-\frac{1}{2}n}\, |\mathbf{V}|^{-\frac{1}{2}}\int_{-\infty}^{\infty}\cdots\int_{-\infty}^{\infty} \exp[\mathbf{t}'\mathbf{x} - \tfrac{1}{2}(\mathbf{x} - \boldsymbol{\mu})'\mathbf{V}^{-1}(\mathbf{x} - \boldsymbol{\mu})]\, dx_1 \ldots dx_n.$$

On rearranging the exponent this becomes

$$M_{\mathbf{x}}(\mathbf{t}) = (2\pi)^{-\frac{1}{2}n}\, |\mathbf{V}|^{-\frac{1}{2}}\int_{-\infty}^{\infty}\cdots\int_{-\infty}^{\infty}$$
$$\exp[-\tfrac{1}{2}(\mathbf{x} - \boldsymbol{\mu} - \mathbf{V}\mathbf{t})'\mathbf{V}^{-1}(\mathbf{x} - \boldsymbol{\mu} - \mathbf{V}\mathbf{t}) + \mathbf{t}'\boldsymbol{\mu} + \tfrac{1}{2}\mathbf{t}'\mathbf{V}\mathbf{t}]\, dx_1 \ldots dx_n$$

$$= \frac{e^{\mathbf{t}'\boldsymbol{\mu}+\frac{1}{2}\mathbf{t}'\mathbf{V}\mathbf{t}}}{(2\pi)^{\frac{1}{2}n}\, |\mathbf{V}|^{\frac{1}{2}}}\int_{-\infty}^{\infty}\cdots\int_{-\infty}^{\infty}$$
$$\exp[-\tfrac{1}{2}(\mathbf{x} - \boldsymbol{\mu} - \mathbf{V}\mathbf{t})'\mathbf{V}^{-1}(\mathbf{x} - \boldsymbol{\mu} - \mathbf{V}\mathbf{t})]\, dx_1 \ldots dx_n.$$

Making the transformation $\mathbf{y} = \mathbf{x} - \boldsymbol{\mu} - \mathbf{V}\mathbf{t}$ from $\mathbf{x}$ to $\mathbf{y}$, for which the Jacobian is unity, the integral then reduces to Aitken's integral with matrix $\mathbf{V}^{-1}$. Hence

$$M_{\mathbf{x}}(t) = \frac{e^{\mathbf{t}'\boldsymbol{\mu}+\frac{1}{2}\mathbf{t}'\mathbf{V}\mathbf{t}}(2\pi)^{\frac{1}{2}n}\, |\mathbf{V}^{-1}|^{-\frac{1}{2}}}{(2\pi)^{\frac{1}{2}n}\, |\mathbf{V}|^{\frac{1}{2}}} = e^{\mathbf{t}'\boldsymbol{\mu}+\frac{1}{2}\mathbf{t}'\mathbf{V}\mathbf{t}}. \tag{25}$$

Differentiating this in the manner of (18) shows that the vector of means is $\boldsymbol{\mu}$ and the variance-covariance matrix is $\mathbf{V}$.

(*iv*) *Marginal distributions.* The definition of the marginal distribution of $x_1, x_2, \ldots, x_k$, namely the first k x's, is, in accord with (7),

$$g(x_1, \ldots, x_k) = \int_{-\infty}^{\infty} \cdots \int_{-\infty}^{\infty} f(x_1, x_2, \ldots, x_n)\, dx_{k+1} \ldots dx_n.$$

The m.g.f. of this distribution is, by (21),

$$M_{x_1,\ldots,x_k}(\mathbf{t}) = \int_{-\infty}^{\infty} \cdots \int_{-\infty}^{\infty} e^{t_1 x_1 + \ldots + t_k x_k} g(x_1, \ldots, x_k)\, dx_1 \ldots dx_k$$

and on substituting for $g(x_1, \ldots, x_k)$ this becomes

$$\begin{aligned} M_{x_1,\ldots,x_k}(\mathbf{t}) &= \int_{-\infty}^{\infty} \cdots \int_{-\infty}^{\infty} e^{t_1 x_1 + \ldots + t_k x_k} f(x_1, \ldots, x_n)\, dx_1 \ldots dx_n \\ &= \text{m.g.f. of } x_1, x_2, \ldots, x_n, \text{ with } t_{k+1} = \ldots = t_n = 0 \\ &= e^{\mathbf{t}'\boldsymbol{\mu} + \frac{1}{2}\mathbf{t}'\mathbf{V}\mathbf{t}}, \text{ with } t_{k+1} = \ldots = t_n = 0. \end{aligned} \tag{26}$$

To make the substitutions $t_{k+1} = \ldots = t_n = 0$ we partition $\mathbf{x}$, $\boldsymbol{\mu}$, $\mathbf{V}$ and $\mathbf{t}$, by defining

$$\mathbf{x}_1' = [x_1 \quad x_2 \quad \cdots \quad x_k] \qquad \text{and} \qquad \mathbf{x}_2' = [x_{k+1} \quad \cdots \quad x_n]$$

so that

$$\mathbf{x}' = [\mathbf{x}_1' \quad \mathbf{x}_2'];$$

then, conformable with this,

$$\boldsymbol{\mu}' = [\boldsymbol{\mu}_1' \quad \boldsymbol{\mu}_2'], \qquad \mathbf{t}' = [\mathbf{t}_1' \quad \mathbf{t}_2']$$

and

$$\mathbf{V} = \begin{bmatrix} \mathbf{V}_{11} & \mathbf{V}_{12} \\ \mathbf{V}_{12}' & \mathbf{V}_{22} \end{bmatrix}.$$

Now putting $\mathbf{t}_2 = \mathbf{0}$ in (26) gives

$$M_{x_1,\ldots,x_k}(\mathbf{t}_1) = e^{\mathbf{t}_1'\boldsymbol{\mu}_1 + \frac{1}{2}\mathbf{t}_1'\mathbf{V}_{11}\mathbf{t}_1}.$$

By analogy with (25) and (21) we therefore have the marginal density function as

$$g(\mathbf{x}_1) = g(x_1, \ldots, x_k) = \frac{\exp[-\frac{1}{2}(\mathbf{x}_1 - \boldsymbol{\mu}_1)'\mathbf{V}_{11}^{-1}(\mathbf{x}_1 - \boldsymbol{\mu}_1)]}{(2\pi)^{\frac{1}{2}k}\,|\mathbf{V}_{11}|^{\frac{1}{2}}}.$$

On comparison with (23) we see that $g(\mathbf{x}_1)$ is a multivariate normal distribution. Similarly, so is

$$g(\mathbf{x}_2) = g(x_{k+1}, \ldots, x_n) = \frac{\exp[-\frac{1}{2}(\mathbf{x}_2 - \boldsymbol{\mu}_2)'\mathbf{V}_{22}^{-1}(\mathbf{x}_2 - \boldsymbol{\mu}_2)]}{(2\pi)^{\frac{1}{2}(n-k)}\,|\mathbf{V}_{22}|^{\frac{1}{2}}}. \tag{27}$$

Thus we see that marginal densities of the multivariate normal distribution are themselves multivariate normal.

Since $\mathbf{V}$ is taken as being positive definite so are $\mathbf{V}_{11}$ and $\mathbf{V}_{22}$. Furthermore, in these expressions use can be made of the partitioned form of $\mathbf{V}$ [see equation (47), Sec. 1.7]. Thus if

$$\mathbf{V}^{-1} = \begin{bmatrix} \mathbf{V}_{11} & \mathbf{V}_{12} \\ \mathbf{V}'_{12} & \mathbf{V}_{22} \end{bmatrix}^{-1} = \begin{bmatrix} \mathbf{W}_{11} & \mathbf{W}_{12} \\ \mathbf{W}'_{12} & \mathbf{W}_{22} \end{bmatrix},$$

then $\quad \mathbf{V}_{11}^{-1} = \mathbf{W}_{11} - \mathbf{W}_{12}\mathbf{W}_{22}^{-1}\mathbf{W}'_{12} \quad$ and $\quad \mathbf{V}_{22}^{-1} = \mathbf{W}_{22} - \mathbf{W}'_{12}\mathbf{W}_{11}^{-1}\mathbf{W}_{12}.$

(*v*) *Conditional distributions.* Let $f(\mathbf{x})$ denote the density function of all n x's. Then equation (8) gives the conditional distribution of the first k x's as

$$f(\mathbf{x}_1 \mid \mathbf{x}_2) = f(\mathbf{x})/g(\mathbf{x}_2)$$

and on substituting from (23) and (27)

$$f(\mathbf{x}_1 \mid \mathbf{x}_2) = \frac{\exp\{-\frac{1}{2}[(\mathbf{x} - \boldsymbol{\mu})'\mathbf{V}^{-1}(\mathbf{x} - \boldsymbol{\mu}) - (\mathbf{x}_2 - \boldsymbol{\mu}_2)'\mathbf{V}_{22}^{-1}(\mathbf{x}_2 - \boldsymbol{\mu}_2)]\}}{(2\pi)^{\frac{1}{2}k}(|\mathbf{V}|/|\mathbf{V}_{22}|)^{\frac{1}{2}}}. \tag{28}$$

Now, in terms of the partitioned form of $\mathbf{V}$ and its inverse given above, we have

$$\mathbf{W}_{11} = (\mathbf{V}_{11} - \mathbf{V}_{12}\mathbf{V}_{22}^{-1}\mathbf{V}'_{12})^{-1} \tag{29}$$

and $$\mathbf{V}^{-1} = \begin{bmatrix} \mathbf{W}_{11} & -\mathbf{W}_{11}\mathbf{V}_{12}\mathbf{V}_{22}^{-1} \\ -\mathbf{V}_{22}^{-1}\mathbf{V}'_{12}\mathbf{W}_{11} & \mathbf{V}_{22}^{-1} + \mathbf{V}_{22}^{-1}\mathbf{V}'_{12}\mathbf{W}_{11}\mathbf{V}_{12}\mathbf{V}_{22}^{-1} \end{bmatrix}.$$

Therefore the exponent in (28) becomes

$$[(\mathbf{x}_1 - \boldsymbol{\mu}_1)' \quad (\mathbf{x}_2 - \boldsymbol{\mu}_2)'] \begin{bmatrix} \mathbf{W}_{11} & -\mathbf{W}_{11}\mathbf{V}_{12}\mathbf{V}_{22}^{-1} \\ -\mathbf{V}_{22}^{-1}\mathbf{V}'_{12}\mathbf{W}_{11} & \mathbf{V}_{22}^{-1} + \mathbf{V}_{22}^{-1}\mathbf{V}'_{12}\mathbf{W}_{11}\mathbf{V}_{12}\mathbf{V}_{22}^{-1} \end{bmatrix}$$
$$\times \begin{bmatrix} (\mathbf{x}_1 - \boldsymbol{\mu}_1) \\ (\mathbf{x}_2 - \boldsymbol{\mu}_2) \end{bmatrix} - (\mathbf{x}_2 - \boldsymbol{\mu}_2)'\mathbf{V}_{22}^{-1}(\mathbf{x}_2 - \boldsymbol{\mu}_2)$$

which simplifies to

$$[(\mathbf{x}_1 - \boldsymbol{\mu}_1)' \quad (\mathbf{x}_2 - \boldsymbol{\mu}_2)'] \begin{bmatrix} \mathbf{I} \\ -\mathbf{V}_{22}^{-1}\mathbf{V}'_{12} \end{bmatrix} \mathbf{W}_{11}[\mathbf{I} \quad -\mathbf{V}_{12}\mathbf{V}_{22}^{-1}] \begin{bmatrix} (\mathbf{x}_1 - \boldsymbol{\mu}_1) \\ (\mathbf{x}_2 - \boldsymbol{\mu}_2] \end{bmatrix}$$

$$= [(\mathbf{x}_1 - \boldsymbol{\mu}_1) - \mathbf{V}_{12}\mathbf{V}_{22}^{-1}(\mathbf{x}_2 - \boldsymbol{\mu}_2)]'\mathbf{W}_{11}[(\mathbf{x}_1 - \boldsymbol{\mu}_1) - \mathbf{V}_{12}\mathbf{V}_{22}^{-1}(\mathbf{x}_2 - \boldsymbol{\mu}_2)]. \tag{30}$$

Furthermore, using the result for the determinant of a partitioned matrix [e.g., Searle (1966, p. 96)],

$$|\mathbf{V}| = |\mathbf{V}_{22}|\,|\mathbf{V}_{11} - \mathbf{V}_{12}\mathbf{V}_{22}^{-1}\mathbf{V}'_{12}| = |\mathbf{V}_{22}|\,|\mathbf{W}_{11}^{-1}|, \qquad \text{from (29)}.$$

Hence

$$|\mathbf{V}|/|\mathbf{V}_{22}| = |\mathbf{W}_{11}^{-1}|. \tag{31}$$

Substituting (30) and (31) in (28) gives

$$f(\mathbf{x}_1 \mid \mathbf{x}_2) = \frac{\begin{aligned}\exp\{-\tfrac{1}{2}[(\mathbf{x}_1 - \boldsymbol{\mu}_1) - \mathbf{V}_{12}\mathbf{V}_{22}^{-1}(\mathbf{x}_2 - \boldsymbol{\mu}_2)]'\mathbf{W}_{11}[(\mathbf{x}_1 - \boldsymbol{\mu}_1) \\ - \mathbf{V}_{12}\mathbf{V}_{22}^{-1}(\mathbf{x}_2 - \boldsymbol{\mu}_2)]\}\end{aligned}}{(2\pi)^{\frac{1}{2}k}\,|\mathbf{W}_{11}^{-1}|^{\frac{1}{2}}}, \tag{32}$$

showing, on comparison with (23), that the conditional distribution is also normal:

$$\mathbf{x}_1 \mid \mathbf{x}_2 \sim N[\boldsymbol{\mu}_1 + \mathbf{V}_{12}\mathbf{V}_{22}^{-1}(\mathbf{x}_2 - \boldsymbol{\mu}_2), \mathbf{W}_{11}^{-1}].$$

(*vi*) *Independence*. Suppose that the vector $\mathbf{x}' = [x_1 \quad x_2 \quad \ldots \quad x_n]$ is partitioned into p sub-vectors $\mathbf{x}' = [\mathbf{x}_1' \quad \mathbf{x}_2' \quad \ldots \quad \mathbf{x}_p']$. Then a necessary and sufficient condition for the vectors to be mutually independent is, in the corresponding partitioning of $\mathbf{V} = \{\mathbf{V}_{ij}\}$ for $i, j = 1, 2, \ldots, p$, that $\mathbf{V}_{ij} = \mathbf{0}$, for $i \neq j$.

Proof of this is established as follows. The m.g.f. of $\mathbf{x}$ is, by (25),

$$M_{\mathbf{x}}(\mathbf{t}) = e^{\mathbf{t}'\boldsymbol{\mu} + \frac{1}{2}\mathbf{t}'\mathbf{V}\mathbf{t}} = \exp\left(\sum_{i=1}^{p} \mathbf{t}_i'\boldsymbol{\mu}_i + \tfrac{1}{2}\sum_{i=j=1}^{p}\sum^{p} \mathbf{t}_i'\mathbf{V}_{ij}\mathbf{t}_j\right)$$

and if $\mathbf{V}_{ij} = \mathbf{0}$ for $i \neq j$ this reduces to

$$M_{\mathbf{x}}(\mathbf{t}) = \exp\sum_{i=1}^{p} (\mathbf{t}_i'\boldsymbol{\mu}_i + \tfrac{1}{2}\mathbf{t}_i'\mathbf{V}_{ii}\mathbf{t}_i) = \prod_{i=1}^{p} \exp(\mathbf{t}_i'\boldsymbol{\mu}_i + \tfrac{1}{2}\mathbf{t}_i'\mathbf{V}_{ii}\mathbf{t}_i).$$

Invoking the property that the m.g.f. of the joint distribution of independent sets of variables is the product of their several m.g.f.'s, we conclude that the $\mathbf{x}_i$'s are independent. Conversely, if they are independent, each with its variance-covariance $\mathbf{K}_{ii}$ say, then the m.g.f. of the joint distribution is

$$\prod_{i=1}^{p} \exp(\mathbf{t}_i'\boldsymbol{\mu}_i + \tfrac{1}{2}\mathbf{t}_i'\mathbf{K}_{ii}\mathbf{t}_i) = \exp\sum_{i=1}^{p} (\mathbf{t}_i'\boldsymbol{\mu}_i + \tfrac{1}{2}\mathbf{t}_i'\mathbf{K}_{ii}\mathbf{t}_i) = \exp(\mathbf{t}'\boldsymbol{\mu} + \tfrac{1}{2}\mathbf{t}'\mathbf{V}\mathbf{t})$$

where $\mathbf{V} = \text{diag}\{\mathbf{K}_{11}, \mathbf{K}_{22}, \ldots, \mathbf{K}_{pp}\}$. Hence $\mathbf{V}_{ij} = \mathbf{0}$ for $i \neq j$.

g. Central χ^2, F and t

When $\mathbf{x}$ is $N(\mathbf{0}, \mathbf{I})$ then $\sum_{i=1}^{n} x_i^2$ has the central χ^2-distribution with n degrees of freedom. Thus, when

$$\mathbf{x} \text{ is } N(\mathbf{0}, \mathbf{I}) \quad \text{and} \quad u = \sum_{i=1}^{n} x_i^2 = \mathbf{x}'\mathbf{x} \quad \text{then} \quad u \sim \chi_n^2.$$

The density function is

$$f(u) = \frac{u^{\frac{1}{2}n-1}e^{-\frac{1}{2}u}}{2^{\frac{1}{2}n}\Gamma(\frac{1}{2}n)} \quad \text{for } u > 0 \tag{33}$$

where $\Gamma(\frac{1}{2}n)$ is the gamma function with argument $\frac{1}{2}n$. [For a positive integer n, $\Gamma(n) = (n-1)!$]. The m.g.f. corresponding to (33) is

$$M_u(t) = (1 - 2t)^{-\frac{1}{2}n} \tag{34}$$

as can be obtained directly from (17) using (33) or as $M_{\mathbf{x}'\mathbf{x}}(t)$ using the $N(\mathbf{0}, \mathbf{I})$ density function in (22). The mean and variance of u are n and $2n$ respectively.

The commonest application of the χ^2-distribution is that when $\mathbf{x}$ is $N(\mu\mathbf{1}, \sigma^2\mathbf{I})$ then $\sum_{i=1}^{n}(x_i - \bar{x})^2/\sigma^2$ is χ^2_{n-1}. This, as we shall see, is a special case of Theorem 2. The same result can also be established using the transformation $\mathbf{y} = \mathbf{H}_0\mathbf{x}$ where $\mathbf{H}_0$ is the last $n - 1$ rows of the Helmert matrix discussed in Sec. 1.

Two independent variables each having central χ^2-distributions form the basis of the F-distribution. Thus if

$$u_1 \text{ is } \chi^2_{n_1} \quad \text{and} \quad u_2 \text{ is } \chi^2_{n_2} \quad \text{then} \quad v = \frac{u_1/n_1}{u_2/n_2} \sim F_{n_1,n_2},$$

the F-distribution with n_1 and n_2 degrees of freedom. The density function is

$$f(v) = \frac{\Gamma(\frac{1}{2}n_1 + \frac{1}{2}n_2)n_1^{\frac{1}{2}n_1}n_2^{\frac{1}{2}n_2}v^{\frac{1}{2}n_1-1}}{\Gamma(\frac{1}{2}n_1)\Gamma(\frac{1}{2}n_2)(n_2 + n_1 v)^{\frac{1}{2}n_1+\frac{1}{2}n_2}} \quad \text{for} \quad v > 0. \tag{35}$$

The mean of the distribution is $n_2/(n_2 - 2)$ and the variance is

$$2n_2^2[1 + (n_2 - 2)/n_1]/(n_2 - 2)^2(n_2 - 4).$$

Finally, the ratio of a normally distributed variable to one that has a χ^2-distribution is the basis of Student's t-distribution. Thus when x is $N(0, 1)$ and u is χ^2_n, independent of x, then

$$z = x/\sqrt{u/n} \text{ is distributed as } t_n,$$

the t-distribution with n degrees of freedom. Its density function is

$$f(z) = \frac{\Gamma(\frac{1}{2}n + \frac{1}{2})}{\sqrt{n\pi}\,\Gamma(\frac{1}{2}n)}\left(1 + \frac{z^2}{n}\right)^{-\frac{1}{2}(n+1)}, \quad \text{for} \quad -\infty < z < \infty, \tag{36}$$

with zero mean and variance $n/(n - 2)$.

A frequent application of this distribution is that if $\mathbf{x}$ is $N(\mu\mathbf{1}, \sigma^2\mathbf{I})$ then

$$\frac{(\bar{x} - \mu)}{1/\sqrt{n}}\sqrt{\frac{n-1}{\sum_{i=1}^{n}(x_i - \bar{x})^2}} \text{ has the } t_{n-1} \text{ distribution.}$$

The relationship between t_n and $F_{1,n}$ can also be demonstrated. For z as described above consider

$$z^2 = \frac{x^2}{u/n}.$$

x^2 is clearly χ^2_1 and u is χ^2_n. Therefore z^2 is $F_{1,n}$; i.e., when a variable is distributed as t_n its square is distributed as $F_{1,n}$.

h. Non-central χ^2

We have already seen that when $\mathbf{x}$ is $N(\mathbf{0}, \mathbf{I}_n)$ the distribution of $\mathbf{x}'\mathbf{x} = \sum x_i^2$ is what is known as a central χ^2-distribution. We now consider the distribution of $\mathbf{u} = \mathbf{x}'\mathbf{x}$ when $\mathbf{x}$ is $N(\boldsymbol{\mu}, \mathbf{I})$. The sole difference is that the mean of $\mathbf{x}$ is $\boldsymbol{\mu}$ and not $\mathbf{0}$. The resulting distribution of $\mathbf{x}'\mathbf{x}$ is known as the non-central χ^2. As with the central χ^2, the non-central χ^2 involves the degrees of freedom, n. It also involves the parameter $\frac{1}{2}\boldsymbol{\mu}'\boldsymbol{\mu} = \frac{1}{2}\sum \mu_i^2$, known as the non-centrality parameter, for which the symbol λ is used; i.e.,

$$\lambda = \tfrac{1}{2}\boldsymbol{\mu}'\boldsymbol{\mu}.$$

Reference to the distribution is by means of the symbol $\chi^{2\prime}(n, \lambda)$, the non-central χ^2 with n degrees of freedom and non-centrality parameter λ. When $\boldsymbol{\mu} = 0$, $\lambda = 0$ and it reduces to the central χ^2-distribution.

The density function of the non-central χ^2-distribution $\chi^{2\prime}(n, \lambda)$ is

$$f(u) = e^{-\lambda}\sum_{k=0}^{\infty}\frac{\lambda^k}{k!}\frac{u^{\frac{1}{2}n+k-1}e^{-\frac{1}{2}u}}{2^{\frac{1}{2}n+k}\Gamma(\frac{1}{2}n + k)}. \tag{37}$$

We observe that this is an infinite weighted sum of density functions of central χ^2's, because the term $(u^{\frac{1}{2}n+k-1}e^{-\frac{1}{2}u})/2^{\frac{1}{2}n+k}\Gamma(\frac{1}{2}n + k)$ in (37) is, by (33), the density function of the $\chi^2_{\frac{1}{2}n+k}$ distribution.

The m.g.f. of $\chi^{2\prime}(n, \lambda)$ can be derived using (37) in (17); because the $\chi^2_{\frac{1}{2}n+k}$ density function occurs in (37) this procedure yields

$$M_u(t) = e^{-\lambda}\sum_{k=0}^{\infty}(\lambda^k/k!)(\text{m.g.f. of } \chi^2_{\frac{1}{2}n+k})$$

and on using (34) this is

$$\begin{aligned} M_u(t) &= e^{-\lambda}\sum_{k=0}^{\infty}(\lambda^k/k!)(1 - 2t)^{-(\frac{1}{2}n+k)} \\ &= e^{-\lambda}e^{\lambda(1-2t)^{-1}}(1 - 2t)^{-\frac{1}{2}n} \\ &= (1 - 2t)^{-\frac{1}{2}n}e^{-\lambda[1-(1-2t)^{-1}]}. \end{aligned} \tag{38}$$

The same result can also be obtained as $M_{\mathbf{x}'\mathbf{x}}(t)$ using the $N(\boldsymbol{\mu}, \mathbf{I})$ density function in (22). This proceeds as follows.

$$\begin{aligned} M_{\mathbf{x}'\mathbf{x}}(t) &= E(e^{t\mathbf{x}'\mathbf{x}}) = E\prod_{i=1}^{n}e^{tx_i^2}, \text{ because the } x_i\text{'s are independent} \\ &= \prod_{i=1}^{n}\int_{-\infty}^{\infty}(2\pi)^{-\frac{1}{2}}\exp[tx_i^2 - \tfrac{1}{2}(x_i - \mu_i)^2]\,dx_i, \end{aligned}$$

and on rearranging the exponent this is

$$M_{\mathbf{x}'\mathbf{x}}(t) = \prod_{i=1}^{n} \int_{-\infty}^{\infty} (2\pi)^{-\frac{1}{2}} \exp -\tfrac{1}{2}\{[x_i(1-2t) - \mu_i]^2(1-2t)^{-1} + \mu_i^2[1-(1-2t)^{-1}]\}\, dx_i$$

$$= \exp -\tfrac{1}{2}(\textstyle\sum \mu_i^2)[1-(1-2t)^{-1}]$$

$$\times \prod_{i=1}^{n} \int_{-\infty}^{\infty} (2\pi)^{-\frac{1}{2}} \exp\left\{\frac{[x_i - \mu_i(1-2t)^{-1}]^2}{2(1-2t)^{-1}}\right\} dx_i$$

$$= e^{-\lambda[1-(1-2t)^{-1}]} \prod_{i=1}^{n} (1-2t)^{-\frac{1}{2}}$$

$$= (1-2t)^{-\frac{1}{2}n} e^{-\lambda[1-(1-2t)^{-1}]},$$

as in (38).

The mean and variance of the $\chi^{2\prime}(n, \lambda)$ distribution are $n + 2\lambda$ and $2n + 8\lambda$ respectively. They may be derived from differentiating the m.g.f., or directly from the independence of the x_i's. Thus, for summation over i,

$$E \sum x_i^2 = \sum E(x_i^2) = \sum (\sigma_x^2 + \mu_i^2) = \sum 1 + \sum \mu_i^2 = n + 2\lambda;$$

and

$$\begin{aligned} v(\textstyle\sum x_i^2) &= \textstyle\sum v(x_i^2) \\ &= \textstyle\sum v[(x_i - \mu_i)^2 + 2\mu_i(x_i - \mu_i) + \mu_i^2] \\ &= \textstyle\sum v(x_i - \mu_i)^2 + 4 \sum \mu_i^2\, v(x_i - \mu_i) \\ &= \textstyle\sum [E(x_i - \mu_i)^4 - \{E(x_i - \mu_i)^2\}^2] + 4 \sum \mu_i^2 \sigma_x^2 \\ &= \textstyle\sum [3\sigma_x^4 - (\sigma_x^2)^2] + 8\lambda \\ &= 2n + 8\lambda. \end{aligned}$$

Notice that properties of the non-central χ^2-distribution reduce to those of the central χ^2 when $\lambda = 0$, as one would expect. A further property is also to be noted: if variables having non-central χ^2-distributions are jointly independent their sum also has a non-central χ^2. Thus if, for $i = 1, 2, \ldots, k$, the

$$u_i \text{ are } \chi^{2\prime}(n_i, \lambda_i) \text{ and independent}$$

then

$$\textstyle\sum u_i \text{ is } \chi^{2\prime}(\sum n_i, \sum \lambda_i).$$

Proof of this is readily established through using moment generating functions and the independence of the u_i's:

$$M_{(u_1, \ldots, u_k)}(\mathbf{t}) = \Pi M_{u_i}(t_i) = \Pi E(e^{t_i u_i})$$

and on putting $t_i = t$ for all i this becomes

$$\Pi M_{u_i}(t) = \Pi E(e^{t u_i}) = E(e^{t \Sigma u_i}) = M_{\Sigma u_i}(t),$$

where the products and sums are over $i = 1, 2, \ldots, k$. Hence

$$\begin{aligned} M_{\Sigma u_i}(t) &= \Pi M_{u_i}(t) \\ &= \Pi(1-2t)^{-\frac{1}{2}n_i}e^{-\lambda_i[1-(1-2t)^{-1}]} \\ &= (1-2t)^{-\frac{1}{2}\Sigma n_i}e^{-\Sigma\lambda\,[1-(1-2t)^{-1}]}. \end{aligned}$$

Comparison with (38) indicates that $\sum u_i \sim \chi^{2\prime}(\sum n_i, \sum \lambda_i)$.

i. Non-central F

Just as there is a non-central analogy of the central χ^2-distribution so also is there a non-central F-distribution. It is specified as follows. If u_1 and u_2 are independent and

$$u_1 \text{ is } \chi^{2\prime}(n_1, \lambda) \qquad \text{and} \qquad u_2 \text{ is } \chi^2_{n_2}$$

then

$$v = \frac{u_1/n_1}{u_2/n_2} \text{ is distributed as } F'(n_1, n_2, \lambda),$$

the non-central F-distribution with n_1 and n_2 degrees of freedom and non-centrality parameter λ. Its density function is

$$f(v) = \sum_{k=0}^{\infty} \frac{e^{-\lambda}\lambda^k}{k!} \frac{n_1^{\frac{1}{2}n_1+k} n_2^{\frac{1}{2}n_2}\Gamma(\frac{1}{2}n_1 + \frac{1}{2}n_2 + k)}{\Gamma(\frac{1}{2}n_1 + k)\Gamma(\frac{1}{2}n_2)} \cdot \frac{v^{\frac{1}{2}n_1+k-1}}{(n_2 + n_1 v)^{\frac{1}{2}n_1+\frac{1}{2}n_2+k}}$$

When $\lambda = 0$ this reduces to (35), the density function of the central F-distribution (when $\lambda = 0, k = 0$). The mean and variance of the distribution are

$$E(v) = \frac{n_2}{n_2 - 2}\left(1 + \frac{2\lambda}{n_1}\right)$$

and

$$\text{variance of } v \text{ is } \frac{2n_2^2}{n_1^2(n_2 - 2)}\left[\frac{(n_1 + 2\lambda)^2}{(n_2 - 2)(n_2 - 4)} + \frac{n_1 + 4\lambda}{n_2 - 4}\right].$$

When $\lambda = 0$ these reduce, of course, to the mean and variance of the central F_{n_1,n_2}-distribution.

Derivation of $f(v)$ is established as follows. Since u_1 and u_2 are independent their joint density function is the product of their individual densities:

$$\begin{aligned} f(u_1, u_2) &= f(u_1)f(u_2) \\ &= \sum_{k=0}^{\infty} \frac{e^{-\lambda}\lambda^k}{k!} \frac{u_1^{\frac{1}{2}n_1+k-1} e^{-\frac{1}{2}u_1}}{2^{\frac{1}{2}n_1+k}\Gamma(\frac{1}{2}n_1 + k)} \frac{u_2^{\frac{1}{2}n_2-1} e^{-\frac{1}{2}u_2}}{2^{\frac{1}{2}n_2}\,\Gamma(\frac{1}{2}n_2)} \end{aligned}$$

For the terms not involving u_1 and u_2 write

$$\alpha_k = \frac{e^{-\lambda}\lambda^k}{k!} \frac{1}{2^{\frac{1}{2}n_1+\frac{1}{2}n_2+k}\Gamma(\frac{1}{2}n_1 + k)\Gamma(\frac{1}{2}n_2)},$$

so that

$$f(u_1, u_2) = \sum_{k=0}^{\infty} \alpha_k u_1^{\frac{1}{2}n_1+k-1} u_2^{\frac{1}{2}n_2-1} e^{-\frac{1}{2}(u_1+u_2)}.$$

Now make a transformation of variables from u_1 and u_2 to v and z where

$$v = \frac{n_2 u_1}{n_1 u_2} \quad \text{and} \quad z = u_1 + u_2.$$

The Jacobian of this transformation is

$$\|\mathscr{J}\| = \begin{vmatrix} \partial v/\partial u_1 & \partial v/\partial u_2 \\ \partial z/\partial u_1 & \partial z/\partial u_2 \end{vmatrix} = \begin{vmatrix} n_2/n_1u_2 & -n_2u_1/n_1u_2^2 \\ 1 & 1 \end{vmatrix} = \frac{n_2(u_1+u_2)}{n_1u_2^2}.$$

Then, after the transformation,

$$f(u_1, u_2)\, du_1\, du_2 \text{ becomes } [f(u_1, u_2)/\|\mathscr{J}\|]\, dv\, dz$$

and so

$$f(v)\, dv = \int_0^\infty [f(u_1, u_2)/\|\mathscr{J}\|]\, dv\, dz.$$

Now the transformations are equivalent to

$$u_1 = \frac{n_1 v z}{n_1 v + n_2} \quad \text{and} \quad u_2 = \frac{n_2 z}{n_1 v + n_2}, \quad \text{and give} \quad \|\mathscr{J}\| = \frac{(n_1 v + n_2)^2}{n_1 n_2 z},$$

and so

$$\begin{aligned} f(v) &= \sum_{k=0}^{\infty} \alpha_k \int_0^\infty \left(\frac{n_1 v z}{n_1 v + n_2}\right)^{\frac{1}{2}n_1+k-1} \left(\frac{n_2 z}{n_1 v + n_2}\right)^{\frac{1}{2}n_2-1} e^{-\frac{1}{2}z} \frac{(n_1 n_2 z)}{(n_1 v + n_2)^2}\, dz \\ &= \sum_{k=0}^{\infty} \alpha_k n_1^{\frac{1}{2}n_1+k} n_2^{\frac{1}{2}n_2} \frac{v^{\frac{1}{2}n_1+k-1}}{(n_1 v + n_2)^{\frac{1}{2}n_1+\frac{1}{2}n_2+k}} \int_0^\infty z^{\frac{1}{2}n_1+\frac{1}{2}n_2+k-1} e^{-\frac{1}{2}z}\, dz \end{aligned}$$

which, on substituting for α_k and evaluating the integral as

$$2^{\frac{1}{2}n_1+\frac{1}{2}n_2+k}\, \Gamma(\tfrac{1}{2}n_1 + \tfrac{1}{2}n_2 + k),$$

becomes the form shown above.

Because of the relative complexity of the density function there would be convenience in having an approximation to it. Consider, as above, $v = (n_2u_1)/(n_1u_2)$ where u_2 is $\chi^2_{n_2}$ and u_1 is $\chi^2(n_1, \lambda)$: the distribution of v is $F'(n_1, n_2, \lambda)$. Suppose some value c exists such that cu_1 is χ^2_m for some value m; i.e., cu_1 has a central χ^2-distribution. Then

$$\frac{v}{m/cn_1} = \frac{cn_1 v}{m} = \frac{cu_1/m}{u_2/n_2}$$

would have a central F-distribution, F_{m,n_2}. No values c and m exist such that this is true; but, as indicated in Scheffé (1959) and shown in Patnaik (1949), approximation to it can be made by choosing c and m so that cu_1, where u_1 is $\chi^{2\prime}(n_1, \lambda)$, has the same mean and variance as χ^2_m. This leads to

$$E(cu_1) = c(n_1 + 2\lambda) = m$$

and $$\text{v}(cu_1) = c^2(2n_1 + 8\lambda) = 2m$$

giving $$c = \frac{n_1 + 2\lambda}{n_1 + 4\lambda} \quad \text{and} \quad m = \frac{(n_1 + 2\lambda)^2}{n_1 + 4\lambda}$$

with $m/cn_1 = (1 + 2\lambda/n_1)$. Hence

$$\frac{v}{m/cn_1} = \frac{v}{1 + 2\lambda/n_1}$$

is approximately distributed as F_{m,n_2}.

j. Other non-central distributions

Two other distributions can be mentioned in the context of non-central distributions: the non-central t-distribution and the doubly non-central F-distribution. If x is $N(\mu, 1)$ and if, independently of x, u is χ^2_n then $x/\sqrt{u/n}$ has the non-central t-distribution, $t'(n, \mu)$, with n degrees of freedom and non-centrality parameter μ. The density function is

$$f(t) = \frac{n^{\frac{1}{2}n}}{\Gamma(\frac{1}{2}n)} \frac{e^{-\frac{1}{2}\mu^2}}{(n + t^2)^{\frac{1}{2}(n+1)}} \sum_{k=0}^{\infty} \frac{\Gamma(\frac{1}{2}n + \frac{1}{2}k + \frac{1}{2})\mu^k 2^{\frac{1}{2}k} t^k}{k!(n + t^2)^{\frac{1}{2}k}}$$

Its derivation is given in Rao (1965, p. 139).

The doubly non-central F-distribution is based on the ratio of two independent non-centrally χ^2-distributed variables. Thus if u_1 is $\chi^{2\prime}(n_1, \lambda_1)$ and u_2 is $\chi^{2\prime}(n_2, \lambda_2)$ then $v = n_2u_1/n_1u_2$ is distributed as $F''(n_1, n_2, \lambda_1, \lambda_2)$, the doubly non-central F-distribution with degrees of freedom n_1 and n_2 and non-centrality parameters λ_1 and λ_2. Scheffé (1959, pp. 135, 415) discusses an application of this distribution and a procedure for approximating it by a central F-distribution. The density function is derived in exactly the same manner as is that of the non-central F shown above, giving

$$f(v) = \sum_{\substack{k_1=0\\k_2=0}}^{\infty} \frac{[\exp(-\lambda_1 - \lambda_2)]\lambda_1^{k_1}\lambda_2^{k_2}\Gamma(\frac{1}{2}n_1 + \frac{1}{2}n_2 + k_1 + k_2)n_1^{\frac{1}{2}n_1+k_1}n_2^{\frac{1}{2}n_2+k_2}v^{\frac{1}{2}n_1+k_1-1}}{k_1!k_2!\Gamma(\frac{1}{2}n_1 + k_1)\Gamma(\frac{1}{2}n_2 + k_2)(n_1v + n_2)^{\frac{1}{2}n_1+\frac{1}{2}n_2+k_1+k_2}}$$

5. DISTRIBUTION OF QUADRATIC FORMS

We discuss here the distribution of a quadratic form $\mathbf{x}'\mathbf{A}\mathbf{x}$ when $\mathbf{x}$ is $N(\boldsymbol{\mu}, \mathbf{V})$. For the most part the discussion is confined to the case of $\mathbf{V}$ being non-singular, although some results pertinent to singular $\mathbf{V}$ are also given. In dealing with just the general case of $\mathbf{x}$ being $N(\boldsymbol{\mu}, \mathbf{V})$ we can readily consider special cases of interest such as $\mathbf{x}$ being $N(\mathbf{0}, \mathbf{I})$ or $N(\mu\mathbf{1}, \mathbf{I})$ or $N(\boldsymbol{\mu}, \mathbf{I})$. But theorems concerning just these alone are not needed. The main results are presented in a series of five theorems. The first relates to cumulants of quadratic forms, the second to the distribution of quadratic forms and the last three to independence properties of quadratic forms.

In all the theorems considerable use is made of the trace of a matrix, $\text{tr}(\mathbf{A})$, the sum of the diagonal elements of $\mathbf{A}$. We recall that $\text{tr}(\mathbf{A})$ equals the sum of the latent roots of $\mathbf{A}$ and that when $\mathbf{A}$ is idempotent $\text{tr}(\mathbf{A}) = r(\mathbf{A})$. Furthermore, under the operation of taking the trace, matrix products are cyclically commutative; e.g., $\text{tr}(\mathbf{ABC}) = \text{tr}(\mathbf{BCA}) = \text{tr}(\mathbf{CAB})$. Also, since a quadratic form is a scalar, it equals its own trace and hence

$$\mathbf{x}'\mathbf{A}\mathbf{x} = \text{tr}(\mathbf{x}'\mathbf{A}\mathbf{x}) = \text{tr}(\mathbf{A}\mathbf{x}\mathbf{x}').$$

These properties of the trace operation are used many times in what follows, without explicit reference thereto. The reader is therefore warned to be familiar with them.

All the theorems relate to $\mathbf{x}$ being $N(\boldsymbol{\mu}, \mathbf{V})$–with one exception, the first part of Theorem 1, which is true for $\mathbf{x}$ being $(\boldsymbol{\mu}, \mathbf{V})$, normal or otherwise. In proving one result for the normal case use is made of the following lemma.

Lemma 10. For any vector $\mathbf{g}$ and any positive definite symmetric matrix $\mathbf{W}$

$$(2\pi)^{\frac{1}{2}n} |\mathbf{W}|^{\frac{1}{2}} e^{\frac{1}{2}\mathbf{g}'\mathbf{W}\mathbf{g}} = \int_{-\infty}^{\infty} \cdots \int_{-\infty}^{\infty} \exp(-\tfrac{1}{2}\mathbf{x}'\mathbf{W}^{-1}\mathbf{x} + \mathbf{g}'\mathbf{x})\, dx_1 \ldots dx_n. \quad (39)$$

Proof. From the integral of a multivariate normal density $N(\boldsymbol{\mu}, \mathbf{W})$ we have

$$(2\pi)^{\frac{1}{2}n} |\mathbf{W}|^{\frac{1}{2}} = \int_{-\infty}^{\infty} \cdots \int_{-\infty}^{\infty} \exp[-\tfrac{1}{2}(\mathbf{x}-\boldsymbol{\mu})'\mathbf{W}^{-1}(\mathbf{x}-\boldsymbol{\mu})]\, dx_1 \ldots dx_n$$

$$= \int_{-\infty}^{\infty} \cdots \int_{-\infty}^{\infty} \exp(-\tfrac{1}{2}\mathbf{x}'\mathbf{W}^{-1}\mathbf{x} + \boldsymbol{\mu}'\mathbf{W}^{-1}\mathbf{x} - \tfrac{1}{2}\boldsymbol{\mu}'\mathbf{W}^{-1}\boldsymbol{\mu})\, dx_1 \ldots dx_n.$$

On writing $\mathbf{g}'$ for $\boldsymbol{\mu}'\mathbf{W}^{-1}$ this gives (39).

a. Cumulants

Theorem 1. When $\mathbf{x}$ is $N(\boldsymbol{\mu}, \mathbf{V})$

(i)
$$E(\mathbf{x}'\mathbf{A}\mathbf{x}) = \text{tr}(\mathbf{AV}) + \boldsymbol{\mu}'\mathbf{A}\boldsymbol{\mu}; \tag{40}$$
(true also when $\mathbf{x}$ is non-normal);

(ii) the rth cumulant of $\mathbf{x}'\mathbf{A}\mathbf{x}$ is
$$K_r(\mathbf{x}'\mathbf{A}\mathbf{x}) = 2^{r-1}(r-1)!\,[\text{tr}(\mathbf{AV})^r + r\boldsymbol{\mu}'\mathbf{A}(\mathbf{VA})^{r-1}\boldsymbol{\mu}];$$

and (iii) the covariance of $\mathbf{x}$ with $\mathbf{x}'\mathbf{A}\mathbf{x}$ is
$$\text{cov}\,(\mathbf{x}, \mathbf{x}'\mathbf{A}\mathbf{x}) = 2\mathbf{VA}\boldsymbol{\mu}.$$

Proof. (i) With $E(\mathbf{x}) = \boldsymbol{\mu}$ and var $(\mathbf{x}) = \mathbf{V}$ we have
$$E(\mathbf{xx}') = \mathbf{V} + \boldsymbol{\mu}\boldsymbol{\mu}'.$$
Hence
$$\begin{aligned} E(\mathbf{x}'\mathbf{A}\mathbf{x}) &= E\,\text{tr}(\mathbf{Axx}') = \text{tr}[\mathbf{A}E(\mathbf{xx}')] \\ &= \text{tr}(\mathbf{AV} + \mathbf{A}\boldsymbol{\mu}\boldsymbol{\mu}') \\ &= \text{tr}(\mathbf{AV}) + \boldsymbol{\mu}'\mathbf{A}\boldsymbol{\mu}. \end{aligned}$$

It is clear from the proof that this part of the theorem holds whether $\mathbf{x}$ is normal or not.

(ii) The m.g.f. of $\mathbf{x}'\mathbf{A}\mathbf{x}$ is
$$M_{\mathbf{x}'\mathbf{A}\mathbf{x}}(t) = (2\pi)^{-\frac{1}{2}n}\,|\mathbf{V}|^{-\frac{1}{2}}\int_{-\infty}^{\infty}\cdots\int_{-\infty}^{\infty}$$
$$\exp[t\mathbf{x}'\mathbf{A}\mathbf{x} - \tfrac{1}{2}(\mathbf{x}-\boldsymbol{\mu})'\mathbf{V}^{-1}(\mathbf{x}-\boldsymbol{\mu})]\,dx_1\cdots dx_n$$

and on rearranging the exponent this becomes
$$M_{\mathbf{x}'\mathbf{A}\mathbf{x}}(t) = \frac{e^{-\frac{1}{2}\boldsymbol{\mu}'\mathbf{V}^{-1}\boldsymbol{\mu}}}{(2\pi)^{\frac{1}{2}n}\,|\mathbf{V}|^{\frac{1}{2}}}\int_{-\infty}^{\infty}\cdots\int_{-\infty}^{\infty}$$
$$\exp[-\tfrac{1}{2}\mathbf{x}'(\mathbf{I} - 2t\mathbf{AV})\mathbf{V}^{-1}\mathbf{x} + \boldsymbol{\mu}'\mathbf{V}^{-1}\mathbf{x}]\,dx_1\ldots dx_n. \tag{41}$$

Now in Lemma 10 put $\mathbf{g}' = \boldsymbol{\mu}'\mathbf{V}^{-1}$ and $\mathbf{W} = [(\mathbf{I} - 2t\mathbf{AV})\mathbf{V}^{-1}]^{-1} = \mathbf{V}(\mathbf{I} - 2t\mathbf{AV})^{-1}$. The right-hand side of (39) then equals the multiple integral in (41) and so (41) becomes
$$M_{\mathbf{x}'\mathbf{A}\mathbf{x}}(t) = e^{-\frac{1}{2}\boldsymbol{\mu}'\mathbf{V}^{-1}\boldsymbol{\mu}}|\mathbf{V}|^{-\frac{1}{2}}\,|\mathbf{V}(\mathbf{I} - 2t\mathbf{AV})^{-1}|^{\frac{1}{2}}\exp[\tfrac{1}{2}\boldsymbol{\mu}'\mathbf{V}^{-1}\mathbf{V}(\mathbf{I} - 2t\mathbf{AV})^{-1}\mathbf{V}^{-1}\boldsymbol{\mu}]$$
which simplifies to
$$M_{\mathbf{x}'\mathbf{A}\mathbf{x}}(t) = |\mathbf{I} - 2t\mathbf{AV}|^{-\frac{1}{2}}\exp\{-\tfrac{1}{2}\boldsymbol{\mu}'[\mathbf{I} - (\mathbf{I} - 2t\mathbf{AV})^{-1}]\mathbf{V}^{-1}\boldsymbol{\mu}\}. \tag{42}$$
The cumulant generating function is the logarithm of the m.g.f. Hence
$$\begin{aligned} \sum_{r=1}^{\infty} K_r t^r/r! &= \log[M_{\mathbf{x}'\mathbf{A}\mathbf{x}}(t)] \\ &= -\tfrac{1}{2}\log|\mathbf{I} - 2t\mathbf{AV}| - \tfrac{1}{2}\boldsymbol{\mu}'[\mathbf{I} - (\mathbf{I} - 2t\mathbf{AV})^{-1}]\mathbf{V}^{-1}\boldsymbol{\mu}. \end{aligned} \tag{43}$$

The two parts of this are evaluated as follows. Use "λ_i of **X**" to denote the "ith latent root of **X**". Then for sufficiently small $|t|$

$$\begin{aligned}
-\tfrac{1}{2}\log|\mathbf{I}-2t\mathbf{AV}| &= -\tfrac{1}{2}\sum_{i=1}^{n}\log[\lambda_i \text{ of } (\mathbf{I}-2t\mathbf{AV})] \\
&= -\tfrac{1}{2}\sum_{i=1}^{n}\log[1-2t(\lambda_i \text{ of } \mathbf{AV})] \\
&= -\tfrac{1}{2}\sum_{i=1}^{n}\sum_{r=1}^{\infty} -[2t(\lambda_i \text{ of } \mathbf{AV})]^r/r \\
&= \sum_{r=1}^{\infty} 2^{r-1}t^r/r \sum_{i=1}^{n}(\lambda_i \text{ of } \mathbf{AV})^r \\
&= \sum_{r=1}^{\infty}(2^{r-1}t^r/r)\operatorname{tr}(\mathbf{AV})^r.
\end{aligned}$$

And, by direct binomial expansion, for sufficiently small $|t|$

$$\mathbf{I}-(\mathbf{I}-2t\mathbf{AV})^{-1} = -\sum_{r=1}^{\infty}2^r t^r(\mathbf{AV})^r.$$

Making these substitutions in (43) and equating the coefficients of t^r gives

$$K_r(\mathbf{x}'\mathbf{A}\mathbf{x}) = 2^{r-1}(r-1)!\,[\operatorname{tr}(\mathbf{AV})^r + r\boldsymbol{\mu}'\mathbf{A}(\mathbf{VA})^{r-1}\boldsymbol{\mu}]. \tag{44}$$

(iii) Finally, the covariance between **x** and **x'Ax** is

$$\begin{aligned}
\operatorname{cov}(\mathbf{x},\mathbf{x}'\mathbf{A}\mathbf{x}) &= E(\mathbf{x}-\boldsymbol{\mu})[\mathbf{x}'\mathbf{A}\mathbf{x}-E(\mathbf{x}'\mathbf{A}\mathbf{x})] \\
&= E(\mathbf{x}-\boldsymbol{\mu})[\mathbf{x}'\mathbf{A}\mathbf{x}-\boldsymbol{\mu}'\mathbf{A}\boldsymbol{\mu}-\operatorname{tr}(\mathbf{AV})] \\
&= E(\mathbf{x}-\boldsymbol{\mu})[(\mathbf{x}-\boldsymbol{\mu})'\mathbf{A}(\mathbf{x}-\boldsymbol{\mu})+2(\mathbf{x}-\boldsymbol{\mu})'\mathbf{A}\boldsymbol{\mu}-\operatorname{tr}(\mathbf{AV})] \\
&= \mathbf{0}+2\mathbf{VA}\boldsymbol{\mu}-\mathbf{0}
\end{aligned}$$

because the first and third moments of $(\mathbf{x}-\boldsymbol{\mu})$ are zero. Hence

$$\operatorname{cov}(\mathbf{x},\mathbf{x}'\mathbf{A}\mathbf{x}) = 2\mathbf{VA}\boldsymbol{\mu}$$

and the theorem is proved.

Corollary 1.1. When $\boldsymbol{\mu} = \mathbf{0}$

$$E(\mathbf{x}'\mathbf{A}\mathbf{x}) = \operatorname{tr}(\mathbf{AV}),$$

and under normality

$$K_r(\mathbf{x}'\mathbf{A}\mathbf{x}) = 2^{r-1}(r-1)!\operatorname{tr}(\mathbf{AV})^r$$

and

$$\operatorname{cov}(\mathbf{x},\mathbf{x}'\mathbf{A}\mathbf{x}) = \mathbf{0}.$$

These are the results given by Lancaster (1954) and others.

Corollary 1.2. An important application of the theorem is the value of its second part when $r = 2$, for then it gives the variance of $\mathbf{x}'\mathbf{A}\mathbf{x}$:

$$\begin{aligned} v(\mathbf{x}'\mathbf{A}\mathbf{x}) &= 2\,\text{tr}(\mathbf{AV})^2 + 4\boldsymbol{\mu}'\mathbf{A}(\mathbf{VA})\boldsymbol{\mu} \\ &= 2\,\text{tr}(\mathbf{AV})^2 + 4\boldsymbol{\mu}'\mathbf{AVA}\boldsymbol{\mu}. \end{aligned} \tag{45}$$

Corollary 1.3. When $\mathbf{x} \sim N(\mathbf{0}, \mathbf{V})$

$$v(\mathbf{x}'\mathbf{A}\mathbf{x}) = 2\,\text{tr}(\mathbf{AV})^2.$$

b. Distributions

Theorem 2. When $\mathbf{x}$ is $N(\boldsymbol{\mu}, \mathbf{V})$ then $\mathbf{x}'\mathbf{A}\mathbf{x}$ is $\chi^{2'}[r(\mathbf{A}), \frac{1}{2}\boldsymbol{\mu}'\mathbf{A}\boldsymbol{\mu}]$ if and only if $\mathbf{AV}$ is idempotent.

Proof (sufficiency). Given that $\mathbf{AV}$ is idempotent to show that $\mathbf{x}'\mathbf{A}\mathbf{x}$ is $\chi^{2'}[r(\mathbf{A}), \frac{1}{2}\boldsymbol{\mu}'\mathbf{A}\boldsymbol{\mu}]$.

From (42) the m.g.f. of $\mathbf{x}'\mathbf{A}\mathbf{x}$ is

$$\begin{aligned} M_{\mathbf{x}'\mathbf{A}\mathbf{x}}(t) &= |\mathbf{I} - 2t\mathbf{AV}|^{-\frac{1}{2}} \exp\{-\tfrac{1}{2}\boldsymbol{\mu}'[\mathbf{I} - (\mathbf{I} - 2t\mathbf{AV})^{-1}]\mathbf{V}^{-1}\boldsymbol{\mu}\} \\ &= \prod_{i=1}^{n}(1 - 2t\lambda_i)^{-\frac{1}{2}} \exp\left\{-\tfrac{1}{2}\boldsymbol{\mu}'\left[-\sum_{k=1}^{\infty}(2t)^k(\mathbf{AV})^k\right]\mathbf{V}^{-1}\boldsymbol{\mu}\right\} \end{aligned}$$

where the λ_i, for $i = 1, 2, \ldots, n$, are the latent roots of $\mathbf{AV}$. Now if $\mathbf{AV}$ is idempotent and r is its rank, r values of the λ_i are unity and $n - r$ are zero; and $(\mathbf{AV})^r = \mathbf{AV}$, so that

$$\begin{aligned} M_{\mathbf{x}'\mathbf{A}\mathbf{x}} &= \prod_{i=1}^{r}(1 - 2t)^{-\frac{1}{2}} \exp\left\{-\tfrac{1}{2}\boldsymbol{\mu}'\left[-\sum_{k=1}^{\infty}(2t)^k\right]\mathbf{AVV}^{-1}\boldsymbol{\mu}\right\} \\ &= (1 - 2t)^{-\frac{1}{2}r} \exp\{-\tfrac{1}{2}\boldsymbol{\mu}'[1 - (1 - 2t)^{-1}]\mathbf{A}\boldsymbol{\mu}\} \\ &= (1 - 2t)^{-\frac{1}{2}r} \exp\{-\tfrac{1}{2}\boldsymbol{\mu}'\mathbf{A}\boldsymbol{\mu}[1 - (1 - 2t)^{-1}]\}. \end{aligned} \tag{46}$$

By comparison with (38) we see that $\mathbf{x}'\mathbf{A}\mathbf{x}$ is $\chi^{2'}(r, \frac{1}{2}\boldsymbol{\mu}'\mathbf{A}\boldsymbol{\mu})$ where $r = r(\mathbf{AV})$. And, since $\mathbf{V}$ is non-singular, $r(\mathbf{AV}) = r(\mathbf{A})$. Hence $\mathbf{x}'\mathbf{A}\mathbf{x}$ is $\chi^{2'}[r(\mathbf{A}), \frac{1}{2}\boldsymbol{\mu}'\mathbf{A}\boldsymbol{\mu}]$.

Proof (necessity). Given that $\mathbf{x}'\mathbf{A}\mathbf{x}$ is $\chi^{2'}(r, \frac{1}{2}\boldsymbol{\mu}'\mathbf{A}\boldsymbol{\mu})$ to show that $\mathbf{AV}$ is idempotent of rank r.

In this case, knowing the distribution of $\mathbf{x}'\mathbf{A}\mathbf{x}$ we have the m.g.f. of $\mathbf{x}'\mathbf{A}\mathbf{x}$ as given in (46), and it is also the form shown in (42). These two forms must be equal—and equal for all values of $\boldsymbol{\mu}$, in particular for $\boldsymbol{\mu} = \mathbf{0}$. Substituting $\boldsymbol{\mu} = \mathbf{0}$ into (42) and (46) and equating gives

$$(1 - 2t)^{-\frac{1}{2}r} = |\mathbf{I} - 2t\mathbf{AV}|^{-\frac{1}{2}}.$$

Writing u for $2t$ and rearranging gives

$$(1 - u)^r = |\mathbf{I} - u\mathbf{AV}|.$$

Letting $\lambda_1, \lambda_2, \ldots, \lambda_n$ be the latent roots of **AV** we then have

$$(1-u)^r = \prod_{i=1}^{n}(1-u\lambda_i).$$

This being an identity in u its right-hand side has no powers of u exceeding r. Hence at least one λ_i is zero. Repeated use of this argument shows that $(n-r)$ of the λ_i's are zero, and so we can write

$$(1-u)^r = \prod_{i=1}^{r}(1-u\lambda_i).$$

Taking logarithms of both sides and equating coefficients gives r equations in the r unknown λ's, namely, all sums of powers of the λ's equal r. These have a solution $\lambda_i = 1$ for $i = 1, 2, \ldots, r$. Thus $n - r$ latent roots of **AV** are zero and r of them are unity. Therefore, by Lemma 9, **AV** is idempotent and the theorem is proved.

Operationally the most important part of this theorem is the sufficiency condition, namely that if **AV** is idempotent then $\mathbf{x'Ax}$ has a non-central χ^2-distribution. However, there are also occasions when the necessity condition is useful.

The theorem does of course have an endless variety of corollaries, depending on the values of $\boldsymbol{\mu}$ and **V** and choice of **A**. For example, consider $\sum_{i=1}^{n}(x_i-\bar{x})^2 = \mathbf{x'H_0'H_0x}$, where $\mathbf{H_0}$ is the last $n-1$ rows of the n-order Helmert matrix discussed in Sec. 1 and exemplified in equation (4) for $n = 4$. Then $\mathbf{H_0H_0'} = \mathbf{I}$ and $\mathbf{H_0'H_0}$ is idempotent. Hence, if **x** is $N(\mu\mathbf{1}, \sigma^2\mathbf{I})$, Theorem 2 tells us that $\sum_{i=1}^{n}(x_i-\bar{x})^2/\sigma^2$ is $\chi^{2\prime}(n-1, \frac{1}{2}\mu\mathbf{1'H_0'H_01}\mu/\sigma^2)$, which is $\chi^2(n-1,0)$ because $\mathbf{1'H_0'H_01} = 0$. Certain more direct corollaries of special interest can be stated as follows.

Corollary 2.1. If **x** is $N(\mathbf{0}, \mathbf{I})$, then $\mathbf{x'Ax}$ is χ^2_r if and only if **A** is idempotent of rank r.

Corollary 2.2. If **x** is $N(\mathbf{0}, \mathbf{V})$ then $\mathbf{x'Ax}$ is χ^2_r if and only if **AV** is idempotent of rank r.

Corollary 2.3. If **x** is $N(\boldsymbol{\mu}, \sigma^2\mathbf{I})$ then $\mathbf{x'x}/\sigma^2$ is $\chi^{2\prime}(n, \frac{1}{2}\boldsymbol{\mu}'\boldsymbol{\mu}/\sigma^2)$.

Corollary 2.4. If **x** is $N(\boldsymbol{\mu}, \mathbf{I})$, then $\mathbf{x'Ax}$ is $\chi^{2\prime}(r, \frac{1}{2}\boldsymbol{\mu}'\mathbf{A}\boldsymbol{\mu})$ if and only if **A** is idempotent of rank r.

Additional special cases are easily established.

The proof of Theorem 2 is based upon moment generating functions. The expression for the cumulants of $\mathbf{x'Ax}$ is given in (44). It shows that when

$\mathbf{x}'\mathbf{A}\mathbf{x}$ has a non-central χ^2-distribution, i.e., when $\mathbf{AV}$ is idempotent of rank r, the kth cumulant of $\mathbf{x}'\mathbf{A}\mathbf{x}$ (with $\mathbf{A}$ being symmetric) is

$$K_k(\mathbf{x}'\mathbf{A}\mathbf{x}) = 2^{k-1}(k-1)![r(\mathbf{A}) + k\boldsymbol{\mu}'\mathbf{A}\boldsymbol{\mu}]. \tag{47}$$

c. Independence

Under this heading we consider the independence of: 1. a quadratic form and a linear form, 2. two quadratic forms, and 3. sets of quadratic forms. There is a theorem for each case. In considering independence let us remember that when two random variables are distributed independently their covariance is always zero. But the fact of two variables having a zero covariance does not always imply independence; it does under normality assumptions.

Theorem 3. When $\mathbf{x} \sim N(\boldsymbol{\mu}, \mathbf{V})$, then $\mathbf{x}'\mathbf{A}\mathbf{x}$ and $\mathbf{Bx}$ are distributed independently if and only if $\mathbf{BVA} = \mathbf{0}$.

Two facets of the theorem are worth noting before proving it: $\mathbf{x}'\mathbf{A}\mathbf{x}$ does not have to have a non-central χ^2-distribution for the theorem to apply; and the theorem does not involve $\mathbf{AVB}$, a product that does not necessarily exist.

Proof of sufficiency: that $\mathbf{BVA} = \mathbf{0}$ implies independence.

From Lemma 7, because $\mathbf{A}$ is symmetric, we have that $\mathbf{A} = \mathbf{LL}'$ for some $\mathbf{L}$ of full column rank. Therefore, if $\mathbf{BVA} = \mathbf{0}$, $\mathbf{BVLL}' = \mathbf{0}$. Since $\mathbf{L}$ has full column rank, $(\mathbf{L}'\mathbf{L})^{-1}$ exists (Corollary to Lemma 9, Chapter 1) and so

$$\mathbf{BVLL}' = \mathbf{0} \quad \text{implies} \quad \mathbf{BVLL}'\mathbf{L}(\mathbf{L}'\mathbf{L})^{-1} = \mathbf{0}, \quad \text{i.e.,} \quad \mathbf{BVL} = \mathbf{0}.$$

Therefore $$\operatorname{cov}(\mathbf{Bx}, \mathbf{x}'\mathbf{L}) = \mathbf{BVL} = \mathbf{0}.$$

Hence, because $\mathbf{x}$ is a vector of normally distributed variables, $\mathbf{Bx}$ and $\mathbf{x}'\mathbf{L}$ are distributed independently. Consequently $\mathbf{Bx}$ and $\mathbf{x}'\mathbf{A}\mathbf{x} = \mathbf{x}'\mathbf{LL}'\mathbf{x}$ are distributed independently.

Proof of necessity: that independence of $\mathbf{x}'\mathbf{A}\mathbf{x}$ and $\mathbf{Bx}$ implies $\mathbf{BVA} = \mathbf{0}$.

The independence property gives $\operatorname{cov}(\mathbf{Bx}, \mathbf{x}'\mathbf{A}\mathbf{x}) = \mathbf{0}$; and Theorem 1(iii) gives $\operatorname{cov}(\mathbf{Bx}, \mathbf{x}'\mathbf{A}\mathbf{x}) = 2\mathbf{BVA}\boldsymbol{\mu}$. Hence $2\mathbf{BVA}\boldsymbol{\mu} = \mathbf{0}$, and since this is true for all $\boldsymbol{\mu}$, $\mathbf{BVA} = \mathbf{0}$, and so the proof is complete.

The next theorem, dealing with the independence of two quadratic forms, is similar to Theorem 3 just considered and its proof follows the same pattern.

Theorem 4. When $\mathbf{x} \sim N(\boldsymbol{\mu}, \mathbf{V})$, the quadratic forms $\mathbf{x}'\mathbf{A}\mathbf{x}$ and $\mathbf{x}'\mathbf{B}\mathbf{x}$ are distributed independently if and only if $\mathbf{AVB} = \mathbf{0}$ (or, equivalently, $\mathbf{BVA} = \mathbf{0}$).

Note that the form of the distributions of $\mathbf{x}'\mathbf{A}\mathbf{x}$ and $\mathbf{x}'\mathbf{B}\mathbf{x}$ is not specified in this theorem. It applies no matter what distributions these quadratics follow, provided only that $\mathbf{x}$ is a vector of normal variables. In practice, the theorem is usually applied in situations where the quadratic forms have χ^2-distributions, as determined by Theorem 2, but this is not a prerequisite of Theorem 4.

Proof. The condition $\mathbf{AVB} = \mathbf{0}$ is equivalent to $\mathbf{BVA} = \mathbf{0}$ because $\mathbf{A}$, $\mathbf{B}$ and $\mathbf{V}$ are symmetric. Each condition therefore implies the other.

Sufficiency: that $\mathbf{AVB} = \mathbf{0}$ implies independence.

By Lemma 7 we can write $\mathbf{A} = \mathbf{LL}'$ and $\mathbf{B} = \mathbf{MM}'$, where each of $\mathbf{L}$ and $\mathbf{M}$ have full column rank. Therefore, if $\mathbf{AVB} = \mathbf{0}$, $\mathbf{LL}'\mathbf{VMM}' = \mathbf{0}$, and because $(\mathbf{L}'\mathbf{L})^{-1}$ and $(\mathbf{M}'\mathbf{M})^{-1}$ exist this means $\mathbf{L}'\mathbf{VM} = \mathbf{0}$. Therefore

$$\text{cov}\,(\mathbf{L}'\mathbf{x}, \mathbf{x}'\mathbf{M}) = \mathbf{L}'\mathbf{VM} = \mathbf{0}.$$

Hence, because $\mathbf{x}$ is a vector of normally distributed variables, $\mathbf{L}'\mathbf{x}$ and $\mathbf{x}'\mathbf{M}$ are distributed independently. Consequently $\mathbf{x}'\mathbf{Ax} = \mathbf{x}'\mathbf{LL}'\mathbf{x}$ and $\mathbf{Bx} = \mathbf{x}'\mathbf{MMx}'$ are distributed independently.[1]

Necessity: that independence implies $\mathbf{AVB} = \mathbf{0}$.

When $\mathbf{x}'\mathbf{Ax}$ and $\mathbf{x}'\mathbf{Bx}$ are distributed independently, $\text{cov}(\mathbf{x}'\mathbf{Ax}, \mathbf{x}'\mathbf{Bx}) = 0$ so that

$$v(\mathbf{x}'\mathbf{Ax} + \mathbf{x}'\mathbf{Bx}) = v(\mathbf{x}'\mathbf{Ax}) + v(\mathbf{x}'\mathbf{Bx}),$$

i.e.,

$$v[\mathbf{x}'(\mathbf{A} + \mathbf{B})\mathbf{x}] = v(\mathbf{x}'\mathbf{Ax}) + v(\mathbf{x}'\mathbf{Bx}).$$

Applying equation (45) to all three terms in this result leads, after a little simplification, to

$$\text{tr}(\mathbf{VAVB}) + 2\boldsymbol{\mu}'\mathbf{AVB}\boldsymbol{\mu} = 0. \tag{48}$$

This is true for all $\boldsymbol{\mu}$, including $\boldsymbol{\mu} = \mathbf{0}$, so that $\text{tr}(\mathbf{VAVB}) = 0$ and on substituting back in (48) this gives $2\boldsymbol{\mu}'\mathbf{AVB}\boldsymbol{\mu} = 0$. This in turn is true for all $\boldsymbol{\mu}$, and so $\mathbf{AVB} = \mathbf{0}$. Thus the theorem is proved.

Before turning to the final theorem concerning independence, Theorem 5, recall that Theorems 3 and 4 are concerned with independence properties only, and apply whether or not the quadratic forms have χ^2-distributions. This is not the case with Theorem 5. It relates to the independence of quadratic forms in a sum of quadratics and is concerned with conditions under which such forms have non-central χ^2-distributions. As such it involves idempotent matrices. The theorem follows; it is lengthy.

Theorem 5. Let the following be given:

$$\mathbf{x}, \text{ order } n \times 1, \text{ distributed as } N(\boldsymbol{\mu}, \mathbf{V});$$

$$\mathbf{A}_i,\ n \times n, \text{ symmetric, of rank } k_i, \text{ for } i = 1, 2, \ldots, p;$$

---

[1] For the proofs of sufficiency in Theorems 3 and 4, I am grateful for discussions with D. L. Solomon and N. S. Urquhart. Proofs can also be established, very tediously, using moment generating functions.

and $\qquad \mathbf{A} = \sum_{i=1}^{p} \mathbf{A}_i$, which is symmetric, with rank k.

Then $\qquad \mathbf{x}'\mathbf{A}_i\mathbf{x}$ is $\chi^{2\prime}(k_i, \tfrac{1}{2}\boldsymbol{\mu}'\mathbf{A}_i\boldsymbol{\mu})$,

and $\qquad$ the $\mathbf{x}'\mathbf{A}_i\mathbf{x}$ are pairwise independent

and $\qquad \mathbf{x}'\mathbf{A}\mathbf{x}$ is $\chi^{2\prime}(k, \tfrac{1}{2}\boldsymbol{\mu}'\mathbf{A}\boldsymbol{\mu})$

if and only if

I: any 2 of (*a*) $\mathbf{A}_i\mathbf{V}$ idempotent, for all i,

(*b*) $\mathbf{A}_i\mathbf{V}\mathbf{A}_j = \mathbf{0}$ for all $i < j$,

(*c*) $\mathbf{A}\mathbf{V}$ idempotent,

are true;

or II: $\qquad$ (*c*) is true and (*d*), $k = \sum_{i=1}^{p} k_i$;

or III: $\qquad$ (*c*) is true and (*e*), $\mathbf{A}_1\mathbf{V}, \ldots, \mathbf{A}_{(p-1)}\mathbf{V}$ are idempotent and $\mathbf{A}_p\mathbf{V}$ is non-negative definite.

Proof of this theorem in statistics rests upon a theorem in matrices, which in turn depends upon a lemma. The matrix theorem, given below as Theorem 5a, is an extension of Graybill (1961, Theorems 1.68 and 1.69). The proof given by Graybill and Marsaglia (1957) is lengthy; that given here follows the much shorter proof of Banerjee (1964) as improved by Loynes (1966), based upon a lemma. Accordingly we first state and prove the lemma given by Loynes.

Loynes' Lemma. If $\mathbf{B}$ is symmetric and idempotent, if $\mathbf{Q}$ is symmetric and non-negative definite, and if $\mathbf{I} - \mathbf{B} - \mathbf{Q}$ is non-negative definite, then $\mathbf{BQ} = \mathbf{QB} = \mathbf{0}$.

Proof of Loynes' Lemma. Let $\mathbf{x}$ be any vector and let $\mathbf{y} = \mathbf{Bx}$. Then

$$\mathbf{y}'\mathbf{B}\mathbf{y} = \mathbf{y}'\mathbf{B}^2\mathbf{x} = \mathbf{y}'\mathbf{B}\mathbf{x} = \mathbf{y}'\mathbf{y},$$

and so

$$\mathbf{y}'(\mathbf{I} - \mathbf{B} - \mathbf{Q})\mathbf{y} = -\mathbf{y}'\mathbf{Q}\mathbf{y}.$$

Furthermore, because $\mathbf{I} - \mathbf{B} - \mathbf{Q}$ is n.n.d.,

$$\mathbf{y}'(\mathbf{I} - \mathbf{B} - \mathbf{Q})\mathbf{y} \geq 0.$$

Hence, $-\mathbf{y}'\mathbf{Q}\mathbf{y} \geq 0$ and so, because $\mathbf{Q}$ is n.n.d. also, $\mathbf{y}'\mathbf{Q}\mathbf{y} = 0$. In addition, since $\mathbf{Q}$ is symmetric, $\mathbf{Q} = \mathbf{L}'\mathbf{L}$ for some $\mathbf{L}$ and therefore $\mathbf{y}'\mathbf{Q}\mathbf{y} \equiv \mathbf{y}'\mathbf{L}'\mathbf{L}\mathbf{y} = 0$ implies $\mathbf{L}\mathbf{y} = \mathbf{0}$ and hence $\mathbf{L}'\mathbf{L}\mathbf{y} = \mathbf{0}$; i.e., $\mathbf{Q}\mathbf{y} = \mathbf{QBx} = \mathbf{0}$. Since this is true for *any* $\mathbf{x}$, $\mathbf{QB} = \mathbf{0}$ and so

$$(\mathbf{QB})' = \mathbf{B}'\mathbf{Q}' = \mathbf{BQ} = \mathbf{0}.$$

Thus is the lemma proved. The matrix theorem follows.

Theorem 5a. Let the following be given:

$$\mathbf{X}_i,\ n \times n,\ \text{symmetric, rank } k_i,\ i = 1, 2, \ldots, p.$$

$$\mathbf{X} = \sum_{i=1}^{p} \mathbf{X}_i, \text{ which is symmetric, with rank } k.$$

Then of the conditions (*a*) $\mathbf{X}_i$, idempotent for all i,

(*b*) $\mathbf{X}_i\mathbf{X}_j = \mathbf{0}$ for $i \neq j$,

(*c*) $\mathbf{X}$ idempotent,

(*d*) $k = \sum_{i=1}^{p} k_i$,

it is true that

I: any 2 of (*a*), (*b*) and (*c*) imply (*a*), (*b*), (*c*) and (*d*);

II: (*c*) and (*d*) imply (*a*) and (*b*);

and III: (*c*) and $\mathbf{X}_1, \mathbf{X}_2, \ldots, \mathbf{X}_{p-1}$ being idempotent with $\mathbf{X}_p$ being non-negative definite, imply that $\mathbf{X}_p$ is idempotent also and hence (*a*); and therefore (*b*) and (*d*).

The analogies between Theorems 5 and 5a are clear; once 5a is proved, the proof of 5 is relatively brief. The part played by Theorem 5a is that it shows that in situations in which any one of sections I, II or III of Theorem 5 hold true, then all of conditions (*a*), (*b*) and (*c*) in section I will hold. The consequences of Theorem 5, the independence of quadratics and their χ^2-distributions, then arise directly from Theorems 2 and 4.

Proof of Theorem 5a. We first prove section I, doing it in four parts.

I(i): Given (*c*), $\mathbf{I} - \mathbf{X}$ is idempotent and hence n.n.d.; and $\mathbf{X} - \mathbf{X}_i - \mathbf{X}_j = \sum_{r \neq i \neq j} \mathbf{X}_r$ is, given (*a*), n.n.d. Therefore $\mathbf{I} - \mathbf{X} + \mathbf{X} - \mathbf{X}_i - \mathbf{X}_j = \mathbf{I} - \mathbf{X}_i - \mathbf{X}_j$ is n.n.d. and so, by Loynes' Lemma, $\mathbf{X}_i\mathbf{X}_j = \mathbf{0}$, which is (*b*). Hence (*a*) and (*c*) imply (*b*).

I(ii): Let λ be a latent root and $\mathbf{u}$ the corresponding latent vector of $\mathbf{X}_1$. Then $\mathbf{X}_1\mathbf{u} = \lambda\mathbf{u}$, and for $\lambda \neq 0$, $\mathbf{u} = \mathbf{X}_1\mathbf{u}/\lambda$. Hence $\mathbf{X}_i\mathbf{u} = \mathbf{X}_i\mathbf{X}_1 u/\lambda$ for $i \neq 1$ is, given (*b*), $\mathbf{0}$; and therefore $\mathbf{X}\mathbf{u} = \mathbf{X}_1\mathbf{u} = \lambda\mathbf{u}$ and so λ is a latent root of $\mathbf{X}$. But, given (*c*), $\mathbf{X}$ is idempotent and hence $\lambda = 0$ or 1. Therefore $\mathbf{X}_1$ is, by Lemma 8, idempotent. Similarly the other $\mathbf{X}_i$'s are idempotent and thus (*a*) is established. Hence (*b*) and (*c*) imply (*a*).

I(iii): Given (*b*) and (*a*) $\mathbf{X}^2 = \sum \mathbf{X}_i^2 = \sum \mathbf{X}_i = \mathbf{X}$, which is (*c*). Thus (*a*) and (*b*) imply (*c*).

I(iv): Given (*c*), $r(\mathbf{X}) = \text{tr}(\mathbf{X})$ and so

$$k = r(\mathbf{X}) = \text{tr}(\mathbf{X}) = \text{tr}(\textstyle\sum \mathbf{X}_i) = \sum \text{tr}(\mathbf{X}_i),$$

and on being given (*a*) $\sum \text{tr}(\mathbf{X}_i) = \sum k_i$. Hence $k = \sum k_i$, which is (*d*). Thus (*a*) and (*c*) imply (*d*).

(II): The proof of this section follows that of Loynes (1966). Given (*c*), $\mathbf{I} - \mathbf{X}$ is idempotent and therefore $\mathbf{X} - \mathbf{I}$ has rank $n - k$; i.e., $\mathbf{X} - \mathbf{I}$ has $n - k$ linearly independent (LIN) rows. Therefore

in $(\mathbf{X} - \mathbf{I})\mathbf{x} = \mathbf{0}$ there are $n - k$ LIN equations;

and in $\mathbf{X}_2\mathbf{x} = \mathbf{0}$ there are k_2 LIN equations;

$\vdots$

and in $\mathbf{X}_p\mathbf{x} = \mathbf{0}$ there are k_p LIN equations.

However, these LIN sets of equations are not all mutually LIN; for example, the k_2 LIN equations in $\mathbf{X}_2\mathbf{x} = \mathbf{0}$ may not be LIN of the k_p LIN equations in $\mathbf{X}_p\mathbf{x} = \mathbf{0}$. Therefore, in

$$\begin{bmatrix} \mathbf{X} - \mathbf{I} \\ \mathbf{X}_2 \\ \vdots \\ \mathbf{X}_p \end{bmatrix} \mathbf{x} = \mathbf{0}$$

the maximum number of LIN equations is, given (*d*),

$$n - k + k_2 + \cdots + k_p = n - k_1;$$

and the equations reduce to $\mathbf{X}_1\mathbf{x} = \mathbf{x}$. Thus the minimum number of LIN solutions to $\mathbf{X}_1\mathbf{x} = \mathbf{x}$ is $n - (n - k_1) = k_1$; that is, for at least k_1 LIN vectors $\mathbf{x}$, $\mathbf{X}_1\mathbf{x} = \mathbf{x} = 1\mathbf{x}$. Hence 1 is a latent root of $\mathbf{X}_1$ with multiplicity at least equal to k_1. But $r(\mathbf{X}_1) = k_1$ and so $\mathbf{X}_1$ has only k_1 non-zero latent roots and so, by Lemma 8, is idempotent; similarly so are the other $\mathbf{X}_i$'s, and thus is (*a*) established. Thus (*c*) and (*d*) imply (*a*) and hence, by I(i), (*b*); and so II is proved.

III: Given (*c*), $\mathbf{X}$ is n.n.d. and then so is $\mathbf{I} - \mathbf{X}$. With $\mathbf{X}_1, \ldots, \mathbf{X}_{p-1}$ being idempotent and hence p.s.d., and $\mathbf{X}_p$ n.n.d. also, then

$$\sum_{r \neq i \neq j}^{p} \mathbf{X}_r = \mathbf{X} - \mathbf{X}_i - \mathbf{X}_j \text{ is n.n.d.}$$

Therefore

$$\mathbf{I} - \mathbf{X} + \mathbf{X} - \mathbf{X}_i - \mathbf{X}_j = \mathbf{I} - \mathbf{X}_i - \mathbf{X}_j \text{ is n.n.d.}$$

and so, by Loynes' Lemma, $\mathbf{X}_i\mathbf{X}_j = \mathbf{0}$; i.e., (*b*) is true. Therefore (*a*) and (*d*) are implied also, and both this section and the whole theorem are proved.

We now have to show how Theorem 5a leads to proving Theorem 5.

Proof of Theorem 5. Since $\mathbf{V}$ is symmetric and positive definite, $\mathbf{V} = \mathbf{T}'\mathbf{T}$ by Lemma 4, for some non-singular $\mathbf{T}$. Then, since $\mathbf{A}_i$ is symmetric, so is $\mathbf{T}\mathbf{A}_i\mathbf{T}'$ and $r(\mathbf{A}_i) = r(\mathbf{T}\mathbf{A}_i\mathbf{T}')$; and $\mathbf{A}_i\mathbf{V}$ is idempotent if and only if $\mathbf{T}\mathbf{A}_i\mathbf{T}'$ is; and $\mathbf{A}_i\mathbf{V}\mathbf{A}_j = \mathbf{0}$ if and only if $\mathbf{T}\mathbf{A}_i\mathbf{T}'\mathbf{T}\mathbf{A}_j\mathbf{T}' = \mathbf{0}$. Hence Theorem 5a holds true using $\mathbf{T}\mathbf{A}_i\mathbf{T}'$ in place of $\mathbf{X}_i$ (and $\mathbf{TAT}'$ in place of $\mathbf{X}$). Then sections I, II and III of Theorem 5a applied to $\mathbf{T}\mathbf{A}_i\mathbf{T}'$ and $\mathbf{TAT}'$ show that when sections I, II or III of Theorem 5 exist conditions (*a*), (*b*) and (*c*) always exist. But, by Theorem 2, $\mathbf{x}'\mathbf{A}_i\mathbf{x}$ is $\chi^{2\prime}(k_i, \frac{1}{2}\boldsymbol{\mu}'\mathbf{A}_i\boldsymbol{\mu})$ if and only if (*a*) is true; also, $\mathbf{x}'\mathbf{A}\mathbf{x}$ is $\chi^{2\prime}(k, \frac{1}{2}\boldsymbol{\mu}'\mathbf{A}\boldsymbol{\mu})$ if and only if (*c*) is true. And by Theorem 4 $\mathbf{x}'\mathbf{A}_i\mathbf{x}$ and $\mathbf{x}'\mathbf{A}_j\mathbf{x}$ are independent if and only if condition (*b*) is true. And so Theorem 5 is proved.

Corollary 5.1. (Cochran's Theorem). When $\mathbf{x}$ is $N(\mathbf{0}, \mathbf{I}_n)$ and $\mathbf{A}_i$ is symmetric of rank r_i for $i = 1, \ldots, p$ with $\sum_{i=1}^{p} \mathbf{A}_i = \mathbf{I}_n$, then the $\mathbf{x}'\mathbf{A}_i\mathbf{x}$, are distributed independently as $\chi^2_{r_i}$ if and only if $\sum_{i=1}^{p} r_i = n$.

Proof. Put $\boldsymbol{\mu} = \mathbf{0}$ and $\mathbf{V} = \mathbf{I}_n = \mathbf{A}$ in Theorem 5. This is the well-known theorem first proved by Cochran (1934).

6. BILINEAR FORMS

Knowing the distributional properties of quadratic forms of normal variables enables us to discuss properties of bilinear forms. We consider the general bilinear form $\mathbf{x}_1'\mathbf{A}_{12}\mathbf{x}_2$ where $\mathbf{x}_1$ and $\mathbf{x}_2$ are of order n_1 and n_2, distributed as $N(\boldsymbol{\mu}_1, \mathbf{C}_{11})$ and as $N(\boldsymbol{\mu}_2, \mathbf{C}_{22})$ respectively, with the matrix of covariances between $\mathbf{x}_1$ and $\mathbf{x}_2$ being $\mathbf{C}_{12}$ of order $n_1 \times n_2$; i.e.,

$$E(\mathbf{x}_1 - \boldsymbol{\mu}_1)(\mathbf{x}_2 - \boldsymbol{\mu}_2)' = \mathbf{C}_{12}.$$

Properties of the bilinear form are readily derived from those of quadratic forms because $\mathbf{x}_1'\mathbf{A}_{12}\mathbf{x}_2$ can be expressed as a quadratic form:

$$\mathbf{x}_1'\mathbf{A}_{12}\mathbf{x}_2 = \tfrac{1}{2}[\mathbf{x}_1' \quad \mathbf{x}_2']\begin{bmatrix} \mathbf{0} & \mathbf{A}_{12} \\ \mathbf{A}_{21} & \mathbf{0} \end{bmatrix}\begin{bmatrix} \mathbf{x}_1 \\ \mathbf{x}_2 \end{bmatrix} \qquad \text{with} \quad \mathbf{A}_{21} = (\mathbf{A}_{12})'.$$

Hence

$$\mathbf{x}_1'\mathbf{A}_{12}\mathbf{x}_2 = \tfrac{1}{2}\mathbf{y}'\mathbf{B}\mathbf{y}$$

where

$$\mathbf{B} = \mathbf{B}' = \begin{bmatrix} \mathbf{0} & \mathbf{A}_{12} \\ \mathbf{A}_{21} & \mathbf{0} \end{bmatrix} \qquad \text{with} \quad \mathbf{A}_{21} = (\mathbf{A}_{12})',$$

and $\quad \mathbf{y}$ is $N(\boldsymbol{\mu}, \mathbf{V})$ with $\boldsymbol{\mu} = \begin{bmatrix} \boldsymbol{\mu}_1 \\ \boldsymbol{\mu}_2 \end{bmatrix}, \quad \mathbf{V} = \begin{bmatrix} \mathbf{C}_{11} & \mathbf{C}_{12} \\ \mathbf{C}_{21} & \mathbf{C}_{22} \end{bmatrix}$

and
$$\mathbf{C}_{21} = (\mathbf{C}_{12})'.$$

Thus properties of $\mathbf{x}_1'\mathbf{A}_{12}\mathbf{x}_2$ are equivalent to those of $\frac{1}{2}(\mathbf{y}'\mathbf{B}\mathbf{y})$ which, for some purposes, is better viewed as $\mathbf{y}'(\frac{1}{2}\mathbf{B})\mathbf{y}$.

Similar to Theorem 1, we have the mean value of $\mathbf{x}_1'\mathbf{A}_{12}\mathbf{x}_2$: whether the distribution of the x's is normal or not,

$$E(\mathbf{x}_1'\mathbf{A}_{12}\mathbf{x}_2) = \text{tr}(\mathbf{A}_{12}\mathbf{C}_{21}) + \boldsymbol{\mu}_1'\mathbf{A}_{12}\boldsymbol{\mu}_2. \tag{49}$$

This is proved in the same manner as is part (i) of Theorem 1. Also, from part (ii) of that theorem we have the rth cumulant of $\mathbf{x}_1'\mathbf{A}_{12}\mathbf{x}_2$ as

$$K_r(\mathbf{x}_1'\mathbf{A}_{12}\mathbf{x}_2) = \tfrac{1}{2}(r-1)![\text{tr}(\mathbf{BV})^r + r\boldsymbol{\mu}'\mathbf{B}(\mathbf{VB})^{r-1}\boldsymbol{\mu}]. \tag{50}$$

And from Theorem 2, $\mathbf{x}_1'\mathbf{A}_{12}\mathbf{x}_2$ is $\chi^{2\prime}[r(\mathbf{B}), \frac{1}{4}\boldsymbol{\mu}'\mathbf{B}\boldsymbol{\mu}]$ if and only if $\frac{1}{2}\mathbf{BV}$ is idempotent. With

$$\mathbf{BV} = \begin{bmatrix} \mathbf{A}_{12}\mathbf{C}_{21} & \mathbf{A}_{12}\mathbf{C}_{22} \\ \mathbf{A}_{21}\mathbf{C}_{11} & \mathbf{A}_{21}\mathbf{C}_{12} \end{bmatrix},$$

notice that, in general, idempotency of $\frac{1}{2}\mathbf{BV}$ does not imply (nor is it implied by) idempotency of $\mathbf{BV}$. In substituting $\mathbf{BV}$ into (50) use is made of $(\mathbf{A}_{21})' = \mathbf{A}_{12}$ and $(\mathbf{C}_{21})' = \mathbf{C}_{12}$ and also of the cyclic commutability of matrix products under the trace operation. In this way

$$\text{tr}(\mathbf{A}_{21}\mathbf{C}_{12}) = \text{tr}(\mathbf{C}_{12}\mathbf{A}_{21}) = \text{tr}(\mathbf{A}_{12}\mathbf{C}_{21})' = \text{tr}(\mathbf{A}_{12}\mathbf{C}_{21}). \tag{51}$$

A special case of (50) is when $r = 2$:

$$v(\mathbf{x}_1'\mathbf{A}_{12}\mathbf{x}_2) = \tfrac{1}{2}[\text{tr}(\mathbf{BV})^2 + 2\boldsymbol{\mu}'\mathbf{BVB}\boldsymbol{\mu}].$$

Substituting for $\mathbf{BV}$ and $\boldsymbol{\mu}$ and using (51) reduces this to

$$\begin{aligned} v(\mathbf{x}_1'\mathbf{A}_{12}\mathbf{x}_2) &= \text{tr}(\mathbf{A}_{12}\mathbf{C}_{21})^2 + \text{tr}(\mathbf{A}_{12}\mathbf{C}_{22}\mathbf{A}_{21}\mathbf{C}_{11}) \\ &\quad + \boldsymbol{\mu}_1'\mathbf{A}_{12}\mathbf{C}_{22}\mathbf{A}_{21}\boldsymbol{\mu}_1 + \boldsymbol{\mu}_2'\mathbf{A}_{21}\mathbf{C}_{11}\mathbf{A}_{12}\boldsymbol{\mu}_2 + 2\boldsymbol{\mu}_1'\mathbf{A}_{12}\mathbf{C}_{21}\mathbf{A}_{12}\boldsymbol{\mu}_2. \end{aligned} \tag{52}$$

We now derive the covariance between two bilinear forms $\mathbf{x}_1'\mathbf{A}_{12}\mathbf{x}_2$ and $\mathbf{x}_3'\mathbf{A}_{34}\mathbf{x}_4$, based on procedures developed by Evans (1969). Let $\mathbf{x}_1$, $\mathbf{x}_2$, $\mathbf{x}_3$ and $\mathbf{x}_4$ have order n_1, n_2, n_3 and n_4 respectively and be normally distributed with respective means $\boldsymbol{\mu}_1$, $\boldsymbol{\mu}_2$, $\boldsymbol{\mu}_3$ and $\boldsymbol{\mu}_4$ and covariance matrices $\mathbf{C}_{ij}$, of order $n_i \times n_j$, for $i, j = 1, 2, 3$ and 4:

$$\mathbf{C}_{ij} = E(\mathbf{x}_i - \boldsymbol{\mu}_i)(\mathbf{x}_j - \boldsymbol{\mu}_j)' = (\mathbf{C}_{ji})'. \tag{53}$$

Also define

$$\mathbf{x}' = [\mathbf{x}_1' \quad \mathbf{x}_2' \quad \mathbf{x}_3' \quad \mathbf{x}_4'] \quad \text{and} \quad \boldsymbol{\mu}' = [\boldsymbol{\mu}_1' \quad \boldsymbol{\mu}_2' \quad \boldsymbol{\mu}_3' \quad \boldsymbol{\mu}_4'] \tag{54}$$

with
$$\mathbf{C} = \{\mathbf{C}_{ij}\} \quad \text{for} \quad i, j = 1, 2, 3, 4 \tag{55}$$
for $\mathbf{C}_{ij}$ of (53); i.e., $\mathbf{x} \sim N(\boldsymbol{\mu}, \mathbf{C})$. Then, with
$$\mathbf{W} = \tfrac{1}{2}\begin{bmatrix} \mathbf{0} & \mathbf{A}_{12} & \mathbf{0} & \mathbf{0} \\ \mathbf{A}_{21} & \mathbf{0} & \mathbf{0} & \mathbf{0} \\ \mathbf{0} & \mathbf{0} & \mathbf{0} & \mathbf{A}_{34} \\ \mathbf{0} & \mathbf{0} & \mathbf{A}_{43} & \mathbf{0} \end{bmatrix}, \tag{56}$$
$$\mathbf{x}'\mathbf{W}\mathbf{x} = \mathbf{x}_1'\mathbf{A}_{12}\mathbf{x}_2 + \mathbf{x}_3'\mathbf{A}_{34}\mathbf{x}_4,$$
so that
$$2\,\mathrm{cov}(\mathbf{x}_1'\mathbf{A}_{12}\mathbf{x}_2, \quad \mathbf{x}_3'\mathbf{A}_{34}\mathbf{x}_4) = v(\mathbf{x}'\mathbf{W}\mathbf{x}) - v(\mathbf{x}_1'\mathbf{A}_{12}\mathbf{x}_2) - v(\mathbf{x}_3'\mathbf{A}_{34}\mathbf{x}_4). \tag{57}$$
Corollary 1.2 applied to the first term of (57) gives
$$v(\mathbf{x}'\mathbf{W}\mathbf{x}) = 2\,\mathrm{tr}(\mathbf{W}\mathbf{C})^2 + 4\boldsymbol{\mu}'\mathbf{W}\mathbf{C}\mathbf{W}\boldsymbol{\mu},$$
for $\boldsymbol{\mu}$, $\mathbf{C}$ and $\mathbf{W}$ of (54), (55) and (56) respectively; and using (52) for $v(\mathbf{x}_1'\mathbf{A}_{12}\mathbf{x}_2)$ and its analogue for $v(\mathbf{x}_3'\mathbf{A}_{34}\mathbf{x}_4)$ we then find that (57) reduces, after repetitive use of properties illustrated in (51), to
$$\begin{aligned} \mathrm{cov}(\mathbf{x}_1'\mathbf{A}_{12}\mathbf{x}_2, \quad \mathbf{x}_3'\mathbf{A}_{34}\mathbf{x}_4) &= \mathrm{tr}(\mathbf{A}_{12}\mathbf{C}_{23}\mathbf{A}_{34}\mathbf{C}_{41} + \mathbf{A}_{12}\mathbf{C}_{24}\mathbf{A}_{43}\mathbf{C}_{31}) \\ &\quad + \boldsymbol{\mu}_1'\mathbf{A}_{12}\mathbf{C}_{23}\mathbf{A}_{34}\boldsymbol{\mu}_4 + \boldsymbol{\mu}_1'\mathbf{A}_{12}\mathbf{C}_{24}\mathbf{A}_{43}\boldsymbol{\mu}_3 \\ &\quad + \boldsymbol{\mu}_2'\mathbf{A}_{21}\mathbf{C}_{13}\mathbf{A}_{34}\boldsymbol{\mu}_4 + \boldsymbol{\mu}_2'\mathbf{A}_{21}\mathbf{C}_{14}\mathbf{A}_{43}\boldsymbol{\mu}_3. \end{aligned} \tag{58}$$

This result does, of course, yield results obtained earlier when used for special cases. For example, to obtain $\mathrm{var}(\mathbf{x}'\mathbf{A}\mathbf{x})$ put all $\mathbf{A}_{ij}$'s equal to $\mathbf{A}$, all $\mathbf{C}_{ij}$'s equal to $\mathbf{V}$ and all $\boldsymbol{\mu}_i$'s equal to $\boldsymbol{\mu}$ and so get the variance of a quadratic form in $\mathbf{x} \sim N(\boldsymbol{\mu}, \mathbf{V})$ as
$$v(\mathbf{x}'\mathbf{A}\mathbf{x}) = 2\,\mathrm{tr}(\mathbf{A}\mathbf{V})^2 + 4\boldsymbol{\mu}'\mathbf{A}\mathbf{V}\mathbf{A}\boldsymbol{\mu}$$
as in (45). Also, to obtain the covariance between two quadratic forms in the same variables, $\mathbf{x}'\mathbf{P}\mathbf{x}$ and $\mathbf{x}'\mathbf{Q}\mathbf{x}$ say, put all the $\boldsymbol{\mu}$'s in (58) equal to $\boldsymbol{\mu}$, all the $\mathbf{C}$'s equal to $\mathbf{V}$, and put $\mathbf{A}_{12} = \mathbf{A}_{21} = \mathbf{P}$ and $\mathbf{A}_{34} = \mathbf{A}_{43} = \mathbf{Q}$ to give
$$\mathrm{cov}(\mathbf{x}'\mathbf{P}\mathbf{x}, \mathbf{x}'\mathbf{Q}\mathbf{x}) = 2\,\mathrm{tr}(\mathbf{P}\mathbf{V}\mathbf{Q}\mathbf{V}) + 4\boldsymbol{\mu}'\mathbf{P}\mathbf{V}\mathbf{Q}\boldsymbol{\mu}.$$

7. THE SINGULAR NORMAL DISTRIBUTION

Up to this point we have assumed that $\mathbf{V}$ is non-singular when $\mathbf{x}$ is $N(\boldsymbol{\mu}, \mathbf{V})$. We now consider the situation when $\mathbf{V}$ is singular. A simple example of this is the variance-covariance matrix of three random variables X_1, X_2 and $X_1 - X_2$.

If

$$\text{var}\begin{bmatrix} X_1 \\ X_2 \end{bmatrix} = \begin{bmatrix} \sigma_1^2 & \sigma_{12} \\ \sigma_{12} & \sigma_2^2 \end{bmatrix},$$

then
$$\mathbf{V} \equiv \text{var}\begin{bmatrix} X_1 \\ X_2 \\ X_1 - X_2 \end{bmatrix} = \begin{bmatrix} \sigma_1^2 & \sigma_{12} & \sigma_1^2 - \sigma_{12} \\ \sigma_{12} & \sigma_2^2 & \sigma_{12} - \sigma_2^2 \\ \sigma_1^2 - \sigma_{12} & \sigma_{12} - \sigma_2^2 & \sigma_1^2 + \sigma_2^2 - 2\sigma_{12} \end{bmatrix}$$

with $\mathbf{V}$ being singular. For such variables being normally distributed we emphasize the singularity of $\mathbf{V}$ by writing, in general, $\mathbf{x} \sim SN(\boldsymbol{\mu}, \mathbf{V})$.

Because $\mathbf{V}^{-1}$ does not exist, the density function of the $SN(\boldsymbol{\mu}, \mathbf{V})$ distribution cannot be written down. However, its characteristic function (m.g.f. using it in place of t) does exist; it is $e^{it'\boldsymbol{\mu} - \frac{1}{2}t'\mathbf{V}t}$. Therefore, by the continuity theorem for characteristic functions [see, for example, Cramer (1951, p. 312) and Anderson (1958, p. 25)], we are guaranteed that the density function exists, even though it cannot be written explicitly.

The general characterization of the $SN(\boldsymbol{\mu}, \mathbf{V})$ distribution given by Anderson (1958, p. 25) is useful. Suppose $\mathbf{y}$ is a vector having the $N(\mathbf{0}, \mathbf{I})$ distribution. Then variables obtained by the transformation $\mathbf{x} = \boldsymbol{\mu} + \mathbf{Ly}$ have the $SN(\boldsymbol{\mu}, \mathbf{LL}')$ distribution, when $\mathbf{LL}'$ is not of full rank. Situations arise in linear models that are similar to this, when we develop equations $\mathbf{X}'\mathbf{X}\mathbf{b}^o = \mathbf{X}'\mathbf{y}$ that have a solution $\mathbf{b}^o = \mathbf{G}\mathbf{X}'\mathbf{y}$ where $\mathbf{X}'\mathbf{X}$ is singular. Then, if $\mathbf{y}$ has a normal distribution, $\mathbf{b}^o$ will also, but its variance-covariance matrix will be singular. Discussion of the singular normal distribution is therefore pertinent. We consider five theorems, 1s–5s, analogues of those for non-singular $\mathbf{V}$ in Sec. 5. Although they are stated as applying to the $SN(\boldsymbol{\mu}, \mathbf{V})$ distribution, we henceforth take this to be either the singular or the non-singular normal distribution; i.e., $\mathbf{V}$ is to be considered as being either *singular or non-singular*. In the case that $\mathbf{V}$ is non-singular, Theorems 1s–5s reduce to Theorems 1–5 respectively.

Theorem 1s. When $\mathbf{x}$ is $SN(\boldsymbol{\mu}, \mathbf{V})$

(i)
$$E(\mathbf{x}'\mathbf{A}\mathbf{x}) = \text{tr}(\mathbf{AV}) + \boldsymbol{\mu}'\mathbf{A}\boldsymbol{\mu}$$

(true also when $\mathbf{x}$ is non-normal)

(ii) the rth cumulant of $\mathbf{x}'\mathbf{A}\mathbf{x}$ is

$$K_r(\mathbf{x}'\mathbf{A}\mathbf{x}) = 2^{r-1}(r-1)![\text{tr}(\mathbf{AV})^r + r\boldsymbol{\mu}'\mathbf{A}(\mathbf{VA})^{r-1}\boldsymbol{\mu}]; \tag{59}$$

and (iii) the covariance of $\mathbf{x}$ with $\mathbf{x}'\mathbf{A}\mathbf{x}$ is

$$\text{cov}(\mathbf{x}, \mathbf{x}'\mathbf{A}\mathbf{x}) = 2\mathbf{VA}\boldsymbol{\mu}.$$

The results in this theorem are identical to those of Theorem 1. Proofs of parts (i) and (iii) are also the same. Proof of part (ii) proceeds as follows, as in Rohde et al. (1966).

Proof of (ii). When $\mathbf{x}$ is $SN(\boldsymbol{\mu}, \mathbf{V})$ with $\mathbf{V}$ singular, there is no loss of generality in supposing that $\mathbf{x} = \boldsymbol{\mu} + \mathbf{Ly}$ where $\mathbf{y}$ is $N(\mathbf{0}, \mathbf{I}_k)$, and $\mathbf{V} = \mathbf{LL}'$ with $\mathbf{L}$ having full column rank k, as in Lemma 7. Then the m.g.f. of $\mathbf{x}'\mathbf{Ax}$ is

$$M_{\mathbf{x}'\mathbf{Ax}}(t) = (2\pi)^{-\frac{1}{2}k} \int_{-\infty}^{\infty} \cdots \int_{-\infty}^{\infty} \exp(t\mathbf{y}'\mathbf{L}'\mathbf{ALy} - \tfrac{1}{2}\mathbf{y}'\mathbf{y} + 2t\boldsymbol{\mu}'\mathbf{ALy} + t\boldsymbol{\mu}'\mathbf{A}\boldsymbol{\mu}) \times dy_1 \ldots dy_k$$

and application of (39) reduces this to

$$M_{\mathbf{x}'\mathbf{Ax}}(t) = |\mathbf{I} - 2t\mathbf{L}'\mathbf{AL}|^{-\frac{1}{2}} \exp[t\boldsymbol{\mu}'\mathbf{A}\boldsymbol{\mu} + 2t^2\boldsymbol{\mu}'\mathbf{AL}(\mathbf{I} - 2t\mathbf{L}'\mathbf{AL})^{-1}\mathbf{L}'\mathbf{A}\boldsymbol{\mu}].$$

Calling the logarithm of this $K_{\mathbf{x}'\mathbf{Ax}}(t)$ and using infinite sums for $-\frac{1}{2}\log|\mathbf{I} - 2t\mathbf{L}'\mathbf{AL}|$ and $(\mathbf{I} - 2t\mathbf{L}'\mathbf{AL})^{-1}$ similar to those used in deriving (44), we get

$$\begin{aligned} K_{\mathbf{x}'\mathbf{Ax}}(t) &= \sum_{r=1}^{\infty} (2^{r-1}t^r/r)\,\text{tr}(\mathbf{L}'\mathbf{AL})^r + t\boldsymbol{\mu}'\mathbf{A}\boldsymbol{\mu} + 2t^2\boldsymbol{\mu}'\mathbf{AL}\sum_{r=0}^{\infty} 2^r t^r (\mathbf{L}'\mathbf{AL})^r\mathbf{L}'\mathbf{AL} \\ &= t[\text{tr}(\mathbf{L}'\mathbf{AL}) + \boldsymbol{\mu}'\mathbf{A}\boldsymbol{\mu}] \\ &\quad + \sum_{r=2}^{\infty} t^r 2^{r-1}[\boldsymbol{\mu}'\mathbf{AL}(\mathbf{L}'\mathbf{AL})^{r-2}\mathbf{L}'\mathbf{A}\boldsymbol{\mu} + \text{tr}(\mathbf{L}'\mathbf{AL})^r/r]. \end{aligned} \tag{60}$$

Now $\mathbf{V} = \mathbf{LL}'$ and so

$$\text{tr}(\mathbf{L}'\mathbf{AL})^r = \text{tr}(\mathbf{VA})^r \qquad \text{for all positive integers } r; \tag{61}$$

also, by induction, it can be shown that

$$\mathbf{AL}(\mathbf{L}'\mathbf{AL})^{r-2}\mathbf{L}'\mathbf{A} = \mathbf{A}(\mathbf{VA})^{r-1}. \tag{62}$$

Hence

$$\begin{aligned} K_{\mathbf{x}'\mathbf{Ax}}(t) &= t[\boldsymbol{\mu}'\mathbf{A}\boldsymbol{\mu} + \text{tr}(\mathbf{AV})] + \sum_{r=2}^{\infty} t^r 2^{r-1}[\boldsymbol{\mu}'\mathbf{A}(\mathbf{VA})^{r-1}\boldsymbol{\mu} + \text{tr}(\mathbf{VA})^r/r] \\ &= \sum_{r=1}^{\infty} t^r 2^{r-1}[\boldsymbol{\mu}'\mathbf{A}(\mathbf{VA})^{r-1}\boldsymbol{\mu} + \text{tr}(\mathbf{VA})^r/r]. \end{aligned}$$

Hence the rth cumulant of $\mathbf{x}'\mathbf{Ax}$, the coefficient of $t^r/r!$ in $K_{\mathbf{x}'\mathbf{Ax}}(t)$, is as given in (59). Note that although the initial definition of the m.g.f. is in terms of $\mathbf{L}$, where $\mathbf{V} = \mathbf{LL}'$, the ultimate expression for the cumulant depends solely on $\mathbf{V}$ and not at all on $\mathbf{L}$, and it is identical to the result for non-singular $\mathbf{V}$ in Theorem 1.

There has recently been a plethora of theorems in the literature on the distribution of quadratic forms in singular normal variables [e.g., Rao (1962),

Khatri (1963), Rayner and Livingstone (1965), Rao (1966), Khatri (1968), Good (1969) and Styan (1969)]. Despite this we give only one here, that which appears to be the most general. It relates to a non-homogeneous form.

Theorem 2s. When $\mathbf{x}$ is $SN(\boldsymbol{\mu}, \mathbf{V})$, the form $\mathbf{x}'\mathbf{A}\mathbf{x} + \mathbf{m}'\mathbf{x} + d$ has a non-central χ^2-distribution with degrees of freedom tr($\mathbf{AV}$) and non-centrality parameter $\frac{1}{2}(\mathbf{A}\boldsymbol{\mu} + \frac{1}{2}\mathbf{m})'\mathbf{V}(\mathbf{A}\boldsymbol{\mu} + \frac{1}{2}\mathbf{m})$ if and only if

(i) $$\mathbf{VAVAV} = \mathbf{VAV},$$

(ii) $$(\mathbf{A}\boldsymbol{\mu} + \tfrac{1}{2}\mathbf{m})'\mathbf{V} = (\mathbf{A}\boldsymbol{\mu} + \tfrac{1}{2}\mathbf{m})'\mathbf{VAV}$$

and (iii) $$\boldsymbol{\mu}'\mathbf{A}\boldsymbol{\mu} + \mathbf{m}'\boldsymbol{\mu} + d = (\mathbf{A}\boldsymbol{\mu} + \tfrac{1}{2}\mathbf{m})'\mathbf{V}(\mathbf{A}\boldsymbol{\mu} + \tfrac{1}{2}\mathbf{m}).$$

This theorem is taken from Rayner and Livingston (1965, Theorem 7.2), who give its proof. Rao (1966) also discusses the topic. Of the many corollaries that can be established we mention but three.

Corollary 2s.1 ($\mathbf{m} = \mathbf{0}$ and $d = 0$.) When $\mathbf{x}$ is $N(\boldsymbol{\mu}, \mathbf{V})$, whether $\mathbf{V}$ be singular or non-singular, $\mathbf{x}'\mathbf{A}\mathbf{x}$ is $\chi^{2\prime}[\text{tr}(\mathbf{AV}), \frac{1}{2}\boldsymbol{\mu}'\mathbf{A}\boldsymbol{\mu}]$ if and only if

(i) $\mathbf{VAVAV} = \mathbf{VAV}$, (ii) $\boldsymbol{\mu}'\mathbf{AV} = \boldsymbol{\mu}'\mathbf{AVAV}$

and (iii) $\boldsymbol{\mu}'\mathbf{A}\boldsymbol{\mu} = \boldsymbol{\mu}'\mathbf{AVA}\boldsymbol{\mu}$.

Corollary 2s.2. ($\mathbf{m} = \mathbf{0}$, $d = 0$ and $\boldsymbol{\mu} = \mathbf{0}$.) When $\mathbf{x}$ is $N(\mathbf{0}, \mathbf{V})$, whether $\mathbf{V}$ be singular or non-singular, $\mathbf{x}'\mathbf{A}\mathbf{x}$ is $\chi^2_{tr(\mathbf{AV})}$ if and only if $\mathbf{VAVAV} = \mathbf{VAV}$.

Corollary 2s.3. (Theorem 2). When $\mathbf{V}$ is non-singular the conditions of Corollary 2s.1 reduce to idempotency of $\mathbf{AV}$.

Despite the condition of idempotency in Corollary 2s.3, when $\mathbf{V}$ is non-singular, one must *not* conclude in the theorem or in Corollaries 2s.1 and 2s.2 that $\mathbf{AV}$ is idempotent, for it is not necessarily so. With $\mathbf{V}$ being p.s.d., $\mathbf{V} = \mathbf{LL}'$ by Lemma 7 and $\mathbf{L}'\mathbf{L}$ is non-singular (by Lemma 9 in Sec. 1.6). Hence on all occasions condition (i) implies, and is implied by, the idempotency of $\mathbf{L}'\mathbf{AL}$, which in turn is equivalent to $\mathbf{AV}$ having all its latent roots equal to 0 or 1. But, by Lemma 9, only when $\mathbf{V}$ is non-singular does this condition imply the idempotency of $\mathbf{AV}$. This is the source of the error in the necessary condition given by Rao (1962), which he later corrected, (1966). The same error occurs in Good (1969), and has been corrected by Styan (1969), who indicates that Good (1969) misquotes Shanbhag (1968) on this point. An example follows.

Example. If

$$\mathbf{V} = \begin{bmatrix} 2 & 0 & -2 \\ 0 & 2 & -2 \\ -2 & -2 & 4 \end{bmatrix} \quad \text{and} \quad \mathbf{A} = (1/16)\begin{bmatrix} 16 & 6 & 5 \\ 6 & 4 & 3 \\ 5 & 3 & 2 \end{bmatrix},$$

then

$$\mathbf{VA} = (1/16)\begin{bmatrix} 22 & 6 & 6 \\ 2 & 2 & 2 \\ -24 & -8 & -8 \end{bmatrix} \text{ and } (\mathbf{VA})^2 = (1/16)\begin{bmatrix} 22 & 6 & 6 \\ 0 & 0 & 0 \\ -22 & -6 & -6 \end{bmatrix}.$$

Clearly, **VA** is not idempotent; but $\text{tr}(\mathbf{VA})^2 = \text{tr}(\mathbf{VA}) = 1$ and the latent roots of **VA** are 1, 0 and 0. Furthermore, condition (i) of Theorem 2s is satisfied, for it will be found that

$$\mathbf{VAVAV} = \begin{bmatrix} 2 & 0 & -2 \\ 0 & 0 & 0 \\ -2 & 0 & 2 \end{bmatrix} = \mathbf{VAV}.$$

The matrix **V** corresponds to $\mathbf{x}' = [x_1 \quad x_2 \quad -(x_1 + x_2)]$ as a result of which

$$\mathbf{x}'\mathbf{A}\mathbf{x} = (8x_1^2 + 2x_2^2 + x_3^2 + 6x_1x_2 + 5x_1x_3 + 3x_2x_3)/8 = \tfrac{1}{2}x_1^2,$$

because $x_3 = -(x_1 + x_2)$. Thus $\mathbf{x}'\mathbf{A}\mathbf{x} = \frac{1}{2}x_1^2$ is clearly distributed as $\chi^{2\prime}[1 = \text{tr}(\mathbf{VA}), \frac{1}{4}\mu_1^2]$ where the degrees of freedom are tr(**VA**); but **VA** is not idempotent.

Theorems relating to independence properties of quadratic forms are based on the work of Khatri (1963) and Good (1963). The one for the independence of a quadratic and a linear form, paralleling Theorem 3, stems from the following result given by Good [1963, Theorem 1C, parts (ii) and (iii)]: when **y** is $SN(\mathbf{0}, \mathbf{W})$, then $\mathbf{y}'\mathbf{P}\mathbf{y}$ and $\mathbf{q}'\mathbf{y}$ are independent if and only if $\mathbf{WPWq} = \mathbf{0}$, and $\mathbf{p}'\mathbf{y}$ and $\mathbf{q}'\mathbf{y}$ are independent if and only if $\mathbf{p}'\mathbf{W}\mathbf{q} = 0$. From this comes

Theorem 3s. When **x** is $SN(\boldsymbol{\mu}, \mathbf{V})$ then $\mathbf{x}'\mathbf{A}\mathbf{x}$ and $\mathbf{B}\mathbf{x}$ are independent if and only if $\mathbf{BVAV} = \mathbf{0}$ and $\mathbf{BVA}\boldsymbol{\mu} = \mathbf{0}$.

Proof. Write $\mathbf{x} = \boldsymbol{\mu} + \mathbf{L}\mathbf{y}$ where $\mathbf{y} \sim N(\mathbf{0}, \mathbf{I})$ and (as in Lemma 7) $\mathbf{V} = \mathbf{L}\mathbf{L}'$, and apply Good's results to

$$\mathbf{x}'\mathbf{A}\mathbf{x} = \mathbf{y}'\mathbf{L}'\mathbf{A}\mathbf{L}\mathbf{y} + 2\boldsymbol{\mu}'\mathbf{A}\mathbf{L}\mathbf{y} + \boldsymbol{\mu}'\mathbf{A}\boldsymbol{\mu}$$

and

$$\mathbf{b}'\mathbf{x} = \mathbf{b}'\mathbf{L}\mathbf{y} + \mathbf{b}'\boldsymbol{\mu}$$

where $\mathbf{b}'$ is any row of **B**. The necessary and sufficient condition for the independence of (i) $\mathbf{y}'\mathbf{L}'\mathbf{A}\mathbf{L}\mathbf{y}$ and $\mathbf{b}'\mathbf{L}\mathbf{y}$, is $\mathbf{I}\mathbf{L}'\mathbf{A}\mathbf{L}\mathbf{I}\mathbf{L}'\mathbf{b} = \mathbf{0}$, which is readily shown to be equivalent to $\mathbf{BVAV} = \mathbf{0}$; and of (ii) $\boldsymbol{\mu}'\mathbf{A}\mathbf{L}\mathbf{y}$ and $\mathbf{b}'\mathbf{L}\mathbf{y}$, is $\boldsymbol{\mu}'\mathbf{A}\mathbf{L}\mathbf{I}\mathbf{L}'\mathbf{b} = 0$, equivalent to $\mathbf{BVA}\boldsymbol{\mu} = \mathbf{0}$. Hence $\mathbf{BVAV} = \mathbf{0}$ and $\mathbf{BVA}\boldsymbol{\mu} = \mathbf{0}$ are the necessary and sufficient conditions for $\mathbf{x}'\mathbf{A}\mathbf{x}$ and $\mathbf{B}\mathbf{x}$ to be independent.

Corollary. $\mathbf{x}'\mathbf{A}\mathbf{x}$ and $\mathbf{B}\mathbf{x}$ are independent if $\mathbf{BVA} = \mathbf{0}$.

Concerning the independence of two quadratic forms, Khatri (1963) proves a theorem pertaining to a Wishart distribution which, in our context, takes the following form.

Theorem 4s. When $\mathbf{x}$ is $SN(\boldsymbol{\mu}, \mathbf{V})$, the quadratic forms $\mathbf{x}'\mathbf{A}\mathbf{x}$ and $\mathbf{x}'\mathbf{B}\mathbf{x}$ are independent if and only if

$$\mathbf{VAVBV} = \mathbf{0},$$

$$\mathbf{VAVB}\boldsymbol{\mu} = \mathbf{VBVA}\boldsymbol{\mu} = \mathbf{0}$$

and

$$\boldsymbol{\mu}'\mathbf{AVB}\boldsymbol{\mu} = 0.$$

Shanbhag (1966) points out that if $\mathbf{A}$ is non-negative definite these conditions reduce to $\mathbf{AVBV} = \mathbf{0}$ and $\mathbf{AVB}\boldsymbol{\mu} = \mathbf{0}$, whereas if both $\mathbf{A}$ and $\mathbf{B}$ are non-negative definite the sole condition is $\mathbf{AVB} = \mathbf{0}$, the same as when $\mathbf{V}$ is non-singular (Theorem 3, above). Good [1963, Theorem 1C, part (i)] considers this situation when $\boldsymbol{\mu} = \mathbf{0}$, erroneously reporting the condition as $\mathbf{AVBV} = \mathbf{0}$ or a cyclic permutation thereof; Shanbhag (1966) points out the error, as acknowledged by Good (1966). The correct condition is $\mathbf{VAVBV} = \mathbf{0}$, as shown above.

Proof. Application of Good's theorem yields a proof to Theorem 4s in the same manner as it does to Theorem 3s (see Exercise 15).

Theorem 5 of Sec. 5 is a generalization of Cochran's theorem. A somewhat similar generalization for the singular normal is given by Styan (1969).

Theorem 5s. Let the following be given:

$\mathbf{x}$, order $n \times 1$, $\sim SN(\boldsymbol{\mu}, \mathbf{V})$;

$\mathbf{A}_i$, $n \times n$, symmetric, rank $(\mathbf{VA}_i\mathbf{V}) = r_i$, $i = 1, \ldots, p$;

and $\mathbf{A} = \sum_{i=1}^{p} \mathbf{A}_i$, rank $\mathbf{VAV} = r$.

If (i) $\mathbf{V}$ is non-singular, or if (ii) $\mathbf{V}$ is singular and $\boldsymbol{\mu} = \mathbf{0}$ or if (iii) $\mathbf{V}$ is singular, $\boldsymbol{\mu}$ is not necessarily null and $\mathbf{A}_i$ is positive semi-definite for $i = 1, 2, \ldots, r$, then the four propositions

(*a*) $\mathbf{x}'\mathbf{A}_i\mathbf{x} \sim \chi^{2\prime}(r_i, \tfrac{1}{2}\boldsymbol{\mu}'\mathbf{A}_i\boldsymbol{\mu})$,

(*b*) the $\mathbf{x}'\mathbf{A}_i\mathbf{x}$ mutually independent,

(*c*) $\mathbf{x}'\mathbf{A}\mathbf{x} \sim \chi^{2\prime}(r, \tfrac{1}{2}\boldsymbol{\mu}'\mathbf{A}\boldsymbol{\mu})$

and (*d*) $r = \sum_{i=1}^{p} r_i$

are implied by any two of (*a*), (*b*) and (*c*) and by (*a*) and (*d*).

Proof. Styan (1969) gives a proof. It follows closely the proof of Theorem 5 in Sec. 5. Because of its length, it is omitted here.

8. EXERCISES

1. Suppose the data for 5 observations in a row-by-column analysis are as follows.

Row	Column 1	Column 2
1	6	4
2	6, 42	12

The analogy for unbalanced data of the interaction sum of squares is

$$\sum_{i=1}^{r}\sum_{j=1}^{c}\frac{y_{ij\cdot}^2}{n_{ij}} - \sum_{i=1}^{r}\frac{y_{i\cdot\cdot}^2}{n_{i\cdot}} - \sum_{j=1}^{c}\frac{y_{\cdot j\cdot}^2}{n_{\cdot j}} + \frac{y_{\cdot\cdot\cdot}^2}{n_{\cdot\cdot}}.$$

Use the above data to show that this expression is not a positive definite form. Why, then, can it not be described as a sum of squares?

2. Derive the moment generating function of the χ_n^2-distribution: (i) from its density function and (ii) using the density function of the $N(0, 1)$ distribution. Use your result to find the mean and variance of the χ_n^2-distribution.

3. Find the first two moments of $1/u$ when u is distributed as χ_n^2.

4. (*a*) Derive the mean and variance of the t_n-distribution and the F_{n_1,n_2}-distribution.

   (*b*) If the random variable r is such that $r/(n\lambda + 1)$ has a central F-distribution with $a - 1$ and $a(n - 1)$ degrees of freedom, show that

$$\hat{\lambda} = \frac{1}{n}\left[r\left(1 - \frac{2}{a(n-1)}\right) - 1\right]$$

   is an unbiased estimator of λ. [*Note:* In certain analysis of variance situations r is a calculated F-statistic and λ is a variance ratio.]

5. Using Helmert's matrix of Sec. 1, show why $\sum_{i=1}^{n}(x_i - \bar{x})^2/\sigma^2$ has a χ_{n-1}^2-distribution when $\mathbf{x}$ is $N(\mu\mathbf{1}, \sigma^2\mathbf{I})$.

6. From the given definition of the t_n- and χ_n^2-distributions show why

$$\frac{\bar{x} - \mu}{1/\sqrt{n}}\sqrt{\frac{n-1}{\sum(x_i - \bar{x})^2}} \sim t_{n-1}$$

when $\mathbf{x}$ is $N(\mu\mathbf{1}, \sigma^2\mathbf{I})$.

7. Show that the variance of a t_n-distribution equals the mean of an $F_{m,n}$-distribution.

8. Using the moment generating function of the $\chi^{2\prime}(n, \lambda)$-distribution derive its mean and variance.

9. Derive the mean and variance of the $F'(n_1, n_2, \lambda)$-distribution.

10. Derive the density function and mean of the doubly non-central F-distribution $F''(n_1, n_2, \lambda_1, \lambda_2)$.

11. When $\mathbf{x} \sim N(\boldsymbol{\mu}, \mathbf{V})$, derive the density function of $\mathbf{Tx}$, proving that it is normal. What conditions must be satisfied for your proof to hold? What is the distribution of $\mathbf{Tx}$ when the conditions are not satisfied? Discuss the case when $\mathbf{V}$ is singular.

12. When $\mathbf{x}$ is $N(\boldsymbol{\mu}_1, \mathbf{I})$ and $\mathbf{y}$ is $N(\boldsymbol{\mu}_2, \mathbf{I})$ and the correlation matrix between $\mathbf{x}$ and $\mathbf{y}$ is $\mathbf{R}$, what are the mean and variance of $\mathbf{x}'\mathbf{A}\mathbf{y}$?

13. When $\mathbf{x}$ is $N(\boldsymbol{\mu}, \mathbf{V})$ show, without using Theorem 2, that if $\mathbf{x}'\mathbf{A}\mathbf{x}$ is $\chi^{2\prime}(r, \frac{1}{2}\boldsymbol{\mu}'\mathbf{A}\boldsymbol{\mu})$ then $(\mathbf{x} - \boldsymbol{\mu})'\mathbf{A}(\mathbf{x} - \boldsymbol{\mu})$ is χ^2_r. Can the converse be proved without the use of Theorem 2?

14. If $\mathbf{y}$ is $N(\mathbf{Xb}, \mathbf{V})$ with $\mathbf{V}^{-1}$ existing, under what conditions is $\mathbf{b}^{o\prime}\mathbf{Q}\mathbf{b}^o$ a χ^2-variable when $\mathbf{b}^o$ is a solution to $\mathbf{X}'\mathbf{X}\mathbf{b}^o = \mathbf{X}'\mathbf{y}$, with $\mathbf{X}'\mathbf{X}$ being singular?

15. In Sec. 7 the two salient features of a theorem from Good (1963) are given. With their aid, prove Theorem 4s.

16. With $\mathbf{x} \sim N(\boldsymbol{\mu}, \mathbf{V})$ what are the necessary and sufficient conditions for $\mathbf{x}'\mathbf{A}_1\mathbf{x} + \mathbf{b}_1'\mathbf{x} + c_1$ and $\mathbf{x}'\mathbf{A}_2\mathbf{x} + \mathbf{b}_2'\mathbf{x} + c_2$ to be independent? What are these conditions when $\mathbf{V}$ is non-singular?

17. (a) From (38) derive the rth cumulant of the $\chi^{2\prime}(q, \lambda)$-distribution.
(b) By equating your result to (59) show that a necessary and sufficient condition for $\mathbf{x}'\mathbf{A}\mathbf{x}$ to be distributed as $\chi^{2\prime}(q, \lambda)$ is

$$\operatorname{tr}(\mathbf{VA})^r + r\boldsymbol{\mu}'\mathbf{A}(\mathbf{VA})^{r-1}\boldsymbol{\mu} = q + 2r\lambda \qquad \text{for all integers } r.$$

(c) Show that this condition is equivalent to (i.e., implies and is implied by) the two conditions

$$\boldsymbol{\mu}'\mathbf{A}(\mathbf{VA})^{r-1}\boldsymbol{\mu} = \boldsymbol{\mu}'\mathbf{A}\boldsymbol{\mu} = 2\lambda \qquad \text{and} \qquad \operatorname{tr}(\mathbf{VA})^r = \operatorname{tr}(\mathbf{VA}) = q$$
$$\text{for all integers } r.$$

(d) Show further that these conditions are also a special case of Theorem 2s.

18. Explain exactly why Cochran's theorem is a corollary of Theorem 5.

19. By writing $\mathbf{x} = \boldsymbol{\mu} + \mathbf{Ly}$ where $\mathbf{y} \sim N(\mathbf{0}, \mathbf{I})$, derive the cumulant generating function of $\mathbf{x}'\mathbf{A}\mathbf{x}$ starting from the density function of $\mathbf{y}$.

20. The non-central χ^2-distribution is defined as the distribution of $\mathbf{x}'\mathbf{x}$ when $\mathbf{x} \sim N(\boldsymbol{\mu}, \mathbf{I}_n)$. Using just this definition prove the sufficiency condition first of Corollary 2.4 and then of Theorem 2.

21. A characterization of the multivariate normal distribution is that $\mathbf{x} \sim N(\boldsymbol{\mu}, \mathbf{V})$ if and only if $\boldsymbol{\lambda}'\mathbf{x}$ has a univariate normal distribution. Using this as a definition of the multivariate normal distribution, derive its moment generating function from that of the univariate normal. [*Hint:* Use $M_{\mathbf{x}}(\mathbf{t}) = M_{\mathbf{t}'\mathbf{x}}(1)$.]

22. Suppose that $x \sim F(n_1, n_2)$ and

$$\Pr\{x \geq F_{n_1,n_2,\alpha}\} = \alpha.$$

Prove that $F_{n_2,n_1,1-\alpha} = 1/F_{n_1,n_2,\alpha}$.

23. If u and v have a bivariate normal distribution with zero means, show that

$$\operatorname{cov}(u^2, v^2) = 2[\operatorname{cov}(u, v)]^2.$$

CHAPTER 3

REGRESSION, OR THE FULL RANK MODEL

1. INTRODUCTION

a. The model

Regression analysis is designed for situations where a variable is thought to be related to one or more other measurements made, usually, on the same object. A purpose of the analysis is to use data (observed values of the variables) to estimate the form of this relationship. An example would be to use information on income and number of years of (formal) schooling to estimate the extent to which a man's annual income is related to his years of schooling. One possibility would be that for a man who had had zero years of school we would anticipate his annual income as being \$$a$; and for every year of schooling he had had we would expect his income to be larger by \$$b$. Thus for a man having x years of schooling we would expect his annual income to be $a + bx$ dollars. In saying that we "expect" him to have an income of $a + bx$ dollars we are thinking of the average of all men who have had x years at school, and if from these men one was picked at random we would expect his income to be $a + bx$. If y denotes income we write $E(y)$ for expected income and thus have

$$E(y) = a + bx. \tag{1}$$

This attempted description of how we think one variable is related to another is an example of what is called *model building*. The model here, that a man's income is expected to be $a + bx$ where x is his number of years of schooling is a *linear model*, linear because we envisage $E(y)$ as being a linear combination of the unknowns, which are called *parameters*, a and b. There are, of course, endless other models, non-linear in a and b, that might be postulated, e.g., that $E(y)$ is a function of x^a or $(\log x)^b$ or perhaps b^x.

However, the linear model is the one that has received greatest attention both in theory and in practice. From the theoretical point of view it is mathematically tractable, and in practical applications of wide variety it has shown itself to be of great value. Furthermore, many models that are apparently non-linear can often be rearranged to be in a linear form. Moreover, while computing demands of linear model analyses can be extensive they are usually not prohibitively excessive, and today's goliath computers are making such analyses ever more readily attainable.

Equation (1) is the equation of our model, in this case the model of how expected income and years of schooling are related. The equation is not the whole model; its other parts have yet to be described. Since the model is something being conjectured, a and b can never be known, and the best that can be done is to obtain estimates of them from data, data which we assume are a random sample from some population to which we conjecture our equation applies. The model is often called a regression model and since its equation is linear the regression is more correctly called *linear regression.* The variable denoted by y is usually called the *dependent variable*, and x is correspondingly called an *independent variable.*

b. Observations

In gathering data, the income of every man with x years of schooling will not be exactly $a + bx$ (with a and b being the same for all men). Indeed this fact is already recognized in the writing of the equation of the model as $E(y) = a + bx$ rather than as $y = a + bx$. Thus if y_i is the income for a man with x_i years of schooling we write

$$E(y_i) = a + bx_i\,, \tag{2}$$

where $E(y_i)$ is not the same as y_i. The difference, $y_i - E(y_i)$, represents the deviation of the observed y_i from its expected value $E(y_i)$ and is written as

$$e_i = y_i - E(y_i) = y_i - a - bx_i\,. \tag{3}$$

Hence

$$y_i = a + bx_i + e_i\,, \tag{4}$$

which we now take as the equation of the model.

The deviation e_i defined in (3) represents the extent to which an observed y_i differs from its expected value $E(y_i) = a + bx_i$. And equations (2), (3) and (4) apply to each of our N observations $y_1, y_2, \ldots, y_N$. Thus the e's include all manner of discrepancies between observed y's and their expected values; for example, they include measurement errors in y_i (its recorded value might not be exactly what the man's income is), and they include deficiencies in the model itself—the extent to which $a + bx_i$ is, in fact, not the man's income (variables other than years of schooling might affect it, the man's age, for example). In this way the e's are considered as random variables, usually called random errors or random residuals.

In order to complete the description of our model in terms of equation (4), characteristics of the e's must be specified. Customary specifications are that the expected value of e_i is zero and its variance is σ^2, for all i; and that covariances between any pairs of e's are zero. Thus

$$E(e_i) = 0 \tag{5}$$

as is obvious from the definition of e_i in (3), and

$$v(e_i) = E[e_i - E(e_i)]^2 = E(e_i^2) = \sigma^2 \qquad \text{for all } i; \tag{6}$$

and $$\operatorname{cov}(e_i e_j) = E[e_i - E(e_i)][e_j - E(e_j)] = E(e_i e_j) = 0 \qquad \text{for } i \neq j. \tag{7}$$

Equations (2)–(7) now constitute the model. They form the basis of the procedure used for estimating a and b.

c. Estimation

There are several well-recognized methods that can be used for estimating a and b (see Sec. 3). The most frequently used is that known as least squares, and it is the one we shall outline here. Its justification as a satisfactory estimation procedure is given in many standard statistical texts.

Least squares estimation involves minimizing the sum of squares of deviations of the observed y_i's from their expected values. In view of (3) this sum of squares is

$$\mathbf{e}'\mathbf{e} = \sum_{i=1}^{N} e_i^2 = \sum_{i=1}^{N} [y_i - E(y_i)]^2 = \sum_{i=1}^{N} (y_i - a - bx_i)^2. \tag{8}$$

Although a and b are fixed (but unknown) values, let us for the moment think of them as mathematical variables. Then those values of them which minimize (8) are the least squares estimators of a and b. They will be denoted by $\hat{a}$ and $\hat{b}$. Minimization of (8) is achieved in the usual manner: differentiate (8) with respect to a and b and equate the differentials to zero. The resulting equations are written in terms of $\hat{a}$ and $\hat{b}$. Their solutions for $\hat{a}$ and $\hat{b}$ are the least squares estimators. Thus from (8)

$$\partial(\mathbf{e}'\mathbf{e})/\partial a = -2\sum (y_i - a - bx_i) = -2(\sum y_i - Na - b\sum x_i) \tag{9}$$

and

$$\partial(\mathbf{e}'\mathbf{e})/\partial b = -2\sum x_i(y_i - a - bx_i) = -2(\sum x_i y_i - a\sum x_i - b\sum x_i^2) \tag{10}$$

where summations are over i, for $i = 1, 2, \ldots, N$. Equating these to zero and writing them in terms of $\hat{a}$ and $\hat{b}$ gives

$$N\hat{a} + \hat{b}\sum x_i = \sum y_i \qquad \text{and} \qquad \hat{a}\sum x_i + \hat{b}\sum x_i^2 = \sum x_i y_i\,. \tag{11}$$

Using the dot notation

$$x_{.} = \sum_i x_i \qquad \text{and} \qquad y_{.} = \sum_i y_i \tag{12}$$

and the corresponding bar notation for observed means,

$$\bar{x}. = x./N \qquad \text{and} \qquad \bar{y}. = y./N, \tag{13}$$

the solution for (11) can be written in the following familiar forms:

$$\hat{b} = \frac{\sum (x_i - \bar{x}.)(y_i - \bar{y}.)}{\sum (x_i - \bar{x}.)^2} = \frac{\sum x_i y_i - N\bar{x}.\bar{y}.}{\sum x_i^2 - N\bar{x}.^2} \tag{14}$$

and

$$\hat{a} = \bar{y}. - \hat{b}\bar{x}. = (y. - \hat{b}x.)/N. \tag{15}$$

d. Example

Suppose in a sample of 5 men that their incomes (in thousands of dollars) and years of schooling are as follows.

i (Man)	y_i (Income, \$1,000)	x_i (Years of Schooling)	
1	10	6	
2	20	12	
3	17	10	
4	12	8	
5	11	9	
$N = 5$	$y. = 70$	$x. = 45$	
	$\bar{y}. = 14$	$\bar{x}. = 9$	
	$\sum y_i^2 = 1054$	$\sum x_i^2 = 425$	$\sum x_i y_i = 665$

From (11) the equations for obtaining $\hat{a}$ and $\hat{b}$ are

$$5\hat{a} + 45\hat{b} = 70 \qquad \text{and} \qquad 45\hat{a} + 425\hat{b} = 665$$

and from (14) and (15) the solutions are

$$\hat{b} = \frac{665 - 5(9)14}{425 - 5(9^2)} = \frac{35}{20} = 1.75$$

and

$$\hat{a} = 14 - 9(1.75) = -1.75.$$

Hence the estimated regression equation, corresponding to (2), is

$$\widehat{E(y_i)} = \hat{a} + \hat{b}x_i = -1.75 + 1.75x_i\,,$$

where the large "hat" over $E(y_i)$ denotes "estimator of" $E(y_i)$ just as does $\hat{a}$ of a.

e. The general case of k x-variables

Suppose that in the study of annual income and years of schooling we also considered the man's age to be a factor affecting income. The model envisaged in (1) is now extended to be

$$E(y) = a + b_1x_1 + b_2x_2$$

where x_1 represents years of schooling and x_2 is age. Thus for the ith man in our data, who has had x_{i1} years of schooling and whose age is x_{i2}, equation (4) could be

$$y_i = a + b_1x_{i1} + b_2x_{i2} + e_i\,.$$

A change in notation is now made: in place of a write b_0, and then for b_0 write b_0x_{i0} with all values of x_{i0} being unity. This gives

$$y_i = b_0x_{i0} + b_1x_{i1} + b_2x_{i2} + e_i\,, \tag{16}$$

for $i = 1, 2, \ldots, N$, with $x_{i0} = 1$ for all i.

Now define the following matrix and vectors:

$$\mathbf{X} = \begin{bmatrix} x_{10} & x_{11} & x_{12} \\ x_{20} & x_{21} & x_{22} \\ \cdot & \cdot & \cdot \\ \cdot & \cdot & \cdot \\ \cdot & \cdot & \cdot \\ x_{N0} & x_{N1} & x_{N2} \end{bmatrix}, \quad \mathbf{y} = \begin{bmatrix} y_1 \\ y_2 \\ \cdot \\ \cdot \\ \cdot \\ y_N \end{bmatrix}, \quad \mathbf{e} = \begin{bmatrix} e_1 \\ e_2 \\ \cdot \\ \cdot \\ \cdot \\ e_N \end{bmatrix} \quad \text{and} \quad \mathbf{b} = \begin{bmatrix} b_0 \\ b_1 \\ b_2 \end{bmatrix}.$$

Then the complete set of equations represented by (16) is

$$\mathbf{y} = \mathbf{Xb} + \mathbf{e}, \qquad \text{with} \quad E(\mathbf{y}) = \mathbf{Xb}. \tag{17}$$

Extension to more than just 2 x-variables (or 3, including x_0) is clear. For k variables

$$\mathbf{X} = \begin{bmatrix} x_{10} & x_{11} & \cdots & x_{1k} \\ x_{20} & x_{21} & \cdots & x_{2k} \\ \cdot & \cdot & & \cdot \\ \cdot & \cdot & & \cdot \\ \cdot & \cdot & & \cdot \\ x_{N0} & x_{N1} & \cdots & x_{Nk} \end{bmatrix}, \qquad \mathbf{b} = \begin{bmatrix} b_0 \\ \cdot \\ \cdot \\ \cdot \\ b_k \end{bmatrix} \tag{18}$$

and $\mathbf{y}$ and $\mathbf{e}$ defined as above are unchanged. Equation (17) is unchanged also, and it represents the model no matter how many x-variables there are, k, so long as they are fewer in number than the number of observations N, i.e., $k < N$. This is the model we now study, dealing with some of its many variations in this and subsequent chapters. (When $k \geq N$, values of the b_i can be derived so that $\mathbf{y} = \mathbf{Xb}$ exactly, and there is no estimation problem.)

Complete specification of the model demands that distributional properties of the vector **e** be defined. For the moment all that is needed are its expected value and variance. These, in accord with (5), (6) and (7) are taken as

$$E(\mathbf{e}) = \mathbf{0}$$

and

$$\text{var}(\mathbf{e}) = E[\mathbf{e} - E(\mathbf{e})][\mathbf{e} - E(\mathbf{e})]' = E(\mathbf{ee}') = \sigma^2\mathbf{I}_N\,. \tag{19}$$

An exact form of the distribution of the elements of **e** will be specified later, when hypothesis testing and confidence intervals are considered.

Derivation of the least squares estimator of **b** follows the same procedure as that used in establishing (11), namely minimization of the sum of squares of the observations from their expected values. Similar to (8) this sum of squares, with $E(\mathbf{e}) = \mathbf{0}$ of (19) and hence $E(\mathbf{y}) = \mathbf{Xb}$, is

$$\begin{aligned}\mathbf{e}'\mathbf{e} &= [\mathbf{y} - E(\mathbf{y})]'[\mathbf{y} - E(\mathbf{y})] = (\mathbf{y} - \mathbf{Xb})'(\mathbf{y} - \mathbf{Xb})\\ &= \mathbf{y}'\mathbf{y} - 2\mathbf{b}'\mathbf{X}'\mathbf{y} + \mathbf{b}'\mathbf{X}'\mathbf{Xb}.\end{aligned}$$

Choosing as the estimator $\hat{\mathbf{b}}$ that value of **b** which minimizes $\mathbf{e}'\mathbf{e}$ involves differentiating $\mathbf{e}'\mathbf{e}$ with respect to the elements of **b** [Searle (1966), Sec. 8.5, for example]. Equating $\partial(\mathbf{e}'\mathbf{e})/\partial\mathbf{b}$ to zero and writing the resulting equations in terms of $\hat{\mathbf{b}}$, we find that these equations are

$$\mathbf{X}'\mathbf{X}\hat{\mathbf{b}} = \mathbf{X}'\mathbf{y}. \tag{20}$$

They are known as the *normal equations*. Provided $(\mathbf{X}'\mathbf{X})^{-1}$ exists they have the unique solution for $\hat{\mathbf{b}}$,

$$\hat{\mathbf{b}} = (\mathbf{X}'\mathbf{X})^{-1}\mathbf{X}'\mathbf{y}. \tag{21}$$

Here is where the description "full rank model" applies. When $\mathbf{X}'\mathbf{X}$ is of full rank the solution of (20) for $\hat{\mathbf{b}}$ can be written as in (21). On the other hand, if $(\mathbf{X}'\mathbf{X})^{-1}$ does not exist, a solution to (20) may be written in terms of a generalized inverse of $\mathbf{X}'\mathbf{X}$. This is the case of models not of full rank, which are taken up in Chapter 5.

By the nature of **X** shown in (18) $\mathbf{X}'\mathbf{X}$ is square of order $k + 1$, with elements that are sums of squares and products, summed over i for $i = 1, 2, \ldots, N$:

$$\mathbf{X}'\mathbf{X} = \begin{bmatrix} \sum x_{i0}^2 & \sum x_{i0}x_{i1} & \cdots & \sum x_{i0}x_{ik} \\ \sum x_{i0}x_{i1} & \sum x_{i1}^2 & \cdots & \sum x_{i1}x_{ik} \\ \cdot & \cdot & & \cdot \\ \cdot & \cdot & & \cdot \\ \cdot & \cdot & & \cdot \\ \sum x_{i0}x_{ik} & \sum x_{i1}x_{ik} & \cdots & \sum x_{ik}^2 \end{bmatrix} \tag{22}$$

and

$$\mathbf{X}'\mathbf{y} = \begin{bmatrix} \sum x_{i0}y_i \\ \sum x_{i1}y_i \\ \cdot \\ \cdot \\ \cdot \\ \sum x_{ik}y_i \end{bmatrix}. \tag{23}$$

Thus $\mathbf{X}'\mathbf{X}$ is the matrix of sum of squares and products of the observed x's and $\mathbf{X}'\mathbf{y}$ is the vector of sums of products of the observed x's and y's. Furthermore, since $x_{i0} = 1$ for all $i = 1, 2, \ldots, N$, and because all summations are over $i = 1, 2, \ldots, N$, $\Sigma x_{i0}^2 = N$, $\Sigma x_{i0}x_{i1} = x_{.1}$ and $\Sigma x_{i0}y_i = y_.$. Hence

$$\mathbf{X}'\mathbf{X} = \begin{bmatrix} N & x_{.1} & x_{.2} & \cdots & x_{.k} \\ x_{.1} & \sum x_{i1}^2 & \sum x_{i1}x_{i2} & \cdots & \sum x_{i1}x_{ik} \\ x_{.2} & \sum x_{i1}x_{i2} & \sum x_{i2}^2 & \cdots & \sum x_{i2}x_{ik} \\ \cdot & \cdot & \cdot & & \cdot \\ \cdot & \cdot & \cdot & & \cdot \\ \cdot & \cdot & \cdot & & \cdot \\ x_{.k} & \sum x_{i1}x_{ik} & \sum x_{i2}x_{ik} & \cdots & \sum x_{ik}^2 \end{bmatrix} \tag{24}$$

f. Example (continued). Suppose in the previous example the ages of the men supplying the data had also been available, as follows.

(Man) i	(Income, \$1000) y_i	(Years of Schooling) x_{i1}	(Age) x_{i2}
1	10	6	28
2	20	12	40
3	17	10	32
4	12	8	36
5	11	9	34
$N = 5$	$y_. = 70$	$x_{.1} = 45$	$x_{.2} = 170$
	$\bar{y}_. = 14$	$\bar{x}_{.1} = 9$	$\bar{x}_{.2} = 34$

$\sum y_i^2 = 1054$ $\quad \sum x_{i1}^2 = 425$ $\quad \sum x_{i2}^2 = 5860$ $\quad \sum x_{i1}x_{i2} = 1562$

$\sum x_{i1}y_i = 665$ $\quad \sum x_{i2}y_i = 2430$

Putting these values into (22) gives

$$\mathbf{X} = \begin{bmatrix} 1 & 6 & 28 \\ 1 & 12 & 40 \\ 1 & 10 & 32 \\ 1 & 8 & 36 \\ 1 & 9 & 34 \end{bmatrix}, \qquad \mathbf{X}'\mathbf{X} = \begin{bmatrix} 5 & 45 & 170 \\ 45 & 425 & 1562 \\ 170 & 1562 & 5860 \end{bmatrix} \tag{25}$$

and

$$(\mathbf{X}'\mathbf{X})^{-1} = \frac{1}{2880}\begin{bmatrix} 50656 & 1840 & -1960 \\ 1840 & 400 & -160 \\ -1960 & -160 & 100 \end{bmatrix}. \tag{26}$$

And with, from (23),

$$\mathbf{X}'\mathbf{y} = \begin{bmatrix} \sum y_i \\ \sum x_{i1}y_i \\ \sum x_{i2}y_i \end{bmatrix} = \begin{bmatrix} 70 \\ 665 \\ 2430 \end{bmatrix}, \tag{27}$$

equation (21) gives

$$\hat{\mathbf{b}} = (\mathbf{X}'\mathbf{X})^{-1}\mathbf{X}'\mathbf{y} = \frac{1}{2880}\begin{bmatrix} 50656 & 1840 & -1960 \\ 1840 & 400 & -160 \\ -1960 & -160 & 100 \end{bmatrix}\begin{bmatrix} 70 \\ 665 \\ 2430 \end{bmatrix}$$

$$= \frac{1}{24}\begin{bmatrix} 56 \\ 50 \\ -5 \end{bmatrix}. \tag{28}$$

Thus from these data the estimated form of the relationship between y and x_1 and x_2 is

$$\widehat{E(y)} = 56/24 + (50/24)x_1 - (5/24)x_2 .$$

g. Intercept and no-intercept models

When all x's are zero in the above models, $E(y) = b_0$ with estimator $\hat{b}_0$. Thus for $x_1 = 0 = x_2$ in the preceding example the estimated value of $E(y)$ is $\hat{b}_0 = 56/24$. Models of this nature are called *intercept models*; the intercept is b_0, the value of $E(y)$ when all x's are zero.

Sometimes it is appropriate to have no term b_0 in the model, in which case the model is called a *no-intercept model*. The matrix $\mathbf{X}$ then has no vector of 1's in it, as does $\mathbf{X}$ of (25) for example, and $\mathbf{X}'\mathbf{X}$ is then the matrix of sums of squares and products of the observations, without the first row and column of totals seen in (24).

Example (*continued*).

For the no-intercept model

$$\mathbf{X}'\mathbf{X} = \begin{bmatrix} 425 & 1562 \\ 1562 & 5860 \end{bmatrix} \quad \text{and} \quad \mathbf{X}'\mathbf{y} = \begin{bmatrix} 665 \\ 2430 \end{bmatrix}$$

for which solution to (20) is

$$\hat{\mathbf{b}} = \begin{bmatrix} \hat{b}_1 \\ \hat{b}_2 \end{bmatrix} = (\mathbf{X}'\mathbf{X})^{-1}\mathbf{X}'\mathbf{y} = \frac{1}{50656}\begin{bmatrix} 5860 & -1562 \\ -1562 & 425 \end{bmatrix}\begin{bmatrix} 665 \\ 2430 \end{bmatrix}$$

$$= \frac{1}{12664}\begin{bmatrix} 25310 \\ -1495 \end{bmatrix}. \tag{29}$$

The no-intercept model thus leads to $E(y)$ being estimated from these data as

$$\widehat{E(y)} = (25310/12664)x_1 - (1495/12664)x_2 = 1.195x_1 - 0.1182x_2\,.$$

2. DEVIATIONS FROM MEANS

The matrix $\mathbf{X}'\mathbf{X}$ and vector $\mathbf{X}'\mathbf{y}$ shown in (22) and (23) have as elements the sums of squares and products of the observations. But it is well known that the regression coefficients $b_1, \ldots, b_k$ can be estimated using a matrix and vector that are just like $\mathbf{X}'\mathbf{X}$ and $\mathbf{X}'\mathbf{y}$ only involving sums of squares and products corrected for their means. Indeed, this is the customary manner in which estimates are calculated. We now establish this formulation. To do so, some additional notation is needed.

Putting $x_{i0} = 1$ in (18) for all i makes the first column of $\mathbf{X}$ all 1's. Therefore, in defining

$$\mathbf{1}_N = \begin{bmatrix} 1 \\ 1 \\ \cdot \\ \cdot \\ \cdot \\ 1 \end{bmatrix} \quad \text{and} \quad \mathbf{X}_1 = \begin{bmatrix} x_{11} & x_{12} & \cdots & x_{1k} \\ x_{21} & x_{22} & \cdots & x_{2k} \\ \cdot & \cdot & & \cdot \\ \cdot & \cdot & & \cdot \\ \cdot & \cdot & & \cdot \\ x_{N1} & x_{N2} & \cdots & x_{Nk} \end{bmatrix}, \tag{30}$$

$\mathbf{X}$ can be written as

$$\mathbf{X} = [\mathbf{1} \quad \mathbf{X}_1], \tag{31}$$

where the order of $\mathbf{1}$ is N; and, as in (30), $\mathbf{X}_1$ is the $N \times k$ matrix of the observed x's. In addition, define

$$\bar{\mathbf{x}}' = [\bar{x}_{.1} \;\; \bar{x}_{.2} \;\; \ldots \;\; \bar{x}_{.k}] \tag{32}$$

as the vector of means of the observed x's. These definitions imply

$$\mathbf{1}'_N\mathbf{1}_N = N, \qquad \mathbf{1}'\mathbf{y} = N\bar{y} \qquad \text{and} \qquad \mathbf{1}'\mathbf{X}_1 = N\bar{\mathbf{x}}', \tag{33}$$

where for convenience we write $\bar{y}$ in place of $\bar{y}.$ for the mean.

The solution $\hat{\mathbf{b}}$ can now be expressed as

$$\begin{aligned}\hat{\mathbf{b}} &= (\mathbf{X}'\mathbf{X})^{-1}\mathbf{X}'\mathbf{y}\\ &= \left[\begin{bmatrix}\mathbf{1}'\\ \mathbf{X}_1'\end{bmatrix}[\mathbf{1} \;\; \mathbf{X}_1]\right]^{-1}\begin{bmatrix}\mathbf{1}'\\ \mathbf{X}_1'\end{bmatrix}\mathbf{y}\\ &= \begin{bmatrix}N & N\bar{\mathbf{x}}'\\ N\bar{\mathbf{x}} & \mathbf{X}_1'\mathbf{X}_1\end{bmatrix}^{-1}\begin{bmatrix}N\bar{y}\\ \mathbf{X}_1'\mathbf{y}\end{bmatrix} \qquad \text{from (33).}\end{aligned}$$

Using the procedure for inverting a partitioned symmetric matrix given in equation (48) of Sec. 1.7, this becomes

$$\hat{\mathbf{b}} = \begin{bmatrix}1/N + \bar{\mathbf{x}}'\mathbf{S}^{-1}\bar{\mathbf{x}} & -\bar{\mathbf{x}}'\mathbf{S}^{-1}\\ -\mathbf{S}^{-1}\bar{\mathbf{x}} & \mathbf{S}^{-1}\end{bmatrix}\begin{bmatrix}N\bar{y}\\ \mathbf{X}_1'\mathbf{y}\end{bmatrix} \tag{34}$$

where

$$\mathbf{S} = \mathbf{X}_1'\mathbf{X}_1 - N\bar{\mathbf{x}}\bar{\mathbf{x}}'. \tag{35}$$

Then, on partitioning

$$\mathbf{b} = \begin{bmatrix}b_0\\ b_1\\ \cdot\\ \cdot\\ \cdot\\ b_k\end{bmatrix} = \begin{bmatrix}b_0\\ \mathscr{b}\end{bmatrix},$$

(34) can be written as

$$\begin{aligned}\begin{bmatrix}\hat{b}_0\\ \hat{\mathscr{b}}\end{bmatrix} &= \left[\begin{bmatrix}1/N & \mathbf{0}\\ \mathbf{0} & \mathbf{0}\end{bmatrix} + \begin{bmatrix}-\bar{\mathbf{x}}'\\ \mathbf{I}\end{bmatrix}\mathbf{S}^{-1}[-\bar{\mathbf{x}} \;\; \mathbf{I}]\right]\begin{bmatrix}N\bar{y}\\ \mathbf{X}_1'\mathbf{y}\end{bmatrix}\\ &= \begin{bmatrix}\bar{y} - \bar{\mathbf{x}}'\mathbf{S}^{-1}(\mathbf{X}_1'\mathbf{y} - N\bar{y}\bar{\mathbf{x}})\\ \mathbf{S}^{-1}(\mathbf{X}_1'\mathbf{y} - N\bar{y}\bar{\mathbf{x}})\end{bmatrix}\end{aligned}$$

so that

$$\hat{\mathscr{b}} = \mathbf{S}^{-1}(\mathbf{X}_1'\mathbf{y} - N\bar{y}\bar{\mathbf{x}}) \tag{36}$$

and
$$\hat{b}_0 = \bar{y} - \bar{\mathbf{x}}'\hat{\boldsymbol{\ell}}. \tag{37}$$

Now consider $\mathbf{S}$ given in (35). First,

$$\mathbf{X}_1'\mathbf{X}_1 = \begin{bmatrix} \sum x_{i1}^2 & \sum x_{i1}x_{i2} & \cdots & \sum x_{i1}x_{ik} \\ \sum x_{i1}x_{i2} & \sum x_{i2}^2 & \cdots & \sum x_{i2}x_{ik} \\ \cdot & \cdot & & \cdot \\ \cdot & \cdot & & \cdot \\ \cdot & \cdot & & \cdot \\ \sum x_{i1}x_{ik} & \sum x_{i2}x_{ik} & \cdots & \sum x_{ik}^2 \end{bmatrix}, \tag{38}$$

the matrix of sums of squares and products of the k observed x-variables. Second, by the nature of $\bar{\mathbf{x}}$ in (32), the matrix $N\bar{\mathbf{x}}\bar{\mathbf{x}}'$ is

$$N\bar{\mathbf{x}}\bar{\mathbf{x}}' = \{N\bar{x}_{.p}\bar{x}_{.q}\} \qquad \text{for} \quad p, q = 1, 2, \ldots, k.$$

Thus

$$\mathbf{S} = \left\{\sum_i x_{ip}x_{iq} - N\bar{x}_{.p}\bar{x}_{.q}\right\} \qquad \text{for} \quad p, q = 1, 2, \ldots, k.$$

Defining

$$\mathscr{X} = \mathbf{X}_1 - \mathbf{1}_N\bar{\mathbf{x}}' \tag{39}$$

as the matrix of observed x's expressed as deviations from their means, it is then easily shown that $\mathbf{S}$ as just derived is

$$\mathbf{S} = \mathscr{X}'\mathscr{X}, \tag{40}$$

i.e., $\mathbf{S}$ in (36) is the matrix of corrected sums of squares and products of the x's. Similarly, the other term in (36) is

$$\begin{aligned} \mathbf{X}_1'\mathbf{y} - N\bar{y}\bar{\mathbf{x}} &= \left\{\sum_i x_{ip}y_i - N\bar{x}_{.p}\bar{y}\right\} \qquad \text{for} \quad p = 1, 2, \ldots, k, \\ &= \mathscr{X}'\mathbf{y}, \end{aligned}$$

the vector of corrected sums of products of the x's and y's. Hence just as $\hat{\mathbf{b}} = (\mathbf{X}'\mathbf{X})^{-1}\mathbf{X}'\mathbf{y}$ in (21) we can now write, from (36),

$$\hat{\boldsymbol{\ell}} = (\mathscr{X}'\mathscr{X})^{-1}\mathscr{X}'\mathbf{y}. \tag{41}$$

This is the inverted matrix of corrected sums of squares and products of the x's pre-multiplying the vector of corrected sums of products of the x's and y's. Then, as in (37), $\hat{b}_0$ is given by

$$\hat{b}_0 = \bar{y} - \hat{\boldsymbol{\ell}}'\bar{\mathbf{x}}. \tag{42}$$

These results, (41) and (42), are the familiar expressions for calculating regression estimators using corrected sums of squares and products.

Example (*continued*) From the data given earlier,

$$\mathscr{X} = \begin{bmatrix} -3 & -6 \\ 3 & 6 \\ 1 & -2 \\ -1 & 2 \\ 0 & 0 \end{bmatrix} \tag{43}$$

and
$$\mathscr{X}'\mathscr{X} = \begin{bmatrix} 425 - 5(9^2) & 1562 - 5(9)34 \\ 1562 - 5(9)34 & 5860 - 5(34)^2 \end{bmatrix} = \begin{bmatrix} 20 & 32 \\ 32 & 80 \end{bmatrix},$$

with

$$(\mathscr{X}'\mathscr{X})^{-1} = \frac{1}{144}\begin{bmatrix} 20 & -8 \\ -8 & 5 \end{bmatrix} \tag{44}$$

and

$$\mathscr{X}'\mathbf{y} = \begin{bmatrix} 665 - 5(14)9 \\ 2430 - 5(14)34 \end{bmatrix} = \begin{bmatrix} 35 \\ 50 \end{bmatrix}. \tag{45}$$

Therefore, on substituting in (41),

$$\hat{\mathscr{b}} = \frac{1}{144}\begin{bmatrix} 20 & -8 \\ -8 & 5 \end{bmatrix}\begin{bmatrix} 35 \\ 50 \end{bmatrix} = \begin{bmatrix} 50/24 \\ -5/24 \end{bmatrix}$$

as in (28). And from (42)

$$\hat{b}_0 = 14 - [50/24 \quad -5/24]\begin{bmatrix} 9 \\ 34 \end{bmatrix} = 56/24$$

as in (28).

Derivation of $\hat{\mathscr{b}}$ in this manner does not apply for the no-intercept model which contains no b_0-term. For then the partitioning of $\mathbf{b}'$ as $[b_0 \quad \mathscr{b}']$ does not exist. $\mathbf{b}'$ is itself the vector of the b's corresponding to the k x-variables and $\hat{\mathbf{b}} = (\mathbf{X}'\mathbf{X})^{-1}\mathbf{X}'\mathbf{y}$ is based on uncorrected sums of squares and products as exemplified in (24).

3. FOUR METHODS OF ESTIMATION

In deriving the estimator $\hat{\mathbf{b}} = (\mathbf{X}'\mathbf{X})^{-1}\mathbf{X}'\mathbf{y}$ in the previous section we blithely adopted the least squares procedure for doing so. This is a well-accepted method of estimation and its rationale will not be discussed here. However, for convenient reference we summarize four common methods of estimation which, although differing in basic concept, all lead to the same estimator under certain frequently-used assumptions. All four procedures are summarized in terms of the full rank model where, in $\mathbf{y} = \mathbf{Xb} + \mathbf{e}$, $\mathbf{X}$ has full column rank, $E(\mathbf{y}) = \mathbf{Xb}$ and $E(\mathbf{e}) = \mathbf{0}$. Reference to their use in models not of full rank is made in Chapter 5.

a. Ordinary least squares

This involves choosing $\hat{\mathbf{b}}$ as the value of $\mathbf{b}$ which minimizes the sum of squares of deviations of the observations from their expected values; i.e., choose $\hat{\mathbf{b}}$ as that $\mathbf{b}$ which minimizes $\sum_{i=1}^{N} [y_i - E(y_i)]^2 = (\mathbf{y} - \mathbf{Xb})'(\mathbf{y} - \mathbf{Xb})$. The resulting estimator is, as we have seen,

$$\hat{\mathbf{b}} = (\mathbf{X}'\mathbf{X})^{-1}\mathbf{X}'\mathbf{y}.$$

b. Generalized least squares

On assuming that the variance-covariance matrix of $\mathbf{e}$ is $\text{var}(\mathbf{e}) = \mathbf{V}$, this method involves minimizing $(\mathbf{y} - \mathbf{Xb})'\mathbf{V}^{-1}(\mathbf{y} - \mathbf{Xb})$ with respect to $\mathbf{b}$. This leads to

$$\tilde{\mathbf{b}} = (\mathbf{X}'\mathbf{V}^{-1}\mathbf{X})^{-1}\mathbf{X}'\mathbf{V}^{-1}\mathbf{y}.$$

Clearly, when $\mathbf{V} = \sigma^2\mathbf{I}$, the generalized and the ordinary least squares estimators are the same: $\tilde{\mathbf{b}} = \hat{\mathbf{b}}$.

c. Maximum likelihood

With least squares estimation no assumption is made about the form of the distribution of the random error terms in the model, the terms represented by $\mathbf{e}$. With maximum likelihood estimation some assumption is made about this distribution (often that it is normal) and the likelihood of the sample of observations represented by the data is then maximized. On assuming that the e's are normally distributed with zero mean and variance-covariance matrix $\mathbf{V}$, i.e., $\mathbf{e} \sim N(\mathbf{0}, \mathbf{V})$, the likelihood is

$$L = (2\pi)^{-\frac{1}{2}N} \, |\mathbf{V}|^{-\frac{1}{2}} \exp\{-\tfrac{1}{2}(\mathbf{y} - \mathbf{Xb})'\mathbf{V}^{-1}(\mathbf{y} - \mathbf{Xb})\}.$$

Maximizing this with respect to $\mathbf{b}$ is equivalent to solving $\partial(\log_e L)/\partial\mathbf{b} = \mathbf{0}$. The solution is the maximum likelihood estimator of $\mathbf{b}$ and turns out to be

$$\tilde{\mathbf{b}} = (\mathbf{X}'\mathbf{V}^{-1}\mathbf{X})^{-1}\mathbf{X}'\mathbf{V}^{-1}\mathbf{y},$$

the same as the generalized least squares estimator. As before, when $\mathbf{V} = \sigma^2\mathbf{I}$, $\tilde{\mathbf{b}}$ simplifies to $\hat{\mathbf{b}}$. Only then, in thinking of $\hat{\mathbf{b}}$ as the maximum likelihood estimator, we do so on the basis of assuming $\mathbf{e} \sim N(\mathbf{0}, \sigma^2\mathbf{I})$.

Two well-known points are worth emphasizing about these estimators. First, least squares estimation does not pre-suppose any distributional properties of the e's other than finite (in our case zero) means and finite variances.

Second, maximum likelihood estimation under normality assumptions leads to the same estimator, $\tilde{\mathbf{b}}$, as generalized least squares; and this reduces to the ordinary least squares estimator $\hat{\mathbf{b}}$ when $\mathbf{V} = \sigma^2\mathbf{I}$.

d. The best linear unbiased estimator (b.l.u.e.)

For any row vector $\mathbf{t}'$ conformable with $\mathbf{b}$ the scalar $\mathbf{t}'\mathbf{b}$ is a linear function of the elements of the parameter vector $\mathbf{b}$. A fourth estimation procedure derives a best, linear, unbiased estimator (b.l.u.e.) of $\mathbf{t}'\mathbf{b}$.

The three characteristics of the estimator inherent in its definition lead to its derivation.

(*i*) *linearity:* it is to be a linear function of the observations $\mathbf{y}$. Let the estimator be $\boldsymbol{\lambda}'\mathbf{y}$, where $\boldsymbol{\lambda}'$ is a row vector of order N. Then $\boldsymbol{\lambda}$ is uniquely determined by the other two characteristics of the definition, as shall be shown.

(*ii*) *unbiasedness:* $\boldsymbol{\lambda}'\mathbf{y}$ is to be an unbiased estimator of $\mathbf{t}'\mathbf{b}$. Therefore $E(\boldsymbol{\lambda}'\mathbf{y})$ must equal $\mathbf{t}'\mathbf{b}$; i.e., $\boldsymbol{\lambda}'\mathbf{X}\mathbf{b} = \mathbf{t}'\mathbf{b}$. Since this is to be true for all $\mathbf{b}$,

$$\boldsymbol{\lambda}'\mathbf{X} = \mathbf{t}'. \tag{46}$$

(*iii*) *a "best" estimator:* "best" means that in the class of linear, unbiased estimators of $\mathbf{t}'\mathbf{b}$, the "best" is to be the one that has minimum variance. This is the criterion for deriving $\boldsymbol{\lambda}'$.

Suppose $\text{var}(\mathbf{y}) = \mathbf{V}$. Then $v(\boldsymbol{\lambda}'\mathbf{y}) = \boldsymbol{\lambda}'\mathbf{V}\boldsymbol{\lambda}$, and for $\boldsymbol{\lambda}'\mathbf{y}$ to be "best" this variance must be a minimum; i.e., $\boldsymbol{\lambda}$ is chosen to minimize $\boldsymbol{\lambda}'\mathbf{V}\boldsymbol{\lambda}$ subject to the limitation that $\boldsymbol{\lambda}'\mathbf{X} = \mathbf{t}'$ derived in (46). Using $2\boldsymbol{\theta}$ as a vector of Lagrange multipliers we therefore minimize

$$w = \boldsymbol{\lambda}'\mathbf{V}\boldsymbol{\lambda} - 2\boldsymbol{\theta}'(\mathbf{X}'\boldsymbol{\lambda} - \mathbf{t})$$

with respect to the elements of $\boldsymbol{\lambda}'$ and $\boldsymbol{\theta}'$. Clearly $\partial w/\partial\boldsymbol{\theta} = \mathbf{0}$ gives (46), and $\partial w/\partial\boldsymbol{\lambda}$ gives

$$\mathbf{V}\boldsymbol{\lambda} = \mathbf{X}\boldsymbol{\theta} \qquad \text{or} \qquad \boldsymbol{\lambda} = \mathbf{V}^{-1}\mathbf{X}\boldsymbol{\theta}$$

since $\mathbf{V}^{-1}$ exists. Substitution in (46) gives $\mathbf{t}' = \boldsymbol{\lambda}'\mathbf{X} = \boldsymbol{\theta}'\mathbf{X}'\mathbf{V}^{-1}\mathbf{X}$ and so $\boldsymbol{\theta}' = \mathbf{t}'(\mathbf{X}'\mathbf{V}^{-1}\mathbf{X})^{-1}$ and hence

$$\boldsymbol{\lambda}' = \boldsymbol{\theta}'\mathbf{X}'\mathbf{V}^{-1} = \mathbf{t}'(\mathbf{X}'\mathbf{V}^{-1}\mathbf{X})^{-1}\mathbf{X}'\mathbf{V}^{-1}. \tag{47}$$

Hence the b.l.u.e. of $\mathbf{t}'\mathbf{b}$ is $\mathbf{t}'(\mathbf{X}'\mathbf{V}^{-1}\mathbf{X})^{-1}\mathbf{X}'\mathbf{V}^{-1}\mathbf{y}$, and its variance is

$$v(\text{b.l.u.e. of } \mathbf{t}'\mathbf{b}) = v(\boldsymbol{\lambda}'\mathbf{y}) = \boldsymbol{\lambda}'\mathbf{V}\boldsymbol{\lambda} = \mathbf{t}'(\mathbf{X}'\mathbf{V}^{-1}\mathbf{X})^{-1}\mathbf{t}, \tag{48}$$

on substituting for $\boldsymbol{\lambda}$ from (47). These results are quite general: from among all estimators of $\mathbf{t}'\mathbf{b}$ that are both linear and unbiased the one having the smallest variance is $\mathbf{t}'(\mathbf{X}'\mathbf{V}^{-1}\mathbf{X})^{-1}\mathbf{X}'\mathbf{V}^{-1}\mathbf{y}$; and the value of this smallest variance is $\mathbf{t}'(\mathbf{X}'\mathbf{V}^{-1}\mathbf{X})^{-1}\mathbf{t}$.

Since (47) is the sole solution to the problem of minimizing $\boldsymbol{\lambda}'\mathbf{V}\boldsymbol{\lambda}$ subject to (46), the b.l.u.e. $\boldsymbol{\lambda}'\mathbf{y}$ of $\mathbf{t}'\mathbf{b}$ is the unique estimator of $\mathbf{t}'\mathbf{b}$ having the properties of linearity, unbiasedness and "bestness"—minimum variance of all linear unbiased estimators. Thus the b.l.u.e. of $\mathbf{t}'\mathbf{b}$ is unique, $\boldsymbol{\lambda}'\mathbf{y}$ for $\boldsymbol{\lambda}'$ given in (47). Furthermore, this result is true for *any* vector $\mathbf{t}'$. Thus for some other vector, $\mathbf{p}'$ say, the b.l.u.e. of $\mathbf{p}'\mathbf{b}$ is $\mathbf{p}'(\mathbf{X}'\mathbf{V}^{-1}\mathbf{X})^{-1}\mathbf{X}'\mathbf{V}^{-1}\mathbf{y}$, and its variance is $\mathbf{p}'(\mathbf{X}'\mathbf{V}^{-1}\mathbf{X})^{-1}\mathbf{p}$; and its covariance with the b.l.u.e. of $\mathbf{t}'\mathbf{b}$ is $\mathbf{p}'(\mathbf{X}'\mathbf{V}^{-1}\mathbf{X})^{-1}\mathbf{t}$, as may be readily shown.

Suppose that $\mathbf{t}'$ takes the value $\mathbf{u}_i'$, the ith row of $\mathbf{I}_k$. Then $\mathbf{u}_i'\mathbf{b}$ is b_i, the ith element of $\mathbf{b}$, and the b.l.u.e. of b_i is $\mathbf{u}_i'(\mathbf{X}'\mathbf{V}^{-1}\mathbf{X})^{-1}\mathbf{X}'\mathbf{V}^{-1}\mathbf{y}$, the ith element of $(\mathbf{X}'\mathbf{V}^{-1}\mathbf{X})^{-1}\mathbf{X}'\mathbf{V}^{-1}\mathbf{y}$; and its variance is $\mathbf{u}_i'(\mathbf{X}'\mathbf{V}^{-1}\mathbf{X})^{-1}\mathbf{u}_i$, the ith diagonal term of $(\mathbf{X}'\mathbf{V}^{-1}\mathbf{X})^{-1}$. Thus by letting $\mathbf{t}'$ be, in turn, each row of $\mathbf{I}_k$, the

$$\text{b.l.u.e. of } \mathbf{b} \text{ is } \tilde{\mathbf{b}} = (\mathbf{X}'\mathbf{V}^{-1}\mathbf{X})^{-1}\mathbf{X}'\mathbf{V}^{-1}\mathbf{y},$$

with

$$\text{var}(\tilde{\mathbf{b}}) = (\mathbf{X}'\mathbf{V}^{-1}\mathbf{X})^{-1}. \tag{49}$$

This expression for $\tilde{\mathbf{b}}$ is identical to that given earlier; i.e., the generalized least squares estimator, the maximum likelihood estimator under normality assumptions and the b.l.u.e. are all the same, $\tilde{\mathbf{b}}$.

It was shown above that $\hat{\mathbf{b}} = (\mathbf{X}'\mathbf{X})^{-1}\mathbf{X}'\mathbf{y}$ is the b.l.u.e. of $\mathbf{b}$ when $\mathbf{V} = \mathbf{I}\sigma^2$. More generally, McElroy (1967) has shown that $\hat{\mathbf{b}}$ is the b.l.u.e. of $\mathbf{b}$ whenever $\mathbf{V} = [(1 - \rho)\mathbf{I} + \mathbf{1}\mathbf{1}'\rho]\sigma^2$ for $0 \leq \rho < 1$. This form of $\mathbf{V}$ demands equality of variances of the e_i's, and equality of all covariances between them, with the correlation between any two e_i's being ρ; clearly $\rho = 0$ is the case $\mathbf{V} = \mathbf{I}\sigma^2$.

4. CONSEQUENCES OF ESTIMATION

Properties of $\hat{\mathbf{b}} = (\mathbf{X}'\mathbf{X})^{-1}\mathbf{X}'\mathbf{y}$ and consequences thereof are now discussed. The topics dealt with in this section are based solely on the two properties so far attributed to $\mathbf{e}$, that $E(\mathbf{e}) = \mathbf{0}$ and $\text{var}(\mathbf{e}) = \sigma^2\mathbf{I}$. In the next section we consider distributional properties, based upon the further assumption of normality of the e's; but this assumption is not made here. The general case of $\text{var}(\mathbf{e}) = \mathbf{V}$ is left largely to the reader (see Sec. 5.8).

a. Unbiasedness

Since $\hat{\mathbf{b}}$ is the b.l.u.e. of $\mathbf{b}$ for $\mathbf{V} = \sigma^2\mathbf{I}$, it is unbiased. This can also be shown directly:

$$E(\hat{\mathbf{b}}) = E(\mathbf{X}'\mathbf{X})^{-1}\mathbf{X}'\mathbf{y} = (\mathbf{X}'\mathbf{X})^{-1}\mathbf{X}'\mathbf{X}\mathbf{b} = \mathbf{b}. \tag{50}$$

Thus the expected value of $\hat{\mathbf{b}}$ is $\mathbf{b}$ and so $\hat{\mathbf{b}}$ is unbiased, implying, of course, that in $\hat{\mathbf{b}}' = [\hat{b}_0 \quad \hat{\mathscr{b}}']$ the estimator $\hat{\mathscr{b}}$ is also unbiased.

b. Variances

With $\hat{\mathbf{b}} = (\mathbf{X}'\mathbf{X})^{-1}\mathbf{X}'\mathbf{y}$ it is clear that the variance-covariance matrix of $\hat{\mathbf{b}}$ is

$$\begin{aligned}\text{var}(\hat{\mathbf{b}}) &= E[\hat{\mathbf{b}} - E(\hat{\mathbf{b}})][\hat{\mathbf{b}} - E(\hat{\mathbf{b}})]' \\ &= E(\mathbf{X}'\mathbf{X})^{-1}\mathbf{X}'[\mathbf{y} - E(\mathbf{y})][\mathbf{y}' - E(\mathbf{y}')]\mathbf{X}(\mathbf{X}'\mathbf{X})^{-1} \\ &= (\mathbf{X}'\mathbf{X})^{-1}\mathbf{X}'E(\mathbf{ee}')\mathbf{X}(\mathbf{X}'\mathbf{X})^{-1} \\ &= (\mathbf{X}'\mathbf{X})^{-1}\sigma^2. \end{aligned} \tag{51}$$

The inverse matrix used for obtaining $\hat{\mathbf{b}}$ therefore also determines the variances and covariances of the elements of $\hat{\mathbf{b}}$.

A similar result holds for $\hat{\boldsymbol{\ell}}$: using the partitioned form of $(\mathbf{X}'\mathbf{X})^{-1}$ shown in (34), with $\mathbf{S} = \mathscr{X}'\mathscr{X}$, result (51) becomes

$$\text{var}\begin{bmatrix}\hat{b}_0 \\ \hat{\boldsymbol{\ell}}\end{bmatrix} = \begin{bmatrix}1/N + \bar{\mathbf{x}}'(\mathscr{X}'\mathscr{X})^{-1}\bar{\mathbf{x}} & -\bar{\mathbf{x}}'(\mathscr{X}'\mathscr{X})^{-1} \\ -(\mathscr{X}'\mathscr{X})^{-1}\bar{\mathbf{x}} & (\mathscr{X}'\mathscr{X})^{-1}\end{bmatrix}\sigma^2.$$

Hence $$\text{var}(\hat{\boldsymbol{\ell}}) = (\mathscr{X}'\mathscr{X})^{-1}\sigma^2, \tag{52}$$

analogous to (51); and

$$v(\hat{b}_0) = \sigma^2/N + \bar{\mathbf{x}}'\,\text{var}(\hat{\boldsymbol{\ell}})\bar{\mathbf{x}} = [1/N + \bar{\mathbf{x}}'(\mathscr{X}'\mathscr{X})^{-1}\bar{\mathbf{x}}]\sigma^2 \tag{53}$$

and $$\text{cov}(\hat{b}_0, \hat{\boldsymbol{\ell}}') = -\bar{\mathbf{x}}'\,\text{var}(\hat{\boldsymbol{\ell}}). \tag{54}$$

c. Estimating $E(y)$

The estimator $\hat{\mathbf{b}}$ can be used for estimating $E(y)$. Analogous to the model

$$E(y) = b_0 + b_1x_1 + \cdots + b_kx_k$$

we have $$\widehat{E(y)} = \hat{b}_0 + \hat{b}_1x_1 + \cdots + \hat{b}_kx_k,$$

as illustrated at the end of each of the examples in Sec. 1. If

$$\mathbf{x}_0' = [x_{00} \quad x_{01} \quad x_{02} \quad \cdots \quad x_{0k}] \tag{55}$$

is a set of x-values (with $x_{00} = 1$) for which we wish to estimate the corresponding value of $E(y)$, that estimator is

$$\widehat{E(y_0)} = \hat{b}_0 + \hat{b}_1x_{01} + \cdots + \hat{b}_kx_{0k} = \mathbf{x}_0'\hat{\mathbf{b}}. \tag{56}$$

We call this the *estimated expected value of y* corresponding to the set of x-values $x_{00}, x_{01}, \ldots, x_{0k}$. When this set of x's is one of those in the data, $\mathbf{x}_0'$ is a row of $\mathbf{X}$ in (18), in which case (56) is an element of $\mathbf{X}\hat{\mathbf{b}}$. Corresponding to $E(\mathbf{y}) = \mathbf{Xb}$ of (17) we therefore have, as N special cases of (56),

$$\widehat{E(\mathbf{y})} = \mathbf{X}\hat{\mathbf{b}}. \tag{57}$$

These are the estimated expected values of y corresponding to the N observed values of y in the data. They are sometimes called *fitted y-values*, or estimated y-values, names which can be misleading because (56) and (57) are both estimates of expected values of y. They correspond, in (56), to any set of predetermined x's in $\mathbf{x}_0'$ of (55), and in (57) to the observed x's in $\mathbf{X}$.

Variances of the estimators (56) and (57) are readily obtained using var($\hat{\mathbf{b}}$) of (51). Thus

$$v[\widehat{E(y_0)}] = \mathbf{x}_0'(\mathbf{X}'\mathbf{X})^{-1}\mathbf{x}_0\sigma^2 \tag{58}$$

and

$$\text{var}[\widehat{E(\mathbf{y})}] = \mathbf{X}(\mathbf{X}'\mathbf{X})^{-1}\mathbf{X}'\sigma^2. \tag{59}$$

On substituting $\mathbf{X} = [\mathbf{1} \quad \mathbf{X}_1]$ from (31) and using $\mathscr{X}$ of (39), this reduces to

$$\begin{aligned}\text{var}[\widehat{E(\mathbf{y})}] &= (\sigma^2/N)\mathbf{11}' + \mathscr{X}\,\text{var}(\hat{\mathscr{b}})\mathscr{X}' \\ &= (\sigma^2/N)\mathbf{11}' + \mathscr{X}(\mathscr{X}'\mathscr{X})^{-1}\mathscr{X}'\sigma^2.\end{aligned} \tag{60}$$

Corresponding to $\mathbf{x}_0'$, the expected y-value is $E(y_0)$, estimated by $\widehat{E(y_0)}$ of (56). In contrast, consider a future observation, y_f say, corresponding to some vector of x-values, $\mathbf{x}_f$ say. Then, by the model, $y_f = \mathbf{x}_f'\mathbf{b} + e_f$ where e_f is a random error term which can be neither observed nor estimated. Hence the best available prediction of y_f, which we shall call $\tilde{y}_f$, is $\tilde{y}_f = \mathbf{x}_f'\hat{\mathbf{b}}$. Thus $\mathbf{x}_f'\hat{\mathbf{b}}$ can be used both as a prediction of a future observation corresponding to $\mathbf{x}_f'$ as well as for its more customary use, that of an estimator of the expected value $E(y_f)$ corresponding to $\mathbf{x}_f'$. The first of these uses prompts inquiring how some future observation y_f varies about its prediction, $\tilde{y}_f = \mathbf{x}_f'\hat{\mathbf{b}}$. To do this we consider the deviation of any y_f from $\tilde{y}_f$:

$$y_f - \tilde{y}_f = y_f - \mathbf{x}_f'\hat{\mathbf{b}} = \mathbf{x}_f'(\mathbf{b} - \hat{\mathbf{b}}) + e_f\,.$$

The variance of this deviation is derived by noting that, because y_f is thought of as an observation obtained independently of those used in deriving $\hat{\mathbf{b}}$, we have $\hat{\mathbf{b}}$ and e_f being independent and so cov($\hat{\mathbf{b}}, e_f) = \mathbf{0}$. Hence

$$v(y_f - \tilde{y}_f) = \mathbf{x}_f'v(\hat{\mathbf{b}} - \mathbf{b})\mathbf{x}_f + v(e_f) = [\mathbf{x}_f'(\mathbf{X}'\mathbf{X})^{-1}\mathbf{x}_f + 1]\sigma^2. \tag{61}$$

Thus the estimated expected value of y corresponding to $\mathbf{x}_f$ is $\widehat{E(y_f)} = \mathbf{x}_f'\hat{\mathbf{b}}$, as in (56), with variance $\mathbf{x}_f'(\mathbf{X}'\mathbf{X})^{-1}\mathbf{x}_f\sigma^2$ similar to (58); and the predicted value of an observation corresponding to $\mathbf{x}_f$ is the same value, $\mathbf{x}_f'\hat{\mathbf{b}} = \tilde{y}_f$, with the variance of deviations of y-values (corresponding to $\mathbf{x}_f$) from this prediction being $[\mathbf{x}_f'(\mathbf{X}'\mathbf{X})^{-1}\mathbf{x}_f + 1]\sigma^2$ of (61). These results are true for any value of $\mathbf{x}_f$. The variance of y_f itself is, of course, σ^2 at all times.

d. Residual error sum of squares

It is convenient to use the symbol $\hat{\mathbf{y}}$ for $\widehat{E(\mathbf{y})}$, the vector of estimated expected values of y corresponding to the vector of observations $\mathbf{y}$; i.e.,

$$\hat{\mathbf{y}} \equiv \widehat{E(\mathbf{y})} = \mathbf{X}\hat{\mathbf{b}}. \tag{62}$$

The vector of deviations of the observed y_i's from their corresponding predicted values is therefore

$$\mathbf{y} - \hat{\mathbf{y}} = \mathbf{y} - \mathbf{X}\hat{\mathbf{b}} = \mathbf{y} - \mathbf{X}(\mathbf{X}'\mathbf{X})^{-1}\mathbf{X}'\mathbf{y} = [\mathbf{I} - \mathbf{X}(\mathbf{X}'\mathbf{X})^{-1}\mathbf{X}']\mathbf{y}. \tag{63}$$

Note that the matrix involved here is idempotent, a fact that gets used repeatedly in the sequel: indeed,

$$\mathbf{I} - \mathbf{X}(\mathbf{X}'\mathbf{X})^{-1}\mathbf{X}' \text{ is symmetric and idempotent} \tag{64}$$

and

$$[\mathbf{I} - \mathbf{X}(\mathbf{X}'\mathbf{X})^{-1}\mathbf{X}']\mathbf{X} = \mathbf{0}. \tag{65}$$

The sum of squares of the deviations of the observed y_i's from their estimated expected values is usually known as the residual, or error sum of squares, for which the symbol SSE will be used. Thus

$$\text{SSE} = \sum_{i=1}^{N} (y_i - \hat{y}_i)^2 = (\mathbf{y} - \hat{\mathbf{y}})'(\mathbf{y} - \hat{\mathbf{y}}). \tag{66}$$

This will be referred to as the *residual error sum of squares*, combining the traditional name "error" with "residual", which is, perhaps, more appropriately descriptive in view of the definition of e_i given in (3).

Computing procedures for SSE are derived from substituting (63) into (66) and using (64) and (65). This gives

$$\begin{aligned} \text{SSE} &= \mathbf{y}'[\mathbf{I} - \mathbf{X}(\mathbf{X}'\mathbf{X})^{-1}\mathbf{X}']\mathbf{y} && (67) \\ &= \mathbf{y}'\mathbf{y} - \mathbf{y}'\mathbf{X}(\mathbf{X}'\mathbf{X})^{-1}\mathbf{X}'\mathbf{y} \\ &= \mathbf{y}'\mathbf{y} - \hat{\mathbf{b}}'\mathbf{X}'\mathbf{y} && (68) \end{aligned}$$

because $\hat{\mathbf{b}}' = \mathbf{y}'\mathbf{X}(\mathbf{X}'\mathbf{X})^{-1}$. This is a convenient form for computing SSE; $\mathbf{y}'\mathbf{y}$ in (68) is the total sum of squares of the observations, and $\hat{\mathbf{b}}'\mathbf{X}'\mathbf{y}$ is the sum of products of the elements of the solution $\hat{\mathbf{b}}$ with their corresponding elements of the right-hand side, $\mathbf{X}'\mathbf{y}$, of the equations from which $\hat{\mathbf{b}}$ is derived, namely $\mathbf{X}'\mathbf{X}\hat{\mathbf{b}} = \mathbf{X}'\mathbf{y}$. Note, however, that in so describing (68) these right-hand side elements must be exactly as they are in the normal equations. Thus if, when solving $\mathbf{X}'\mathbf{X}\hat{\mathbf{b}} = \mathbf{X}'\mathbf{y}$, some or all of the equations are amended by factorizing out some common factors, then it is not the right-hand sides of the equations so amended that are used in $\hat{\mathbf{b}}'\mathbf{X}'\mathbf{y}$ of (68) but the $\mathbf{X}'\mathbf{y}$ of the original normal equations.

An expression for SSE involving $\hat{\ell}$ and $\mathscr{X}'\mathbf{y}$ can also be established. For (68) is equivalent to

$$\begin{aligned} \text{SSE} &= \mathbf{y}'\mathbf{y} - [\bar{y} - \hat{\ell}'\bar{\mathbf{x}} \quad \hat{\ell}'] \begin{bmatrix} N\bar{y} \\ \mathbf{X}_1'\mathbf{y} \end{bmatrix} \\ &= \mathbf{y}'\mathbf{y} - N\bar{y}^2 - \hat{\ell}'(\mathbf{X}_1'\mathbf{y} - N\bar{y}\bar{\mathbf{x}}) \qquad (69) \\ &= \mathscr{y}'\mathscr{y} - \hat{\ell}'\mathscr{X}'\mathbf{y}, \qquad (70) \end{aligned}$$

where $\mathscr{y}'\mathscr{y}$ denotes the corrected sum of squares of the y's. The form of (70) is clearly analogous to that of (68) and it is equally, if not more, useful for computing purposes: $\mathscr{y}'\mathscr{y}$ is the corrected sum of squares of the y's, and $\hat{\ell}'\mathscr{X}'\mathbf{y}$ is the sum of products of elements of the solution with the corresponding elements of the right-hand side, $\mathscr{X}'\mathbf{y}$, of the equations from which $\hat{\ell}$ is derived, namely $\mathscr{X}'\mathscr{X}\hat{\ell} = \mathscr{X}'\mathbf{y}$. Note that $\mathscr{X}'\mathbf{y} = \mathscr{X}'\mathscr{y}$ because $\mathscr{X}'\mathscr{y} = \mathscr{X}'(\mathbf{y} - \bar{y}\mathbf{1})$ and, by (33) and (39), $\mathscr{X}'\mathbf{1} = 0$.

e. Estimating the residual error variance

In (67) SSE is written as a quadratic form in $\mathbf{y}$:

$$\text{SSE} = \mathbf{y}'[\mathbf{I} - \mathbf{X}(\mathbf{X}'\mathbf{X})^{-1}\mathbf{X}']\mathbf{y}.$$

Therefore, with $\mathbf{y}$ being distributed $(\mathbf{Xb}, \mathbf{I}\sigma^2)$ the expected value of SSE is, from Theorem 1 of Sec. 2.5,

$$\begin{aligned} E[\text{SSE}] &= \text{tr}[\mathbf{I} - \mathbf{X}(\mathbf{X}'\mathbf{X})^{-1}\mathbf{X}']\mathbf{I}\sigma^2 + \mathbf{b}'\mathbf{X}'[\mathbf{I} - \mathbf{X}(\mathbf{X}'\mathbf{X})^{-1}\mathbf{X}']\mathbf{Xb} \\ &= r[\mathbf{I} - \mathbf{X}(\mathbf{X}'\mathbf{X})^{-1}\mathbf{X}']\sigma^2 \\ &= [N - r(\mathbf{X})]\sigma^2, \end{aligned}$$

on utilizing (64) and (65) and the fact that the trace of an idempotent matrix equals its rank. Hence an unbiased estimator of σ^2 is

$$\hat{\sigma}^2 = \frac{\text{SSE}}{N - r(\mathbf{X})} = \frac{\text{SSE}}{N - r} \qquad (71)$$

using r for $r(\mathbf{X})$, the rank of $\mathbf{X}$. Even though, in this full model regression situation we know that

$$r = r(\mathbf{X}) = k + 1,$$

the use of r will be retained, to emphasize that it is the rank of $\mathbf{X}$ and not just the number of x-variables plus one. It also makes for easier transition to the non-full rank case, where it is essential to use $r(\mathbf{X})$.

f. Partitioning the total sum of squares

The total sum of squares, which we shall call SST, is

$$SST = \mathbf{y}'\mathbf{y} = \sum_{i=1}^{N} y_i^2.$$

And the sum of squares of deviations of the observed y_i's from their predicted values is

$$\text{SSE} = \mathbf{y}'\mathbf{y} - \hat{\mathbf{b}}'\mathbf{X}'\mathbf{y} = \mathbf{y}'\mathbf{y} - N\bar{y}^2 - \hat{\mathscr{b}}'\mathscr{X}'\mathbf{y}$$

as in (68) and (69). The difference

$$\text{SSR} = \text{SST} - \text{SSE} = \hat{\mathbf{b}}'\mathbf{X}'\mathbf{y} = \hat{\mathbf{b}}'\mathbf{X}'\mathbf{X}\hat{\mathbf{b}} = N\bar{y}^2 + \hat{\mathscr{b}}'\mathscr{X}'\mathbf{y}$$

represents that portion of SST attributable to having fitted the regression, and so is called the *sum of squares due to regression*, SSR. It is also often called the *reduction in sum of squares*. This partitioning of SST can be summarized in a manner that serves as a foundation for developing the traditional analysis of variance table:

$$\begin{array}{lll} \text{SSR} = \hat{\mathbf{b}}'\mathbf{X}'\mathbf{y} & (=\hat{\mathbf{b}}'\mathbf{X}'\mathbf{X}\hat{\mathbf{b}}) & = N\bar{y}^2 + \hat{\mathscr{b}}'\mathscr{X}'\mathbf{y} \\ \text{SSE} = \mathbf{y}'\mathbf{y} - \hat{\mathbf{b}}'\mathbf{X}'\mathbf{y} & & = \mathbf{y}'\mathbf{y} - N\bar{y}^2 - \hat{\mathscr{b}}'\mathscr{X}'\mathbf{y} \\ \hline \text{SST} = \mathbf{y}'\mathbf{y} & & = \mathbf{y}'\mathbf{y} \\ \hline \end{array} \tag{72}$$

Now suppose the model had had no x-variables in it but had simply been $y_i = b_0x_{i0} + e_i$, i.e., $y_i = b_0 + e_i$. Then $\hat{b}_0$ would be $\bar{y}$ and SSR would become $N\bar{y}^2$. This we recognize as the usual correction for the mean, which shall be written

$$\text{SSM} = N\bar{y}^2.$$

Then in (72) we see that

$$\text{SSR} = \text{SSM} + \hat{\mathscr{b}}'\mathscr{X}'\mathbf{y}$$

and so we can call

$$\text{SSR}_m = \text{SSR} - \text{SSM} = \hat{\mathscr{b}}'\mathscr{X}'\mathbf{y} = \hat{\mathscr{b}}'\mathscr{X}'\mathscr{X}\hat{\mathscr{b}}$$

the regression sum of squares corrected for the mean. In this way (72) becomes

$$\begin{array}{ll} \text{SSM} & = N\bar{y}^2 \\ \text{SSR}_m & = \hat{\mathscr{b}}'\mathscr{X}'\mathbf{y} = \hat{\mathscr{b}}'\mathscr{X}'\mathscr{X}\hat{\mathscr{b}} \\ \text{SSE} & = \mathbf{y}'\mathbf{y} - N\bar{y}^2 - \hat{\mathscr{b}}'\mathscr{X}'\mathbf{y} \\ \hline \text{SST} & = \mathbf{y}'\mathbf{y} \\ \hline \end{array} \tag{73}$$

Similar to SSR_m we also have

$$\text{SST}_m = \text{SST} - \text{SSM} = \mathbf{y}'\mathbf{y} - N\bar{y}^2 = \mathscr{y}'\mathscr{y} \tag{74}$$

as the corrected sum of squares of the y's. With it, SSR_m and SSE of (73) can be summarized as

$$\begin{array}{lll} \text{SSR}_m = \hat{\mathscr{b}}'\mathscr{X}'\mathbf{y} & & = \hat{\mathscr{b}}'\mathscr{X}'\mathbf{y} \\ \text{SSE} \;\;= \mathscr{y}'\mathscr{y} - \hat{\mathscr{b}}'\mathscr{X}'\mathbf{y} & = \mathbf{y}'\mathbf{y} - N\bar{y}^2 & - \hat{\mathscr{b}}'\mathscr{X}'\mathbf{y} \\ \hline \text{SST}_m \;= \mathscr{y}'\mathscr{y} & & = \mathbf{y}'\mathbf{y} - N\bar{y}^2 \\ \hline \end{array} \tag{75}$$

This format is identical to that of (72); in the one case, (72), uncorrected sums of squares are used with total SST, and in the other, (75), corrected sums of squares are used with total SST_m. The error terms are the same in the two cases, however, namely SSE.

The summary shown in (75) is the basis of the traditional analysis of variance table for fitting linear regression. Distributional properties of these sums of squares are considered in Sec. 5.

g. Multiple correlation

A measure of the goodness of fit of the regression is the multiple correlation coefficient, estimated as the product moment correlation between the observed y_i's and the predicted $\hat{y}_i$'s. Denoted by R, it can be calculated as

$$\begin{array}{lll} & R^2 = \text{SSR}/\text{SST} & \text{in the no-intercept model} \\ \text{and as} & R^2 = \text{SSR}_m/\text{SST}_m & \text{in the intercept model.} \end{array} \tag{76}$$

This we now show.

In the no-intercept model the mean $\bar{y}$ is ignored, and the product moment correlation between the y_i's and $\hat{y}_i$'s is defined by

$$R^2 = \frac{(\sum y_i\hat{y}_i)^2}{(\sum y_i^2)(\sum \hat{y}_i^2)} = \frac{(\mathbf{y}'\hat{\mathbf{y}})^2}{\mathbf{y}'\mathbf{y}(\hat{\mathbf{y}}'\hat{\mathbf{y}})}\,. \tag{77}$$

With $\hat{\mathbf{y}} = \mathbf{X}\hat{\mathbf{b}} = \mathbf{X}(\mathbf{X}'\mathbf{X})^{-1}\mathbf{X}'\mathbf{y}$ it can be shown (see Exercise 18) that (77) reduces to $R^2 = \text{SSR}/\text{SST}$ as in (76).

In the intercept model the definition of R^2 is

$$R^2 = \frac{[\sum (y_i - \bar{y})(\hat{y}_i - \bar{\hat{y}})]^2}{\sum (y_i - \bar{y})^2 \sum (\hat{y}_i - \bar{\hat{y}})^2}\,. \tag{78}$$

In simplifying this expression we use

$$\bar{y} = \mathbf{1}'\mathbf{y}/N \qquad \text{and} \qquad \mathbf{1}'\mathbf{X}(\mathbf{X}'\mathbf{X})^{-1}\mathbf{X}' = \mathbf{1}' \tag{79}$$

the latter arising from $\mathbf{X}'\mathbf{X}(\mathbf{X}'\mathbf{X})^{-1}\mathbf{X}' = \mathbf{X}'$ because the first row of $\mathbf{X}'$ is $\mathbf{1}'$.

These results, together with (74), lead (see Exercise 18) to (78) reducing to $R^2 = \text{SSR}_m^2/\text{SST}_m(\text{SSR}_m) = \text{SSR}_m/\text{SST}_m$ as in (76).

Intuitively the ratio SSR/SST (or $\text{SSR}_m/\text{SST}_m$) has appeal, since it represents that fraction of the total sum of squares which is accounted for by fitting the model—in this case fitting the regression. Thus, although R has traditionally been thought of and used as a multiple correlation coefficient in some sense, its more frequent use nowadays is in the form of R^2, where it represents the fraction of the total sum of squares accounted for by fitting the model.

Care must be taken in using these formulae for R^2 for, although SSR_m and SST_m in the intercept model have been defined as $\text{SSR} - N\bar{y}^2$ and $\text{SST} - N\bar{y}^2$, the value of SSR used in the intercept model is not the same as its value in the corresponding no-intercept model. This is brought out in the example.

h. Example (continued)

In (28) we found

$$\hat{\mathbf{b}} = \frac{1}{24}\begin{bmatrix} 56 \\ 50 \\ -5 \end{bmatrix}$$

and so the vector $\widehat{E(\mathbf{y})} = \hat{\mathbf{y}}$ is

$$\widehat{E(\mathbf{y})} = \hat{\mathbf{y}} = \mathbf{X}\hat{\mathbf{b}} = \frac{1}{24}\begin{bmatrix} 1 & 6 & 28 \\ 1 & 12 & 40 \\ 1 & 10 & 32 \\ 1 & 8 & 36 \\ 1 & 9 & 34 \end{bmatrix}\begin{bmatrix} 56 \\ 50 \\ -5 \end{bmatrix} = \frac{1}{24}\begin{bmatrix} 216 \\ 456 \\ 396 \\ 276 \\ 336 \end{bmatrix} = \begin{bmatrix} 9 \\ 19 \\ 16\frac{1}{2} \\ 11\frac{1}{2} \\ 14 \end{bmatrix}.$$

Hence from (59), using $(\mathbf{X}'\mathbf{X})^{-1}$ of (26),

$$\text{var}[\widehat{E(\mathbf{y})}] = \mathbf{X}(\mathbf{X}'\mathbf{X})^{-1}\mathbf{X}'\sigma^2$$

$$= \begin{bmatrix} 1 & 6 & 28 \\ 1 & 12 & 40 \\ 1 & 10 & 32 \\ 1 & 8 & 36 \\ 1 & 9 & 34 \end{bmatrix} \frac{\sigma^2}{2880}\begin{bmatrix} 50656 & 1840 & -1960 \\ 1840 & 400 & -160 \\ -1960 & -160 & 100 \end{bmatrix}\begin{bmatrix} 1 & 1 & 1 & 1 & 1 \\ 6 & 12 & 10 & 8 & 9 \\ 28 & 40 & 32 & 36 & 34 \end{bmatrix}.$$

Also, from (60), using $\mathscr{X}$ of (43) and $(\mathscr{X}'\mathscr{X})^{-1}$ of (44),

$$\widehat{\text{var}[E(\bar{\mathbf{y}})]} = \tfrac{1}{5}\mathbf{1}\mathbf{1}'\sigma^2 + \mathscr{X}(\mathscr{X}'\mathscr{X})^{-1}\mathscr{X}'\sigma^2$$

$$= \frac{1}{5}\begin{bmatrix} 1 & 1 & 1 & 1 & 1 \\ 1 & 1 & 1 & 1 & 1 \\ 1 & 1 & 1 & 1 & 1 \\ 1 & 1 & 1 & 1 & 1 \\ 1 & 1 & 1 & 1 & 1 \end{bmatrix}\sigma^2$$

$$+ \begin{bmatrix} -3 & -6 \\ 3 & 6 \\ 1 & -2 \\ -1 & 2 \\ 0 & 0 \end{bmatrix} \frac{\sigma^2}{144}\begin{bmatrix} 20 & -8 \\ -8 & 5 \end{bmatrix}\begin{bmatrix} -3 & 3 & 1 & -1 & 0 \\ -6 & 6 & -2 & 2 & 0 \end{bmatrix},$$

and on carrying out the arithmetic it will be found that both forms reduce to

$$\widehat{\text{var}[E(\mathbf{y})]} = \begin{bmatrix} .7 & -.3 & .2 & .2 & .2 \\ -.3 & .7 & .2 & .2 & .2 \\ .2 & .2 & .7 & -.3 & .2 \\ .2 & .2 & -.3 & .7 & .2 \\ .2 & .2 & .2 & .2 & .2 \end{bmatrix}\sigma^2.$$

An estimate of this is obtained by replacing σ^2 by $\hat{\sigma}^2$, derived below.

From $\mathbf{y}$ and $\hat{\mathbf{y}}$ we get

$$(\mathbf{y} - \hat{\mathbf{y}}) = \begin{bmatrix} 10 \\ 20 \\ 17 \\ 12 \\ 11 \end{bmatrix} - \begin{bmatrix} 9 \\ 19 \\ 16\frac{1}{2} \\ 11\frac{1}{2} \\ 14 \end{bmatrix} = \begin{bmatrix} 1 \\ 1 \\ \frac{1}{2} \\ \frac{1}{2} \\ -3 \end{bmatrix}$$

and hence, from its definition,

$$\text{SSE} = 1^2 + 1^2 + (\tfrac{1}{2})^2 + (\tfrac{1}{2})^2 + 3^2 = 11\tfrac{1}{2}.$$

The alternative form for SSE, given in (68), is

$$\text{SSE} = \mathbf{y}'\mathbf{y} - \hat{\mathbf{b}}'\mathbf{X}'\mathbf{y}$$

and with $\mathbf{y}'\mathbf{y} = \Sigma y_i^2 = 1{,}054$ in the basic data, $\hat{\mathbf{b}}'$ from (28) and $\mathbf{X}'\mathbf{y}$ from (27), this gives

$$\text{SSE} = 1{,}054 - (1/24)[56 \quad 50 \quad -5]\begin{bmatrix} 70 \\ 665 \\ 2{,}430 \end{bmatrix}$$

$$= 1{,}054 - 1{,}042\tfrac{1}{2} = 11\tfrac{1}{2} \text{ as before.}$$

Likewise, using the form given in (70),

$$\text{SSE} = 1{,}054 - 5(14^2) - (1/24)[50 \quad -5]\begin{bmatrix} 35 \\ 50 \end{bmatrix}$$

$$= 1{,}054 - 980 - 62\tfrac{1}{2} = 11\tfrac{1}{2} \text{ again.}$$

Hence in (71)

$$\hat{\sigma}^2 = 11\tfrac{1}{2}/(5 - 3) = 5.75 .$$

From the calculations for SSE the summaries in (72), (73) and (75) are as shown in Table 3.1. From the last of these R^2 is $\text{SSR}_m/\text{SST}_m = 62\frac{1}{2}/74 = .84$,

TABLE 3.1. PARTITIONING OF SUM OF SQUARES: INTERCEPT MODEL

Eqs. (72)	Eqs. (73)	Eqs. (75)
	SSM = 980	
SSR = $1{,}042\frac{1}{2}$	SSR_m = $62\frac{1}{2}$	SSR_m = $62\frac{1}{2}$
SSE = $11\frac{1}{2}$	SSE = $11\frac{1}{2}$	SSE = $11\frac{1}{2}$
SST = 1,054	SST = 1,054	SST_m = 74

since the model being used is the intercept model. Were a no-intercept model to be used on these data the formal expression for R^2 would be SSR/SST, although not with the SSR shown above, for that is the value of SSR with the intercept model. Thus for the no-intercept model for these data the normal equations for $\hat{\mathbf{b}}$ and $\mathbf{X}'\mathbf{y}$ are given in (29) and so

$$\text{SSR} = \hat{\mathbf{b}}'\mathbf{X}'\mathbf{y} = (1/12{,}664)[25{,}310 \quad -1{,}495]\begin{bmatrix} 665 \\ 2{,}430 \end{bmatrix} = 1{,}038.24,$$

different from the value of SSR for the intercept model given under (72) in Table 3.1. The corresponding value of R^2 is $1{,}038.24/1{,}054 = .98$.

5. DISTRIBUTIONAL PROPERTIES

The normality assumption is now introduced. We assume that $\mathbf{e}$ is normally distributed:

$$\mathbf{e} \sim N(\mathbf{0}, \sigma^2\mathbf{I}).$$

Distributional properties of $\mathbf{y}$ and functions of $\mathbf{y}$ follow at once. In particular, the distributions of $\hat{\mathbf{b}}$, of $\hat{\sigma}^2$ and of various sums of squares are derived, using the results in Chapter 2.

a. y is normal

From $\mathbf{y} = \mathbf{Xb} + \mathbf{e}$ we have $\mathbf{y} - \mathbf{Xb} = \mathbf{e}$ and therefore

$$\mathbf{y} \sim N(\mathbf{Xb}, \sigma^2\mathbf{I}_N).$$

b. $\hat{\mathbf{b}}$ is normal

$\hat{\mathbf{b}}$ is a linear function of $\mathbf{y}$. Therefore (see Exercise 11 of Chapter 2), it is normally distributed:

$$\hat{\mathbf{b}} = (\mathbf{X}'\mathbf{X})^{-1}\mathbf{X}'\mathbf{y} \sim N[\mathbf{b}, (\mathbf{X}'\mathbf{X})^{-1}\sigma^2].$$

Its mean and variance are as already derived in (50) and (51). The same reasoning shows that $\hat{\mathscr{b}}$ is also normally distributed, with mean and variance given in (50) and (52):

$$\hat{\mathscr{b}} = (\mathscr{X}'\mathscr{X})^{-1}\mathscr{X}'\mathbf{y} \sim N[\mathscr{b}, (\mathscr{X}'\mathscr{X})^{-1}\sigma^2].$$

c. $\hat{\mathbf{b}}$ and $\hat{\sigma}^2$ are independent

We have

$$\hat{\mathbf{b}} = (\mathbf{X}'\mathbf{X})^{-1}\mathbf{X}'\mathbf{y}$$

and

$$\text{SSE} = \mathbf{y}'[\mathbf{I} - \mathbf{X}(\mathbf{X}'\mathbf{X})^{-1}\mathbf{X}']\mathbf{y}.$$

But, by (65)

$$(\mathbf{X}'\mathbf{X})^{-1}\mathbf{X}'[\mathbf{I} - \mathbf{X}(\mathbf{X}'\mathbf{X})^{-1}\mathbf{X}'] = \mathbf{0}$$

and, so by Theorem 3 of Sec. 2.5b, the statistics $\hat{\mathbf{b}}$ and $\hat{\sigma}^2$ are distributed independently.

d. SSE/σ^2 has a χ^2-distribution

From (67), SSE is a quadratic in $\mathbf{y}$:

$$\text{SSE} = \mathbf{y}'[\mathbf{I} - \mathbf{X}(\mathbf{X}'\mathbf{X})^{-1}\mathbf{X}']\mathbf{y} = \mathbf{y}'\mathbf{P}\mathbf{y} \text{ say,}$$

defining $\mathbf{P}$ as

$$\mathbf{P} = \mathbf{I} - \mathbf{X}(\mathbf{X}'\mathbf{X})^{-1}\mathbf{X}'. \tag{80}$$

Now $\text{SSE}/\sigma^2 = \mathbf{y}'(1/\sigma^2)\mathbf{P}\mathbf{y}$, and by (64) $\mathbf{P}$ is idempotent; and $\text{var}(\mathbf{y}) = \sigma^2\mathbf{I}$. Therefore $(1/\sigma^2)\mathbf{P}\sigma^2\mathbf{I}$ is idempotent and so, from Theorem 2 of Chapter 2,

$$\text{SSE}/\sigma^2 \sim \chi^{2'}\{r[\mathbf{I} - \mathbf{X}(\mathbf{X}'\mathbf{X})^{-1}\mathbf{X}'], \mathbf{b}'\mathbf{X}'[\mathbf{I} - \mathbf{X}(\mathbf{X}'\mathbf{X})^{-1}\mathbf{X}']\mathbf{Xb}/2\sigma^2\}$$

which, because of (64) and (65), reduces to

$$\text{SSE}/\sigma^2 \sim \chi^2_{N-r}, \qquad \text{where} \quad r = r(\mathbf{X}).$$

Hence

$$(N - r)\hat{\sigma}^2/\sigma^2 \sim \chi^2_{N-r}.$$

e. Non-central χ^2's

Having shown that SSE/σ^2 has a central χ^2-distribution, we now show that SSR, SSM and SSR_m, the other terms in the partitioning of the total sum of squares in (72), (73) and (75), have non-central χ^2-distributions. Furthermore, these terms are independent of SSE. Thus we are led to F-statistics that have non-central F-distributions. And these in turn are central F-distributions under certain null hypotheses, and so tests of these hypotheses are established.

From (72) we have

$$\text{SSR} = \hat{\mathbf{b}}'\mathbf{X}'\mathbf{y} = \mathbf{y}'\mathbf{X}(\mathbf{X}'\mathbf{X})^{-1}\mathbf{X}'\mathbf{y}.$$

The matrix $\mathbf{X}(\mathbf{X}'\mathbf{X})^{-1}\mathbf{X}'$ involved here is idempotent and its products with $\mathbf{I} - \mathbf{X}(\mathbf{X}'\mathbf{X})^{-1}\mathbf{X}'$ are null. Therefore, by Theorem 4 of Chapter 2, SSR is independent of SSE and by Theorem 2 of the same chapter

$$\text{SSR}/\sigma^2 \sim \chi^{2\prime}\{r[\mathbf{X}(\mathbf{X}'\mathbf{X})^{-1}\mathbf{X}'], \mathbf{b}'\mathbf{X}'\mathbf{X}(\mathbf{X}'\mathbf{X})^{-1}\mathbf{X}'\mathbf{X}\mathbf{b}/2\sigma^2\},$$

i.e.,

$$\text{SSR}/\sigma^2 \sim \chi^{2\prime}(r, \mathbf{b}'\mathbf{X}'\mathbf{X}\mathbf{b}/2\sigma^2).$$

Similarly, in (73)

$$\text{SSM} = N\bar{y}^2 = \mathbf{y}'N^{-1}\mathbf{1}\mathbf{1}'\mathbf{y}$$

where $N^{-1}\mathbf{1}\mathbf{1}'$ is idempotent and its products with $\mathbf{I} - \mathbf{X}(\mathbf{X}'\mathbf{X})^{-1}\mathbf{X}'$ are null. Therefore SSM is distributed independently of SSE and

$$\text{SSM}/\sigma^2 \sim \chi^{2\prime}[r(N^{-1}\mathbf{1}\mathbf{1}'), \mathbf{b}'\mathbf{X}'N^{-1}\mathbf{1}\mathbf{1}'\mathbf{X}\mathbf{b}/2\sigma^2],$$

i.e.,

$$\text{SSM}/\sigma^2 \sim \chi^{2\prime}[1, (\mathbf{1}'\mathbf{X}\mathbf{b})^2/2N\sigma^2].$$

Also, in (75)

$$\text{SSR}_m = \hat{\mathscr{b}}'\mathscr{X}'\mathbf{y} = \hat{\mathscr{b}}'\mathscr{X}'\mathscr{X}\hat{\mathscr{b}}.$$

Hence, because $\hat{\mathscr{b}} \sim N[\mathscr{b}, (\mathscr{X}'\mathscr{X})^{-1}\sigma^2]$

$$\text{SSR}_m/\sigma^2 \sim \chi^{2\prime}[r(\mathscr{X}'\mathscr{X}), \mathscr{b}'\mathscr{X}'\mathscr{X}\mathscr{b}/2\sigma^2],$$

i.e.,

$$\text{SSR}_m/\sigma^2 \sim \chi^{2\prime}[r - 1, \mathscr{b}'\mathscr{X}'\mathscr{X}\mathscr{b}/2\sigma^2].$$

Furthermore, SSR_m can be expressed as $\mathbf{y}'\mathbf{Q}\mathbf{y}$ where not only is $\mathbf{Q}$ idempotent but its products with $\mathbf{I} - \mathbf{X}(\mathbf{X}'\mathbf{X})^{-1}\mathbf{X}'$ and $N^{-1}\mathbf{1}\mathbf{1}'$ are null. Hence, by Theorem 4 of Chapter 2, SSR_m is independent of both SSE and SSM.

Finally, of course,

$$\mathbf{y}'\mathbf{y}/\sigma^2 \sim \chi^{2\prime}(N, \mathbf{b}'\mathbf{X}'\mathbf{X}\mathbf{b}/2\sigma^2).$$

f. *F*-distributions

Applying the definition of the non-central *F*-distribution to the foregoing results it is clear that the *F*-statistic

$$F(R) = \frac{\text{SSR}/r}{\text{SSE}/(N-r)} \sim F'(r, N-r, \mathbf{b}'\mathbf{X}'\mathbf{X}\mathbf{b}/2\sigma^2). \tag{81}$$

Similarly,

$$F(M) = \frac{\text{SSM}/1}{\text{SSE}/(N-r)} \sim F'[1, N-r, (\mathbf{1}'\mathbf{X}\mathbf{b})^2/2N\sigma^2] \tag{82}$$

and

$$F(R_m) = \frac{\text{SSR}_m/(r-1)}{\text{SSE}/(N-r)} \sim F'[r-1, N-r, \ell'\mathscr{X}'\mathscr{X}\ell/2\sigma^2]. \tag{83}$$

Under certain null hypotheses the non-centrality parameters in (81)–(83) are zero, and these non-central *F*'s then become central *F*'s, so providing us with statistics for testing those hypotheses. This is discussed subsequently.

g. Analyses of variance

Calculation of the above *F*-statistics can be summarized in analyses of variance tables. An outline of such tables is given in (72), (73) and (75). For example, (72) and the calculation of (81) are summarized in Table 3.2.

TABLE 3.2. ANALYSIS OF VARIANCE FOR FITTING REGRESSION

Source of Variation	d.f.[1]	Sum of Squares	Mean Square	*F*-statistic
Regression	r	$\text{SSR} = \hat{\mathbf{b}}'\mathbf{X}'\mathbf{y}$	$\text{MSR} = \text{SSR}/r$	$F(R) = \dfrac{\text{MSR}}{\text{MSE}}$
Residual error	$N-r$	$\text{SSE} = \mathbf{y}'\mathbf{y} - \hat{\mathbf{b}}'\mathbf{X}'\mathbf{y}$	$\text{MSE} = \dfrac{\text{SSE}}{N-r}$	
Total	N	$\text{SST} = \mathbf{y}'\mathbf{y}$		

[1] $r = r(\mathbf{X}) = k + 1$ when there are k regression variables (x's).

This table summarizes not only the sums of squares—already summarized in (72)—but also the degrees of freedom (d.f.) of the associated χ^2-distributions. In the mean squares, which are sums of squares divided by degrees of freedom, it also shows calculation of the numerator and denominator of *F*. And then the calculation of *F* itself is shown. Thus the analysis of variance table is simply a convenient summary of the steps involved in calculating an *F*-statistic.

TABLE 3.3. ANALYSIS OF VARIANCE, SHOWING A TERM FOR THE MEAN

Source of Variation[1]	d.f.[2]	Sum of Squares	Mean Square	F-statistics
Mean	1	$\text{SSM} = N\bar{y}^2$	$\text{MSM} = \text{SSM}/1$	$F(M) = \dfrac{\text{MSM}}{\text{MSE}}$
Regression (c.f.m.)	$r - 1$	$\text{SSR}_m = \hat{\mathbf{b}}'\mathfrak{X}'\mathbf{y}$	$\text{MSR}_m = \dfrac{\text{SSR}_m}{r-1}$	$F(R_m) = \dfrac{\text{MSR}_m}{\text{MSE}}$
Residual error	$N - r$	$\text{SSE} = \mathbf{y}'\mathbf{y} - N\bar{y}^2 - \hat{\mathbf{b}}'\mathfrak{X}'\mathbf{y}$	$\text{MSE} = \dfrac{\text{SSE}}{N-r}$	
Total	N	$\text{SST} = \mathbf{y}'\mathbf{y}$		

[1] c.f.m. = corrected for the mean.
[2] $r = r(\mathbf{X}) = k + 1$ when there are k regression variables (x's).

In a manner similar to Table 3.2, (73) and the F-ratios of (82) and (83) are summarized in Table 3.3.

And the abbreviated form of this, based on (75) and showing only the calculation of (83), is as shown in Table 3.4.

Tables 3.2, 3.3 and 3.4 all are summarizing the same thing. They show development of the customary form of this analysis, namely Table 3.4. Although it *is* the form customarily seen it is not necessarily the most informative. Credit on that account goes to Table 3.3 which shows how SSR of Table 3.2 (the basic form) is partitioned into SSM and SSR_m the regression

TABLE 3.4. ANALYSIS OF VARIANCE (CORRECTED FOR THE MEAN)

Source of Variation[1]	d.f.[2]	Sum of Squares	Mean Square	F-statistic
Regression (c.f.m.)	$r - 1$	$\text{SSR}_m = \hat{\mathbf{b}}'\mathfrak{X}'\mathbf{y}$	$\text{MSR}_m = \dfrac{\text{SSR}_m}{r-1}$	$F(R_m) = \dfrac{\text{MSR}_m}{\text{MSE}}$
Residual error	$N - r$	$\text{SSE} = \mathbf{y}'\mathbf{y} - N\bar{y}^2 - \hat{\mathbf{b}}'\mathfrak{X}'\mathbf{y}$	$\text{MSE} = \dfrac{\text{SSE}}{N-r}$	
Total (c.f.m.)	$N - 1$	$\text{SST}_m = \mathbf{y}'\mathbf{y} - N\bar{y}^2$		

[1] c.f.m. = corrected for the mean.
[2] $r = r(\mathbf{X}) = k + 1$ when there are k regression variables (x's).

sum of squares corrected for the mean (c.f.m.). Following that, Table 3.4 is simply an abbreviated version of Table 3.3, with SSM removed from the body of the table and subtracted from SST to give $\text{SST}_m = \text{SST} - \text{SSM} = \mathbf{y}'\mathbf{y} - N\bar{y}^2$, the corrected sum of squares of the y-observations. Thus, although Table 3.4 does not show $F(M) = \text{MSM/MSE}$, it is identical to Table 3.3 insofar as $F(R_m) = \text{MSR}_m/\text{MSE}$ is concerned.

h. Pure error

Data sometimes have the characteristic that the sets of x's corresponding to several y's are the same. For example, in the case of a laboratory experiment involving temperature, the x-observations (temperature in degrees Centigrade) for 9 different y-observations might be 62, 78, 69, 62, 69, 58, 78, 75 and 62; i.e., 3 x-observations of 62, 2 of 69, 2 of 78 and 1 each of 58 and 75. These are called *repeated* x's. Their presence provides a partitioning of SSE into two terms, one of which represents "lack of fit" of the model and the other represents "pure error". Description is given in terms of simple regression (involving one x-variable); extension to several x's is straightforward.

Suppose $x_1, x_2, \ldots, x_p$ are the p distinct values of the x's, where x_i occurs in the data n_i times, i.e., with n_i y-values, y_{ij} for $j = 1, 2, \ldots, n_i$ and for $i = 1, 2, \ldots, p$. For all i, $n_i \geq 1$, and we will write

$$n. = \sum_{i=1}^{p} n_i = N \quad .$$

Then

$$\begin{aligned} \text{SSE} &= \sum_{i=1}^{p}\sum_{j=1}^{n_i} y_{ij}^2 - \hat{\mathbf{b}}'\mathbf{X}'\mathbf{y} \\ &= \sum_{i=1}^{p}\sum_{j=1}^{n_i} y_{ij}^2 - N\bar{y}_{..}^2 - \hat{\ell}\left(\sum_{i=1}^{p}\sum_{j=1}^{n_i} x_{ij}y_{ij} - N\bar{x}_{..}\bar{y}_{..}\right), \end{aligned}$$

with $N - 2$ degrees of freedom, can be partitioned into

$$\text{SSPE} = \sum_{i=1}^{p}\left[\sum_{j=1}^{n_i} y_{ij}^2 - n_i(\bar{y}_{i.})^2\right] \qquad \text{with } N - p \text{ degrees of freedom}$$

and

$$\text{SSLF} = \text{SSE} - \text{SSPE} \qquad \text{with } p - 2 \text{ degrees of freedom.}$$

In this form $\text{SSPE}/(N - p)$, known as the mean square due to *pure error*, is an estimator of σ^2. $\text{SSLF}/(p - 2)$ is a mean square due to the *lack of fit* of the model. It provides a test of the lack of fit by comparing

$$F(\text{LF}) = \frac{\text{SSLF}/(p - 2)}{\text{SSPE}/(N - p)}$$

against $F_{p-2,N-p}$. Significance indicates inadequacy of the model. Lack of significance indicates, as Draper and Smith (1966) so aptly put it, "that there appears to be no reason for doubting the adequacy of the model", in which case SSE/$(N - 2)$ provides a pooled estimator of σ^2. Full discussion of pure error and lack of fit are to be found in Draper and Smith (1966).

i. Tests of hypotheses

Immediately after (81)–(83) the comment was made that those results provide us with statistics for testing hypotheses. This we now illustrate, prior to consideration of the general linear hypothesis in Sec. 6.

In Table 3.2 the statistic $F(R)$ is, as shown in (81), distributed as a non-central F with non-centrality parameter $\mathbf{b}'\mathbf{X}'\mathbf{X}\mathbf{b}/2\sigma^2$. This is zero under the null hypothesis H: $\mathbf{b} = \mathbf{0}$, when $F(R)$ then has a central F-distribution, $F_{r,N-r}$, and can be compared to tabulated values thereof to test that hypothesis. When

$$F(R) \geq \text{tabulated } F_{r,N-r} \text{ at the } 100\alpha\% \text{ level,}$$

we reject the null hypothesis H: $\mathbf{b} = \mathbf{0}$; otherwise we do not reject it. Apropos assuming the model $E(\mathbf{y}) = \mathbf{X}\mathbf{b}$ we might then say, borrowing a phrase from Williams (1959), that when $F(R)$ is significant there is "concordance of the data with this assumption" of the model; i.e., the model accounts for a significant portion of the variation in the y-variable. But this does not mean that this model, for the particular set of x's used, is necessarily the most suitable model: there may be a subset of those x's which are as significant as the whole, or there may be further x's which, when used alone, or in combination with some or all of the x's already used, are significantly better than those already used; or there may be non-linear functions of these x's that are at least as suitable as the x's as used. None of these contingencies is inconsistent with $F(R)$ being significant and the ensuing conclusion that the data are in concordance with the model $E(\mathbf{y}) = \mathbf{X}\mathbf{b}$.

The non-centrality parameter of the F-statistic $F(M)$ of Table 3.3 is, as in (82), $(\mathbf{1}'\mathbf{X}\mathbf{b})^2/2N\sigma^2$. For the numerator of this expression,

$$\mathbf{1}'\mathbf{X}\mathbf{b} = \mathbf{1}'E(\mathbf{y}) = E(\mathbf{1}'\mathbf{y}) = E(N\bar{y}) = NE(\bar{y}).$$

Hence the non-centrality parameter in (82) is $N[E(\bar{y})]^2/2\sigma^2$, which is zero under the hypothesis H: $E(\bar{y}) = 0$. The statistic $F(M)$ is then distributed as $F_{1,N-r}$ and so can be used to test H: $E(\bar{y}) = 0$; i.e., $F(M)$ can be used to test the hypothesis that the expected value of the mean of the observed y's is zero. This is an interpretation for the phrase "testing the mean" sometimes used for describing the test based on $F(M)$. Equivalently, $\sqrt{F(M)}$ has the t-distribution with $N - r$ degrees of freedom, because

$$F(M) = N\bar{y}^2/\hat{\sigma}^2 = [\bar{y}/(\hat{\sigma}/\sqrt{N})]^2$$

is the square of a t-variable.

Another way of looking at the test provided by $F(M)$ is based on the model $E(y_i) = b_0$. The reduction in sum of squares for fitting this model is SSM, and the non-centrality parameter in (82) is then $Nb_0^2/2\sigma^2$. Hence $F(M)$ can be used to test whether the model $E(y_i) = b_0$ accounts for variation in the y-variable.

In using a test based on $F(R)$ we are testing the hypothesis that all b_i's, including b_0, are simultaneously zero. However, for the null hypothesis H: $\mathbf{b} = \mathbf{0}$, i.e., that just the b's corresponding to the x-variables are zero, then the test is based on $F(R_m)$ in Tables 3.3 and 3.4. This is so because, from (83), we see that the non-centrality parameter in the non-central F-distribution of $F(R_m)$ is zero under the null hypothesis $\mathbf{b} = \mathbf{0}$, in which case $F(R_m)$ has a central F-distribution on $r - 1$ and $N - r$ degrees of freedom. Thus $F(R_m)$ provides a test of the hypothesis $\mathbf{b} = \mathbf{0}$. If $F(R_m)$ is significant the hypothesis is rejected. This is not to be taken as evidence that all elements of $\mathbf{b}$ are non-zero, but only that at least one of them may be. If $F(M)$ has first been found significant then $F(R_m)$ being significant indicates that a model with the x's in it explains significantly more of the variation in the y-variable than does the model $E(y) = b_0$.

Tests using $F(M)$ and $F(R_m)$ are based on numerators SSM and SSR_m that are, as shown earlier in Sec. 3.5e, statistically independent. Therefore the significance or otherwise of $F(M)$ and/or $F(R_m)$ is independent of the numerator sum of squares of the other, although the F's themselves are not independent because they have the same denominator mean square. Consideration of significance levels for both $F(M)$ and $F(R_m)$ is therefore a case of simultaneous statistical inference, a topic beyond the scope of this book. It is dealt with most fully by Miller (1966). We continue to discuss the significance of $F(M)$, $F(R_m)$ and extensions thereof, however, ignoring the simultaneity problem, having here tacitly acknowledged its existence.

The case of both $F(M)$ and $F(R_m)$ being significant is discussed above; as a further possibility suppose $F(M)$ is not significant and $F(R_m)$ is.[1] This would be evidence that even though $E(\bar{y})$ might be zero, fitting the x's does explain variation in the y-variable; a situation when this might occur is when the y-variable can have both positive and negative values, such as weight gain in beef cattle, where some gains may in fact be losses, i.e., negative gains.

j. Example (continued)

Using the summaries shown in Table 3.1, the analyses of variance in Tables 3.2, 3.3 and 3.4 are shown in Table 3.5. The first part of Table 3.5 shows $F(R) = 60.4$, with 3 and 2 degrees of freedom; and since the tabulated value of the $F_{3,2}$-distribution is 19.15 at the 5% level, and $F(R) = 60.4 > 19.15$, we conclude that the model accounts for a significant (at the 5% level) portion

---

[1] I am grateful to N. S. Urquhart for emphasizing this possibility.

TABLE 3.5. TABLES 3.2, 3.3 AND 3.4 FOR THE EXAMPLE

Source of Variation	d.f.	Sum of Squares	Mean Square	F-statistic
	Table 3.2			
Regression	3	SSR $= 1{,}042\frac{1}{2}$	347.5	$F(R) = 347.5/5.75 = 60.4$
Residual error	2	SSE $= 11\frac{1}{2}$	5.75	
Total	5	SST $= 1{,}054$		
	Table 3.3			
Mean	1	SSM $=$ 980	980	$F(M) = 980/5.75 = 170.5$
Regression (c.f.m.)	2	$\text{SSR}_m =$ $62\frac{1}{2}$	31.25	$F(R_m) = 31.25/5.75 = 5.6$
Residual error	2	SSE $=$ $11\frac{1}{2}$	5.75	
Total	5	SST $= 1{,}054$		
	Table 3.4			
Regression (c.f.m.)	2	$\text{SSR}_m =$ $62\frac{1}{2}$	31.25	$F(R_m) = 31.25/5.75 = 5.6$
Residual error	2	SSE $=$ $11\frac{1}{2}$	5.75	
Total	4	$\text{SST}_m =$ 74		

of the variation in the y-variable. Similarly $F(M)$ of the Table 3.3 portion of Table 3.5 has 1 and 2 degrees of freedom; and since $F(M) = 170.5 > 18.51$, the tabulated value of the $F_{1,2}$-distribution at the 5% level, we reject the hypothesis that $E(\bar{y})$ is zero. And finally, since $F(R_m) = 5.6 < 19.00$, the tabulated value of the $F_{2,2}$ distribution at the 5% level, we do not reject the hypothesis that $b_1 = b_2 = 0$; this test provides evidence that the x's are contributing little in terms of accounting for the variation in the y-variable. Most of it is accounted for by the mean, as is evident from the sums of squares values in the Table 3.3 section of Table 3.5. As is true generally, the Table 3.4 section is simply an abbreviated form of the Table 3.3 section, omitting the line for the mean. Just how much of the total sum of squares has been accounted for by the mean is, of course, not evident in the Table 3.4 section. This is a disadvantage to Table 3.4, traditional though its usage is.

k. Confidence intervals

On the basis of normality assumptions we have seen earlier, in Sec. 3.5b, that $\hat{\mathbf{b}}$ has a normal distribution. From that,

$$\frac{\hat{b}_i - b_i}{\sqrt{a^{ii}\sigma^2}} \sim N(0, 1) \tag{84}$$

for $i = 0, 1, 2, \ldots,$ or k where, in accord with (51), a^{ii} is the ith diagonal element of $(\mathbf{X}'\mathbf{X})^{-1}$; i.e., from the development of (52) and (53)

$$a^{00} = 1/N + \bar{\mathbf{x}}'(\mathscr{X}'\mathscr{X})^{-1}\bar{\mathbf{x}}; \tag{85}$$

and for $i = 1, 2, \ldots, k$

$$a^{ii} = i\text{th diagonal element of } (\mathscr{X}'\mathscr{X})^{-1}. \tag{86}$$

With these values of a^{ii}, and in (84) replacing σ^2 by $\hat{\sigma}^2$ of (71), we also have

$$\frac{\hat{b}_i - b_i}{\sqrt{a^{ii}\hat{\sigma}^2}} \sim t_{N-r} \tag{87}$$

where t_{N-r} represents the t-distribution on $N - r$ degrees of freedom.

Let us define $t_{N-r,\alpha,L}$ and $t_{N-r,\alpha,U}$ as a pair of lower and upper limits respectively of the t_{N-r}-distribution such that

$$\Pr\{t \le t_{N-r,\alpha,L}\} + \Pr\{t \ge t_{N-r,\alpha,U}\} = \alpha$$

and so

$$\Pr\{t_{N-r,\alpha,L} \le t \le t_{N-r,\alpha,U}\} = 1 - \alpha \tag{88}$$

for $t \sim t_{N-r}$. Then by (87)

$$\Pr\left\{t_{N-r,\alpha,L} \le \frac{\hat{b}_i - b_i}{\sqrt{a^{ii}\hat{\sigma}^2}} \le t_{N-r,\alpha,U}\right\} = 1 - \alpha$$

and rearrangement of this probability statement in the form

$$\Pr\{\hat{b}_i - \hat{\sigma}t_{N-r,\alpha,U}\sqrt{a^{ii}} \le b_i \le \hat{b}_i - \hat{\sigma}t_{N-r,\alpha,L}\sqrt{a^{ii}}\} = 1 - \alpha$$

provides

$$(\hat{b}_i - \hat{\sigma}t_{N-r,\alpha,U}\sqrt{a^{ii}}), \qquad (\hat{b}_i - \hat{\sigma}t_{N-r,\alpha,L}\sqrt{a^{ii}}) \tag{89}$$

as a $100(1 - \alpha)\%$ confidence interval for b_i. For this confidence interval to be symmetric with respect to b_i, as is often required, we need

$$-t_{N-r,\alpha,L} = t_{N-r,\alpha,U} = t_{N-r,\frac{1}{2}\alpha} \qquad \text{where} \quad \Pr\{t \ge t_{N-r,\frac{1}{2}\alpha}\} = \tfrac{1}{2}\alpha \tag{90}$$

and the interval (89) becomes

$$\hat{b}_i \pm \hat{\sigma}t_{N-r,\frac{1}{2}\alpha}\sqrt{a^{ii}}, \tag{91}$$

of width $2\hat{\sigma}t_{N-r,\frac{1}{2}\alpha}\sqrt{a^{ii}}$.

When the degrees of freedom are large ($N - r > 100$, say), the distribution in (87) is approximately $N(0, 1)$ and on defining $\nu_{\alpha,L}$ and $\nu_{\alpha,U}$ such that

$$\Pr\{\nu_{\alpha,L} \leq \nu \leq \nu_{\alpha,U}\} = 1 - \alpha \qquad \text{for} \quad \nu \sim N(0, 1) \tag{92}$$

$\nu_{\alpha,L}$ and $\nu_{\alpha,U}$ can be used in (89) in place of $t_{N-r,\alpha,L}$ and $t_{N-r,\alpha,U}$. In particular, for a symmetric confidence interval,

$$\nu_{\alpha,L} = -\nu_{\alpha,U} = z_{\frac{1}{2}\alpha}, \qquad \text{where} \quad (2\pi)^{-\frac{1}{2}}\int_{z_{\frac{1}{2}\alpha}}^{\infty} e^{-\frac{1}{2}x^2}\, dx = \tfrac{1}{2}\alpha,$$

and the interval is

$$\hat{b}_i \pm \hat{\sigma} z_{\frac{1}{2}\alpha}\sqrt{a^{ii}}. \tag{93}$$

Tabulated values of $z_{\frac{1}{2}\alpha}$ for a variety of values of $\frac{1}{2}\alpha$ are available in Table 1 of the Appendix.

Confidence intervals for any linear combination of the b's, $\mathbf{q}'\mathbf{b}$ say, can be established in like manner. The argument is unchanged, except that at all stages b_i and $\hat{b}_i$ are replaced by $\mathbf{q}'\mathbf{b}$ and $\mathbf{q}'\hat{\mathbf{b}}$ respectively, and $a^{ii}\hat{\sigma}^2$ is replaced by $\mathbf{q}'(\mathbf{X}'\mathbf{X})^{-1}\mathbf{q}\hat{\sigma}^2$. Thus the symmetric confidence interval for $\mathbf{q}'\mathbf{b}$ is, from (93),

$$\mathbf{q}'\hat{\mathbf{b}} \pm \hat{\sigma} t_{N-r,\frac{1}{2}\alpha}\sqrt{\mathbf{q}'(\mathbf{X}'\mathbf{X})^{-1}\mathbf{q}} \tag{94}$$

with $z_{\frac{1}{2}\alpha}$ replacing $t_{N-r,\frac{1}{2}\alpha}$ when $N - r$ is large.

In equation (56) we developed $\mathbf{x}_0'\hat{\mathbf{b}}$ as the estimator of $E(y_0)$ corresponding to the set of x's in $\mathbf{x}_0'$. Result (94) now provides a confidence interval on $\mathbf{x}_0'\mathbf{b}$, namely

$$\mathbf{x}_0'\hat{\mathbf{b}} \pm \hat{\sigma} t_{N-r,\frac{1}{2}\alpha}\sqrt{\mathbf{x}_0'(\mathbf{X}'\mathbf{X})^{-1}\mathbf{x}_0}\,. \tag{95}$$

In the case of simple regression involving only one x-variable (where $k = 1$ and $r = 2$ as in the footnote to Table 3.4), $\mathbf{x}_0' = [1 \quad x_0]$ and (95) becomes

$$[1 \quad x_0]\begin{bmatrix} \bar{y} - \hat{b}\bar{x} \\ \hat{b} \end{bmatrix} \pm \hat{\sigma} t_{N-2,\frac{1}{2}\alpha}\sqrt{[1 \quad x_0]\begin{bmatrix} N & N\bar{x} \\ N\bar{x} & \sum_{i=1}^{N} x_i^2 \end{bmatrix}^{-1}\begin{bmatrix} 1 \\ x_0 \end{bmatrix}}$$

which simplifies to

$$\bar{y} + \hat{b}(x_0 - \bar{x}) \pm \hat{\sigma} t_{N-2,\frac{1}{2}\alpha}\sqrt{\frac{1}{N} + \frac{(\bar{x} - x_0)^2}{\sum_{i=1}^{N} x_i^2 - N\bar{x}^2}}\,, \tag{96}$$

the familiar expression [e.g., Steel and Torrie (1960), p. 170] for the confidence interval on $E(y)$ in a simple regression model. Plotting the values of this interval for a series of values of x_0 provides the customary confidence belt for the regression line $y = b_0 + bx$.

A confidence interval on an estimated observation is sometimes called a *tolerance interval.* In keeping with the variance given in (61) the tolerance interval comes from using $\mathbf{x}_0'(\mathbf{X}'\mathbf{X})^{-1}\mathbf{x}_0 + 1$ instead of $\mathbf{x}_0'(\mathbf{X}'\mathbf{X})^{-1}\mathbf{x}_0$ in (95). In line with (95) it reduces for simple regression to

$$\bar{y} + \hat{b}(x_0 - \bar{x}) \pm \hat{\sigma} t_{N-r,\frac{1}{2}\alpha} \sqrt{1 + \frac{1}{N} + \frac{(\bar{x} - x_0)^2}{\sum_{i=1}^{N} x_i^2 - N\bar{x}^2}}\,.$$

l. Example (continued)

Confidence intervals on b_1 will be calculated for the example used earlier, a non-symmetric interval from (89) and a symmetric interval from (91). For both we use

$$\hat{b}_1 = 50/24 = 2.08$$

from (28),

$$\hat{\sigma} = \sqrt{5.75} = 2.40 \qquad \text{and} \qquad N - r = 2$$

from Table 3.5, and

$$a^{11} = 20/144 = 0.139$$

from (86) and (44). Then in (89) a non-symmetric confidence interval for b_1 is

$$\begin{aligned} &2.08 - 2.40\, t_{2,\alpha,U}\sqrt{0.139} \text{ to } 2.08 + 2.40\, t_{2,\alpha,L}\sqrt{0.139} \\ &\qquad = 2.08 - 0.89\, t_{2,\alpha,U} \text{ to } 2.08 + 0.89\, t_{2,\alpha,L}\,. \end{aligned} \tag{97}$$

From tabulated values of the t_2-distribution, [e.g., Vogler and Norton (1957)] we find that

$$\Pr(t \leq -3.6) = 0.04 \qquad \text{and} \qquad \Pr(t \geq 7.1) = 0.01,$$

so that by (88), for $\alpha = 0.05$,

$$t_{2,\cdot 05,L} = -3.6 \qquad \text{and} \qquad t_{2,\cdot 05,U} = 7.1$$

and so in (97) the confidence interval becomes

$$2.08 - 0.89(7.1) \text{ to } 2.08 - 0.89(-3.6) = (-4.23, 5.08). \tag{98}$$

It is questionable, of course, as to what kind of situation would reasonably lead to needing a non-symmetric confidence interval with the t-distribution. The example illustrates, however, how such intervals can be calculated and doing so emphasizes the important fact that there are many such intervals—because there are many values $t_{N-r,\alpha,L}$ and $t_{N-r,\alpha,U}$ that satisfy (88). In contrast, there is only one symmetric confidence interval, the interval which has the optimal property that for given $N - r$ and α it is the interval of shortest length. This is the interval given in (91) for which, for the example, (90) is

$$\Pr\{t \geq 4.30\} = 0.025$$

for $t \sim t_2$. Hence the symmetric interval on b_1 is, from (91),

$$\begin{aligned} 2.08 \pm 2.40 t_{2,\frac{1}{2}\alpha}\sqrt{0.139} &= 2.08 \pm 0.89 t_{2,0.025} \\ &= 2.08 \pm 0.89(4.30) \\ &= (-1.75, 5.91). \end{aligned}$$

The length of this interval is $1.75 + 5.91 = 7.66$, shorter than the length of the non-symmetric interval in (98), namely $4.23 + 5.08 = 9.31$.

6. THE GENERAL LINEAR HYPOTHESIS

a. Testing linear hypotheses

The literature of linear models abounds with discussions of different kinds of hypotheses that can be of interest in widely differing fields of application. Four hypotheses of particular interest are: (i) $H\colon\ \mathbf{b} = \mathbf{0}$, the hypothesis that all elements of $\mathbf{b}$ are zero. (ii) $H\colon\ \mathbf{b} = \mathbf{b}_0$, the hypothesis that $b_i = b_{i0}$ for $i = 0, 1, 2, \ldots, k$, i.e., that each b_i is equal to some specified value b_{i0}. (iii) $H\colon\ \boldsymbol{\lambda}'\mathbf{b} = m$, that some linear combination of the elements of $\mathbf{b}$ equals a specified constant. (iv) $H\colon\ \mathbf{b}_q = \mathbf{0}$, that some of the b_i's, q of them where $q < k$, are zero. Although the calculations for the F-statistic for these hypotheses and variants of them appear, on the surface, to differ markedly from one kind of hypothesis to another, we will show that *all* linear hypotheses can be handled by one universal procedure. Specific hypotheses such as those listed above are then just special cases of the general procedure.

The general hypothesis we consider is

$$H\colon\quad \mathbf{K}'\mathbf{b} = \mathbf{m}$$

where $\mathbf{b}$, of course, is the $(k + 1)$-order vector of parameters of the model; $\mathbf{K}'$ is any matrix of s rows and $k + 1$ columns; and $\mathbf{m}$ is a vector, of order s, of specified constants. There is only one limitation on $\mathbf{K}'$: that it have full row rank, i.e., $r(\mathbf{K}') = s$. This simply means that the linear functions of $\mathbf{b}$ which form the hypothesis must be linearly independent; that is, the hypothesis must be made up of linearly independent functions of $\mathbf{b}$ and must contain no functions which are linear combinations of others therein. This is quite reasonable because it means, for example, that if the hypothesis relates to $b_1 - b_2$ and $b_2 - b_3$ then there is no point in having it also relate, explicitly, to $b_1 - b_3$. Clearly, this condition on $\mathbf{K}'$ is not at all restrictive in limiting the application of the hypothesis H: $\mathbf{K'b} = \mathbf{m}$ to real problems. Furthermore, although it might seem necessary to also require that $\mathbf{m}$ be such that the equations $\mathbf{K'b} = \mathbf{m}$ be consistent, this is automatically achieved by demanding that $\mathbf{K'}$ have full row rank, for the equations $\mathbf{K'b} = \mathbf{m}$ are then consistent for any vector $\mathbf{m}$.

We now develop the F-statistic to test the hypothesis H: $\mathbf{K'b} = \mathbf{m}$. We already have the following:

$$\mathbf{y} \sim N(\mathbf{Xb}, \sigma^2\mathbf{I}),$$

$$\hat{\mathbf{b}} = (\mathbf{X'X})^{-1}\mathbf{X'y}$$

and

$$\hat{\mathbf{b}} \sim N[\mathbf{b}, (\mathbf{X'X})^{-1}\sigma^2].$$

Therefore $$\mathbf{K'\hat{b}} - \mathbf{m} \sim N[\mathbf{K'b} - \mathbf{m}, \mathbf{K'(X'X)^{-1}K}\sigma^2].$$

Hence, by an application of Theorem 2 in Chapter 2, the following quadratic in $\mathbf{K'\hat{b}} - \mathbf{m}$, using $[\mathbf{K'(X'X)^{-1}K}]^{-1}$ as the matrix of the quadratic, has a χ^2-distribution: if

$$Q = (\mathbf{K'\hat{b}} - \mathbf{m})'[\mathbf{K'(X'X)^{-1}K}]^{-1}(\mathbf{K'\hat{b}} - \mathbf{m})$$

then $$Q/\sigma^2 \sim \chi^{2\prime}\{s, (\mathbf{K'b} - \mathbf{m})'[\mathbf{K'(X'X)^{-1}K}]^{-1}(\mathbf{K'b} - \mathbf{m})/2\sigma^2\}. \quad (99)$$

The independence of Q and SSE is now shown, using Theorem 4 of Chapter 2. To do this we first express Q and SSE as quadratic forms of the same normally distributed random variable, noting initially that the inverse of $\mathbf{K'(X'X)^{-1}K}$ used in (99) exists because $\mathbf{K'}$ has full row rank and $\mathbf{X'X}$ is symmetric. Then, on replacing $\hat{\mathbf{b}}$ by $(\mathbf{X'X})^{-1}\mathbf{X'y}$, equation (99) for Q becomes

$$Q = [\mathbf{K'(X'X)^{-1}X'y} - \mathbf{m}]'[\mathbf{K'(X'X)^{-1}K}]^{-1}[\mathbf{K'(X'X)^{-1}X'y} - \mathbf{m}].$$

But because $\mathbf{K'}$ has full row rank, $(\mathbf{K'K})^{-1}$ exists—see corollary of Lemma 5, Sec. 2.2. Therefore

$$\mathbf{K'(X'X)^{-1}X'y} - \mathbf{m} = \mathbf{K'(X'X)^{-1}X'}[\mathbf{y} - \mathbf{XK(K'K)^{-1}m}],$$

and so

$$Q = [\mathbf{y} - \mathbf{XK(K'K)^{-1}m}]'\mathbf{X(X'X)^{-1}K}[\mathbf{K'(X'X)^{-1}K}]^{-1} \times \mathbf{K'(X'X)^{-1}X'}[\mathbf{y} - \mathbf{XK(K'K)^{-1}m}].$$

Now consider the error sum of squares

$$\text{SSE} = \mathbf{y'}[\mathbf{I} - \mathbf{X(X'X)^{-1}X'}]\mathbf{y}.$$

Because the products $\mathbf{X'}[\mathbf{I} - \mathbf{X(X'X)^{-1}X'}]$ and $[\mathbf{I} - \mathbf{X(X'X)^{-1}X'}]\mathbf{X}$ are both null, SSE can be rewritten as

$$\text{SSE} = [\mathbf{y} - \mathbf{XK(K'K)^{-1}m}]'[\mathbf{I} - \mathbf{X(X'X)^{-1}X'}][\mathbf{y} - \mathbf{XK(K'K)^{-1}m}].$$

Both Q and SSE have now been expressed as quadratics in the vector $\mathbf{y} - \mathbf{XK(K'K)^{-1}m}$. And although we already know that Q/σ^2 and SSE/σ^2

have $\chi^{2\prime}$-distributions, this is further seen from their being quadratics in $\mathbf{y} - \mathbf{XK}(\mathbf{K}'\mathbf{K})^{-1}\mathbf{m}$ which is a normally distributed vector; and the matrix in each quadratic is idempotent. But, more importantly, the product of the two matrices is null:

$$[\mathbf{I} - \mathbf{X}(\mathbf{X}'\mathbf{X})^{-1}\mathbf{X}']\mathbf{X}(\mathbf{X}'\mathbf{X})^{-1}\mathbf{K}[\mathbf{K}'(\mathbf{X}'\mathbf{X})^{-1}\mathbf{K}]^{-1}\mathbf{K}'(\mathbf{X}'\mathbf{X})^{-1}\mathbf{X}' = \mathbf{0}.$$

Therefore by Theorem 4 of Chapter 2, Q and SSE are distributed independently. Hence

$$F(H) = \frac{Q/s}{\text{SSE}/[N - r(\mathbf{X})]} = Q/s\hat{\sigma}^2$$
$$\sim F'\{s, N - r(\mathbf{X}), (\mathbf{K}'\mathbf{b} - \mathbf{m})'[\mathbf{K}'(\mathbf{X}'\mathbf{X})^{-1}\mathbf{K}]^{-1}(\mathbf{K}'\mathbf{b} - \mathbf{m})/2\sigma^2\} \tag{100}$$

and under the null hypothesis $H\colon\ \mathbf{K}'\mathbf{b} = \mathbf{m}$

$$F(H) \sim F_{s, N-r(\mathbf{X})}.$$

Hence $F(H)$ provides a test of the hypothesis $H\colon\ \mathbf{K}'\mathbf{b} = \mathbf{m}$. Thus the F-statistic for testing the hypothesis $H\colon\ \mathbf{K}'\mathbf{b} = \mathbf{m}$ is

$$F(H) = \frac{Q}{s\hat{\sigma}^2} = \frac{(\mathbf{K}'\hat{\mathbf{b}} - \mathbf{m})'[\mathbf{K}'(\mathbf{X}'\mathbf{X})^{-1}\mathbf{K}]^{-1}(\mathbf{K}'\hat{\mathbf{b}} - \mathbf{m})}{s\hat{\sigma}^2} \tag{101}$$

with s and $N - r$ degrees of freedom, s being the number of rows in $\mathbf{K}'$, it being of full row rank.

The generality of this result merits emphasis: it applies for *any* linear hypothesis $\mathbf{K}'\mathbf{b} = \mathbf{m}$, the only limitation being that $\mathbf{K}'$ have full row rank. Other than this, $F(H)$ can be used to test any linear hypothesis whatever. No matter what the hypothesis is, it has only to be written in the form $\mathbf{K}'\mathbf{b} = \mathbf{m}$ and $F(H)$ of (101) provides the test. Having once solved the normal equations for the model $\mathbf{y} = \mathbf{Xb} + \mathbf{e}$ and so obtained $(\mathbf{X}'\mathbf{X})^{-1}$, $\hat{\mathbf{b}} = (\mathbf{X}'\mathbf{X})^{-1}\mathbf{X}'\mathbf{y}$ and $\hat{\sigma}^2$, the testing of $H\colon\ \mathbf{K}'\mathbf{b} = \mathbf{m}$ can be achieved by immediate application of $F(H)$. The appeal of this result is illustrated below in subsection c, for the four hypotheses listed at the beginning of this section. Note that $\hat{\sigma}^2$ is universal to every application of $F(H)$. Thus, in considering different hypotheses the only term in $F(H)$ that alters is Q/s.

b. Estimation under the null hypothesis

When considering the hypothesis $H\colon\ \mathbf{K}'\mathbf{b} = \mathbf{m}$ it is natural to ask, "What is the estimator of $\mathbf{b}$ under the null hypothesis?" This might be especially pertinent following non-rejection of the hypothesis by the preceding F-test. The desired estimator, $\tilde{\mathbf{b}}$ say, is readily obtainable using constrained least squares. Thus $\tilde{\mathbf{b}}$ is derived so as to minimize $(\mathbf{y} - \mathbf{X}\tilde{\mathbf{b}})'(\mathbf{y} - \mathbf{X}\tilde{\mathbf{b}})$ subject to the constraint $\mathbf{K}'\tilde{\mathbf{b}} = \mathbf{m}$.

With $2\boldsymbol{\theta}'$ as a vector of Lagrange multipliers we minimize

$$(\mathbf{y} - \mathbf{X}\tilde{\mathbf{b}})'(\mathbf{y} - \mathbf{X}\tilde{\mathbf{b}}) + 2\boldsymbol{\theta}'(\mathbf{K}'\tilde{\mathbf{b}} - \mathbf{m})$$

with respect to the elements of $\tilde{\mathbf{b}}$ and $\boldsymbol{\theta}$. Differentiation with respect to these elements leads to the equations

$$\mathbf{X}'\mathbf{X}\tilde{\mathbf{b}} + \mathbf{K}\boldsymbol{\theta} = \mathbf{X}'\mathbf{y}$$

and

$$\mathbf{K}'\tilde{\mathbf{b}} = \mathbf{m}. \tag{102}$$

These equations are solved as follows: from the first,

$$\tilde{\mathbf{b}} = (\mathbf{X}'\mathbf{X})^{-1}(\mathbf{X}'\mathbf{y} - \mathbf{K}\boldsymbol{\theta}) = \hat{\mathbf{b}} - (\mathbf{X}'\mathbf{X})^{-1}\mathbf{K}\boldsymbol{\theta},$$

and in the second

$$\mathbf{K}'\tilde{\mathbf{b}} = \mathbf{K}'\hat{\mathbf{b}} - \mathbf{K}'(\mathbf{X}'\mathbf{X})^{-1}\mathbf{K}\boldsymbol{\theta} = \mathbf{m}.$$

Hence

$$\boldsymbol{\theta} = [\mathbf{K}'(\mathbf{X}'\mathbf{X})^{-1}\mathbf{K}]^{-1}(\mathbf{K}'\hat{\mathbf{b}} - \mathbf{m})$$

and so

$$\tilde{\mathbf{b}} = \hat{\mathbf{b}} - (\mathbf{X}'\mathbf{X})^{-1}\mathbf{K}[\mathbf{K}'(\mathbf{X}'\mathbf{X})^{-1}\mathbf{K}]^{-1}(\mathbf{K}'\hat{\mathbf{b}} - \mathbf{m}). \tag{103}$$

This expression and (101) apply directly to $\tilde{\boldsymbol{\ell}}$ when the hypothesis is $\mathbf{L}'\boldsymbol{\ell} = \mathbf{m}$ (see Exercise 8).

Having thus estimated $\mathbf{b}$ under the hypothesis we now show that the corresponding residual sum of squares is SSE + Q where Q is the numerator sum of squares of $F(H)$, the F-statistic used in testing the hypothesis in (101). The residual is

$$\begin{aligned}(\mathbf{y} - \mathbf{X}\tilde{\mathbf{b}})'(\mathbf{y} - \mathbf{X}\tilde{\mathbf{b}}) &= [\mathbf{y} - \mathbf{X}\hat{\mathbf{b}} + \mathbf{X}(\hat{\mathbf{b}} - \tilde{\mathbf{b}})]'[\mathbf{y} - \mathbf{X}\hat{\mathbf{b}} + \mathbf{X}(\hat{\mathbf{b}} - \tilde{\mathbf{b}})] \\ &= (\mathbf{y} - \mathbf{X}\hat{\mathbf{b}})'(\mathbf{y} - \mathbf{X}\hat{\mathbf{b}}) + (\hat{\mathbf{b}} - \tilde{\mathbf{b}})'\mathbf{X}'\mathbf{X}(\hat{\mathbf{b}} - \tilde{\mathbf{b}}) \end{aligned} \tag{104}$$

with the other terms vanishing because $\mathbf{X}'(\mathbf{y} - \mathbf{X}\hat{\mathbf{b}}) = \mathbf{0}$. Now from (103)

$$\hat{\mathbf{b}} - \tilde{\mathbf{b}} = (\mathbf{X}'\mathbf{X})^{-1}\mathbf{K}[\mathbf{K}'(\mathbf{X}'\mathbf{X})^{-1}\mathbf{K}]^{-1}(\mathbf{K}'\hat{\mathbf{b}} - \mathbf{m})$$

and so on substituting in (104)

$$\begin{aligned}(\mathbf{y} - \mathbf{X}\tilde{\mathbf{b}})'(\mathbf{y} - \mathbf{X}\tilde{\mathbf{b}}) \\ &= \text{SSE} + (\mathbf{K}'\hat{\mathbf{b}} - \mathbf{m})'[\mathbf{K}'(\mathbf{X}'\mathbf{X})^{-1}\mathbf{K}]^{-1}\mathbf{K}'(\mathbf{X}'\mathbf{X})^{-1}\mathbf{X}'\mathbf{X}(\mathbf{X}'\mathbf{X})^{-1} \\ &\qquad \times \mathbf{K}[\mathbf{K}'(\mathbf{X}'\mathbf{X})^{-1}\mathbf{K}]^{-1}(\mathbf{K}'\hat{\mathbf{b}} - \mathbf{m}) \\ &= \text{SSE} + (\mathbf{K}'\hat{\mathbf{b}} - \mathbf{m})'[\mathbf{K}'(\mathbf{X}'\mathbf{X})^{-1}\mathbf{K}]^{-1}(\mathbf{K}'\hat{\mathbf{b}} - \mathbf{m}) \\ &= \text{SSE} + Q \end{aligned} \tag{105}$$

from (99).

c. Four common hypotheses

The preceding expressions for $F(H)$ and $\tilde{\mathbf{b}}$, namely (101) and (103), are here illustrated for four commonly occurring hypotheses.

(*i*) H: $\mathbf{b} = \mathbf{0}$. Testing this hypothesis has already been considered earlier in the analysis of variance tables. However, it illustrates the reduction of $F(H)$ to the F-statistic of the analysis of variance tables. To apply $F(H)$ the equations $\mathbf{b} = \mathbf{0}$ have to be written as $\mathbf{K}'\mathbf{b} = \mathbf{m}$: hence $\mathbf{K}' = \mathbf{I}$, $s = k + 1$ and $\mathbf{m} = \mathbf{0}$. Thus $[\mathbf{K}'(\mathbf{X}'\mathbf{X})^{-1}\mathbf{K}]^{-1}$ becomes $\mathbf{X}'\mathbf{X}$ and so

$$F(H) = \frac{\hat{\mathbf{b}}'\mathbf{X}'\mathbf{X}\hat{\mathbf{b}}}{(k+1)\hat{\sigma}^2} = \frac{\text{SSR}}{r}\frac{N-r}{\text{SSE}}$$

as before. Under the null hypothesis $F(H)$ is $F_{r,N-r}$, where $r = k + 1$.

The corresponding value of $\tilde{\mathbf{b}}$ is, of course,

$$\tilde{\mathbf{b}} = \hat{\mathbf{b}} - (\mathbf{X}'\mathbf{X})^{-1}[(\mathbf{X}'\mathbf{X})^{-1}]^{-1}\hat{\mathbf{b}} = \mathbf{0}.$$

(*ii*) H: $\mathbf{b} = \mathbf{b}_0$, *i.e.*, $b_i = b_{i0}$ *for all* i. Rewriting $\mathbf{b} = \mathbf{b}_0$ as $\mathbf{K}'\mathbf{b} = \mathbf{m}$ gives

$$\mathbf{K}' = \mathbf{I}, \qquad s = k + 1, \qquad \mathbf{m} = \mathbf{b}_0 \quad \text{and} \quad [\mathbf{K}'(\mathbf{X}'\mathbf{X})^{-1}\mathbf{K}]^{-1} = \mathbf{X}'\mathbf{X}$$

and so

$$F(H) = \frac{(\hat{\mathbf{b}} - \mathbf{b}_0)'\mathbf{X}'\mathbf{X}(\hat{\mathbf{b}} - \mathbf{b}_0)}{(k+1)\hat{\sigma}^2}. \tag{106}$$

The numerator can be expressed alternatively as

$$\begin{aligned}(\hat{\mathbf{b}} - \mathbf{b}_0)'\mathbf{X}'\mathbf{X}(\hat{\mathbf{b}} - \mathbf{b}_0) &= (\mathbf{y} - \mathbf{X}\mathbf{b}_0)'\mathbf{X}(\mathbf{X}'\mathbf{X})^{-1}\mathbf{X}'\mathbf{X}(\mathbf{X}'\mathbf{X})^{-1}\mathbf{X}'(\mathbf{y} - \mathbf{X}\mathbf{b}_0)\\ &= (\mathbf{y} - \mathbf{X}\mathbf{b}_0)'\mathbf{X}(\mathbf{X}'\mathbf{X})^{-1}\mathbf{X}'(\mathbf{y} - \mathbf{X}\mathbf{b}_0),\end{aligned}$$

although the form shown in (106) is probably the most suitable for computing purposes. Under the null hypothesis $F(H)$ is distributed as $F_{r,N-r}$, where $r = k + 1$.

In this case the estimator of $\mathbf{b}$ under the hypothesis is

$$\tilde{\mathbf{b}} = \hat{\mathbf{b}} - (\mathbf{X}'\mathbf{X})^{-1}[(\mathbf{X}'\mathbf{X})^{-1}]^{-1}(\hat{\mathbf{b}} - \mathbf{b}_0) = \mathbf{b}_0\,.$$

(*iii*) H: $\boldsymbol{\lambda}'\mathbf{b} = m$. Here

$$\mathbf{K}' = \boldsymbol{\lambda}', \qquad s = 1, \qquad \mathbf{m} = m$$

and

$$F(H) = \frac{(\boldsymbol{\lambda}'\hat{\mathbf{b}} - m)'[\boldsymbol{\lambda}'(\mathbf{X}'\mathbf{X})^{-1}\boldsymbol{\lambda}]^{-1}(\boldsymbol{\lambda}'\hat{\mathbf{b}} - m)}{\hat{\sigma}^2}$$

and because $\boldsymbol{\lambda}'$ is a vector this can be rewritten as

$$F(H) = \frac{(\boldsymbol{\lambda}'\hat{\mathbf{b}} - m)^2}{\boldsymbol{\lambda}'(\mathbf{X}'\mathbf{X})^{-1}\boldsymbol{\lambda}\hat{\sigma}^2}.$$

Under the null hypothesis $F(H)$ has the $F_{1,N-r}$-distribution. Hence

$$\sqrt{F(H)} = \frac{\boldsymbol{\lambda}'\hat{\mathbf{b}} - m}{\hat{\sigma}\sqrt{\boldsymbol{\lambda}'(\mathbf{X}'\mathbf{X})^{-1}\boldsymbol{\lambda}}} \sim t_{N-r}\,.$$

This is as one would expect, since $\boldsymbol{\lambda}'\hat{\mathbf{b}}$ is normally distributed with variance $\boldsymbol{\lambda}'(\mathbf{X}'\mathbf{X})^{-1}\boldsymbol{\lambda}\sigma^2$.

For this hypothesis the value of $\tilde{\mathbf{b}}$ is

$$\tilde{\mathbf{b}} = \hat{\mathbf{b}} - (\mathbf{X}'\mathbf{X})^{-1}\boldsymbol{\lambda}[\boldsymbol{\lambda}'(\mathbf{X}'\mathbf{X})^{-1}\boldsymbol{\lambda}]^{-1}(\boldsymbol{\lambda}'\hat{\mathbf{b}} - m)$$

$$= \hat{\mathbf{b}} - \left\{\frac{\boldsymbol{\lambda}'\mathbf{b} - m}{\boldsymbol{\lambda}'(\mathbf{X}'\mathbf{X})^{-1}\boldsymbol{\lambda}}\right\}(\mathbf{X}'\mathbf{X})^{-1}\boldsymbol{\lambda}.$$

At this point it is appropriate to comment on the lack of emphasis being given to the t-test in hypothesis testing. This is because the equivalence of t-statistics with F-statistics that have 1 degree of freedom in the numerator term makes it unnecessary to consider t-tests. Whenever a t-test might be proposed, the hypothesis to be tested can be put in the form H: $\boldsymbol{\lambda}'\mathbf{b} = m$ and the F-statistic $F(H)$ derived as here. If the t-statistic is insisted upon it is then obtained as $\sqrt{F(H)}$. No further discussion of using the t-test is therefore necessary.

(iv) H: $\mathbf{b}_q = \mathbf{0}$, *i.e.*, $b_i = 0$ *for* $i = 0, 1, 2, \ldots, q - 1$, *for* $q < k$. In this case

$$\mathbf{K}' = [\mathbf{I}_q \quad \mathbf{0}] \quad \text{and} \quad \mathbf{m} = \mathbf{0} \quad \text{so that } s = q.$$

We write

$$\mathbf{b}_q' = [b_0 \quad b_1 \quad \cdots \quad b_{q-1}]$$

and partition $\mathbf{b}$, $\hat{\mathbf{b}}$ and $(\mathbf{X}'\mathbf{X})^{-1}$ accordingly:

$$\mathbf{b} = \begin{bmatrix}\mathbf{b}_q \\ \mathbf{b}_p\end{bmatrix}, \quad \hat{\mathbf{b}} = \begin{bmatrix}\hat{\mathbf{b}}_q \\ \hat{\mathbf{b}}_p\end{bmatrix} \quad \text{and} \quad (\mathbf{X}'\mathbf{X})^{-1} = \begin{bmatrix}\mathbf{T}_{qq} & \mathbf{T}_{qp} \\ \mathbf{T}_{pq} & \mathbf{T}_{pp}\end{bmatrix}$$

where $p + q =$ the order of $\mathbf{b} = k + 1$. Then in $F(H)$ of (101)

$$\mathbf{K}'\hat{\mathbf{b}} = \hat{\mathbf{b}}_q$$

and

$$[\mathbf{K}'(\mathbf{X}'\mathbf{X})^{-1}\mathbf{K}]^{-1} = \mathbf{T}_{qq}^{-1},$$

giving

$$F(H) = \hat{\mathbf{b}}_q'\mathbf{T}_{qq}^{-1}\hat{\mathbf{b}}_q/q\hat{\sigma}^2. \tag{107}$$

In the numerator we recognize the result [e.g., Searle (1966), Sec. 9.11] of "invert part of the inverse"; i.e., take the inverse of $\mathbf{X}'\mathbf{X}$ and invert that part of it which corresponds to $\mathbf{b}_q$ of the hypothesis H: $\mathbf{b}_q = \mathbf{0}$. Although demonstrated here for a $\mathbf{b}_q$ that consists of the first q b's in $\mathbf{b}$, it clearly applies for any subset of q b's. In particular, for just one b, it leads to the usual F-test on 1 degree of freedom, equivalent to a t-test (see Exercise 15).

The estimator of $\mathbf{b}$ under this hypothesis is

$$\begin{aligned}\tilde{\mathbf{b}} &= \hat{\mathbf{b}} - (\mathbf{X}'\mathbf{X})^{-1}\begin{bmatrix}\mathbf{I}_q\\ 0\end{bmatrix}\mathbf{T}_{qq}^{-1}(\hat{\mathbf{b}}_q - \mathbf{0})\\
&= \hat{\mathbf{b}} - \begin{bmatrix}\mathbf{T}_{qq}\\ \mathbf{T}_{pq}\end{bmatrix}\mathbf{T}_{qq}^{-1}\hat{\mathbf{b}}_q = \begin{bmatrix}\hat{\mathbf{b}}_q\\ \hat{\mathbf{b}}_p\end{bmatrix} - \begin{bmatrix}\hat{\mathbf{b}}_q\\ \mathbf{T}_{pq}\mathbf{T}_{qq}^{-1}\hat{\mathbf{b}}_q\end{bmatrix}\\
&= \begin{bmatrix}\mathbf{0}\\ \hat{\mathbf{b}}_p - \mathbf{T}_{pq}\mathbf{T}_{qq}^{-1}\hat{\mathbf{b}}_q\end{bmatrix};\end{aligned}$$

i.e., the estimator of the b's not in the hypothesis is $\hat{\mathbf{b}}_p - \mathbf{T}_{pq}\mathbf{T}_{qq}^{-1}\hat{\mathbf{b}}_q$.

The expressions obtained for $F(H)$ and $\tilde{\mathbf{b}}$ for these four hypotheses concerning $\mathbf{b}$ are in terms of $\hat{\mathbf{b}}$. They also apply to similar hypotheses in terms of $\boldsymbol{\ell}$ (see Exercise 7), as do analogous results for any hypothesis $L'\boldsymbol{\ell} = \mathbf{m}$ (see Exercise 8).

d. Reduced models

We now consider, in turn, the effect on the model $\mathbf{y} = \mathbf{Xb} + \mathbf{e}$ of the hypotheses $\mathbf{K}'\mathbf{b} = \mathbf{m}$, $\mathbf{K}'\mathbf{b} = \mathbf{0}$ and $\mathbf{b}_q = \mathbf{0}$.

(*i*) $\mathbf{K}'\mathbf{b} = \mathbf{m}$. In estimating $\mathbf{b}$ subject to $\mathbf{K}'\mathbf{b} = \mathbf{m}$ it could be said that we are dealing with a model $\mathbf{y} = \mathbf{Xb} + \mathbf{e}$ on which has been imposed the limitation $\mathbf{K}'\mathbf{b} = \mathbf{m}$. We refer to the model that we start with, $\mathbf{y} = \mathbf{Xb} + \mathbf{e}$ without the limitation, as the *full model*; and the model with the limitation imposed, $\mathbf{y} = \mathbf{Xb} + \mathbf{e}$ with $\mathbf{K}'\mathbf{b} = \mathbf{m}$, is called the *reduced model*. For example, if the full model is

$$y_i = b_0 + b_1x_{i1} + b_2x_{i2} + b_3x_{i3} + e_i$$

and the hypothesis is H: $b_1 = b_2$, the reduced model is

$$y_i = b_0 + b_1(x_{i1} + x_{i2}) + b_3x_{i3} + e_i\,.$$

The meaning of Q and of SSE + Q is now investigated in terms of sums of squares associated with the full and reduced models. To aid description we introduce the terms reduction(full) and residual(full) for the reduction and residual sums of squares after fitting the full model:

$$\text{reduction(full)} = \text{SSR} \qquad \text{and} \qquad \text{residual(full)} = \text{SSE}.$$

Similarly

$$\text{SSE} + Q = \text{residual(reduced)}, \tag{108}$$

as established in (105). Hence

$$\begin{aligned}Q &= \text{SSE} + Q - \text{SSE}\\ &= \text{residual(reduced)} - \text{residual(full)}\end{aligned} \tag{109}$$

and also

$$\begin{aligned} Q &= \mathbf{y}'\mathbf{y} - \text{SSE} - [\mathbf{y}'\mathbf{y} - (\text{SSE} + Q)] \\ &= \text{SSR} \quad - [\mathbf{y}'\mathbf{y} - (\text{SSE} + Q)] \\ &= \text{reduction(full)} - [\mathbf{y}'\mathbf{y} - (\text{SSE} + Q)]. \end{aligned} \tag{110}$$

Comparison of (110) with (109) tempts one to conclude that $\mathbf{y}'\mathbf{y} - (\text{SSE} + Q)$ is reduction(reduced), the reduction in sum of squares due to fitting the reduced model. The temptation to do this is heightened by the fact that SSE + Q is residual(reduced) as in (108). However, we shall show that only in special cases is $\mathbf{y}'\mathbf{y} - (\text{SSE} + Q)$ the reduction in sum of squares due to fitting the reduced model. It is not always so. The circumstances of these special cases are quite wide, as well as useful, but they are not universal. First we show that $\mathbf{y}'\mathbf{y} - (\text{SSE} + Q)$ is not generally a sum of squares, since it can be negative: for, in

$$\begin{aligned} \mathbf{y}'\mathbf{y} - \text{SSE} - Q &= \text{SSR} - Q \\ &= \hat{\mathbf{b}}'\mathbf{X}'\mathbf{y} - (\mathbf{K}'\hat{\mathbf{b}} - \mathbf{m})'[\mathbf{K}'(\mathbf{X}'\mathbf{X})^{-1}\mathbf{K}]^{-1}(\mathbf{K}'\hat{\mathbf{b}} - \mathbf{m}) \end{aligned} \tag{111}$$

the second term is a positive semi-definite form. Therefore it is never negative, and if one or more of the elements of $\mathbf{m}$ are sufficiently large that term will exceed $\hat{\mathbf{b}}'\mathbf{X}'\mathbf{y}$ and (111) will be negative. Hence $\mathbf{y}'\mathbf{y} - (\text{SSE} + Q)$ is not a sum of squares.

The reason that $\mathbf{y}'\mathbf{y} - (\text{SSE} + Q)$ is not necessarily a reduction in sum of squares due to fitting the reduced model is that $\mathbf{y}'\mathbf{y}$ is not always the total sum of squares for the reduced model. For example, if the full model is

$$y_i = b_0 + b_1x_{i1} + b_2x_{i2} + e_i$$

and the hypothesis is $b_1 = b_2 + 4$, then the reduced model would be

$$y_i = b_0 + (b_2 + 4)x_{i1} + b_2x_{i2} + e_i;$$

i.e.,

$$y_i - 4x_{i1} = b_0 + b_2(x_{i1} + x_{i2}) + e_i\,. \tag{112}$$

The total sum of squares for this reduced model is $(\mathbf{y} - 4\mathbf{x}_1)'(\mathbf{y} - 4\mathbf{x}_1)$ and not $\mathbf{y}'\mathbf{y}$, and so $\mathbf{y}'\mathbf{y} - (\text{SSE} + Q)$ is not the reduction in sum of squares. Furthermore, (112) is not the only reduced model, because the hypothesis $b_1 = b_2 + 4$ could just as well be used to amend the model to be

$$y_i = b_0 + b_1x_{i1} + (b_1 - 4)x_{i2} + e_i;$$

i.e.,

$$y_i + 4x_{i2} = b_0 + b_1(x_{i1} + x_{i2}) + e_i\,. \tag{113}$$

The total sum of squares will now be $(\mathbf{y} + 4\mathbf{x}_2)'(\mathbf{y} + 4\mathbf{x}_2)$. So in this case there are two reduced models, (112) and (113), and they are not identical. Hence neither are their total sums of squares, neither of which equal $\mathbf{y}'\mathbf{y}$. Therefore $\mathbf{y}'\mathbf{y} - (\text{SSE} + Q)$ is not the reduction in sum of squares due to fitting the reduced model. Indeed, by the existence of (112) and (113) there is no unique reduced model. And yet, despite this, SSE + Q is the residual sum of squares for all possible reduced models—their total sums of squares and reductions in sums of squares differ from model to model but their residual sums of squares are all the same.

The situation just described is true in general for the hypothesis $\mathbf{K}'\mathbf{b} = \mathbf{m}$. Suppose $\mathbf{L}'$ is such that $\mathbf{R} = \begin{bmatrix} \mathbf{K}' \\ \mathbf{L}' \end{bmatrix}$ has full rank and $\mathbf{R}^{-1} = [\mathbf{P} \quad \mathbf{S}]$ is its inverse. Then the model $\mathbf{y} = \mathbf{X}\mathbf{b} + \mathbf{e}$ can be written as

$$\begin{aligned} \mathbf{y} &= \mathbf{X}\mathbf{R}^{-1}\mathbf{R}\mathbf{b} + \mathbf{e} \\ &= \mathbf{X}[\mathbf{P} \quad \mathbf{S}]\begin{bmatrix} \mathbf{K}'\mathbf{b} \\ \mathbf{L}'\mathbf{b} \end{bmatrix} + \mathbf{e}, \\ &= \mathbf{X}\mathbf{P}\mathbf{m} + \mathbf{X}\mathbf{S}\mathbf{L}'\mathbf{b} + \mathbf{e}; \end{aligned}$$

i.e.,
$$\mathbf{y} - \mathbf{X}\mathbf{P}\mathbf{m} = \mathbf{X}\mathbf{S}\mathbf{L}'\mathbf{b} + \mathbf{e}. \tag{114}$$

This is a model in the elements of $\mathbf{L}'\mathbf{b}$, which represents $r - s$ LIN functions of the elements of $\mathbf{b}$. But since $\mathbf{L}'$ is arbitrary, chosen to make $\mathbf{R}$ non-singular, the model (114) is not unique. Despite this, it can be shown that the residual sum of squares after fitting any one of the models implicit in (114) is SSE + Q. And the corresponding value of the estimator of $\mathbf{b}$ is $\tilde{\mathbf{b}}$ given in (103) (see Exercise 10).

(*ii*) $\mathbf{K}'\mathbf{b} = \mathbf{0}$. One case in which $\mathbf{y}'\mathbf{y} - (\text{SSE} + Q)$ *is* a reduction in sum of squares due to fitting the reduced model is when $\mathbf{m} = \mathbf{0}$. For then (114) becomes

$$\mathbf{y} = \mathbf{X}\mathbf{S}\mathbf{L}'\mathbf{b} + \mathbf{e}$$

and so the total sum of squares for the reduced model is $\mathbf{y}'\mathbf{y}$, the same as that of the full model. Hence in this case

$$\mathbf{y}'\mathbf{y} - (\text{SSE} + Q) = \text{reduction(reduced)}. \tag{115}$$

That it is a sum of squares, i.e., is positive semi-definite, is seen from (111) wherein putting $\mathbf{m} = \mathbf{0}$ gives

$$\begin{aligned} \mathbf{y}'\mathbf{y} - (\text{SSE} + Q) &= \hat{\mathbf{b}}'\mathbf{X}'\mathbf{y} - \hat{\mathbf{b}}'\mathbf{K}[\mathbf{K}'(\mathbf{X}'\mathbf{X})^{-1}\mathbf{K}]^{-1}\mathbf{K}'\hat{\mathbf{b}} \\ &= \mathbf{y}'\{\mathbf{X}(\mathbf{X}'\mathbf{X})^{-1}\mathbf{X}' - \mathbf{X}(\mathbf{X}'\mathbf{X})^{-1}\mathbf{K}[\mathbf{K}'(\mathbf{X}'\mathbf{X})^{-1}\mathbf{K}]^{-1} \\ &\qquad \times \mathbf{K}'(\mathbf{X}'\mathbf{X})^{-1}\mathbf{X}'\}\mathbf{y}\,. \end{aligned} \tag{116}$$

Since the matrix enclosed in curly brackets is idempotent it is positive semi-definite. Therefore so is $\mathbf{y}'\mathbf{y} - (\text{SSE} + Q)$; i.e., it is a sum of squares. From (115)

$$Q = \mathbf{y}'\mathbf{y} - \text{SSE} - \text{reduction(reduced)}.$$

But
$$\mathbf{y}'\mathbf{y} - \text{SSE} = \text{SSR} = \text{reduction(full)}$$

and so
$$Q = \text{reduction(full)} - \text{reduction(reduced)}.$$

Therefore, since the sole difference between the full and reduced models is just the hypothesis, it is logical to describe

Q as the reduction in a sum of squares due to the hypothesis.

With this description we insert the partitioning of SSR as the sum of Q and SSR $- Q$ into the analysis of variance of Table 3.2 to yield Table 3.6. In

TABLE 3.6. ANALYSIS OF VARIANCE FOR TESTING THE HYPOTHESIS $\mathbf{K}'\mathbf{b} = \mathbf{0}$

Source of Variation	Degrees of Freedom		Sum of Squares	
Regression (full model)	r		SSR	
Hypothesis		s		Q
Reduced model		$r - s$		SSR $- Q$
Residual error	$N - r$		SSE	
Total	N		SST	

doing so we utilize (99), that when $\mathbf{m} = \mathbf{0}$,

$$Q/\sigma^2 \sim \chi^{2\prime}\{s, \mathbf{b}'\mathbf{K}[\mathbf{K}'(\mathbf{X}'\mathbf{X})^{-1}\mathbf{K}]^{-1}\mathbf{K}'\mathbf{b}/2\sigma^2\}.$$

Then, because

$$(\mathbf{y}'\mathbf{y} - \text{SSE})/\sigma^2 \sim \chi^{2\prime}\{r, \mathbf{b}'\mathbf{X}'\mathbf{X}\mathbf{b}/2\sigma^2\},$$

an application of Theorem 5 of Chapter 2 shows that

$$(\text{SSR} - Q)/\sigma^2 \sim \chi^{2\prime}\{r - s, \mathbf{b}'[\mathbf{X}'\mathbf{X} - \mathbf{K}\{\mathbf{K}'(\mathbf{X}'\mathbf{X})^{-1}\mathbf{K}\}^{-1}\mathbf{K}']\mathbf{b}/2\sigma^2\}$$

and is independent of SSE$/\sigma^2$. This, of course, can also be derived directly from (116). Furthermore, the non-centrality parameter in the distribution of SSR $- Q$ can, in terms of (114), be shown to be equal to $\mathbf{b}'\mathbf{L}(\mathbf{S}'\mathbf{X}'\mathbf{X}\mathbf{S})\mathbf{L}'\mathbf{b}/2\sigma^2$ (see Exercise 11). Hence, under the null hypothesis, this non-centrality parameter is zero when $\mathbf{L}'\mathbf{b} = \mathbf{0}$. Thus SSR $- Q$ forms the basis of an F-test for the sub-hypothesis $\mathbf{L}'\mathbf{b} = \mathbf{0}$ under the null hypothesis $\mathbf{K}'\mathbf{b} = \mathbf{0}$.

We now have the following F-tests:

$$\frac{\text{SSR}/r}{\text{SSE}/(N-r)} \quad \text{tests the full model,}$$

$$\frac{Q/s}{\text{SSE}/(N-r)} \quad \text{tests the hypothesis } \mathbf{K'b} = \mathbf{0}$$

and, under the null hypothesis,

$$\frac{(\text{SSR} - Q)/(r-s)}{\text{SSE}/(N-r)} \quad \text{tests the sub-hypothesis } \mathbf{L'b} = \mathbf{0}.$$

(*iii*) $\mathbf{b}_q = \mathbf{0}$. The most useful case of the reduced model when $\mathbf{m} = \mathbf{0}$ is when $\mathbf{K}' = [\mathbf{I}_q \quad \mathbf{0}]$ for some $q \leq k$. The null hypothesis $\mathbf{K'b} = \mathbf{m}$ is then $\mathbf{b}_q = \mathbf{0}$, where $\mathbf{b}'_q = [b_0 \quad b_1 \quad \cdots \quad b_{q-1}]$ say, a subset of q of the b's. This situation was discussed earlier where we found, in (107),

$$F(H) = Q/q\hat{\sigma}^2, \qquad \text{with} \quad Q = \hat{\mathbf{b}}'_q\mathbf{T}_{qq}^{-1}\hat{\mathbf{b}}_q,$$

involving the "invert part of the inverse" rule. Hence a special case of Table 3.6 is the analysis of variance table for testing the hypothesis H: $\mathbf{b}_q = \mathbf{0}$, shown in Table 3.7.

TABLE 3.7. ANALYSIS OF VARIANCE FOR TESTING THE HYPOTHESIS $\mathbf{b}_q = \mathbf{0}$

Source of Variation	Degrees of Freedom	Sum of Squares
Full model ($\mathbf{b}$)	r	$\text{SSR} = \hat{\mathbf{b}}'\mathbf{X}'\mathbf{y}$
Hypothesis: $\mathbf{b}_q = \mathbf{0}$	q	$Q = \hat{\mathbf{b}}'_q\mathbf{T}_{qq}^{-1}\hat{\mathbf{b}}_q$
Reduced model ($\mathbf{b}_p$)	$r - q$	$\text{SSR} - Q$
Residual error	$N - r$	$\text{SSE} = \text{SST} - \text{SSR}$
Total	N	$\text{SST} = \mathbf{y}'\mathbf{y}$

Shown in Table 3.7 is the most direct way of computing its parts: $\text{SSR} = \hat{\mathbf{b}}'\mathbf{X}'\mathbf{y}$, $Q = \hat{\mathbf{b}}'_q\mathbf{T}_{qq}^{-1}\hat{\mathbf{b}}_q$, $\text{SSR} - Q$ by differencing, $\text{SST} = \mathbf{y}'\mathbf{y}$ and SSE by differencing. Although $\text{SSR} - Q$ is obtained most readily by differencing it can also be expressed as $\tilde{\mathbf{b}}'_p\mathbf{X}'_p\mathbf{X}_p\tilde{\mathbf{b}}_p$ (see Exercise 12). The estimator $\tilde{\mathbf{b}}_p$ is derived from (103) as

$$\tilde{\mathbf{b}}_p = \hat{\mathbf{b}}_p - \mathbf{T}_{pq}\mathbf{T}_{qq}^{-1}\hat{\mathbf{b}}_q \tag{117}$$

using $\mathbf{K}'(\mathbf{X}'\mathbf{X})^{-1}\mathbf{K} = \mathbf{T}_{qq}$ as in (107).

Example. For the following data

y	x_1	x_2	x_3
8	2	1	4
10	−1	2	1
9	1	−3	4
6	2	1	2
12	1	4	6

$$(\mathbf{X}'\mathbf{X})^{-1} = \begin{bmatrix} 11 & 3 & 21 \\ 3 & 31 & 20 \\ 21 & 20 & 73 \end{bmatrix}^{-1} = \begin{bmatrix} .2145 & .0231 & -.0680 \\ .0231 & .0417 & -.0181 \\ -.0680 & -.0181 & .0382 \end{bmatrix},$$

$$\mathbf{y}'\mathbf{y} = 425 \quad \text{and} \quad \mathbf{X}'\mathbf{y} = \begin{bmatrix} 39 \\ 55 \\ 162 \end{bmatrix}.$$

We consider no-intercept models only. Then

$$\hat{\mathbf{b}}' = [-1.39 \quad 0.27 \quad 2.54]$$

and the analysis of variance is

Source	Degrees of Freedom	Sum of Squares
Full model	3	SSR = 372.9
Residual error	2	SSE = 52.1
Total	5	SST = 425.0

For testing the hypothesis H: $b_1 = b_2 + 4$ the reduction Q is, from (99),

$$Q = (\hat{b}_1 - \hat{b}_2 - 4)\left[[1 \quad -1 \quad 0](\mathbf{X}'\mathbf{X})^{-1}\begin{bmatrix} 1 \\ -1 \\ 0 \end{bmatrix}\right]^{-1}(\hat{b}_1 - \hat{b}_2 - 4)$$

$$= \frac{(-1.39 - 0.27 - 4.0)^2}{.2145 + 0.417 - 2(.0231)} = \frac{(-5.66)^2}{.21} = 152.55.$$

Hence the F-statistic for testing the hypothesis is $152.2/(52.1/2) = 5.8$.

Were a reduced model to be derived by replacing b_1 by $b_2 + 4$ it would be

$$y - 4x_1 = b_2(x_1 + x_2) + b_3x_3 + e \tag{118}$$

for which the data are

$y - 4x_1$	$x_1 + x_2$	x_3
0	3	4
14	1	1
5	−2	4
−2	3	2
8	5	6

The total sum of squares is now $0^2 + 14^2 + 5^2 + 2^2 + 8^2 = 289$; and the residual sum of squares, using SSE from the analysis of variance and Q from the F-statistic, is

$$\text{SSE} + Q = 52.1 + 152.2 = 204.3.$$

Therefore the analysis of variance for the reduced model is

Source	Degrees of Freedom	Sum of Squares
Regression (reduced model)	2	84.7
Residual error	3	204.3
Total	5	289.0

The value of 84.7 for the reduction in sum of squares for the reduced model can be verified by deriving the normal equations for the model (118) directly. From the data they are

$$\begin{bmatrix} 48 & 41 \\ 41 & 73 \end{bmatrix} \begin{bmatrix} \tilde{b}_2 \\ \tilde{b}_3 \end{bmatrix} = \begin{bmatrix} 38 \\ 78 \end{bmatrix}$$

and hence $$\begin{bmatrix} \tilde{b}_2 \\ \tilde{b}_3 \end{bmatrix} = (1/1823) \begin{bmatrix} 73 & -41 \\ -41 & 48 \end{bmatrix} \begin{bmatrix} 38 \\ 78 \end{bmatrix} = \begin{bmatrix} -0.23 \\ 1.20 \end{bmatrix}.$$

Then the reduction in the sum of squares is

$$[-0.23 \quad 1.20] \begin{bmatrix} 38 \\ 78 \end{bmatrix} = 93.6 - 8.9 = 84.7$$

as in the analysis of variance.

These calculations are, of course, shown here purely to illustrate the sum of squares in the analysis of variance. They are not needed specifically because for the reduced model the residual is always SSE + Q. And the estimator of **b** can be found from (103) as

$$\tilde{\mathbf{b}} = \begin{bmatrix} -1.39 \\ 0.27 \\ 2.54 \end{bmatrix} - (\mathbf{X}'\mathbf{X})^{-1} \begin{bmatrix} 1 \\ -1 \\ 0 \end{bmatrix} \frac{1}{.21}(-5.66)$$

$$= \begin{bmatrix} -1.39 \\ 0.27 \\ 2.54 \end{bmatrix} + \begin{bmatrix} .2145 - .0231 \\ .0231 - .0417 \\ -.0680 + .0181 \end{bmatrix} (26.95) = \begin{bmatrix} 3.77 \\ -0.23 \\ 1.20 \end{bmatrix}$$

wherein $\tilde{b}_1 - \tilde{b}_2 = 4$, of course, and $\tilde{b}_2$ and $\tilde{b}_3$ are as before.

For testing the hypothesis $b_1 = 0$, $Q = (-1.39)^2/(.2145) = 8.9$ and the analysis of variance of Table 3.6 is

Source	Degrees of Freedom	Sum of Squares
Full model	3	372.9
Hypothesis	1	8.9
Reduced model	2	364.0
Residual error	2	52.1
Total	5	425.0

with

$$\tilde{\mathbf{b}} = \begin{bmatrix} -1.39 \\ .27 \\ 2.54 \end{bmatrix} - \begin{bmatrix} .2145 \\ .0231 \\ -.0680 \end{bmatrix} (-1.39)/.2145 = \begin{bmatrix} 0 \\ 0.42 \\ 2.10 \end{bmatrix}.$$

Again these results can be verified from the normal equations of the reduced model, in this case

$$\begin{bmatrix} 31 & 20 \\ 20 & 73 \end{bmatrix} \begin{bmatrix} \tilde{b}_2 \\ \tilde{b}_3 \end{bmatrix} = \begin{bmatrix} 55 \\ 162 \end{bmatrix}$$

They give

$$\begin{bmatrix} \tilde{b}_2 \\ \tilde{b}_3 \end{bmatrix} = (1/1863) \begin{bmatrix} 73 & -20 \\ -20 & 31 \end{bmatrix} \begin{bmatrix} 55 \\ 162 \end{bmatrix} = \begin{bmatrix} 0.42 \\ 2.10 \end{bmatrix}$$

as above; and the reduction in sum of squares is

$$[0.42 \quad 2.10] \begin{bmatrix} 31 & 20 \\ 20 & 73 \end{bmatrix} \begin{bmatrix} 0.42 \\ 2.10 \end{bmatrix} = 364.0.$$

7. RELATED TOPICS

It is appropriate to briefly mention certain topics related to the preceding development that are customarily associated with testing hypotheses. The treatment of these topics will do no more than act as an outline to the reader, showing him their application to the linear models situation. As with the discussion of distribution functions in Chapter 2, the reader will have to look elsewhere for a complete discussion of these topics.

a. The likelihood ratio test

Tests of linear hypotheses $\mathbf{K}'\mathbf{b} = \mathbf{m}$ have been developed from the starting point of the F-statistic. This, in turn, can be shown to stem from the likelihood ratio test.

For a sample of N observations $\mathbf{y}$, where $\mathbf{y}$ is $N(\mathbf{Xb}, \sigma^2\mathbf{I})$ the likelihood function is

$$L(\mathbf{b}, \sigma^2) = (2\pi\sigma^2)^{-\frac{1}{2}N} \exp\{-[(\mathbf{y} - \mathbf{Xb})'(\mathbf{y} - \mathbf{Xb})/2\sigma^2]\}.$$

The likelihood ratio test utilizes two values of $L(\mathbf{b}, \sigma^2)$:

(i) $\text{Max}(L_w)$, the maximum value of $L(\mathbf{b}, \sigma^2)$ maximized over the complete range of parameters, namely $0 < \sigma^2 < \infty$, and $-\infty < b_i < \infty$ for all i.

(ii) $\text{Max}(L_H)$, the maximum value of $L(\mathbf{b}, \sigma^2)$ maximized over the range of parameters limited (restricted or defined) by the hypothesis H.

The likelihood ratio is the ratio of these two maxima:

$$L = \frac{\max(L_H)}{\max(L_w)}.$$

Each maximum is found in the usual manner: differentiate $L(\mathbf{b}, \sigma^2)$ with respect to σ^2 and the elements of $\mathbf{b}$, equate the differentials to zero, solve the resulting equations for $\mathbf{b}$ and σ^2 and use these solutions in the place of $\mathbf{b}$ and σ^2 in $L(\mathbf{b}, \sigma^2)$. In the case of $\max(L_H)$ the maximization procedure is carried out within the limitations of the hypothesis. We demonstrate for the case of the hypothesis H: $\mathbf{b} = \mathbf{0}$. First, $\partial L(\mathbf{b}, \sigma^2)/\partial\mathbf{b} = \mathbf{0}$ gives, as we have seen, $\hat{\mathbf{b}} = (\mathbf{X}'\mathbf{X})^{-1}\mathbf{X}'\mathbf{y}$; and $\partial L(\mathbf{b}, \sigma^2)/\partial\sigma^2 = 0$ gives $\hat{\sigma}^2 = (\mathbf{y} - \mathbf{X}\hat{\mathbf{b}})'(\mathbf{y} - \mathbf{X}\hat{\mathbf{b}})/N$. Thus

$$\begin{aligned}\max(L_w) = L(\hat{\mathbf{b}}, \hat{\sigma}^2) &= (2\pi\hat{\sigma}^2)^{-\frac{1}{2}N} \exp\{-[(\mathbf{y} - \mathbf{X}\hat{\mathbf{b}})'(\mathbf{y} - \mathbf{X}\hat{\mathbf{b}})/2\hat{\sigma}^2]\} \\ &= \frac{e^{-\frac{1}{2}N}N^{\frac{1}{2}N}}{(2\pi)^{\frac{1}{2}N}[(\mathbf{y} - \mathbf{X}\hat{\mathbf{b}})'(\mathbf{y} - \mathbf{X}\hat{\mathbf{b}})]^{\frac{1}{2}N}}.\end{aligned}$$

This is the denominator of L. The numerator comes from amending L by the hypothesis $\mathbf{b} = \mathbf{0}$, so giving

$$L(\mathbf{0}, \sigma^2) = (2\pi\sigma^2)^{-\frac{1}{2}N} \exp -(\mathbf{y}'\mathbf{y}/2\sigma^2).$$

Maximizing this with respect to σ^2 by using the equation $\partial L(\mathbf{0}, \sigma^2)/\partial\sigma^2 = 0$ gives $\tilde{\sigma}^2 = \mathbf{y}'\mathbf{y}/N$ and so

$$\max(L_H) = L(\mathbf{0}, \tilde{\sigma}^2) = (2\pi\tilde{\sigma}^2)^{-\frac{1}{2}N} \exp -(\mathbf{y}'\mathbf{y}/2\,\tilde{\sigma}^2)$$
$$= \frac{e^{-\frac{1}{2}N} N^{\frac{1}{2}N}}{(2\pi)^{\frac{1}{2}N}(\mathbf{y}'\mathbf{y})^{\frac{1}{2}N}}.$$

With these values for the maxima, the likelihood ratio is

$$L = \frac{\max(L_H)}{\max(L_w)} = \left[\frac{(\mathbf{y} - \mathbf{X}\hat{\mathbf{b}})'(\mathbf{y} - \mathbf{X}\hat{\mathbf{b}})}{\mathbf{y}'\mathbf{y}}\right]^{\frac{1}{2}N} = \left[\frac{1}{1 + \text{SSR}/\text{SSE}}\right]^{\frac{1}{2}N}.$$

Clearly L is a single-valued function of SSR/SSE, monotonic decreasing when SSR/SSE increases. Therefore SSR/SSE can be used as a test statistic in place of L. By the same reasoning so can (SSR/SSE)$[(N - r)/r]$ whose use as the F-statistic has already been discussed. Thus is the use of the F-statistic established as an outcome of the likelihood ratio test. The basis of $F(H)$ can be established similarly.

b. Type I and II errors

Under the null hypothesis H: $\mathbf{K}'\mathbf{b} = \mathbf{m}$, $F(H) = (N - r)Q/s\text{SSE}$ has the $F_{s,N-r}$ distribution. For a significance test at the $100\alpha\%$ level the rule of the test is to not reject H whenever $F(H) \leq F_{\alpha,s,N-r}$, the tabulated value of the $F_{s,N-r}$ distribution, at the $100\alpha\%$ point. This means $F_{\alpha,s,N-r}$ is defined as follows: if u is any variable having the $F_{s,N-r}$ distribution then

$$\Pr\{u \geq F_{\alpha,s,N-r}\} = \alpha.$$

The probability α is the (significance) level of the significance test. An oft-used value for it is 0.05, but there is nothing sacrosanct about this; any value between 0 and 1 can be used for α. Other frequently used values are 0.01 and 0.10.

The rule of whether or not to reject the hypothesis H is to reject it whenever $F(H) > F_{\alpha,s,N-r}$ and to not reject it whenever $F(H) \leq F_{\alpha,s,N-r}$. By the nature of the statistic $F(H)$ we know that over repeated sampling $F(H)$ will exceed $F_{\alpha,s,N-r}$ on $100\alpha\%$ (5%, say) of the time; and when it does we will reject H. Therefore, in situations in which the null hypothesis H is actually true, this rejection will constitute an error of judgment. It is the error known as a Type I error, or rejection error. It consists of wrongly rejecting the null hypothesis H when it is true; the probability of its occurrence is α.

Now consider the situation when not H but some other hypothesis. H_a: $\mathbf{K}_a'\mathbf{b} = \mathbf{m}_a$, is true. Then, as in (100),

$$F(H) \sim F'(s, N - r, \lambda) \tag{119}$$

with non-centrality parameter

$$\begin{aligned}\lambda &= (\mathbf{K}'\mathbf{b} - \mathbf{m})'[\mathbf{K}'(\mathbf{X}'\mathbf{X})^{-1}\mathbf{K}]^{-1}(\mathbf{K}'\mathbf{b} - \mathbf{m})/2\sigma^2 \\ &= \tfrac{1}{2}(\mathbf{K}'\mathbf{b} - \mathbf{m})'[\text{var}(\mathbf{K}'\hat{\mathbf{b}})]^{-1}(\mathbf{K}'\mathbf{b} - \mathbf{m})\end{aligned} \tag{120}$$

using (51) for var($\hat{\mathbf{b}}$). Note that $\lambda \neq 0$ because $\mathbf{K}'\mathbf{b} \neq \mathbf{m}$ but $\mathbf{K}_a'\mathbf{b}_a = \mathbf{m}_a$. Suppose that, without our knowing it, this alternative hypothesis H_a had been true at the time the data were collected. And suppose that, with those data, the hypothesis H: $\mathbf{K}'\mathbf{b} = \mathbf{m}$ is tested using $F(H)$ as already described. When $F(H) \leq F_{\alpha,s,N-r}$, we do not reject H. But in doing this an error is made—an error of not rejecting H when (even though we did not know it) H_a was true and hence H was not true; i.e., we fail to reject H when it is false. This is called a Type II error, where we fail to reject the null hypothesis H when the alternative, H_a, is true. The probability of this, to be denoted by P(II), is

$$\begin{aligned}\text{P(II)} &= \Pr\{\text{Type II error occurring}\} \\ &= \Pr\{\text{not rejecting } H \text{ when } H \text{ is not true}\} \\ &= \Pr\{F(H) \leq F_{\alpha,s,N-r} \text{ where } F(H) \sim F'(s, N - r, \lambda)\}\end{aligned} \tag{121}$$

which we shall write as

$$\text{P(II)} = \Pr\{F'(s, N - r, \lambda) \leq F_{\alpha,s,N-r}\} \tag{122}$$

from (119) and (120). By the right-hand side of (122) we mean the probability that a random variable distributed as $F'(s, N - r, \lambda)$ is less than $F_{\alpha,s,N-r}$, the $100\alpha\%$ point in the central $F_{s,N-r}$-distribution. The two types of errors are summarized in Table 3.8.

TABLE 3.8. TYPE I AND TYPE II ERRORS IN HYPOTHESIS TESTING

	Result of Test of Hypothesis	
Null Hypothesis	$F(H) \leq F_{\alpha,s,N-r}$	$F(H) > F_{\alpha,s,N-r}$
(H: $\mathbf{K}'\mathbf{b} = \mathbf{m}$)	Conclusion	
	Do not reject H	Reject H
True	No error	Type I error[1]
False (H_a: $\mathbf{K}_a\mathbf{b} = \mathbf{m}_a$ is true)	Type II error[2]	No error

[1] Pr{Type I error} = α = Pr{$F(H) > F_{\alpha,s,N-r}$ when H: $\mathbf{K}'\mathbf{b} = \mathbf{m}$ is true}.
[2] Pr{Type II error} = P(II) = Pr{$F(H) \leq F_{\alpha,s,N-r}$ when H_a: $\mathbf{K}_a'\mathbf{b} = \mathbf{m}_a$ is true}.

The probability of a Type II error, P(II) of (122) using λ of (120), is not readily available from tables because tables of the non-central F-distribution are limited in extent. However, tabulations given by Tang (1938) circumvent this difficulty through considering a function E^2, dependent on s, $N - r$ and λ, such that

$$\text{P(II)} = \Pr\{F'(s, N - r, \lambda) \leq F_{\alpha,s,N-r}\}$$

of (122) is equivalent to

$$\text{P(II)} = \Pr\{E^2[s, N - r, \sqrt{2\lambda/(s + 1)}] \leq sF_{\alpha,s,N-r}/(N - r + sF_{\alpha,s,N-r})\}. \tag{123}$$

For $\alpha = 0.05$ and 0.01, Tang (1938) tabulates this probability for a variety of values of s, $N - r$ and $\sqrt{2\lambda/(s + 1)}$, denoted more generally by f_1, f_2 and φ, respectively. Similar tables are also available in Kempthorne (1952) and Graybill (1961).

Example. Consider testing the hypothesis H: $b = 1$ in the case of simple regression involving just a single x-variable. Then $s = 1$ and $r = 2$, and, on supposing there are 22 observations, $N - r = 20$. For $\alpha = 0.05$, the tabulated value of $F_{\alpha,s,N-r}$ is $F_{0\cdot05,1,20} = 4.35$. Therefore in (122)

$$\text{P(II)} = \Pr\{F'(1, 20, \lambda) \leq 4.35\}. \tag{124}$$

Suppose that $v(\hat{b}) = \frac{1}{18}$ and that the alternative hypothesis is H_a: $b = b_a \neq 1$. Then from (120)

$$\lambda = \tfrac{1}{2}(b_a - 1)(\tfrac{1}{18})^{-1}(b_a - 1) = 9(b_a - 1)^2 \tag{125}$$

for $b_a \neq 1$. Substitution of (125) in (124) gives

$$\text{P(II)} = \Pr\{F'[1, 20, 9(b_a - 1)^2] \leq 4.35\}$$

and so in (123)

$$\begin{aligned} \text{P(II)} &= \Pr\{E^2[1, 20, \sqrt{18(b_a - 1)^2/2}] \leq 4.35/(20 + 4.35)\} \\ &= \Pr\{E^2[1, 20, 3(b_a - 1)] \leq 0.179\}. \end{aligned} \tag{126}$$

From tables of this probability, e.g., Graybill (1961, p. 444), we find the following values of P(II) and 1 − P(II) for different values of b_a:

b_a:	$1\frac{1}{3}$	$1\frac{1}{2}$	$1\frac{2}{3}$	$1\frac{5}{6}$	2	
P(II) :	.730	.477	.233	.081	.019	for $N = 22, N - r = 20$.
1 − P(II) :	.270	.523	.767	.919	.981	

Note here that P(II) decreases as b_a increases—and, correspondingly, 1 − P(II) increases. Further reference to the tables also shows that a larger (smaller) value of N leads to smaller (larger) values of P(II), with correspondingly larger (smaller) values of 1 − P(II): e.g., for $N > 22$,

b_a:	$1\frac{1}{3}$	$1\frac{1}{2}$	$1\frac{2}{3}$	$1\frac{5}{6}$	2	
P(II):	.722	.463	.219	.072	.016	for $N = 32, N - r = 30$
1 − P(II):	.278	.537	.781	.928	.984	

and for $N < 22$,

P(II):	.751	.517	.278	.111	.032	for $N = 12, N - r = 10.$
1 − P(II):	.259	.483	.722	.889	.968	

In these cases the expressions analogous to (126) are

$$\text{P(II)} = \Pr\{E^2[1, 30, 3(b_a - 1)] \leq 0.122\} \qquad \text{for} \quad N = 32, N - r = 30$$

and

$$\text{P(II)} = \Pr\{E^2[1, 10, 3(b_a - 1)] \leq 0.332\} \qquad \text{for} \quad N = 12, N - r = 10,$$

corresponding respectively to $F_{0\cdot05,1,30} = 4.17$ and $F_{0\cdot05,1,10} = 4.96$.

c. The power of a test

Through the expression for λ in (120) it can be seen that P(II) of (122) depends upon $\mathbf{K}_a'$ and $\mathbf{m}_a$ of the alternative hypothesis H_a: $\mathbf{K}_a'\mathbf{b} = \mathbf{m}_a$. The probability 1 − P(II) is similarly dependent. It is known as the *power of the test* with respect to the alternative hypothesis H_a. From (121) it is

$$\begin{aligned} \text{Power} &= 1 - \text{P(II)} \\ &= 1 - \Pr\{\text{not rejecting } H \text{ when } H \text{ is not true}\} \\ &= \Pr\{\text{rejecting } H \text{ when } H \text{ is not true}\}. \end{aligned} \tag{127}$$

Tests of hypotheses as described in this chapter are based on assigning a small value α to the probability of a Type I error—the probability of rejecting H when H is true. In addition, whatever the test procedure is, rejecting H when H is *not* true is usually something we want a test to achieve with as high a probability as possible. In terms of (127) we therefore want the power of a test, for a given value of α, to be as large as possible. In the preceding example it can be seen that, for given values of α and b_a, increasing the amount of data, i.e., increasing N, is one way of increasing the power, 1 − P(II). This kind of result is true fairly generally. Also, for given α and N, the alternative hypothesis for which λ is largest (in the example, the largest value of b_a) is that which has the largest power. Extensive discussion of the power of a test and the part it plays in the theory of hypothesis testing in general is beyond the scope of this book. Scheffé (1959), Graybill (1961) and Rao (1965) are three of the many places where such discussion may be found.

d. Examining residuals

The estimated error vector

$$\hat{\mathbf{e}} = \mathbf{y} - \mathbf{X}\hat{\mathbf{b}}$$

is customarily referred to as the *vector of residuals*. In a variety of ways, its elements can be plotted and otherwise investigated to see if they suggest that assumptions inherent in the assumed model are not being upheld. The general problem entailed here—of examining residuals—is a large one; we do no more than hint at some of the available analyses and refer the reader elsewhere for more complete discussions.

Several elementary, but important, properties of the residuals are worth noting. With $\mathbf{P} = \mathbf{I} - \mathbf{X}(\mathbf{X}'\mathbf{X})^{-1}\mathbf{X}'$ of (80), which is symmetric and idempotent, as noted in (64),

$$\hat{\mathbf{e}} = \mathbf{y} - \mathbf{X}\hat{\mathbf{b}} = \mathbf{y} - \mathbf{X}(\mathbf{X}'\mathbf{X})^{-1}\mathbf{X}'\mathbf{y} = [\mathbf{I} - \mathbf{X}(\mathbf{X}'\mathbf{X})^{-1}\mathbf{X}']\mathbf{y} = \mathbf{P}\mathbf{y}$$

where, from (65), $\mathbf{PX} = \mathbf{0}$. A first property of the residuals is that they sum to zero:

$$\sum_{i=1}^{N} \hat{e}_i = \mathbf{1}'\hat{\mathbf{e}} = \mathbf{1}'\mathbf{P}\mathbf{y} = 0$$

using $\mathbf{1}'\mathbf{P} = \mathbf{0}'$ given by (79). Second, their sum of squares is SSE—as is evident in (66):

$$\sum_{i=1}^{N} \hat{e}_i^2 = \hat{\mathbf{e}}'\hat{\mathbf{e}} = \mathbf{y}'\mathbf{P}'\mathbf{P}\mathbf{y} = \mathbf{y}'\mathbf{P}\mathbf{y} = \mathbf{y}'\mathbf{y} - \mathbf{y}'\mathbf{X}(\mathbf{X}'\mathbf{X})^{-1}\mathbf{X}'\mathbf{y} = \text{SSE}.$$

Concerning distributional properties of residuals, their expected values are zero and their variance-covariance matrix is $\mathbf{P}\sigma^2$:

$$E(\hat{\mathbf{e}}) = E(\mathbf{P}\mathbf{y}) = \mathbf{P}\mathbf{X}\mathbf{b} = \mathbf{0}$$

and

$$\text{var}(\hat{\mathbf{e}}) = \text{var}(\mathbf{P}\mathbf{y}) = \mathbf{P}^2\sigma^2 = \mathbf{P}\sigma^2.$$

Additional results are shown in Exercise 19.

The properties just described hold true for the residuals of any intercept model. Consideration of the extent to which they satisfy other conditions is the means whereby assumptions of the model can be investigated. For example, in assuming normality of the error terms in the model we have $\hat{\mathbf{e}} \sim N(\mathbf{0}, \mathbf{P}\sigma^2)$. Plotting the values of $\hat{e}_i$ to see if they appear normally distributed therefore provides a means of seeing if the assumption $\mathbf{e} \sim N(\mathbf{0}, \sigma^2\mathbf{I})$ might be wrong. In doing this we ignore the fact that because $\text{var}(\hat{\mathbf{e}}) = \mathbf{P}\sigma^2$ the $\hat{e}_i$'s are correlated since, as Anscombe and Tukey (1963) indicate, for at least a two-way table with more than three rows and columns, "the effect of correlation [among residuals] upon graphical procedures is usually negligible". Draper and Smith (1966) provide further discussion of this point.

Other graphical procedures that may provide evidence of inappropriate assumptions in the model involve plotting the residuals against $\widehat{E(y)}$ and against the observed x's. The latter can be especially meaningful when the x's are time—and even when this is not the case, if time has been a factor involved in collecting the data the plot of residuals against time may be quite revealing. Two conclusions that these kinds of plots might suggest are that the variance of the error terms may not be constant or that additional terms are needed in the model. Draper and Smith (1966) give a most readable account of using these procedures, including reference to appropriate research papers. Three more recent publications are those of Theil (1968), Cox and Snell (1968) and Loynes (1969).

8. SUMMARY OF REGRESSION CALCULATIONS

The more frequently used of the general expressions developed in this chapter for estimating the linear regression on k x-variables are summarized and listed below.

N: number of observations on each variable.

k: number of x-variables.

$\mathbf{y}$: $N \times 1$ vector of observed y-values.

$\mathbf{X}_1$: $N \times k$ matrix of observed x-values.

$\mathbf{X} = [\mathbf{1} \quad \mathbf{X}_1]$.

$\bar{y}$: mean of the observed y's.

$\bar{\mathbf{x}}' = (1/N)\mathbf{1}'\mathbf{X}_1$: vector of means of observed x's.

$\mathbf{b} = \begin{bmatrix} b_0 \\ \mathscr{b} \end{bmatrix}$: b_0 is the intercept; : $\mathscr{b}$ is vector of regression coefficients.

$\mathscr{X} = \mathbf{X}_1 - \mathbf{1}\bar{\mathbf{x}}'$: matrix of observed x's expressed as deviations from their means.

$\mathscr{X}'\mathscr{X}$: matrix of corrected sums of squares and products of observed x's.

$\mathscr{X}'\mathbf{y}$: vector of corrected sums of products of observed x's and y's.

$r = k + 1$: rank of $\mathbf{X}$.

$\text{SST}_m = \mathbf{y}'\mathbf{y} - N\bar{y}^2$: total sum of squares (c.f.m.).

$\hat{\mathbf{b}} = (\mathscr{X}'\mathscr{X})^{-1}\mathscr{X}'\mathbf{y}$: estimated regression coefficients.

$\text{SSE} = \text{SST}_m - \hat{\mathbf{b}}'\mathscr{X}'\mathbf{y}$: error sum of squares.

$\hat{\sigma}^2 = \text{SSE}/(N - r)$: estimated residual error variance.

$\widehat{\text{var}}(\hat{\mathbf{b}}) = (\mathscr{X}'\mathscr{X})^{-1}\hat{\sigma}^2$: estimated covariance matrix of $\hat{\mathbf{b}}$.

$\text{SSR}_m = \hat{\mathbf{b}}'\mathscr{X}'\mathbf{y}$: sum of squares due to fitting model over and above the mean.

$R^2 = \text{SSR}/\text{SST}$: coefficient of determination.

$F_{r-1,N-r} = \text{SSR}_m/(r - 1)\hat{\sigma}^2$: F-statistic for testing H: $\mathbf{b} = \mathbf{0}$.

$a^{ii} =$ ith diagonal element of $(\mathscr{X}'\mathscr{X})^{-1}$.

$t_i = \hat{b}_i/\hat{\sigma}\sqrt{a^{ii}}$: t-statistic, on $N - r$ degrees of freedom, for testing hypothesis $b_i = 0$.

$\hat{b}_i \pm t_{N-r,\frac{1}{2}\alpha}\sqrt{a^{ii}\hat{\sigma}^2}$: symmetric $100(1 - \alpha)\%$ confidence interval for b_i.

$F_{q,N-r} = \hat{\mathbf{b}}_q'\mathscr{T}_{qq}^{-1}\hat{\mathbf{b}}_q/q\hat{\sigma}^2$: F-statistic for testing H: $\mathbf{b}_q = 0$.

$\hat{b}_0 = \bar{y} - \bar{\mathbf{x}}'\hat{\mathbf{b}}$: estimated intercept.

$\text{cov}(\hat{b}_0,\hat{\mathbf{b}}) = -(\mathscr{X}'\mathscr{X})^{-1}\bar{\mathbf{x}}'\hat{\sigma}^2$: estimated vector of covariances of $\hat{b}_0$ with $\hat{\mathbf{b}}$.

$\hat{v}(\hat{b}_0) = [1/N + \bar{\mathbf{x}}'(\mathscr{X}'\mathscr{X})^{-1}\bar{\mathbf{x}}]\hat{\sigma}^2$: estimated variance of $\hat{b}_0$.

$t_0 = \hat{b}_0/\sqrt{\hat{v}(\hat{b}_0)}$: t-statistic, on $N - r$ degrees of freedom, for testing hypothesis $b_0 = 0$.

$\hat{b}_0 \pm t_{N-r,\frac{1}{2}\alpha}\sqrt{\hat{v}(\hat{b}_0)}$: symmetric $100(1 - \alpha)\%$ confidence interval for b_0.

No-intercept model. Modify the above expressions as follows.

Use $\mathbf{X}_1$ in place of $\mathscr{X}$:

$\mathbf{X}_1'\mathbf{X}_1 =$ matrix of uncorrected sums of squares and products of observed x's.

$\mathbf{X}_1'\mathbf{y} =$ vector of uncorrected sums of products of observed x's and y's.

Put $r = k$ (instead of $k + 1$).

Use $\text{SST} = \mathbf{y}'\mathbf{y}$ (instead of $\text{SST}_m = \mathbf{y}'\mathbf{y} - N\bar{y}^2$).

Ignore b_0 and $\hat{b}_0$.

9. EXERCISES

1. For the following data

i:	1	2	3	4	5	6	7	8	9	10
y_i:	12	32	36	18	17	20	21	40	30	24
x_i:	65	43	44	59	60	50	52	38	42	40

   (*a*) Calculate the normal equations (11).
   (*b*) Calculate $\hat{b}$ and $\hat{a}$ as in (14) and (15).

2. When $k = 1$ show that (41) and (42) are equivalent to (14) and (15) and also equivalent to (21).

3. When $\mathbf{y}$ has variance-covariance matrix $\mathbf{V}$, prove that the covariance of the b.l.u.e.'s of $\mathbf{p}'\mathbf{b}$ and $\mathbf{q}'\mathbf{b}$ is $\mathbf{p}'(\mathbf{X}'\mathbf{V}^{-1}\mathbf{X})^{-1}\mathbf{q}$.

4. Since $\text{SSM} = \mathbf{y}'N^{-1}\mathbf{1}\mathbf{1}'\mathbf{y}$, show that $N^{-1}\mathbf{1}\mathbf{1}'$ is idempotent and has null products with $\mathbf{I} - \mathbf{X}(\mathbf{X}'\mathbf{X})^{-1}\mathbf{X}'$. What are the consequences of these properties of $N^{-1}\mathbf{1}\mathbf{1}'$?

5. Derive the matrix $\mathbf{Q}$ such that $\text{SSR}_m = \mathbf{y}'\mathbf{Q}\mathbf{y}$; show that $\mathbf{Q}$ is idempotent and has null products with $\mathbf{I} - \mathbf{X}(\mathbf{X}'\mathbf{X})^{-1}\mathbf{X}'$. What are the consequences of these properties of $\mathbf{Q}$? Show that SSR_m and SSM are independent.

6. Show that the non-centrality parameters of the non-central χ^2-distributions of SSM, SSR_m and SSE add to that of SST.

7. With the notation of this chapter, derive the *F*-statistics and values of $\tilde{\mathbf{b}}$ shown below.

	Hypothesis	*F*-statistic	$\tilde{\mathbf{b}}$
(i)	$\mathscr{b} = 0$	$\text{SSR}_m/k\hat{\sigma}^2$	$\tilde{\mathbf{b}}' = [\bar{y} \quad \mathbf{0}']$
(ii)	$\mathscr{b} = \mathscr{b}_0$	$(\hat{\mathscr{b}} - \mathscr{b}_0)'\mathscr{X}'\mathscr{X}(\hat{\mathscr{b}} - \mathscr{b}_0)/k\hat{\sigma}^2$	$\tilde{\mathbf{b}}' = [\bar{y} - \bar{x}'\mathscr{b}_0 \quad \mathscr{b}_0']$
(iii)	$\boldsymbol{\lambda}'\mathscr{b} = m$	$(\boldsymbol{\lambda}'\hat{\mathscr{b}} - m)^2/\boldsymbol{\lambda}'(\mathscr{X}'\mathscr{X})^{-1}\boldsymbol{\lambda}\hat{\sigma}^2$	$\tilde{\mathbf{b}} = \hat{\mathbf{b}} + \left(\dfrac{\boldsymbol{\lambda}'\hat{\mathscr{b}} - m}{\boldsymbol{\lambda}'(\mathscr{X}'\mathscr{X})^{-1}\boldsymbol{\lambda}}\right) \times \begin{bmatrix}\bar{\mathbf{x}}' \\ -\mathbf{I}\end{bmatrix}(\mathscr{X}'\mathscr{X})^{-1}\boldsymbol{\lambda}$
(iv)	$\mathscr{b}_q = \mathbf{0}$	$\hat{\mathscr{b}}_q'\mathscr{T}_{qq}^{-1}\hat{\mathscr{b}}_q/q\hat{\sigma}^2$	$\tilde{\mathbf{b}} = \begin{bmatrix}\bar{y} - \bar{\mathbf{x}}_p'\tilde{\mathscr{b}}_p \\ \mathbf{0} \\ \tilde{\mathscr{b}}_p = \hat{\mathscr{b}}_p - \mathscr{T}_{pq}\mathscr{T}_{qq}^{-1}\hat{\mathbf{b}}_q\end{bmatrix}$

In each case state the distribution of the *F*-statistic under the null hypothesis.

8. Show that the F-statistic for testing the hypothesis $\mathbf{L}'\mathscr{b} = \mathbf{m}$ takes essentially the same form as $F(H)$. Derive the estimator of $\mathscr{b}$ under the null hypothesis $\mathbf{L}'\mathscr{b} = \mathbf{m}$, showing that $\tilde{b}_0 = \hat{b}_0 + \bar{\mathbf{x}}'(\hat{\mathscr{b}} - \tilde{\mathscr{b}})$.

9. Suppose $\hat{\sigma}^2 = 200$ and $\hat{\mathbf{b}}' = [3 \quad 5 \quad 2]$ where

$$\hat{v}(\hat{b}_1) = 28 \qquad \hat{v}(\hat{b}_2) = 24 \qquad \hat{v}(\hat{b}_3) = 18$$

$$\widehat{\text{cov}}(\hat{b}_1, \hat{b}_2) = -16 \qquad \widehat{\text{cov}}(\hat{b}_1, \hat{b}_3) = 14 \qquad \widehat{\text{cov}}(\hat{b}_2, \hat{b}_3) = -12.$$

Show that the F-statistic for testing the hypothesis $b_1 = b_2 + 4 = b_3 + 7$ has a value of 1.0. Calculate the estimate of $\mathbf{b}$ under the null hypothesis.

10. By writing $\boldsymbol{\gamma}$ for $\mathbf{L}'\mathbf{b}$ in equation (114) write down the estimator of $\boldsymbol{\gamma}$ and residual sum of squares for fitting that model. Show that the estimator of $\boldsymbol{\gamma}$ corresponds to the estimator $\tilde{\mathbf{b}}$ given in equation (103) and that the residual sum of squares is identical to SSE + Q.

11. By using expression (116) prove directly that $[\mathbf{y}'\mathbf{y} - (\text{SSE} + Q)]/\sigma^2$ has a non-central χ^2-distribution, independent of SSE, when $\mathbf{m} = \mathbf{0}$; and show that, under the null hypothesis, the non-centrality parameter is $\mathbf{b}'\mathbf{L}(\mathbf{S}'\mathbf{X}'\mathbf{X}\mathbf{S})\mathbf{L}'\mathbf{b}/2\sigma^2$.

12. Prove that in Table 3.7 SSR $- Q = \tilde{\mathbf{b}}_p'\mathbf{X}_p'\mathbf{X}_p\tilde{\mathbf{b}}_p$. [*Hint:* Use (117) and $(\mathbf{X}'\mathbf{X})^{-1}$ defined before (107).]

13. Show that if in the example of Sec. 6d the reduced model is derived by replacing b_2 by $b_1 - 4$ then the analysis of variance is as follows:

Source	Degrees of Freedom	Sum of Squares
Reduced model	2	1156.7
Error	3	204.65
Total	5	1361.0

14. Show that Table 3.6 reduces to the customary analysis of variance table for testing the hypothesis $b_1 = 0$ in the model $E(y_i) = a + b_1x_{i1} + b_2x_{i2}$; i.e., show that $\text{SSR}_m - Q$ reduces to the SSR_m when fitting $a + b_2x_{i2}$. Use the intercept model.

15. If $\hat{b}_{k+1}$ is the estimated regression coefficient for the $(k + 1)$th independent variable in a model having just $k + 1$ such variables, the corresponding t-statistic for testing the hypothesis $b_{k+1} = 0$ is $t = \hat{b}_{k+1}/\sqrt{\widehat{\text{var}}(\hat{b}_{k+1})}$ where $\widehat{\text{var}}(\hat{b}_{k+1})$ is the estimated variance of $\hat{b}_{k+1}$. Prove that the F-statistic for testing the same hypothesis is identical to t^2.

16. For $\boldsymbol{\lambda}'$ and $\tilde{\mathbf{b}}$ of (48) and (49), $\mathbf{t}'\tilde{\mathbf{b}} = \boldsymbol{\lambda}'\mathbf{y}$ is the unique b.l.u.e. of $\mathbf{t}'\mathbf{b}$. Prove this by assuming that $\mathbf{t}'\tilde{\mathbf{b}} + \mathbf{q}'\mathbf{y}$ is a b.l.u.e. of $\mathbf{t}'\mathbf{b}$ different from $\mathbf{t}'\tilde{\mathbf{b}}$ and showing that $\mathbf{q}'$ is null.

17. Prove that $\boldsymbol{\lambda}' = \mathbf{t}'(\mathbf{X}'\mathbf{V}^{-1}\mathbf{X})^{-1}\mathbf{X}'\mathbf{V}^{-1}$ of (47) minimizes rather than maximizes $w = \boldsymbol{\lambda}'\mathbf{V}\boldsymbol{\lambda} - 2\boldsymbol{\theta}'(\mathbf{X}'\boldsymbol{\lambda} - \mathbf{t})$.

18. Prove that the definitions in (77) and (78) are equivalent to the computing formulae given in (76).

19. Prove the following results for $\hat{\mathbf{e}}$ of an intercept model. What are the analogous results in a no-intercept model?

$$\text{cov}(\hat{\mathbf{e}}, \mathbf{y}) = \mathbf{P}\sigma^2 \quad \text{and} \quad \text{cov}(\hat{\mathbf{e}}, \hat{\mathbf{y}}) = \mathbf{0}_{N\times N};$$

$$\text{cov}(\hat{\mathbf{e}}, \hat{\mathbf{b}}) = \mathbf{0}_{N\times(k+1)}, \quad \text{but} \quad \text{cov}(\mathbf{e}, \hat{\mathbf{b}}) = \mathbf{X}(\mathbf{X}'\mathbf{X})^{-1}\sigma^2;$$

$$\sum_{i=1}^{N} \hat{e}_i y_i = \text{SSE} \quad \text{and} \quad \sum_{i=1}^{N} \hat{e}_i \hat{y}_i = 0.$$

CHAPTER 4

INTRODUCING LINEAR MODELS: REGRESSION ON DUMMY VARIABLES

This chapter begins by describing, in terms of an example, a type of regression analysis that is not recommended. It highlights, however, the advantages of an alternative analysis known as regression on dummy (0, 1) variables. This in turn is a useful precursor to linear models that are not of full rank—the subject of the next chapter.

1. REGRESSION ON ALLOCATED CODES

a. Allocated Codes

The Bureau of Labor Statistics' Consumer Survey 1960–61 reports detailed data about household expenditure habits and the characteristics of each household sampled. Of the many questions that could be asked of such data one is, "To what extent is a household's investment in consumer durables associated with the occupation of the head of household?" Investment behavior is, of course, related to many factors other than occupation, but for purposes of illustration we consider this question just as it stands.

The survey data contain figures on investment in consumer durables (hereafter referred to simply as investment) for some 9,000 families; and for each, the occupation of the head of the household is also recorded, in one of 14 different classes. Suppose the 14 classes are further grouped into 4 categories:

1. Laborer
2. Artisan
3. Professional
4. Self-employed

Also, suppose a regression analysis has been proposed, of investment on occupation, as a means of answering the question posed. A problem immediately arises: how can occupation be measured? One possibility is to "measure" it by the code numbers 1, 2, 3 and 4 listed above. In some sense one might rationalize that these numbers correspond to a measure of occupational status, and how else, it might be asked, can one "measure" occupation recorded in this way in order to investigate the effect of occupation on investment? Accepting these numbers 1, 2, 3 and 4, the procedure would be to carry out a regression analysis of y, investment, on x, which would be 1, 2, 3 or 4 depending on which occupational category the head of the household belonged to. Details of the regression analysis would proceed in the usual fashion using a model

$$E(y_i) = b_0 + b_1 x_i \tag{1}$$

and a test of the hypothesis $b_1 = 0$ could easily be made.

b. Difficulties and criticism

As an analysis procedure, what we have just described is permissible. An inherent difficulty, however, occurs with the definition of x, the independent variable occupational status. Although the 4 categories of occupation represent different kinds of occupation, allocation of the numbers 1, 2, 3 and 4 to these categories as "measures" of occupational status may not accurately correspond to the underlying measure of whatever is meant by occupational status. The allocation of the numbers is, in this sense, quite arbitrary. For example, does a professional man have 3 times as much status as a laborer? If the answer is "no", and a different set of numbers is allocated to the categories, the same kind of criticism can be leveled: whatever the allocation may be it is essentially arbitrary.

This criticism of allocating codes to the categories is not entirely justified so far as the suggested model, namely (1), is concerned. By giving a self-employed person an x-value of 4 we are not really saying he has twice as much status as an artisan (for whom $x = 2$). But, in terms of the model, what we *are* saying is that

$$
\begin{aligned}
&E(\text{investment of a laborer}) &&= b_0 + b_1,\\
&E(\text{investment of an artisan}) &&= b_0 + 2b_1,\\
&E(\text{investment of a professional}) &&= b_0 + 3b_1,\\
\text{and}\quad &E(\text{investment of a self-employed}) &&= b_0 + 4b_1.
\end{aligned}
$$

This means, for example, that

$$
\begin{aligned}
&E(\text{investment of a self-employed}) - E(\text{investment of an artisan})\\
&\quad = E(\text{investment of a professsional}) - E(\text{investment of a laborer}) \qquad (2)\\
&\quad = 2[E(\text{investment of a professional}) - E(\text{investment of an artisan})]\\
&\quad = 2b_1.
\end{aligned}
$$

This, in terms of the real world, may be quite unrealistic. And yet, even without data, allocation of the numbers 1, 2, 3 and 4 forces this consequence on the analysis. The only estimation the analysis will yield will be that of b_1 (and b_0). This will also be the case even if a set of numbers different from 1, 2, 3 and 4 is allocated to the categories: relationships akin to (2) will still apply and, so far as they are concerned, estimation of b_1 will be the only achievement from a regression analysis.

The inherent difficulty with the analysis suggested above is the allocation of codes to non-quantitative variables such as "occupation". Yet such variables are frequently of interest: religion and nationality in the behavioral sciences; species, fertilizer and soil type in agriculture; source of raw material, treatment and plant location in an industrial process; and so on. Allocating codes to these variables involves at least two difficulties: often it cannot be made a reasonable procedure (e.g., allocating codes to "measure" geographical regions of the United States), and in making any such allocation we automatically impose value differences on the categories of the variables in the manner illustrated in equation (2).

c. Grouped variables

These same difficulties also arise with variables that are more measurable than those just considered. Education is an example. It can be measured as the number of years of formal education but then an immediate question is, When does formal education start? Measurement difficulties of this nature can, of course, be avoided by defining education as a series of categories, such as high school incomplete, high school graduate, college graduate, and advanced degree. These are not unlike the categories of occupation discussed earlier although they do have a clear-cut sense of ordinality about them and hence some sense of "measure". However, this would disappear at once were a fifth category "foreign education" to be added. The matter is also further complicated by the subjectivity of decisions that have to be made in classifying people within such categories. For example, how would a man with a foreign education but an American doctorate be classified; or what would be the classification of a college dropout who had subsequently passed the Institute of Flycatchers' examination?

Many instances could be cited where variables are grouped into categories in a manner similar to the education example just given. Income is a common example, with such categories as high, medium, low and poor; city size is another, such as metropolis, large city, city, town and village; and so on. In all these cases it is possible but, for the reasons described, not very rational to impose codes on the categories of independent variables of this nature. This problem is avoided by using the technique of regression on dummy (0, 1) variables. As an analysis procedure it is also more informative than regression on allocated codes because it leads to a larger multiple correlation coefficient (R^2, as defined in Sec. 3.4g; see also Exercise 9 of Chapter 6). Furthermore, it provides from the data estimated values to be associated with categories of the independent variables, rather than allocating codes arbitrarily, regardless of the data. [Illustration of these estimated values and of the larger R^2-values can be found, for example, in Searle and Udell (1970) where regression on allocated codes and regression on dummy variables have both been carried out on the same set of data.]

d. Unbalanced data

Despite the limitations of using allocated codes, an investigator with data to analyze and who has limited training and experience in statistics might well be tempted to use these codes. Armed with a knowledge of regression and of analysis of variance as depicted from the point of view of carefully designed experiments (albeit a good knowledge of these topics), an investigator could easily feel that regression on allocated codes was an appropriate analysis. For example, for 100 people in a (pilot) survey designed to investigate the effect of both occupation and education on investment suppose that the number of people reporting data were distributed as in Table 4.1. Faced with

TABLE 4.1. NUMBER OF PEOPLE, CLASSIFIED ACCORDING TO OCCUPATION AND EDUCATION, WHO REPORTED INVESTMENT DATA

	Education		
Occupation	High School Incomplete	High School Graduate	College Graduate
Laborer	14	8	7
Artisan	10	—	—
Professional	—	17	22
Self-employed	3	9	10

data from people so classified, the choice of an analysis procedure may not, for some investigators, be easy. A patent difficulty with such data is that the numbers of observations in the subclasses of the data are not all the same. Data where these numbers are the same are known as *equal-numbers data* or, more frequently, as *balanced data*. In contrast, those like Table 4.1 with unequal numbers of observations in the subclasses, including perhaps some that contain no observations at all (empty subclasses, or empty cells), are called *unequal-numbers data* or, more usually, *unbalanced data*, or sometimes "messy" data.

Traditional analysis of variance methods, in terms of well-designed experiments, are generally applicable only to balanced data. (Exceptions are the specified patterns of Latin square designs, balanced incomplete block designs and derivatives thereof.) Hence for unbalanced data like those of Table 4.1, analysis of variance in its traditional framework is inapplicable. On the other hand, regression can be used with some degree of propriety by allocating codes to "education" and "occupation". Disadvantages implicit in doing this are incurred, as has just been described, but at least *some* analysis can be conducted, a computer can do the arithmetic and interpretation is straightforward. The possibility that regression on allocated codes may be used must therefore not be ignored. Indeed, in the presence of powerful computer programs for regression analysis, the possibility of its being used has greatly increased.

The preferred analysis is regression on dummy (0, 1) variables. This is so not only because of the advantages already discussed but also because it is identical to established analysis of variance procedures that are available for unbalanced data. As well as being called regression on dummy variables, or analysis of variance for unbalanced data, it is also known as the *method of fitting constants*—fitting the constants, or terms, of a linear model. The calculations involved in this method of analysis are, for unbalanced data, usually more complicated than those of traditional analysis of variance for balanced data, so that prior to the present era of computers there has been limited demand for analyzing unbalanced data. Nowadays, however, in view of the availability of vast computer storage and editing of data we are witnessing a great increase in the demand for analysis of unbalanced data, analysis which *cannot* be made merely by means of minor adjustments to traditional analyses of variance of balanced data. Indeed, the situation is just the opposite: unbalanced data have their own analysis of variance techniques, and those for balanced data are merely special cases of the techniques for unbalanced data. The position is that unbalanced data analyses can be couched in matrix expressions, many of which simplify very little in terms of summation formulas. In contrast, when the numbers of observations in the subclasses are all the same, these matrix expressions simplify considerably. They reduce, in fact, to the well-known summation formulae of traditional analysis of variance of designed experiments, such as randomized complete blocks, factorial experiment designs and others. One can therefore think of such analyses simply as special cases of the more basic analyses of variance for unbalanced data. This is the attitude taken in this book. General analysis procedures are developed in Chapter 5 and applied to specific situations in Chapters 6, 7 and 8—but at all times for unbalanced data. Passing reference is made to simplification of the results in the case of balanced data, but there is little detailed discussion of such cases.

The remainder of this chapter acts as a preface to the development of general linear model theory in the chapter that follows. Since regression on dummy variables is identical to a wide class of linear models, the one serves to introduce the other. Furthermore, although there is widespread use of regression on dummy variables in many fields of application, its equivalence with linear models is not always appreciated, and ramifications of linear model theory are not always adopted by the users of regression on dummy (0, 1) variables. We therefore discuss this first, demonstrate its equivalence to linear models, characterize the description of linear models and thereafter confine our attention to them.

2. REGRESSION ON DUMMY (0, 1) VARIABLES

a. Factors and levels

Discussion of regression on dummy variables is enhanced by the notion of factors and levels, a descriptive terminology that can be usefully adapted from the literature of experimental design.

In studying the effect of the variables "occupation" and "education" on investment behavior, as in Table 4.1, we are interested in the extent to which each category of each variable is associated with investment. Thus we are interested in seeing to what extent a person's being an artisan affects his investment and to what extent someone else's being self-employed affects *his* investment. More particularly, we are interested in investigating the difference between the effects of these two categories in the population of people of whom our data are considered to be a random sample. To acknowledge the immeasurability of the variables and the associated arbitrariness or subjectivity in deciding on their categories (as discussed in the previous section), we introduce the terms "factor" and "level". The word *factor* denotes what has heretofore been called a variable. Thus occupation is one factor, and education is another. The categories into which each factor (variable) has been divided are called *levels* of that factor. Thus laborer is one level of the factor occupation, and professional is another level of that factor. This use of "factor" in place of "variable" emphasizes that what is being called a factor cannot be measured precisely by cardinal values: the word "variable" is reserved for that which can be so measured. Given this interpretation of "variable", investment is the only variable in our investigation. Other elements of the investigation are factors, each with a number of levels. The term "levels" emphasizes that the groupings of a factor are just arbitrary divisions with no imposition of allocated values. It is these that we seek to estimate from data. In this context the ordinal numbers 1, 2, 3 and 4 shown in the list of occupations are no longer values given to the categories of a variable but are used solely to identify levels of factors. For example, level 2 of the occupation factor is artisan.

Thinking in terms of levels of factors rather than groupings of variables overcomes many difficulties inherent in using allocated codes. Even when groupings of a non-quantitative variable have no sense of ordinality, they can still be thought of as levels of a factor; and whenever value differences cannot be placed rationally on the groupings, the concept of levels enables us to estimate differences between the effects that the levels of a factor have on the variable being studied, without any *a priori* imposition of values. This estimation of differences is brought about by regression on dummy (0, 1) variables.

b. The regression

Our aim is to consider the effects of the levels of each factor on investment. We begin by estimating just the effect of education on investment, more particularly, the effect on investment of each of three levels of the factor education shown in Table 4.1. To do this we set up a regression on three independent variables x_1, x_2 and x_3:

$$y_i = b_0 + b_1x_{i1} + b_2x_{i2} + b_3x_{i3} + e_i\,. \qquad (3)$$

In this context y_i is investment and b_0 and e_i are, respectively, the customary constant and error terms found in regression analysis. Corresponding to the x's (the independent variables), which have yet to be defined, are the regression coefficients b_1, b_2 and b_3. Through the manner in which the x's will be defined these b's turn out to be terms that lead to estimates of the differences between the effects on investment of the levels of the factor education.

To define the x's, we note that each person falls into one and only one educational level. Whichever level he is in, let the corresponding x take the value unity and let all other x's for that person have a value of zero. Thus a high school graduate is in level 2 of the education factor, and for him $x_{i2} = 1$, with $x_{i1} = 0$ and $x_{i3} = 0$. In this way numerical values (0's and 1's) can be assigned to all three x's for each person in the data. On these values a regression analysis is carried out.

It is because each x-value is unity when someone belongs to the corresponding level of education, and zero otherwise, that the x's are described as (0, 1) variables: and because they are not true variables in the sense previously defined they are often called "dummy" variables. Despite this, the formal procedures of regression can be carried out, with consequences of great interest.

Example. We suppose that we have investment data on 3 people who did not complete high school, on 2 who did, and on 1 college graduate. These

TABLE 4.2. INVESTMENT INDICES OF 6 PEOPLE

Educational Status	Investment Index
1. (High school incomplete)	y_{11}, y_{12}, y_{13}
2. (High school graduate)	y_{21}, y_{22}
3. (College graduate)	y_{31}

6 observations (investment indices) are shown in Table 4.2, where y_{ij} is the observation on the jth person in the ith level of educational status. Then, with $e_{ij} = y_{ij} - E(y_{ij})$ just as in regression (except for having two subscripts rather than one), we write the observations in terms of (3) as follows:

$$\begin{aligned}
y_{11} &= b_0 + b_1(1) + b_2(0) + b_3(0) + e_{11} \\
y_{12} &= b_0 + b_1(1) + b_2(0) + b_3(0) + e_{12} \\
y_{13} &= b_0 + b_1(1) + b_2(0) + b_3(0) + e_{13} \\
y_{21} &= b_0 + b_1(0) + b_2(1) + b_3(0) + e_{21} \\
y_{22} &= b_0 + b_1(0) + b_2(1) + b_3(0) + e_{22} \\
y_{31} &= b_0 + b_1(0) + b_2(0) + b_3(1) + e_{31}.
\end{aligned}$$

The 1's and 0's in parentheses are the values of the dummy (0, 1) variables. Their pattern can be seen more clearly when the equations are written as

$$\begin{bmatrix} y_{11} \\ y_{12} \\ y_{13} \\ y_{21} \\ y_{22} \\ y_{31} \end{bmatrix} = \begin{bmatrix} 1 & 1 & 0 & 0 \\ 1 & 1 & 0 & 0 \\ 1 & 1 & 0 & 0 \\ 1 & 0 & 1 & 0 \\ 1 & 0 & 1 & 0 \\ 1 & 0 & 0 & 1 \end{bmatrix} \begin{bmatrix} b_0 \\ b_1 \\ b_2 \\ b_3 \end{bmatrix} + \begin{bmatrix} e_{11} \\ e_{12} \\ e_{13} \\ e_{21} \\ e_{22} \\ e_{31} \end{bmatrix} \tag{4}$$

and, by writing

$$\mathbf{y} = \begin{bmatrix} y_{11} \\ y_{12} \\ y_{13} \\ y_{21} \\ y_{22} \\ y_{31} \end{bmatrix}, \quad \mathbf{e} = \begin{bmatrix} e_{11} \\ e_{12} \\ e_{13} \\ e_{21} \\ e_{22} \\ e_{31} \end{bmatrix}, \quad \mathbf{b} = \begin{bmatrix} b_0 \\ b_1 \\ b_2 \\ b_3 \end{bmatrix} \quad \text{and} \quad \mathbf{X} = \begin{bmatrix} 1 & 1 & 0 & 0 \\ 1 & 1 & 0 & 0 \\ 1 & 1 & 0 & 0 \\ 1 & 0 & 1 & 0 \\ 1 & 0 & 1 & 0 \\ 1 & 0 & 0 & 1 \end{bmatrix}, \tag{5}$$

the equations become the familiar form

$$\mathbf{y} = \mathbf{Xb} + \mathbf{e} \tag{6}$$

that has been dealt with so fully in the preceding chapter. On defining the properties of the e-terms in (6) exactly as in regression, namely $\mathbf{e} \sim (\mathbf{0}, \sigma^2\mathbf{I})$, least squares applied to (6) yields the same normal equations as before, $\mathbf{X'X\hat{b}} = \mathbf{X'y}$. Now, however, $\mathbf{X}$ does not have full column rank—as seen in (5), the sum of its last 3 columns equals its first. Thus is a model described as a "*model not of full rank*". Its property is that $\mathbf{X}$ does not have full column rank, with the important consequence that $(\mathbf{X'X})^{-1}$ does not exist and so $\mathbf{X'X\hat{b}} = \mathbf{X'y}$ cannot be solved as $\mathbf{\hat{b}} = (\mathbf{X'X})^{-1}\mathbf{X'y}$. However, by using a generalized inverse of $\mathbf{X'X}$ solutions can be found; but before discussing them, in Chapter 5, we give another example and then describe other aspects of linear models.

Example. Countless experiments are undertaken each year in agriculture and the plant sciences to investigate the effect on growth and yield of various fertilizer treatments applied to different varieties of a species. Suppose we have data from 6 plants, representing 3 varieties being tested in combination with 2 fertilizer treatments. Although the experiment would not necessarily be conducted by growing the plants in varietal rows, it is convenient to visualize the data as in Table 4.3. The entries in the table are such that y_{ijk}

TABLE 4.3. YIELDS OF 6 PLANTS

	Treatment	
Variety	1	2
1	y_{111}, y_{112}	y_{121}
2	y_{211}	y_{221}
3	y_{311}	

represents the yield of the kth plant of variety i that received treatment j. We will now write these out, using 5 dummy (0, 1) variables and 5 regression coefficients corresponding to the 3 varieties and the 2 treatments. The regression coefficients for the 3 varieties will be denoted by α_1, α_2 and α_3 and those for the treatments will be β_1 and β_2. Furthermore, the intercept term in the regression, previously denoted by b_0, will now be written as μ. Thus the vector of parameters $\mathbf{b}'$ will be

$$\mathbf{b}' = [\mu \quad \alpha_1 \quad \alpha_2 \quad \alpha_3 \quad \beta_1 \quad \beta_2].$$

[This notation clearly distinguishes between regression coefficients for varieties (α's) and those for treatments (β's) and, in contrast to using b's as elements of **b**, avoids double subscripting which could then provide that clarity.] With this notation the regression equation for y_{ijk} is

$$y_{ijk} = \mu + \alpha_1 x_{ijk,1} + \alpha_2 x_{ijk,2} + \alpha_3 x_{ijk,3} + \beta_1 x^*_{ijk,1} + \beta_2 x^*_{ijk,2} + e_{ij}$$

where the x's and x^*'s are dummy (0, 1) variables. Thus for the observation on variety 1 and treatment 2, $x_{121,1} = 1$, $x_{121,2} = 0$ and $x_{121,3} = 0$; and $x^*_{121.1} = 0$ with $x^*_{121,2} = 1$. In this way the regression equations for the yields in Table 4.3 are

$$\begin{aligned}
y_{111} &= \mu + \alpha_1(1) + \alpha_2(0) + \alpha_3(0) + \beta_1(1) + \beta_2(0) + e_{111} \\
y_{112} &= \mu + \alpha_1(1) + \alpha_2(0) + \alpha_3(0) + \beta_1(1) + \beta_2(0) + e_{112} \\
y_{121} &= \mu + \alpha_1(1) + \alpha_2(0) + \alpha_3(0) + \beta_1(0) + \beta_2(1) + e_{121} \\
y_{211} &= \mu + \alpha_1(0) + \alpha_2(1) + \alpha_3(0) + \beta_1(1) + \beta_2(0) + e_{211} \\
y_{221} &= \mu + \alpha_1(0) + \alpha_2(1) + \alpha_3(0) + \beta_1(0) + \beta_2(1) + e_{221} \\
y_{311} &= \mu + \alpha_1(0) + \alpha_2(0) + \alpha_3(1) + \beta_1(1) + \beta_2(0) + e_{311}\,.
\end{aligned} \tag{7}$$

Using **y** and **e** to denote the vectors of observations and error terms in the usual way, these equations become

$$\mathbf{y} = \begin{bmatrix} 1 & 1 & 0 & 0 & 1 & 0 \\ 1 & 1 & 0 & 0 & 1 & 0 \\ 1 & 1 & 0 & 0 & 0 & 1 \\ 1 & 0 & 1 & 0 & 1 & 0 \\ 1 & 0 & 1 & 0 & 0 & 1 \\ 1 & 0 & 0 & 1 & 1 & 0 \end{bmatrix} \begin{bmatrix} \mu \\ \alpha_1 \\ \alpha_2 \\ \alpha_3 \\ \beta_1 \\ \beta_2 \end{bmatrix} + \mathbf{e}. \tag{8}$$

Now write

$$\mathbf{X} = \begin{bmatrix} 1 & 1 & 0 & 0 & 1 & 0 \\ 1 & 1 & 0 & 0 & 1 & 0 \\ 1 & 1 & 0 & 0 & 0 & 1 \\ 1 & 0 & 1 & 0 & 1 & 0 \\ 1 & 0 & 1 & 0 & 0 & 1 \\ 1 & 0 & 0 & 1 & 1 & 0 \end{bmatrix} \quad \text{and} \quad \mathbf{b} = \begin{bmatrix} \mu \\ \alpha_1 \\ \alpha_2 \\ \alpha_3 \\ \beta_1 \\ \beta_2 \end{bmatrix}, \tag{9}$$

where **X** is not of full column rank: the sum of columns 2, 3 and 4 equals column 1, as does that of columns 5 and 6. With this proviso equations (8) are $\mathbf{y} = \mathbf{Xb} + \mathbf{e}$ just as before, the equation of a model that is not of full rank. In general, the matrix **X**, having elements that are all 0 or 1, is called an *incidence matrix*, because the presence of the 1's among its elements describes the incidence of the terms of the model (μ, the α's and the β's) in the data.

3. DESCRIBING LINEAR MODELS

a. A 1-way classification

Equations (4) and (8) in the above examples have been developed from the point of view of regression on dummy (0, 1) variables. Consider equations (4) again. They relate to investment indices of 6 people in 3 different levels of educational status, as shown in Table 4.2. Suppose that equations (4) are rewritten as

$$\begin{aligned} y_{11} &= \mu + b_1 + e_{11}, \\ y_{12} &= \mu + b_1 + e_{12}, \\ y_{13} &= \mu + b_1 + e_{13}, \\ y_{21} &= \mu + b_2 + e_{21}, \\ y_{22} &= \mu + b_2 + e_{22}, \\ \text{and} \quad y_{31} &= \mu + b_3 + e_{31}, \end{aligned} \tag{10}$$

where the x's are no longer explicitly shown and μ is written for b_0. Then we see that in each equation of (10) the subscript on the b corresponds exactly to the first subscript on the y; e.g., b_1 is found in y_{11}, y_{12} and y_{13} and b_2 is in y_{21} and y_{22}. Hence each equation of (10) can be written as

$$y_{ij} = \mu + b_i + e_{ij} \tag{11}$$

for the various values that i and j take in the data. In this case $i = 1, 2, 3$ and the upper limit on j in the ith class is the number of observations in that ith class. Denoting this by n_i we have $j = 1, 2, \ldots, n_i$ where $n_1 = 3$, $n_2 = 2$ and $n_3 = 1$. Thus have we developed (11) as the equation of the general linear model for 3 classes; for a classes it applies for $i = 1, 2, \ldots, a$.

Although (11) is the general form of a linear model equation, its specific values are still as shown in (4), exactly as developed in the regression context. Now, however, there is no need to view the elements of **b** as regression coefficients, nor the 0's and 1's of **X** as dummy variables. The elements of **b** can be given meanings in their own rights, and the 0's and 1's of **X** relate to "absence" and "presence" of levels of factors.

Since μ enters into every equation in (10) it is described as the general mean of the population of investment indices. It represents some overall mean regardless of educational status.

To give meaning to the b's consider b_1: in equations (10) [or (4), they are equivalent] b_1 occurs in only those equations pertaining to investment indices of people of educational status 1 (high school incomplete), namely y_{11}, y_{12} and y_{13}. Similarly for b_2: in (10) it occurs only in the equations for people of educational status 2, y_{21} and y_{22}. Likewise b_3 is in the equation for y_{31} and nowhere else. Thus b_1 gets described as the effect on investment of a person's being of educational status 1: similar descriptions apply to b_2 and b_3. In general, in terms of (11), b_i is described as the *effect* on investment due to educational status i.

Description of a linear model is incomplete without specifying distributional properties of the random error terms, the e_{ij}'s evident in equations (4), (10) and (11). This is usually done by attributing to them the same kind of properties as in regression analysis [see equations (5), (6) and (7) in Sec. 3.1b]. Thus e_{ij} is defined as $e_{ij} = y_{ij} - E(y_{ij})$ and so $E(e_{ij}) = 0$, giving

$$E(y_{ij}) = \mu + b_i .$$

The variance of each e_{ij} is defined as σ^2 and so

$$v(e_{ij}) = E[e_{ij} - E(e_{ij})]^2 = E(e_{ij}^2) = \sigma^2 \qquad \text{for all } i \text{ and } j.$$

Furthermore, covariances between all pairs of different e's are taken to be zero, so that $\text{cov}(e_{ij}, e_{i'j'}) = 0$ unless $i = i'$ and $j = j'$ in which case the covariance becomes the variance σ^2. Thus

$$\text{var}(\mathbf{e}) = \sigma^2\mathbf{I}.$$

The general description of the 1-way classification model can therefore be summarized as follows. For y_{ij} being the jth observation in the ith class, the equation of the model is (11):

$$y_{ij} = \mu + b_i + e_{ij} .$$

μ is the general mean, b_i is the effect on y_{ij} due to the ith class and e_{ij} is a random error term peculiar to y_{ij} with

$$\mathbf{e} \sim (\mathbf{0}, \sigma^2\mathbf{I}).$$

For a classes, $i = 1, 2, \ldots, a$ and $j = 1, 2, \ldots, n_i$ for the ith class. The additional assumption of normality is made when hypothesis testing and confidence intervals are considered: i.e., we then assume

$$\mathbf{e} \sim N(\mathbf{0}, \sigma^2\mathbf{I}).$$

b. A 2-way classification

Suppose equations (7) are rewritten with the x's no longer explicitly shown, just as were equations (4) in the preceding example, in (10). Then (7) becomes

$$\begin{aligned} y_{111} &= \mu + \alpha_1 + \beta_1 + e_{111} \\ y_{112} &= \mu + \alpha_1 + \beta_1 + e_{112} \\ y_{121} &= \mu + \alpha_1 + \beta_2 + e_{121} \\ y_{211} &= \mu + \alpha_2 + \beta_1 + e_{211} \\ y_{221} &= \mu + \alpha_2 + \beta_2 + e_{221} \\ y_{311} &= \mu + \alpha_3 + \beta_1 + e_{311}. \end{aligned} \tag{12}$$

Here, in each equation, the subscripts on α and β correspond respectively to the first two on y: α_1 and β_1 are found in y_{111} and y_{112}, and α_2 and β_1 are in y_{211}. Hence each equation in (12) can be written as

$$y_{ijk} = \mu + \alpha_i + \beta_j + e_{ijk}\,. \tag{13}$$

The values taken by i, j and k in the data are, in this case, $i = 1, 2, 3$ and $j = 1, 2$ with the upper limit of k being the number of observations of the ith variety receiving the jth treatment. Denoting this by n_{ij} we have $k = 1, 2, \ldots, n_{ij}$ where $n_{11} = 2$, $n_{12} = 1$, $n_{21} = 1$, $n_{22} = 1$, $n_{31} = 1$ and $n_{32} = 0$. Thus (13) is the equation of the general linear model involving varieties and treatments.

As with the 1-way classification of the preceding section so here, the elements of **b** (in this case μ, the α's and β's) do not need to be viewed as regression coefficients but can be given meanings in their own rights. First μ: it is described as the mean of the whole population of yields, representing some overall mean yield regardless of variety or treatment. Second, the α's: in equations (12) [or (7), they are equivalent] α_1 occurs in only those equations pertaining to yields of variety 1, namely y_{111}, y_{112} and y_{121}. Similarly for α_2: in (12) it occurs only in the equations of yields of variety 2, y_{211} and y_{221}. Likewise α_3 is in the equation for y_{311} and nowhere else. Thus α_1 gets described as the effect on yield of a plant's being of variety 1; similar descriptions apply to α_2 and α_3. In general, α_i is described as the effect on yield due to variety i. Likewise the β's: β_1 occurs only in equations of yields that received treatment 1, y_{111}, y_{112}, y_{211} and y_{311}; and β_2 is in only the equations pertaining to treatment 2, those for y_{121} and y_{221}. Thus β_j is described as the effect on yield due to treatment j. Hence general description of the β's is similar to that of the α's; both are effects on yield, but whereas the α's are effects due to variety, the β's are effects due to treatment.

The error terms in this model, the e_{ijk}, are assumed to have exactly the same properties as before: i.e., if **e** is the vector of the e_{ijk}, then we assume that

$\mathbf{e} \sim (\mathbf{0}, \sigma^2\mathbf{I})$, with the additional assumption of normality for hypothesis testing and confidence intervals.

Apart from μ and e_{ijk}, equation (13) has terms for just two factors, which can be referred to generally as an α-factor and a β-factor. The model for which (13) is the equation could therefore be called a 2-factor model, although the name 2-way classification is more firmly established. Its general description is as follows. For y_{ijk} being the kth observation on the ith level of the α-factor and the jth level of the β-factor the equation of the model is (13):

$$y_{ijk} = \mu + \alpha_i + \beta_j + e_{ijk}.$$

μ is the general mean, α_i is the effect on y_{ijk} due to the ith level of the α-factor, β_j is the effect due to the jth level of the β-factor and e_{ijk} is a random error term peculiar to y_{ijk} with

$$\mathbf{e} \sim (\mathbf{0}, \sigma^2\mathbf{I}).$$

When the α-factor has a levels, $i = 1, 2, \ldots, a$; and for the β-factor having b levels $j = 1, 2, \ldots, b$; and $k = 1, 2, \ldots, n_{ij}$ for n_{ij} observations in the (i, j) cell or subclass—the "intersection" of the ith level of the α-factor and the jth level of the β-factor. And, of course, for hypothesis testing and confidence intervals we further assume

$$\mathbf{e} \sim N(\mathbf{0}, \sigma^2\mathbf{I}).$$

The example of this model described here is from agriculture, but the same kind of model can apply to other situations involving 2 factors. Thus for the example of Table 4.1 concerning the effect of occupation and education on investment, equation (13) could act equally as well as it could for the agricultural example. α_i would then be the effect on investment of the ith occupation category (the ith level of the occupation factor), and β_j would be the effect of the jth level of the education factor.

Similarities between the above description of the 2-way classification and that of the 1-way classification at the end of the preceding section will be clearly apparent. They extend quite naturally to many-factored models. The following outline of a 3-way classification illustrates this.

c. A 3-way classification

Suppose that for the data of Table 4.1 the hometown region of the United States (Northeast, South, Midwest, Southwest, Rockies or West Coast) was also recorded for each person. Then a study of the effects on investment of occupation, education and region could be made using a model whose equation is

$$y_{ijkh} = \mu + \alpha_i + \beta_j + \gamma_k + e_{ijkh} \tag{14}$$

where y_{ijkh} is the investment index of the hth person in the ith occupation and jth level of education living in the kth region. μ is the general mean, α_i is the effect on investment due to the ith occupation, β_j is the effect due to the jth level of education and γ_k is the effect due to the kth region. As usual, e_{ijkh} is an error term peculiar to y_{ijkh}, and we assume $\mathbf{e} \sim (\mathbf{0}\ \sigma^2\mathbf{I})$. If in the data there are a levels of occupation, then $i = 1, 2, \ldots, a$; for b levels of education, $j = 1, 2, \ldots, b$; and for c regions, $k = 1, 2, \ldots, c$. And $h = 1, 2, \ldots, n_{ijk}$, for n_{ijk} observations in the subclass of the data represented by the ith occupation, the jth level of education and the kth region.

Extension of models of this nature to 4-way and higher-ordered classifications is clear.

d. Main effects and interactions

(i) *Main effects.* The α's, β's and γ's of the preceding examples each represent the effect on y of one level of one factor. Thus, in the 2-way classification of Table 4.3, α_i of equation (13) refers to the effect on yield of the ith level of the factor variety: i.e., of variety i. And β_j in the same equation refers to the effect on yield of treatment j. Effects of this nature that pertain to a single level of a factor are called *main effects*. This is logical: the effects of variety and treatment on yield are the effects in which our main interest lies. Hence the elements of the model that correspond to them are called the main effects of the model.

By its very nature, the equation of the model implies that the effect α_i is added to the effect β_j in conjecturing the expected value of y_{ijk} as being

$$E(y_{ijk}) = \mu + \alpha_i + \beta_j \,. \tag{15}$$

This means that the total effect of variety i and treatment j on expected yield is considered as being the sum of the two individual effects α_i and β_j. For this reason the effects are described as being *additive*. The model also means that the effect of variety i on expected yield is considered as being the same, no matter what treatment is used on it. For all treatments the effect of variety i is assumed to be α_i, and the combined effect of variety i and treatment j (over and above μ) is taken to be $\alpha_i + \beta_j$.

Suppose values of μ, the α_i's and β_j's are as follows:

$$\begin{array}{llll} \mu = 4, & \alpha_1 = 1 & \text{and} & \beta_1 = 4 \\ & \alpha_2 = 3 & & \beta_2 = 7. \\ & \alpha_3 = 2 & & \end{array} \tag{16}$$

These are hypothetical values of the elements (μ, the α's and β's) of the model (15), introduced for the sake of illustration. They are not observed values. Indeed, these elements can never be observed, and in practice they are never known, for they are population values which can only be estimated from available data. However, for purposes of illustrating certain aspects of linear models it is instructive to give arithmetic values to these elements so that consequences thereof may be portrayed graphically. For example, with the assumed values of (16)

$$E(y_{11k}) = \mu + \alpha_1 + \beta_1 = 4 + 1 + 4 = 9. \tag{17}$$

This is not an observed value of $E(y_{11k})$ or of y_{11k} itself: it is an assumed value of $E(y_{11k})$ based on the assumed values of the parameters given in (16).

First note that (15) for a given i and j is the same for all k. Since the subscript k is merely the identifier of individual observations in the (i, j) subclass (15) means that the expected value of every observation in that subclass is the same. Thus, by (17), the expected value of every observation in the (1, 1) cell is, in our hypothetical example, 9; i.e., for all $k = 1, 2, \ldots, n_{11}$, $E(y_{11k}) = 9$. With this interpretation the expected values for other subclasses derived from (16) are those shown in Table 4.4 and plotted in Figure 4.1.

TABLE 4.4. EXPECTED VALUES OF A NO-INTERACTION MODEL. EQUATIONS (16) SUBSTITUTED IN (15). (SEE FIGURES 4.1 AND 4.3.)

	Treatment	
Variety	1	2
1	$E(y_{11k}) = 4 + 1 + 4 = 9$	$E(y_{12k}) = 4 + 1 + 7 = 12$
2	$E(y_{21k}) = 4 + 3 + 4 = 11$	$E(y_{22k}) = 4 + 3 + 7 = 14$
3	$E(y_{31k}) = 4 + 2 + 4 = 10$	$E(y_{32k}) = 4 + 2 + 7 = 13$

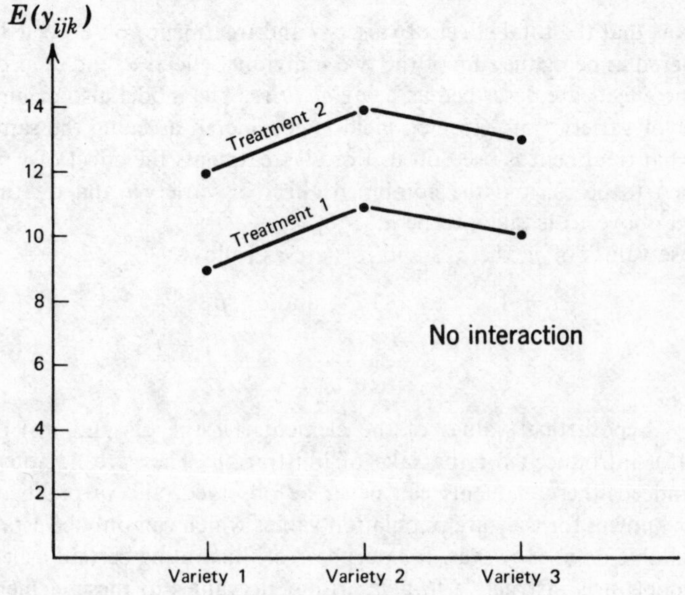

Figure 4.1. Expected values of Table 4.4.

In Figure 4.1 it is to be noted that the "variable" of the abscissa, variety number, is not a continuous variable. Therefore lines joining values of $E(y_{ijk})$ in no way indicate a continuous change in $E(y_{ijk})$ from one variety to the next. The lines are shown merely to emphasize the trend in the change, and they are used in similar fashion in Figures 4.2, 4.3 and 4.4. Furthermore, the ordinates plotted in these figures are values of $E(y_{ijk})$ and not of actual observations y_{ijk}. With this in mind, it is clear from Figure 4.1 that in the hypothetical example of the model given in (15) the effect of variety is the same regardless of treatment. For *both* treatments, variety 2 has an expected yield two units larger than does variety 1; and for both treatments variety 3 is one unit lower than variety 2.

(*ii*) *Interactions.* In some other hypothetical example suppose that the plots of expected yields are those shown in Figure 4.2. The difference between

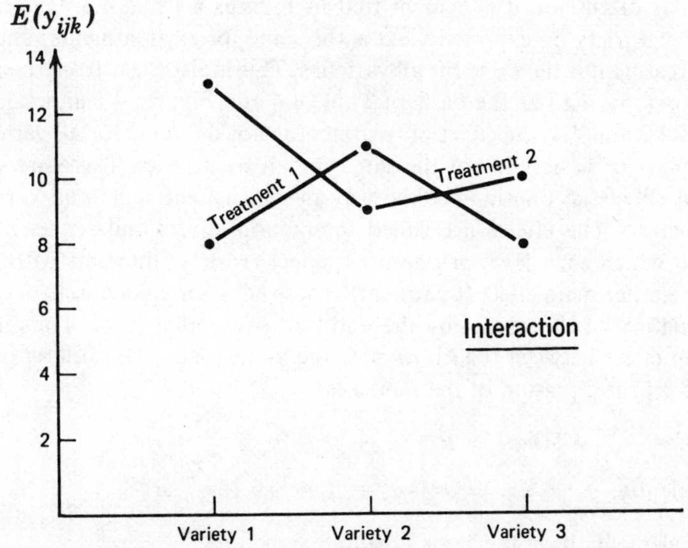

Figure 4.2. Expected values for an interaction model (see Table 4.5).

this and Figure 4.1 is obvious: the lines for the two treatments are not parallel. This indicates that the effect of variety is different for the different treatments. With treatment 1, variety 2 is three units *larger* (in expected yield) than is variety 1 with the same treatment, but for treatment 2, variety 2 is four units *smaller* than variety 1. Thus in this second hypothetical example the varieties are acting differently according to which treatment is used. We say that varieties are "interacting" with treatments. The extent to which they are not acting in the same manner for each treatment is termed an "interaction".

This discussion can also be put another way. In Figure 4.1 the difference between treatments is the same for each variety; it does not change from variety to variety but is constant over all varieties. This is evidenced by parallelism of the lines in Figure 4.1. On the other hand, the lack of parallelism in Figure 4.2 indicates that the differences between the treatments differ from variety to variety. Thus the difference "treatment 1 minus treatment 2" is -5, $+2$ and -2 for the three varieties respectively, whereas in Figure 4.1 it is -3 for every variety. This difference between the two hypothetical examples is well illustrated when they are plotted as in Figures 4.3 and 4.4.

The parallel lines of Figure 4.3 (corresponding to those of Figure 4.1) illustrate, for the first hypothetical example (Table 4.4), the uniform difference between treatments of all varieties. But in Figure 4.4 the non-parallel lines illustrate, for the second hypothetical example, the lack of uniformity in the differences between treatments over all varieties.

From this discussion it is evident that in Figures 4.1 and 4.3 (Table 4.4) the effect of variety on expected yield is the same for all treatments; and the effect of treatment is the same for all varieties. This is also clear from the form of equation (15), used as the basis of Table 4.4 and Figures 4.1 and 4.3. But in Figures 4.2 and 4.4, the effect of treatment is not the same for all varieties, and the effect of variety is not the same for all treatments. There are some additional effects accounting for the way in which treatments and varieties are interacting. The effects are called interaction effects and represent the manner in which each level of one main effect (variety) interacts with each level of the other main effect (treatment). These effects are taken into account in the equation of the model by the addition of another term. Thus if the interaction effect between the ith level of the α-effect and the jth level of the β-effect is γ_{ij} the equation of the model is

$$E(y_{ijk}) = \mu + \alpha_i + \beta_j + \gamma_{ij} \tag{18}$$

or, equivalently,

$$y_{ijk} = \mu + \alpha_i + \beta_j + \gamma_{ij} + e_{ijk}\,. \tag{19}$$

All other elements have the same meaning as before.

The second hypothetical example (plotted in Figures 4.2 and 4.4) is based on the same hypothetical values for μ, the α's and β's given in (16) together with the following hypothetical values for the interaction effects γ_{ij}:

$$\begin{aligned} \gamma_{11} &= -1 & \gamma_{21} &= 1 \\ \gamma_{12} &= 0 & \gamma_{22} &= -5 \\ \gamma_{13} &= -2 & \gamma_{31} &= -3. \end{aligned} \tag{20}$$

In this way the expected values derived from (18) are those shown in Table 4.5 and plotted in Figures 4.2 and 4.4.

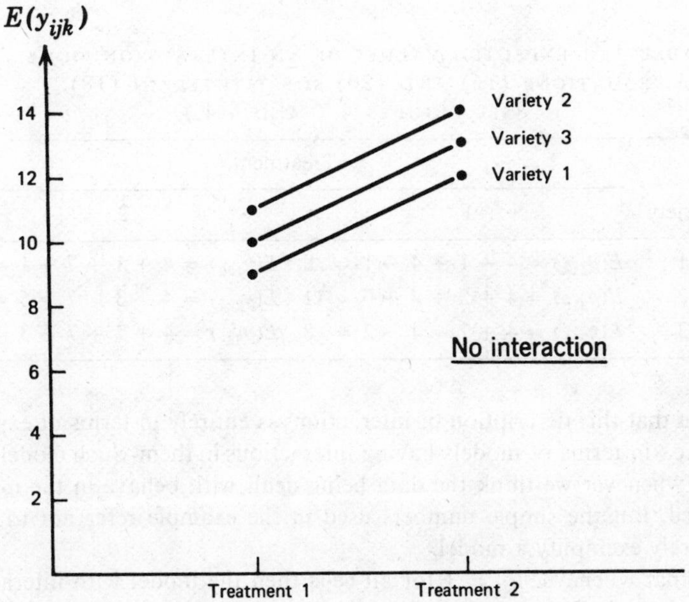

Figure 4.3. Expected values of Table 4.4 (see also Figure 4.1).

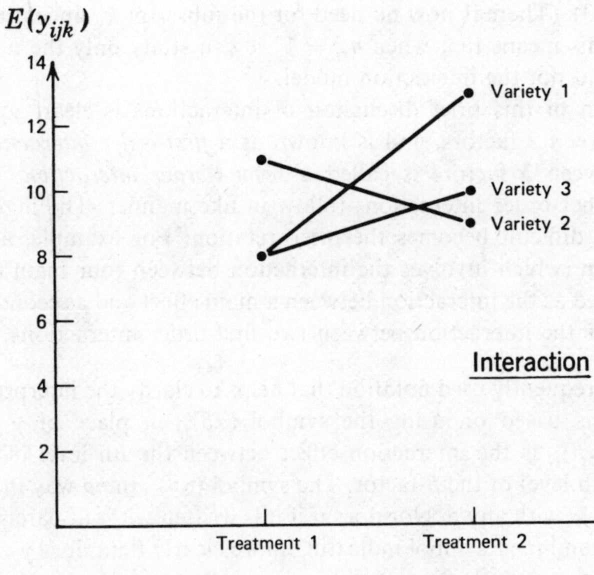

Figure 4.4. Expected values for an interaction model (see Table 4.5 and also Figure 4.2).

TABLE 4.5. EXPECTED VALUES OF AN INTERACTION MODEL. EQUATIONS (16) AND (20) SUBSTITUTED IN (18). (SEE FIGURES 4.2 AND 4.4.)

	Treatment	
Variety	1	2
1	$E(y_{11k}) = 4 + 1 + 4 - 1 = 8$	$E(y_{12k}) = 4 + 1 + 7 + 1 = 13$
2	$E(y_{21k}) = 4 + 3 + 4 + 0 = 11$	$E(y_{22k}) = 4 + 3 + 7 - 5 = 9$
3	$E(y_{31k}) = 4 + 2 + 4 - 2 = 8$	$E(y_{32k}) = 4 + 2 + 7 - 3 = 10$

Notice that this description of interactions is entirely in terms of expected yields, i.e., in terms of models having interactions in them. Such models may be used whenever we think the data being dealt with behave in the manner illustrated. But the simple numbers used in the example refer not to data; they merely exemplify a model.

Note that whenever $n_{ij} = 1$ for all cells then the model with interaction, (19), becomes indistinguishable from the model without interaction, (13). The γ_{ij} and e_{ijk} terms of (19) get combined, $\gamma_{ij} + e_{ijk} = \epsilon_{ij}$ say, and so (19) becomes

$$y_{ij} = \mu + \alpha_i + \beta_j + \epsilon_{ij},$$

equivalent to (13). (There is now no need for the subscript k, since it is unity for all cells.) This means that when $n_{ij} = 1$ we can study only the no-interaction model and not the interaction model.

Generalization of this brief discussion of interactions is clear. γ_{ij} is an interaction between 2 factors, and is known as a *first-order interaction*. An interaction between 3 factors is called a *second-order interaction*. Third-, fourth- and higher-order interactions follow in like manner. The higher the order the more difficult becomes the interpretation. For example, a third-order interaction (which involves the interaction between four main effects) can be interpreted as the interaction between a main effect and a second-order interaction or as the interaction between two first-order interactions.

Notation. A frequently used notation that helps to clarify the interpretation of interactions is based on using the symbol $(\alpha\beta)_{ij}$ in place of γ_{ij}. This indicates that $(\alpha\beta)_{ij}$ is the interaction effect between the ith level of the α-factor and the jth level of the β-factor. The symbol $(\alpha\beta)_{ij}$ in *no* way indicates a product of any α with any β. Nor does it if it is written without parentheses, as $\alpha\beta_{ij}$. It is a combined symbol indicating more clearly than does γ_{ij} that it represents an interaction between levels of an α-factor and a β-factor. By this means a high-order interaction, $(\alpha\beta\gamma\delta)_{hijk}$ for example, is readily interpreted: as the interaction between α_h and $(\beta\gamma\delta)_{ijk}$, or as the interaction between $(\alpha\beta)_{hi}$ and $(\gamma\delta)_{jk}$, or as one of many other interpretations. This notation also clarifies the writing of a model: e.g.,

$$y_{ijkm} = \mu + \alpha_i + \beta_j + \gamma_k + \tau_{ij} + \rho_{ik} + \theta_{jk} + \varphi_{ijk} + e_{ijkm}$$

is not as readily comprehended as is

$$y_{ijkm} = \mu + \alpha_i + \beta_j + \gamma_k + (\alpha\beta)_{ij} + (\alpha\gamma)_{ik} + (\beta\gamma)_{jk} + (\alpha\beta\gamma)_{ijk} + e_{ijkm}.$$

Finally, even when a model has interactions its order is still described by the number of main effect factors it has. Thus (18) is an equation for a 2-way classification, just as is (13). But (18) includes interactions and (13) does not.

e. Nested and crossed classifications

In the example of Table 4.3 every treatment is applied to every variety. True, there are no observations on the use of treatment 2 with variety 3, but the feasibility of this combination is not precluded by the absence of data. A situation of this nature is described as a *crossed classification*. This means that every level of every factor could be used in combination with every level of every other factor: in this way the factors "cross" each other; their "intersections" are the subclasses or cells of the situation, wherein data arise. Absence of data from a cell does not imply non-existence of that cell, only that it has no data. The total number of cells in a crossed classification is the product of the number of levels of the various factors; i.e., ab in a 2-way classification. Not all of these may have observations in them; suppose s of them do. Then the total number of observations is the sum of the numbers of observations in these s cells.

Example. Suppose at a university a student survey is carried out to ascertain the reaction to instructors' usage of a new computing facility that provides typewriter terminals in the classroom. We suppose that all freshmen have to take English or Geology or Chemistry in their first semester (and one other of these courses in their second semester). All three courses in the first semester are large and are divided into sections, each section with a different instructor and not all sections necessarily having the same number of students. In the survey, the response provided by each student is opinion (measured on a scale of 1 through 10) of his instructor's use of the computer. Based on these data the questions of interest are, Do the instructors differ in their use and is the use of the computer affected by the subject matter being taught?

A possible model for this situation would include a general mean μ and main effects α_1, α_2 and α_3 for the three types of courses. It would also include terms for the sections of each course. Suppose for the moment there are 10 sections for each course, and that we try to use a model $y_{ijk} = \mu + \alpha_i + \beta_j + e_{ijk}$ for $i = 1, 2, 3$, $j = 1, 2, \ldots, 10$ and $k = 1, 2, \ldots, n_{ij}$ where n_{ij}

is the number of students in section j of course i. Consider β_j. It represents the effect of section j: and for $j = 1$ say, it would be the effect for section 1 of the English course, of the Geology course and of the Chemistry course. This is meaningless because these three sections, composed of different groups of students, have nothing in common other than that they are all numbered 1 in their respective courses. But, assuming students in all courses have been allocated randomly to their sections, this numbering is purely for identification purposes; it indicates nothing in common about the three sections that are numbered 1. Neither is there anything in common about the three sections that are numbered 5, or 6, or any other number. They are not like the treatments in the agricultural example, where treatment 1 on variety 1 was the same as treatment 1 on variety 2 and on variety 3. The sections are not related in this way; they are identities within their own courses. Thus we refer to them as sections within courses, and describe them as being *nested* within courses. Thus sections is a nested factor, or a *nested classification*, sometimes also referred to as a *hierarchical* classification.

The difference between a crossed classification and a nested classification is exemplified in Table 4.6, in terms of the variety and treatment example described earlier and the sections-within-courses example just discussed

TABLE 4.6. SCHEMATIC REPRESENTATION OF A CROSSED CLASSIFICATION AND A NESTED CLASSIFICATION

A Crossed Classification			A Nested Classification		
Treatment			Course		
Variety	1	2	English	Geology	Chemistry
1			Sec. 1 of English	Sec. 1 of Geology	Sec. 1 of Chemistry
2			Sec. 2 of English	Sec. 2 of Geology	Sec. 2 of Chemistry
3			Sec. 3 of English		Sec. 3 of Chemistry
					Sec. 4 of Chemistry

In the crossed classification variety 1 is used in combination with both treatment 1 and treatment 2 and it is the same variety on both occasions. In the nested classification section 1 of English is in no way related to section 1 of Geology. The only thing in common between the two is the number 1, which is purely an identifier. In the crossed classification every level of the one factor is used in combination with every level of the other factor, but in the nested classification the levels of the nested factor (sections) are unrelated to one another and are nested within a level of the other factor. Further, as seen here, there may be different numbers of levels of the nested factor within each level of the other factor (different numbers of sections in the different courses).

The equation of the model accounts for the nesting of sections within courses by giving to the effect β_j for the jth section the subscript i, for course, so that β_{ij} is then the effect for the jth section nested within the ith course. This signifies that the jth section cannot be defined alone but only in the context of which course it belongs to. Thus the model is

$$y_{ijk} = \mu + \alpha_i + \beta_{ij} + e_{ijk} \tag{21}$$

where y_{ijk} is the opinion of student k in the jth section of course i. The limits of k are $k = 1, 2, \ldots, n_{ij}$ where there are n_{ij} students in the jth section of the ith course, and $j = 1, 2, \ldots, b_i$ where there are b_i sections in course i, and $i = 1, 2, 3$. Table 4.7 summarizes a situation of a total of 263 students in 3 sections in English, 2 in Geology and 4 in Chemistry.

TABLE 4.7. A NESTED CLASSIFICATION

English ($i = 1$) 3 sections, $b_1 = 3$	Geology ($i = 2$) 2 sections, $b_2 = 2$	Chemistry ($i = 3$) 4 sections, $b_3 = 4$
$n_{11} = 28$	$n_{21} = 31$	$n_{31} = 27$
$n_{12} = 27$	$n_{22} = 29$	$n_{32} = 32$
$n_{13} = 30$		$n_{33} = 29$
		$n_{34} = 30$

The situation illustrated in Table 4.7 is described as a 2-way nested classification: sections within courses. Now consider asking a student his opinion on two different occasions. If y_{ijkh} is the hth reply ($h = 1$ or 2) of the kth student in section j of course i, a suitable model might be

$$y_{ijkh} = \mu + \alpha_i + \beta_{ij} + \gamma_{ijk} + e_{jjkh}\,. \tag{22}$$

Now we have not only sections nested within courses but also students nested within sections. For exactly the same reason that sections are nested within courses so also are students nested within sections. They cannot be a crossed classification. This is an example of a 3-way nested classification: students nested within sections within courses. In general there is no limit to the degree of nesting that can be handled: the extent of its use depends entirely on the data and the environment from which they came.

Notation. The meaning of the term γ_{ij} in (19) might, at first sight, be confused with the meaning of β_{ij} in (21), although the context of each does make their meaning clear. By the presence of α_i and β_j in (19), γ_{ij} is clearly an interaction effect; and by the lack of a term with just a j-subscript from (21), it is clear that β_{ij} is an effect for a nested factor. However, additional clarity can be brought to the situation by using the $(\alpha\beta)_{ij}$-notation for interactions (as already described), for then β_{ij} is clearly a main effect. Similar clarity is also gained by using $\beta_{(i)j}$ instead of β_{ij} for the nested effect. This makes it clear that $\beta_{(i)j}$ is not an interaction effect like γ_{ij}. Either (or both) of these notations can be used to insure against confusion.

Interaction effects are effects peculiar to specific combinations of the factors involved. Thus $(\alpha\beta)_{ij}$ is the interaction effect peculiar to the combination of the ith level of the α-factor with the jth level of the β-factor. Interactions between a factor and one nested within it cannot, therefore, exist. This is so because, for example, when sections are nested within courses they are defined only within that context—there is no such thing as a section factor in which the identically same level occurs in combination with the levels of the course factor; e.g., section 1 as defined for English never occurs in combination with Chemistry (which has its own section 1), and so there is no such thing as an interaction between course and sections nested within courses. The notation of (21) makes this quite clear: the interaction between α_i and β_{ij} would be $(\alpha\beta)_{ij}$ which cannot be identified separately from β_{ij}. Therefore there is no interaction.

Nested and crossed classifications are by no means mutually exclusive. Both can occur in the same model. For example, in using (22) as the model for the repeated surveying of the students we are ignoring the fact that the two surveys (assumed to be conducted with the same questionnaire) will have been made at different times. If the time element is to be included a suitable model could be

$$y_{ijkh} = \mu + \delta_h + \alpha_i + \beta_{ij} + \gamma_{ijk} + e_{ijkh} \tag{23}$$

where all terms are the same as previously, with the addition of δ_h, the effect of time h. The δ-factor (time) and the α-factor (courses) are crossed factors; each level of the one occurs with every level of the other. And, as before, the β-factor (sections within courses) and the γ-factor (students within sections) are nested factors. Interactions could be included in a model for this situation, too; thus the model

$$y_{ijkh} = \mu + \delta_h + \alpha_i + (\alpha\delta)_{ih} + \beta_{ij} + \gamma_{ijk} + e_{ijkh} \tag{24}$$

includes a term for the interaction between time and course.

Clearly the variations that may be rung on the same theme are very numerous. Just exactly what goes into a model depends, of course, on the nature of the data to be analyzed, the things of interest to the researcher and the assumptions he is prepared to make. For example, if time is to be ignored, either by assumption or because it is known to be of no importance, then (22) would be an acceptable model. Even so, it might be questioned whether or not we truly know that something is of no importance, and in this light maybe model (23) or (24) should be used. On the other hand, if the second student survey had been carried out following a major modification to the computer system designed to improve its efficiency and attractiveness to instructors, then there is no question that (22) would be an unsuitable model compared to (23) or (24). δ_1 and δ_2 would then represent the effects of the two computer systems, unmodified and modified. On all occasions the environment in which the data were gathered determines the model.

In conclusion it is to be emphasized that all these kinds of models can be written as $\mathbf{y} = \mathbf{Xb} + \mathbf{e}$ just as they were in equations (4) and (8). For all of them $\mathbf{X}$ will have 0's and 1's for its elements and not be of full column rank. But for all these models the estimation procedures of the previous chapter can be used to derive normal equations $\mathbf{X}'\mathbf{X}\hat{\mathbf{b}} = \mathbf{X}'\mathbf{y}$. In these, $\mathbf{X}'\mathbf{X}$ does not have full rank, but the equations can be solved using a generalized inverse of $\mathbf{X}'\mathbf{X}$. This and its consequences are discussed in detail in the next chapter. As prelude we consider a numerical example to illustrate some points involved.

4. THE NORMAL EQUATIONS

The equation of the general linear model is $\mathbf{y} = \mathbf{Xb} + \mathbf{e}$, identical to that used for regression analysis in the preceding chapter. There, the normal equations for estimating $\mathbf{b}$ were written as $\mathbf{X}'\mathbf{X}\hat{\mathbf{b}} = \mathbf{X}'\mathbf{y}$, where $\hat{\mathbf{b}}$ was the estimator of $\mathbf{b}$. The same kind of normal equations can be used here. However, we now write them as $\mathbf{X}'\mathbf{X}\mathbf{b}^o = \mathbf{X}'\mathbf{y}$. This is done because, as will be shown, these equations have no single solution for $\mathbf{b}^o$. $\mathbf{X}'\mathbf{X}$ is singular and so there are infinitely many solutions. No one of them is an estimator of $\mathbf{b}$ in the sense that $\hat{\mathbf{b}}$ is in regression analysis, and so we introduce the symbol $\mathbf{b}^o$. It represents a solution to the normal equations, but it is not an estimator of $\mathbf{b}$. This point is emphasized repeatedly in the next chapter and is illustrated here as introduction thereto.

Suppose the values of the 6 observations in Table 4.2 are, in some appropriate unit of measurement,

$$\mathbf{y}' = [16 \quad 10 \quad 19 \quad 11 \quad 13 \quad 27].$$

Comparable to $\mathbf{b}$ in (5) we now use

$$\mathbf{b}' = [\mu \quad \alpha_1 \quad \alpha_2 \quad \alpha_3]$$

where μ is a general mean and the α's are effects due to educational status (see Table 4.2). Then, with $\mathbf{X}$ of (5), the normal equations are

$$\begin{bmatrix} 6 & 3 & 2 & 1 \\ 3 & 3 & 0 & 0 \\ 2 & 0 & 2 & 0 \\ 1 & 0 & 0 & 1 \end{bmatrix} \begin{bmatrix} \mu^o \\ \alpha_1^o \\ \alpha_2^o \\ \alpha_3^o \end{bmatrix} = \begin{bmatrix} 96 \\ 45 \\ 24 \\ 27 \end{bmatrix} \tag{25}$$

equivalent to

$$\begin{aligned} 6\mu^o + 3\alpha_1^o + 2\alpha_2^o + \alpha_3^o &= 96 \\ 3\mu^o + 3\alpha_1^o \qquad\qquad\quad &= 45 \\ 2\mu^o \qquad\quad + 2\alpha_2^o \qquad &= 24 \\ \mu^o \qquad\qquad\qquad + \alpha_3^o &= 27. \end{aligned}$$

The next chapter discusses the derivation of equations such as these. All we note here is that the sum of the last three equals the first and hence they have infinitely many solutions. Four are shown in Table 4.8.

TABLE 4.8. FOUR SOLUTIONS, $\mathbf{b}_1^o$, $\mathbf{b}_2^o$, $\mathbf{b}_3^o$ AND $\mathbf{b}_4^o$, TO EQUATIONS (25)

Element of Solution	Solution			
	$\mathbf{b}_1^o$	$\mathbf{b}_2^o$	$\mathbf{b}_3^o$	$\mathbf{b}_4^o$
μ^o	16	14	27	−2982
α_1^o	−1	1	−12	2997
α_2^o	−4	−2	−15	2994
α_3^o	11	13	0	3009

The differences between the same elements of the four solutions shown in Table 4.8 make it crystal clear why no solution $\mathbf{b}^o$ can be considered an estimator of $\mathbf{b}$. For this reason $\mathbf{b}^o$ is always referred to as a solution of normal equations and never as an estimator. The notation $\mathbf{b}^o$ emphasizes this, distinguishing it from $\hat{\mathbf{b}}$ and $\tilde{\mathbf{b}}$ of equations (21) and (103) in the preceding chapter.

An investigator having data to be analyzed will clearly have no use for any $\mathbf{b}^o$ as it stands, whatever its numerical value. But what about linear functions of the elements of $\mathbf{b}^o$? Suppose, for example, there is interest in estimating the mean effect on investment of high school and of college education. (It will be remembered, see Table 4.2, that corresponding to α_1, α_2 and α_3 in the model are the three levels of educational status, high school incomplete, high school graduate and college graduate; and the y-variable is investment in consumer durables.) Thus the question is: Even if $\mathbf{b}^o$ is of no use in itself, what does it do for values such as $\frac{1}{2}(\alpha_2^o + \alpha_3^o)$—or for $(\mu^o + \alpha_1^o + \alpha_2^o + \alpha_3^o)/3$? The answer is seen in Table 4.9. Exactly as with the elements of $\mathbf{b}^o$ itself in Table 4.8, the

TABLE 4.9. VALUES OF $\frac{1}{2}(\alpha_2^o + \alpha_3^o)$ AND $(\mu^o + \alpha_1^o + \alpha_2^o + \alpha_3^o)/3$

	Solution (See Table 4.8)			
Linear Function	$\mathbf{b}_1^o$	$\mathbf{b}_2^o$	$\mathbf{b}_3^o$	$\mathbf{b}_4^o$
$\frac{1}{2}(\alpha_2^o + \alpha_3^o)$	$3\frac{1}{2}$	$5\frac{1}{2}$	$-7\frac{1}{2}$	$3{,}001\frac{1}{2}$
$(\mu^o + \alpha_1^o + \alpha_2^o + \alpha_3^o)/3$	22/3	26/3	0	2,006

values of the functions in Table 4.9 vary greatly from solution to solution.

However, this situation is not true of all linear functions. Consider those in Table 4.10. We see at once that the value of each of these expressions is

TABLE 4.10. ESTIMATES OF FOUR ESTIMABLE FUNCTIONS

	Solution (See Table 4.8)			
Linear Function	$\mathbf{b}_1^o$	$\mathbf{b}_2^o$	$\mathbf{b}_3^o$	$\mathbf{b}_4^o$
$\alpha_1^o - \alpha_2^o$	3	3	3	3
$\mu^o + \alpha_1^o$	15	15	15	15
$\mu^o + \frac{1}{2}(\alpha_2^o + \alpha_3^o)$	$19\frac{1}{2}$	$19\frac{1}{2}$	$19\frac{1}{2}$	$19\frac{1}{2}$
$\frac{1}{2}(\alpha_2^o + \alpha_3^o) - \alpha_1^o$	$4\frac{1}{2}$	$4\frac{1}{2}$	$4\frac{1}{2}$	$4\frac{1}{2}$

invariant to whatever solution $\mathbf{b}^o$ is used. Since this is so for all of the infinitely many solutions $\mathbf{b}^o$, these expressions are of value to the investigator whose data they are. And, by their nature, these expressions are often those of specific interest to the investigator for they can be described as follows:

$\alpha_1^o - \alpha_2^o$: estimator of difference between effects of 2 levels.

$\mu^o + \alpha_1^o$: estimator of general mean plus effect of a level.

$\mu^o + \frac{1}{2}(\alpha_2^o + \alpha_3^o)$: estimator of general mean plus mean effect of two levels.

$\frac{1}{2}(\alpha_2^o + \alpha_3^o) - \alpha_1^o$: estimator of superiority of mean effect of two levels over effect of another level.

These are, of course, only four of the many such linear functions of elements of $\mathbf{b}^o$ having the property demonstrated in Table 4.10. Others similar to them are, for example, $\alpha_3^o - \alpha_1^o$, $\mu^o + \alpha_2^o$, $\mu^o + \frac{1}{2}(\alpha_1^o + \alpha_2^o)$ and so on.

Functions such as these are known as *estimators of estimable functions*. They all have the property that they are invariant to whatever solution is obtained to the normal equations. Because of this invariance property they are the only functions that can be of interest, so far as estimation of the parameters of a linear model is concerned. Distinguishing this class of functions from functions such as those illustrated in Table 4.9, which do not have the invariance property, is important—as is deriving their other properties. This is done in the next chapter.

5. EXERCISES

Suppose an oil company gets its crude oil from 4 different sources, refines it in 3 different refineries, using the same 2 processes in each refinery. In one part of the refining process a measurement of efficiency is taken as a percentage and recorded as an integer between 0 and 100. Table 4.11 shows the available measurements of efficiency for different samples of oil.

TABLE 4.11. RESULTS OF EFFICIENCY TESTS

		Source			
Refinery	Process	Texas	Oklahoma	Gulf of Mexico	Iran
Galveston	1	31, 33, 44, 36	38	26	—
	2	37, 59	42	—	—
Newark	1	—	—	42	34, 42, 28
	2	39	36	32, 38	—
Savannah	1	42	36	—	22
	2	—	42, 46	26	37, 43

1. (*a*) For the 8 observations on Texas oil write out the equations for a regression on dummy variables for considering the effect of refinery and process on efficiency.

   (*b*) Rewrite the equations in terms of a linear model.

   (*c*) Write down the equation of the general linear model for this situation.

2. Repeat Exercise 1 for the Oklahoma data.

3. Repeat Exercise 1 for the Gulf of Mexico data.

4. Repeat Exercise 1 for the Iran data.

5. Repeat Exercises 1–4 with interactions between refinery and process included.

6. (*a*) For all 25 observations in Table 4.11 write down the equations of the linear model for considering the effect of source, refinery and process on efficiency. Do not include interactions.
   (*b*) Write down the equation of the general model for this situation.
   (*c*) Write down the normal equations.

7. Repeat Exercise 6 with interactions between source and refinery and between refinery and process.

8. Repeat Exercise 6 with all possible interactions included.

9. In any of the above exercises derive two solutions of the normal equations and investigate functions of the elements of the solutions that might be invariant to whatever solution is used.

10. Repeat the above exercises assuming that processes are nested within refineries, suitably modifying the interactions where necessary.

CHAPTER 5

MODELS NOT OF FULL RANK

Chapter 3 discusses regression analysis in terms of a model having equation $\mathbf{y} = \mathbf{Xb} + \mathbf{e}$ where $\mathbf{X}$ has full column rank; Chapter 4 illustrates how the same equation can apply to linear models generally, where $\mathbf{X}$ does not have full column rank. Estimation and hypothesis testing for this case are now considered, following the same sequence of development as in Chapter 3. Discussion of estimable functions, demonstrated in a simple way at the end of Chapter 4, forms part of the chapter.

1. THE NORMAL EQUATIONS

The model we deal with is

$$\mathbf{y} = \mathbf{Xb} + \mathbf{e}$$

where $\mathbf{y}$ is an $N \times 1$ vector of observations y_i, $\mathbf{b}$ is a $p \times 1$ vector of parameters, $\mathbf{X}$ is an $N \times p$ matrix of known values (in most cases 0's and 1's) and $\mathbf{e}$ is a vector of random error terms. As before, $\mathbf{e}$ can be considered defined as

$$\mathbf{e} = \mathbf{y} - E(\mathbf{y})$$

so that $E(\mathbf{e}) = \mathbf{0}$ and $E(\mathbf{y}) = \mathbf{Xb}$. Every element in $\mathbf{e}$ is assumed to have variance σ^2 and zero covariance with every other element; i.e.,

$$\text{var}(\mathbf{e}) = E(\mathbf{ee}') = \sigma^2\mathbf{I}_N.$$

Thus $\qquad \mathbf{e} \sim (\mathbf{0}, \sigma^2\mathbf{I}) \quad$ and $\quad \mathbf{y} \sim (\mathbf{Xb}, \sigma^2\mathbf{I}),$

with normality being introduced subsequently, when needed for hypothesis testing and derivation of confidence intervals.

a. The equations

Just as in Chapter 3, the normal equations corresponding to the model $\mathbf{y} = \mathbf{Xb} + \mathbf{e}$ can be derived by least squares. And, as before, for var($\mathbf{e}$) $= \sigma^2\mathbf{I}$ they turn out to be

$$\mathbf{X}'\mathbf{X}\hat{\mathbf{b}} = \mathbf{X}'\mathbf{y}. \tag{1}$$

The more general cases where var($\mathbf{e}$) $= \mathbf{V}$, whether $\mathbf{V}$ be non-singular or singular, are discussed in Sec. 8.

Before solving equations (1) we look at their form, initially in terms of an example.

b. Example

Federer (1955) reports an analysis of rubber-producing plants called guayule, for which the plant weights were available for 54 plants of three different kinds, 27 of them normal, 15 off-types and 12 aberrants. We will consider just 6 plants for purposes of illustration, 3 normals, 2 off-types and 1 aberrant, as shown in Table 5.1.

TABLE 5.1. WEIGHTS OF SIX PLANTS

	Type of Plant		
	Normal	Off-Type	Aberrant
	101	84	32
	105	88	
	94		
Totals	300	172	32

For the entries in this table let y_{ij} denote the weight of the jth plant of the ith type, i taking values 1, 2, 3 for normal, off-type and aberrant respectively, and $j = 1, 2, \ldots, n_i$, where n_i is the number of observations in the ith type. The problem is to estimate the effect of type of plant on weight of plant. To do this we assume that the observation y_{ij} is the sum of three parts

$$y_{ij} = \mu + \alpha_i + e_{ij},$$

where μ represents the population mean of the weight of plant, α_i is the effect of type i on weight, and e_{ij} is a random error term peculiar to the observation y_{ij}.

To develop the normal equations we write down the 6 observations in terms of the equation of the model:

$$\begin{aligned}
101 = y_{11} &= \mu + \alpha_1 && + e_{11} \\
105 = y_{12} &= \mu + \alpha_1 && + e_{12} \\
94 = y_{13} &= \mu + \alpha_1 && + e_{13} \\
84 = y_{21} &= \mu \qquad + \alpha_2 && + e_{21} \\
88 = y_{22} &= \mu \qquad + \alpha_2 && + e_{22} \\
32 = y_{31} &= \mu \qquad\qquad + \alpha_3 && + e_{31}\,.
\end{aligned}$$

They are easily rewritten in the form $\mathbf{y} = \mathbf{Xb} + \mathbf{e}$ as

$$\begin{bmatrix} 101 \\ 105 \\ 94 \\ 84 \\ 88 \\ 32 \end{bmatrix} = \begin{bmatrix} y_{11} \\ y_{12} \\ y_{13} \\ y_{21} \\ y_{22} \\ y_{31} \end{bmatrix} = \begin{bmatrix} 1 & 1 & 0 & 0 \\ 1 & 1 & 0 & 0 \\ 1 & 1 & 0 & 0 \\ 1 & 0 & 1 & 0 \\ 1 & 0 & 1 & 0 \\ 1 & 0 & 0 & 1 \end{bmatrix} \begin{bmatrix} \mu \\ \alpha_1 \\ \alpha_2 \\ \alpha_3 \end{bmatrix} + \begin{bmatrix} e_{11} \\ e_{12} \\ e_{13} \\ e_{21} \\ e_{22} \\ e_{31} \end{bmatrix} \tag{2}$$

where $\mathbf{y}$ is the vector of observations, $\mathbf{X}$ is the matrix of 0's and 1's, $\mathbf{b}$ is the vector of parameters to be considered,

$$\mathbf{b}' = [\mu \quad \alpha_1 \quad \alpha_2 \quad \alpha_3],$$

and $\mathbf{e}$ is the vector of error terms.

The vector $\mathbf{b}$ in $\mathbf{y} = \mathbf{Xb} + \mathbf{e}$ is the vector of parameters; it is the vector of all the elements of the model, in this case the elements μ, α_1, α_2 and α_3. And this is so in general; for example, if data can be arranged in rows and columns according to two different classifications, the vector $\mathbf{b}$ will have as its elements the term μ, the terms representing row effects, those representing column effects and those representing interaction effects between rows and columns; for r rows and c columns it can have as many as $1 + r + c + rc$ elements.

The matrix $\mathbf{X}$ in $\mathbf{y} = \mathbf{Xb} + \mathbf{e}$ is called the *incidence matrix*, or sometimes the *design matrix*, because the location of the 0's and 1's throughout its elements represents the incidence of terms of the model among the observations—and hence of the classifications in which the observations lie. This is particularly evident if one writes $\mathbf{X}$ as a 2-way table with the parameters as headings to the columns and the observations as labels for the rows, as illustrated in Table 5.2.

TABLE 5.2. X AS A 2-WAY TABLE

	Parameters of Model			
Observations	μ	α_1	α_2	α_3
y_{11}	1	1	0	0
y_{12}	1	1	0	0
y_{13}	1	1	0	0
y_{21}	1	0	1	0
y_{22}	1	0	1	0
y_{31}	1	0	0	1

In Table 5.2, as in equations (2), it is clear that the sum of the last 3 columns equals the first column. (This is so because every y_{ij} observation contains μ and so the first column of **X** is all 1's; and every y_{ij} also contains just one α and so the sum of the last 3 columns is also all 1's.) Thus **X** is not of full column rank.

Now consider the normal equations (1). They involve **X′X**, which is obviously square and symmetric. Its elements are the inner products of the columns of **X** with each other; e.g.,

$$\mathbf{X}'\mathbf{X} = \begin{bmatrix} 6 & 3 & 2 & 1 \\ 3 & 3 & 0 & 0 \\ 2 & 0 & 2 & 0 \\ 1 & 0 & 0 & 1 \end{bmatrix}. \tag{3}$$

Furthermore, because **X** does not have full column rank, **X′X** is not of full rank.

The normal equations also involve the vector **X′y**; its elements are the inner products of the columns of **X** with the vector **y**, and since the only non-zero elements of **X** are ones, the elements of **X′y** are certain sums of elements of **y**; e.g., from (2)

$$\mathbf{X}'\mathbf{y} = \begin{bmatrix} 1 & 1 & 1 & 1 & 1 & 1 \\ 1 & 1 & 1 & 0 & 0 & 0 \\ 0 & 0 & 0 & 1 & 1 & 0 \\ 0 & 0 & 0 & 0 & 0 & 1 \end{bmatrix} \begin{bmatrix} y_{11} \\ y_{12} \\ y_{13} \\ y_{21} \\ y_{22} \\ y_{31} \end{bmatrix}$$

$$= \begin{bmatrix} y_{11} + y_{12} + y_{13} + y_{21} + y_{22} + y_{31} \\ y_{11} + y_{12} + y_{13} \\ y_{21} + y_{22} \\ y_{31} \end{bmatrix} = \begin{bmatrix} y_{..} \\ y_{1.} \\ y_{2.} \\ y_{3.} \end{bmatrix} = \begin{bmatrix} 504 \\ 300 \\ 172 \\ 32 \end{bmatrix}. \tag{4}$$

This is often the nature of $\mathbf{X}'\mathbf{y}$ in linear models—a vector of various subtotals of the y-observations.

Whenever $\mathbf{X}'\mathbf{X}$ is not of full rank, as in (3), the normal equations (1) cannot be solved with one solitary solution $\hat{\mathbf{b}} = (\mathbf{X}'\mathbf{X})^{-1}\mathbf{X}'\mathbf{y}$ as in Chapter 3. Many solutions are available. To emphasize this we write the normal equations as

$$\mathbf{X}'\mathbf{X}\mathbf{b}^o = \mathbf{X}'\mathbf{y}, \tag{5}$$

using the symbol $\mathbf{b}^o$ to distinguish the many solutions of (5) from the solitary solution that exists when $\mathbf{X}'\mathbf{X}$ has full rank. We shall also use $\mathbf{b}^o$ to denote a solution $\mathbf{G}\mathbf{X}'\mathbf{y}$ to (5), where $\mathbf{G}$ is a generalized inverse of $\mathbf{X}'\mathbf{X}$.

The normal equations of the example are, from (3) and (4),

$$\begin{bmatrix} 6 & 3 & 2 & 1 \\ 3 & 3 & 0 & 0 \\ 2 & 0 & 2 & 0 \\ 1 & 0 & 0 & 1 \end{bmatrix} \begin{bmatrix} \mu^o \\ \alpha_1^o \\ \alpha_2^o \\ \alpha_3^o \end{bmatrix} = \begin{bmatrix} y_{..} \\ y_{1.} \\ y_{2.} \\ y_{3.} \end{bmatrix} = \begin{bmatrix} 504 \\ 300 \\ 172 \\ 32 \end{bmatrix}. \tag{6}$$

By retaining the algebraic form of $\mathbf{X}'\mathbf{y}$ as well as its arithmetic form, it can be seen that if $\mathbf{X}'\mathbf{X}$ is written in a 2-way table the row headings of the table will be the totals in $\mathbf{X}'\mathbf{y}$ and its column headings the parameters. Indeed, the elements of $\mathbf{X}'\mathbf{X}$ are the numbers of times that a parameter of the model occurs in a total; for example, μ occurs 6 times in $y_{..}$ and α_1 occurs 3 times; likewise α_2 does not occur at all in $y_{1.}$; and so on. Another way of looking at $\mathbf{X}'\mathbf{X}$ is that its elements are the coefficients of the parameters of the model in the expected values of the totals in $\mathbf{X}'\mathbf{y}$. In this sense we might write the normal equations as $E(\widehat{\mathbf{X}'\mathbf{y}}) = \mathbf{X}'\mathbf{y}$ replacing $\mathbf{b}$ implicit in the left-hand side by $\mathbf{b}^o$. However, the easiest way of deriving $\mathbf{X}'\mathbf{y}$ and $\mathbf{X}'\mathbf{X}$ other than carrying out the matrix products explicitly is to form $\mathbf{X}'\mathbf{y}$ as the vector of all class and subclass totals of the observations (including the grand total), and to form $\mathbf{X}'\mathbf{X}$ as the matrix of the number of times that each parameter arises in each total that occurs in $\mathbf{X}'\mathbf{y}$.

c. Solutions

Since $\mathbf{X}$ does not have full column rank, $\mathbf{X}'\mathbf{X}$ has no inverse and the normal equations (5) have no unique solution. They have many solutions. To get any one of them we find any generalized inverse $\mathbf{G}$ of $\mathbf{X}'\mathbf{X}$ and write the corresponding solution as

$$\mathbf{b}^o = \mathbf{G}\mathbf{X}'\mathbf{y}. \tag{7}$$

The ability to do this comes directly from Theorem 8 of Chapter 1. The results of that chapter are used repeatedly here, especially those of Sec. 1.5.

The notation $\mathbf{b}^o$ in equation (7) for a solution to the normal equations (5) emphasizes that what is derived by solving (5) is *only* a solution to the equations and *not* an estimator of $\mathbf{b}$. This point cannot be over-emphasized. In a general discussion of linear models that are not of full rank, it is essential to realize that what is obtained as a solution of the normal equations is just that, a solution and *nothing more*. It is misleading and in most cases quite wrong for $\mathbf{b}^o$ to be termed an estimator, particularly an estimator of $\mathbf{b}$. It is true that $\mathbf{b}^o$ is, as shall be shown, an estimator of something, but not of $\mathbf{b}$, and indeed the expression it estimates depends entirely upon which generalized inverse of $\mathbf{X}'\mathbf{X}$ is used in obtaining $\mathbf{b}^o$. For this reason $\mathbf{b}^o$ is always referred to as a solution and not an estimator.

2. CONSEQUENCES OF A SOLUTION

$\mathbf{b}^o$ is clearly a function of the observations $\mathbf{y}$, even though it is not an estimator of $\mathbf{b}$. The expected value and variance and ensuing consequences of $\mathbf{b}^o$ are therefore not identical to those of $\hat{\mathbf{b}}$ of Chapter 3.

a. Expected values

For the generalized inverse $\mathbf{G}$,

$$E(\mathbf{b}^o) = \mathbf{G}\mathbf{X}'E(\mathbf{y}) = \mathbf{G}\mathbf{X}'\mathbf{X}\mathbf{b} = \mathbf{H}\mathbf{b}; \tag{8}$$

i.e., $\mathbf{b}^o$ has expected value $\mathbf{Hb}$ where $\mathbf{H} = \mathbf{G}\mathbf{X}'\mathbf{X}$. Hence $\mathbf{b}^o$ is an unbiased estimator of $\mathbf{Hb}$, but *not* of $\mathbf{b}$.

b. Variances

From (7)

$$\begin{aligned} \text{var}(\mathbf{b}^o) &= \text{var}(\mathbf{G}\mathbf{X}'\mathbf{y}) \\ &= \mathbf{G}\mathbf{X}' \,\text{var}(\mathbf{y})\, \mathbf{X}\mathbf{G}' \\ &= \mathbf{G}\mathbf{X}'\mathbf{X}\mathbf{G}'\sigma^2. \end{aligned} \tag{9}$$

Although this is no analogue of its counterpart $(\mathbf{X}'\mathbf{X})^{-1}\sigma^2$ in the full model, as would be $\mathbf{G}\sigma^2$, we shall see that the result (9) causes no difficulties in applications. Of course, by appropriate choice of $\mathbf{G}$, (9) can be expressed as $\mathbf{G}\sigma^2$. For, since $\mathbf{X}'\mathbf{X}$ is symmetric, there exists an (orthogonal) permutation matrix $\mathbf{P}$ such that

$$\mathbf{P}'\mathbf{X}'\mathbf{X}\mathbf{P} = \begin{bmatrix} \mathbf{A}_{11} & \mathbf{A}_{12} \\ \mathbf{A}_{12}' & \mathbf{A}_{22} \end{bmatrix}$$

where $\mathbf{A}_{11}$ is square, of full rank, equal to the rank of $\mathbf{X}'\mathbf{X}$. Then

$$\mathbf{G} = \mathbf{P}\begin{bmatrix} \mathbf{A}_{11}^{-1} & \mathbf{0} \\ \mathbf{0} & \mathbf{0} \end{bmatrix}\mathbf{P}'$$

is a symmetric generalized inverse of $\mathbf{X}'\mathbf{X}$ with $\mathbf{G}\mathbf{X}'\mathbf{X}\mathbf{G}' = \mathbf{G}$. Hence, with this $\mathbf{G}$, $\text{var}(\mathbf{b}^o) = \mathbf{G}\sigma^2$. There is, however, no particular advantage to this form.

c. Estimating $E(\mathbf{y})$

Corresponding to the vector of observations $\mathbf{y}$ we have the vector of estimated expected values $\widehat{E(\mathbf{y})}$, just as in Sec. 3.4c:

$$\widehat{E(\mathbf{y})} \equiv \hat{\mathbf{y}} = \mathbf{X}\mathbf{b}^o = \mathbf{X}\mathbf{G}\mathbf{X}'\mathbf{y}. \tag{10}$$

This vector is invariant to the choice of whatever generalized inverse of $\mathbf{X}'\mathbf{X}$ is used for $\mathbf{G}$, because $\mathbf{X}\mathbf{G}\mathbf{X}'$ is invariant, as in Theorem 7 of Sec. 1.5a. Hence (10) is *the* vector of estimated expected values corresponding to the vector of observations. This means that no matter what solution of the normal equations is used for $\mathbf{b}^o$ the vector $\hat{\mathbf{y}} = \mathbf{X}\mathbf{G}\mathbf{X}'\mathbf{y}$ will always be the same.

This result, and others of similar nature that will be developed, are of great importance. It means that we can get a solution to the normal equations in any way we please, call it $\mathbf{b}^o$, and no matter which solution it is, $\hat{\mathbf{y}} = \mathbf{X}\mathbf{b}^o$ will be the correct value of $\hat{\mathbf{y}}$.

d. Residual error sum of squares

As before, the residual error sum of squares is defined as

$$\begin{aligned} \text{SSE} &= (\mathbf{y} - \mathbf{X}\mathbf{b}^o)'(\mathbf{y} - \mathbf{X}\mathbf{b}^o) \\ &= \mathbf{y}'(\mathbf{I} - \mathbf{X}\mathbf{G}\mathbf{X}')(\mathbf{I} - \mathbf{X}\mathbf{G}\mathbf{X}')\mathbf{y} \\ &= \mathbf{y}'(\mathbf{I} - \mathbf{X}\mathbf{G}\mathbf{X}')\mathbf{y} \end{aligned} \tag{11}$$

because $\mathbf{I} - \mathbf{X}\mathbf{G}\mathbf{X}'$ is idempotent and, by Theorem 7 of Sec. 1.5a, it is symmetric. Further, because $\mathbf{X}\mathbf{G}\mathbf{X}'$ is invariant to $\mathbf{G}$, so is SSE. Thus SSE is invariant to whatever solution of the normal equations is used for $\mathbf{b}^o$. This is another result invariant to the many solutions there are to the normal equations.

A computing form for SSE can be derived exactly as in regression:

$$\begin{aligned} \text{SSE} &= \mathbf{y}'(\mathbf{I} - \mathbf{X}\mathbf{G}\mathbf{X}')\mathbf{y} = \mathbf{y}'\mathbf{y} - \mathbf{y}'\mathbf{X}\mathbf{G}\mathbf{X}'\mathbf{y} \\ &= \mathbf{y}'\mathbf{y} - \mathbf{b}^{o\prime}\mathbf{X}'\mathbf{y}. \end{aligned} \tag{12}$$

This, we see, is exactly the same result as in the full rank case: $\mathbf{y}'\mathbf{y}$ is the total sum of squares of the observed y's; and $\mathbf{b}^{o\prime}\mathbf{X}'\mathbf{y}$ is the sum of products of the solutions in $\mathbf{b}^{o\prime}$ multiplied by the corresponding elements of the right-hand sides of the equations $\mathbf{X}'\mathbf{X}\mathbf{b}^o = \mathbf{X}'\mathbf{y}$ from which $\mathbf{b}^o$ is derived.

e. Estimating the residual error variance

Since $\mathbf{y}$ is distributed with mean $\mathbf{X}\mathbf{b}$ and variance matrix $\sigma^2\mathbf{I}$, equation (40) of Chapter 2 yields

$$E(\text{SSE}) = E[\mathbf{y}'(\mathbf{I} - \mathbf{X}\mathbf{G}\mathbf{X}')\mathbf{y}] = \text{tr}[(\mathbf{I} - \mathbf{X}\mathbf{G}\mathbf{X}')\mathbf{I}\sigma^2] + \mathbf{b}'\mathbf{X}'(\mathbf{I} - \mathbf{X}\mathbf{G}\mathbf{X}')\mathbf{X}\mathbf{b}.$$

Through the properties of $\mathbf{XGX}'$ in Theorem 7 of Sec. 1.5a this reduces to

$$E(\text{SSE}) = \sigma^2 r(\mathbf{I} - \mathbf{XGX}') = [N - r(\mathbf{X})]\sigma^2.$$

Hence an unbiased estimator of σ^2 is

$$\hat{\sigma}^2 = \frac{\text{SSE}}{N - r(\mathbf{X})}. \tag{13}$$

Again we see a similarity with the full rank case: only now, the importance of using $r(\mathbf{X})$ in the expectation is clear, because $\mathbf{X}$ is not of full column rank and its rank is therefore not equal to its number of columns. In fact, the rank of $\mathbf{X}$ depends on the nature of the data available.

f. Partitioning the total sum of squares

Partitioning the total sum of squares as shown in Sec. 3.4f for the full rank model occurs in exactly similar fashion for the model not of full rank. The only difference is that there is no utility in corrected sums of squares and products of the x-variables. Matrices such as $\mathscr{X}'\mathscr{X}$ do not arise, therefore. However, use is still made of $\text{SST}_m = \mathbf{y}'\mathbf{y} - N\bar{y}^2$, the corrected sum of squares of the y-observations. The three forms of partitioning the sums of squares are shown in Table 5.3.

TABLE 5.3. PARTITIONING SUMS OF SQUARES

	$\text{SSM} = N\bar{y}^2 = \mathbf{y}'N^{-1}\mathbf{11}'\mathbf{y}$	
$\text{SSR} = \mathbf{y}'\mathbf{XGX}'\mathbf{y}$	$\text{SSR}_m = \mathbf{y}'(\mathbf{XGX}' - N^{-1}\mathbf{11}')\mathbf{y}$	$\text{SSR}_m = \mathbf{y}'(\mathbf{XGX}' - N^{-1}\mathbf{11}')\mathbf{y}$
$\text{SSE} = \mathbf{y}'(\mathbf{I} - \mathbf{XGX}')\mathbf{y}$	$\text{SSE} = \mathbf{y}'(\mathbf{I} - \mathbf{XGX}')\mathbf{y}$	$\text{SSE} = \mathbf{y}'(\mathbf{I} - \mathbf{XGX}')\mathbf{y}$
$\text{SST} = \mathbf{y}'\mathbf{y}$	$\text{SST} = \mathbf{y}'\mathbf{y}$	$\text{SST}_m = \mathbf{y}'\mathbf{y} - N\bar{y}^2$

The three columns in Table 5.3 correspond to the three partitionings shown in (72), (73) and (75) of Chapter 3. The first column shows

$$\text{SSR} = \text{SST} - \text{SSE} = \mathbf{y}'\mathbf{XGX}'\mathbf{y} = \mathbf{b}^{\circ\prime}\mathbf{X}'\mathbf{y}, \tag{14}$$

the sum of squares attributable to fitting the model $\mathbf{y} = \mathbf{Xb} + \mathbf{e}$, similar to the sum of squares for regression of Chapter 3. In the second column,

$$\text{SSM} = N\bar{y}^2 \tag{15}$$

is the sum of squares due to fitting a general mean, and

$$\text{SSR}_m = \text{SSR} - \text{SSM} = \text{SSR} - N\bar{y}^2 \tag{16}$$

is the sum of squares for fitting the model, corrected for the mean. The third column is identical to the second except that SSM has been deleted from the body of the table and subtracted from SST to give

$$\text{SST}_m = \text{SST} - \text{SSM} = \sum y_i^2 - N\bar{y}^2 \tag{17}$$

as the total sum of squares corrected for the mean. In all three columns the residual error sum of squares is the same, SSE of (12).

Table 5.3 forms the basis of traditional analysis of variance tables, as is shown in Sec. 3.

g. Coefficient of determination

The estimated expected values of y corresponding to the observations $\mathbf{y}$ are the elements of $\hat{\mathbf{y}}$ given in (10). The product-moment correlation between the observed y's and the corresponding elements of $\hat{\mathbf{y}}$ is, when squared, commonly referred to as the coefficient of determination. Since the usual linear model has a mean in it we define

$$\begin{aligned} R^2 &= \text{coefficient of determination} \\ &= \frac{[\sum (y_i - \bar{y})(\hat{y}_i - \bar{\hat{y}})]^2}{\sum (y_i - \bar{y})^2 \sum (\hat{y}_i - \bar{\hat{y}})^2}. \end{aligned} \tag{18}$$

In simplifying this we utilize $\mathbf{X'XGX'} = \mathbf{X'}$ (Theorem 7, Sec. 1.5a) and so, because $\mathbf{1'}$ is the first row of $\mathbf{X'}$,

$$\mathbf{1'XGX'} = \mathbf{1'}. \tag{19}$$

By this means we can show that $\bar{\hat{y}} = \bar{y}$, and hence, as in equations (76) of Chapter 3,

$$R^2 = \frac{(\text{SSR}_m)^2}{\text{SST}_m(\text{SSR}_m)} = \frac{\text{SSR}_m}{\text{SST}_m}. \tag{20}$$

h. Example (continued)

The normal equations for the example of Sec. 1b are given in (6). For $\mathbf{X'X}$ given there, a generalized inverse is

$$\mathbf{G} = \begin{bmatrix} 0 & 0 & 0 & 0 \\ 0 & \frac{1}{3} & 0 & 0 \\ 0 & 0 & \frac{1}{2} & 0 \\ 0 & 0 & 0 & 1 \end{bmatrix} \quad \text{with} \quad \mathbf{H} = \begin{bmatrix} 0 & 0 & 0 & 0 \\ 1 & 1 & 0 & 0 \\ 1 & 0 & 1 & 0 \\ 1 & 0 & 0 & 1 \end{bmatrix}. \tag{21}$$

Then, from (7)

$$(\mathbf{b}^o)' = (\mathbf{GX'y})' = [0 \quad 100 \quad 86 \quad 32] \tag{22}$$

for which, from (10),

$$(\hat{\mathbf{y}})' = (\mathbf{Xb}^o)' = [100 \quad 100 \quad 100 \quad 86 \quad 86 \quad 32] \tag{23}$$

and

$$\text{SSR} = \mathbf{b}^{o\prime}\mathbf{X'y} = 100(300) + 86(172) + 32(32) = 45{,}816. \tag{24}$$

Demonstration of the invariance of $\hat{\mathbf{y}}$ and SSR to the choice of $\mathbf{G}$ is seen from taking the generalized inverse as

$$\mathbf{G}_1 = \begin{bmatrix} 1 & -1 & -1 & 0 \\ -1 & \frac{4}{3} & 1 & 0 \\ -1 & 1 & 1\frac{1}{2} & 0 \\ 0 & 0 & 0 & 0 \end{bmatrix}$$

with $$\mathbf{H}_1 = \begin{bmatrix} 1 & 0 & 0 & 1 \\ 0 & 1 & 0 & -1 \\ 0 & 0 & 1 & -1 \\ 0 & 0 & 0 & 0 \end{bmatrix} \quad \text{and} \quad \mathbf{b}_1^o = \begin{bmatrix} 32 \\ 68 \\ 54 \\ 0 \end{bmatrix}. \tag{25}$$

Then again

$$\hat{\mathbf{y}}' = \mathbf{b}_1^{o\prime}\mathbf{X}' = [100 \quad 100 \quad 100 \quad 86 \quad 86 \quad 32]$$

and $$\text{SSR} = \mathbf{b}_1^{o\prime}\mathbf{X}'\mathbf{y} = 32(504) + 68(300) + 54(172) = 45,816.$$

The vector of observations is, from (2),

$$\mathbf{y}' = [101 \quad 105 \quad 94 \quad 84 \quad 88 \quad 32]$$

and so $$\text{SST} = \sum y^2 = \mathbf{y}'\mathbf{y} = 45{,}886 \tag{26}$$

and $$\text{SSM} = N\bar{y}^2 = 42{,}336. \tag{27}$$

Hence the partitioning of sums of squares shown in Table 5.3 is, for the example, as given in Table 5.4.

TABLE 5.4. PARTITIONING SUMS OF SQUARES. (DATA OF TABLE 5.1)

	SSM = 42,336	
SSR = 45,816	SSR_m = 3,480	SSR_m = 3,480
SSE = 70	SSE = 70	SSE = 70
SST = 45,886	SST = 45,886	SST_m = 3,550

The value of R^2, calculated from (20), is $R^2 = 3480/3550 = 0.98$.

3. DISTRIBUTIONAL PROPERTIES

We now introduce normality for the error terms,

$$\mathbf{e} \sim N(\mathbf{0}, \sigma^2\mathbf{I}_N),$$

and derive distributional properties of $\mathbf{y}$ and functions of $\mathbf{y}$ in a manner similar to the full rank case (Sec. 3.5).

a. y is normal

Because $\mathbf{y} = \mathbf{Xb} + \mathbf{e}$ and $E(\mathbf{y}) = \mathbf{Xb}$ we have $\mathbf{y} \sim N(\mathbf{Xb}, \sigma^2\mathbf{I})$.

b. $\mathbf{b}^o$ is normal

Since $\mathbf{b}^o$ is a linear function of $\mathbf{y}$ it is also normally distributed, with mean and variance derived in (8) and (9):

$$\mathbf{b}^o = \mathbf{GX'y} \sim N(\mathbf{Hb}, \mathbf{GX'XG'}\sigma^2).$$

Notice that the variance-covariance matrix of $\mathbf{b}^o$ is singular.

c. $\mathbf{b}^o$ and $\hat{\sigma}^2$ are independent

In applying Theorem 3 of Sec. 2.5c to

$$\mathbf{b}^o = \mathbf{GX'y} \qquad \text{and} \qquad \text{SSE} = \mathbf{y'(I - XGX')y}$$

we see that

$$\mathbf{GX'I}\sigma^2\mathbf{(I - XGX')} = \mathbf{G(X' - X'XGX')}\sigma^2 = \mathbf{0}$$

because $\mathbf{X'} = \mathbf{X'XGX'}$ (Theorem 7, Sec. 1.5a). Therefore $\mathbf{b}^o$ and $\hat{\sigma}^2$ are independent.

d. SSE/σ^2 is χ^2

$$\text{SSE}/\sigma^2 = \mathbf{y'(I - XGX')y}/\sigma^2,$$

and in applying Theorem 2 of Sec. 2.5b we have

$$\mathbf{I}\sigma^2\mathbf{(I - XGX')}/\sigma^2 = \mathbf{I - XGX'},$$

which is idempotent. Therefore, by Theorem 2 of Chapter 2

$$\text{SSE}/\sigma^2 \sim \chi^{2\prime}[r(\mathbf{I - XGX'}), \mathbf{b'X'(I - XGX')Xb}/2\sigma^2]$$

which, because of properties of $\mathbf{XGX'}$ and with $r(\mathbf{X}) = r$, reduces to

$$\text{SSE}/\sigma^2 \sim \chi^2_{N-r}. \tag{28}$$

e. Non-central χ^2's

With SSE being χ^2_{N-r} we now show that the other terms in Table 5.3 have non-central χ^2-distributions independent of SSE. This leads to F-statistics that have non-central F-distributions which in turn are central F-distributions under certain null hypotheses. Thus are tests of hypotheses established.

First, from (14), SSR $= \mathbf{y}'\mathbf{XGX}'\mathbf{y}$, in which $\mathbf{XGX}'$ is idempotent and its products with $\mathbf{I} - \mathbf{XGX}'$ are null. Therefore by Theorems 4 and 2 of Chapter 2, SSR$/\sigma^2$ is distributed independently of SSE, with

$$\begin{aligned} \text{SSR}/\sigma^2 &\sim \chi^{2\prime}[r(\mathbf{XGX}'), \mathbf{b}'\mathbf{X}'\mathbf{XGX}'\mathbf{Xb}/2\sigma^2] \\ &\sim \chi^{2\prime}(r, \mathbf{b}'\mathbf{X}'\mathbf{Xb}/2\sigma^2). \end{aligned} \tag{29}$$

Similarly

$$\text{SSM}/\sigma^2 = \mathbf{y}'N^{-1}\mathbf{11}'\mathbf{y}/\sigma^2$$

where $N^{-1}\mathbf{11}'$ is idempotent; further, using (19) gives

$$N^{-1}\mathbf{11}'\mathbf{XGX}' = N^{-1}\mathbf{11}', \tag{30}$$

so that the products of $N^{-1}\mathbf{11}'$ and $(\mathbf{I} - \mathbf{XGX}')$ are null. Hence SSM is distributed independently of SSE and

$$\begin{aligned} \text{SSM}/\sigma^2 &\sim \chi^{2\prime}[r(N^{-1}\mathbf{11}'), \mathbf{b}'\mathbf{X}'N^{-1}\mathbf{11}'\mathbf{Xb}/2\sigma^2] \\ &\sim \chi^{2\prime}[1, (\mathbf{1}'\mathbf{Xb})^2/2N\sigma^2], \end{aligned} \tag{31}$$

just as in the full rank case.

Similar argument applies to

$$\text{SSR}_m = \mathbf{y}'(\mathbf{XGX}' - N^{-1}\mathbf{11}')\mathbf{y}.$$

$\mathbf{XGX}' - N^{-1}\mathbf{11}'$ is idempotent [using (30)] and has null products with both $N^{-1}\mathbf{11}'$ and $(\mathbf{I} - \mathbf{XGX}')$. Hence, by Theorems 2 and 4 of Chapter 2, SSR$_m$ is independent of SSM and SSE, and

$$\begin{aligned} \text{SSR}_m/\sigma^2 &\sim \chi^{2\prime}[r(\mathbf{XGX}' - N^{-1}\mathbf{11}'), \mathbf{b}'\mathbf{X}'(\mathbf{XGX}' - N^{-1}\mathbf{11}')\mathbf{Xb}/2\sigma^2] \\ &\sim \chi^{2\prime}[r - 1, \mathbf{b}'\mathbf{X}'(\mathbf{I} - N^{-1}\mathbf{11}')\mathbf{Xb}/2\sigma^2]. \end{aligned} \tag{32}$$

Now, whatever $\mathbf{X}$ is, so long as its first column is $\mathbf{1}$ we can write $\mathbf{X} = [\mathbf{1} \quad \mathbf{X}_1]$ and then

$$\mathbf{X}'(\mathbf{I} - N^{-1}\mathbf{11}')\mathbf{X} = \begin{bmatrix} 0 & \mathbf{0}' \\ \mathbf{0} & \mathbf{X}_1'(\mathbf{I} - N^{-1}\mathbf{11}')\mathbf{X}_1 \end{bmatrix} = \begin{bmatrix} 0 & \mathbf{0}' \\ \mathbf{0} & \mathscr{X}'\mathscr{X} \end{bmatrix} \tag{33}$$

where $\mathscr{X}'\mathscr{X}$ represents the same kind of matrix that it does in Chapter 3: the sums of squares and products of the deviations of the elements of the columns of $\mathbf{X}$ (other than the first column) from their means. Symbolically it is simpler than its equivalent form $\mathbf{X}_1'(\mathbf{I} - N^{-1}\mathbf{11}')\mathbf{X}_1$ but computationally it offers little advantage, in distinction to the full rank model where it is advantageous.

Nevertheless, writing

$$\mathbf{b} = \begin{bmatrix} b_0 \\ \boldsymbol{\ell} \end{bmatrix} \tag{34}$$

just as in the full rank case, with b_0 representing a general mean, we have from (32) and (33)

$$\begin{aligned} \text{SSR}_m/\sigma^2 &\sim \chi^{2\prime}[r - 1, \boldsymbol{\ell}'\mathbf{X}_1'(\mathbf{I} - N^{-1}\mathbf{1}\mathbf{1}')\mathbf{X}_1\boldsymbol{\ell}/2\sigma^2] \\ &\sim \chi^{2\prime}[r - 1, \boldsymbol{\ell}'\mathcal{X}'\mathcal{X}\boldsymbol{\ell}/2\sigma^2]. \end{aligned} \tag{35}$$

f. *F*-distributions

From the foregoing results the definition of the non-central *F*-distribution leads to the following *F*-statistics:

$$F(R) = \frac{\text{SSR}/r}{\text{SSE}/(N - r)} \sim F'(r, N - r, \mathbf{b}'\mathbf{X}'\mathbf{X}\mathbf{b}/2\sigma^2) \tag{36}$$

$$F(M) = \frac{\text{SSM}/1}{\text{SSE}/(N - r)} \sim F'[1, N - r, (\mathbf{1}'\mathbf{X}\mathbf{b})^2/2N\sigma^2] \tag{37}$$

$$F(R_m) = \frac{\text{SSR}_m/(r - 1)}{\text{SSE}/(N - r)} \sim F'[r - 1, N - r, \boldsymbol{\ell}'\mathbf{X}_1'(\mathbf{I} - N^{-1}\mathbf{1}\mathbf{1}')\mathbf{X}_1\boldsymbol{\ell}/2\sigma^2] \tag{38}$$

$$\sim F'(r - 1, N - r, \boldsymbol{\ell}'\mathcal{X}'\mathcal{X}\boldsymbol{\ell}/2\sigma^2) \tag{39}$$

Under certain null hypotheses these non-central *F*'s become central *F*'s, and so provide us with tests of those hypotheses. These are discussed subsequently in subsection h, and again in Sec. 5.

g. Analyses of variance

Calculation of the above *F*-statistics can be summarized in analysis of variance tables just as is done in Tables 3.2, 3.3 and 3.4 of Sec. 3.5h. Similar tables are shown in Tables 5.5 and 5.6. The sums of squares are those of Table 5.3.

TABLE 5.5. ANALYSIS OF VARIANCE FOR FITTING THE MODEL $\mathbf{y} = \mathbf{Xb} + \mathbf{e}$

Source of Variation	d.f.	Sum of Squares	Mean Square	*F*-statistic
Model	$r = r(\mathbf{X})$	$\text{SSR} = \mathbf{b}^{o\prime}\mathbf{X}'\mathbf{y}$	$\text{MSR} = \text{SSR}/r$	$F(R) = \dfrac{\text{MSR}}{\text{MSE}}$
Residual error	$N - r$	$\text{SSE} = \mathbf{y}'\mathbf{y} - \mathbf{b}^{o\prime}\mathbf{X}'\mathbf{y}$	$\text{MSE} = \dfrac{\text{SSE}}{N - r}$	
Total	N	$\text{SST} = \mathbf{y}'\mathbf{y}$		

Table 5.5 summarizes not only the sums of squares but also the degrees of freedom associated with the χ^2-distributions. It also shows, in the mean squares, the calculation of the numerator and denominator of $F(R)$ of equation (36), as well as $F(R)$ itself. The table is therefore a convenient summary of these calculations.

TABLE 5.6. ANALYSIS OF VARIANCE FOR FITTING THE MODEL $\mathbf{y} = \mathbf{Xb} + \mathbf{e}$

Table 5.6a. Complete form

Source of Variation[2]	d.f.[1]	Sum of Squares	Mean Square	F-Statistics
Mean	1	$\text{SSM} = N\bar{y}^2$	$\text{MSM} = \dfrac{\text{SSM}}{1}$	$F(M) = \dfrac{\text{MSM}}{\text{MSE}}$
Model (a.f.m.)	$r-1$	$\text{SSR}_m = \mathbf{b}^{o\prime}\mathbf{X}'\mathbf{y} - N\bar{y}^2$	$\text{MSR}_m = \dfrac{\text{SSR}_m}{r-1}$	$F(R_m) = \dfrac{\text{MSR}_m}{\text{MSE}}$
Residual error	$N-r$	$\text{SSE} = \mathbf{y}'\mathbf{y} - \mathbf{b}^{o\prime}\mathbf{X}'\mathbf{y}$	$\text{MSE} = \dfrac{\text{SSE}}{N-r}$	
Total	N	$\text{SST} = \mathbf{y}'\mathbf{y}$		

Table 5.6b. Abbreviated form

Model (a.f.m.)	$r-1$	$\text{SSR}_m = \mathbf{b}^{o\prime}\mathbf{X}'\mathbf{y} - N\bar{y}^2$	$\text{MSR}_m = \dfrac{\text{SSR}_m}{r-1}$	$F(R_m) = \dfrac{\text{MSR}_m}{\text{MSE}}$
Residual error	$N-r$	$\text{SSE} = \mathbf{y}'\mathbf{y} - \mathbf{b}^{o\prime}\mathbf{X}'\mathbf{y}$	$\text{MSE} = \dfrac{\text{SSE}}{N-r}$	
Total (a.f.m.)	$N-1$	$\text{SST}_m = \mathbf{y}'\mathbf{y} - N\bar{y}^2$		

[1] $r = r(\mathbf{X})$.

[2] a.f.m. = after fitting the mean.

Table 5.6 shows the same thing for $F(M)$ and $F(R_m)$ of (37) and (38). Table 5.6b shows the abbreviated form of the complete analysis of variance table shown in Table 5.6a. This abbreviated form is derived by removing SSM from the body of the table and subtracting it from SST to give SST_m, as in Table 5.3. Thus Table 5.6b does not contain $F(M)$, but it is identical to

Table 5.6a insofar as $F(R_m) = \text{MSR}_m/\text{MSE}$ is concerned. Thus the two sections of Table 5.6 are similar to Tables 3.3 and 3.4 of Sec. 3.5h.

Although Table 5.6b is the form in which this analysis of variance is most usually seen, it is not necessarily the most informative. Credit on that account goes to Table 5.6a, which shows how SSR of Table 5.4 gets partitioned into SSM and SSR_m, so summarizing both $F(M)$ and $F(R_m)$.

h. Tests of hypotheses

Following equations (36)–(39) we indicated that those results provide statistics suitable for testing certain hypotheses. This we now discuss, prior to considering the general linear hypothesis in Sec. 5.

The $F(R)$ statistic of (36), whose calculation is summarized in Table 5.5, has a non-central F-distribution with non-centrality parameter $\mathbf{b}'\mathbf{X}'\mathbf{X}\mathbf{b}/2\sigma^2$, which is zero under the null hypothesis H: $\mathbf{X}\mathbf{b} = \mathbf{0}$. $F(R)$ then has a central $F_{r,N-r}$-distribution and can be compared to tabulated values thereof to test that hypothesis. When $F(R)$ is significant we might say, just as we did in Sec. 3.5i, that there is concordance of the data with the model $E(\mathbf{y}) = \mathbf{X}\mathbf{b}$; i.e., the model accounts for a significant portion of the variation in the y-variable. This does not mean that the model used is necessarily the most suitable model: there may be a subset of the elements used that is as significant as the whole set; or there may be other elements (factors) which, when used alone, or in combination with some or all of those already used, are significantly better than those used; or there may be non-linear models that are at least as suitable as the model used. None of these contingencies is inconsistent with $F(R)$ being significant and the ensuing conclusion that the data are in concordance with the model $E(\mathbf{y}) = \mathbf{X}\mathbf{b}$.

Notice, in contrast to the full model case in Sec. 3.5i, that the test based on $F(R)$ cannot be described formally as testing H: $\mathbf{b} = \mathbf{0}$ because, as shown in Secs. 4 and 5 that follow, $\mathbf{b}$ is not what is called an "estimable function" and this means H: $\mathbf{b} = \mathbf{0}$ cannot be tested. However, H: $\mathbf{X}\mathbf{b} = \mathbf{0}$ can be tested and $F(R)$ is the appropriate statistic, as is soon discussed.

The non-centrality parameter of $F(M)$ in Table 5.6a is, by (37), $(\mathbf{1}'\mathbf{X}\mathbf{b})^2/2N\sigma^2$ and, just as in the full rank case (Sec. 3.5i), this parameter equals $N[E(\bar{y})]^2/2\sigma^2$. It is zero under the hypothesis H: $E(\bar{y}) = 0$ whereupon the statistic $F(M)$ is then distributed as $F_{1,N-r}$, and hence $F(M)$ provides a test of the hypothesis H: $E(\bar{y}) = 0$. The test is based on comparing $F(M)$ with tabulated values of the $F_{1,N-r}$-distribution. An equivalent test is to compare $\sqrt{F(M)}$ against tabulations of the t-distribution having $N - r$ degrees of freedom. This hypothesis, H: $E(\bar{y}) = 0$, is one interpretation of what is meant by "testing the mean". Another interpretation, just as in the full rank case, is that $F(M)$ can be used to test whether the model $E(y_{ij}) = b_0$ accounts for variation in the y-variable.

Just as $F(R)$ provides a test of the model $E(\mathbf{y}) = \mathbf{Xb}$, so does $F(R_m)$ provide a test of the model over and above the mean. For the same reason that $F(R)$ cannot be described as testing H: $\mathbf{b} = \mathbf{0}$, so also $F(R_m)$ cannot be described as testing H: $\mathscr{b} = \mathbf{0}$; $\mathscr{b}$ is not, in general, what is called an "estimable function" and so H: $\mathscr{b} = \mathbf{0}$ cannot be tested (see Secs. 4 and 5). In general, therefore, $F(R_m)$ must be looked on as providing a test of the model $E(\mathbf{y}) = \mathbf{Xb}$ over and above the model $E(y) = b_0$. Since the latter can be considered as fitting a general mean we look upon $F(R_m)$ as providing a test of the model $E(\mathbf{y}) = \mathbf{Xb}$ over and above the mean. When $F(R_m)$ is significant we conclude that the model satisfactorily accounts for variation in the y-variable. This is not to be taken as evidence that all elements of $\mathscr{b}$ are non-zero, but only that at least one of them, or one linear combination of them, may be. If $F(M)$ has first been found significant, then $F(R_m)$ being significant indicates that a model with terms in it additional to a mean explains significantly more of the variation in the y-variable than does the model $E(y) = b_0$.

Similar to regression, the tests using $F(M)$ and $F(R_m)$ are based on numerators that are statistically independent although their denominators, the residual mean square, are identical. The F-statistics are therefore not independent.

TABLE 5.7. TABLES 5.5 AND 5.6 FOR THE EXAMPLE

Source of Variation	d.f.	Sum of Squares	Mean Square	F-Statistic
Table 5.5				
Model	3	SSR = 45,816	15,272	$F(R) = 654.51$
Residual error	3	SSE = 70	$23\frac{1}{3}$	
Total	6	SST = 45,886		
Table 5.6a				
Mean	1	SSM = 42,336	42,336	$F(M) = 1814.4$
Model (a.f.m)	2	SSR_m = 3,480	1,740	$F(R_m) = 74.3$
Residual error	3	SSE = 70	$23\frac{1}{3}$	
Total	6	SST = 45,886		
Table 5.6b				
Model (a.f.m)	2	SSR_m = 3,480	1,740	$F(R_m) = 74.3$
Residual error	3	SSE = 70	$23\frac{1}{3}$	
Total (a.f.m)	5	SST_m = 3,550		

The case of both $F(M)$ and $F(R_m)$ being significant has just been discussed; a further possibility is that $F(M)$ is not significant but $F(R_m)$ is. This is evidence of the mean's being zero but that fitting the rest of the model explains variation in the y-variable. As in regression, a likely situation when this might occur is when the y-variable can have both positive and negative values.

i. Example (continued)

Using the sums of squares in Table 5.4, the analyses of variance of Tables 5.5 and 5.6 are shown in Table 5.7. This is a case where all three F-statistics are significant, $F(R)$, $F(M)$ and $F(R_m)$ indicating, respectively, that the model accounts for a significant portion of the variation in y, that the mean is unlikely to be zero and that the model needs in it something more than the mean to explain variation in y.

4. ESTIMABLE FUNCTIONS

The underlying idea of an estimable function was introduced at the end of Chapter 4. Basically it is a linear function of the parameters for which an estimator can be found from $\mathbf{b}^o$ that is invariant to whatever solution of the normal equations is used for $\mathbf{b}^o$. We now discuss such functions in detail, confining ourselves to linear functions of the form $\mathbf{q}'\mathbf{b}$ where $\mathbf{q}'$ is a row vector.

a. Definition

A (linear) function of the parameters is defined as *estimable* if it is identically equal to some linear function of the expected value of the vector of observations $\mathbf{y}$. This means that $\mathbf{q}'\mathbf{b}$ is estimable if $\mathbf{q}'\mathbf{b} = \mathbf{t}'E(\mathbf{y})$ for some vector $\mathbf{t}'$; i.e., if there exists a vector $\mathbf{t}'$ such that $\mathbf{t}'E(\mathbf{y}) = \mathbf{q}'\mathbf{b}$ then $\mathbf{q}'\mathbf{b}$ is said to be estimable. Note that in no way is there any sense of uniqueness about $\mathbf{t}'$ for a given $\mathbf{q}'\mathbf{b}$; $\mathbf{t}'$ simply has to exist. Thus in the example of the preceding section

$$E(y_{1j}) = \mu + \alpha_1 \qquad \text{and} \qquad E(y_{2k}) = \mu + \alpha_2 \,.$$

Hence $E(y_{1j} - y_{2k}) = \alpha_1 - \alpha_2$ and therefore $\alpha_1 - \alpha_2$ is an estimable function. In this case $\mathbf{t}'$ is a row vector of zeros except for $+1$ and -1 in the elements corresponding, respectively, to y_{1j} and y_{2k}.

The value of $\mathbf{t}'$ is not as important as its existence, and in this sense all that need be done to establish estimability of $\mathbf{q}'\mathbf{b}$ is to be satisfied that there is at least one linear function of the expected values of the y's, $\mathbf{t}'E(\mathbf{y})$, whose value is $\mathbf{q}'\mathbf{b}$. Since $\mathbf{t}'E(\mathbf{y}) = E(\mathbf{t}'\mathbf{y})$ this is equivalent to establishing some linear function of the y's, $\mathbf{t}'\mathbf{y}$, whose expected value is $\mathbf{q}'\mathbf{b}$. There are usually many such functions of the y's; establishing the existence of any one of them is sufficient for establishing estimability.

b. Properties

Five important properties arise from the definition of an estimable function.

(*i*) *The expected value of any observation is estimable.* The definition of an estimable function is that $\mathbf{q}'\mathbf{b}$ is estimable if $\mathbf{q}'\mathbf{b} = \mathbf{t}'E(\mathbf{y})$ for some vector $\mathbf{t}'$. Consider a $\mathbf{t}'$ which has one element unity and the others zero: $\mathbf{t}'E(\mathbf{y})$ will be estimable and it will be an element of $E(\mathbf{y})$; i.e., the expected value of an observation. Hence the expected value of any observation is estimable. In the example, $E(y_{1j}) = \mu + \alpha_1$ and so $\mu + \alpha_1$ is estimable.

(*ii*) *Linear combinations of estimable functions.* Any linear combination of estimable functions is estimable. This is so because any (and every) estimable function is a linear combination of the elements of $E(\mathbf{y})$. So, therefore, is a linear combination of estimable functions: and therefore it is estimable. Thus, if $\mathbf{q}_1'\mathbf{b}$ and $\mathbf{q}_2'\mathbf{b}$ are estimable, then $\mathbf{q}_1'\mathbf{b} = \mathbf{t}_1'E(\mathbf{y})$ and $\mathbf{q}_2'\mathbf{b} = \mathbf{t}_2'E(\mathbf{y})$ for some $\mathbf{t}_1$ and $\mathbf{t}_2$. Hence $c_1\mathbf{q}_1'\mathbf{b} + c_2\mathbf{q}_2'\mathbf{b} = (c_1\mathbf{t}_1' + c_2\mathbf{t}_2')E(\mathbf{y})$ and so it is estimable.

(*iii*) *The form of an estimable function.* If $\mathbf{q}'\mathbf{b}$ is estimable, $\mathbf{q}'\mathbf{b} = \mathbf{t}'E(\mathbf{y})$ for some $\mathbf{t}'$, by definition, and so $\mathbf{q}'\mathbf{b} = \mathbf{t}'\mathbf{X}\mathbf{b}$. Since estimability is a concept that does not depend on the value of $\mathbf{b}$, this last result must be true for all values of $\mathbf{b}$. Therefore

$$\mathbf{q}' = \mathbf{t}'\mathbf{X} \tag{40}$$

for some vector $\mathbf{t}'$. For any estimable function $\mathbf{q}'\mathbf{b}$ it is not any specific value of $\mathbf{t}'$ that is so important; it is the *existence* of some $\mathbf{t}'$ that satisfies (40) that is important. In this context (40) gets used repeatedly; i.e., $\mathbf{q}'\mathbf{b}$ is estimable whenever $\mathbf{q}' = \mathbf{t}'\mathbf{X}$ and, conversely, estimability of $\mathbf{q}'\mathbf{b}$ implies $\mathbf{q}' = \mathbf{t}'\mathbf{X}$ for some $\mathbf{t}'$.

(*iv*) *Invariance to the solution* $\mathbf{b}^o$. When $\mathbf{q}'\mathbf{b}$ is estimable, $\mathbf{q}'\mathbf{b}^o$ is invariant to whatever solution of $\mathbf{X}'\mathbf{X}\mathbf{b}^o = \mathbf{X}'\mathbf{y}$ is used for $\mathbf{b}^o$. This is so because, by (40),

$$\mathbf{q}'\mathbf{b}^o = \mathbf{t}'\mathbf{X}\mathbf{b}^o = \mathbf{t}'\mathbf{X}\mathbf{G}\mathbf{X}'\mathbf{y}$$

and $\mathbf{X}\mathbf{G}\mathbf{X}'$ is invariant to $\mathbf{G}$ (Theorem 7, Sec. 1.5a). Therefore $\mathbf{q}'\mathbf{b}^o$ is invariant to $\mathbf{G}$ and hence to $\mathbf{b}^o$, when $\mathbf{q}'\mathbf{b}$ is estimable. This is the importance of estimability. If $\mathbf{q}'\mathbf{b}$ is estimable, $\mathbf{q}'\mathbf{b}^o$ has the same value for all solutions $\mathbf{b}^o$ to the normal equations.

(*v*) *The b.l.u.e.* The b.l.u.e. of the estimable function $\mathbf{q}'\mathbf{b}$ is $\mathbf{q}'\mathbf{b}^o$; i.e.,

$$\widehat{\mathbf{q}'\mathbf{b}} = \mathbf{q}'\mathbf{b}^o, \tag{41}$$

where by the "hat" notation we mean "b.l.u.e. of".

To prove (41) we demonstrate properties of linearity, unbiasedness and "bestness". First, $\mathbf{q}'\mathbf{b}^o$ is clearly a linear function of the observations, because $\mathbf{q}'\mathbf{b}^o = \mathbf{q}'\mathbf{GX}'\mathbf{y}$. Second, $\mathbf{q}'\mathbf{b}^o$ is an unbiased estimator of $\mathbf{q}'\mathbf{b}$ because

$$E(\mathbf{q}'\mathbf{b}^o) = \mathbf{q}'E(\mathbf{b}^o) = \mathbf{q}'\mathbf{Hb} = \mathbf{t}'\mathbf{XHb} = \mathbf{t}'\mathbf{Xb} = \mathbf{q}'\mathbf{b},$$

invoking both (40) and, from Theorem 7 of Sec. 1.5a,

$$\mathbf{X} = \mathbf{XH} = \mathbf{XGX}'\mathbf{X} \quad \text{which also implies} \quad \mathbf{X}' = \mathbf{X}'\mathbf{XG}'\mathbf{X}'. \tag{42}$$

To demonstrate that $\mathbf{q}'\mathbf{b}^o$ is a best estimator we need its variance:

$$\begin{aligned} v(\mathbf{q}'\mathbf{b}^o) &= \mathbf{q}'\mathbf{GX}'\mathbf{XG}'\mathbf{q}\sigma^2, && \text{from (9)} \\ &= \mathbf{q}'\mathbf{GX}'\mathbf{XG}'\mathbf{X}'\mathbf{t}\sigma^2, && \text{from (40)} \\ &= \mathbf{q}'\mathbf{GX}'\mathbf{t}\sigma^2, && \text{from (42)} \\ &= \mathbf{q}'\mathbf{Gq}\sigma^2, && \text{from (40).} \end{aligned} \tag{43}$$

This illustrates the implication that the result $\text{var}(\mathbf{b}^o) = \mathbf{GX}'\mathbf{XG}'\sigma^2$ of (9), involving $\mathbf{G}$ and $\mathbf{G}'$, leads to no difficulties in application. Estimable functions $\mathbf{q}'\mathbf{b}$ are the only ones of interest and their b.l.u.e.'s have variance $\mathbf{q}'\mathbf{Gq}\sigma^2$ even though $\mathbf{GX}'\mathbf{XG}'\sigma^2$ for the variance of $\mathbf{b}^o$ is not an exact analogue of $(\mathbf{X}'\mathbf{X})^{-1}\sigma^2$ of the full rank case.

In the light of (43) we now show, following Rao (1962), that $\mathbf{q}'\mathbf{b}^o$ has minimum variance among all linear unbiased estimators of $\mathbf{q}'\mathbf{b}$ and hence is best. Suppose $\mathbf{k}'\mathbf{y}$ is some other linear unbiased estimator of $\mathbf{q}'\mathbf{b}$ different from $\mathbf{q}'\mathbf{b}^o$. Then, because $\mathbf{k}'\mathbf{y}$ is unbiased $E(\mathbf{k}'\mathbf{y}) = \mathbf{q}'\mathbf{b}$ and so $\mathbf{k}'\mathbf{X} = \mathbf{q}'$. Therefore

$$\text{cov}(\mathbf{q}'\mathbf{b}^o, \mathbf{k}'\mathbf{y}) = \text{cov}(\mathbf{q}'\mathbf{GX}'\mathbf{y}, \mathbf{k}'\mathbf{y}) = \mathbf{q}'\mathbf{GX}'\mathbf{k}\sigma^2 = \mathbf{q}'\mathbf{Gq}\sigma^2.$$

Consequently

$$\begin{aligned} v(\mathbf{q}'\mathbf{b}^o - \mathbf{k}'\mathbf{y}) &= v(\mathbf{q}'\mathbf{b}^o) + v(\mathbf{k}'\mathbf{y}) - 2\,\text{cov}(\mathbf{q}'\mathbf{b}^o, \mathbf{k}'\mathbf{y}) \\ &= v(\mathbf{k}'\mathbf{y}) - \mathbf{q}'\mathbf{Gq}\sigma^2 \\ &= v(\mathbf{k}'\mathbf{y}) - v(\mathbf{q}'\mathbf{b}^o). \end{aligned} \tag{44}$$

But $v(\mathbf{q}'\mathbf{b}^o - \mathbf{k}'\mathbf{y})$ is positive and so therefore, from (44), $v(\mathbf{k}'\mathbf{y})$ exceeds $v(\mathbf{q}'\mathbf{b}^o)$; i.e., $\mathbf{q}'\mathbf{b}^o$ has a smaller variance than any other linear unbiased estimator of $\mathbf{q}'\mathbf{b}$, and so is "best".

The importance of this result must not be overlooked. If $\mathbf{q}'\mathbf{b}$ is an estimable function its b.l.u.e. is $\mathbf{q}'\mathbf{b}^o$ with variance $\mathbf{q}'\mathbf{Gq}\sigma^2$; and this is so for any solution $\mathbf{b}^o$ to the normal equations using any generalized inverse $\mathbf{G}$. Both the estimator and its variance are invariant to the choice of $\mathbf{G}$ (and $\mathbf{b}^o$); but this is so *only* for estimable functions and not for non-estimable functions.

The covariance between the b.l.u.e.'s of two estimable functions is derived in a manner similar to (43):

$$\text{cov}(\mathbf{q}_1'\mathbf{b}^o, \mathbf{q}_2'\mathbf{b}^o) = \mathbf{q}_1'\mathbf{Gq}_2\sigma^2, \tag{45}$$

and hence, if $\mathbf{Q}'\mathbf{b}^o$ represents the b.l.u.e.'s of several estimable functions, the variance-covariance matrix of those b.l.u.e.'s is

$$\text{var}(\mathbf{Q}'\mathbf{b}^o) = \mathbf{Q}'\mathbf{G}\mathbf{Q}\sigma^2. \tag{46}$$

c. Confidence intervals

Since it is only estimable functions that have estimators (b.l.u.e.'s) that are invariant to the solution of the normal equations, they are the only functions for which establishing confidence intervals is valid. Similar to equation (94) of Sec. 3.5k we have, on the basis of normality, that the symmetric $100(1 - \alpha)\%$ confidence interval on the estimable function $\mathbf{q}'\mathbf{b}$ is

$$\mathbf{q}'\mathbf{b}^o \pm \hat{\sigma} t_{N-r,\frac{1}{2}\alpha}\sqrt{\mathbf{q}'\mathbf{G}\mathbf{q}} \tag{47}$$

where $t_{N-r,\frac{1}{2}\alpha}$ is defined by the probability statement $\Pr\{t \geq t_{N-r,\frac{1}{2}\alpha}\} = \frac{1}{2}\alpha$ for t having the t-distribution with $N - r$ degrees of freedom. As before, when $N - r$ is large ($N - r \geq 100$, say) $z_{\frac{1}{2}\alpha}$ may be used in place of $t_{N-r,\frac{1}{2}\alpha}$ where $(2\pi)^{-\frac{1}{2}} \int_{z_{\frac{1}{2}\alpha}}^{\infty} e^{-\frac{1}{2}x^2}\, dx = \frac{1}{2}\alpha$.

d. Example (continued)

When defining an estimable function we showed that $\alpha_1 - \alpha_2$ is estimable; and

$$\alpha_1 - \alpha_2 = [0 \quad 1 \quad -1 \quad 0]\mathbf{b} = \mathbf{q}'\mathbf{b}$$

has

$$\mathbf{q}' = [0 \quad 1 \quad -1 \quad 0].$$

Then, using $\mathbf{b}^o$ of (22), the b.l.u.e. of $\alpha_1 - \alpha_2$ is

$$\widehat{\alpha_1 - \alpha_2} = \mathbf{q}'\mathbf{b}^o = [0 \quad 1 \quad -1 \quad 0]\mathbf{b}^o = 100 - 86 = 14$$

with variance

$$v(\widehat{\alpha_1 - \alpha_2}) = \mathbf{q}'\mathbf{G}\mathbf{q}\sigma^2 = (\tfrac{1}{3} + \tfrac{1}{2})\sigma^2 = \tfrac{5}{6}\sigma^2.$$

Using $\mathbf{G}_1$ and $\mathbf{b}_1^o$ of (25) gives the same results: the b.l.u.e. of $\alpha_1 - \alpha_2$ is

$$\widehat{\alpha_1 - \alpha_2} = [0 \quad 1 \quad -1 \quad 0]\mathbf{b}_1^o = 68 - 54 = 14,$$

with

$$v(\widehat{\alpha_1 - \alpha_2}) = \mathbf{q}'\mathbf{G}_1\mathbf{q} = (\tfrac{4}{3} + 1\tfrac{1}{2} - 2)\sigma^2 = \tfrac{5}{6}\sigma^2.$$

The reader should verify that these same properties hold true for other estimable functions, such as $\alpha_2 - \alpha_3$, $\mu + \frac{1}{2}(\alpha_1 + \alpha_2)$, $\alpha_2 - 2\alpha_1 + \alpha_3$ and so on.

From these results, and using $\hat{\sigma}^2 = 23\frac{1}{3}$ from Table 5.7, the symmetric $100(1 - \alpha)\%$ confidence interval on $\alpha_1 - \alpha_2$ is, from (47),

$$14 \pm \sqrt{23\tfrac{1}{3}}\, t_{6-3,\frac{1}{2}\alpha}\sqrt{\tfrac{5}{6}} = 14 \pm 4.41 t_{3,\frac{1}{2}\alpha}$$

and with $t_{3,\frac{1}{2}\alpha} = 3.18$ for $\alpha = 0.05$ this becomes

$$14 \pm 4.41(3.18) = 14 \pm 14 = 0 \text{ to } 28.$$

e. What functions are estimable?

Whenever $\mathbf{q}' = \mathbf{t}'\mathbf{X}$ for some $\mathbf{t}$, then $\mathbf{q}'\mathbf{b}$ is estimable, with b.l.u.e. $\mathbf{q}'\mathbf{b}^o$ having variance $\mathbf{q}'\mathbf{G}\mathbf{q}\sigma^2$. Some special cases are now considered.

Any linear function of $\mathbf{Xb}$ is estimable: $\mathbf{m}'\mathbf{Xb}$, say, for any vector $\mathbf{m}'$. Its b.l.u.e. is

$$\widehat{\mathbf{m}'\mathbf{Xb}} = \mathbf{m}'\mathbf{Xb}^o = \mathbf{m}'\mathbf{XGX}'\mathbf{y}$$

with variance $$v(\widehat{\mathbf{m}'\mathbf{Xb}}) = \mathbf{m}'\mathbf{XGX}'\mathbf{m}\sigma^2. \tag{48}$$

Also estimable is any linear function of $\mathbf{X}'\mathbf{Xb}$: it is a linear function of $\mathbf{Xb}$, $\mathbf{s}'\mathbf{X}'\mathbf{Xb}$ say. Replacing $\mathbf{m}'$ in (48) by $\mathbf{s}'\mathbf{X}'$ gives

$$\widehat{\mathbf{s}'\mathbf{X}'\mathbf{Xb}} = \mathbf{s}'\mathbf{X}'\mathbf{y}$$

and $$v(\widehat{\mathbf{s}'\mathbf{X}'\mathbf{Xb}}) = \mathbf{s}'\mathbf{X}'\mathbf{Xs}\sigma^2. \tag{49}$$

Since $\mathbf{X}'\mathbf{Xb}$ is the same as the left-hand side of the normal equations with $\mathbf{b}^o$ replaced by $\mathbf{b}$ and the b.l.u.e. of $\mathbf{s}'\mathbf{X}'\mathbf{Xb}$ is $\mathbf{s}'\mathbf{X}'\mathbf{y}$ where $\mathbf{X}'\mathbf{y}$ is the right-hand side of the normal equations, we might in this sense say that the b.l.u.e. of any linear function of the left-hand sides of the normal equations is the same function of the right-hand sides.

Linear functions of $E(\mathbf{b}^o)$ are also estimable, because $\mathbf{u}'E(\mathbf{b}^o) = \mathbf{u}'\mathbf{Hb} = \mathbf{u}'\mathbf{GX}'\mathbf{Xb}$. Using $\mathbf{u}'\mathbf{G}$ in place of $\mathbf{s}'$ in (49) shows that

$$\widehat{\mathbf{u}'E(\mathbf{b}^o)} = \mathbf{u}'\mathbf{GX}'\mathbf{y} = \mathbf{u}'\mathbf{b}^o$$

and $$v[\widehat{\mathbf{u}'E(\mathbf{b}^o)}] = v(\mathbf{u}'\mathbf{b}^o) = \mathbf{u}'\mathbf{GX}'\mathbf{XG}'\mathbf{u}\sigma^2 \quad \text{from (9)}. \tag{50}$$

A special case of this result is when $\mathbf{u}'$ takes, in turn, the values of the rows of $\mathbf{I}$: then $\mathbf{b}^o$ is the b.l.u.e. of $\mathbf{Hb}$. These results are summarized in Table 5.8.

In view of the discussion of the F-statistics $F(R)$ and $F(R_m)$ in Sec. 3, it is worth emphasizing two vectors that are not estimable, namely $\mathbf{b}$ and its sub-vector $\boldsymbol{\ell}$. They are not estimable because no value of $\mathbf{q}' = \mathbf{t}'\mathbf{X}$ can be found such that $\mathbf{q}'\mathbf{b}$ reduces to an element of $\mathbf{b}$; i.e., no individual element of $\mathbf{b}$ is estimable. Therefore neither $\mathbf{b}$ nor $\boldsymbol{\ell}$ is estimable.

TABLE 5.8. ESTIMABLE FUNCTIONS AND THEIR B.L.U.E.'S

Estimable Function			
Description	Function	b.l.u.e.	Variance of b.l.u.e.
General case: $\mathbf{q}' = \mathbf{t}'\mathbf{X}$	$\mathbf{q}'\mathbf{b}$	$\mathbf{q}'\mathbf{b}^o$	$\mathbf{q}'\mathbf{G}\mathbf{q}\sigma^2$
Linear function of $\mathbf{Xb}$ ($\mathbf{m}'$ arbitrary)	$\mathbf{m}'\mathbf{Xb}$	$\mathbf{m}'\mathbf{Xb}^o$	$\mathbf{m}'\mathbf{XGX}'\mathbf{m}\sigma^2$
Linear function of $\mathbf{X}'\mathbf{Xb}$ ($\mathbf{s}'$ arbitrary)	$\mathbf{s}'\mathbf{X}'\mathbf{Xb}$	$\mathbf{s}'\mathbf{X}'\mathbf{Xb}^o = \mathbf{s}'\mathbf{X}'\mathbf{y}$	$\mathbf{s}'\mathbf{X}'\mathbf{Xs}\sigma^2$
Linear function of $E(\mathbf{b}^o)$ ($\mathbf{u}'$ arbitrary)	$\mathbf{u}'E(\mathbf{b}^o)$	$\mathbf{u}'\mathbf{b}^o$	$\mathbf{u}'\text{var}(\mathbf{b}^o)\mathbf{u}$
Vector $\mathbf{Hb}$ having $\mathbf{b}^o$ as b.l.u.e.	$\mathbf{Hb}$	$\mathbf{b}^o$	$\text{var}(\mathbf{b}^o) = \mathbf{GX}'\mathbf{XG}'\sigma^2$

f. Linearly independent estimable functions

From Table 5.8 it is evident that there are infinitely many estimable functions. If we ask "How many linearly independent estimable functions are there?" the answer is r, the rank of $\mathbf{X}$; i.e., there are $r(\mathbf{X})$ LIN estimable functions.

Since $\mathbf{q}'\mathbf{b}$ with $\mathbf{q}' = \mathbf{t}'\mathbf{X}$ is estimable for any $\mathbf{t}'$, let $\mathbf{T}'_{N\times N}$ be a matrix of full rank. Then, with $\mathbf{Q}' = \mathbf{T}'\mathbf{X}$, the functions $\mathbf{Q}'\mathbf{b}$ are N estimable functions. But $r(\mathbf{Q}) = r(\mathbf{X})$. Therefore there are only $r(\mathbf{X})$ LIN rows in $\mathbf{Q}'$ and hence only $r(\mathbf{X})$ LIN terms in $\mathbf{Q}'\mathbf{b}$; i.e., only $r(\mathbf{X})$ LIN estimable functions. Thus any set of estimable functions cannot contain more than r LIN such functions.

g. Testing for estimability

A given function $\mathbf{q}'\mathbf{b}$ is estimable if some vector $\mathbf{t}'$ can be found such that $\mathbf{t}'\mathbf{X} = \mathbf{q}'$. However, for $\mathbf{q}'$ known, derivation of a $\mathbf{t}'$ satisfying $\mathbf{t}'\mathbf{X} = \mathbf{q}'$ may not always be easy, especially when $\mathbf{X}$ has large dimensions. Alternative to deriving $\mathbf{t}'$, the estimability of $\mathbf{q}'\mathbf{b}$ can be investigated by seeing whether $\mathbf{q}'$ is such that the equation $\mathbf{q}'\mathbf{H} = \mathbf{q}'$ is satisfied: $\mathbf{q}'\mathbf{b}$ is estimable if and only if $\mathbf{q}'\mathbf{H} = \mathbf{q}'$. This is easily proved: if $\mathbf{q}'\mathbf{b}$ is estimable, $\mathbf{q}' = \mathbf{t}'\mathbf{X}$ and $\mathbf{q}'\mathbf{H} = \mathbf{t}'\mathbf{XH} = \mathbf{t}'\mathbf{X} = \mathbf{q}'$; and if $\mathbf{q}'\mathbf{H} = \mathbf{q}'$, then $\mathbf{q}' = \mathbf{q}'\mathbf{GX}'\mathbf{X} = \mathbf{t}'\mathbf{X}$ for $\mathbf{t}' = \mathbf{q}'\mathbf{GX}'$.

"Is $\mathbf{q}'\mathbf{b}$ estimable?" is now easily answered. It is estimable if $\mathbf{q}'$ satisfies the equation $\mathbf{q}'\mathbf{H} = \mathbf{q}'$. Otherwise it is not. Thus we have a direct procedure for testing the estimability of $\mathbf{q}'\mathbf{b}$: ascertain whether or not $\mathbf{q}'\mathbf{H}$ equals $\mathbf{q}'$. When $\mathbf{q}'\mathbf{H}$ does equal $\mathbf{q}'$, not only is $\mathbf{q}'\mathbf{b}$ estimable but, from the last line of Table 5.8, the b.l.u.e. of $\mathbf{q}'\mathbf{b} = \mathbf{q}'\mathbf{Hb}$ is $\mathbf{q}'\mathbf{b}^o$. This corresponds to the invariance property of $\mathbf{q}'\mathbf{b}^o$ for $\mathbf{q}'\mathbf{H} = \mathbf{q}'$ derived in Theorem 6 of Chapter 1.

h. General expressions

In Table 5.8 and equations (48), $\mathbf{m}'\mathbf{Xb}$ is estimable with b.l.u.e. $\mathbf{m}'\mathbf{Xb}^o$ for any vector $\mathbf{m}'$ of order N. Thus, if we define $\mathbf{x}_j$ as the jth column of $\mathbf{X}$, then

$$\mathbf{X} \equiv [\mathbf{x}_1 \quad \mathbf{x}_2 \quad \cdots \quad \mathbf{x}_p]$$

and

$$\mathbf{m}'\mathbf{Xb} = (\mathbf{m}'\mathbf{x}_1)b_1 + (\mathbf{m}'\mathbf{x}_2)b_2 + \cdots + (\mathbf{m}'\mathbf{x}_p)b_p \tag{51}$$

with b.l.u.e.

$$\widehat{\mathbf{m}'\mathbf{Xb}} = \mathbf{m}'\mathbf{Xb}^o = (\mathbf{m}'\mathbf{x}_1)b_1^o + (\mathbf{m}'\mathbf{x}_2)b_2^o + \cdots + (\mathbf{m}'\mathbf{x}_p)b_p^o. \tag{52}$$

For any values given to the m_i's, the elements of $\mathbf{m}$, those same values used in (51) yield an estimable function, and used in (52) they yield the b.l.u.e. of that estimable function. Hence (51) and (52) constitute general expressions for an estimable function and its b.l.u.e.

Similar results hold for $\mathbf{s}'\mathbf{X}'\mathbf{Xb}$ of (49) where $\mathbf{s}'$ is any vector of order p, in distinction to $\mathbf{m}'$ of (51) and (52) which has order N. Defining $\mathbf{z}_j$ as the jth column of $\mathbf{X}'\mathbf{X}$,

$$\mathbf{X}'\mathbf{X} = [\mathbf{z}_1 \quad \mathbf{z}_2 \quad \cdots \quad \mathbf{z}_p],$$

we have the estimable function

$$\mathbf{s}'\mathbf{X}'\mathbf{Xb} = (\mathbf{s}'\mathbf{z}_1)b_1 + (\mathbf{s}'\mathbf{z}_2)b_2 + \cdots + (\mathbf{s}'\mathbf{z}_p)b_p \tag{53}$$

with b.l.u.e.

$$\widehat{\mathbf{s}'\mathbf{X}'\mathbf{Xb}} = \mathbf{s}'\mathbf{X}'\mathbf{Xb}^o = (\mathbf{s}'\mathbf{z}_1)b_1^o + (\mathbf{s}'\mathbf{z}_2)b_2^o + \cdots + (\mathbf{s}'\mathbf{z}_p)b_p^o. \tag{54}$$

These expressions hold for any elements in $\mathbf{s}'$ of order p, just as (51) and (52) hold for any elements of $\mathbf{m}'$ in order N.

From the last line of Table 5.8 we also have that $\mathbf{w}'\mathbf{Hb}$ is estimable with b.l.u.e. $\mathbf{w}'\mathbf{b}^o$. Thus if

$$\mathbf{w}' = [w_1 \quad w_2 \quad \cdots \quad w_p]$$

and

$$\mathbf{H} = [\mathbf{h}_1 \quad \mathbf{h}_2 \quad \cdots \quad \mathbf{h}_p]$$

then an estimable function is

$$\mathbf{w}'\mathbf{Hb} = (\mathbf{w}'\mathbf{h}_1)b_1 + (\mathbf{w}'\mathbf{h}_2)b_2 + \cdots + (\mathbf{w}'\mathbf{h}_p)b_p \tag{55}$$

and its b.l.u.e. is

$$\widehat{\mathbf{w}'\mathbf{Hb}} = \mathbf{w}'\mathbf{b}^o = w_1b_1^o + w_2b_2^o + \cdots + w_pb_p^o. \tag{56}$$

Expressions (55) and (56) have advantages over (51) and (52) based on $\mathbf{m}'\mathbf{Xb}$ because of fewer arbitrary elements, p instead of N; and over (53) and (54) because of greater simplicity. This is evident in (56), which is just a linear combination of the elements of $\mathbf{b}^o$, each element multiplied by a single arbitrary w. And (55) often has a simple form also, because when $\mathbf{X}'\mathbf{X}$ is a design matrix $\mathbf{H}$ often has $p - r$ null rows $[r = r(\mathbf{X})]$, with its other r rows having elements that are either 0, 1 or -1. The estimable function in (55) accordingly takes on a simple form and involves only r elements of $\mathbf{w}$. Furthermore, $\mathbf{b}^o$ in such cases can have only r non-zero elements too, and so the b.l.u.e. in (56) then involves only r terms.

That $\mathbf{H}$ can often be obtained as a matrix of 0's, 1's and -1's when $\mathbf{X}'\mathbf{X}$ is a design matrix is established as follows. Suppose that

$$\mathbf{X}'\mathbf{X} = \begin{bmatrix} \mathbf{X}_1'\mathbf{X}_1 & \mathbf{X}_1'\mathbf{X}_2 \\ \mathbf{X}_2'\mathbf{X}_1 & \mathbf{X}_2'\mathbf{X}_2 \end{bmatrix} \quad \text{and} \quad \mathbf{G} = \begin{bmatrix} (\mathbf{X}_1'\mathbf{X}_1)^{-1} & \mathbf{0} \\ \mathbf{0} & \mathbf{0} \end{bmatrix}$$

where $\mathbf{X}_1'\mathbf{X}_1$ has full rank, equal to $r(\mathbf{X})$, and $\mathbf{G}$ is a generalized inverse of $\mathbf{X}'\mathbf{X}$. Since $\mathbf{X} = [\mathbf{X}_1 \quad \mathbf{X}_2]$ where $\mathbf{X}_1$ has full column rank, $\mathbf{X}_2 = \mathbf{X}_1\mathbf{M}$ for some matrix $\mathbf{M}$, and because all elements of $\mathbf{X}$ are 0 or 1, those of $\mathbf{M}$ can often be 0, 1, or -1. Hence

$$\mathbf{H} = \mathbf{G}\mathbf{X}'\mathbf{X} = \begin{bmatrix} \mathbf{I} & (\mathbf{X}_1'\mathbf{X}_1)^{-1}\mathbf{X}_1'\mathbf{X}_2 \\ \mathbf{0} & \mathbf{0} \end{bmatrix} = \begin{bmatrix} \mathbf{I} & \mathbf{M} \\ \mathbf{0} & \mathbf{0} \end{bmatrix}$$

and so $p - r$ rows of $\mathbf{H}$ are null and elements in the r non-null rows are often 0, 1 or -1.

i. Example (continued)

From (2), (6), and (21) the values of $\mathbf{X}$, $\mathbf{X}'\mathbf{X}$ and $\mathbf{H}$ are

$$\mathbf{X} = \begin{bmatrix} 1 & 1 & 0 & 0 \\ 1 & 1 & 0 & 0 \\ 1 & 1 & 0 & 0 \\ 1 & 0 & 1 & 0 \\ 1 & 0 & 1 & 0 \\ 1 & 0 & 0 & 1 \end{bmatrix}, \quad \mathbf{X}'\mathbf{X} = \begin{bmatrix} 6 & 3 & 2 & 1 \\ 3 & 3 & 0 & 0 \\ 2 & 0 & 2 & 0 \\ 1 & 0 & 0 & 1 \end{bmatrix} \quad \text{and} \quad \mathbf{H} = \begin{bmatrix} 0 & 0 & 0 & 0 \\ 1 & 1 & 0 & 0 \\ 1 & 0 & 1 & 0 \\ 1 & 0 & 0 & 1 \end{bmatrix}$$

with

$$\mathbf{b} = \begin{bmatrix} \mu \\ \alpha_1 \\ \alpha_2 \\ \alpha_3 \end{bmatrix} \quad \text{and} \quad \mathbf{b}^o = \begin{bmatrix} 0 \\ 100 \\ 86 \\ 32 \end{bmatrix}$$

from (22). With these values, $\mathbf{m}'\mathbf{X}\mathbf{b}$ of (51) is

$$\mathbf{m}'\mathbf{X}\mathbf{b} = (m_1 + m_2 + m_3 + m_4 + m_5 + m_6)\mu + (m_1 + m_2 + m_3)\alpha_1 + (m_4 + m_5)\alpha_2 + m_6\alpha_3 \quad (57)$$

with b.l.u.e., from (52),

$$\widehat{\mathbf{m}'\mathbf{X}\mathbf{b}} = \mathbf{m}'\mathbf{X}\mathbf{b}^o = (m_1 + m_2 + m_3)100 + (m_4 + m_5)86 + m_6 32. \quad (58)$$

Thus for any values of $m_1, \ldots, m_6$, (57) is an estimable function and (58) is its b.l.u.e. Similarly, from (53) and (54) and using $\mathbf{X'X}$

$$\mathbf{s'X'Xb} = (6s_1 + 3s_2 + 2s_3 + s_4)\mu + 3(s_1 + s_2)\alpha_1 + 2(s_1 + s_3)\alpha_2 + (s_1 + s_4)\alpha_3 \quad (59)$$

is estimable with b.l.u.e.

$$\widehat{\mathbf{s'X'Xb}} = \mathbf{s'X'Xb}^o = 300(s_1 + s_2) + 172(s_1 + s_3) + 32(s_1 + s_4). \quad (60)$$

These expressions hold for any values given to the arbitrary s's, of which there are only $p = 4$, compared to $N = 6$ arbitrary m's in (57) and (58). Those with the fewer arbitrary values seem preferable. Likewise, from (55) and (56) and using $\mathbf{H}$, an estimable function is

$$\mathbf{w'Hb} = (w_2 + w_3 + w_4)\mu + w_2\alpha_1 + w_3\alpha_2 + w_4\alpha_3 \quad (61)$$

having b.l.u.e.

$$\mathbf{w'Hb}^o = \mathbf{w'b}^o = 100w_2 + 86w_3 + 32w_4. \quad (62)$$

For any values of w_2, w_3 and w_4, (61) is estimable and (62) is its b.l.u.e.

Note that in using (55) and (56), of which (61) and (62) are examples, the $\mathbf{H}$ used in $\mathbf{w'Hb}$ of (55) must correspond to the $\mathbf{b}^o$ used in $\mathbf{w'b}^o$ of (56). In (55) one cannot use an $\mathbf{H}$ based on a generalized inverse that is different from the one used in deriving $\mathbf{b}^o = \mathbf{GX'y}$. This point is obvious, but important. Of course, (55) and (56) apply for *any* $\mathbf{b}^o$ and its corresponding $\mathbf{H}$. Thus for $\mathbf{b}_1^o$ and $\mathbf{H}_1$ of (25), equations (55) and (56) indicate that

$$\mathbf{w'H_1b} = w_{11}\mu + w_{12}\alpha_1 + w_{13}\alpha_2 + (w_{11} - w_{12} - w_{13})\alpha_3 \quad (63)$$

is estimable with b.l.u.e.

$$\widehat{\mathbf{w_1'H_1b}} = \mathbf{w_1'b_1^o} = 32w_{11} + 68w_{12} + 54w_{13}\,, \quad (64)$$

these results holding for any values w_{11}, w_{12} and w_{13}. Clearly, (63) and (64) are not identical to (61) and (62) but for different sets of values of w_2, w_3 and w_4 in (61) and (62) and of w_{11}, w_{12} and w_{13} in (63) and (64), both pairs of expressions will generate the same set of estimable functions and their b.l.u.e.'s. For example, with $w_2 = \frac{1}{2}$, $w_3 = 0$ and $w_4 = \frac{1}{2}$ equations (61) and (62) give $\mu + \frac{1}{2}(\alpha_1 + \alpha_2)$ estimable with b.l.u.e. $100(\frac{1}{2}) + 86(0) + 32(\frac{1}{2}) = 66$; and with $w_{11} = 1$, $w_{12} = \frac{1}{2}$ and $w_{13} = 0$, equations (63) and (64) give $\mu + \frac{1}{2}(\alpha_1 + \alpha_2)$ as estimable with b.l.u.e. $32(1) + 68(\frac{1}{2}) + 54(0) = 66$.

5. THE GENERAL LINEAR HYPOTHESIS

We here develop the theory for the general linear hypothesis written as H: $\mathbf{K'b} = \mathbf{m}$, just as in Sec. 3.6 for the full rank case. Before considering a test of this hypothesis we first establish its "testability"; some hypotheses can be tested and others cannot.

a. Testable hypotheses

The definition that a testable hypothesis is one which can be tested tells us little. By it we mean that a testable hypothesis is a hypothesis that can be expressed in terms of estimable functions. Of itself this is not to say that hypotheses composed of non-estimable functions cannot be tested—although this *is* the case, as is proved in subsection d. Nevertheless, it seems reasonable that a testable hypothesis be one made up of estimable functions for the following reason: if $\mathbf{K}'\mathbf{b} = \mathbf{m}$ is to be tested, then results for the full rank case suggest that $\mathbf{K}'\mathbf{b}^o - \mathbf{m}$ will be part of the test statistic which, of course, will need to be invariant to $\mathbf{b}^o$. And it will be invariant only if $\mathbf{K}'\mathbf{b}$ is estimable.

A testable hypothesis $\mathbf{H}$: $\mathbf{K}'\mathbf{b} = \mathbf{m}$ is therefore taken as one where

$$\mathbf{K}'\mathbf{b} \equiv \{\mathbf{k}_i'\mathbf{b}\} \qquad \text{for} \quad i = 1, 2, \ldots, s$$

such that $\mathbf{k}_i'\mathbf{b}$ is estimable for all i. Hence $\mathbf{k}_i' = \mathbf{t}_i'\mathbf{X}$ for some $\mathbf{t}_i'$ and so

$$\mathbf{K}' = \mathbf{T}'\mathbf{X} \tag{65}$$

for some matrix $(\mathbf{T}')_{s\times N}$. Furthermore, any hypothesis is considered only in terms of its linearly independent components, and so $(\mathbf{K}')_{s\times p}$ is always of full row rank, s.

Since $\mathbf{K}'\mathbf{b}$ is taken to be a set of estimable functions their b.l.u.e.'s are

$$\widehat{\mathbf{K}'\mathbf{b}} = \mathbf{K}'\mathbf{b}^o \tag{66}$$

with
$$E(\mathbf{K}'\mathbf{b}^o) = \mathbf{K}'\mathbf{b} \tag{67}$$

and
$$\begin{aligned} \text{var}(\widehat{\mathbf{K}'\mathbf{b}}) &= \mathbf{K}' \,\text{var}(\mathbf{b}^o)\mathbf{K} \\ &= \mathbf{K}'\mathbf{G}\mathbf{X}'\mathbf{X}\mathbf{G}'\mathbf{K}\sigma^2, \qquad \text{from (9)} \\ &= \mathbf{K}'\mathbf{G}\mathbf{X}'\mathbf{X}\mathbf{G}'\mathbf{X}'\mathbf{T}\sigma^2, \qquad \text{from (65)} \\ &= \mathbf{K}'\mathbf{G}\mathbf{K}\sigma^2, \end{aligned} \tag{68}$$

using Theorem 7 of Chapter 1, and (65) again. This matrix is non-singular, as is now shown. Because the functions $\mathbf{K}'\mathbf{b}$ are estimable $\mathbf{K}'$ can be represented not only as $\mathbf{T}'\mathbf{X}$ for some $\mathbf{T}'$ but also as $\mathbf{S}'\mathbf{X}'\mathbf{X}$ for some $\mathbf{S}'$ of full row rank s. Then with

$$\mathbf{K}' = \mathbf{S}'\mathbf{X}'\mathbf{X}, \text{ of order } s \times p, \qquad \text{and} \qquad r(\mathbf{K}') = s$$

where $s \leq r$, it is readily shown that

$$\mathbf{S}' \text{ and } \mathbf{S}'\mathbf{X}' \text{ have full row rank } s.$$

[For example, $r(\mathbf{S}') \geq r(\mathbf{K}')$ and so $r(\mathbf{S}') \geq s$; but $\mathbf{S}'$ is $s \times p$ and so $r(\mathbf{S}') \leq s$; hence $r(\mathbf{S}') = s$.] Furthermore

$$\mathbf{K}'\mathbf{G}\mathbf{K} = \mathbf{S}'\mathbf{X}'\mathbf{X}\mathbf{G}\mathbf{X}'\mathbf{X}\mathbf{S} = \mathbf{S}'\mathbf{X}'\mathbf{X}\mathbf{S}$$

and so $r(\mathbf{K}'\mathbf{G}\mathbf{K}) = r(\mathbf{S}'\mathbf{X}') = s =$ the order of $\mathbf{K}'\mathbf{G}\mathbf{K}$. Hence $\mathbf{K}'\mathbf{G}\mathbf{K}$ is non-singular.

b. Testing testable hypotheses

The test for the testable hypothesis H: $\mathbf{K}'\mathbf{b} = \mathbf{m}$ is developed just as in the full rank case (Sec. 3.6a). Normality of $\mathbf{e}$ is assumed and we have, from Secs. 3a and 3b,

$$\mathbf{y} \sim N(\mathbf{X}\mathbf{b}, \sigma^2\mathbf{I}),$$

$$\mathbf{b}^o \sim N(\mathbf{G}\mathbf{X}'\mathbf{X}\mathbf{b}, \mathbf{G}\mathbf{X}'\mathbf{X}\mathbf{G}'\sigma^2)$$

and

$$\mathbf{K}'\mathbf{b}^o - \mathbf{m} \sim N(\mathbf{K}'\mathbf{b} - \mathbf{m}, \mathbf{K}'\mathbf{G}\mathbf{K}\sigma^2)$$

from (67) and (68). Therefore, using Theorem 2 of Chapter 2, the quadratic form

$$Q = (\mathbf{K}'\mathbf{b}^o - \mathbf{m})'(\mathbf{K}'\mathbf{G}\mathbf{K})^{-1}(\mathbf{K}'\mathbf{b}^o - \mathbf{m}) \tag{69}$$

is such that

$$Q/\sigma^2 \sim \chi^{2\prime}[s, (\mathbf{K}'\mathbf{b} - \mathbf{m})'(\mathbf{K}'\mathbf{G}\mathbf{K})^{-1}(\mathbf{K}'\mathbf{b} - \mathbf{m})/2\sigma^2]$$

Furthermore,

$$Q = [\mathbf{y} - \mathbf{X}\mathbf{K}(\mathbf{K}'\mathbf{K})^{-1}\mathbf{m}]'\mathbf{X}\mathbf{G}'\mathbf{K}(\mathbf{K}'\mathbf{G}\mathbf{K})^{-1}\mathbf{K}'\mathbf{G}\mathbf{X}'[\mathbf{y} - \mathbf{X}\mathbf{K}(\mathbf{K}'\mathbf{K})^{-1}\mathbf{m}],$$

with $(\mathbf{K}'\mathbf{K})^{-1}$ existing because $\mathbf{K}'$ has full row rank, and

$$\begin{aligned}\mathbf{K}'\mathbf{G}\mathbf{X}'\mathbf{X}\mathbf{K}(\mathbf{K}'\mathbf{K})^{-1}\mathbf{m} &= \mathbf{T}'\mathbf{X}\mathbf{G}\mathbf{X}'\mathbf{X}\mathbf{K}(\mathbf{K}'\mathbf{K})^{-1}\mathbf{m} \\ &= \mathbf{T}'\mathbf{X}\mathbf{K}(\mathbf{K}'\mathbf{K})^{-1}\mathbf{m} = \mathbf{K}'\mathbf{K}(\mathbf{K}'\mathbf{K})^{-1}\mathbf{m} = \mathbf{m}.\end{aligned}$$

Also,

$$\text{SSE} = [\mathbf{y} - \mathbf{X}\mathbf{K}(\mathbf{K}'\mathbf{K})^{-1}\mathbf{m}]'(\mathbf{I} - \mathbf{X}\mathbf{G}\mathbf{X}')[\mathbf{y} - \mathbf{X}\mathbf{K}(\mathbf{K}'\mathbf{K})^{-1}\mathbf{m}],$$

this being so because $\mathbf{X}'(\mathbf{I} - \mathbf{X}\mathbf{G}\mathbf{X}') = \mathbf{0}$. For the same reason, the matrices of the quadratic forms in these expressions have null products and so Q and SSE are distributed independently. Therefore

$$F(H) = \frac{Q/s}{\text{SSE}/(N - r)} \sim F'[s, N - r, (\mathbf{K}'\mathbf{b} - \mathbf{m})'(\mathbf{K}'\mathbf{G}\mathbf{K})^{-1}(\mathbf{K}'\mathbf{b} - \mathbf{m})/2\sigma^2]$$

and under the null hypothesis H: $\mathbf{K}'\mathbf{b} = \mathbf{m}$, the non-centrality parameter is zero and so $F(H) \sim F_{s,N-r}$. Thus $F(H)$ provides a test of the hypothesis H: $\mathbf{K}'\mathbf{b} = \mathbf{m}$, with

$$F(H) = (\mathbf{K}'\mathbf{b}^o - \mathbf{m})'(\mathbf{K}'\mathbf{G}\mathbf{K})^{-1}(\mathbf{K}'\mathbf{b}^o - \mathbf{m})/s\hat{\sigma}^2 \tag{70}$$

with s and $N - r$ degrees of freedom.

Suppose we now seek a solution for $\mathbf{b}^o$ under the hypothesis H: $\mathbf{K}'\mathbf{b} = \mathbf{m}$. Denote it by $\mathbf{b}^o_H$. Then it will come from minimizing $(\mathbf{y} - \mathbf{X}\mathbf{b}^o_H)'(\mathbf{y} - \mathbf{X}\mathbf{b}^o_H)$ subject to $\mathbf{K}'\mathbf{b}^o_H = \mathbf{m}$. Using a Lagrange multiplier $2\boldsymbol{\theta}'$ this leads, exactly as in equation (102) of Chapter 3, to

$$\mathbf{X}'\mathbf{X}\mathbf{b}^o_H + \mathbf{K}\boldsymbol{\theta} = \mathbf{X}'\mathbf{y} \quad \text{and} \quad \mathbf{K}'\mathbf{b}^o_H = \mathbf{m}. \tag{71}$$

From the first of these a solution is

$$\mathbf{b}^o_H = \mathbf{G}\mathbf{X}'\mathbf{y} - \mathbf{G}\mathbf{K}\boldsymbol{\theta} = \mathbf{b}^o - \mathbf{G}\mathbf{K}\boldsymbol{\theta}$$

and substitution in the second leads, as in (103) of Chapter 3, to

$$\mathbf{b}^o_H = \mathbf{b}^o - \mathbf{G}\mathbf{K}(\mathbf{K}'\mathbf{G}\mathbf{K})^{-1}(\mathbf{K}'\mathbf{b}^o - \mathbf{m}). \tag{72}$$

The error sum of squares after fitting this, to be denoted by SSE_H, is

$$\begin{aligned}\text{SSE}_H &= (\mathbf{y} - \mathbf{X}\mathbf{b}^o_H)'(\mathbf{y} - \mathbf{X}\mathbf{b}^o_H)\\ &= [\mathbf{y} - \mathbf{X}\mathbf{b}^o + \mathbf{X}(\mathbf{b}^o - \mathbf{b}^o_H)]'[\mathbf{y} - \mathbf{X}\mathbf{b}^o + \mathbf{X}(\mathbf{b}^o - \mathbf{b}^o_H)]\\ &= (\mathbf{y} - \mathbf{X}\mathbf{b}^o)'(\mathbf{y} - \mathbf{X}\mathbf{b}^o) + (\mathbf{b}^o - \mathbf{b}^o_H)'\mathbf{X}'\mathbf{X}(\mathbf{b}^o - \mathbf{b}^o_H),\end{aligned} \tag{73}$$

the cross-product term vanishing because $\mathbf{X}'(\mathbf{y} - \mathbf{X}\mathbf{b}^o) = \mathbf{0}$. Substituting from (72) for $\mathbf{b}^o - \mathbf{b}^o_H$ this gives

$$\text{SSE}_H = \text{SSE} + (\mathbf{K}'\mathbf{b}^o - \mathbf{m})'(\mathbf{K}'\mathbf{G}'\mathbf{K})^{-1}\mathbf{K}'\mathbf{G}'\mathbf{X}'\mathbf{X}\mathbf{G}\mathbf{K}(\mathbf{K}'\mathbf{G}\mathbf{K})^{-1}(\mathbf{K}'\mathbf{b}^o - \mathbf{m}).$$

Now $\mathbf{K}' = \mathbf{T}'\mathbf{X}$ and so

$$\mathbf{K}'\mathbf{G}'\mathbf{X}'\mathbf{X}\mathbf{G}\mathbf{K}(\mathbf{K}'\mathbf{G}\mathbf{K})^{-1} = \mathbf{T}'\mathbf{X}\mathbf{G}'\mathbf{X}'\mathbf{X}\mathbf{G}\mathbf{K}(\mathbf{K}'\mathbf{G}\mathbf{K})^{-1} = \mathbf{T}'\mathbf{X}\mathbf{G}\mathbf{K}(\mathbf{K}'\mathbf{G}\mathbf{K})^{-1} = \mathbf{I};$$

and

$$\mathbf{K}'\mathbf{G}'\mathbf{K} = \mathbf{T}'\mathbf{X}\mathbf{G}'\mathbf{X}'\mathbf{T} = \mathbf{T}'\mathbf{X}\mathbf{G}\mathbf{X}'\mathbf{T} = \mathbf{K}'\mathbf{G}\mathbf{K}.$$

Hence

$$\begin{aligned}\text{SSE}_H &= \text{SSE} + (\mathbf{K}'\mathbf{b}^o - \mathbf{m})'(\mathbf{K}'\mathbf{G}\mathbf{K})^{-1}(\mathbf{K}'\mathbf{b}^o - \mathbf{m})\\ &= \text{SSE} + Q\end{aligned} \tag{74}$$

for Q of (69).

c. The hypothesis $\mathbf{K}'\mathbf{b} = \mathbf{0}$

Application of the above results to certain special cases, as was done for the full rank case in Sec. 3.6c, cannot be undertaken here because (74) is limited to cases where $\mathbf{K}'\mathbf{b}$ is estimable. For example, the hypotheses H: $\mathbf{b} = \mathbf{b}_0$ and H: $\mathbf{b}_q = \mathbf{0}$ cannot be tested because $\mathbf{b}$ and $\mathbf{b}_q$ are not estimable. Neither is $\boldsymbol{\ell}$. This is why, as indicated in Sec. 3, tests based on $F(R)$ and $F(R_m)$ cannot be described as testing hypotheses of this nature. Nevertheless, as discussed in Sec. 6.2f(iii), the test based on $F(R_m)$ can sometimes be thought of as appearing equivalent to testing $\boldsymbol{\ell} = \mathbf{0}$.

One special case of the general hypothesis $\mathbf{K}'\mathbf{b} = \mathbf{m}$ is when $\mathbf{m}$ is null. In this situation Q and $\mathbf{b}^o_H$ of (69) and (72) become

$$Q = \mathbf{b}^{o\prime}\mathbf{K}(\mathbf{K}'\mathbf{G}\mathbf{K})^{-1}\mathbf{K}'\mathbf{b}^o \quad \text{and} \quad \mathbf{b}^o_H = \mathbf{b}^o - \mathbf{G}\mathbf{K}(\mathbf{K}'\mathbf{G}\mathbf{K})^{-1}\mathbf{K}'\mathbf{b}^o \tag{75}$$

with

$Q = \text{SSR} -$ reduction in sum of squares due to fitting the reduced model.

Hence, corresponding to Table 3.6 we have the analysis of variance shown in Table 5.9.

TABLE 5.9. ANALYSIS OF VARIANCE FOR TESTING THE HYPOTHESIS $\mathbf{K}'\mathbf{b} = \mathbf{0}$

Source of Variation	d.f.[1]	Sum of Squares
Full model	r	$\text{SSR} = \mathbf{b}^{o\prime}\mathbf{X}'\mathbf{y}$
Hypothesis	s	$Q = \mathbf{b}^{o\prime}\mathbf{K}(\mathbf{K}'\mathbf{G}\mathbf{K})^{-1}\mathbf{K}'\mathbf{b}^o$
Reduced model	$r - s$	$\text{SSR} - Q$
Residual error	$N - r$	SSE
Total	N	$\text{SST} = \mathbf{y}'\mathbf{y}$

[1] $r = r(\mathbf{X})$, and $s = r(\mathbf{K}')$, with $\mathbf{K}'$ having full row rank.

As before, we have three tests of hypotheses:

$$\frac{\text{SSR}/r}{\text{SSE}/(N - r)} \text{ tests the full model,}$$

$$\frac{Q/s}{\text{SSE}/(N - r)} \text{ tests the hypothesis } H: \quad \mathbf{K}'\mathbf{b} = \mathbf{0}$$

and, under the null hypothesis,

$$\frac{(\text{SSR} - Q)/(r-s)}{\text{SSE}/(N - r)} \text{ tests the reduced model.}$$

The first and last of these tests are not to be construed as testing the fit of the models concerned but rather as testing their adequacy in terms of accounting for variation in the y-variable.

Table 5.9 can, of course, be rewritten to make it in terms of "after fitting the mean" (a.f.m.). This is done by subtracting $N\bar{y}^2$ from SSR and SST to get SSR_m and SST_m, as shown in Table 5.10.

TABLE 5.10. ANALYSIS OF VARIANCE FOR TESTING THE HYPOTHESIS $\mathbf{K}'\mathbf{b} = \mathbf{0}$ AFTER FITTING THE MEAN

Source of Variation[1]	d.f.[2]	Sum of Squares
Full model (a.f.m.)	$r - 1$	$SSR_m = SSR - N\bar{y}^2$
Hypothesis	s	$Q = \mathbf{b}^{o\prime}\mathbf{K}(\mathbf{K}'\mathbf{G}\mathbf{K})^{-1}\mathbf{K}'\mathbf{b}^o$
Reduced model (a.f.m.)	$r - s - 1$	$SSR_m - Q$
Residual error	$N - r$	SSE
Total (a.f.m.)	$N - 1$	$SST_m = \mathbf{y}'\mathbf{y} - N\bar{y}^2$

[1] a.f.m. = after fitting the mean.
[2] $r = r(\mathbf{X})$, and $s = r(\mathbf{K}')$, with $\mathbf{K}'$ having full row rank.

The tests of hypotheses are then

$$\frac{SSR_m/(r-1)}{SSE/(N-r)} \text{ tests the full model (a.f.m.),}$$

$$\frac{Q/s}{SSE/(N-r)} \text{ tests the hypothesis } H\colon \mathbf{K}'\mathbf{b} = \mathbf{0}$$

and, under the null hypothesis,

$$\frac{(SSR_m - Q)/(r-s-1)}{SSE/(N-r)} \text{ tests the reduced model (a.f.m.).}$$

As was stated below Table 5.9, the first and last of these tests relate to the adequacy of the models concerned in explaining variation in the y-variable.

The analogy of all these results with the full rank case is clear. In the non-full rank case $\mathbf{G}$ and $\mathbf{b}^o$ are used in place of $(\mathbf{X}'\mathbf{X})^{-1}$ and $\hat{\mathbf{b}}$ of the full rank case. In fact, of course, the full rank model is just a special case of the non-full rank model. For, when $\mathbf{X}'\mathbf{X}$ is non-singular, $\mathbf{G} = (\mathbf{X}'\mathbf{X})^{-1}$ and $\mathbf{b}^o = \hat{\mathbf{b}}$ and all results for the full rank model follow from those of the non-full rank model.

d. Non-testable hypotheses

In stating earlier that a testable hypothesis is one composed of estimable functions we appealed to the intuitive need for having $\mathbf{K}'\mathbf{b}^o$ invariant to $\mathbf{b}^o$ in order to be able to test H: $\mathbf{K}'\mathbf{b} = \mathbf{m}$. We now show explicitly that if $\mathbf{K}'\mathbf{b}$ is not estimable the corresponding value of SSE_H is SSE and so there is no test of H: $\mathbf{K}'\mathbf{b} = \mathbf{m}$.

The equations that result from minimizing $(\mathbf{y} - \mathbf{X}\mathbf{b}^o)'(\mathbf{y} - \mathbf{X}\mathbf{b}^o)$ subject to $\mathbf{K}'\mathbf{b}^o = \mathbf{m}$ are, just as in (71),

$$\mathbf{X}'\mathbf{X}\mathbf{b}^o_H + \mathbf{K}\boldsymbol{\theta} = \mathbf{X}'\mathbf{y} \quad \text{and} \quad \mathbf{K}'\mathbf{b}^o_H = \mathbf{m}, \tag{76}$$

where $2\boldsymbol{\theta}'$ is a vector of Lagrange multipliers. Consider the equations

$$\mathbf{K}'(\mathbf{H}-\mathbf{I})\mathbf{z}_1 = \mathbf{m} - \mathbf{K}'\mathbf{G}\mathbf{X}'\mathbf{y} \tag{77}$$

in $\mathbf{z}_1$. As indicated in the proof of Theorem 4 of Sec. 1.2c, $(\mathbf{H}-\mathbf{I})\mathbf{z}_1$ contains $p - r$ arbitrary elements. Now because $\mathbf{K}'\mathbf{b}$ is not estimable, $\mathbf{K}' \neq \mathbf{T}'\mathbf{X}$ for any $\mathbf{T}'$ and so, because $\mathbf{X} = \mathbf{X}\mathbf{G}\mathbf{X}'\mathbf{X}$ (Theorem 7, Chapter 1), $\mathbf{K}' \neq (\mathbf{T}'\mathbf{X}\mathbf{G})\mathbf{X}'\mathbf{X}$ for any $\mathbf{T}'$; i.e., the rows of $\mathbf{K}'$ are LIN of those of $\mathbf{X}'\mathbf{X}$. But $\mathbf{X}'\mathbf{X}$ has order p and rank r. Also, the rows of $\mathbf{K}'$ have order p and are to be LIN of each other. Therefore, if they are also to be LIN of the rows of $\mathbf{X}'\mathbf{X}$ there can be no more than $p - r$ of them; i.e., $\mathbf{K}'$ has no more than $p - r$ rows. Hence (77) represents no more than $p - r$ equations in the $p - r$ unknowns of $(\mathbf{H}-\mathbf{I})\mathbf{z}_1$ and so always has at least one solution for $\mathbf{z}_1$. Using it for $\mathbf{z}$ in

$$\mathbf{b}^o = \mathbf{G}\mathbf{X}'\mathbf{y} + (\mathbf{H}-\mathbf{I})\mathbf{z} \tag{78}$$

to obtain

$$\mathbf{b}^o_H = \mathbf{G}\mathbf{X}'\mathbf{y} + (\mathbf{H}-\mathbf{I})\mathbf{z}_1, \tag{79}$$

we find that $\boldsymbol{\theta} = \mathbf{0}$ and $\mathbf{b}^o_H$ of (79) satisfy (76). Consequently, because (79) is just a subset of the solutions (78) to $\mathbf{X}'\mathbf{X}\mathbf{b}^o = \mathbf{X}'\mathbf{y}$,

$$\text{SSE}_H = (\mathbf{y} - \mathbf{X}\mathbf{b}^o_H)'(\mathbf{y} - \mathbf{X}\mathbf{b}^o_H) = \text{SSE}$$

and so there is no test of the hypothesis. Thus when $\mathbf{K}'\mathbf{b}$ is not estimable, there is no test of the hypothesis H: $\mathbf{K}'\mathbf{b} = \mathbf{m}$.

In comparing equations (71) and (76) the sole difference between them is that $\mathbf{K}'\mathbf{b}$ is estimable in (71) whereas in (76) it is not. In solving (71) it is the estimability condition ($\mathbf{K}' = \mathbf{T}'\mathbf{X}$ for some $\mathbf{T}'$) that leads to the solution (72). On the other hand, in solving (76) the solution for $\mathbf{b}^o_H$ is also a solution of $\mathbf{X}'\mathbf{X}\mathbf{b} = \mathbf{X}'\mathbf{y}$, as shown in (77) and (79). It is the lack of estimability of $\mathbf{K}'\mathbf{b}$ that allows this. In contrast, in (71) where $\mathbf{K}'\mathbf{b}$ is estimable, $\mathbf{K}' = \mathbf{S}'\mathbf{X}'\mathbf{X}$ for some $\mathbf{S}'$ and so, for $\mathbf{b}^o$ of (78), $\mathbf{K}'\mathbf{b}^o = \mathbf{S}'\mathbf{X}'\mathbf{X}\mathbf{b}^o_H = \mathbf{S}'\mathbf{X}'\mathbf{y}$ for all values of $\mathbf{z}$. Therefore no value of $\mathbf{z}$ in (78) can be found such that $\mathbf{K}'\mathbf{b}^o = \mathbf{m}$, and so no value of (78) exists which satisfies (71).

A further extension is that of trying to test a hypothesis which consists partly of estimable functions and partly of non-estimable functions. Suppose H: $\mathbf{K}'\mathbf{b} = \mathbf{m}$ can be written as

$$H: \quad \begin{bmatrix} \mathbf{K}_1'\mathbf{b} \\ \mathbf{k}'\mathbf{b} \end{bmatrix} = \begin{bmatrix} \mathbf{m}_1 \\ m_2 \end{bmatrix} \tag{80}$$

where $\mathbf{K}_1'\mathbf{b}$ is estimable but $\mathbf{k}'\mathbf{b}$ is not. Then, using two Lagrange multipliers, the same development as above will lead to the conclusion that testing (80) is indistinguishable from testing just H: $\mathbf{K}_1'\mathbf{b} = \mathbf{m}_1$. Hence in carrying out the test of a hypothesis which consists partly of estimable functions and partly of non-estimable functions, all we are doing is testing the hypothesis made up of just the estimable functions. (See Exercise 7.)

e. Checking for testability

The logic of deriving

$$Q = (\mathbf{K}'\mathbf{b}^\circ - \mathbf{m})'(\mathbf{K}'\mathbf{G}\mathbf{K})^{-1}(\mathbf{K}'\mathbf{b}^\circ - \mathbf{m})$$

depends upon $\mathbf{K}'\mathbf{b}$ being estimable. Nevertheless, when $\mathbf{K}'\mathbf{b}$ is not estimable Q can be calculated so long as $\mathbf{K}'\mathbf{G}\mathbf{K}$ is non-singular. This is so because estimability of $\mathbf{K}'\mathbf{b}$ is a sufficient condition for the existence of Q, in particular for the existence of $(\mathbf{K}'\mathbf{G}\mathbf{K})^{-1}$, but it is not a necessary condition. Hence whenever $(\mathbf{K}'\mathbf{G}\mathbf{K})^{-1}$ exists Q can be calculated even if $\mathbf{K}'\mathbf{b}$ is not estimable. Checking to see that $\mathbf{K}'\mathbf{b}$ is estimable is therefore essential before calculating Q and $F(H)$. This can be done by ascertaining the existence of some $\mathbf{T}'$ such that $\mathbf{K}' = \mathbf{T}'\mathbf{X}$, or by seeing if $\mathbf{K}'$ satisfies $\mathbf{K}' = \mathbf{K}'\mathbf{H}$.

Suppose, however, that checking the estimability of $\mathbf{K}'\mathbf{b}$ in this manner is overlooked and Q is calculated. Then, if in fact $\mathbf{K}'\mathbf{b}$ is not estimable, what hypothesis, if any, is $F(H)$ testing? The answer is H: $\mathbf{K}'\mathbf{H}\mathbf{b} = \mathbf{m}$. We show this as follows. Since H: $\mathbf{K}'\mathbf{H}\mathbf{b} = \mathbf{m}$ is always testable the value of Q for testing this hypothesis (call it Q_1) is, from (69),

$$Q_1 = (\mathbf{K}'\mathbf{H}\mathbf{b}^\circ - \mathbf{m})'(\mathbf{K}'\mathbf{H}\mathbf{G}\mathbf{H}'\mathbf{K})^{-1}(\mathbf{K}'\mathbf{H}\mathbf{b}^\circ - \mathbf{m}). \qquad (81)$$

In this expression

$$\mathbf{K}'\mathbf{H}\mathbf{b}^\circ = \mathbf{K}'\mathbf{G}\mathbf{X}'\mathbf{X}\mathbf{G}\mathbf{X}'\mathbf{y} = \mathbf{K}'\mathbf{G}\mathbf{X}'\mathbf{X}\mathbf{G}'\mathbf{X}'\mathbf{y} = \mathbf{K}'\mathbf{G}_1\mathbf{X}'\mathbf{y}$$

because $\mathbf{X}\mathbf{G}\mathbf{X}' = \mathbf{X}\mathbf{G}'\mathbf{X}'$ (Theorem 7, Sec. 1.5a), and where

$$\mathbf{G}_1 = \mathbf{G}\mathbf{X}'\mathbf{X}\mathbf{G}'$$

is a generalized inverse of $\mathbf{X}'\mathbf{X}$. Therefore

$$\mathbf{K}'\mathbf{H}\mathbf{b}^\circ = \mathbf{K}'\mathbf{G}_1\mathbf{X}'\mathbf{y} = \mathbf{K}'\mathbf{b}_1^\circ$$

where

$$\mathbf{b}_1^\circ = \mathbf{G}_1\mathbf{X}'\mathbf{y}$$

is a solution of $\mathbf{X}'\mathbf{X}\mathbf{b}^\circ = \mathbf{X}'\mathbf{y}$. Also,

$$\mathbf{K}'\mathbf{H}\mathbf{G}\mathbf{H}'\mathbf{K} = \mathbf{K}'\mathbf{G}\mathbf{X}'\mathbf{X}\mathbf{G}\mathbf{X}'\mathbf{X}\mathbf{G}'\mathbf{K} = \mathbf{K}'\mathbf{G}\mathbf{X}'\mathbf{X}\mathbf{G}'\mathbf{K} = \mathbf{K}'\mathbf{G}_1\mathbf{K}$$

and so in (81)

$$Q_1 = (\mathbf{K}'\mathbf{b}_1^\circ - \mathbf{m})'(\mathbf{K}'\mathbf{G}_1\mathbf{K})^{-1}(\mathbf{K}'\mathbf{b}_1^\circ - \mathbf{m}).$$

Thus Q_1 is identical to the numerator sum of squares that would be calculated from (69) for testing the non-testable hypothesis $\mathbf{K}'\mathbf{b} = \mathbf{m}$ using the solution $\mathbf{b}_1^\circ = \mathbf{G}_1\mathbf{X}'\mathbf{y}$. Hence the calculations that might be made when trying to test the non-testable hypothesis $\mathbf{K}'\mathbf{b} = \mathbf{m}$ are indistinguishable from those entailed in testing the testable hypothesis $\mathbf{K'Hb} = \mathbf{m}$; i.e., if $F(H)$ of (70) is calculated for a hypothesis $\mathbf{K'b} = \mathbf{m}$ that is non-testable, the hypothesis actually being tested is $\mathbf{K'Hb} = \mathbf{m}$.

f. Example (continued)

From (21) and (22)

$$\mathbf{G} = \begin{bmatrix} 0 & 0 & 0 & 0 \\ 0 & \frac{1}{3} & 0 & 0 \\ 0 & 0 & \frac{1}{2} & 0 \\ 0 & 0 & 0 & 1 \end{bmatrix}, \qquad \mathbf{H} = \begin{bmatrix} 0 & 0 & 0 & 0 \\ 1 & 1 & 0 & 0 \\ 1 & 0 & 1 & 0 \\ 1 & 0 & 0 & 1 \end{bmatrix} \quad \text{and} \quad \mathbf{b}^o = \begin{bmatrix} 0 \\ 100 \\ 86 \\ 32 \end{bmatrix} \tag{82}$$

with (24), (26) and (27) being

$$\text{SSR} = 45{,}816, \qquad \text{SST} = \mathbf{y'y} = 45{,}886 \quad \text{and} \quad \text{SSM} = 42{,}336, \tag{83}$$

and hence

$$\hat{\sigma}^2 = (45{,}886 - 45{,}816)/3 = 70/3. \tag{84}$$

Consider H: $\alpha_1 = \alpha_2 + 10$. It can be written as $[0 \quad 1 \quad -1 \quad 0]\mathbf{b} = 10$ where $\mathbf{k'H} = \mathbf{k'} = [0 \quad 1 \quad -1 \quad 0]$. Therefore it is a testable hypothesis and the F-statistic for testing it, (70), is derived as follows:

$$\begin{aligned} \mathbf{k'b}^o - m &= 100 - 86 - 10 = 4, \\ \mathbf{k'Gk} &= \tfrac{1}{3} + \tfrac{1}{2} = \tfrac{5}{6} \\ \text{and} \qquad F(H) &= \frac{4(5/6)^{-1}4}{1(70/3)} = \frac{3(6)16}{5(70)} = \frac{144}{175}. \end{aligned} \tag{85}$$

Or again, consider H: $\mu + \alpha_1 = \mu + \alpha_2 = 90$, written as

$$\mathbf{K'b} = \begin{bmatrix} 1 & 1 & 0 & 0 \\ 1 & 0 & 1 & 0 \end{bmatrix}\mathbf{b} = \begin{bmatrix} 90 \\ 90 \end{bmatrix}$$

where $\mathbf{K'H} = \mathbf{K'}$. Hence the hypothesis is testable and

$$\mathbf{K'b}^o - \mathbf{m} = \begin{bmatrix} 100 \\ 86 \end{bmatrix} - \begin{bmatrix} 90 \\ 90 \end{bmatrix} = \begin{bmatrix} 10 \\ -4 \end{bmatrix},$$

$$\mathbf{K'GK} = \begin{bmatrix} 0 & \frac{1}{3} & 0 & 0 \\ 0 & 0 & \frac{1}{2} & 0 \end{bmatrix} \begin{bmatrix} 1 & 1 \\ 1 & 0 \\ 0 & 1 \\ 0 & 0 \end{bmatrix} = \begin{bmatrix} \frac{1}{3} & 0 \\ 0 & \frac{1}{2} \end{bmatrix},$$

and

$$F(H) = \frac{[10 \quad -4]\begin{bmatrix} 3 & 0 \\ 0 & 2 \end{bmatrix}\begin{bmatrix} 10 \\ -4 \end{bmatrix}}{2(70/3)} = \frac{332(3)}{2(70)} = \frac{249}{35}.$$

This same hypothesis could also be written as

$$\mathbf{K}'\mathbf{b} = \begin{bmatrix} 1 & 1 & 0 & 0 \\ 0 & 1 & -1 & 0 \end{bmatrix}\mathbf{b} = \begin{bmatrix} 90 \\ 0 \end{bmatrix}$$

for which

$$\mathbf{K}'\mathbf{b}^o - \mathbf{m} = \begin{bmatrix} 100 \\ 14 \end{bmatrix} - \begin{bmatrix} 90 \\ 0 \end{bmatrix} = \begin{bmatrix} 10 \\ 14 \end{bmatrix}$$

and
$$\mathbf{K}'\mathbf{G}\mathbf{K} = \begin{bmatrix} 0 & \frac{1}{3} & 0 & 0 \\ 0 & \frac{1}{3} & -\frac{1}{2} & 0 \end{bmatrix}\begin{bmatrix} 1 & 0 \\ 1 & 1 \\ 0 & -1 \\ 0 & 0 \end{bmatrix} = \begin{bmatrix} \frac{1}{3} & \frac{1}{3} \\ \frac{1}{3} & \frac{5}{6} \end{bmatrix}.$$

Hence
$$F(H) = \frac{[10 \quad 14]6\begin{bmatrix} \frac{5}{6} & -\frac{1}{3} \\ -\frac{1}{3} & \frac{1}{3} \end{bmatrix}\begin{bmatrix} 10 \\ 14 \end{bmatrix}}{2(70/3)} = \frac{500 - 560 + 392}{2(70/3)} = \frac{249}{35},$$

the same result as before.

To test the hypothesis H: $\alpha_1 = \alpha_2$, it is written as $[0 \quad 1 \quad -1 \quad 0]\mathbf{b} = 0$, and is seen to be testable. As in (85), $\mathbf{k}'\mathbf{G}\mathbf{k} = \frac{5}{6}$, and now $\mathbf{k}'\mathbf{b}^o - m = 14$. Hence $Q = 14^2(\frac{5}{6})^{-1} = 235.2$. Table 5.9 then has the values shown in Table 5.11. If fitting the mean is to be taken into account, as in Table 5.10, SSM = 42,336 is subtracted from SSR and SST to get SSR_m and SST_m, as shown in Table 5.12.

As an example of a non-testable hypothesis consider H: $\alpha_1 + \alpha_2 = 220$ which written as $\mathbf{k}'\mathbf{b} = m$ gives $\mathbf{k}' = [0 \quad 1 \quad 1 \quad 0]$. Equation (77) is then

$$[0 \quad 1 \quad 1 \quad 0]\begin{bmatrix} -1 & 0 & 0 & 0 \\ 1 & 0 & 0 & 0 \\ 1 & 0 & 0 & 0 \\ 1 & 0 & 0 & 0 \end{bmatrix}\mathbf{z}_1 = 220 - [0 \quad 1 \quad 1 \quad 0]\begin{bmatrix} 0 \\ 100 \\ 86 \\ 32 \end{bmatrix}$$

giving $2\mathbf{z}_1 = 220 - 186$, i.e., $\mathbf{z}_1 = 17$. Therefore in (79)

$$\mathbf{b}_H^o = \begin{bmatrix} 0 \\ 100 \\ 86 \\ 32 \end{bmatrix} + \begin{bmatrix} -1 \\ 1 \\ 1 \\ 1 \end{bmatrix} 17 = \begin{bmatrix} -17 \\ 117 \\ 103 \\ 49 \end{bmatrix}$$

TABLE 5.11. EXAMPLE OF TABLE 5.9

Source	d.f.	Sum of Squares
Full model	3	SSR = 45,816
Hypothesis	1	$Q =$ 235.2
Reduced model	2	SSR $- Q =$ 45,580.8
Residual error	3	SSE = 70
Total	6	SST = 45,886

TABLE 5.12. EXAMPLE OF TABLE 5.10

Source	d.f.	Sum of Squares
Full model (a.f.m.)	2	$\text{SSR}_m = 3480$
Hypothesis	1	$Q = 235.2$
Reduced model (a.f.m.)	1	$\text{SSR}_m - Q = 3244.8$
Residual error	3	SSE = 70
Total (a.f.m.)	5	$\text{SST}_m = 3550$

and the error sum of squares after fitting this is, by (12),

$$\begin{aligned}\text{SSE}_H &= 45{,}886 - [-17(504) + 117(300) + 103(172) + 49(32)]\\ &= 45{,}886 - 45{,}816 = 70\end{aligned}$$

identical to SSE. Hence there is no test for H: $\alpha_1 + \alpha_2 = 220$.

Suppose the testability had not been checked. The formula for Q could nevertheless be calculated:

$$\mathbf{k}'\mathbf{b}^o - m = 186 - 220 = -34 \quad \text{and} \quad \mathbf{k}'\mathbf{G}\mathbf{k} = 5/6$$

so that

$$F(H) = (-34)(5/6)^{-1}(-34)/(70/3) = (18)34^2/350 = 10{,}404/175.$$

We show that this is the same as testing H: $\mathbf{k}'\mathbf{H}\mathbf{b} = 220$, namely

$$H: \quad \begin{bmatrix} 0 & 1 & 1 & 0 \end{bmatrix} \begin{bmatrix} 0 & 0 & 0 & 0 \\ 1 & 1 & 0 & 0 \\ 1 & 0 & 1 & 0 \\ 1 & 0 & 0 & 1 \end{bmatrix} \begin{bmatrix} \mu \\ \alpha_1 \\ \alpha_2 \\ \alpha_3 \end{bmatrix} = 220$$

which is H: $2\mu + \alpha_1 + \alpha_2 = 220$. For this, $\mathbf{k}' = [2 \quad 1 \quad 1 \quad 0]$, $\mathbf{k}'\mathbf{H} = \mathbf{k}'$, $\mathbf{k}'\mathbf{b}^o - m = -34$ and $\mathbf{k}'\mathbf{G}\mathbf{k} = 5/6$, and so

$$F(H) = (-34)(5/6)^{-1}(-34)/(70/3) = 18(34^2)/350 = 10{,}404/175.$$

g. Independent and orthogonal contrasts

The numerator sum of squares for testing H: $\mathbf{K}'\mathbf{b} = \mathbf{0}$ is, as in (75),

$$Q = \mathbf{b}^{o\prime}\mathbf{K}(\mathbf{K}'\mathbf{G}\mathbf{K})^{-1}\mathbf{K}'\mathbf{b}^o. \tag{86}$$

For $\mathbf{K}'\mathbf{b}$ being estimable, $\mathbf{K}' = \mathbf{S}'\mathbf{X}'\mathbf{X}$ for some $\mathbf{S}'$, as discussed following equation (68), and so with $\mathbf{b}^o = \mathbf{G}\mathbf{X}'\mathbf{y}$

$$\begin{aligned} Q &= \mathbf{y}'\mathbf{X}\mathbf{G}'\mathbf{X}'\mathbf{X}\mathbf{S}(\mathbf{S}'\mathbf{X}'\mathbf{X}\mathbf{G}\mathbf{X}'\mathbf{X}\mathbf{S})^{-1}\mathbf{S}'\mathbf{X}'\mathbf{X}\mathbf{G}\mathbf{X}'\mathbf{y} \\ &= \mathbf{y}'\mathbf{X}\mathbf{S}(\mathbf{S}'\mathbf{X}'\mathbf{X}\mathbf{S})^{-1}\mathbf{S}'\mathbf{X}'\mathbf{y}, \end{aligned}$$

on using Theorem 7 of Sec. 1.5a. Furthermore, $\mathbf{K}'$ has full row rank s, and when $s = r = r(\mathbf{X})$, it can be shown that $\mathbf{X}\mathbf{S} = \mathbf{X}_1\mathbf{P}\mathbf{X}_1'\mathbf{X}$ where $\mathbf{X}_1$, a sub-matrix of $\mathbf{X}$, is $N \times r$ of full column rank, with $\mathbf{P}$ and $\mathbf{X}_1'\mathbf{X}_1$ both non-singular. This leads to $\mathbf{S}(\mathbf{S}'\mathbf{X}'\mathbf{X}\mathbf{S})^{-1}\mathbf{S}'$ being a generalized inverse of $\mathbf{X}'\mathbf{X}$ (see Exercise 4) and so

$$Q = \mathbf{y}'\mathbf{X}\mathbf{G}\mathbf{X}'\mathbf{y} = \text{SSR} \qquad \text{when} \quad s = r = r(\mathbf{X}). \tag{87}$$

Now $r = r(\mathbf{X})$ is the maximum number of LIN estimable functions (see Sec. 4f). Hence (87) shows that the sum of squares SSR due to fitting the model $E(\mathbf{y}) = \mathbf{X}\mathbf{b}$ is exactly equivalent to the numerator sum of squares for testing the hypothesis $\mathbf{K}'\mathbf{b} = \mathbf{0}$ when $\mathbf{K}'\mathbf{b}$ represents the maximum number of LIN estimable functions, namely $r = r(\mathbf{X})$. This means that if $\mathbf{k}_i'$ is a row of $\mathbf{K}'$, then the numerator sum of squares for simultaneously testing $\mathbf{k}_i'\mathbf{b} = 0$ for $i = 1, 2, \ldots, r$ equals SSR. But it does not necessarily mean that for testing the r hypotheses $\mathbf{k}_i'\mathbf{b} = 0$ individually the sums of squares add up to SSR. This will be so only in certain cases, which are now discussed.

Suppose that $\mathbf{k}_i'$ and $\mathbf{k}_j'$ are two rows of $\mathbf{K}'$. Then

$$q_i = \mathbf{b}^{o\prime}\mathbf{k}_i(\mathbf{k}_i'\mathbf{G}\mathbf{k}_i)^{-1}\mathbf{k}_i'\mathbf{b}^o = \mathbf{y}'\mathbf{X}\mathbf{G}'\mathbf{k}_i(\mathbf{k}_i'\mathbf{G}\mathbf{k}_i)^{-1}\mathbf{k}_i'\mathbf{G}\mathbf{X}'\mathbf{y}$$

and

$$q_j = \mathbf{b}^{o\prime}\mathbf{k}_j(\mathbf{k}_j'\mathbf{G}\mathbf{k}_j)^{-1}\mathbf{k}_j'\mathbf{b}^o = \mathbf{y}'\mathbf{X}\mathbf{G}'\mathbf{k}_j(\mathbf{k}_j'\mathbf{G}\mathbf{k}_j)^{-1}\mathbf{k}_j'\mathbf{G}\mathbf{X}'\mathbf{y} \tag{88}$$

are the numerator sums of squares for testing the hypotheses $\mathbf{k}_i'\mathbf{b} = 0$ and $\mathbf{k}_j'\mathbf{b} = 0$ respectively. By Theorem 4 of Chapter 2 these sums of squares, viewed as quadratics in $\mathbf{y}$ which we are assuming has the $N(\mathbf{X}\mathbf{b}, \sigma^2\mathbf{I})$ distribution, will be independent when

$$\mathbf{X}\mathbf{G}'\mathbf{k}_i(\mathbf{k}_i'\mathbf{G}\mathbf{k}_i)^{-1}\mathbf{k}_i'\mathbf{G}\mathbf{X}'\mathbf{X}\mathbf{G}'\mathbf{k}_j(\mathbf{k}_j'\mathbf{G}\mathbf{k}_j)^{-1}\mathbf{k}_j'\mathbf{G}\mathbf{X}' = \mathbf{0}.$$

A necessary and sufficient condition for this is

$$\mathbf{k}_i'\mathbf{G}\mathbf{X}'\mathbf{X}\mathbf{G}'\mathbf{k}_j = 0.$$

Since $\mathbf{k}_j'\mathbf{b}$ is estimable, $\mathbf{k}_j' = \mathbf{t}_j'\mathbf{X}$ for some $\mathbf{t}_j'$, and so the condition becomes

$$\mathbf{k}_i'\mathbf{G}\mathbf{X}'\mathbf{X}\mathbf{G}'\mathbf{X}'\mathbf{t}_j = \mathbf{k}_i'\mathbf{G}\mathbf{X}'\mathbf{t}_j = \mathbf{k}_i'\mathbf{G}\mathbf{k}_j = 0. \tag{89}$$

Thus (89) is a condition that makes q_i and q_j of (88) independent. It also makes

$$(\mathbf{K}'\mathbf{G}\mathbf{K})^{-1} = \text{diag}\{(\mathbf{k}_i'\mathbf{G}\mathbf{k}_i)^{-1}\} \quad \text{for } i = 1, 2, \ldots, r$$

and so (86) becomes

$$\begin{aligned} Q &= \sum_{i=1}^{r} \mathbf{b}^{o\prime}\mathbf{k}_i(\mathbf{k}_i'\mathbf{G}\mathbf{k}_i)^{-1}\mathbf{k}_i'\mathbf{b}^o \\ &= \sum_{i=1}^{r} \frac{(\mathbf{k}_i'\mathbf{b}^o)^2}{\mathbf{k}_i'\mathbf{G}\mathbf{k}_i} = \sum_{i=1}^{r} q_i. \end{aligned} \tag{90}$$

Condition (89) is also, by (45), the condition that $\mathbf{k}_i'\mathbf{b}^o$ and $\mathbf{k}_j'\mathbf{b}^o$ are independent. Hence when $\mathbf{K}'\mathbf{b}$ consists of $r = r(\mathbf{X})$ LIN functions $\mathbf{k}_i'\mathbf{b}$ for $i = 1, 2, \ldots, r$, and when the $\mathbf{k}_i'\mathbf{b}^o$ are distributed independently, then the numerator sum of squares Q for testing $\mathbf{K}'\mathbf{b} = \mathbf{0}$ not only equals SSR by (87) but also equals, in (90), the sum of the numerator sums of squares q_i for testing the hypotheses $\mathbf{k}_i'\mathbf{b} = 0$ for $i = 1, 2, \ldots, r$. This can be summarized as follows. When, for $i = 1, 2, \ldots, r$,

$$\mathbf{k}_i' = \mathbf{k}_i'\mathbf{H}, \tag{91}$$

$$\mathbf{k}_i'\mathbf{G}\mathbf{k}_j = 0 \quad \text{for} \quad i \neq j \tag{92}$$

$$\text{the } \mathbf{k}_i' \text{ are LIN}, \tag{93}$$

then $\qquad F(H) = Q/r\hat{\sigma}^2$ tests H: $\mathbf{K}'\mathbf{b} = \mathbf{0}$

and $\qquad F(H_i) = q_i/\hat{\sigma}^2$ tests H_i: $\mathbf{k}_i'\mathbf{b} = 0$

and

$$Q = \text{SSR} = \sum_{i=1}^{r} q_i, \tag{94}$$

and the q_i's are mutually independent, with

$$q_i = \frac{(\mathbf{k}_i'\mathbf{b}^o)^2}{\mathbf{k}_i'\mathbf{G}\mathbf{k}_i}.$$

Under their respective null hypotheses, $F(H)$ is distributed as $F_{r,N-r}$ and $F(H_i)$ as $F_{1,N-r}$, the latter being equivalent to t-tests with $N - r$ degrees of freedom using

$$\sqrt{\frac{q_i}{\hat{\sigma}^2}} = \frac{\mathbf{k}_i'\mathbf{b}^o}{\sqrt{\mathbf{k}_i'\mathbf{G}\mathbf{k}_i\hat{\sigma}^2}}$$

as the t-statistic to test H_i.

In the case of balanced data these conditions lead to sets of values for the $\mathbf{k}_i'$ such that the $\mathbf{k}_i'\mathbf{b}$ are often called orthogonal contrasts; "orthogonal" because $\mathbf{G}$ is then such that (92) reduces to $\mathbf{k}_i'\mathbf{k}_j = 0$ and "contrasts" because the $\mathbf{k}_i'\mathbf{b}$ can be expressed as sums of differences between elements of $\mathbf{b}$. The name "orthogonal contrasts" is retained here (for unbalanced data), meaning orthogonal in the sense of (92). Examples are given below and in Chapter 6.

h. Example (continued)

In the example, $r(\mathbf{X}) = r = 3$. Illustration of (87) is given by considering the hypothesis H: $\mathbf{K}'\mathbf{b} = \mathbf{0}$ for

$$\mathbf{K}' = \begin{bmatrix} 3 & 1 & 1 & 1 \\ 0 & 2 & -1 & -1 \\ 0 & 0 & 1 & -1 \end{bmatrix}.$$

The rows of $\mathbf{K}'$ are LIN and, because $\mathbf{K}'\mathbf{H} = \mathbf{K}'$, the elements of $\mathbf{K}'\mathbf{b}$ are estimable. Using $\mathbf{b}^\circ$ and $\mathbf{G}$ of (82) the numerator sum of squares for testing the hypothesis is, from (86),

$$Q = [218 \quad 82 \quad 54] \left[\tfrac{1}{6} \begin{bmatrix} 11 & -5 & -3 \\ -5 & 17 & 3 \\ -3 & 3 & 9 \end{bmatrix} \right]^{-1} \begin{bmatrix} 218 \\ 82 \\ 54 \end{bmatrix}$$

$$= [218 \quad 82 \quad 54] \tfrac{1}{12} \begin{bmatrix} 8 & 2 & 2 \\ 2 & 5 & -1 \\ 2 & -1 & 9 \end{bmatrix} \begin{bmatrix} 218 \\ 82 \\ 54 \end{bmatrix}$$

$$= [8(218^2) + 5(82^2) + 9(54^2) + 2(2)218(82) + 2(2)218(54) - 2(1)82(54)]/12$$

$$= 45{,}816 = \text{SSR in Table 5.11}.$$

Thus simultaneous testing of

$$H_1: \quad 3\mu + \alpha_1 + \alpha_2 + \alpha_3 = 0,$$
$$H_2: \quad 2\alpha_1 - \alpha_2 - \alpha_3 = 0$$

and

$$H_3: \quad \alpha_2 - \alpha_3 = 0$$

utilizes a numerator sum of squares equal to SSR. But adding the numerator sums of squares for testing these hypotheses individually does not give SSR:

Hypothesis	Numerator Sum of Squares
$3\mu + \alpha_1 + \alpha_2 + \alpha_3 = 0$	$218^2/(11/6) = 25{,}922.2$
$2\alpha_1 - \alpha_2 - \alpha_3 = 0$	$82^2/(17/6) = 2{,}373.2$
$\alpha_2 - \alpha_3 = 0$	$54^2/(9/6) = 1{,}944.0$
	Total $30{,}239.4 \neq 45{,}816$

With balanced data, the individual hypotheses of $\mathbf{K}'\mathbf{b} = \mathbf{0}$ given above would be considered orthogonal contrasts. Not so with unbalanced data, however, for the b.l.u.e.'s of the estimable functions involved in the hypotheses are not distributed independently. This is so because the covariance matrix of the b.l.u.e.'s [see (46)],

$$\text{var}\,(\mathbf{K}'\mathbf{b}^\circ) = \mathbf{K}'\mathbf{G}\mathbf{K}\sigma^2 = \tfrac{1}{6}\begin{bmatrix} 11 & -5 & -3 \\ -5 & 17 & 3 \\ -3 & 3 & 9 \end{bmatrix}\sigma^2,$$

does not have its off-diagonal elements zero. With balanced data $\mathbf{K}'\mathbf{G}\mathbf{K}$ would be diagonal, so giving rise to independence.

To derive a set of orthogonal contrasts in the manner of (92) we need $\mathbf{K}'$ such that its rows satisfy (91)–(93). Suppose one contrast of interest is $\alpha_1 - \alpha_3$, and in seeking two others orthogonal to it we take $\mathbf{K}'$ to have the form

$$\mathbf{K}' = \begin{bmatrix} a & b & c & d \\ 0 & 1 & 0 & -1 \\ 0 & f & g & h \end{bmatrix},$$

Then (91), using $\mathbf{H}$ of (82), demands that

$$b + c + d = a \qquad \text{and} \qquad f + g + h = 0;$$

and (92), using $\mathbf{G}$ of (82), requires that

$$\tfrac{1}{3}b - d = 0, \qquad \tfrac{1}{3}f - h = 0 \qquad \text{and} \qquad \tfrac{1}{3}bf + \tfrac{1}{2}cg + dh = 0.$$

Solutions to these two sets of equations are

$$\tfrac{1}{6}a = \tfrac{1}{3}b = \tfrac{1}{2}c = d \qquad \text{and} \qquad \tfrac{1}{3}f = -\tfrac{1}{4}g = h$$

for any values of d and h. For example, putting $d = 1$ and $h = 1$ gives

$$\mathbf{K}' = \begin{bmatrix} 6 & 3 & 2 & 1 \\ 0 & 1 & 0 & -1 \\ 0 & 3 & -4 & 1 \end{bmatrix}$$

for which

$$\mathbf{K}'\mathbf{b}^o = \begin{bmatrix} 504 \\ 68 \\ -12 \end{bmatrix} \qquad \text{and} \qquad \mathbf{K}'\mathbf{G}\mathbf{K} = \begin{bmatrix} 6 & 0 & 0 \\ 0 & \tfrac{4}{3} & 0 \\ 0 & 0 & 12 \end{bmatrix},$$

the latter having its off-diagonal elements zero, in accord with having chosen $\mathbf{K}$ so as to satisfy (92). Furthermore, it is readily seen that the rows of $\mathbf{K}'$ are LIN and so satisfy (93). The hypothesis $\mathbf{K}'\mathbf{b} = \mathbf{0}$ is then tested using, from (86),

$$Q = [504 \quad 68 \quad -12]\begin{bmatrix} \frac{1}{6} & 0 & 0 \\ 0 & \frac{3}{4} & 0 \\ 0 & 0 & \frac{1}{12} \end{bmatrix}\begin{bmatrix} 504 \\ 68 \\ -12 \end{bmatrix}$$

$$= 504^2/6 + 68^2(\tfrac{3}{4}) + (-12)^2/12$$

$$= 42{,}336 + 3{,}468 + 12$$

$$= 45{,}816 = \text{SSR of Table 5.11.}$$

Thus the contrasts

$$6\mu + 3\alpha_1 + 2\alpha_2 + \alpha_3,$$

$$\alpha_1 \qquad - \alpha_3$$

and

$$3\alpha_1 - 4\alpha_2 + \alpha_3$$

which are estimable and LIN are also orthogonal in the manner of (92), and the numerator sums of squares for testing the hypotheses that each of them is zero add to that for testing them simultaneously, namely SSR. Thus is (94) illustrated.

Notice that for testing H: $6\mu + 3\alpha_1 + 2\alpha_2 + \alpha_3 = 0$ the numerator sum of squares is $504^2/6 = 42{,}336 = N\bar{y}^2 = \text{SSM}$; and the sums of squares for the contrasts orthogonal to this, 3468 and 12, sum to 3480, SSR_m, the sum of squares due to fitting the model, after correcting for the mean (see Table 5.12). In general, consider any contrast $\mathbf{k}'\mathbf{b}$ that is to be orthogonal to $6\mu + 3\alpha_1 + 2\alpha_2 + \alpha_3$. By (91), with $\mathbf{H}$ of (82) the form of $\mathbf{k}'$ must be

$$\mathbf{k}' = [k_2 + k_3 + k_4 \quad k_2 \quad k_3 \quad k_4].$$

And (92) requires that $\mathbf{k}'$ must satisfy

$$\mathbf{k}'\mathbf{G}\begin{bmatrix} 6 \\ 3 \\ 2 \\ 1 \end{bmatrix} = \mathbf{k}'\begin{bmatrix} 0 & 0 & 0 & 0 \\ 0 & \frac{1}{3} & 0 & 0 \\ 0 & 0 & \frac{1}{2} & 0 \\ 0 & 0 & 0 & 1 \end{bmatrix}\begin{bmatrix} 6 \\ 3 \\ 2 \\ 1 \end{bmatrix} = \mathbf{k}'\begin{bmatrix} 0 \\ 1 \\ 1 \\ 1 \end{bmatrix} = 0.$$

By the form of $\mathbf{k}'$ this means that $k_2 + k_3 + k_4 = 0$, so that

$$\mathbf{k}' = [0 \quad k_2 \quad k_3 \quad k_4]$$

with $k_2 + k_3 + k_4 = 0$. Thus any contrast $\mathbf{k}'\mathbf{b}$ having $k_2 + k_3 + k_4 = 0$ satisfies (91) and (92), is orthogonal in the manner of (92) to $6\mu + 3\alpha_1 + 2\alpha_2 + \alpha_3$ and, because the first term of $\mathbf{k}'$ is zero, does not involve μ.

$2\alpha_1 - \alpha_2 - \alpha_3$ is one such contrast. Any $r - 1$ such contrasts orthogonal to each other will have numerator sums of squares that sum to SSR_m. For example, for

$$\mathbf{K}' = \begin{bmatrix} 0 & 2 & -1 & -1 \\ 0 & a & b & c \end{bmatrix}$$

$\mathbf{K}'\mathbf{b}$ will be a pair of orthogonal contrasts, orthogonal to each other and to $6\mu + 3\alpha_1 + 2\alpha_2 + \alpha_3$ if

$$a + b + c = 0$$

and

$$[0 \quad 2 \quad -1 \quad -1]\mathbf{G}\begin{bmatrix} 0 \\ a \\ b \\ c \end{bmatrix} = \tfrac{2}{3}a - \tfrac{1}{2}b - c = 0.$$

Solutions to these equations are

$$\tfrac{7}{3} = -\tfrac{7}{10}b = c$$

for any c. Thus for $c = 7$

$$\mathbf{K}' = \begin{bmatrix} 0 & 2 & -1 & -1 \\ 0 & 3 & -10 & 7 \end{bmatrix}$$

for which

$$\mathbf{K}'\mathbf{b}^o = \begin{bmatrix} 82 \\ -336 \end{bmatrix} \quad \text{and} \quad \mathbf{K}'\mathbf{G}\mathbf{K} = \begin{bmatrix} 17/6 & 0 \\ 0 & 102 \end{bmatrix}$$

so that in (86)

$$\begin{aligned} Q &= 82^2(6)/17 + 336^2/102 = 2{,}373\tfrac{3}{17} + 1{,}106\tfrac{14}{17} \\ &= 3{,}480 = \text{SSR}_m \text{ of Table 5.12.} \end{aligned}$$

These few examples illustrate the several ways in which (91)–(94) can be used for establishing independent and orthogonal contrasts for unbalanced data and testing hypotheses about them. Other examples are shown in Chapter 6.

6. RESTRICTED MODELS

Reference has been made to the fact that sometimes a linear model may include restrictions on the elements of the parameter vector. Such restrictions are quite different from the "usual constraints" frequently introduced for the sole purpose of getting a solution to the normal equations. These are discussed in Sec. 7. In contrast, the restrictions envisaged here are considered to be an integral part of the model and as such must be taken into account in the estimation and testing processes.

The discussion so far has been in terms of models whose parameters have been very loosely defined. Indeed, no formal definition has been made. In writing the equation of the model as $\mathbf{y} = \mathbf{Xb} + \mathbf{e}$ we simply described $\mathbf{b}$ as being the vector of the "parameters of the model" and left it at that; thus in the example, μ is described simply as a general mean and α_1, α_2 and α_3 as effects on yield arising from three different plant varieties. No further definition is implied. Sometimes, however, more explicit definitions inherent in the model result in relationships (or restrictions) existing among the parameters of the model. These are considered part and parcel of the model. For example, the situation may be such that the parameters of the model satisfy the relation $\alpha_1 + \alpha_2 + \alpha_3 = 0$; that is, we take this not as a hypothesis to be tested but as a fact, without question. Relationships of this nature, existing as an integral part of a model, will be called *restrictions on the model*. Their origin and concept is not the same as that of relationships that sometimes get imposed on the solutions of normal equations in order to simplify obtaining those solutions; those relationships will be called *constraints on the solutions*. They are discussed in Sec. 7. But here we are concerned with an aspect of the model, that it includes relationships among its parameters. A simple example might be a model involving the three angles of a triangle; or one involving a total weight and its components, such as fat, bone, muscle and lean in a dressed beef carcass.

The models already discussed, those that contain no restrictions of the kind just referred to, will be referred to as *unrestricted models*. And models that do include restrictions of this nature will be called *restricted models*. The question then arises as to how the estimation and hypothesis testing processes developed for unrestricted models apply to restricted models. In general, we consider the set of restrictions

$$\mathbf{P}'\mathbf{b} = \boldsymbol{\delta} \tag{95}$$

as part of the model, where $\mathbf{P}'$ has full row rank q. The restricted model is then $\mathbf{y} = \mathbf{Xb} + \mathbf{e}$ restricted by $\mathbf{P}'\mathbf{b} = \boldsymbol{\delta}$. Fitting this restricted model leads, just as in (71), to

$$\mathbf{X}'\mathbf{X}\mathbf{b}_r^o + \mathbf{P}\boldsymbol{\theta} = \mathbf{X}'\mathbf{y}$$

and (96)

$$\mathbf{P}'\mathbf{b}_r^o = \boldsymbol{\delta}$$

where $2\boldsymbol{\theta}$ is a vector of Lagrange multipliers, and the subscript r on $\mathbf{b}_r^o$ denotes that $\mathbf{b}_r^o$ is a solution to the normal equations of the restricted model. To solve (96) distinction must be made as to whether, in the unrestricted model, $\mathbf{P}'\mathbf{b}$ is estimable or not estimable, because the solution is not the same in the two cases. We first consider the case of $\mathbf{P}'\mathbf{b}$ being estimable.

a. Restrictions involving estimable functions

When $\mathbf{P}'\mathbf{b}$ is estimable we have, by analogy with (72), that a solution to (96) is

$$\mathbf{b}_r^o = \mathbf{b}^o - \mathbf{GP}(\mathbf{P}'\mathbf{GP})^{-1}(\mathbf{P}'\mathbf{b}^o - \boldsymbol{\delta}). \tag{97}$$

Its expected value is

$$E(\mathbf{b}_r^o) = \mathbf{Hb} - \mathbf{GP}(\mathbf{P}'\mathbf{GP})^{-1}(\mathbf{P}'\mathbf{Hb} - \boldsymbol{\delta}) = \mathbf{Hb},$$

using $E(\mathbf{b}^o) = \mathbf{Hb}$ of (8), $\mathbf{P}'\mathbf{H} = \mathbf{P}'$ because $\mathbf{P}'\mathbf{b}$ is estimable, and (95). Similarly the variance of $\mathbf{b}_r^o$ is, after a little simplification,

$$\text{var}(\mathbf{b}_r^o) = \text{var}\{[\mathbf{I} - \mathbf{GP}(\mathbf{P}'\mathbf{GP})^{-1}\mathbf{P}']\mathbf{b}^o\} = \mathbf{G}[\mathbf{X}'\mathbf{X} - \mathbf{P}(\mathbf{P}'\mathbf{GP})^{-1}\mathbf{P}']\mathbf{G}'\sigma^2.$$

The error sum of squares after fitting this restricted model is

$$\text{SSE}_r = (\mathbf{y} - \mathbf{Xb}_r^o)'(\mathbf{y} - \mathbf{Xb}_r^o)$$

and from (73) and (74) this is seen to be

$$\text{SSE}_r = \text{SSE} + (\mathbf{P}'\mathbf{b}^o - \boldsymbol{\delta})'(\mathbf{P}'\mathbf{GP})^{-1}(\mathbf{P}'\mathbf{b}^o - \boldsymbol{\delta}) \tag{98}$$

with

$$E(\text{SSE}_r) = (N - r)\sigma^2 + E\mathbf{b}^{o\prime}\mathbf{P}(\mathbf{P}'\mathbf{GP})^{-1}\mathbf{P}'\mathbf{b}^o - \boldsymbol{\delta}'(\mathbf{P}'\mathbf{GP})^{-1}\boldsymbol{\delta}.$$

On applying Theorem 1 of Chapter 2 to the middle term and using (8) and (95) again, this reduces to

$$E(\text{SSE}_r) = (N - r + q)\sigma^2.$$

Hence, in the restricted model, an unbiased estimator of the error variance is

$$\hat{\sigma}_r^2 = \frac{\text{SSE}_r}{N - r + q}. \tag{99}$$

(There should be no confusion over the letter r used as the rank of $\mathbf{X}$ and as a subscript to denote "restricted".)

$\mathbf{b}_r^o$ and SSE_r of (97) and (98) not being the same as $\mathbf{b}^o$ and SSE indicates that estimable restrictions on the parameters of a model affect the estimation process. However, this does not affect the estimability of any function that is estimable in the unrestricted model. Thus if $\mathbf{k}'\mathbf{b}$ is estimable in the unrestricted model it is still estimable in the restricted model. The condition for estimability—that for some $\mathbf{t}'$, $E(\mathbf{t}'\mathbf{y}) = \mathbf{k}'\mathbf{b}$—remains unaltered. However, although the function is still estimable, it *is* a function of the parameters and is therefore subject to the restrictions $\mathbf{P}'\mathbf{b} = \boldsymbol{\delta}$. These may change the form of $\mathbf{k}'\mathbf{b}$. Thus, in the example, the function $\mathbf{k}'\mathbf{b} = \mu + \frac{1}{2}(\alpha_1 + \alpha_2)$ is estimable, but in a restricted model having $\alpha_1 - \alpha_2 = 0$ as a restriction $\mathbf{k}'\mathbf{b}$ becomes $\mu + \alpha_1$, or equivalently $\mu + \alpha_2$.

In general, the estimable function $\mathbf{k}'\mathbf{b}$ is changed to $\mathbf{k}'\mathbf{b} + \boldsymbol{\lambda}'(\mathbf{P}'\mathbf{b} - \boldsymbol{\delta})$ where, in order that this be just a function of the b's, $\boldsymbol{\lambda}'$ must be such that $\boldsymbol{\lambda}'\boldsymbol{\delta} = 0$. (When $\boldsymbol{\delta}$ is null $\boldsymbol{\lambda}'$ can be any vector.) Then $\mathbf{k}'\mathbf{b}$ becomes $\mathbf{k}'\mathbf{b} + \boldsymbol{\lambda}'\mathbf{P}'\mathbf{b} = (\mathbf{k}' + \boldsymbol{\lambda}'\mathbf{P}')\mathbf{b}$. This, of course, is also estimable under the unrestricted model because both $\mathbf{k}'\mathbf{b}$ and $\mathbf{P}'\mathbf{b}$ are.

In the restricted model the hypothesis H: $\mathbf{K}'\mathbf{b} = \mathbf{m}$ can be considered only if it is consistent with $\mathbf{P}'\mathbf{b} = \boldsymbol{\delta}$; for example, if $\mathbf{P}'\mathbf{b} = \boldsymbol{\delta}$ is $\alpha_1 - \alpha_2 = 0$ one cannot consider the hypothesis $\alpha_1 - \alpha_2 = 4$. Within this limitation of consistency the hypothesis $\mathbf{K}'\mathbf{b} = \mathbf{m}$ is tested in the restricted model by considering the unrestricted model $\mathbf{y} = \mathbf{X}\mathbf{b} + \mathbf{e}$ subject to both the restrictions $\mathbf{P}'\mathbf{b} = \boldsymbol{\delta}$ and the testable hypothesis $\mathbf{K}'\mathbf{b} = \mathbf{m}$. The restricted model reduced by the hypothesis $\mathbf{K}'\mathbf{b} = \mathbf{m}$ can be called the *reduced restricted model*. On writing

$$\mathbf{Q}' = \begin{bmatrix} \mathbf{P}' \\ \mathbf{K}' \end{bmatrix} \quad \text{and} \quad \ell = \begin{bmatrix} \boldsymbol{\delta} \\ \mathbf{m} \end{bmatrix}$$

we minimize $(\mathbf{y} - \mathbf{X}\mathbf{b})'(\mathbf{y} - \mathbf{X}\mathbf{b})$ subject to

$$\mathbf{Q}'\mathbf{b} = \ell.$$

Since both $\mathbf{P}'$ and $\mathbf{K}'$ have full row rank and their rows are mutually LIN, $\mathbf{Q}'$ has full row rank and $\mathbf{Q}'\mathbf{b}$ is estimable. The minimization, leading to the solution $\mathbf{b}^o_{r,H}$, therefore gives in accord with (72)

$$\mathbf{b}^o_{r,H} = \mathbf{b}^o - \mathbf{G}\mathbf{Q}(\mathbf{Q}'\mathbf{G}\mathbf{Q})^{-1}(\mathbf{Q}'\mathbf{b}^o - \ell).$$

The corresponding residual error sum of squares is

$$\text{SSE}_{r,H} = \text{SSE} + (\mathbf{Q}'\mathbf{b}^o - \ell)'(\mathbf{Q}'\mathbf{G}\mathbf{Q})^{-1}(\mathbf{Q}'\mathbf{b}^o - \ell)$$

and the test of the hypothesis $\mathbf{K}'\mathbf{b} = \mathbf{m}$ is based on

$$F(H_r) = (\text{SSE}_{r,H} - \text{SSE}_r)/s\,\hat{\sigma}^2_r \tag{100}$$

where $\hat{\sigma}^2_r = \text{SSE}_r/(N - r + q)$ as in (99).

Just as with estimable functions, a hypothesis that is testable in the unrestricted model is also testable in the restricted model. Modification of the hypothesis by the restrictions may change the form of the hypothesis, but its modified form will be testable not only under the restricted model but also in the unrestricted model.

Example. The hypothesis H: $\mu + \frac{1}{2}(\alpha_1 + \alpha_2) = 20$ is testable in the unrestricted model. In a restricted model having $\alpha_1 - \alpha_2 = 4$ as a restriction the hypothesis is modified to be H: $\mu + \alpha_2 = 18$ or, equivalently, H: $\mu + \alpha_1 = 22$. These are testable in the restricted model: they are also testable in the unrestricted model.

In general, if $\mathbf{K}'\mathbf{b} = \mathbf{m}$ is testable in the unrestricted model then, for $\mathbf{L}_{s\times q}$ being any matrix, $(\mathbf{K}' + \mathbf{LP}')\mathbf{b} = \mathbf{m} + \mathbf{L}\boldsymbol{\delta}$ will be testable in the restricted model: it will also be testable in the unrestricted model.

b. Restrictions involving non-estimable functions

When the restrictions are $\mathbf{P}'\mathbf{b} = \boldsymbol{\delta}$ and $\mathbf{P}'\mathbf{b}$ is not estimable, the solutions to (96) are, similar to (79),

$$\mathbf{b}_r^o = \mathbf{b}^o + (\mathbf{H} - \mathbf{I})\mathbf{z}_1 \tag{101}$$

where, following (77), $\mathbf{z}_1$ satisfies

$$\mathbf{P}'(\mathbf{H} - \mathbf{I})\mathbf{z}_1 = \boldsymbol{\delta} - \mathbf{P}'\mathbf{G}\mathbf{X}'\mathbf{y}. \tag{102}$$

Hence $\mathbf{b}_r^o$ is just one of the solutions to the normal equations $\mathbf{X}'\mathbf{X}\mathbf{b} = \mathbf{X}'\mathbf{y}$. Therefore, in this case $\text{SSE}_r = \text{SSE}$; i.e., the restrictions do not affect the residual error sum of squares.

Just as before, the inclusion of restrictions in the model does not alter the estimability of a function that is estimable in the unrestricted model. It is still estimable in the restricted model. But, because of the restrictions it will be amended; and since the restrictions do not involve estimable functions the amended form of an estimable function may be a function which, although estimable in the restricted model, is not estimable in the unrestricted model. For example, the function $\mu + \frac{1}{2}(\alpha_1 + \alpha_2)$ is estimable in the unrestricted model, but in a restricted model that includes the restriction $\alpha_1 = 0$, it is amended to be $\mu + \frac{1}{2}\alpha_2$, and this, although estimable in the restricted model, would not be estimable in the unrestricted model.

Thus it is that functions which are not estimable in unrestricted models may be estimable in restricted models. In general, if $\mathbf{k}'\mathbf{b}$ is estimable in the unrestricted model then $\mathbf{k}'\mathbf{b} + \boldsymbol{\lambda}'(\mathbf{P}'\mathbf{b} - \boldsymbol{\delta})$ is estimable in the restricted model; to eliminate $\boldsymbol{\delta}$ when it is non-null, $\boldsymbol{\lambda}'$ must be such that $\boldsymbol{\lambda}'\boldsymbol{\delta} = 0$; and the function $\mathbf{k}'\mathbf{b} + \boldsymbol{\lambda}'\mathbf{P}'\mathbf{b}$ is then estimable in the restricted model.

Just as $\text{SSE}_r = \text{SSE}$ when the restrictions involve non-estimable functions so too, when testing the hypothesis $\mathbf{K}'\mathbf{b} = \mathbf{m}$, will $\text{SSE}_{r,H} = \text{SSE}_H$. Hence the F-statistic for testing the hypothesis is identical to that of the unrestricted model. Thus, so far as calculation of the F-statistic is concerned, the imposition of restrictions involving non-estimable functions makes no difference at all. SSE and SSE_H are calculated in the usual manner. Thus the F-statistic is calculated just as in (70).

However, although calculation of the F-statistic is not affected by the model having restrictions on its parameters—restrictions involving non-estimable functions—these restrictions *do* apply to the hypotheses being tested, just as they do to estimable functions, discussed above. Thus hypotheses that are testable in the unrestricted model are also testable in the restricted model; but application of the restrictions may change their form so that although they are testable in the restricted model they may not be testable in the unrestricted model. For example, H: $\alpha_1 - 2\alpha_2 + \alpha_3 = 17$ is testable in the unrestricted model, but in a restricted model having the restriction $\alpha_1 + \alpha_2 = 3$ the hypothesis becomes H: $3\alpha_1 + \alpha_3 = 23$; this is testable in the restricted model, but would not be testable in the unrestricted model.

In general, if $\mathbf{K}'\mathbf{b} = \mathbf{m}$ is testable in the unrestricted model then, for $\mathbf{L}_{s\times q}$ being any matrix, $(\mathbf{K}' + \mathbf{LP}')\mathbf{b} = \mathbf{m} + \mathbf{L\delta}$ is testable in the restricted model; it would not be testable in the unrestricted model.

Table 5.13 summarizes the results of this section so far as estimable functions and tests of hypotheses are concerned.

7. THE "USUAL CONSTRAINTS"

The source of difficulties with the model not of full rank is that the normal equations $\mathbf{X}'\mathbf{X}\mathbf{b}^{\circ} = \mathbf{X}'\mathbf{y}$ have no unique solution. Our discussions have skirted this situation by using a generalized inverse of $\mathbf{X}'\mathbf{X}$. Other presentations impose "the usual constraints" or the "usual restrictions". By this is meant, for example, solving normal equations such as

$$\begin{aligned} 6\mu^{\circ} + 2\alpha_1^{\circ} + 2\alpha_2^{\circ} + 2\alpha_3^{\circ} &= y_{..} \\ 2\mu^{\circ} + 2\alpha_1^{\circ} \qquad\qquad\qquad &= y_{1.} \\ 2\mu^{\circ} \qquad\quad + 2\alpha_2^{\circ} \qquad\quad &= y_{2.} \\ 2\mu^{\circ} \qquad\qquad\qquad + 2\alpha_3^{\circ} &= y_{3.} \end{aligned} \tag{103}$$

by imposing the "usual constraint" of $\alpha_1^{\circ} + \alpha_2^{\circ} + \alpha_3^{\circ} = 0$. This leads to a solution immediately.

The use of these "usual constraints" is described in many ways. For example, Kempthorne (1952, p. 80) writes, "To obtain a unique solution we may impose any condition, the simplest one (generally) being $\sum \hat{t}_j = 0$"; and Federer (1955, p. 159) has, "Now in order to obtain a unique solution the following restrictions are necessary: $\sum r_i = \sum c_j = \sum t_h = 0$"; while in Steel and Torrie (1960, p. 115) we find, "... impose a restriction that $\sum \hat{\tau}_j = 0$ where $\hat{\tau}_j$ is our estimate of τ_j; in the population either $\sum \tau_j = 0$ (fixed effects) or, $\mu_\tau = 0$ (random effects) .... In our illustration $\hat{\alpha}_1 + 2\hat{\alpha}_2 = 0 \cdots$". Constraints (as we shall call them) of this kind can be perfectly permissible, so long as the implications of their use are well understood by the user. The above quotations illustrate the difficulties a student might have in comprehending the "usual constraints", for each of them errs in a manner all too commonly found. First, these constraints cannot be "any" conditions. Second, in situations of unbalanced data those of the form

TABLE 5.13. SUMMARY OF ESTIMATION AND HYPOTHESIS TESTING IN UNRESTRICTED AND RESTRICTED MODELS

Property	Unrestricted model, and restricted model with restrictions $\mathbf{P}'\mathbf{b} = \boldsymbol{\delta}$ where $\mathbf{P}'\mathbf{b}$ is non-estimable and $\mathbf{P}'$ has full row rank q	Restricted model with restrictions $\mathbf{P}'\mathbf{b} = \boldsymbol{\delta}$ where $\mathbf{P}'\mathbf{b}$ is estimable and $\mathbf{P}'$ has full row rank q
Solutions to normal equations	$\mathbf{b}^o = \mathbf{GX}'\mathbf{y}$	$\mathbf{b}^o_r = \mathbf{b}^o - \mathbf{GP}(\mathbf{P}'\mathbf{GP})^{-1}(\mathbf{P}'\mathbf{b}^o - \boldsymbol{\delta})$
Error sum of squares	$\text{SSE} = \mathbf{y}'\mathbf{y} - \mathbf{b}^{o\prime}\mathbf{X}'\mathbf{y}$	$\text{SSE}_r = \text{SSE} + \mathbf{t}'(\mathbf{P}'\mathbf{GP})^{-1}\mathbf{t}$ where $\mathbf{t} = \mathbf{P}'\mathbf{b}^o - \boldsymbol{\delta}$
Estimated error variance	$\hat{\sigma}^2 = \text{SSE}/(N - r)$	$\hat{\sigma}^2_r = \text{SSE}_r/(N - r + q)$
Estimable functions	$\mathbf{k}'\mathbf{b}$ for $\mathbf{k}'\mathbf{H} = \mathbf{k}'$ and, in restricted models, $\mathbf{k}'\mathbf{b} + \boldsymbol{\lambda}'\mathbf{P}'\mathbf{b}$ for $\boldsymbol{\lambda}'$ such that $\boldsymbol{\lambda}'\boldsymbol{\delta} = 0$ (any $\boldsymbol{\lambda}'$ when $\boldsymbol{\delta} = \mathbf{0}$)	
Is a function that is estimable in the restricted model always estimable in the unrestricted model?	No	Yes

Testable hypotheses	$\mathbf{K}'\mathbf{b} = \mathbf{m}$ for $\mathbf{K}'\mathbf{H} = \mathbf{K}'$, of full row rank s and, in restricted models, $(\mathbf{K}' + \mathbf{L}\mathbf{P}')\mathbf{b} = \mathbf{m} + \mathbf{L}\boldsymbol{\delta}$ for any $\mathbf{L}$	
Is a hypothesis that is testable in the restricted model always testable in the unrestricted model?	No	Yes
F-statistic for testing testable hypotheses	$F(H) = \dfrac{(\mathbf{K}'\mathbf{b}^o - \mathbf{m})'(\mathbf{K}'\mathbf{G}\mathbf{K})^{-1}(\mathbf{K}'\mathbf{b}^o - \mathbf{m})}{s\hat{\sigma}^2}$ with $(s, N - r)$ degrees of freedom.	With $\mathbf{Q} = \begin{bmatrix} \mathbf{P}' \\ \mathbf{K}' \end{bmatrix}$ and $\ell = \begin{bmatrix} \boldsymbol{\delta} \\ \mathbf{m} \end{bmatrix}$ $F(H_r) = \dfrac{\text{SSE} - \text{SSE}_r + (\mathbf{Q}'\mathbf{b}^o - \ell)'(\mathbf{Q}'\mathbf{G}\mathbf{Q})^{-1}(\mathbf{Q}'\mathbf{b}^o - \ell)}{s\hat{\sigma}_r^2}$ with $(s, N - r + q)$ degrees of freedom.
Solution for $\mathbf{b}$ under null hypothesis	$\mathbf{b}_H^o = \mathbf{b}^o - \mathbf{G}\mathbf{K}(\mathbf{K}'\mathbf{G}\mathbf{K})^{-1}(\mathbf{K}'\mathbf{b}^o - \mathbf{m})$	$\mathbf{b}_{r,H}^o = \mathbf{b}^o - \mathbf{G}\mathbf{Q}(\mathbf{Q}'\mathbf{G}\mathbf{Q})^{-1}(\mathbf{Q}'\mathbf{b}^o - \ell)$

$\sum \alpha_i^o = 0$ are generally not the simplest. Third, such constraints are not necessary for solving normal equations; they are only sufficient. Any constraints that lead to a solution suffice. Fourth, they can be used whether or not a similar relationship holds for the elements of the model; and only if it does, with enough such relationships in the model to make it a full rank model, will the solutions of the normal equations then be estimates of the parameters of the model. These points we now expand on.

We have seen that with any solution $\mathbf{b}^o$ to the normal equations we can derive most things of interest in linear model estimation: SSE $= \mathbf{y}'\mathbf{y} - \mathbf{b}^{o\prime}\mathbf{X}'\mathbf{y}$, the analysis of variance, the error variance estimate $\hat{\sigma}^2 = \text{SSE}/(N - r)$, and the b.l.u.e. of any estimable function $\mathbf{k}'\mathbf{b}$ as $\widehat{\mathbf{k}'\mathbf{b}} = \mathbf{k}'\mathbf{b}^o$. These things can be obtained provided we have a solution $\mathbf{b}^o$, no matter how it has been derived. However, for some things, the generalized inverse of $\mathbf{X}'\mathbf{X}$ that yielded $\mathbf{b}^o$ is needed.* For example, to ascertain the estimability of a function or to test a testable hypothesis the generalized inverse is, if not absolutely necessary, certainly very useful. Yet applying constraints to the solutions is, as will be shown, probably the easiest way of getting a solution to the normal equations; but if we also want the generalized inverse corresponding to that solution, the constraints must be imposed in a way that readily yields the generalized inverse, *and* the implications of doing this must be recognized.

a. Limitations on constraints

First, the constraints need apply *only* to the elements of the solution vector $\mathbf{b}^o$. They are imposed solely for deriving a solution and need have nothing to do with the model. They do not apply to parameters of the model. Second, if the constraints are of the form $\mathbf{C}'\mathbf{b}^o = \boldsymbol{\gamma}$, we know from (71) that minimizing $(\mathbf{y} - \mathbf{X}\mathbf{b}^o)'(\mathbf{y} - \mathbf{X}\mathbf{b}^o)$ subject to $\mathbf{C}'\mathbf{b}^o = \boldsymbol{\gamma}$ leads to the equations

$$\mathbf{X}'\mathbf{X}\mathbf{b}^o + \mathbf{C}\boldsymbol{\lambda} = \mathbf{X}'\mathbf{y}$$

and

$$\mathbf{C}'\mathbf{b}^o = \boldsymbol{\gamma},$$

equivalent to

$$\begin{bmatrix} \mathbf{X}'\mathbf{X} & \mathbf{C} \\ \mathbf{C}' & \mathbf{0} \end{bmatrix} \begin{bmatrix} \mathbf{b}^o \\ \boldsymbol{\lambda} \end{bmatrix} = \begin{bmatrix} \mathbf{X}'\mathbf{y} \\ \boldsymbol{\gamma} \end{bmatrix} \tag{104}$$

where $\boldsymbol{\lambda}$ is a vector of Lagrange multipliers. For these equations to have but one solitary solution for $\mathbf{b}^o$ (and $\boldsymbol{\lambda}$) it is clear that $\mathbf{C}'$ must have full row rank of sufficient rows to make $\begin{bmatrix} \mathbf{X}'\mathbf{X} & \mathbf{C} \\ \mathbf{C}' & \mathbf{0} \end{bmatrix}$ non-singular; and by applying Lemma 6 to (30) in Chapter 1, the rows of $\mathbf{C}'$ must be LIN of those of $\mathbf{X}$; i.e., $\mathbf{C}'$ cannot be of the form $\mathbf{C}' = \mathbf{L}'\mathbf{X}$. Thus we see that the constraints $\mathbf{C}'\mathbf{b}^o = \boldsymbol{\gamma}$ must be such that $\mathbf{C}'\mathbf{b}$ is not estimable. They cannot, therefore, be "any" constraints. They

*Use of "the" generalized inverse is correct here because it refers to the particular generalized inverse used to get a particular solution $\mathbf{b}^o$.

must be constraints for which $\mathbf{C}'\mathbf{b}$ is non-estimable, and there must be $p - r$ of them where $\mathbf{X}$ has p columns and rank r. Under these conditions the inverse given in Sec. 1.5b can be used to obtain the unique solution to (104), and this can be shown to be equivalent to the solution obtainable by the methods of (101) and (102) (see Exercise 11).

b. Constraints of the form $b_i^o = 0$

In balanced data that lead to normal equations like (103), for example, constraints of the type $\sum \alpha_i^o = 0$ are indeed the easiest to use. But they are not the easiest for unbalanced data. The constraints easiest to use with unbalanced data are the simple ones of putting $p - r$ elements of $\mathbf{b}^o$ equal to zero. They cannot be *any* $p - r$ elements, of course, for they must be judiciously chosen so as to make $\begin{bmatrix} \mathbf{X}'\mathbf{X} & \mathbf{C} \\ \mathbf{C}' & \mathbf{0} \end{bmatrix}$ non-singular. Ways of doing this are discussed in the chapters on applications (Chapters 6 and 7).

Using constraints that make some of the elements of $\mathbf{b}^o$ be zero is equivalent to putting those elements equal to zero in the normal equations or, more exactly, in $(\mathbf{y} - \mathbf{X}\mathbf{b}^o)'(\mathbf{y} - \mathbf{X}\mathbf{b}^o)$ which is minimized subject to such constraints. This has the effect of eliminating from the normal equations all those terms involving the zeroed b_i^o's, and also the equations corresponding to the same b_i^o's. And this, in turn, is equivalent to eliminating from $\mathbf{X}'\mathbf{X}$ the rows and columns corresponding to those b_i^o's and eliminating from $\mathbf{X}'\mathbf{y}$ the corresponding elements. What remains of $\mathbf{X}'\mathbf{X}$ is a symmetric matrix of order r that is non-singular. Hence these "modified" equations—modified by the constraints of putting some b_i^o's zero—can be solved. The solutions, together with the zeroed b_i^o's of the constraints, then constitute $\mathbf{b}^o$, a solution to the normal equations. Details of this procedure are now described, including derivation of the corresponding generalized inverse of $\mathbf{X}'\mathbf{X}$.

Putting $(p - r)$ b_i^o's equal to zero is equivalent to $\mathbf{C}'\mathbf{b}^o = \mathbf{0}$ with $\mathbf{C}'$ having $p - r$ rows each of which is null except for a single unity element. Suppose that $\mathbf{R}$ is a permutation matrix [equation (6), Sec. 1.1b] of order p such that

$$\mathbf{C}'\mathbf{R} = [\mathbf{0}_{(p-r)\times r} \quad \mathbf{I}_{p-r}]. \tag{105}$$

Then, remembering that $\mathbf{R}$ is orthogonal, the equations to be solved, (104), can be rewritten as

$$\begin{bmatrix} \mathbf{R}' & \mathbf{0} \\ \mathbf{0} & \mathbf{I} \end{bmatrix} \begin{bmatrix} \mathbf{X}'\mathbf{X} & \mathbf{C} \\ \mathbf{C}' & \mathbf{0} \end{bmatrix} \begin{bmatrix} \mathbf{R} & \mathbf{0} \\ \mathbf{0} & \mathbf{I} \end{bmatrix} \begin{bmatrix} \mathbf{R}' & \mathbf{0} \\ \mathbf{0} & \mathbf{I} \end{bmatrix} \begin{bmatrix} \mathbf{b}^o \\ \boldsymbol{\lambda} \end{bmatrix} = \begin{bmatrix} \mathbf{R}' & \mathbf{0} \\ \mathbf{0} & \mathbf{I} \end{bmatrix} \begin{bmatrix} \mathbf{X}'\mathbf{y} \\ \boldsymbol{\gamma} \end{bmatrix}$$

which reduce to

$$\begin{bmatrix} \mathbf{R}'\mathbf{X}'\mathbf{X}\mathbf{R} & \mathbf{R}'\mathbf{C} \\ \mathbf{C}'\mathbf{R} & \mathbf{0} \end{bmatrix} \begin{bmatrix} \mathbf{R}'\mathbf{b}^o \\ \boldsymbol{\lambda} \end{bmatrix} = \begin{bmatrix} \mathbf{R}'\mathbf{X}'\mathbf{y} \\ \mathbf{0} \end{bmatrix}. \tag{106}$$

Partitioning $\mathbf{R'X'XR}$, $\mathbf{R'b}^o$ and $\mathbf{R'X'y}$ to conform with $\mathbf{C'R}$ in (105), namely,

$$\mathbf{R'X'XR} = \begin{bmatrix} \mathbf{Z}_{11} & \mathbf{Z}_{12} \\ \mathbf{Z}_{21} & \mathbf{Z}_{22} \end{bmatrix}, \quad \mathbf{R'b}^o = \begin{bmatrix} \mathbf{b}_1^o \\ \mathbf{b}_2^o \end{bmatrix} \quad \text{and} \quad \mathbf{R'X'y} = \begin{bmatrix} (\mathbf{X'y})_1 \\ (\mathbf{X'y})_2 \end{bmatrix}, \tag{107}$$

we then have

$$\mathbf{Z}_{11}, \text{ of full rank, } = (\mathbf{X'X})_m, \tag{108}$$

the $\mathbf{X'X}$ matrix modified by deletion of rows and columns,

$$\mathbf{b}_1^o = \text{solutions of modified equations}$$

and $$\mathbf{b}_2^o = \text{zeroed } b_i^o\text{'s}.$$

Then equations (106) become

$$\begin{bmatrix} \mathbf{Z}_{11} & \mathbf{Z}_{12} & \mathbf{0} \\ \mathbf{Z}_{21} & \mathbf{Z}_{22} & \mathbf{I} \\ \mathbf{0} & \mathbf{I} & \mathbf{0} \end{bmatrix} \begin{bmatrix} \mathbf{b}_1^o \\ \mathbf{b}_2^o \\ \boldsymbol{\lambda} \end{bmatrix} = \begin{bmatrix} (\mathbf{X'y})_1 \\ (\mathbf{X'y})_2 \\ \mathbf{0} \end{bmatrix}$$

with the solution

$$\begin{bmatrix} \mathbf{b}_1^o \\ \mathbf{b}_2^o \\ \boldsymbol{\lambda} \end{bmatrix} = \begin{bmatrix} \mathbf{Z}_{11}^{-1} & \mathbf{0} & -\mathbf{Z}_{11}^{-1}\mathbf{Z}_{12} \\ \mathbf{0} & \mathbf{0} & \mathbf{I} \\ -\mathbf{Z}_{21}\mathbf{Z}_{11}^{-1} & \mathbf{I} & \mathbf{0} \end{bmatrix} \begin{bmatrix} (\mathbf{X'y})_1 \\ (\mathbf{X'y})_2 \\ \mathbf{0} \end{bmatrix}. \tag{109}$$

The important part of this solution is

$$\mathbf{b}_1^o = \mathbf{Z}_{11}^{-1}(\mathbf{X'y})_1. \tag{110}$$

It is easily derived: the inverse of the modified $\mathbf{X'X}$ matrix post-multiplied by the modified $\mathbf{X'y}$ vector. Then $\mathbf{b}_1^o$ and the b_i^o's zeroed by the constraints constitute a complete solution $\mathbf{b}^o$.

The generalized inverse of $\mathbf{X'X}$ corresponding to the solution (110) is derived as follows. From (109)

$$\begin{bmatrix} \mathbf{b}_1^o \\ \mathbf{b}_2^o \end{bmatrix} = \begin{bmatrix} \mathbf{Z}_{11}^{-1} & \mathbf{0} \\ \mathbf{0} & \mathbf{0} \end{bmatrix} \begin{bmatrix} (\mathbf{X'y})_1 \\ (\mathbf{X'y})_2 \end{bmatrix} = \begin{bmatrix} \mathbf{Z}_{11}^{-1} & \mathbf{0} \\ \mathbf{0} & \mathbf{0} \end{bmatrix} \mathbf{R'X'y}$$

from (107); and using the orthogonality of $\mathbf{R}$ and (107) again this gives

$$\mathbf{b}^o = \mathbf{R}(\mathbf{R'b}^o) = \mathbf{R} \begin{bmatrix} \mathbf{b}_1^o \\ \mathbf{b}_2^o \end{bmatrix} \tag{111}$$

$$= \mathbf{R} \begin{bmatrix} \mathbf{Z}_{11}^{-1} & \mathbf{0} \\ \mathbf{0} & \mathbf{0} \end{bmatrix} \mathbf{R'X'y}. \tag{112}$$

But, from Sec. 1.1b with the definition of $\mathbf{Z}_{11}$ given in (108),

$$\mathbf{G} = \mathbf{R}\begin{bmatrix} \mathbf{Z}_{11}^{-1} & \mathbf{0} \\ \mathbf{0} & \mathbf{0} \end{bmatrix}\mathbf{R}' \tag{113}$$

is a generalized inverse of $\mathbf{X}'\mathbf{X}$. And so, from (112), $\mathbf{G}$ of (113) is the generalized inverse of $\mathbf{X}'\mathbf{X}$ corresponding to the solution $\mathbf{b}^o$ found by using (110) and (111). We therefore have the following procedure.

* **c. Procedure for deriving $\mathbf{b}^o$ and G**

1. For $\mathbf{X}'\mathbf{X}$ of order p, find its rank; call it r.
2. Delete $p - r$ rows and corresponding columns from $\mathbf{X}'\mathbf{X}$, to leave a symmetric sub-matrix of full rank r. Call that modified matrix $(\mathbf{X}'\mathbf{X})_m$.
3. Corresponding to the rows deleted from $\mathbf{X}'\mathbf{X}$ delete elements from $\mathbf{X}'\mathbf{y}$. Call the modified vector $(\mathbf{X}'\mathbf{y})_m$.
4. Calculate $\mathbf{b}_m^o = [(\mathbf{X}'\mathbf{X})_m]^{-1}(\mathbf{X}'\mathbf{y})_m$.
5. In $\mathbf{b}^o$, all elements corresponding to rows deleted from $\mathbf{X}'\mathbf{X}$ are zero; other elements are those of $\mathbf{b}_m^o$, in sequence.
6. In $\mathbf{X}'\mathbf{X}$ replace all elements of $(\mathbf{X}'\mathbf{X})_m$ by those of its inverse; and put all other elements zero. The resulting matrix is $\mathbf{G}$, the generalized inverse corresponding to the solution $\mathbf{b}^o$. Its derivation is in line with the algorithm (for symmetric matrices) of Sec. 1.1b.

d. Restrictions on the model

Throughout the preceding discussion of constraints no mention has been made of restrictions on the parameters of the model corresponding to constraints imposed on a solution. This is because constraints on the solution are used solely for obtaining a solution and need have no bearing on the model whatever. But if the model is such that there are restrictions on its parameters, these same restrictions can be used as constraints on the solutions, provided they relate to non-estimable functions, i.e., if restrictions $\mathbf{P}'\mathbf{b} = \boldsymbol{\delta}$ have $\mathbf{P}'\mathbf{b}$ not estimable. If $\mathbf{P}'$ were of full row rank, $p - r$, then the solutions would be given by

$$\begin{bmatrix} \mathbf{X}'\mathbf{X} & \mathbf{P} \\ \mathbf{P}' & \mathbf{0} \end{bmatrix}\begin{bmatrix} \mathbf{b}^o \\ \boldsymbol{\lambda} \end{bmatrix} = \begin{bmatrix} \mathbf{X}'\mathbf{y} \\ \boldsymbol{\delta} \end{bmatrix} \tag{114}$$

and the solution would in fact be the b.l.u.e. of $\mathbf{b}$. Of course, the solution to (114) could also be obtained by using the solution derived from simple constraints of the form $b_i^o = 0$ discussed in the preceding section, namely (112). This can be amended in accord with (101) and (102) to give a solution satisfying (114). It will be, from 101,

$$\mathbf{b}_{r,0}^o = \mathbf{b}_0^o + (\mathbf{H} - \mathbf{I})\mathbf{z}_1, \tag{115}$$

using $\mathbf{b}^o$ of (111) as $\mathbf{b}_0^o$, $\mathbf{G}$ of (113) and $\mathbf{H} = \mathbf{GX'X}$, in the usual way. From (102), the $\mathbf{z}_1$ of (115) will be such that

$$\mathbf{P'(H - I)z_1} = \boldsymbol{\delta} - \mathbf{P'GX'y} \tag{116}$$

as in (102). This procedure will be especially useful when the restrictions in the model, $\mathbf{P'b} = \boldsymbol{\delta}$, involve $\mathbf{P}'$ of less than $p - r$ rows.

The important thing about restrictions in the model is their effect on estimable functions and testable hypotheses, as has already been pointed out. Equally as important is the fact that constraints on the solutions do not necessarily imply restrictions in the model and therefore constraints do not affect estimable functions or testable hypotheses. Furthermore, since constraints are only a means to obtaining a solution $\mathbf{b}^o$ they do not affect sums of squares. Confusion on these points often arises because of certain kinds of restrictions that often occur; these same restrictions applied as constraints to the solution also greatly aid in obtaining a solution. For example, the model equation $y_{ij} = \mu + \alpha_i + e_{ij}$ is often written as $y_{ij} = \mu_i + e_{ij}$ with μ and α_i defined as $\mu = \sum_{i=1}^{c} \mu_i/c$ and $\alpha_i = \mu_i - \mu$ respectively. In this way a restriction in the model is $\sum_{i=1}^{c} \alpha_i = 0$. Suppose for $c = 3$ the normal equations were, for such a model,

$$\begin{aligned}
6\mu^o + 2\alpha_1^o + 2\alpha_2^o + 2\alpha_3^o &= y_{..}\\
2\mu^o + 2\alpha_1^o \qquad\qquad\qquad &= y_{1.}\\
2\mu^o \qquad + 2\alpha_2^o \qquad\qquad &= y_{2.}\\
2\mu^o \qquad\qquad\qquad + 2\alpha_3^o &= y_{3.}
\end{aligned}$$

Because $\alpha_1 + \alpha_2 + \alpha_3 = 0$ in the model and in order to help in solving the equations, we impose the constraint

$$\alpha_1^o + \alpha_2^o + \alpha_3^o = 0. \tag{117}$$

But suppose the normal equations were

$$\begin{aligned}
6\mu^o + 3\alpha_1^o + 2\alpha_2^o + \alpha_3^o &= y_{..}\\
3\mu^o + 3\alpha_1^o \qquad\qquad\qquad &= y_{1.}\\
2\mu^o \qquad + 2\alpha_2^o \qquad\qquad &= y_{2.}\\
\mu^o \qquad\qquad\qquad + \alpha_3^o &= y_{3.}
\end{aligned} \tag{118}$$

Then the constraint (117) is of no particular help in solving equations (118). On the other hand,

$$3\alpha_1^o + 2\alpha_2^o + \alpha_3^o = 0 \tag{119}$$

is of help. But this is no reason at all for making $3\alpha_1 + 2\alpha_2 + \alpha_3 = 0$ be part of the model. Not only might it be quite inappropriate but there is no need for it. Suppose in fact that $\alpha_1 + \alpha_2 + \alpha_3 = 0$ is a meaningful restriction in the model. Then (119) could still be used for solving equations (118) and, provided the corresponding generalized inverse of $\mathbf{X}'\mathbf{X}$ was found, the solution could be amended to satisfy (117) by using (115) and (116). Thus if $\mathbf{b}_0^o$ is the solution satisfying (119) then that satisfying (117) is (115) with (116) using $\mathbf{P}' = [0 \quad 1 \quad 1 \quad 1]$, $\boldsymbol{\delta} = \mathbf{0}$ and $\mathbf{G}$ corresponding to $\mathbf{b}_0^o$.

e. Example (continued)

The normal equations are, from (6):

$$\begin{aligned} 6\mu^o + 3\alpha_1^o + 2\alpha_2^o + \alpha_3^o &= 504 \\ 3\mu^o + 3\alpha_1^o \qquad\qquad\quad &= 300 \\ 2\mu^o \qquad\quad + 2\alpha_2^o \qquad &= 172 \\ \mu^o \qquad\qquad\qquad + \alpha_3^o &= 32. \end{aligned}$$

The procedure described in subsection c goes as follows:

Step 1: $p = 4$ and $r = 3$.

Steps 2 and 3: $(\mathbf{X}'\mathbf{X})_m = \begin{bmatrix} 6 & 3 & 2 \\ 3 & 3 & 0 \\ 2 & 0 & 2 \end{bmatrix}$ and $(\mathbf{X}'\mathbf{y})_m = \begin{bmatrix} 504 \\ 300 \\ 172 \end{bmatrix}$.

Step 4: $\mathbf{b}_m^o = \begin{bmatrix} 1 & -1 & -1 \\ -1 & \frac{4}{3} & 1 \\ -1 & 1 & 1\frac{1}{2} \end{bmatrix} \begin{bmatrix} 504 \\ 300 \\ 172 \end{bmatrix} = \begin{bmatrix} 32 \\ 68 \\ 54 \end{bmatrix}$.

Step 5: $\mathbf{b}^{o\prime} = [32 \quad 68 \quad 54 \quad 0]$. (120)

Step 6: $\mathbf{G} = \begin{bmatrix} 1 & -1 & -1 & 0 \\ -1 & \frac{4}{3} & 1 & 0 \\ -1 & 1 & 1\frac{1}{2} & 0 \\ 0 & 0 & 0 & 0 \end{bmatrix}$.

These results, it will be noted, are identical to those shown in (25). Another way of carrying out the same procedure would be as follows:

Steps 2 and 3: $(\mathbf{X}'\mathbf{X})_m = \begin{bmatrix} 6 & 2 & 1 \\ 2 & 2 & 0 \\ 1 & 0 & 1 \end{bmatrix}$ and $(\mathbf{X}'\mathbf{y})_m = \begin{bmatrix} 504 \\ 172 \\ 32 \end{bmatrix}$.

Step 4: $\mathbf{b}_m^o = (1/6)\begin{bmatrix} 2 & -2 & -2 \\ -2 & 5 & 2 \\ -2 & 2 & 8 \end{bmatrix}\begin{bmatrix} 504 \\ 172 \\ 32 \end{bmatrix} = \begin{bmatrix} 100 \\ -14 \\ -68 \end{bmatrix}.$

Step 5: $\mathbf{b}^{o\prime} = [100 \quad 0 \quad -14 \quad -68].$

Step 6: $\mathbf{G} = (1/6)\begin{bmatrix} 2 & 0 & -2 & -2 \\ 0 & 0 & 0 & 0 \\ -2 & 0 & 5 & 2 \\ -2 & 0 & 2 & 8 \end{bmatrix}.$

One check on this result is

$$\text{SSR} = \mathbf{b}^{o\prime}\mathbf{X}'\mathbf{y} = 100(504) - 14(172) - 68(32) = 45{,}816$$

as before.

Suppose now that restrictions on the model are $\alpha_1 + \alpha_2 + \alpha_3 = 0$. Then equations (114) are

$$\begin{bmatrix} 6 & 3 & 2 & 1 & 0 \\ 3 & 3 & 0 & 0 & 1 \\ 2 & 0 & 2 & 0 & 1 \\ 1 & 0 & 0 & 1 & 1 \\ 0 & 1 & 1 & 1 & 0 \end{bmatrix}\begin{bmatrix} \mu^o \\ \alpha_1^o \\ \alpha_2^o \\ \alpha_3^o \\ \lambda \end{bmatrix} = \begin{bmatrix} 504 \\ 300 \\ 172 \\ 32 \\ 0 \end{bmatrix}. \tag{121}$$

The solution is

$$\begin{bmatrix} \mathbf{b}^o \\ \lambda \end{bmatrix} = \begin{bmatrix} \mu^o \\ \alpha_1^o \\ \alpha_2^o \\ \alpha_3^o \\ \lambda \end{bmatrix} = (1/54)\begin{bmatrix} 11 & -5 & -2 & 7 & -18 \\ -5 & 17 & -4 & -13 & 18 \\ -2 & -4 & 20 & -16 & 18 \\ 7 & -13 & -16 & 29 & 18 \\ -18 & 18 & 18 & 18 & 0 \end{bmatrix}\begin{bmatrix} 504 \\ 300 \\ 172 \\ 32 \\ 0 \end{bmatrix}$$

so that $\mathbf{b}^o$ is

$$\mathbf{b}^{o\prime} = [\mu^o \quad \alpha_1^o \quad \alpha_2^o \quad \alpha_3^o] = [72\tfrac{2}{3} \quad 27\tfrac{1}{3} \quad 13\tfrac{1}{3} \quad -40\tfrac{2}{3}]. \tag{122}$$

The alternative way of getting this solution is that of (115); use a solution based on the constraint $\alpha_3^o = 0$ and amend it to satisfy $\alpha_1^o + \alpha_2^o + \alpha_3^o = 0$.

To do this we use $\mathbf{b}^{o\prime} = [32 \quad 68 \quad 54 \quad 0]$ of (120), where the corresponding **H**-matrix is

$$\mathbf{H} = \mathbf{GX'X} = \begin{bmatrix} 1 & 0 & 0 & 1 \\ 0 & 1 & 0 & -1 \\ 0 & 0 & 1 & -1 \\ 0 & 0 & 0 & 0 \end{bmatrix}.$$

Hence, as in (115), the solution to (121) is

$$\mathbf{b}^o_{r,0} = \begin{bmatrix} 32 \\ 68 \\ 54 \\ 0 \end{bmatrix} + \begin{bmatrix} 0 & 0 & 0 & 1 \\ 0 & 0 & 0 & -1 \\ 0 & 0 & 0 & -1 \\ 0 & 0 & 0 & -1 \end{bmatrix} \mathbf{z}_1 \tag{123}$$

for which (116) is

$$[0 \quad 1 \quad 1 \quad 1] \begin{bmatrix} 0 & 0 & 0 & 1 \\ 0 & 0 & 0 & -1 \\ 0 & 0 & 0 & -1 \\ 0 & 0 & 0 & -1 \end{bmatrix} \mathbf{z}_1 = -[0 \quad 1 \quad 1 \quad 1] \begin{bmatrix} 32 \\ 68 \\ 54 \\ 0 \end{bmatrix}$$

Therefore $\mathbf{z}_1' = [z_1 \quad z_2 \quad z_3 \quad 40\frac{2}{3}]$

and substitution in (123) gives

$$\mathbf{b}^o_{r,0} = \begin{bmatrix} 32 \\ 68 \\ 54 \\ 0 \end{bmatrix} + 40\tfrac{2}{3} \begin{bmatrix} 1 \\ -1 \\ -1 \\ -1 \end{bmatrix} = \begin{bmatrix} 72\frac{2}{3} \\ 27\frac{1}{3} \\ 13\frac{1}{3} \\ -40\frac{2}{3} \end{bmatrix}$$

as in (122).

Finally, suppose we use $3\alpha_1^o + 2\alpha_2^o + \alpha_3^o = 0$ to solve the equations. The solution is

$$\mathbf{b}^{o\prime} = [84 \quad 16 \quad 2 \quad -52], \tag{124}$$

and the generalized inverse of **X'X** corresponding to this is

$$\mathbf{G} = \tfrac{1}{6} \begin{bmatrix} 1 & 0 & 0 & 0 \\ -1 & 2 & 0 & 0 \\ -1 & 0 & 3 & 0 \\ -1 & 0 & 0 & 6 \end{bmatrix} \quad \text{and} \quad \mathbf{H} = \tfrac{1}{6} \begin{bmatrix} 6 & 3 & 2 & 1 \\ 0 & 3 & -2 & -1 \\ 0 & -3 & 4 & -1 \\ 0 & -3 & -2 & 5 \end{bmatrix}.$$

Then, to amend this solution to satisfy $\alpha_1^o + \alpha_2^o + \alpha_3^o = 0$ we solve (116), namely

$$[0 \quad 1 \quad 1 \quad 1]\tfrac{1}{6}\begin{bmatrix} 0 & 3 & 2 & 1 \\ 0 & -3 & -2 & -1 \\ 0 & -3 & -2 & -1 \\ 0 & -3 & -2 & -1 \end{bmatrix}\mathbf{z}_1 = -[0 \quad 1 \quad 1 \quad 1]\begin{bmatrix} 84 \\ 16 \\ 2 \\ -52 \end{bmatrix}$$

i.e.,

$$-(3z_2 + 2z_3 + z_4) = -2(18 - 52) = 68$$

and then, using (124) for $\mathbf{b}_0^o$ in (115) the solution satisfying $\alpha_1^o + \alpha_2^o + \alpha_3^o = 0$ is

$$\begin{bmatrix} 84 \\ 16 \\ 2 \\ -52 \end{bmatrix} + \tfrac{1}{6}\begin{bmatrix} 0 & 3 & 2 & 1 \\ 0 & -3 & -2 & -1 \\ 0 & -3 & -2 & -1 \\ 0 & -3 & -2 & -1 \end{bmatrix}\begin{bmatrix} z_1 \\ z_2 \\ z_3 \\ z_4 \end{bmatrix} = \begin{bmatrix} 84 \\ 16 \\ 2 \\ -52 \end{bmatrix} + \tfrac{1}{6}\begin{bmatrix} -68 \\ 68 \\ 68 \\ 68 \end{bmatrix} = \begin{bmatrix} 72\frac{2}{3} \\ 27\frac{1}{3} \\ 13\frac{1}{3} \\ -40\frac{2}{3} \end{bmatrix}$$

as in (122).

8. GENERALIZATIONS

We have now discussed both the full rank model (Chapter 3) and the model not of full rank. The latter is, if course, just a generalization of the former. More specifically, the full rank model is a special case of the non-full rank model with $\mathbf{G}$ and $\mathbf{b}^o$ taking the forms $(\mathbf{X}'\mathbf{X})^{-1}$ and $\hat{\mathbf{b}}$ respectively. In general, therefore, the non-full rank model covers all cases.

Estimability and testability, however, enter into only the non-full rank model. All linear functions are estimable and all linear hypotheses are testable in the full rank case. There is therefore merit in dealing with the two models separately, as has been done. But in both models only one special case has been considered, namely that where the error terms have $\text{var}(\mathbf{e}) = \sigma^2\mathbf{I}$. We now briefly discuss the general case of $\text{var}(\mathbf{e}) = \sigma^2\mathbf{V}$, both where $\mathbf{V}$ is non-singular and where it is singular.

a. Non-singular V

When $\text{var}(\mathbf{e}) = \sigma^2\mathbf{V}$ with $\mathbf{V}$ non-singular the normal equations are, as indicated in Sec. 3.3,

$$\mathbf{X}'\mathbf{V}^{-1}\mathbf{X}\mathbf{b}^o = \mathbf{X}'\mathbf{V}^{-1}\mathbf{y}. \tag{125}$$

For the full rank model this has the single solution

$$\hat{\mathbf{b}} = (\mathbf{X}'\mathbf{V}^{-1}\mathbf{X})^{-1}\mathbf{X}'\mathbf{V}^{-1}\mathbf{y} \tag{126}$$

as given in Sec. 3.3. With the non-full rank model a generalized inverse of $\mathbf{X}'\mathbf{V}^{-1}\mathbf{X}$ must be used to solve (125). Denoting this by $\mathbf{F}$ gives

$$\mathbf{b}^{\circ} = \mathbf{F}\mathbf{X}'\mathbf{V}^{-1}\mathbf{y} \quad \text{with} \quad \mathbf{X}'\mathbf{V}^{-1}\mathbf{X}\mathbf{F}\mathbf{X}'\mathbf{V}^{-1}\mathbf{X} = \mathbf{X}'\mathbf{V}^{-1}\mathbf{X}, \tag{127}$$

of which (126) is a special case. Thus we see that estimation in the model having $\text{var}(\mathbf{e}) = \sigma^2\mathbf{V}$ for non-singular $\mathbf{V}$ is identical to that when $\text{var}(\mathbf{e}) = \sigma^2\mathbf{I}$, except for using a generalized inverse of $\mathbf{X}'\mathbf{V}^{-1}\mathbf{X}$; and $\mathbf{X}'\mathbf{V}^{-1}\mathbf{y}$ is used in place of $\mathbf{X}'\mathbf{y}$.

Furthermore, since $\mathbf{V}$ is a symmetric positive definite matrix $\mathbf{V}^{-1} = \mathbf{L}\mathbf{L}'$ for some non-singular $\mathbf{L}$. Putting $\mathbf{x} = \mathbf{L}'\mathbf{y}$ transforms the model $\mathbf{y} = \mathbf{X}\mathbf{b} + \mathbf{e}$ into $\mathbf{x} = \mathbf{L}'\mathbf{X}\mathbf{b} + \boldsymbol{\epsilon}$ where $\boldsymbol{\epsilon} = \mathbf{L}'\mathbf{e}$ and $\text{var}(\boldsymbol{\epsilon}) = \sigma^2\mathbf{I}$. Estimating $\mathbf{b}$ from this model for $\mathbf{x}$ gives $\hat{\mathbf{b}}$ or $\mathbf{b}^{\circ}$ of (126) or (127) respectively, and the corresponding error sum of squares is

$$\mathbf{x}'\mathbf{x} - \mathbf{b}^{\circ\prime}\mathbf{X}'\mathbf{L}\mathbf{x} = \mathbf{y}'\mathbf{V}^{-1}\mathbf{y} - \mathbf{b}^{\circ\prime}\mathbf{X}'\mathbf{V}^{-1}\mathbf{y} \tag{128}$$

using $\hat{\mathbf{b}}$ for $\mathbf{b}^{\circ}$ in the full-rank case. Thus the weighted sum of squares, $\mathbf{y}'\mathbf{V}^{-1}\mathbf{y}$, is used in place of $\mathbf{y}'\mathbf{y}$ in the corresponding analyses of variance.

b. Singular V

At least two conditions among data can lead to $\text{var}(\mathbf{y}) = \mathbf{V}$ being singular: if any elements of $\mathbf{y}$ are linear functions of other elements or if any elements of $\mathbf{y}$ are a constant plus a linear function of other elements. For example, if $v(y_1) = v(y_2) = \sigma^2$ and $\text{cov}(y_1, y_2) = 0$ then

$$\text{var}\begin{bmatrix} y_1 \\ y_2 \\ y_1 + y_2 \end{bmatrix} = \begin{bmatrix} 1 & 0 & 1 \\ 0 & 1 & 1 \\ 1 & 1 & 2 \end{bmatrix}\sigma^2; \tag{129}$$

and for any constant θ

$$\text{var}\begin{bmatrix} y_1 \\ y_2 \\ y_1 + \theta \end{bmatrix} = \begin{bmatrix} 1 & 0 & 1 \\ 0 & 1 & 0 \\ 1 & 0 & 1 \end{bmatrix}\sigma^2. \tag{130}$$

Suppose we write $\mathbf{w}'$ for the vector $[y_1 \quad y_2]$, and let the equation of the model for $\mathbf{w}$ be

$$\mathbf{w} = \begin{bmatrix} y_1 \\ y_2 \end{bmatrix} = \mathbf{T}\mathbf{b} + \boldsymbol{\epsilon}. \tag{131}$$

Then the equation for

$$\mathbf{y} = \begin{bmatrix} y_1 \\ y_2 \\ y_1 + y_2 \end{bmatrix}$$

of (129) can be written as

$$\mathbf{y} = \begin{bmatrix} 1 & 0 \\ 0 & 1 \\ 1 & 1 \end{bmatrix} \begin{bmatrix} y_1 \\ y_2 \end{bmatrix} = \begin{bmatrix} 1 & 0 \\ 0 & 1 \\ 1 & 1 \end{bmatrix} \mathbf{w} = \begin{bmatrix} 1 & 0 \\ 0 & 1 \\ 1 & 1 \end{bmatrix} (\mathbf{Tb} + \boldsymbol{\epsilon}),$$

i.e., as $\mathbf{y} = \mathbf{Mw}$ for some matrix $\mathbf{M}$. On the other hand the equation for

$$\mathbf{y} = \begin{bmatrix} y_1 \\ y_2 \\ y_1 + \theta \end{bmatrix}$$

of (130) cannot be written in the form $\mathbf{y} = \mathbf{Mw}$ for $\mathbf{w}$ of (131). Of these two possible ways in which a vector of observations $\mathbf{y}$ can have a singular variance-covariance matrix we consider only the first, like (129), in which $\mathbf{y}$ can be written as $\mathbf{y} = \mathbf{Mw}$. The more general case, where $\text{var}(\mathbf{y}) = \mathbf{V}$ is singular but $\mathbf{y}$ cannot necessarily be written as $\mathbf{y} = \mathbf{Mw}$, is considered by Zyskind and Martin (1969). The situation of $\mathbf{y} = \mathbf{Mw}$ is a special case of the results given there. However, because it represents the way in which a singular $\mathbf{V}$ does frequently arise, brief discussion is given to it here. It is also the case for which the normal equations, their solution and ensuing results are most easily described.

Whenever some elements of $\mathbf{y}$ can be expressed as linear functions of other elements, $\mathbf{y}$ can be written as

$$\mathbf{y} = \mathbf{Mw} \tag{132}$$

where no element of $\mathbf{w}$ is a linear function of the others. Thus $\mathbf{M}$ has full column rank. Furthermore, on taking the equation of the model for $\mathbf{w}$ as being

$$\mathbf{w} = \mathbf{Tb} + \boldsymbol{\epsilon} \tag{133}$$

we have

$$\mathbf{y} = \mathbf{Mw} = \mathbf{MTb} + \mathbf{M}\boldsymbol{\epsilon}$$

so that if the model for $\mathbf{y}$ is

$$\mathbf{y} = \mathbf{Xb} + \mathbf{e}$$

we can take

$$\mathbf{X} = \mathbf{MT} \tag{134}$$

and $\mathbf{e} = \mathbf{M}\boldsymbol{\epsilon}$. Furthermore, if $\text{var}(\boldsymbol{\epsilon}) = \sigma^2\mathbf{I}$, and $\text{var}(\mathbf{y}) = \mathbf{V}\sigma^2$ we have

$$\mathbf{V}\sigma^2 = \text{var}(\mathbf{y}) = \text{var}(\mathbf{e}) = \text{var}(\mathbf{M}\boldsymbol{\epsilon}) = \mathbf{MM}'\sigma^2$$

so that
$$\mathbf{V} = \mathbf{MM}'. \tag{135}$$

Now from (131), the normal equations for $\mathbf{b}^o$ are
$$\mathbf{T}'\mathbf{T}\mathbf{b}^o = \mathbf{T}'\mathbf{w}. \tag{136}$$

But with $\mathbf{M}$ having full column rank it can be readily shown that $\mathbf{M}(\mathbf{M}'\mathbf{M})^{-2}\mathbf{M}'$ is the unique Penrose generalized inverse (see Sec. 1.3) of $\mathbf{V}$ of (135). Furthermore, by Theorem 7 of Sec. 1.5a, $\mathbf{M}'(\mathbf{MM}')^{-}\mathbf{M}$ is unique for all generalized inverses $\mathbf{V}^{-} = (\mathbf{MM}')^{-}$ of $\mathbf{V} = \mathbf{MM}'$, and using the Penrose inverse for this shows that
$$\mathbf{M}'\mathbf{V}^{-}\mathbf{M} = \mathbf{M}'(\mathbf{MM}')^{-}\mathbf{M} = \mathbf{M}'\mathbf{M}(\mathbf{M}'\mathbf{M})^{-2}\mathbf{M}'\mathbf{M} = \mathbf{I}. \tag{137}$$

Therefore (136), which is
$$\mathbf{T}'\mathbf{I}\mathbf{T}\mathbf{b}^o = \mathbf{T}'\mathbf{I}\mathbf{w},$$
becomes
$$\mathbf{T}'\mathbf{M}'\mathbf{V}^{-}\mathbf{M}\mathbf{T}\mathbf{b}^o = \mathbf{T}'\mathbf{M}'\mathbf{V}^{-}\mathbf{M}\mathbf{w}$$
which, with (132) and (134), is equivalent to
$$\mathbf{X}'\mathbf{V}^{-}\mathbf{X}\mathbf{b}^o = \mathbf{X}'\mathbf{V}^{-}\mathbf{y}. \tag{138}$$
Hence
$$\mathbf{b}^o = (\mathbf{X}'\mathbf{V}^{-}\mathbf{X})^{-}\mathbf{X}'\mathbf{V}^{-}\mathbf{y}$$
where $(\mathbf{X}'\mathbf{V}^{-}\mathbf{X})^{-}$ is any generalized inverse of $\mathbf{X}'\mathbf{V}^{-}\mathbf{X}$ and $\mathbf{V}^{-}$ is any generalized inverse of $\mathbf{V}$. These results for the normal equations and their solution are identical to those for non-singular $\mathbf{V}$, (125) and (126), only with a generalized inverse $\mathbf{V}^{-}$ of $\mathbf{V}$ used in place of $\mathbf{V}^{-1}$.

The residual error sum of squares in the singular case is, from fitting (133),
$$\text{SSE} = \mathbf{w}'\mathbf{w} - \mathbf{b}^{o\prime}\mathbf{T}'\mathbf{w} \tag{139}$$
and with the aid of (132), (134) and (137) this reduces, in the same way that (138) was derived, to
$$\text{SSE} = \mathbf{y}'\mathbf{V}^{-}\mathbf{y} - \mathbf{b}^{o\prime}\mathbf{X}'\mathbf{V}^{-}\mathbf{y}.$$
This is the same result as (128) using $\mathbf{V}^{-}$ in place of $\mathbf{V}^{-1}$. Its expected value from (139), is
$$\begin{aligned} E(\text{SSE}) &= E(\mathbf{w}'\mathbf{w} - \mathbf{b}^{o\prime}\mathbf{T}'\mathbf{w}) \\ &= [(\text{number of elements in } \mathbf{w}) - r(\mathbf{T})]\sigma^2 \\ &= [r(\mathbf{M}) - r(\mathbf{T})]\sigma^2 \end{aligned}$$
and because of (135) and (134) and $\mathbf{M}$ having full column rank this is equivalent to
$$E(\text{SSE}) = [r(\mathbf{V}) - r(\mathbf{X})]\sigma^2.$$
Hence an unbiased estimator of σ^2 is
$$\hat{\sigma}^2 = \frac{\text{SSE}}{r(\mathbf{V}) - r(\mathbf{X})} = \frac{\mathbf{y}'\mathbf{V}^{-}\mathbf{y} - \mathbf{b}^{o\prime}\mathbf{X}'\mathbf{V}^{-}\mathbf{y}}{r(\mathbf{V}) - r(\mathbf{X})}.$$

9. SUMMARY

The basic results of this chapter are summarized at the beginning of the next, prior to using them on applications in that and succeeding chapters. Additional summaries are to be found as follows:

Procedure for deriving $\mathbf{G}$	:	Sec. 5.7c
Analysis of variance for fitting model	:	Tables 5.5 and 5.6, Sec. 5.3g
Estimable functions	:	Table 5.8, Sec. 5.4e
Analysis of variance for testing hypothesis $\mathbf{K}'\mathbf{b} = \mathbf{0}$	:	Tables 5.9 and 5.10, Sec. 5.5c
Restricted models	:	Tables 5.13, Sec. 5.6

10. EXERCISES

1. Use part or all of the data from the Exercise of Chapter 4 to fit a model and establish analyses of variance such as those in Table 5.7. Derive estimable functions and orthogonal contrasts, find their estimators and test hypotheses about them.

2. Rework the example of this chapter using $\mathbf{G}_1$, $\mathbf{H}_1$ and $\mathbf{b}_1^o$ of equation (25).

3. Show that R^2 of equation (18) reduces to (20).

4. For $\mathbf{X}$ of order $N \times p$ and rank r, and $\mathbf{S}'$ and $\mathbf{S}'\mathbf{X}'$ of full row rank r, show that $\mathbf{S}(\mathbf{S}'\mathbf{X}'\mathbf{X}\mathbf{S})^{-1}\mathbf{S}'$ is a generalized inverse of $\mathbf{X}'\mathbf{X}$.

5. If $\mathbf{T}$ has full row rank prove that $\mathbf{T}(\mathbf{T}'\mathbf{T})^{-}\mathbf{T}' = \mathbf{I}$.

6. Using

$$\begin{bmatrix} \mathbf{X}'\mathbf{X} & \mathbf{K} \\ \mathbf{K}' & \mathbf{0} \end{bmatrix}^{-1} = \begin{bmatrix} \mathbf{B}_{11} & \mathbf{B}_{12} \\ \mathbf{B}_{21} & \mathbf{0} \end{bmatrix}$$

given in Sec. 1.5b show that the resulting solutions of equations (76) are $\boldsymbol{\theta} = \mathbf{0}$ and $\mathbf{b}_H^o$ of (79) as obtained in this chapter. [$\mathbf{H}$ of Sec. 1.5b is now represented by $\mathbf{K}'$ of the non-testable hypothesis $\mathbf{K}'\mathbf{b} = \mathbf{m}$ with $\mathbf{K}'$ of full row rank $p - r$; and m of Sec. 1.5b is $p - r$ here.]

7. (*a*) Suppose $\mathbf{S}'\mathbf{b}$ is estimable and $\mathbf{q}'\mathbf{b}$ is not. Prove that testing the hypothesis

$$H: \quad \begin{bmatrix} \mathbf{S}'\mathbf{b} \\ \mathbf{q}'\mathbf{b} \end{bmatrix} = \begin{bmatrix} \mathbf{m}_1 \\ m_2 \end{bmatrix}$$

is indistinguishable from testing the hypothesis $H: \quad \mathbf{S}'\mathbf{b} = \mathbf{m}_1$.
(*b*) In terms of the example of Sec. 5f, demonstrate (*a*) for the hypothesis

$$H: \alpha_1 = \alpha_2 = 110.$$

8. For the example of this chapter derive the contrasts specified below and find the numerator sums of squares for testing the hypotheses that these contrasts are zero.
(*a*) A contrast orthogonal to both $6\mu + 3\alpha_1 + 2\alpha_2 + \alpha_3$ and $\alpha_1 - 2\alpha_2 + \alpha_3$.
(*b*) Two contrasts orthogonal to one another and to $\alpha_1 - \alpha_2$. Relate the results in (*b*) to Table 5.11. [Define "orthogonal" as in (92).]

9. Show, formally, that testing the hypothesis $\lambda\mathbf{K}'\mathbf{b} = \mathbf{0}$ is identical to testing $\mathbf{K}'\mathbf{b} = \mathbf{0}$, for λ being a scalar.

10. Suppose a model can be expressed as

$$y_{ijk} = \alpha_i + \epsilon_{ijk}$$

where y_{ijk} is an observation and $i = 1, \ldots, c$, $j = 1, \ldots, N_i$, and $k = 1, \ldots, n_{ij}$. The vector of observations can be written as

$$\mathbf{y}' = [y_{111} \quad y_{112} \quad y_{11n_{11}} \quad \cdots \quad y_{1N_11} \quad \cdots \quad y_{1,N_1,n_{1N_1}} \quad \cdots \quad y_{c,N_c,1} \quad \cdots \quad y_{c,N_c,n_{c,N_c}}]$$

where the observations are ordered by k, within j within i. If $\mathbf{V}$ is the variance-covariance matrix of $\mathbf{y}$, it is a diagonal matrix of matrices $\mathbf{A}_{ij}$, for $i = 1, \ldots, c$ and $j = 1, \ldots, N_i$, where $\mathbf{A}_{ij} = e\mathbf{I}_{n_{ij}} + b\mathbf{J}_{n_{ij}}$, and $\mathbf{1}'_{n_{ij}}$ is a vector of n_{ij}1's and $\mathbf{J}_{n_{ij}} = \mathbf{1}_{n_{ij}}\mathbf{1}'_{n_{ij}}$. The normal equations for estimating $\boldsymbol{\alpha}$, the vector of the α_i's, are then

$$\mathbf{X}'\mathbf{V}^{-1}\mathbf{X}\hat{\boldsymbol{\alpha}} = \mathbf{X}'\mathbf{V}^{-1}\mathbf{y}$$

where $(\mathbf{X}'\mathbf{V}^{-1}\mathbf{X})^{-1}$ exists.
(*a*) For $c = 2$, with $n_{11} = 2$, $n_{12} = 3$, $n_{21} = 4$, $n_{22} = 1$ and $n_{23} = 2$ write down $\mathbf{y}'$ and $\mathbf{V}$ in full.
(*b*) For the general case, write down $\mathbf{X}$ and $\mathbf{V}$.
(*c*) Solve the normal equations for $\hat{\boldsymbol{\alpha}}$, showing that

$$\hat{\alpha}_i = \frac{\displaystyle\sum_{j=1}^{N_i} \frac{\bar{y}_{ij\cdot}}{b + e/n_{ij}}}{\displaystyle\sum_{j=1}^{N_i} \frac{1}{b + e/n_{ij}}}.$$

11. The solution to (104) can be obtained in two different ways: either in the manner of (101) and (102), or using equation (30) and the ensuing results in Sec. 1.5b. Show the equivalence of the two solutions.

CHAPTER 6

TWO ELEMENTARY MODELS

The methods of the preceding chapter are now demonstrated for specific applications. Only unbalanced data are considered in detail, with but passing reference to the simpler cases of balanced data. The applications discussed do by no means exhaust the great variety available, but they cover a sufficiently wide spectrum for the reader to gain an adequate understanding of the methodology, so that he can then apply it to other situations.

Throughout this chapter and the next two it is assumed that the individual error terms have zero mean and variance σ^2 and are pairwise uncorrelated; i.e., it is assumed that $E(\mathbf{e}) = \mathbf{0}$ and $\text{var}(\mathbf{e}) = \sigma^2\mathbf{I}$. These are the only assumptions made for purposes of point estimation whereas, for hypothesis testing and confidence interval estimation, normality of error terms is additionally assumed. Thus for point estimation $\mathbf{e} \sim (\mathbf{0}, \sigma^2\mathbf{I})$ is assumed, and for hypothesis testing and confidence intervals $\mathbf{e} \sim N(\mathbf{0}, \sigma^2\mathbf{I})$ is assumed. A more general assumption would be $\text{var}(\mathbf{e}) = \mathbf{V}\sigma^2$ for $\mathbf{V}$ symmetric and positive definite (or perhaps positive semi-definite). Although there is brief discussion of this in Sec. 5.8, examples of it are postponed to Chapters 9 and 10, under the heading of "mixed models".

Numerical illustrations continue to be based on hypothetical data, using numbers that have been chosen with an eye to simplifying the arithmetic. The implausibility of such numbers as real data is, it is felt, more than compensated for by improved readability of the ensuing arithmetic. Bearing in mind that the *raison d'être* of the arithmetic is to illustrate techniques, emphasis on readability seems more to be desired than mimicry of real life. The latter inevitably involves numbers that become as difficult to follow as the algebra they purport to illustrate.

1. SUMMARY OF GENERAL RESULTS

It is appropriate to summarize the main results of Chapter 5 that are used in this and the next two chapters.

The equation of the model is

$$\mathbf{y} = \mathbf{Xb} + \mathbf{e}, \tag{1}$$

with normal equations

$$\mathbf{X'Xb}^o = \mathbf{X'y}, \tag{2}$$

whose solution is

$$\mathbf{b}^o = \mathbf{GX'y} \tag{3}$$

where $\mathbf{G}$ is a generalized inverse of $\mathbf{X'X}$, meaning that it satisfies $\mathbf{X'XGX'X} = \mathbf{X'X}$.

Development of the general theory in Chapter 5 has, as its starting point, the finding of a matrix $\mathbf{G}$. However, Sec. 5.7b describes a procedure for solving the normal equations by putting some elements of $\mathbf{b}^o$ equal to zero and then finding the $\mathbf{G}$ that corresponds to this solution. In certain cases this is an easy procedure because putting some elements of $\mathbf{b}^o$ equal to zero so greatly simplifies the normal equations that their solution becomes "obvious", and the corresponding $\mathbf{G}$ (by the methods of Sec. 5.7c) equally so. The basis of this procedure when $\mathbf{X'X}$ has order p and rank r is to

$$\text{set } p - r \text{ elements of } \mathbf{b}^o \text{ equal to zero} \tag{4}$$

and to strike out corresponding equations from the normal equations, leaving a set of r equations of full rank. Details are given in Sec. 5.7.

Having obtained a value for $\mathbf{b}^o$ the predicted value of $\mathbf{y}$ corresponding to its observed value is

$$\hat{\mathbf{y}} = \mathbf{XGX'y} \tag{5}$$

and the residual sum of squares is

$$\text{SSE} = \mathbf{y'y} - \mathbf{b}^{o\prime}\mathbf{X'y}$$

with the estimated error variance being

$$\hat{\sigma}^2 = \text{MSE} = \text{SSE}/(N - r), \qquad \text{where} \quad r = r(\mathbf{X}). \tag{6}$$

The sum of squares due to fitting the mean is

$$\text{SSM} = N\bar{y}^2 \tag{7}$$

where $\bar{y}$ is the mean of all observations; and the sum of squares due to fitting the model is

$$\text{SSR} = \mathbf{b}^{o\prime}\mathbf{X'y}, \tag{8}$$

while the total sum of squares is

$$\text{SST} = \mathbf{y}'\mathbf{y} = \sum y^2 \quad (9)$$

where $\sum y^2$ represents the sum of squares of the individual observation[s]. Hence

$$\text{SSE} = \text{SST} - \text{SSR}. \quad (10)$$

Furthermore, SSR and SST both corrected for the mean are

$$\text{SSR}_m = \text{SSR} - \text{SSM} \quad (11)$$

and

$$\text{SST}_m = \text{SST} - \text{SSM} \quad (12)$$

with

$$\text{MSR}_m = \text{SSR}_m/(r-1).$$

Analysis of variance tables summarizing these calculations are Tables 5.[5] and 5.6 of Sec. 5.3. From them comes the coefficient of determination

$$R^2 = \text{SSR}_m/\text{SST}_m. \quad (13)$$

Also, on the basis of normality,

$$F(R_m) = \text{MSR}_m/\text{MSE} \quad (14)$$

compared to tabulated values of the $F_{r-1,N-r}$-distribution tests whether th[e] model $E(\mathbf{y}) = \mathbf{Xb}$, over and above a general mean, accounts for variation i[n] the y-variable. Similarly

$$F(M) = \text{SSM}/\text{MSE} = N\bar{y}^2/\hat{\sigma}^2 \quad (15)$$

compared to tabulated values of $F_{1,N-r}$ tests the hypothesis H: $E(\bar{y}) = 0$. Comparing $\sqrt{F(M)}$ against the t_{N-r}-distribution is an identical test.

As indicated in Sec. 5.4, the expected value of any observation is estim[-]able: i.e., any (and every) element of **Xb** is estimable, and its correspondin[g] b.l.u.e. (best linear unbiased estimator) is the same element of $\mathbf{Xb}^o$; and an[y] linear combination of elements of **Xb** is estimable with its b.l.u.e. being th[e] same linear combination of elements of $\mathbf{Xb}^o$. More generally,

$$\mathbf{q}'\mathbf{b} \text{ is estimable when } \mathbf{q}' = \mathbf{t}'\mathbf{X} \text{ for any } \mathbf{t}', \quad (16)$$

and then

$$\widehat{\mathbf{q}'\mathbf{b}} = \mathbf{q}'\mathbf{b}^o \text{ is the b.l.u.e. of } \mathbf{q}'\mathbf{b} \quad (17)$$

with

$$v(\widehat{\mathbf{q}'\mathbf{b}}) = \mathbf{q}'\mathbf{G}\mathbf{q}\sigma^2; \quad (18)$$

and the $100(1-\alpha)\%$ symmetric confidence interval on $\mathbf{q}'\mathbf{b}$ is

$$\mathbf{q}'\mathbf{b}^o \pm \hat{\sigma} t_{N-r,\frac{1}{2}\alpha}\sqrt{\mathbf{q}'\mathbf{G}\mathbf{q}}. \quad (19)$$

Table 5.7 in Sec. 5.4c shows a variety of special cases of estimable functions.

A test of the general linear hypothesis

$$H: \ \mathbf{K}'\mathbf{b} = \mathbf{m}, \text{ for } \mathbf{K}'\mathbf{b} \text{ estimable and } \mathbf{K}' \text{ having full row rank } s \quad (20)$$

s to compare
$$F(H) = Q/s\hat{\sigma}^2,$$
n which
$$Q = (\mathbf{K}'\mathbf{b}^\circ - \mathbf{m})'(\mathbf{K}'\mathbf{G}\mathbf{K})^{-1}(\mathbf{K}'\mathbf{b}^\circ - \mathbf{m}), \tag{21}$$
gainst tabulated values of the $F_{s,N-r}$-distribution. The solution of the ormal equations under the null hypothesis is then, if needed,

$$\mathbf{b}_H^\circ = \mathbf{b}^\circ - \mathbf{G}\mathbf{K}(\mathbf{K}'\mathbf{G}\mathbf{K})^{-1}(\mathbf{K}'\mathbf{b}^\circ - \mathbf{m}).$$

'articular interest attaches to hypotheses of the form $\mathbf{K}'\mathbf{b} = \mathbf{0}$ in which $\mathbf{m}$ f the general case (20) is null. These are discussed in Sec. 5.5c, with the nalysis of variance shown in Table 5.9 therein, and appropriate F-tests ollowing it. Section 5.5g deals with orthogonal contrasts $\mathbf{k}_i'\mathbf{b}$ among the lements of $\mathbf{b}$. These contrasts are such that

$$\mathbf{k}_i'\mathbf{G}\mathbf{k}_j = 0 \qquad \text{for} \quad i \neq j. \tag{22}$$

Vhen (22) is true for $i, j = 1, 2, \ldots, r$, the test of the hypothesis H: $\mathbf{K}'\mathbf{b} =$ $\mathbf{0}$ has a numerator sum of squares which not only equals SSR but also equals he sum of the numerator sums of squares for testing the r hypotheses H_i: $\mathbf{k}_i'\mathbf{b} = \mathbf{0}$ where $\mathbf{K}' = \{\mathbf{k}_i'\}$ for $i = 1, 2, \ldots, r$.

Models that include restrictions on the parameters are also dealt with in Chapter 5. Their analyses are summarized in Table 5.13 of Sec. 5.6.

2. THE 1-WAY CLASSIFICATION

Chapter 4 contains discussion of data concerning the investment on consumer durables of people with different levels of education. Assuming that nvestment is measured by an index number, suppose that available data consist of values of this index for 7 people, as shown in Table 6.1. It is a very small example but adequate for illustrative purposes.

a. Model

A suitable model for these data suggested in Sec. 4.3a is

$$y_{ij} = \mu + \alpha_i + e_{ij} \tag{23}$$

TABLE 6.1. INVESTMENT INDICES OF 7 PEOPLE

Level of Education	No. of People	Indices	Total
1 (High school incomplete)	3	74, 68, 77	219
2 (High school graduate)	2	76, 80	156
3 (College graduate)	2	85, 93	178
Totals	7		553

where y_{ij} is the investment index of the jth person in the ith education level, μ is a general mean, α_i is the effect on investment of the ith level of education and e_{ij} is the random error term peculiar to y_{ij}. For the data of Table 6.1 there are 3 education levels, and i takes values $i = 1, 2, 3$; and for a given i, subscript j takes values $j = 1, 2, \ldots, n_i$ where n_i is the number of observations in the ith education level: $n_1 = 3$, $n_2 = 2$ and $n_3 = 2$ in Table 6.1.

The model (23) is the model for the 1-way classification. In general, the groupings such as education levels are called classes and in (23) y_{ij} is the response of the jth observation in the ith class, μ is a general mean, α_i is the effect on the response of the ith class and e_{ij} is the error term. When the number of classes in the data is a, $i = 1, 2, \ldots, a$, with $j = 1, 2, \ldots, n_i$. Although described here in terms of investment as the response and levels of education as the classes, this is a model that can apply to many situations. For example, the classes may be varieties of a plant, makes of a machine, or different levels of income in a community. Analysis of this model has already been used as an illustration in Chapter 5, interspersed with the development of the general methods in that chapter; we give a further example here and indicate some results that apply to the model generally.

The normal equations come from writing the data of Table 6.1 in terms of equation (23):

$$\begin{bmatrix} 74 \\ 68 \\ 77 \\ 76 \\ 80 \\ 85 \\ 93 \end{bmatrix} = \begin{bmatrix} y_{11} \\ y_{12} \\ y_{13} \\ y_{21} \\ y_{22} \\ y_{31} \\ y_{32} \end{bmatrix} = \begin{bmatrix} \mu + \alpha_1 + e_{11} \\ \mu + \alpha_1 + e_{12} \\ \mu + \alpha_1 + e_{13} \\ \mu + \alpha_2 + e_{21} \\ \mu + \alpha_2 + e_{22} \\ \mu + \alpha_3 + e_{31} \\ \mu + \alpha_3 + e_{32} \end{bmatrix}, \tag{24}$$

i.e.,

$$\begin{bmatrix} 74 \\ 68 \\ 77 \\ 76 \\ 80 \\ 85 \\ 93 \end{bmatrix} = \mathbf{y} = \begin{bmatrix} 1 & 1 & 0 & 0 \\ 1 & 1 & 0 & 0 \\ 1 & 1 & 0 & 0 \\ 1 & 0 & 1 & 0 \\ 1 & 0 & 1 & 0 \\ 1 & 0 & 0 & 1 \\ 1 & 0 & 0 & 1 \end{bmatrix} \begin{bmatrix} \mu \\ \alpha_1 \\ \alpha_2 \\ \alpha_3 \end{bmatrix} + \begin{bmatrix} e_{11} \\ e_{12} \\ e_{13} \\ e_{21} \\ e_{22} \\ e_{31} \\ e_{32} \end{bmatrix} = \mathbf{Xb} + \mathbf{e}.$$

Thus

$$\mathbf{X} = \begin{bmatrix} 1 & 1 & 0 & 0 \\ 1 & 1 & 0 & 0 \\ 1 & 1 & 0 & 0 \\ 1 & 0 & 1 & 0 \\ 1 & 0 & 1 & 0 \\ 1 & 0 & 0 & 1 \\ 1 & 0 & 0 & 1 \end{bmatrix} \quad \text{and} \quad \mathbf{b} = \begin{bmatrix} \mu \\ \alpha_1 \\ \alpha_2 \\ \alpha_3 \end{bmatrix}, \tag{25}$$

with **y** being the vector of observations and **e** the vector of corresponding error terms.

General formulation of the model (1) for the 1-way classification is achieved by writing the vector of responses as

$$\mathbf{y}' = [y_{11} \quad y_{12} \quad \cdots \quad y_{1n_1} \quad \cdots \quad y_{i1} \quad y_{i2} \quad \cdots \quad y_{in_i} \quad \cdots \quad y_{a1} \quad y_{a2} \quad \cdots \quad y_{an_a}] \tag{26}$$

and the vector of parameters as

$$\mathbf{b}' = [\mu \quad \alpha_1 \quad \alpha_2 \quad \cdots \quad \alpha_a], \tag{27}$$

in which case the matrix **X** has order $N \times (a + 1)$, where

$$N = n. = \sum_{i=1}^{a} n_i$$

and the symbols N and $n.$ are used interchangeably.

The form of **X** in (25) is typical of its general form. Its first column is $\mathbf{1}_N$ and of its other columns the ith one has $\mathbf{1}_{n_i}$ in its $(\sum_{k=1}^{i-1} n_k + 1)$th to $(\sum_{k=1}^{i} n_k)$th rows, and zeros elsewhere. Thus in these a columns the $\mathbf{1}_{n_i}$-vectors lie down the "diagonal", as in (25), and so can be written as a direct sum, using the following notation.

Notation. The direct sum of three matrices $\mathbf{A}_1$, $\mathbf{A}_2$ and $\mathbf{A}_3$ is defined [e.g., Searle (1966, Sec. 8.9)] as

$$\sum_{i=1}^{3}{}^{+} \mathbf{A}_i = \begin{bmatrix} \mathbf{A}_1 & \mathbf{0} & \mathbf{0} \\ \mathbf{0} & \mathbf{A}_2 & \mathbf{0} \\ \mathbf{0} & \mathbf{0} & \mathbf{A}_3 \end{bmatrix}.$$

The symbol Σ^+ for a direct sum is introduced here for subsequent convenience. Using Σ^+, the form of $\mathbf{X}$ in the general 1-way classification is, as in (25),

$$\mathbf{X} = \begin{bmatrix} \mathbf{1}_N & \sum_{i=1}^{a}{}^{+}\,\mathbf{1}_{n_i} \end{bmatrix}. \tag{28}$$

b. Normal equations

The normal equations $\mathbf{X}'\mathbf{X}\mathbf{b}^o = \mathbf{X}'\mathbf{y}$ of (2) are, from (26) and (28),

$$\mathbf{X}'\mathbf{X}\mathbf{b} = \begin{bmatrix} n. & n_1 & n_2 & n_3 & \cdots & n_a \\ n_1 & n_1 & 0 & 0 & \cdots & 0 \\ n_2 & 0 & n_2 & 0 & \cdots & 0 \\ n_3 & 0 & 0 & n_3 & \cdots & 0 \\ \cdot & \cdot & \cdot & \cdot & & \cdot \\ \cdot & \cdot & \cdot & \cdot & & \cdot \\ \cdot & \cdot & \cdot & \cdot & & \cdot \\ n_a & 0 & 0 & 0 & \cdots & n_a \end{bmatrix} \begin{bmatrix} \mu^o \\ \alpha_1^o \\ \alpha_2^o \\ \alpha_3^o \\ \cdot \\ \cdot \\ \cdot \\ \alpha_a^o \end{bmatrix} = \begin{bmatrix} y_{..} \\ y_{1\cdot} \\ y_{2\cdot} \\ y_{3\cdot} \\ \cdot \\ \cdot \\ \cdot \\ y_{a\cdot} \end{bmatrix} = \mathbf{X}'\mathbf{y}. \tag{29}$$

It can be seen that $\mathbf{X}'\mathbf{X}$ has $n_. = N$ as its leading element and the rest of its first row (and column) consists of the n_i's, which are also the remaining elements of the diagonal. $\mathbf{X}'\mathbf{y}$ is the vector of the response totals, totals of the y_{ij}'s: the first is the grand total and the others are the class totals. This is evident in the normal equations for the example, which from (24) and (25) are

$$\begin{bmatrix} 7 & 3 & 2 & 2 \\ 3 & 3 & 0 & 0 \\ 2 & 0 & 2 & 0 \\ 2 & 0 & 0 & 2 \end{bmatrix} \begin{bmatrix} \mu^o \\ \alpha_1^o \\ \alpha_2^o \\ \alpha_3^o \end{bmatrix} = \begin{bmatrix} y_{..} \\ y_{1\cdot} \\ y_{2\cdot} \\ y_{3\cdot} \end{bmatrix} = \begin{bmatrix} 553 \\ 219 \\ 156 \\ 178 \end{bmatrix}. \tag{30}$$

These clearly have the form of (29), with the right-hand vector $\mathbf{X}'\mathbf{y}$ having as elements the totals shown in Table 6.1.

c. Solving the normal equations

Solving the normal equations (29) by means of (4) demands ascertaining the rank of $\mathbf{X}$ (or equivalently of $\mathbf{X}'\mathbf{X}$). In both (25) and (28) it is clear that in $\mathbf{X}$ the first column equals the sum of the others, as in $\mathbf{X}'\mathbf{X}$ also, of (29) and (30). Therefore with $p = a + 1$, the order of $\mathbf{X}$, we have $r = r(\mathbf{X}) = a + 1 - 1 = a$, and so $p - r = 1$. Hence by (4) we can solve the normal equations by putting one element of $\mathbf{b}^o$ equal to zero, and crossing out one equation. In (29) and (30) the obvious element to equate to zero is μ^o, deleting the first equation in so doing. As a result, the solution of (29) is

$$\mathbf{b}^o = \begin{bmatrix} \mu^o \\ \alpha_1^o \\ \alpha_2^o \\ \vdots \\ \alpha_a^o \end{bmatrix} = \begin{bmatrix} 0 \\ \bar{y}_{1\cdot} \\ \bar{y}_{2\cdot} \\ \vdots \\ \bar{y}_{a\cdot} \end{bmatrix}; \tag{31}$$

i.e., the solutions to the normal equations are $\mu^o = 0$ and $\alpha_i^o = \bar{y}_{i\cdot}$ for $i = 1, 2, \ldots, a$. The corresponding generalized inverse of $\mathbf{X}'\mathbf{X}$ is

$$\mathbf{G} = \begin{bmatrix} 0 & \mathbf{0} \\ \mathbf{0} & \mathbf{D}\{1/n_i\} \end{bmatrix} \tag{32}$$

where $\mathbf{D}\{1/n_i\}$ is the diagonal matrix of elements $1/n_i$ for $i = 1, 2, \ldots, a$, and

$$\mathbf{H} = \mathbf{GX}'\mathbf{X} = \begin{bmatrix} 0 & \mathbf{0}' \\ \mathbf{1}_a & \mathbf{I}_a \end{bmatrix}, \tag{33}$$

on multiplying (32) and $\mathbf{X}'\mathbf{X}$ of (29).

For the example, $\mathbf{b}^o$ of (31) is

$$\mathbf{b}^o = \begin{bmatrix} 0 \\ 219/3 \\ 156/2 \\ 178/2 \end{bmatrix} = \begin{bmatrix} 0 \\ 73 \\ 78 \\ 89 \end{bmatrix} \tag{34}$$

and from (32)

$$\mathbf{G} = \begin{bmatrix} 0 & 0 & 0 & 0 \\ 0 & \frac{1}{3} & 0 & 0 \\ 0 & 0 & \frac{1}{2} & 0 \\ 0 & 0 & 0 & \frac{1}{2} \end{bmatrix} \tag{35}$$

and

$$\mathbf{H} = \begin{bmatrix} 0 & 0 & 0 & 0 \\ 0 & \frac{1}{3} & 0 & 0 \\ 0 & 0 & \frac{1}{2} & 0 \\ 0 & 0 & 0 & \frac{1}{2} \end{bmatrix} \begin{bmatrix} 7 & 3 & 2 & 2 \\ 3 & 3 & 0 & 0 \\ 2 & 0 & 2 & 0 \\ 2 & 0 & 0 & 2 \end{bmatrix} = \begin{bmatrix} 0 & 0 & 0 & 0 \\ 1 & 1 & 0 & 0 \\ 1 & 0 & 1 & 0 \\ 1 & 0 & 0 & 1 \end{bmatrix}. \tag{36}$$

d. Analysis of variance

In all cases, SSM and SST of (7) and (9) are easily computed. The other term basic to the analysis of variance is SSR of (8), and by (29) and (31) this is

$$\text{SSR} = \mathbf{b}^{o\prime}\mathbf{X}'\mathbf{y} = \sum_{i=1}^{a} \bar{y}_{i\cdot}y_{i\cdot} = \sum_{i=1}^{a} y_{i\cdot}^2/n_i\,. \tag{37}$$

With the data of Table 6.1, calculation of these terms proceeds as follows.

$$\text{SSM} = N\bar{y}^2 = 7(553/7)^2 = 43{,}687, \tag{38}$$

$$\text{SST} = \sum y^2 = 74^2 + 68^2 + 77^2 + 76^2 + 80^2 + 85^2 + 93^2 = 44{,}079 \tag{39}$$

and, from (37),

$$\text{SSR} = 219^2/3 + 156^2/2 + 178^2/2 = 43{,}997. \tag{40}$$

Hence from (10)

$$\begin{aligned}\text{SSE} &= \text{SST} - \text{SSR}\\ &= \sum y^2 - \sum_{i=1}^{a} y_{i\cdot}^2/n_i = 44{,}079 - 43{,}997 = 82,\end{aligned}$$

and from (11) and (12)

$$\text{SSR}_m = \text{SSR} - \text{SSM} = 43{,}997 - 43{,}687 = 310$$

and

$$\text{SST}_m = \text{SST} - \text{SSM} = 44{,}079 - 43{,}687 = 392.$$

From these values, the analysis of variance (based on Table 5.6b of Sec. 5.3g) is that shown in Table 6.2. From this, as in (6), the estimated error

TABLE 6.2. ANALYSIS OF VARIANCE OF DATA IN TABLE 6.1

Term	Degrees of Freedom	Sum of Squares	Mean Square	F-statistic
Model (after mean)	$a - 1 = 2$	$\text{SSR}_m = 310$	$\text{MSR}_m = 155$	$F(R_m) = 7.56$
Residual error	$N - a = 4$	$\text{SSE} = 82$	$\text{MSE} = 20.5$	
Total (after mean)	$N - 1 = 6$	$\text{SST}_m = 392$		

variance is

$$\hat{\sigma}^2 = \text{MSE} = 20.5 \tag{41}$$

and the coefficient of determination, as in (13), is

$$R^2 = \text{SSR}_m/\text{SST}_m = 310/392 = 0.79;$$

i.e., fitting the model $y_{ij} = \mu + \alpha_i + e_{ij}$ accounts for 79% of the total sum of squares.

The statistic $F(R_m)$ of (14) is

$$F(R_m) = \text{MSR}_m/\text{MSE} = 155/20.5 = 7.56$$

with $r - 1 = 2$ and $N - r = 4$ degrees of freedom. On the basis of normality, comparison of this with tabulated values of the $F_{2,4}$-distribution provides a test of whether the model, over and above a mean, accounts for the variation in y. Since the 5% critical value of the $F_{2,4}$-distribution is 6.94 and this is exceeded by $F(R_m) = 7.56$, we conclude that the model does account for variation in y. Equivalently, we could conclude that the α-effects in the model contribute significantly to the interpretive value of the model over and above μ. Similarly, calculating (15) from (38) and (41) gives

$$F(M) = 43{,}687/20.5 = 2131.1.$$

Since the 5% critical point of the $F_{1,4}$-distribution is 7.71 we reject the hypothesis H: $E(\bar{y}) = 0$. This can also be construed as rejecting the hypothesis H: $\mu = 0$ when ignoring the α's.

e. Estimable functions

The expected value of any observation is estimable: thus $\mu + \alpha_i$ is estimable and, correspondingly,

$$\text{the b.l.u.e. of } \mu + \alpha_i \text{ is } \mu^o + \alpha_i^o .$$

Using ^ over an expression to denote the b.l.u.e. of that expression, and noting the values of μ^o and α_i^o from (31), gives

$$\widehat{\mu + \alpha_i} = \mu^o + \alpha_i^o = \bar{y}_{i\cdot} . \tag{42}$$

The variance of the b.l.u.e. of an estimable function comes from expressing that function as $\mathbf{q}'\mathbf{b}$: $\mathbf{q}'\mathbf{b}^o$ is its b.l.u.e. and $\mathbf{q}'\mathbf{G}\mathbf{q}\sigma^2$ is the variance of the b.l.u.e. For example, with $\mathbf{b}'$ of (27)

$$\mu + \alpha_1 = [1 \quad 1 \quad 0 \quad \cdots \quad 0]\mathbf{b}$$

and so with

$$\mathbf{q}' = [1 \quad 1 \quad 0 \quad \cdots \quad 0] \tag{43}$$

$\widehat{\mu + \alpha_1}$ of (42) is $\mathbf{q}'\mathbf{b}^o$:

$$\widehat{\mu + \alpha_1} = \bar{y}_{1\cdot} = \mathbf{q}'\mathbf{b}^o$$

and hence

$$v(\widehat{\mu + \alpha_1}) = v(\bar{y}_{1\cdot}) = \sigma^2/n_1 = \mathbf{q}'\mathbf{G}\mathbf{q}\sigma^2. \tag{44}$$

The equality $\mathbf{q}'\mathbf{G}\mathbf{q} = 1/n_1$ for $\mathbf{q}'$ of (43) is easily verified, using $\mathbf{G}$ of (32).

The basic result concerning estimable functions is (42). It provides b.l.u.e.'s of all other estimable functions. Any linear combination of the

$\mu + \alpha_i$ is estimable and its b.l.u.e. is the same linear combination of b.l.u.e.'s of the $\mu + \alpha_i$, i.e. of the $\bar{y}_{i\cdot}$. Thus for scalars λ_i

$$\sum_{i=1}^{a} \lambda_i(\mu + \alpha_i) \text{ is estimable,} \quad \text{with b.l.u.e. } \sum_{i=1}^{a} \lambda_i \bar{y}_{i\cdot}; \tag{45}$$

i.e.,

$$\widehat{\sum_{i=1}^{a} \lambda_i(\mu + \alpha_i)} = \sum_{i=1}^{a} \lambda_i(\widehat{\mu + \alpha_i}) = \sum_{i=1}^{a} \lambda_i \bar{y}_{i\cdot}\,. \tag{46}$$

Furthermore, although the variance of this b.l.u.e. can be obtained as $\mathbf{q}'\mathbf{G}\mathbf{q}\sigma^2$ by expressing the estimable function as $\mathbf{q}'\mathbf{b}$, it is clear from (46) that the variance depends solely on variances and covariances of the $\bar{y}_{i\cdot}$. These are

$$v(\bar{y}_{i\cdot}) = \sigma^2/n_i \qquad \text{and} \qquad \text{cov}(\bar{y}_{i\cdot}, \bar{y}_{k\cdot}) = 0 \qquad \text{for } i \neq k,$$

and so from (46)

$$v[\widehat{\sum \lambda_i(\mu + \alpha_i)}] = v(\sum \lambda_i \bar{y}_{i\cdot}) = (\sum \lambda_i^2/n_i)\sigma^2, \tag{47}$$

where summation is over $i = 1, 2, \ldots, a$. From this, the $100(1 - \alpha)\%$ symmetric confidence interval on $\sum q_i(\mu + \alpha_i)$ is, from (19),

$$\sum q_i(\widehat{\mu + \alpha_i}) \pm \hat{\sigma} t_{N-r,\frac{1}{2}\alpha}\sqrt{\sum q_i^2/n_i} = \sum q_i \bar{y}_{i\cdot} \pm \hat{\sigma} t_{N-r,\frac{1}{2}\alpha}\sqrt{\sum q_i^2/n_i}\,. \tag{48}$$

For example,

$$\alpha_1 - \alpha_2 = (\mu + \alpha_1) - (\mu + \alpha_2) \tag{49}$$

is estimable, with $\lambda_1 = 1$ and $\lambda_2 = -1$ and $\lambda_3 = 0$ in (45). Hence using (34) in (46)

$$\widehat{\alpha_1 - \alpha_2} = (\widehat{\mu + \alpha_1}) - (\widehat{\mu + \alpha_2}) = \bar{y}_{1\cdot} - \bar{y}_{2\cdot} = 73 - 78 = -5;$$

and in (47)

$$v(\widehat{\alpha_1 - \alpha_2}) = [1^2/3 + (-1)^2/2]\sigma^2 = 5\sigma^2/6;$$

and from (48) the $100(1 - \alpha)\%$ symmetric confidence interval on $\alpha_1 - \alpha_2$ is

$$-5 \pm \hat{\sigma} t_{4,\frac{1}{2}\alpha}\sqrt{5/6} = -5 \pm t_{4,\frac{1}{2}\alpha}\sqrt{205/12}\,,$$

using $\hat{\sigma}^2 = 20.5$ from (41). Similarly

$$3\alpha_1 + 2\alpha_2 - 5\alpha_3 = 3(\mu + \alpha_1) + 2(\mu + \alpha_3) - 5(\mu + \alpha_3) \tag{50}$$

is estimable, with $\lambda_1 = 3$, $\lambda_2 = 2$ and $\lambda_3 = -5$, and so

$$\begin{aligned} \widehat{3\alpha_1 + 2\alpha_2 - 5\alpha_3} &= 3(\widehat{\mu + \alpha_1}) + 2(\widehat{\mu + \alpha_2}) - 5(\widehat{\mu + \alpha_3}) \\ &= 3(73) + 2(78) - 5(89) \\ &= -70 \end{aligned}$$

and from (47)

$$\widehat{v(3\alpha_1 + 2\alpha_2 - 5\alpha_3)} = (3^2/3 + 2^2/2 + 5^2/2)\sigma^2 = 17\tfrac{1}{2}\sigma^2.$$

With estimates of variances of this nature being obtained by replacing σ^2 by $\hat{\sigma}^2$ [$=20.5$ in (41)], the $100(1-\alpha)\%$ symmetric confidence interval on $3\alpha_1 + 2\alpha_2 - 5\alpha_3$ is

$$-70 \pm \sqrt{20.5}\, t_{4,\frac{1}{2}\alpha}\sqrt{17.5} = -70 \pm t_{4,\frac{1}{2}\alpha}\sqrt{358\tfrac{3}{4}}.$$

Certain implications of (45), which can be written as

$$\sum \lambda_i(\mu + \alpha_i) = \mu \sum \lambda_i + \sum \lambda_i \alpha_i \text{ is estimable,} \tag{51}$$

are worth noting. First, observe that $r(\mathbf{X}) = a$ and so, from Sec. 5.4f, the maximum number of LIN estimable functions is a. Since there are a functions $\mu + \alpha_i$, which are estimable, they therefore constitute a LIN set of estimable functions. Hence all other estimable functions are linear combinations of the $\mu + \alpha_i$, i.e., are of the form (51). Specific results follow from this.

(*i*) *μ is not estimable.* Proof: Suppose that μ is estimable. Then for some set of λ_i-values (51) must reduce to μ, and for these λ_i we would then have

$$\mu = \mu \sum \lambda_i + \sum \lambda_i \alpha_i, \text{ identically.}$$

For this to be so the λ_i must satisfy two conditions:

$$\sum \lambda_i = 1 \qquad \text{and} \qquad \sum \lambda_i \alpha_i = 0 \qquad \text{for all } \alpha_i\,.$$

The second of these conditions can be true only if $\lambda_i = 0$ for all i, in which case $\sum \lambda_i \neq 1$ and the first is not true. Hence no λ_i exist such that (51) reduces to μ; i.e., μ is not estimable. Q.E.D.

(*ii*) *α_i is not estimable.* Proof: Suppose α_k is estimable for some subscript k. Then in the second term of (51) we must have $\lambda_k = 1$ and $\lambda_i = 0$ for all $i \neq k$. But then (51) becomes $\mu + \alpha_k$. Hence α_k is not estimable. Q.E.D.

(*iii*) *$(\sum \lambda_i)\mu + \sum \lambda_i \alpha_i$ is estimable for any λ_i.* This is simply a restatement of (51), made for purposes of emphasizing the estimability of any linear combination of μ and the α_i's in which the coefficient of μ is the sum of the coefficients of the α_i. From (46) its b.l.u.e. is

$$\widehat{(\sum \lambda_i)\mu + \sum \lambda_i \alpha_i} = \sum \lambda_i \bar{y}_{i\cdot}\,.$$

For example, $13.7\mu + 6.8\alpha_1 + 2.3\alpha_2 + 4.6\alpha_3$ is estimable and its b.l.u.e. is $6.8\bar{y}_{1\cdot} + 2.3\bar{y}_{2\cdot} + 4.6\bar{y}_{3\cdot}$. Two others, of more likely interest are

$$\mu + \frac{1}{N}\sum n_i \alpha_i \qquad \text{with b.l.u.e. } \bar{y}_{\cdot\cdot} \tag{52}$$

and

$$\mu + \frac{1}{a}\sum \alpha_i \qquad \text{with b.l.u.e. } \sum \bar{y}_{i\cdot}/a\,. \tag{53}$$

These are (45)—or, equivalently, (51)—and (46) with $\lambda_i = n_i/n.$ in (52) and $\lambda_i = 1/a$ in (53). For balanced data $n_i = n$ for all i and then (52) and (53) are the same.

(*iv*) $\sum \lambda_i \alpha_i$ for $\sum \lambda_i = 0$ *is estimable*. This is just a special case of (51) in which $\sum \lambda_i = 0$, so eliminating the term in μ from (51). Thus is demonstrated the estimability of any linear combination of the α_i in which the sum of the coefficients is zero. From (46) its b.l.u.e. is

$$\widehat{\sum \lambda_i \alpha_i} = \sum \lambda_i \bar{y}_{i.} \qquad \text{with } \sum \lambda_i = 0. \tag{54}$$

$3.6\alpha_1 + 2.7\alpha_2 - 6.3\alpha_3$ with b.l.u.e. $3.6\bar{y}_{1.} + 2.7\bar{y}_{2.} - 6.3\bar{y}_{3.}$ is an example; another is $\alpha_1 + \alpha_2 - 2\alpha_3$, or $\frac{1}{2}\alpha_1 + \frac{1}{2}\alpha_2 - \alpha_3$.

(*v*) $\alpha_i - \alpha_k$ *for* $i \neq k$ *is estimable*. This arises as a special case of the preceding result: putting $\lambda_i = 1$ and $\lambda_k = -1$ and all other λ's zero shows that

$$\alpha_i - \alpha_k \text{ is estimable for every } i \neq k; \tag{55}$$

i.e., the difference between any pair of α's is estimable. Its estimator, by (46), is

$$\widehat{\alpha_i - \alpha_k} = \bar{y}_{i.} - \bar{y}_{k.}$$

with

$$v(\widehat{\alpha_i - \alpha_k}) = (1/n_i + 1/n_k)\sigma^2$$

and the $100(1 - \alpha)\%$ symmetric confidence interval on $\alpha_i - \alpha_k$ is

$$\bar{y}_{i.} - \bar{y}_{k.} \pm \hat{\sigma} t_{N-r, \frac{1}{2}\alpha} \sqrt{1/n_i + 1/n_k}\,.$$

The differences $\alpha_i - \alpha_k$ are frequently called *contrasts* (see subsection *g* that follows); they, and all linear combinations of them, often called contrasts also, are estimable, in accord with the principles of (46), (47) and (48); e.g., $\alpha_1 + \alpha_2 - 2\alpha_3 = \alpha_1 - \alpha_3 + \alpha_2 - \alpha_3$ is estimable, as above.

Estimability of the above functions could, of course, be established from the basic property common to all estimable functions, that they are functions of expected values of observations; e.g.,

$$\alpha_1 - \alpha_2 = E(y_{1j}) - E(y_{2j'}) = (\mu + \alpha_1) - (\mu + \alpha_2).$$

But the detailed derivations show how particular cases are all part of the general result (42) to which estimable functions belong.

f. Tests of linear hypotheses

(*i*) *General hypotheses*. The only hypotheses that can be tested are those that involve estimable functions, various forms of which have just been discussed. In all cases they are tested in accord with (21) of Sec 6.1.

For example, $\alpha_2 - \alpha_1$ and $2\alpha_3 - \alpha_1 - \alpha_2$ are estimable and, for example, the hypothesis

$$H: \begin{array}{c} \alpha_2 - \alpha_1 = 9 \\ 2\alpha_3 - \alpha_1 - \alpha_2 = 30 \end{array},$$

equivalent to

$$\begin{bmatrix} 0 & -1 & 1 & 0 \\ 0 & -1 & -1 & 2 \end{bmatrix} \mathbf{b} = \begin{bmatrix} 9 \\ 30 \end{bmatrix},$$

is tested by using

$$\mathbf{K}' = \begin{bmatrix} 0 & -1 & 1 & 0 \\ 0 & -1 & -1 & 2 \end{bmatrix}, \qquad \mathbf{K}'\mathbf{b}^o = \begin{bmatrix} 5 \\ 27 \end{bmatrix}$$

and

$$(\mathbf{K}'\mathbf{G}\mathbf{K})^{-1} = \begin{bmatrix} \frac{5}{6} & -\frac{1}{6} \\ -\frac{1}{6} & \frac{17}{6} \end{bmatrix}^{-1} = \frac{1}{14}\begin{bmatrix} 17 & 1 \\ 1 & 5 \end{bmatrix}. \tag{56}$$

Then

$$\mathbf{K}'\mathbf{b}^o - \mathbf{m} = \begin{bmatrix} 5 \\ 27 \end{bmatrix} - \begin{bmatrix} 9 \\ 30 \end{bmatrix} = \begin{bmatrix} -4 \\ -3 \end{bmatrix}$$

so that Q of (21) is

$$Q = [-4 \quad -3]\tfrac{1}{14}\begin{bmatrix} 17 & 1 \\ 1 & 5 \end{bmatrix}\begin{bmatrix} -4 \\ -3 \end{bmatrix} = 341/14.$$

Then, using $s = r(\mathbf{K}') = 2$, and $\hat{\sigma}^2 = 20.5$ of (41)

$$F(H) = (341/14)/2(20.5) = 341/574 < 1.$$

Comparison with tabulated values of the $F_{2,4}$-distribution indicates non-rejection of the hypothesis. (The 5% value of the $F_{2,4}$-distribution is 6.94.)

(ii) *The test based on F(M)*. The hypothesis H: $E(\bar{y}) = 0$ tested by using $F(M)$ of (15) is identical to H: $N\mu + \sum n_i\alpha_i = 0$ because, from the model (23), $NE(\bar{y}) = N\mu + \sum n_i\alpha_i$. That H: $N\mu + \sum n_i\alpha_i = 0$ *is* a testable hypothesis is readily evident by rewriting it as

$$H: \quad \boldsymbol{\lambda}'\mathbf{b} = 0 \quad \text{with} \quad \boldsymbol{\lambda}' = [N \quad n_1 \quad n_2 \quad \cdots \quad n_a]. \tag{57}$$

Clearly $\boldsymbol{\lambda}'\mathbf{b}$ is estimable, as in (52). To show that (21) reduces to SSM for (57) we use (31) and (32) to derive

$$\boldsymbol{\lambda}'\mathbf{b}^o = \sum n_i\bar{y}_{i.} = N\bar{y}_{..}, \qquad \boldsymbol{\lambda}'\mathbf{G} = [0 \quad \mathbf{1}'] \qquad \text{and} \qquad \boldsymbol{\lambda}'\mathbf{G}\boldsymbol{\lambda} = \sum n_i = N.$$

Hence the numerator sum of squares for testing H is, from (21),

$$Q = \mathbf{b}^{o\prime}\boldsymbol{\lambda}(\boldsymbol{\lambda}'\mathbf{G}\boldsymbol{\lambda})^{-1}\boldsymbol{\lambda}'\mathbf{b}^o = N\bar{y}_{..}^2 = \text{SSM}.$$

Furthermore, in (21) s is defined as $s = r(\mathbf{K}')$ and so here $s = r(\boldsymbol{\lambda}') = 1$ and so (21) is

$$F(H) = Q/s\hat{\sigma}^2 = \text{SSM}/\hat{\sigma}^2 = \text{SSM}/\text{MSE} = F(M)$$

of (15). Hence the F-test using $F(M)$ does test $H\colon\ n.\mu + \sum n_i\alpha_i = 0$ or, equivalently, $H\colon\ \mu + \sum n_i\alpha_i/n. = 0$. ($n. = N$)

Example. To test $H\colon\ 7\mu + 3\alpha_1 + 2\alpha_2 + 2\alpha_3 = 0$ in our example,

$$\boldsymbol{\lambda}'\mathbf{b}^o = [7 \quad 3 \quad 2 \quad 2]\begin{bmatrix} 0 \\ 73 \\ 78 \\ 89 \end{bmatrix} = 553 \qquad \text{and} \qquad \boldsymbol{\lambda}'\mathbf{G}\boldsymbol{\lambda} = [0 \quad 1 \quad 1 \quad 1]\begin{bmatrix} 7 \\ 3 \\ 2 \\ 2 \end{bmatrix} = 7$$

so that

$$Q = 553(7^{-1})553 = 553^2/7 = 43{,}687 = \text{SSM of (38).}$$

Hence
$$F(H) = Q/s\hat{\sigma}^2 = 43{,}687/20.5 = 2131.1 = F(M)$$

calculated earlier.

(*iii*) *The test based on $F(R_m)$.* The test based on $F(R_m)$ shown in (14) is equivalent (for the 1-way classification) to testing

$$H\colon\ \text{all } \alpha\text{'s equal,}$$

and this in turn appears equivalent to testing that all the α's are zero.

First consider the example, where there are only three α's. Then the above hypothesis $H\colon\ \alpha_1 = \alpha_2 = \alpha_3$ is identical to $H\colon\ \alpha_1 - \alpha_2 = \alpha_1 - \alpha_3 = 0$ which can be written as

$$H\colon \quad \begin{bmatrix} 0 & 1 & -1 & 0 \\ 0 & 1 & 0 & -1 \end{bmatrix}\begin{bmatrix} \mu \\ \alpha_1 \\ \alpha_2 \\ \alpha_3 \end{bmatrix} = \begin{bmatrix} 0 \\ 0 \end{bmatrix}. \tag{58}$$

Writing this as $\mathbf{K}'\mathbf{b} = \mathbf{0}$ we have

$$\mathbf{K}' = \begin{bmatrix} 0 & 1 & -1 & 0 \\ 0 & 1 & 0 & -1 \end{bmatrix}, \qquad \mathbf{K}'\mathbf{b}^o = \begin{bmatrix} -5 \\ -16 \end{bmatrix}$$

and
$$(\mathbf{K}'\mathbf{G}\mathbf{K})^{-1} = \begin{bmatrix} \frac{5}{6} & \frac{1}{3} \\ \frac{1}{3} & \frac{5}{6} \end{bmatrix}^{-1} = \tfrac{1}{7}\begin{bmatrix} 10 & -4 \\ -4 & 10 \end{bmatrix}$$

using $\mathbf{b}^o$ and $\mathbf{G}$ of (34) and (35). Hence in (21), where $s = r(\mathbf{K}') = 2$,

$$Q = [-5 \quad -16]\,\tfrac{1}{7}\begin{bmatrix} 10 & -4 \\ -4 & 10 \end{bmatrix}\begin{bmatrix} -5 \\ -16 \end{bmatrix} = 2170/7 = 310 = \text{SSR}_m$$

of Table 6.2. Therefore

$$F(H) = Q/s\hat{\sigma}^2 = 310/2\hat{\sigma}^2 = 155/20.5 = 7.56 = F(R_m)$$

of Table 6.2.

Generalization of this result follows. The hypothesis of equality of all the α's can be written as

$$H:\quad \mathbf{K'b} = \mathbf{0} \qquad \text{with} \qquad \mathbf{K'} = [\mathbf{0}\mathbf{1}_{a-1} \quad \mathbf{1}_{a-1} \quad -\mathbf{I}_{a-1}] \tag{59}$$

where $\mathbf{K'}$ has full row rank $s = a - 1$. It can then be shown (see Exercise 7) that Q of (21) reduces to

$$Q = \sum n_i\bar{y}_{i.}^2 - N\bar{y}^2 = \text{SSR} - \text{SSM} = \text{SSR}_m,$$

using SSR defined in (37). Thus,

$$F(H) = Q/s\hat{\sigma}^2 = \text{SSR}_m/(a-1)\text{MSE} = \text{MSR}_m/\text{MSE} = F(R_m)$$

as illustrated above; i.e., the test statistic $F(R_m)$ provides a test of the hypothesis H: all α's equal.

The apparent equivalence of the preceding hypothesis to one in which all α_i's are zero is now considered. First we note that because α_i is not estimable the hypothesis H: $\alpha_i = 0$ cannot be tested. Therefore H: all α_i's $= 0$ cannot, formally, be tested. However, it can be shown that there is apparent equivalence of the two hypotheses. Consider Q, the numerator sum of squares for testing H: $\mathbf{K'b} = \mathbf{0}$. The identity

$$Q = \text{SSR} - (\text{SSR} - Q)$$

is, from Tables 3.8 and 5.9, equivalent to

$$Q = \text{SSR} - \text{sum of squares due to fitting the reduced model.}$$

Now for the 1-way classification based on

$$y_{ij} = \mu + \alpha_i + e_{ij}, \tag{60}$$

we have just seen that the hypothesis H: all α_i's equal can be expressed in the form H: $\mathbf{K'b} = \mathbf{0}$ and tested. In carrying out this test the underlying reduced model is derived by putting all α_i's equal (to α say) in (60) and so getting

$$y_{ij} = \mu + \alpha + e_{ij} = \mu' + e_{ij}$$

as the reduced model (with $\mu' = \mu + \alpha$). The sum of squares for fitting this model is clearly the same as that for fitting

$$y_{ij} = \mu + e_{ij}$$

derived from putting $\alpha_i = 0$ in (60). Thus the reduced model for H: all α_i's equal appears indistinguishable from that for H: all α_i's zero. Hence the test based on $F(R_m)$ sometimes gets referred to as testing H: all α_i's zero. More correctly it is testing H: all α_i's equal.

g. Independent and orthogonal contrasts

The general form of a contrast among effects α_i is a linear combination $\sum k_i\alpha_i$ such that $\sum k_i = 0$. It can be written as

$$\sum k_i\alpha_i = \mathbf{k}'\mathbf{b} \quad \text{with} \quad \mathbf{k}' = [0 \quad k_1 \quad \cdots \quad k_a] \text{ and } \sum k_i = 0. \tag{61}$$

All such contrasts are orthogonal to $N\mu + \sum n_i\alpha_i$ considered in (57) because (22) is then satisfied: i.e., with $\boldsymbol{\lambda}'$ of (57), $\mathbf{G}$ of (32) and $\sum k_i$ of (61), equation (22) is satisfied:

$$\boldsymbol{\lambda}'\mathbf{G}\mathbf{k} = [0 \quad \mathbf{1}']\mathbf{k} = \sum k_i = 0. \tag{62}$$

Furthermore, when testing a hypothesis that $(a - 1)$ LIN such contrasts are zero, $Q = \text{SSR}_m$. For example, in testing

$$H: \begin{bmatrix} 0 & -1 & 1 & 0 \\ 0 & -1 & -1 & 2 \end{bmatrix} \mathbf{b} = \mathbf{0}$$

the values in (56) give Q of (21) as

$$Q = [5 \quad 27]\tfrac{1}{14}\begin{bmatrix} 17 & 1 \\ 1 & 5 \end{bmatrix}\begin{bmatrix} 5 \\ 27 \end{bmatrix} = 4340/14 = 310 = \text{SSR}_m$$

of Table 6.2.

Differences between pairs of α_i's are the simplest forms of contrast. Such differences are the basis of the hypotheses considered in (58) and (59), which also satisfy (62). Hence the numerators of $F(M)$ and $F(R_m)$ are independent—as already established in Sec. 5.3 for the general case.

Although, for example, $\alpha_1 - \alpha_2$ and $\alpha_1 - \alpha_3$ are both orthogonal to $7\mu + 3\alpha_1 + 2\alpha_2 + 2\alpha_3$ they are not orthogonal to each other. To find a contrast $\Sigma k_i\alpha_i$ orthogonal to $\alpha_1 - \alpha_2$ it is necessary that (22) be satisfied for $\Sigma k_i\alpha_i$ and $\alpha_1 - \alpha_2$, i.e., that

$$[0 \quad k_1 \quad k_2 \quad k_3]\mathbf{G}\begin{bmatrix} 0 \\ 1 \\ -1 \\ 0 \end{bmatrix} = k_1/3 - k_2/2 = 0$$

as well as having

$$\sum k_i = 0.$$

Any k's of the form $k_1 = -0.6k_3$, $k_2 = -0.4k_3$ and $k_3 = k_3$ will suffice. For example, $\mathbf{k}' = [0 \quad -3 \quad -2 \quad 5]$ gives $\mathbf{k}'\mathbf{b} = -3\alpha_1 - 2\alpha_2 + 5\alpha_3$ which is orthogonal to $\alpha_1 - \alpha_2$ and, of course, to $7\mu + 3\alpha_1 + 2\alpha_2 + 2\alpha_3$. Testing

$$\begin{bmatrix} 0 & 1 & -1 & 0 \\ 0 & -3 & -2 & 5 \end{bmatrix} \mathbf{b} = 0$$

then involves

$$\mathbf{K}'\mathbf{b}^o = \begin{bmatrix} -5 \\ 70 \end{bmatrix} \quad \text{and} \quad (\mathbf{K}'\mathbf{G}\mathbf{K})^{-1} = \begin{bmatrix} \frac{5}{6} & 0 \\ 0 & 17\frac{1}{2} \end{bmatrix}^{-1} = \begin{bmatrix} \frac{6}{5} & 0 \\ 0 & \frac{2}{35} \end{bmatrix}$$

and so Q of (21) is

$$Q = [-5 \quad 70] \begin{bmatrix} \frac{6}{5} & 0 \\ 0 & \frac{2}{35} \end{bmatrix} \begin{bmatrix} -5 \\ 70 \end{bmatrix} = 30 + 280 = 310 = \text{SSR}_m.$$

The terms that make up this sum, namely 30 and 280, are the numerator sums of squares for testing $\alpha_1 - \alpha_2 = 0$ and $-3\alpha_1 - 2\alpha_2 + 5\alpha_3 = 0$ respectively, as can be easily verified. Evidence that this is so is seen in the zero off-diagonal elements of $\mathbf{K}'\mathbf{G}\mathbf{K}$, showing the independence of the elements of $\mathbf{K}'\mathbf{b}^o$.

h. Models that include restrictions

It has been emphasized in Sec. 5.6 that linear models do not need to include restrictions on their elements. But if they do, estimable functions and testable hypotheses may take different forms from those they have in the unrestricted model. In particular, functions of interest that are not estimable in the unrestricted model may be estimable in the restricted model—the model that includes restrictions.

In considering restrictions we confine ourselves to those relating to non-estimable functions. This is because restrictions relating to estimable functions do not alter the form of estimable functions and testable hypotheses available in the unrestricted model. This is shown in Table 5.13 where we also see that the only changes from an unrestricted model incurred by having a restricted model are those wrought in estimable functions by the restriction that is part of the restricted model. These are particularly interesting in the 1-way classification, some of which we now illustrate.

Suppose the restricted model has the restriction $\Sigma n_i\alpha_i = 0$. Then the function $\mu + \Sigma n_i\alpha_i/n_.$, which is estimable in the unrestricted model [as in (52)], becomes μ in the restricted model, with its b.l.u.e. by (52) being $\bar{y}_{..}$. Thus in the model having $\Sigma n_i\alpha_i = 0$ as a restriction, μ is estimable with b.l.u.e. $\bar{y}_{..}$. Furthermore, the hypothesis considered in (57) and tested by means of $F(M)$ then becomes H: $\mu = 0$; i.e., under the restriction $\Sigma n_i\alpha_i = 0$ the F-statistic $F(M)$ can be used to test the hypothesis H: $\mu = 0$.

Suppose the model included the restriction $\Sigma\alpha_i = 0$. In the unrestricted model $\mu + \Sigma\alpha_i/a$ is estimable with b.l.u.e. $\Sigma\bar{y}_{i.}/a$—as in (53). In the restricted model with $\Sigma\alpha_i = 0$ this means μ is estimable with b.l.u.e. $\Sigma\bar{y}_{i.}/a$. In this case the hypothesis $\mu = 0$ is tested by the F-statistic derived in the unrestricted model for testing H: $\mu + \Sigma\alpha_i/a = 0$. This is H: $\mathbf{k}'\mathbf{b} = 0$ for $k' = [1 \quad a^{-1}\mathbf{1}']$, for which $\mathbf{k}'\mathbf{b}^o = \Sigma\bar{y}_{i.}/a$ and $\mathbf{k}'\mathbf{Gk} = a^{-2}\,\Sigma(1/n_i)$. Hence the F-statistic for testing H: $\mu = 0$ in this restricted model is

$$F(H) = (\Sigma\bar{y}_{i.})^2/(\hat{\sigma}^2\,\Sigma n_i^{-1}).$$

The preceding two paragraphs illustrate how different restrictions can lead to the same parameter being estimable in different restricted models even though that parameter may not be estimable in the unrestricted model. Furthermore, even though it is formally the same parameter in the different restricted models (i.e., the same symbol), its b.l.u.e. in those models may not be the same. Its b.l.u.e. is the b.l.u.e. of the estimable function in the unrestricted model from which the estimable function in the restricted model has been derived by application of the restriction. Thus in a model having $\Sigma n_i\alpha_i = 0$ the b.l.u.e. of μ is $\bar{y}_{..}$, the b.l.u.e. of $\mu + \Sigma n_i\alpha_i/n_.$ in the unrestricted model; but in a model having $\Sigma\alpha_i = 0$ the b.l.u.e. of μ is $\Sigma\bar{y}_{i.}/a$, the b.l.u.e. of $\mu + \Sigma\alpha_i/a$ in the unrestricted model. A third example is that, in a model having $\Sigma w_i\alpha_i = 0$ for some weights w_i, μ is estimable with b.l.u.e. $\Sigma w_i\bar{y}_{i.}/\Sigma w_i$, this being the b.l.u.e. of $\mu + \Sigma w_i\alpha_i/\Sigma w_i$ in the unrestricted model. Here the F-statistic for testing H: $\mu = 0$ comes from testing

$$H\colon\ \mathbf{k}'\mathbf{b} = 0 \quad \text{with} \quad \mathbf{k}' = [1 \quad w_1/w. \quad \cdots \quad w_a/w.]$$

for $w_. = \Sigma\, w_i$. Thus

$$\mathbf{k}'\mathbf{b}^o = \Sigma w_i\bar{y}_{i.}/w. \qquad \text{and} \qquad \mathbf{k}'\mathbf{Gk} = (\Sigma w_i^2/n_i)/w_.^2$$

and so the F-statistic for testing H: $\mu = 0$ is

$$F(H) = \frac{(\Sigma w_i\bar{y}_{i.})^2}{\hat{\sigma}^2\,\Sigma w_i^2/n_i}.$$

These three cases are summarized in Table 6.3. The first two rows of Table 6.3 are, of course, just special cases of the last row: $w_i = n_i$ for the first row, and $w_i = 1$ for the second. In all three rows μ is estimable, and because $\mu + \alpha_i$ is estimable too (anything estimable in the unrestricted model is estimable in a restricted model) it follows that, in the restricted models, α_i is estimable with its b.l.u.e. being $\bar{y}_{i.}$ minus the b.l.u.e. of μ.

The choice of what model to use, the unrestricted model, one of those in Table 6.3 or some other, depends upon the nature of one's data. For unbalanced data we often find $\Sigma n_i\alpha_i = 0$ used. Having the same restrictions on the solutions, $\Sigma n_i\alpha_i^o = 0$, leads to an easy procedure for solving the normal equations, as is evident from (29): $\mu^o = \bar{y}_{..}$ and $\alpha_i^o = \bar{y}_{i.} - \bar{y}_{..}$.

TABLE 6.3. ESTIMATORS OF μ, AND F-STATISTICS FOR TESTING H: $\mu = 0$, IN THREE DIFFERENT RESTRICTED MODELS

Restriction on model	Estimable function in unrestricted model which reduces to μ in restricted model	b.l.u.e. of μ in restricted model (= b.l.u.e. of function in preceding column in unrestricted model)	F-statistic for testing H: $\mu = 0$
$\Sigma n_i\alpha_i = 0$	$\mu + \Sigma n_i\alpha_i/n.$	$\bar{y}_{..}$	$F(M) = n.\bar{y}^2_{..}/\hat{\sigma}^2$
$\Sigma\alpha_i = 0$	$\mu + \Sigma\alpha_i/a$	$\Sigma\bar{y}_{i.}/a$	$(\Sigma\bar{y}_{i.})^2/(\hat{\sigma}^2\,\Sigma n_i^{-1})$
$\Sigma w_i\alpha_i = 0$	$\mu + \Sigma w_i\alpha_i/w.$	$\Sigma w_i\bar{y}_{i.}/w.$	$(\Sigma w_i\bar{y}_{i.})^2/(\hat{\sigma}^2\,\Sigma w_i^2 n_i^{-1})$

This is perfectly permissible for finding a solution $\mathbf{b}^o$, it being of course the oft-referred-to method of applying the "usual constraints" as discussed in Sec. 5.7. But although $\Sigma n_i\alpha_i^o = 0$ provides an easy solution for $\mathbf{b}^o$, the same restriction applied to parameters of the model, $\Sigma n_i\alpha_i = 0$, may not always be appropriate. For example, suppose an experiment to estimate the efficacy of a feed additive for dairy cows was made on 7 Holsteins, 5 Jerseys and 2 Guernseys; the "constraint" $7\alpha_1^o + 5\alpha_2^o + 2\alpha_3^o = 0$ would lead very easily to solutions for μ^o, α_1^o, α_2^o and α_3^o. But if the proportions of these three breeds in the whole population of dairy cows (assumed to consist of just these three breeds and no others) was 6:2:2 it would be more meaningful to use $6\alpha_1 + 2\alpha_2 + 2\alpha_3 = 0$ rather than $7\alpha_1 + 5\alpha_2 + 2\alpha_3 = 0$, if any such restriction was desired. In this case we would use the third row of Table 6.3 rather than the first.

i. Balanced data

With balanced data $n_i = n$ for all i, and the first two rows of Table 6.3 are then equal. $\Sigma\alpha_i^o = 0$ as a "constraint" provides an easy solution to the normal equations, $\mu^o = \bar{y}_{..}$ and $\alpha_i^o = \bar{y}_{i.} - \bar{y}_{..}$, familiarly found in the literature. Apart from this, all other results stand fast: for example, $\mu + \alpha_i$ and $\alpha_i - \alpha_k$ are estimable, with b.l.u.e.'s $\bar{y}_{i.}$ and $\bar{y}_{i.} - \bar{y}_{k.}$ respectively; and SSE $= \Sigma y^2 - \Sigma y_{i.}^2/n$ as usual.

Sometimes the restriction $\Sigma\alpha_i = 0$ is also used as part of the model. This is in accord with the "constraint" $\Sigma\alpha_i^o = 0$ useful for solving the normal equations. As a restriction it can also be opportunely rationalized in terms of defining the α_i's as deviations from their mean, and hence having their mean be zero, i.e., $\Sigma\alpha_i = 0$. The effect of the restriction is to make μ and α_i estimable with b.l.u.e.'s $\hat{\mu} = \bar{y}_{..}$ and $\hat{\alpha}_i = \bar{y}_{i.} - \bar{y}_{..}$, and hypotheses about individual values of μ and α_i are then testable.

3. REDUCTIONS IN SUMS OF SQUARES

a. The R( ) notation

Consideration of models more complex than that for the 1-way classification will lead us to comparing the adequacy of different models for the same set of data. Since in the identity SSE = SST − SSR we have SSR as the reduction in total sum of squares due to fitting any particular model it, SSR, is a measure of the variation in y accounted for by that model. Comparison of different models in terms of a given set of data can therefore be made by comparing the different values of SSR that result from fitting the different models. To facilitate discussion of these comparisons we refer, as previously, to SSR as a reduction in sum of squares, and now denote it by $R(\quad)$, with the contents of the brackets indicating the model fitted. For example, in fitting $y_{ij} = \mu + \alpha_i + e_{ij}$ the reduction in sum of squares is $R(\mu, \alpha)$, the μ and α indicating a model that has parameters μ and those of an α-factor. Similarly, $R(\mu, \alpha, \beta)$ is the reduction in sum of squares for fitting $y_{ijk} = \mu + \alpha_i + \beta_j + e_{ijk}$; and $R(\mu, \alpha, \beta{:}\alpha)$ is the reduction due to fitting the nested model $y_{ijk} = \mu + \alpha_i + \beta_{ij} + e_{ijk}$, the symbol $\beta{:}\alpha$ in $R(\mu, \alpha, \beta{:}\alpha)$ indicating that the β-factor is nested within the α-factor. Extension to more complex models is clear, and at all times the letter R is mnemonic for "reduction" in sum of squares and not for "residual", as used by some writers. In this book $R(\quad)$ is always a reduction in the sum of squares.

The model $y_i = \mu + e_i$ has the normal equation $N\mu = y_.$, and the corresponding reduction in sum of squares, $R(\mu)$, is readily found to be $N\bar{y}^2$. But $N\bar{y}^2$ is, for all models, SSM. Therefore

$$R(\mu) = N\bar{y}^2 = \text{SSM}.$$

With the 1-way classification model $y_{ij} = \mu + \alpha_i + e_{ij}$, the reduction in sum of squares, which we now write as $R(\mu, \alpha)$ is, by (37),

$$\text{SSR} \equiv R(\mu, \alpha) = \sum y_{i.}^2/n_i\,.$$

Therefore, from (11)

$$\text{SSR}_m = \text{SSR} - \text{SSM} = R(\mu, \alpha) - R(\mu). \tag{63}$$

Thus for the 1-way classification SSR_m is the difference between the reductions in sums of squares due to fitting two different models, one containing μ and an α-factor and the other containing just μ. SSR_m of (63) can therefore be viewed as the additional reduction in sum of squares due to fitting a model containing μ and an α-factor over and above fitting one containing just μ. Hence $R(\mu, \alpha) - R(\mu)$ is the additional reduction due to fitting μ and α, over and above fitting just μ; or, more succinctly, it is the reduction due to fitting α over and above μ. An equivalent interpretation is that, having once fitted μ, the difference $R(\mu, \alpha) - R(\mu)$ represents the reduction in sum of squares due to fitting an α-factor additional to μ. In this way $R(\mu, \alpha) - R(\mu)$ is the reduction due to fitting "α, having already fitted μ", or to fitting "α after μ". In view of this we use the symbol $R(\alpha \mid \mu)$ for (63) and write

$$R(\alpha \mid \mu) = R(\mu, \alpha) - R(\mu). \tag{64}$$

This notation readily admits of extension. For example,

$$R(\alpha \mid \mu, \beta) = R(\mu, \alpha, \beta) - R(\mu, \beta)$$

is the reduction in sum of squares due to fitting "α, after μ and β"; i.e., the reduction due to fitting a model containing μ, an α-factor and a β-factor having already fitted one having μ and a β-factor. It is a measure of the extent to which a model can explain more of the variation in y by having in it, in a specified manner, something more than just μ and a β-factor.

Every $R(\ )$-term is, by definition, the SSR of some model. Its form is therefore $\mathbf{y}'\mathbf{X}(\mathbf{X}'\mathbf{X})^{-}\mathbf{X}'\mathbf{y}$ for $\mathbf{X}$ appropriate to that model, and with $\mathbf{X}(\mathbf{X}'\mathbf{X})^{-}\mathbf{X}'$ idempotent. Therefore, for $\mathbf{y} \sim N(\boldsymbol{\mu}, \sigma^2\mathbf{I})$ for any vector $\boldsymbol{\mu}$, the distribution of $R(\ )/\sigma^2$ is a non-central χ^2 independent of SSE. Suppose $R(\mathbf{b}_1, \mathbf{b}_2)$ is the reduction for fitting $\mathbf{y} = \mathbf{X}\mathbf{b}_1 + \mathbf{Z}\mathbf{b}_2 + \mathbf{e}$, and $R(\mathbf{b}_1)$ is the reduction for fitting $\mathbf{y} = \mathbf{X}\mathbf{b}_1 + \mathbf{e}$. Then it can be shown (see Exercise 12) that $R(\mathbf{b}_2 \mid \mathbf{b}_1)/\sigma^2$ has a non-central χ^2-distribution, independent of $R(\mathbf{b}_1)$ and of SSE. Hence whenever the reduction in sum of squares $R(\mathbf{b}_1, \mathbf{b}_2)$ for fitting a model is partitioned as $R(\mathbf{b}_1, \mathbf{b}_2) = R(\mathbf{b}_2 \mid \mathbf{b}_1) + R(\mathbf{b}_1)$, we know that both $R(\mathbf{b}_2 \mid \mathbf{b}_1)$ and $R(\mathbf{b}_1)$ have non-central χ^2-distributions and that they are independent of each other and of SSE.

The succinctness of the $R(\ )$ notation and its identifiability with its corresponding model are readily apparent. This and the distributional properties just discussed provide great convenience for considering the effectiveness of different models. As such it is used extensively in what follows.

b. Analyses of variance

Table 6.2 is an example of the analysis of variance given in Table 5.6b of Sec. 5.3g. Its underlying sums of squares can be expressed in terms of the $R(\ )$ notation as follows:

$$\text{SSM} = R(\mu) = 43{,}687, \qquad \text{SSR} = R(\mu, \alpha) = 43{,}997,$$
$$\text{SSR}_m = R(\alpha \mid \mu) = 310, \qquad \text{SSE} = \text{SST} - R(\mu, \alpha) = 82.$$

These are summarized in Table 6.4 where the aptness of the $R(\ )$ notation for highlighting the meaning of sums of squares is evident: $R(\mu)$ is the reduction due to fitting the mean μ; $R(\alpha \mid \mu)$ is that due to fitting the α-factor after μ; and with $R(\mu, \alpha)$ being the reduction due to fitting the model, which consists of an α-factor and μ, SSE $=$ SST $- R(\mu, \alpha)$ is the attendant residual sum of squares. $R(\mu, \alpha)$, of course, equals $R(\mu) + R(\alpha \mid \mu)$ as in (64).

The clarity provided by the $R(\ )$ notation is even more evident for models that involve several factors. The notation is therefore used universally in all analysis of variance tables that follow. Also, all such tables have a format similar to that of Table 6.4: they show a line for the mean, $R(\mu)$, and a total

TABLE 6.4. ANALYSIS OF VARIANCE USING $R(\ )$ NOTATION. (SEE ALSO TABLES 6.2 AND 5.5)

Source of Variation	d.f.	Sum of Squares	Mean Square	F-statistic
Mean	$1 = 1$	$R(\mu) = 43{,}687$	43,687	2131.1
α-factor after mean	$a - 1 = 2$	$R(\alpha \mid \mu) = 310$	155	7.56
Residual error	$N - a = 4$	SSE $=$ SST $- R(\mu, \alpha) = 82$	$20\frac{1}{2}$	
Total	$N = 7$	SST $= 44{,}079$		

sum of squares SST $= \sum y^2$, not corrected for the mean. The only term which such a table does not yield at a glance is the coefficient of determination $R^2 = \text{SSR}_m/\text{SST}_m$, of equation (13). However, since this can always be expressed as

$$R^2 = 1 - \frac{\text{SSE}}{\text{SST} - R(\mu)} \tag{65}$$

it too can then be readily derived from analysis of variance tables such as Table 6.4.

c. Tests of hypotheses

In Sec. 2f(iii) we saw how $F(M)$ is a suitable statistic for testing H: $n_{.}\mu + \sum n_i\alpha_i = 0$. But

$$F(M) = \text{SSM}/\text{MSE} = R(\mu)/\hat{\sigma}^2$$

and so we see that $R(\mu)$ is the numerator sum of squares for testing H: $n_{.}\mu + \sum n_i\alpha_i = 0$ as well as being the reduction in sum of squares due to fitting the model $y_{ij} = \mu + e_{ij}$. This dual interpretation of $R(\mu)$ should be noted.

Section 2f(iii) describes why $F(R_m)$ is referred to as testing H: all α_i's equal. But

$$F(R_m) = \frac{\text{MSR}_m}{\text{MSE}} = \frac{\text{SSR}_m}{(a-1)\text{MSE}} = \frac{R(\alpha \mid \mu)}{(a-1)\hat{\sigma}^2}$$

d so we see that $R(\alpha \mid \mu)$ is the numerator sum of squares for testing : all α_i's equal, as well as being the reduction in sum of squares due to ting α after μ. These two interpretations of $R(\alpha \mid \mu)$ should be borne in ind. The association of $R(\alpha \mid \mu) = R(\mu, \alpha) - R(\mu)$ with the effective sting of H: all α_i's zero is particularly convenient: in the symbol $R(\mu, \alpha)$, utting $\alpha = 0$ reduces the symbol to $R(\mu)$ and the difference between these vo, $R(\mu, \alpha) - R(\mu)$, is the required numerator sum of squares.

In terms of $R(\alpha \mid \mu)$ being the numerator sum of squares for testing the ypothesis H: all α_i's equal, Table 6.4 is an application of Table 5.9. /riting H: all α_i's equal in the form H: $\mathbf{K'b} = \mathbf{0}$, as in (59), we see in able 5.9 that $R(\alpha \mid \mu)$ is the numerator sum of squares for testing H: $\mathbf{K'b} =$, and $R(\mu)$ is the sum of squares for the reduced model $y_{ij} = \mu + \alpha + e_{ij} =$ $' + e_{ij}$.

4. THE 2-WAY NESTED CLASSIFICATION

Chapter 4 describes a student opinion poll of instructors' classroom use of computing facility in courses in English, Geology and Chemistry. Partial ata from such a poll are shown in Table 6.5. These are data from a 2-way ested classification, the analysis of which is now described.

TABLE 6.5. STUDENT OPINION POLL OF INSTRUCTORS' CLASSROOM USE OF COMPUTER FACILITY

Course	Section of Course	Observations: Individual	Total	Number[1]	Mean
English	1	5	5	(1)	5
	2	8, 10, 9	27	(3)	9
			Total 32	(4)	8
Geology	1	8, 10	18	(2)	9
	2	6, 2, 1, 3	12	(4)	3
	3	3, 7	10	(2)	5
			Total 40	(8)	5
			Grand total 72	(12)	6

For clarity, number of observations are in parenthesis.

a. Model

As suggested in Chapter 4, a suitable model is

$$y_{ijk} = \mu + \alpha_i + \beta_{ij} + e_{ijk} \tag{66}$$

where y_{ijk} is the kth observation in the jth section of the ith course, μ is a general mean, α_i is the effect due to the ith course, β_{ij} is the effect due to the jth section in the ith course and e_{ijk} is the usual error term. Having a levels of the α-factor (courses), $i = 1, 2, \ldots, a$ with $a = 2$ in the data of Table 6.5. For b_i levels of the β-factor nested within the α-factor (sections nested within courses) $j = 1, 2, \ldots, b_i$, with $b_1 = 2$ and $b_2 = 3$ in the example. And for n_{ij} observations in the jth section of the ith course, $k = 1, 2, \ldots, n_{ij}$; values of the n_{ij} in Table 5 are those in the penultimate column thereof. Shown there also are values of $n_{i.} = \sum_{j=1}^{b_i} n_{ij}$ and $n_{..} = \sum_{i=1}^{a} n_{i.}$; e.g., $n_{11} = 1$, $n_{12} = 3$ and $n_{1.} = 4$; and $n_{..} = 12$. Corresponding totals and means of the y_{ijk}'s are also shown in the table.

b. Normal equations

For the 12 observations of Table 6.5 the equations of the model (66) are

$$\begin{bmatrix} 5 \\ 8 \\ 10 \\ 9 \\ 8 \\ 10 \\ 6 \\ 2 \\ 1 \\ 3 \\ 3 \\ 7 \end{bmatrix} = \begin{bmatrix} y_{111} \\ y_{121} \\ y_{122} \\ y_{123} \\ y_{211} \\ y_{212} \\ y_{221} \\ y_{222} \\ y_{223} \\ y_{224} \\ y_{231} \\ y_{232} \end{bmatrix} = \begin{bmatrix} 1 & 1 & 0 & 1 & 0 & 0 & 0 & 0 \\ 1 & 1 & 0 & 0 & 1 & 0 & 0 & 0 \\ 1 & 1 & 0 & 0 & 1 & 0 & 0 & 0 \\ 1 & 1 & 0 & 0 & 1 & 0 & 0 & 0 \\ 1 & 0 & 1 & 0 & 0 & 1 & 0 & 0 \\ 1 & 0 & 1 & 0 & 0 & 1 & 0 & 0 \\ 1 & 0 & 1 & 0 & 0 & 0 & 1 & 0 \\ 1 & 0 & 1 & 0 & 0 & 0 & 1 & 0 \\ 1 & 0 & 1 & 0 & 0 & 0 & 1 & 0 \\ 1 & 0 & 1 & 0 & 0 & 0 & 1 & 0 \\ 1 & 0 & 1 & 0 & 0 & 0 & 0 & 1 \\ 1 & 0 & 1 & 0 & 0 & 0 & 0 & 1 \end{bmatrix} \begin{bmatrix} \mu \\ \alpha_1 \\ \alpha_2 \\ \beta_{11} \\ \beta_{12} \\ \beta_{21} \\ \beta_{22} \\ \beta_{23} \end{bmatrix} + \begin{bmatrix} e_{111} \\ e_{121} \\ e_{122} \\ e_{123} \\ e_{211} \\ e_{212} \\ e_{221} \\ e_{222} \\ e_{223} \\ e_{224} \\ e_{231} \\ e_{232} \end{bmatrix} \tag{67}$$

Writing $\mathbf{X}$ for the 12×8 matrix of 0's and 1's it is readily evident that the normal equations $\mathbf{X}'\mathbf{X}\mathbf{b}^o = \mathbf{X}'\mathbf{y}$ are

$$\left[\begin{array}{c|cc|cc|ccc} 12 & 4 & 8 & 1 & 3 & 2 & 4 & 2 \\ \hline 4 & 4 & \cdot & 1 & 3 & \cdot & \cdot & \cdot \\ 8 & \cdot & 8 & \cdot & \cdot & 2 & 4 & 2 \\ \hline 1 & 1 & \cdot & 1 & \cdot & \cdot & \cdot & \cdot \\ 3 & 3 & \cdot & \cdot & 3 & \cdot & \cdot & \cdot \\ \hline 2 & \cdot & 2 & \cdot & \cdot & 2 & \cdot & \cdot \\ 4 & \cdot & 4 & \cdot & \cdot & \cdot & 4 & \cdot \\ 2 & \cdot & 2 & \cdot & \cdot & \cdot & \cdot & 2 \end{array}\right] \left[\begin{array}{c} \mu^o \\ \hline \alpha_1^o \\ \alpha_2^o \\ \hline \beta_{11}^o \\ \beta_{12}^o \\ \hline \beta_{21}^o \\ \beta_{22}^o \\ \beta_{23}^o \end{array}\right] = \left[\begin{array}{c} 72 \\ \hline 32 \\ 40 \\ \hline 5 \\ 27 \\ \hline 18 \\ 12 \\ 10 \end{array}\right] = \left[\begin{array}{c} y_{\cdot\cdot\cdot} \\ \hline y_{1\cdot\cdot} \\ y_{2\cdot\cdot} \\ \hline y_{11\cdot} \\ y_{12\cdot} \\ \hline y_{21\cdot} \\ y_{22\cdot} \\ y_{23\cdot} \end{array}\right] \tag{68}$$

where periods represent zeros. The general form of these equations is

$$\left[\begin{array}{c|cc|cc|ccc} n_{\cdot\cdot} & n_{1\cdot} & n_{2\cdot} & n_{11} & n_{12} & n_{21} & n_{22} & n_{23} \\ \hline n_{1\cdot} & n_{1\cdot} & \cdot & n_{11} & n_{12} & \cdot & \cdot & \cdot \\ n_{2\cdot} & \cdot & n_{2\cdot} & \cdot & \cdot & n_{21} & n_{22} & n_{23} \\ \hline n_{11} & n_{11} & \cdot & n_{11} & \cdot & \cdot & \cdot & \cdot \\ n_{12} & n_{12} & \cdot & \cdot & n_{12} & \cdot & \cdot & \cdot \\ \hline n_{21} & \cdot & n_{21} & \cdot & \cdot & n_{21} & \cdot & \cdot \\ n_{22} & \cdot & n_{22} & \cdot & \cdot & \cdot & n_{22} & \cdot \\ n_{23} & \cdot & n_{23} & \cdot & \cdot & \cdot & \cdot & n_{23} \end{array}\right] \left[\begin{array}{c} \mu^o \\ \hline \alpha_1^o \\ \alpha_2^o \\ \hline \beta_{11}^o \\ \beta_{12}^o \\ \hline \beta_{21}^o \\ \beta_{22}^o \\ \beta_{23}^o \end{array}\right] = \left[\begin{array}{c} y_{\cdot\cdot\cdot} \\ \hline y_{1\cdot\cdot} \\ y_{2\cdot\cdot} \\ \hline y_{11\cdot} \\ y_{12\cdot} \\ \hline y_{21\cdot} \\ y_{22\cdot} \\ y_{23\cdot} \end{array}\right] \tag{69}$$

The partitioning shown in (69) suggests how more levels of the factors would be incorporated in the normal equations.

c. Solving the normal equations

$\mathbf{X}'\mathbf{X}$ in equations (68) and (69) has order 8 and rank 5: rows 2 and 3 sum to the first row, rows 4 and 5 sum to row 2 and rows 6, 7 and 8 sum to row 3. Hence $r(\mathbf{X}'\mathbf{X}) = 8 - 3 = 5$. For the general 2-way nested classification $\mathbf{X}'\mathbf{X}$ has rank $b_{\cdot}$, the number of subclasses. This is so because its order p is, for a levels of the main classification (courses in our example), $p = 1 + a + b_{\cdot}$; but the rows corresponding to the α-equations add to that of the μ-equation (1 dependency) and the rows corresponding to the β-equations in each α-level add to the row for that α-equation (a dependencies, linearly independent of the first one). Therefore $r = r(\mathbf{X}'\mathbf{X}) = 1 + a + b_{.} - (1 + a) = b_{.}$. Hence by (4) the normal equations can be solved by putting $p - r = 1 + a$ elements of $\mathbf{b}^o$ equal to zero. From the nature of (68) and (69) it is clear that the easiest $1 + a$ elements of $\mathbf{b}^o$ to set equal to zero are μ^o and $\alpha_1^o, \alpha_2^o, \ldots, \alpha_a^o$. Doing this gives the other elements of $\mathbf{b}^o$ as

$$\beta_{ij}^o = \bar{y}_{ij\cdot} \qquad \text{for all } i \text{ and } j \tag{70}$$

so that

$$\mathbf{b}^{o\prime} = [\mathbf{0}'_{1\times(1+a)} \quad \bar{\mathbf{y}}'] \tag{71}$$

where $\bar{\mathbf{y}}'$ is the row vector of cell means. In the case of the example, we see from Table 6.5 that

$$\bar{\mathbf{y}}' = [5 \quad 9 \quad 9 \quad 3 \quad 5]$$

and so

$$\mathbf{b}^{o\prime} = [0 \quad 0 \quad 0 \quad 5 \quad 9 \quad 9 \quad 3 \quad 5]. \tag{72}$$

The corresponding generalized inverse of $\mathbf{X}'\mathbf{X}$ is

$$\mathbf{G} = \begin{bmatrix} \mathbf{0} & \mathbf{0} \\ \mathbf{0} & \mathbf{D}(1/n_{ij}) \end{bmatrix} \qquad \text{for} \quad i = 1, \ldots, a, j = 1, 2, \ldots, b_i \tag{73}$$

where $\mathbf{D}(1/n_{ij})$ for the example is diagonal, with non-zero elements 1, $\frac{1}{3}$, $\frac{1}{2}$, $\frac{1}{4}$ and $\frac{1}{2}$.

d. Analysis of variance

Sums of squares for the analysis of variance of this model, with their calculated values for the example of Table 6.5, are as follows.

$$R(\mu) = \text{SSM} = n_{..}\bar{y}_{...}^2 = 12(6^2) \qquad = 432;$$

$$R(\mu, \alpha, \beta:\alpha) = \text{SSR} = \mathbf{b}^{o\prime}\mathbf{X}'\mathbf{y} = \sum_{i=1}^{a}\sum_{j=1}^{b_i} y_{ij\cdot}^2/n_{ij}$$

$$= 5^2/1 + 27^2/3 + 18^2/2 + 12^2/4 + 10^2/2 \qquad = 516;$$

$$R(\alpha, \beta:\alpha \mid \mu) = R(\mu, \alpha, \beta:\alpha) - R(\mu) = 516 - 432 \qquad = 84;$$

$$\text{SST} = \sum_{i=1}^{a}\sum_{j=1}^{b_i}\sum_{k=1}^{n_{ij}} y_{ijk}^2 = 5^2 + 8^2 + \cdots + 3^2 + 7^2 = 542;$$

$$\text{SSE} = \text{SST} - R(\mu, \alpha, \beta:\alpha) = 542 - 516 \qquad = 26.$$

TABLE 6.6. ANALYSIS OF VARIANCE FOR THE DATA OF TABLE 6.5

Source of Variation	d.f.	Sum of Squares	Mean Square	F-statistic
Mean	$1 = 1$	$R(\mu) = 432$	432	$F(M) = 116.3$
Model, after mean	$b_{.} - 1 = 4$	$R(\alpha, \beta:\alpha \mid \mu) = 84$	21	$F(R_m) = 5.7$
Residual	$N - b_{.} = 7$	$\text{SSE} = 26$	$3\frac{5}{7}$	
Total	$N = 12$	$\text{SST} = 542$		

Hence the analysis of variance table, in the style of Table 6.4, is that shown in Table 6.6. From $F(M) = 116.3$ we reject the hypothesis H: $E(\bar{y}) = 0$, because $F(M)$ exceeds the 5% value of the $F_{1,7}$-distribution, namely 5.59. Also, comparing $F(R_m) = 5.7$ with the 5% value of the $F_{4,7}$-distribution, 4.12, we reject the hypothesis (at the 5% significance level) that the model $E(y_{ijk}) = \mu + \alpha_i + \beta_{ij}$ of (66) does not account for more variation in the y-variable than does the model $E(y_{ijk}) = \mu$.

Suppose to the data of Table 6.5 we fit the 1-way classification model

$$y_{ijk} = \mu + \alpha_i + e_{ijk}\,.$$

Then, as in (37) and (63) the reduction for fitting this model is

$$R(\mu, \alpha) = \sum_{i=1}^{a} y_{i\cdot\cdot}^2/n_{i\cdot} = 32^2/4 + 40^2/8 = 456.$$

Hence

$$R(\beta{:}\alpha \mid \mu, \alpha) = R(\mu, \alpha, \beta{:}\alpha) - R(\mu, \alpha) = 516 - 456 = 60,$$

and

$$R(\alpha \mid \mu) = R(\mu, \alpha) - R(\mu) = 456 - 432 = 24.$$

In this way $R(\alpha, \beta{:}\alpha \mid \mu)$ of Table 6.6 can be divided into two portions:

$$\begin{aligned} 84 = R(\alpha, \beta{:}\alpha \mid \mu) &= R(\mu, \alpha, \beta{:}\alpha) - R(\mu) \\ &= R(\mu, \alpha, \beta{:}\mu) - R(\mu, \alpha) + R(\mu, \alpha) - R(\mu) \\ &= R(\beta{:}\alpha \mid \mu, \alpha) + R(\alpha \mid \mu) \\ &= 60 + 24. \end{aligned}$$

The result of doing this is seen in Table 6.7. There, the F-statistic

$$F(\alpha \mid \mu) = \frac{R(\alpha \mid \mu)}{(a - 1)\text{MSE}} = 6.5 \tag{74}$$

TABLE 6.7. ANALYSIS OF VARIANCE FOR DATA OF TABLE 6.5 (2-WAY NESTED CLASSIFICATION)

Source of Variation	d.f.	Sum of Squares	Mean Square	F-statistic
Mean, μ	$1 = 1$	$R(\mu) = 432$	432	116.3
α after μ	$a - 1 = 1$	$R(\alpha \mid \mu) = 24$	24	6.5
$\beta{:}\alpha$ after μ and α	$b_{\cdot} - a = 3$	$R(\beta{:}\alpha \mid \mu, \alpha) = 60$	20	5.4
Residual	$N - b_{\cdot} = 7$	SSE $= 26$	$3\frac{5}{7}$	
Total	$N = 12$	SST $= 542$		

tests the significance of fitting α after μ; and

$$F(\beta:\alpha \mid \mu, \alpha) = \frac{R(\beta:\alpha \mid \mu, \alpha)}{(b. - a)\text{MSE}} = 5.4 \tag{75}$$

tests the significance of fitting $\beta:\alpha$ after μ and α. The 5% critical values of the $F_{1,7}$- and $F_{3,7}$-distributions are 5.59 and 4.35 respectively, and are exceeded by (74) and (75). Hence we conclude that fitting α after μ as well as $\beta:\alpha$ after μ and α accounts for variation in the y-variable.

e. Estimable functions

Applying the general theory of estimability to any design models involves many of the points detailed in Sec. 2e for the 1-way classification. These details are not repeated in what follows.

The expected value of any observation is estimable, and so $\mu + \alpha_i + \beta_{ij}$ is estimable, with b.l.u.e. $\mu^o + \alpha_i^o + \beta_{ij}^o = \bar{y}_{ij.}$ from (70) and (72). This result and linear combinations thereof are shown in Table 6.8. An example of one

TABLE 6.8. ESTIMABLE FUNCTIONS IN THE 2-WAY NESTED CLASSIFICATION $y_{ij} = \mu + \alpha_i + \beta_{ij} + e_{ijk}$

Estimable Function	b.l.u.e.	Variance of b.l.u.e.
$\mu + \alpha_i + \beta_{ij}$	$\bar{y}_{ij.}$	σ^2/n_{ij}
$\beta_{ij} - \beta_{ij'}$, for $j \neq j'$	$\bar{y}_{ij.} - \bar{y}_{ij'.}$	$\sigma^2(1/n_{ij} + 1/n_{ij'})$
$\mu + \alpha_i + \sum_{j=1}^{b_i} w_{ij}\beta_{ij}$, for $\sum_{j=1}^{b_i} w_{ij} = 1$	$\sum_{j=1}^{b_i} w_{ij}\bar{y}_{ij.}$	$\sigma^2\left(\sum_{j=1}^{b_i} w_{ij}^2/n_{ij}\right)$
$\alpha_i - \alpha_{i'} + \sum_{j=1}^{b_i} w_{ij}\beta_{ij} - \sum_{j=1}^{b_{i'}} w_{i'j}\beta_{i'j}$, for $\sum_{j=1}^{b_i} w_{ij} = 1 = \sum_{j=1}^{b_{i'}} w_{i'j}$	$\sum_{j=1}^{b_i} w_{ij}\bar{y}_{ij.} - \sum_{j=1}^{b_{i'}} w_{i'j}\bar{y}_{i'j.}$	$\sigma^2\left(\sum_{j=1}^{b_i} w_{ij}^2/n_{ij} + \sum_{j=1}^{b_{i'}} w_{i'j}^2/n_{i'j}\right)$

of them is, using (72),

$$\widehat{\beta_{11} - \beta_{12}} = 5 - 9 = -4$$

with

$$v\left(\widehat{\beta_{11} - \beta_{12}}\right) = \sigma^2(\tfrac{1}{1} + \tfrac{1}{3}) = 4\sigma^2/3,$$

an unbiased estimate of this variance being, from Table 6.6, $4\hat{\sigma}^2/3 = 4(\text{MSE})/3 = 104/21$. Values typically used for w_{ij} in the last two rows of Table 6.8 are either $1/b_i$ or $n_{ij}/n_{i.}$. With the former and using (72) again we have, for example

$$\alpha_1 - \alpha_2 + \tfrac{1}{2}(\beta_{11} + \beta_{12}) - \tfrac{1}{3}(\beta_{21} + \beta_{22} + \beta_{23}) \tag{76}$$

has b.l.u.e. $\quad \tfrac{1}{2}(5 + 9) - \tfrac{1}{3}(9 + 3 + 5) = 1\tfrac{1}{3}$

and the variance of this b.l.u.e. is

$$\sigma^2[(\tfrac{1}{2})^2(\tfrac{1}{1} + \tfrac{1}{3}) + (\tfrac{1}{3})^2(\tfrac{1}{2} + \tfrac{1}{4} + \tfrac{1}{2})] = 35\sigma^2/72.$$

It is to be noted in Table 6.8 that μ is not estimable; neither is $\mu + \alpha_i$, nor is α_i.

. Tests of hypotheses

The estimable functions of Table 6.8 form the basis of testable hypotheses. The F-statistic for testing the null hypothesis that any one of the functions in Table 6.8 is zero is the square of its b.l.u.e. divided by that b.l.u.e.'s variance with $\hat{\sigma}^2$ replacing σ^2. Such a statistic has the $F_{1,N-b.}$-distribution under the null hypothesis and its square root has the $t_{N-b.}$ distribution; e.g.,

$$F = \frac{(\bar{y}_{ij\cdot} - \bar{y}_{ij'\cdot})^2}{\hat{\sigma}^2(1/n_{ij} + 1/n_{ij'})}$$

or, equivalently $\sqrt{F}$, can be used to test the hypothesis that $\beta_{ij} = \beta_{ij'}$.

The hypothesis H: $\beta_{i1} = \beta_{i2} = \cdots = \beta_{ib_i}$ for all i is of especial interest. It is the hypothesis of equal β's within each α-level. By writing it in the form H: $\mathbf{K'b} = \mathbf{0}$ it can be shown that the resulting F-statistic of (21) is $F(\beta:\alpha \mid \mu, \alpha)$ given in (75) and used in Table 6.8. Thus, whereas in (75) $F(\beta:\alpha \mid \mu, \alpha)$ is described as being used for testing the significance of fitting $\beta:\alpha$ after μ and α, we now see that, equivalently, $F(\beta:\alpha \mid \mu, \alpha)$ can also be used for testing the hypothesis of equality of the β's within each α-level.

Example. Carrying out this test for the data of Table 6.5 involves

$$\mathbf{K}' = \begin{bmatrix} 0 & 0 & 0 & 1 & -1 & 0 & 0 & 0 \\ 0 & 0 & 0 & 0 & 0 & 1 & -1 & 0 \\ 0 & 0 & 0 & 0 & 0 & 1 & 0 & -1 \end{bmatrix} \tag{77}$$

for which, using $\mathbf{b}^{o\prime}$ of (72) and $\mathbf{G}$ implicit in (73), gives

$$\mathbf{K'b}^o = \begin{bmatrix} -4 \\ 6 \\ 4 \end{bmatrix} \quad \text{and} \quad (\mathbf{K'GK})^{-1} = \begin{bmatrix} \frac{4}{3} & 0 & 0 \\ 0 & \frac{3}{4} & \frac{1}{2} \\ 0 & \frac{1}{2} & 1 \end{bmatrix}^{-1} = \begin{bmatrix} \frac{3}{4} & 0 & 0 \\ 0 & 2 & -1 \\ 0 & -1 & \frac{3}{2} \end{bmatrix}$$

so that Q of (21) is

$$Q = [-4 \quad 6 \quad 4]\begin{bmatrix} \frac{3}{4} & 0 & 0 \\ 0 & 2 & -1 \\ 0 & -1 & \frac{3}{2} \end{bmatrix}\begin{bmatrix} -4 \\ 6 \\ 4 \end{bmatrix}$$

$$= 60 = R(\beta:\alpha \mid \mu, \alpha) \text{ of Table 6.7.}$$

Thus the F-value is $60/3\hat{\sigma}^2 = 20/3\frac{5}{7} = 5.4$ as in (75) and Table 6.7.

Consider the hypothesis

$$H:\quad \mathbf{k}'\mathbf{b} = 0 \qquad \text{for } \mathbf{k}' = [0 \quad 1 \quad -1 \quad \tfrac{1}{4} \quad \tfrac{3}{4} \quad -\tfrac{2}{8} \quad -\tfrac{4}{8} \quad -\tfrac{2}{8}]. \tag{78}$$

Here we have

$$\mathbf{k}'\mathbf{b} = \alpha_1 - \alpha_2 + (\beta_{11} + 3\beta_{12})/4 - (2\beta_{21} + 4\beta_{22} + 2\beta_{23})/8$$

which is the estimable function typified in the last line of Table 6.8, with $w_{ij} = n_{ij}/n_{i.}$. From (78), (72) and (73)

$$\mathbf{k}'\mathbf{b}^o = 3 \qquad \text{and} \qquad \mathbf{k}'\mathbf{G}\mathbf{k} = \tfrac{3}{8}$$

so that by (21) the numerator sum of squares for testing the hypothesis in (78) is

$$Q = 3^2(\tfrac{8}{3}) = 24 = R(\alpha \mid \mu) \text{ of Table 6.7.} \tag{79}$$

This is no accident. Although $R(\alpha \mid \mu)$ is, as indicated in (74), the numerator sum of squares for testing the fit of α after μ, it is also the numerator sum of squares for testing

$$H:\quad \alpha_i + \sum_{j=1}^{b_i} n_{ij}\beta_{ij}/n_{i\cdot} = \alpha_{i'} + \sum_{j=1}^{b_{i'}} n_{i'j}\beta_{i'j}/n_{i'\cdot} \qquad \text{for all} \quad i \neq i'. \tag{80}$$

Furthermore, this hypothesis is orthogonal to

$$H:\quad \beta_{ij} = \beta_{ij'} \qquad \text{for} \quad j \neq j', \text{ within each } i, \tag{81}$$

orthogonal in the sense of (62); e.g., $\mathbf{k}'$ of (78) and $\mathbf{K}'$ of (77) are examples of (80) and (81) respectively, and $\mathbf{k}'$ and every row of $\mathbf{K}'$ satisfy (62). And, in testing (80) by using (21) it will be found that $F(H)$ given there reduces to $F(\alpha \mid \mu)$, as exemplified in (79). Hence $F(\alpha \mid \mu)$ tests (80), the numerator sum of squares being $R(\alpha \mid \mu)$; and $F(\beta\!:\!\alpha \mid \mu, \alpha)$ tests (81), its numerator being $R(\beta\!:\!\alpha \mid \mu, \alpha)$. The two numerator sums of squares $R(\alpha \mid \mu)$ and $R(\beta\!:\!\alpha \mid \mu, \alpha)$ are statistically independent, as can be established by expressing each of them as quadratics in $\mathbf{y}$ and applying Theorem 4 of Chapter 2 (see Exercise 9).

The equivalence of the F-statistic for testing (80) and $F(\alpha \mid \mu)$ can also be appreciated by noting in (80) that if the β_{ij} did not exist then (80) would represent

$$H:\quad \text{all } \alpha\text{'s equal (in the absence of } \beta\text{'s)}$$

which is indeed the context of earlier interpreting $F(\alpha \mid \mu)$ as testing α after μ.

g. Models that include restrictions

The general effect of having restrictions as part of the model has been discussed in Sec. 5.6 and illustrated in detail in Sec. 2h of this chapter. The points made there apply equally as well here: restrictions that involve non-estimable functions of the parameters affect the form of functions that are estimable and hypotheses that are testable. Of particular interest here are restrictions $\sum_{j=1}^{b_i} w_{ij}\beta_{ij} = 0$ with $\sum_{j=1}^{b_i} w_{ij} = 1$ for all i, because then from Table 6.8 we see that $\mu + \alpha_i$ and $\alpha_i - \alpha_{i'}$ are estimable, and hypotheses about them are testable. If, further, the w_{ij} of the restrictions are $n_{ij}/n_{i.}$, so that the restrictions are $\sum_{j=1}^{b_i} n_{ij}\beta_{ij} = 0$ for all i, then (80) becomes H: all α_i's equal and (80) is, as we have just shown, tested by $F(\alpha \mid \mu)$, which is independent of $F(\beta:\alpha \mid \mu, \alpha)$ that tests H: all β's equal within each α-level. However, if the w_{ij} of the restrictions are not $n_{ij}/n_{i.}$, but some other form satisfying $\sum_{j=1}^{b_i} w_{ij} = 1$ for all i, e.g., $w_{ij} = 1/b_i$, the hypothesis H: all α_i's equal can still be tested, but the F-statistic will not equal $F(\alpha \mid \mu)$, nor will its numerator be independent of that of $F(\beta:\alpha \mid \mu, \alpha)$.

h. Balanced data

The position with balanced data ($n_{ij} = n$ for all i and j, and $b_i = b$ for all i) is akin to that of the 1-way classification, discussed in Sec. 2i earlier. "Constraints" $\sum_{j}^{b_i} \beta^o_{ij} = 0$ for all i and $\sum_{i=1}^{a} \alpha^o_i = 0$ on the solutions applied to the normal equations lead to easy solutions thereof: $\mu^o = \bar{y}_{...}$, $\alpha^o_i = \bar{y}_{i..} - \bar{y}_{...}$ and $\beta^o_{ij} = \bar{y}_{ij.} - \bar{y}_{i..}$, as found in many texts. Other results are unaffected e.g., the estimable functions and their b.l.u.e's of Table 6.8 are unaltered.

When restrictions paralleling the constraints are taken as part of the model, $\sum_{i=1}^{a} \alpha_i = 0$ and $\sum_{j=1}^{b_i} \beta_{ij} = 0$ for all i, the effect is to make μ, α_i and β_{ij} individually estimable with b.l.u.e.'s $\hat{\mu} = \bar{y}_{...}$, $\hat{\alpha}_i = \bar{y}_{i..} - \bar{y}_{...}$, and $\hat{\beta}_{ij} = \bar{y}_{ij.} - \bar{y}_{i..}$. And, as in the 1-way classification, rationalization of such restrictions is opportune: the α_i's are defined as deviations from their mean as are the β_{ij}'s from their within α-level means.

5. NORMAL EQUATIONS FOR DESIGN MODELS

Models of the nature described here and in Chapter 4 are sometimes called *design* models [e.g., Graybill (1961, Chapters 11–15)]. Several general properties of the normal equations $\mathbf{X}'\mathbf{X}\mathbf{b}^o = \mathbf{X}'\mathbf{y}$ of such models will now be characterized, using (68) for illustration.

First, there is one equation corresponding to each effect of a model. Second, the right-hand side of any equation (the element of $\mathbf{X}'\mathbf{y}$) is the sum of all observations that contain, in their model, a specific effect; e.g., the right-hand side of the first equation in (68) is the sum of all observations that contain μ. Third, the left-hand side of each equation is the expected value of its right-hand side with $\mathbf{b}$ replaced by $\mathbf{b}^o$. Thus the first equation in (68) corresponds to μ: its right-hand side is $y_{...}$ and its left-hand side is $E(y_{...})$ with $\mathbf{b}$ therein replaced by $\mathbf{b}^o$. Hence the equation is, as implied in (68),

$$12\mu^o + 4\alpha_1^o + 8\alpha_2^o + \beta_{11}^o + 3\beta_{12}^o + 2\beta_{21}^o + 4\beta_{22}^o + 2\beta_{23}^o = y_{...} = 72. \quad (82)$$

Similarly, the second equation of (68) relates to α_1: its right-hand side is the sum of all observations that have α_1 in their model, namely $y_{1..}$; and its left-hand side is $E(y_{1..})$ with $\mathbf{b}$ replaced by $\mathbf{b}^o$. Thus the equation is

$$4\mu^o + 4\alpha_1^o + \beta_{11}^o + 3\beta_{12}^o = y_{1..} = 32. \quad (83)$$

Suppose in a design model that θ_i is the effect (parameter) for the ith level of the θ-factor. Let $y_{\theta_i\cdot}$ be the total of the observations in this level of this factor. Then the normal equations are

$$[E(y_{\theta_i\cdot}) \text{ with } \mathbf{b} \text{ replaced by } \mathbf{b}^o] = y_{\theta_i\cdot}\,, \quad (84)$$

with i ranging over all levels of all factors θ, including the solitary level of the μ-factor.

The coefficient of each term in (82) is the number of times that its corresponding parameter occurs in $y_{...}$; e.g., the coefficient of μ^o is 12 because μ occurs 12 times in $y_{...}$; that of α_1^o is 4 because α_1 occurs 4 times; and so on. Similarly, the term in β_{11}^o in (83) is β_{11}^o because β_{11} occurs once in $y_{1..}$; and the term in β_{12}^o is $3\beta_{12}^o$ because β_{12} occurs thrice in $y_{1..}$. In general the coefficients of the terms in the normal equations (i.e., the elements of $\mathbf{X}'\mathbf{X}$) are the n_{ij}'s of the data, determined as follows.

Equation (84) may be called the θ_i-equation, not only because of its form as shown there but also because of its derivation from the least squares procedure when differentiating with respect to θ_i. The coefficient of φ_j^o (corresponding to the parameter φ_j) in (84) is as follows:

$$\left.\begin{matrix}\text{coefficient of } \varphi_j^o \\ \text{in the } \theta_i\text{-equation}\end{matrix}\right\} = \left\{\begin{matrix}\text{no. of observations in the} \\ i\text{th level of the } \theta\text{-factor} \\ \text{and } j\text{th level of the } \varphi\text{-factor}\end{matrix}\right.$$
$$\equiv n(\theta_i\,, \varphi_j);$$

e.g., (83) is the α_1-equation and the coefficient of β_{12}^o is $n(\beta_{12}\,, \alpha_1) = n_{12} = 3$ as shown. These n's are the elements of $\mathbf{X}'\mathbf{X}$. The property

$$n(\theta_i\,, \varphi_j) = n(\varphi_j\,, \theta_i),$$

arising from the definition of $n(\theta_i\,, \varphi_j)$ just given, accords with the symmetry of

X'X. The fact that

$$n(\mu, \theta_i) = n(\theta_i, \mu) = n(\theta_i, \theta_i) = n_{\theta_i} = \begin{array}{l}\text{no. of observations}\\ \text{in the } i\text{th level}\\ \text{of the } \theta\text{-factor}\end{array}$$

is what leads to **X'X** having in its first row, in its first column and in its diagonal all the n's (and their various sums) of the data. This is evident in (68) and will be further apparent in subsequent examples. In addition, partitioning of the form shown in (68) helps to identify the location of the n's and their sums in **X'X**; e.g., the μ-equation is first, followed by the two α-equations and then by the sets of 2 and of 3 β-equations corresponding to the level of the β-factor within each level of the α-factor. Partitioning **X'X** in this manner is always of assistance in identifying its elements.

6. EXERCISES

1. In the model $y_{ij} = \mu_i + e_{ij}$ prove that μ_i is estimable and find its b.l.u.e.
2. Suppose the population of a community consists of 12% who did not complete high school and 68% who did, with the remainder having graduated from college. With the data of Table 6.1, find
(*a*) the estimated population average index;
(*b*) the estimated variance of the estimator in (*a*);
(*c*) the 95% symmetric confidence interval on the population average;
(*d*) the F-statistic for testing the hypothesis H: $\mu + \alpha_1 = 70$ and $\alpha_1 = \alpha_3 - 15$; and
(*e*) a contrast that is orthogonal to $4\alpha_1 - 3\alpha_2 - \alpha_3$; test the hypothesis that it and $4\alpha_1 - 3\alpha_2 - \alpha_3$ are zero.
3. An opinion poll yields the scores of four laborers as 37, 25, 42 and 28 for some attribute; those of two artisans are 23 and 29; of three professionals 38, 30 and 25 and two self-employed people score 23 and 29. If in the population from which these people come the percentages in these four groups are respectively 10%, 20%, 40% and 30%, what are the estimates and estimated variances of
(*a*) the population score?
(*b*) the difference in score between professionals and an average of the other three groups?
(*c*) the difference between a self-employed and a professional?
4. (Exercise 3 continued) Test the hypothesis that a laborer's score equals an artisan's score equals the arithmetic average of a professional's and a self-employed's scores equals the weighted population average of a laborer's and a professional's scores.
5. (Exercise 3 continued) Find two mutually orthogonal contrasts (one not involving self-employed people) that are orthogonal to the difference between a laborer's and an artisan's score, and test the joint hypothesis that all three contrasts are zero.

6. Suppose that in Exercise 3 it is found that the score of 23 recorded as that of an artisan is in fact that of a laborer. What are now the answers to Exercises 3, 4 and 5?

7. In the 1-way classification prove that for the hypothesis H: all α's equal $F(H)$ reduces to $F(R_m)$. [*Hint:* Use $\mathbf{K}'$ of Sec. 6.2f(iii), and derive, explicitly, the inverse of $\mathbf{D}\{a_i\} + b\mathbf{11}'$ for any b, and any a_i for $i = 1, 2, \ldots, a$.]

8. Derive the expression for a non-symmetric $(1 - \alpha)\%$ confidence interval for the contrast $\sum \lambda_i \alpha_i$ of the 1-way classification.

9. If in a 1-way classification the a classes are allocated codes 1 through a in some manner, show that regression of the response variable y on the allocated codes gives rise to a reduction in sum of squares that is less than that arising from fitting the 1-way classification model. [*Hint:* Make use of Lagrange's identity:

$$\sum a_i^2 \sum b_i^2 - \left(\sum a_i b_i\right)^2 = \tfrac{1}{2} \sum_{i \neq i'}\sum (a_i b_{i'} - a_{i'} b_i)^2.]$$

How might this result be generalized to many-factor models?

10. Using Theorems 2 and 4 of Chapter 2 prove that $R(\alpha \mid \mu)/\sigma^2$ and $R(\beta : \alpha \mid \alpha, \mu)/\sigma^2$ of Table 6.7 are independently distributed as non-central χ^2-variables.

11. Suppose the data of a student opinion poll, similar to that of Section 4 of the chapter, are as shown below. (Each column represents a section of a course.)

English			Geology		Chemistry			
2	7	8	2	10	8	6	1	8
5	9	4	6	8		2	3	6
2		3		9		3	2	
		6				1		
		4						

(*a*) Write down the normal equations and a solution of them.
(*b*) Calculate an analysis of variance table similar to Table 6.7.
(*c*) Test the following hypotheses (one at a time):
  (i) Sections within courses have the same opinions.
  (ii) Courses, ignoring sections, have similar opinions.
(*d*) Specify a set of restrictions in the model which enables you to test the following hypotheses (one at a time) independently of the test in (i) of (*c*) above.
  (i) All courses have the same opinion.
  (ii) Geology's opinion is the mean of that of English and Chemistry, and and English's opinion equals Chemistry.
(*e*) Repeat (*d*) without the independence property.

12. Suppose $\mathbf{y} = \mathbf{Xb}_1 + \mathbf{Zb}_2 + \mathbf{e}$, with $\mathbf{y} \sim N(\mathbf{Xb}_1 + \mathbf{Zb}_2, \sigma^2\mathbf{I})$, and that $R(\mathbf{b}_1, \mathbf{b}_2)$ is the reduction in sum of squares for fitting this model. Prove that $R(\mathbf{b}_2 \mid \mathbf{b}_1)/\sigma^2$ has a non-central χ^2-distribution independent of $R(\mathbf{b}_1)$ and of SSE.

CHAPTER 7

THE 2-WAY CROSSED CLASSIFICATION

This chapter continues with applications of Chapter 5, as started in Chapter 6, dealing at length with the 2-way crossed classification (with and without interactions).

1. THE 2-WAY CLASSIFICATION WITHOUT INTERACTION

A sophomore course in Home Economics might include in its laboratory exercises an experiment to illustrate the cooking speed of 3 makes of pan used with 4 brands of stove. Using pans of uniform diameter, but made by different manufacturers, the students collect data on the number of seconds (beyond 3 minutes) that it takes to bring 2 quarts of water to the boil. Although the experiment is designed to use each of the 3 makes of pan with each of the 4 stoves, one student carelessly fails to record 3 of her times. Her resulting data are shown in Table 7.1. The totals for each brand of stove and make of pan are also shown, as well as the number of readings for each and their mean time. As before, the number of readings are shown in parentheses to distinguish them from the readings themselves.

The foregoing description implies that the observations which the student failed to record are, in some sense, "missing observations". This is true, and we could, if we wished, analyze the data using one of the many available "missing observations" techniques [see, for example, Federer (1955, p. 133)]. Most of these techniques involve estimating the missing observations in some manner, putting these estimates into the data and then proceeding more or less as if the data were balanced, except for minor adjustments in the degrees of freedom. Although such procedures can be recommended on many occasions (see Sec. 8.2) they are of greatest use when only very few observations are missing. This might be considered the case with Table 7.1 (although it *is* 25% of the designed experiment that has been lost), but these data serve merely to illustrate techniques involved in situations for which the "missing observation" concept is wholly inappropriate—situations in which large numbers of cells may be empty, not because observations have been lost but because none were obtainable. Data of this nature occur quite frequently (e.g., Table 4.1) and it is to their exact analysis that we turn our attention, using Table 7.1 as illustration.

The data of Table 7.1 are described as coming from a 2-way crossed classification—two factors, with every level of one occurring in combination with

TABLE 7.1. NUMBER OF SECONDS (BEYOND 3 MINUTES) TAKEN TO BOIL 2 QUARTS OF WATER

Brand of Stove	Make of Pan A	B	C	Total	No. of Observations	Mean
X	18	12	24	54	(3)	18
Y	—	—	9	9	(1)	9
Z	3	—	15	18	(2)	9
W	6	3	18	27	(3)	9
Total	27	15	66	108		
No. of observations	(3)	(2)	(4)		(9)	
Mean	9	$7\frac{1}{2}$	$16\frac{1}{2}$			12

every level of the other. Models for such data are discussed in Sec. 4.3, where particular attention is paid to describing the inclusion of interaction effects in such models. However, it is also pointed out there that, when there is only one observation per cell, the usual model with interactions cannot be used. This is also true of the data in Table 7.1 where some cells have not even one observation but are empty.

a. Model

A suitable equation of the model for analyzing the data of Table 7.1 is therefore

$$y_{ij} = \mu + \alpha_i + \beta_j + e_{ij} \tag{1}$$

where y_{ij} is the observation in the ith row (brand of stove) and jth column (make of pan), μ is a mean, α_i is the effect of the ith row, β_j is the effect of the jth column, and e_{ij} is an error term. Outside the context of rows and columns (which is a useful one) α_i is equivalently the effect due to the ith level of the α-factor and β_j is the effect due to the jth level of the β-factor. In general we have a levels of the α-factor with $i = 1, 2, \ldots, a$ and b levels of the β-factor with $j = 1, 2, \ldots, b$; in the example $a = 4$ and $b = 3$.

With balanced data every one of the ab cells of a table like Table 7.1 would have one (or n) observations, and $n(\geq 1)$ would be the only symbol needed to describe the number of observations in each cell. In Table 7.1, however, some cells have zero observations and some have one. We therefore need n_{ij} as the number of observations in the ith row and jth column. Then, in Table 7.1, all $n_{ij} = 0$ or 1, and the numbers of observations shown in that table are then the values of

$$n_{i\cdot} = \sum_{j=1}^{b} n_{ij}, \qquad n_{\cdot j} = \sum_{i=1}^{a} n_{ij} \qquad \text{and} \quad N = n_{\cdot\cdot} = \sum_{i=1}^{a}\sum_{j=1}^{b} n_{ij}.$$

Corresponding totals and means of the observations are shown also. This n_{ij}-notation, convenient here, is also identical to that used for data in which there are none, one or many observations per cell, as discussed in the next section.

For the observations in Table 7.1 the equations $\mathbf{y} = \mathbf{Xb} + \mathbf{e}$ of the model are as shown in equation (2). As well as having the elements of $\mathbf{b}$, namely μ, $\alpha_1, \ldots, \alpha_4, \beta_1, \beta_2$ and β_3 shown in a vector in the usual manner they are also shown as headings to columns of the matrix $\mathbf{X}$. This is purely for convenience in reading the equations: it clarifies the incidence of the elements of the model in the data, as does the partitioning of the matrix according to the different factors μ, α and β. For the same reason dots are used to represent 0's in the $\mathbf{X}$-matrix. Thus the model equations for the data of Table 7.1 are

$$\begin{array}{c} \\ \begin{bmatrix} 18 \\ 12 \\ 24 \\ 9 \\ 3 \\ 15 \\ 6 \\ 3 \\ 18 \end{bmatrix} \end{array} = \begin{array}{c} \\ \begin{bmatrix} y_{11} \\ y_{12} \\ y_{13} \\ y_{23} \\ y_{31} \\ y_{33} \\ y_{41} \\ y_{42} \\ y_{43} \end{bmatrix} \end{array} = \begin{array}{c} \begin{array}{c|cccc|ccc} \mu & \alpha_1 & \alpha_2 & \alpha_3 & \alpha_4 & \beta_1 & \beta_2 & \beta_3 \end{array} \\ \left[\begin{array}{c|cccc|ccc} 1 & 1 & \cdot & \cdot & \cdot & 1 & \cdot & \cdot \\ 1 & 1 & \cdot & \cdot & \cdot & \cdot & 1 & \cdot \\ 1 & 1 & \cdot & \cdot & \cdot & \cdot & \cdot & 1 \\ 1 & \cdot & 1 & \cdot & \cdot & \cdot & \cdot & 1 \\ 1 & \cdot & \cdot & 1 & \cdot & 1 & \cdot & \cdot \\ 1 & \cdot & \cdot & 1 & \cdot & \cdot & \cdot & 1 \\ 1 & \cdot & \cdot & \cdot & 1 & 1 & \cdot & \cdot \\ 1 & \cdot & \cdot & \cdot & 1 & \cdot & 1 & \cdot \\ 1 & \cdot & \cdot & \cdot & 1 & \cdot & \cdot & 1 \end{array}\right] \end{array} \begin{array}{c} \\ \left[\begin{array}{c} \mu \\ \hline \alpha_1 \\ \alpha_2 \\ \alpha_3 \\ \alpha_4 \\ \hline \beta_1 \\ \beta_2 \\ \beta_3 \end{array}\right] \end{array} + \begin{array}{c} \\ \begin{bmatrix} e_{11} \\ e_{12} \\ e_{13} \\ e_{23} \\ e_{31} \\ e_{33} \\ e_{41} \\ e_{42} \\ e_{43} \end{bmatrix} \end{array}. \qquad (2)$$

b. Normal equations

Equations (2) are $\mathbf{y} = \mathbf{X}\mathbf{b} + \mathbf{e}$. The corresponding normal equations, $\mathbf{X}'\mathbf{X}\mathbf{b}^o = \mathbf{X}'\mathbf{y}$, written in a manner similar to (2), are

$$
\begin{array}{c}
\mu^o \ \alpha_1^o \ \alpha_2^o \ \alpha_3^o \ \alpha_4^o \ \beta_1^o \ \beta_2^o \ \beta_3^o \\
\left[\begin{array}{c|cccc|ccc}
9 & 3 & 1 & 2 & 3 & 3 & 2 & 4 \\
\hline
3 & 3 & \cdot & \cdot & \cdot & 1 & 1 & 1 \\
1 & \cdot & 1 & \cdot & \cdot & \cdot & \cdot & 1 \\
2 & \cdot & \cdot & 2 & \cdot & 1 & \cdot & 1 \\
3 & \cdot & \cdot & \cdot & 3 & 1 & 1 & 1 \\
\hline
3 & 1 & \cdot & 1 & 1 & 3 & \cdot & \cdot \\
2 & 1 & \cdot & \cdot & 1 & \cdot & 2 & \cdot \\
4 & 1 & 1 & 1 & 1 & \cdot & \cdot & 4
\end{array}\right]
\end{array}
\left[\begin{array}{c}
\mu^o \\ \hline \alpha_1^o \\ \alpha_2^o \\ \alpha_3^o \\ \alpha_4^o \\ \hline \beta_1^o \\ \beta_2^o \\ \beta_3^o
\end{array}\right]
=
\left[\begin{array}{c}
y_{..} \\ \hline y_{1.} \\ y_{2.} \\ y_{3.} \\ y_{4.} \\ \hline y_{.1} \\ y_{.2} \\ y_{.3}
\end{array}\right]
=
\left[\begin{array}{c}
108 \\ \hline 54 \\ 9 \\ 18 \\ 27 \\ \hline 27 \\ 15 \\ 66
\end{array}\right]. \tag{3}
$$

Properties of such equations, described in general in Sec. 6.4, are further evident here. In this case the first row and column and the diagonal of $\mathbf{X}'\mathbf{X}$ have $n_{..}$, the $n_{i.}$'s and the $n_{.j}$'s in them. The only other non-zero off-diagonal elements are those in an $a \times b$ matrix of 1's and 0's (and its transpose) corresponding to the pattern of observations. The partitioning indicated in (3) highlights the form of $\mathbf{X}'\mathbf{X}$ and suggests how more levels of the factors would be accommodated.

c. Solving the normal equations

In the examples of Secs. 6.2 and 6.4 solutions of the normal equations were easily derived by the procedure indicated in equation (4) of Chapter 6. Now, however, even after making use of that procedure, there is no neat, explicit solution. Numerically, a solution can be readily obtained, but algebraically it cannot be expressed succinctly.

In (3) the sum of the a rows of $\mathbf{X}'\mathbf{X}$ immediately after the first (the α-equations) equals the first row; and the sum of the last b rows (the β-equations) also equals the first row. Hence, with $\mathbf{X}'\mathbf{X}$ having order $q = 1 + a + b$, its rank is $r = r(\mathbf{X}'\mathbf{X}) = 1 + a + b - 2 = a + b - 1$. Thus $p - r = 2$, and we solve (3) by setting an appropriate two elements of $\mathbf{b}^o$ to zero and deleting the corresponding equations. One of the easiest ways to do this is to put $\mu^o = 0$ and either $\alpha_1^o = 0$ or $\beta_b^o = 0$, according to whether $a < b$ or $a > b$ (when $a = b$ it is immaterial); i.e., when there are fewer α-levels than β-levels put $\alpha_1^o = 0$, but when there are fewer β-levels than α-levels put $\beta_b^o = 0$.

The latter is the case in our example and so with $\mu^o = 0 = \beta_3^o$ we get from (3)

$$\left[\begin{array}{cccc:cc} 3 & \cdot & \cdot & \cdot & 1 & 1 \\ \cdot & 1 & \cdot & \cdot & \cdot & \cdot \\ \cdot & \cdot & 2 & \cdot & 1 & \cdot \\ \cdot & \cdot & \cdot & 3 & 1 & 1 \\ \hdashline 1 & \cdot & 1 & 1 & 3 & \cdot \\ 1 & \cdot & \cdot & 1 & \cdot & 2 \end{array}\right] \left[\begin{array}{c} \alpha_1^o \\ \alpha_2^o \\ \alpha_3^o \\ \alpha_4^o \\ \hdashline \beta_1^o \\ \beta_2^o \end{array}\right] = \left[\begin{array}{c} y_{1\cdot} \\ y_{2\cdot} \\ y_{3\cdot} \\ y_{4\cdot} \\ \hdashline y_{\cdot 1} \\ y_{\cdot 2} \end{array}\right] = \left[\begin{array}{c} 54 \\ 9 \\ 18 \\ 27 \\ \hdashline 27 \\ 15 \end{array}\right] \tag{4}$$

Written in full these equations are

$$\begin{array}{rl} 3\alpha_1^o \qquad\qquad\qquad + \beta_1^o + \beta_2^o &= 54 \\ \alpha_2^o \qquad\qquad\qquad\qquad\quad &= 9 \\ 2\alpha_3^o \qquad + \beta_1^o \qquad\quad &= 18 \\ 3\alpha_4^o + \beta_1^o + \beta_2^o &= 27 \end{array} \tag{5}$$

and

$$\begin{array}{rl} \alpha_1^o + \alpha_3^o + \alpha_4^o + 3\beta_1^o \qquad &= 27 \\ \alpha_1^o \qquad + \alpha_4^o \qquad + 2\beta_2^o &= 15. \end{array} \tag{6}$$

From (5) the α^o's are expressed in terms of the β^o's and substitution in (6) then leads to solutions for the β^o's. Thus (5) gives

$$\begin{aligned} \alpha_1^o &= 54/3 - \tfrac{1}{3}(\beta_1^o + \beta_2^o) = 18 - \tfrac{1}{3}[1(\beta_1^o) + 1(\beta_2^o)] \\ \alpha_2^o &= 9/1 \qquad\qquad\qquad\quad = 9 - \tfrac{1}{1}[0(\beta_1^o) + 0(\beta_2^o)] \\ \alpha_3^o &= 18/2 - \tfrac{1}{2}\beta_1^o \qquad\quad\;\; = 9 - \tfrac{1}{2}[1(\beta_1^o) + 0(\beta_2^o)] \\ \alpha_4^o &= 27/3 - \tfrac{1}{3}(\beta_1^o + \beta_2^o) = 9 - \tfrac{1}{3}[1(\beta_1^o) + 1(\beta_2^o)]. \end{aligned} \tag{7}$$

[The reason for including the coefficients 1 and 0 in the right-hand sides of (7) becomes clear when considering the generalization of this procedure (see below). For this reason they are retained.] Substituting (7) into (6) gives

$$\begin{aligned} &\{3 - [1(1)/3 + 0(0)/1 + 1(1)/2 + 1(1)/3]\}\beta_1^o \\ &\qquad - [1(1)/3 + 0(0)/1 + 1(0)/2 + 1(1)/3]\beta_2^o \\ &\qquad\qquad = 27 - [1(18) + 0(9) + 1(9) + 1(9)] \\ &-[1(1)/3 + 0(0)/1 + 0(1)/2 + 1(1)/3]\beta_1^o \\ &\qquad + \{2 - [1(1)/3 + 0(0)/1 + 0(0)/2 + 1(1)/3]\}\beta_2^o \\ &\qquad\qquad = 15 - [1(18) + 0(9) + 0(9) + 1(9)] \end{aligned} \tag{8}$$

which reduce to

$$(11/6)\beta_1^o - (4/6)\beta_2^o = -9 \quad \text{and} \quad (-4/6)\beta_1^o + (8/6)\beta_2^o = -12 \tag{9}$$

with solutions

$$\boldsymbol{\beta}^o = \begin{bmatrix} \beta_1^o \\ \beta_2^o \end{bmatrix} = \begin{bmatrix} -10 \\ -14 \end{bmatrix}. \tag{10}$$

Hence in (7)

$$\alpha_1^o = 26, \quad \alpha_2^o = 9, \quad \alpha_3^o = 14 \quad \text{and} \quad \alpha_4^o = 17$$

so that the solution to the normal equations is

$$\mathbf{b}^{o\prime} = [0 \quad 26 \quad 9 \quad 14 \quad 17 \quad -10 \quad -14 \quad 0]. \tag{11}$$

d. Absorbing equations

Development of (11) as a solution to (3) illustrates what is sometimes called the *absorption process:* in going from (4) to (8) the α-equations of (5) are "absorbed" into the β-equations of (6). Here we see the reason given earlier for the rule about deciding whether to put $\alpha_1^o = 0$ or $\beta_b^o = 0$: the object is for (8) to have as few equations as possible. Hence, if there are fewer β-levels than α-levels we put $\beta_b^o = 0$, absorb the α-equations and have equations (8) in terms of $(b - 1)$ β^o's. But if $a < b$, we put $\alpha_1^o = 0$, absorb the β-equations and have equations like (8) in terms of $(a - 1)$ α^o's. It is of no consequence in using the ultimate solution which one is obtained; the important thing is the number of equations in (8), either $a - 1$ or $b - 1$, whichever is less. In many instances the number of equations is, in fact, of little importance because, even if one of a and b is much larger than the other, the solving of (8) will undoubtedly be a computer operation. However, in Chapter 9 we discuss situations in which one of a and b is considerably larger than the other ($a = 10$ and $b = 2{,}000$, say), and then the method of obtaining (8) is of material importance.

We now describe the absorption process in general terms. Akin to (3) the normal equations are

$$\left[\begin{array}{c|ccc|ccc} n_{..} & n_{1.} & \cdots & n_{a.} & n_{.1} & \cdots & n_{.b} \\ \hline n_{1.} & n_{1.} & & \mathbf{0} & & & \\ \vdots & & \ddots & & & \{n_{ij}\} & \\ n_{a.} & \mathbf{0} & & n_{a.} & & & \\ \hline n_{.1} & & & & n_{.1} & & \mathbf{0} \\ \vdots & & \{n_{ji}\} & & & \ddots & \\ n_{.b} & & & & \mathbf{0} & & n_{.b} \end{array}\right] \left[\begin{array}{c} \mu^o \\ \hline \alpha_1^o \\ \vdots \\ \alpha_a^o \\ \hline \beta_1^o \\ \vdots \\ \beta_b^o \end{array}\right] = \left[\begin{array}{c} y_{..} \\ \hline y_{1.} \\ \vdots \\ y_{a.} \\ \hline y_{.1} \\ \vdots \\ y_{.b} \end{array}\right] \tag{12}$$

and, in putting $\mu^o = 0$ and $\beta_b^o = 0$ these reduce, similar to (4), to

$$\left[\begin{array}{ccc|ccc} n_{1\cdot} & & \mathbf{0} & n_{11} & \cdots & n_{1,b-1} \\ & \ddots & & \vdots & & \vdots \\ \mathbf{0} & & n_{a\cdot} & n_{a1} & \cdots & n_{a,b-1} \\ \hline n_{11} & \cdots & n_{a1} & n_{\cdot 1} & & \mathbf{0} \\ \vdots & & \vdots & & \ddots & \\ n_{1,b-1} & \cdots & n_{a,b-1} & \mathbf{0} & & n_{\cdot b-1} \end{array}\right] \left[\begin{array}{c} \alpha_1^o \\ \vdots \\ \alpha_a^o \\ \hline \beta_1^o \\ \vdots \\ \beta_{b-1}^o \end{array}\right] = \left[\begin{array}{c} y_{1\cdot} \\ \vdots \\ y_{a\cdot} \\ \hline y_{\cdot 1} \\ \vdots \\ y_{\cdot b-1} \end{array}\right]. \tag{13}$$

Solving the first a equations of (13) gives

$$\alpha_i^o = \bar{y}_{i\cdot} - \frac{1}{n_{i\cdot}} \sum_{j=1}^{b-1} n_{ij}\beta_j^o \qquad \text{for} \quad i = 1, 2, \ldots, a, \tag{14}$$

as in (7); and substitution of these values in the last $b - 1$ equations of (13) gives

$$\left(n_{\cdot j} - \sum_{i=1}^{a} \frac{n_{ij}^2}{n_{i\cdot}}\right)\beta_j^o - \sum_{j' \neq j}^{b-1} \left(\sum_{i=1}^{a} \frac{n_{ij}n_{ij'}}{n_{i\cdot}}\right)\beta_{j'}^o = y_{\cdot j} - \sum_{i=1}^{a} n_{ij}\bar{y}_{i\cdot} \tag{15}$$

$$\text{for} \quad j, j' = 1, 2, \ldots, b - 1,$$

as illustrated in (8). Written in vector form these are

$$\mathbf{C}\boldsymbol{\beta}_{b-1}^o = \mathbf{r} \qquad \text{with solution} \quad \boldsymbol{\beta}_{b-1}^o = \mathbf{C}^{-1}\mathbf{r} \tag{16}$$

where $\mathbf{C} = \{c_{jj'}\}$ and $\mathbf{r} = \{r_j\}$ for $j = 1, \ldots, b - 1$

with
$$c_{jj} = n_{\cdot j} - \sum_{i=1}^{a} \frac{n_{ij}^2}{n_{i\cdot}}, \qquad c_{jj'} = -\sum_{i=1}^{a} \frac{n_{ij}n_{ij'}}{n_{i\cdot}} \quad \text{for} \quad j \neq j' \tag{17}$$

and
$$r_j = y_{\cdot j} - \sum_{i=1}^{a} n_{ij}\bar{y}_{i\cdot} \qquad \text{for} \quad j = 1, \ldots, b - 1. \tag{18}$$

A check on these calculations is provided by also calculating c_{bb}, c_{jb} and r_b and confirming that

$$\sum_{j'=1}^{b} c_{jj'} = 0 \quad \text{for all } j, \qquad \text{and} \qquad \sum_{j=1}^{b} r_j = 0.$$

The solution $\boldsymbol{\beta}_{b-1}^o$ in (16) is subscripted to emphasize that it has $b - 1$ and not b elements.

To express the solutions α_i^o in matrix form we write

$$\boldsymbol{\alpha}^o = \begin{bmatrix} \alpha_1^o \\ \vdots \\ \alpha_a^o \end{bmatrix}, \qquad \mathbf{y}_a = \begin{bmatrix} y_{1\cdot} \\ \vdots \\ y_{a\cdot} \end{bmatrix}$$

and $\mathbf{D}_a = \mathbf{D}\{n_{i\cdot}\}, \qquad \text{for} \quad i = 1, 2, \ldots, a,$

a diagonal matrix (see Sec. 1.1) of order a, of the $n_{i\cdot}$-values. We also define

$$\mathbf{N}_{a\times(b-1)} = \begin{bmatrix} n_{11} & \cdots & n_{1,b-1} \\ \vdots & & \vdots \\ n_{a1} & \cdots & n_{a,b-1} \end{bmatrix},$$

$$\mathbf{M}_{a\times(b-1)} = \mathbf{D}_a^{-1}\mathbf{N} = \{n_{ij}/n_{i\cdot}\} \qquad \text{for} \quad i = 1, \ldots, a \text{ and } j = 1, \ldots, b-1,$$

and $$\bar{\mathbf{y}}_a = \mathbf{D}_a^{-1}\mathbf{y}_a = \{\bar{y}_{i\cdot}\} \qquad \text{for} \quad i = 1, \ldots, a. \tag{19}$$

Then from (13)

$$\boldsymbol{\alpha}^o = \mathbf{D}_a^{-1}\mathbf{y}_a - \mathbf{M}\boldsymbol{\beta}_{b-1}^o = \bar{\mathbf{y}}_a - \mathbf{M}\boldsymbol{\beta}_{b-1}^o\,.$$

Thus $$\mathbf{b}^o = \begin{bmatrix} 0 \\ \boldsymbol{\alpha}^o \\ \boldsymbol{\beta}_{b-1}^o \\ 0 \end{bmatrix} = \begin{bmatrix} 0 \\ \bar{\mathbf{y}}_a - \mathbf{M}\mathbf{C}^{-1}\mathbf{r} \\ \mathbf{C}^{-1}\mathbf{r} \\ 0 \end{bmatrix}. \tag{20}$$

Section 4 of this chapter deals with the condition of "connectedness" of unbalanced data. Although most modestly sized sets of data are usually connected, large sets of survey-style data are sometimes not connected. The condition is important because only when data are connected do $\mathbf{C}^{-1}$ and the solution in (20) exist. Further discussion therefore relates solely to data that are connected, a condition that must be satisfied before this analysis can be undertaken. Section 4 indicates how to ascertain if data are connected.

Corresponding to the solution (20), the generalized inverse of $\mathbf{X}'\mathbf{X}$ of (12) is

$$\mathbf{G} = \begin{bmatrix} 0 & \mathbf{0} & \mathbf{0} & 0 \\ \mathbf{0} & \mathbf{D}_a^{-1} + \mathbf{M}\mathbf{C}^{-1}\mathbf{M}' & -\mathbf{M}\mathbf{C}^{-1} & \mathbf{0} \\ \mathbf{0} & -\mathbf{C}^{-1}\mathbf{M}' & \mathbf{C}^{-1} & \mathbf{0} \\ 0 & \mathbf{0} & \mathbf{0} & 0 \end{bmatrix}. \tag{21}$$

The non-null part of this matrix is, of course, the (regular) inverse of the matrix of coefficients in equations (13). Thus **G** is in accord with Sec. 5.7.

Example (*continued*). From (9)

$$\mathbf{C}^{-1} = \begin{bmatrix} 11/6 & -4/6 \\ -4/6 & 8/6 \end{bmatrix}^{-1} = \frac{1}{12}\begin{bmatrix} 8 & 4 \\ 4 & 11 \end{bmatrix}$$

and from (4)

$$\mathbf{D}_a = \begin{bmatrix} 3 & 0 & 0 & 0 \\ 0 & 1 & 0 & 0 \\ 0 & 0 & 2 & 0 \\ 0 & 0 & 0 & 3 \end{bmatrix}$$

and $$\mathbf{N} = \begin{bmatrix} 1 & 1 \\ 0 & 0 \\ 1 & 0 \\ 1 & 1 \end{bmatrix}, \quad \text{so that} \quad \mathbf{M} = \mathbf{D}_a^{-1}\mathbf{N} = \begin{bmatrix} \frac{1}{3} & \frac{1}{3} \\ 0 & 0 \\ \frac{1}{2} & 0 \\ \frac{1}{3} & \frac{1}{3} \end{bmatrix}.$$

Therefore, for use in **G**,

$$\mathbf{MC}^{-1} = \frac{1}{12}\begin{bmatrix} 4 & 5 \\ 0 & 0 \\ 4 & 2 \\ 4 & 5 \end{bmatrix} \quad \text{and} \quad \mathbf{MC}^{-1}\mathbf{M}' = \frac{1}{12}\begin{bmatrix} 3 & 0 & 2 & 3 \\ 0 & 0 & 0 & 0 \\ 2 & 0 & 2 & 2 \\ 3 & 0 & 2 & 3 \end{bmatrix}$$

and so in (21)

$$\mathbf{G} = \frac{1}{12}\begin{bmatrix} 0 & 0 & 0 & 0 & 0 & 0 & 0 & 0 \\ 0 & 7 & 0 & 2 & 3 & -4 & -5 & 0 \\ 0 & 0 & 12 & 0 & 0 & 0 & 0 & 0 \\ 0 & 2 & 0 & 8 & 2 & -4 & -2 & 0 \\ 0 & 3 & 0 & 2 & 7 & -4 & -5 & 0 \\ 0 & -4 & 0 & -4 & -4 & 8 & 4 & 0 \\ 0 & -5 & 0 & -2 & -5 & 4 & 11 & 0 \\ 0 & 0 & 0 & 0 & 0 & 0 & 0 & 0 \end{bmatrix}.$$

Post-multiplication of this by the right-hand side of equation (3) gives the solution $\mathbf{GX}'\mathbf{y} = \mathbf{b}^o$ shown in (11).

e. Analyses of variance

(*i*) *Basic calculations.* The reduction of sum of squares due to fitting the model is $R(\mu, \alpha, \beta) = \mathbf{b}^{o\prime}\mathbf{X}'\mathbf{y}$. In preceding examples $\mathbf{b}^{o\prime}\mathbf{X}'\mathbf{y}$ simplified, but it does not do so here because of the manner in which $\mathbf{b}^o$ has been derived. However, on defining

$$\mathbf{y}'_\beta = [y_{.1} \quad \cdots \quad y_{.,b-1}] \tag{22}$$

it will be found from (20) that

$$R(\mu, \alpha, \beta) = (\bar{\mathbf{y}}_a - \mathbf{M}\mathbf{C}^{-1}\mathbf{r})'\mathbf{y}_a + (\mathbf{C}^{-1}\mathbf{r})'\mathbf{y}_\beta \,.$$

Also, because in (18)

$$\mathbf{r} = \mathbf{y}_\beta - \mathbf{M}'\mathbf{y}_a\,,$$

there is further simplification to

$$R(\mu, \alpha, \beta) = \bar{\mathbf{y}}'_a\mathbf{y}_a + \mathbf{r}'\mathbf{C}^{-1}\mathbf{r}. \tag{23}$$

As usual we have

$$R(\mu) = n_{..}\bar{y}^2_{..} = y^2_{..}/n_{..} \tag{24}$$

and, in line with Sec. 6.3a,

$$\begin{aligned} R(\mu, \alpha) &= \sum_{i=1}^{a} n_{i.}\bar{y}^2_{i.} = \sum_{i=1}^{a} y^2_{i.}/n_{i.} \\ &= \bar{\mathbf{y}}'_a\mathbf{y}_a\,, \quad \text{from (19).} \end{aligned} \tag{25}$$

Hence in (23)

$$\begin{aligned} R(\mu, \alpha, \beta) &= R(\mu, \alpha) + \mathbf{r}'\mathbf{C}^{-1}\mathbf{r} \\ &= \sum_{i=1}^{a} n_{i.}\bar{y}^2_{i.} + \boldsymbol{\beta}^{o\prime}\mathbf{r} \end{aligned} \tag{26}$$

with the terms of $\mathbf{r}'\mathbf{C}^{-1}\mathbf{r} = \boldsymbol{\beta}^{o\prime}\mathbf{r}$ defined as in (16), (17) and (18).

Calculation of these terms for the data of Table 7.1 is as follows.

$$R(\mu) = 108^2/9 \qquad = 1{,}296 \tag{27}$$

$$R(\mu, \alpha) = 54^2/3 + 9^2/1 + 18^2/2 + 27^2/3 \qquad = 1{,}458 \tag{28}$$

$$R(\mu, \alpha, \beta) = 1{,}458 + (-10)(-9) + (-14)(-12) = 1{,}716, \tag{29}$$

using (10) and (9) for $\boldsymbol{\beta}^{o\prime}$ and $\mathbf{r}$ in (26).

(*ii*) *Fitting the model.* The first analysis of variance to be considered is that for fitting the model (1). This partitions $R(\mu, \alpha, \beta)$, the sum of squares for fitting the model, into two parts: $R(\mu)$ for fitting the mean and $R(\alpha, \beta \mid \mu)$ for fitting the α- and β-factors after the mean. The latter is

$$\begin{aligned} R(\alpha, \beta \mid \mu) &= R(\mu, \alpha, \beta) - R(\mu) \\ &= \sum_{i=1}^{a} n_{i.}\bar{y}^2_{i.} + \mathbf{r}'\mathbf{C}^{-1}\mathbf{r} - N\bar{y}^2_{..} \end{aligned} \tag{30}$$

from (24) and (26). We note what is obvious here: that the sum of the two terms concerned is $R(\mu, \alpha, \beta)$, namely

$$R(\mu) + R(\alpha, \beta \mid \mu) = R(\mu, \alpha, \beta)$$

by the definition of $R(\alpha, \beta \mid \mu)$.

The values of the terms for the example are $R(\mu) = 1{,}296$ from (27) and

$$R(\alpha, \beta \mid \mu) = 1{,}716 - 1{,}296 = 420$$

from (27) and (29). These and the other terms of the analysis,

$$\text{SST} = \sum_i \sum_j y_{ij}^2 = 18^2 + \cdots + 18^2 \qquad = 1{,}728$$

and
$$\text{SSE} = \text{SST} - R(\mu, \alpha, \beta) = 1{,}728 - 1{,}716 = \quad 12$$

using (29), are shown in Table 7.2a. The corresponding F-statistics (based on normality of the e's) are also shown, $F(M) = 324$ and $F(R_m) = 21$. Clearly they are significant at the 5% level. (Tabulated values of the $F_{1,3}$- and $F_{5,3}$-distributions are 10.13 and 9.01 respectively, at the 5% level.) Therefore we reject the hypothesis that $E(\bar{y})$ is zero, and we further conclude that the model needs in it something more than just μ in order to satisfactorily explain variation in the y-variable.

(*iii*) *Fitting rows before columns.* The significance of the statistic $F(R_m)$ in Table 7.2a leads us to enquire if it is the α's (rows, or brands of stove) or the β's (columns, or makes of pan) or both that are contributing to this significance. First consider the α's, in terms of fitting the model

$$y_{ij} = \mu + \alpha_i + e_{ij}\,.$$

Since this is just the model for a 1-way classification the sum of squares for fitting it is $R(\mu, \alpha)$ as given in (25). Therefore the sum of squares attributable to fitting α after μ is

$$\begin{aligned} R(\alpha \mid \mu) &= R(\mu, \alpha) - R(\mu) \\ &= \sum_{i=1}^{a} n_{i\cdot}\bar{y}_{i\cdot}^2 - n_{\cdot\cdot}\bar{y}_{\cdot\cdot}^2 \end{aligned} \qquad (31)$$

from (24) and (25). Furthermore, the sum of squares attributable to fitting the β's after μ and the α's is

$$\begin{aligned} R(\beta \mid \mu, \alpha) &= R(\mu, \alpha, \beta) - R(\mu, \alpha) \\ &= \boldsymbol{\beta}^{o\prime}\mathbf{r} = \mathbf{r}'\mathbf{C}^{-1}\mathbf{r} \end{aligned} \qquad (32)$$

from (26). These two sums of squares are shown in Table 7.2b. They are, of course, a partitioning of $R(\alpha, \beta \mid \mu)$ shown in Table 7.2a, since

$$\begin{aligned} R(\alpha \mid \mu) + R(\beta \mid \mu, \alpha) &= R(\mu, \alpha) - R(\mu) + R(\mu, \alpha, \beta) - R(\mu, \alpha) \\ &= R(\mu, \alpha, \beta) - R(\mu) \\ &= R(\alpha, \beta \mid \mu). \end{aligned} \qquad (33)$$

TABLE 7.2. ANALYSES OF VARIANCE FOR 2-WAY CLASSIFICATION, NO INTERACTION (DATA OF TABLE 7.1)

Source of Variation	Degrees of Freedom[1]	Sum of Squares	Mean Square	F-statistic
	Table 7.2a: For fitting μ, and α and β after μ			
Mean	$1 = 1$	$R(\mu) = 1{,}296$	1,296	$F(M) = 324$
α and β after μ	$a+b-2 = 5$	$R(\alpha, \beta \mid \mu) = 420$	84	$F(R_m) = 21$
Residual error	$N' = 3$	$\mathrm{SSE} = 12$	4	
Total	$N = 9$	$\mathrm{SST} = 1{,}728$		
	Table 7.2b: For fitting μ, α after μ, and β after μ and α			
Mean	$1 = 1$	$R(\mu) = 1{,}296$	1,296	$F(M) = 324$
α after μ	$a - 1 = 3$	$R(\alpha \mid \mu) = 162$	54	$F(\alpha \mid \mu) = 13\frac{1}{2}$
β after μ and α	$b - 1 = 2$	$R(\beta \mid \mu, \alpha) = 258$	129	$F(\beta \mid \mu, \alpha) = 32\frac{1}{4}$
Residual error	$N' = 3$	$\mathrm{SSE} = 12$	4	
Total	$N = 9$	$\mathrm{SST} = 1{,}728$		
	Table 7.2c: For fitting μ, β after μ, and α after μ and β			
Mean	$1 = 1$	$R(\mu) = 1{,}296$	1,296	$F(M) = 324$
β after μ	$b - 1 = 2$	$R(\beta \mid \mu) = 148\frac{1}{2}$	$74\frac{1}{4}$	$F(\beta \mid \mu) = 18\frac{9}{16}$
α after μ and β	$a - 1 = 3$	$R(\alpha \mid \mu, \beta) = 271\frac{1}{2}$	$90\frac{1}{2}$	$F(\alpha \mid \mu, \beta) = 22\frac{5}{8}$
Residual error	$N' = 3$	$\mathrm{SSE} = 12$	4	
Total	$N = 9$	$\mathrm{SST} = 1{,}728$		

[1] $N' = N - a - b + 1$.

And, similarly, all three R's shown in Table 7.2b sum to $R(\mu, \alpha, \beta)$ because

$$R(\mu) + R(\alpha \mid \mu) + R(\beta \mid \mu, \alpha) = R(\mu) + R(\alpha, \beta \mid \mu) = R(\mu, \alpha, \beta). \quad (34)$$

Calculation of $R(\alpha \mid \mu)$ and $R(\beta \mid \mu, \alpha)$ for Table 7.2b is as follows. Substituting in (31) from (27) and (28) yields

$$R(\alpha \mid \mu) = 1{,}458 - 1{,}296 = 162;$$

and in (32) using (9) and (10) gives

$$R(\beta \mid \mu, \alpha) = -9(-10) - 12(-14) = 258. \quad (35)$$

The validity of (33) is evident:

$$R(\alpha \mid \mu) + R(\beta \mid \mu, \alpha) = 162 + 258 = 420 = R(\alpha, \beta \mid \mu)$$

of Table 7.2a.

F-statistics corresponding to the R's are also shown in Table 7.2b. Comparing $F(\alpha \mid \mu) = 13\frac{1}{2}$ and $F(\beta \mid \mu, \alpha) = 32\frac{1}{4}$ to tabulated values of the $F_{3,3}$- and $F_{2,3}$-distributions respectively, namely 9.28 and 9.55 at the 5% level, we conclude that having both α- and β-effects in the model adds significantly to its adequacy in terms of explaining variation in y.

(*iv*) *Fitting columns before rows.* Table 7.2b is for fitting μ, then μ and α and then μ, α and β. But we could just as well consider the α's and β's in reverse order and contemplate fitting μ, μ and β and then μ, α and β. To do this we would first fit the model $y_{ij} = \mu + \beta_j + e_{ij}$, which leads to

$$R(\mu, \beta) = \sum_{j=1}^{b} n_{.j}\bar{y}_{.j}^2 \tag{36}$$

similar to (25). Then, analogous to (31) we have

$$\begin{aligned} R(\beta \mid \mu) &= R(\mu, \beta) - R(\mu) \\ &= \sum_{j=1}^{b} n_{.j}\bar{y}_{.j}^2 - n_{..}\bar{y}_{..}^2 . \end{aligned} \tag{37}$$

Also, similar to the first part of (32), we have

$$R(\alpha \mid \mu, \beta) = R(\mu, \alpha, \beta) - R(\mu, \beta) \tag{38}$$

for the sum of squares due to fitting the α after fitting μ and β. Now, however, we do not have an expression for $R(\alpha \mid \mu, \beta)$ analogous to $\boldsymbol{\beta}^{o\prime}\mathbf{r}$ of (32), the only term that presents a little difficulty to calculate. However, by means of (34) such an expression can be avoided, because using (34) in (38) gives

$$\begin{aligned} R(\alpha \mid \mu, \beta) &= R(\mu, \alpha) + R(\beta \mid \mu, \alpha) - R(\mu, \beta) \\ &= \sum_{i=1}^{a} n_{i.}\bar{y}_{i.}^2 + \mathbf{r}'\mathbf{C}^{-1}\mathbf{r} - \sum_{j=1}^{b} n_{.j}\bar{y}_{.j}^2 \end{aligned} \tag{39}$$

on substituting from (25), (32) and (36) respectively. Hence, having once obtained $\mathbf{r}'\mathbf{C}^{-1}\mathbf{r}$, we have $R(\alpha \mid \mu, \beta)$ directly available without further ado. Analogues of (33) and (34) are, of course, true also:

$$\begin{aligned} & R(\beta \mid \mu) + R(\alpha \mid \mu, \beta) = R(\alpha, \beta \mid \mu) \\ \text{and} \quad & R(\mu) + R(\beta \mid \mu) + R(\alpha \mid \mu, \beta) = R(\mu, \alpha, \beta). \end{aligned} \tag{40}$$

With the data of Table 7.1, equation (36) is

$$R(\mu, \beta) = 27^2/3 + 15^2/2 + 66^2/4 = 1{,}444\tfrac{1}{2}. \tag{41}$$

Using this and (27) in (37) gives

$$R(\beta \mid \mu) = 1{,}444\tfrac{1}{2} - 1{,}296 = 148\tfrac{1}{2}.$$

Then in (39)

$$R(\alpha \mid \mu, \beta) = 1{,}458 + 258 - 1{,}444\tfrac{1}{2} = 271\tfrac{1}{2}$$

from (28), (35) and (41) respectively. We note, as indicated in (40), that

$$R(\beta \mid \mu) + R(\alpha \mid \mu, \beta) = 148\tfrac{1}{2} + 271\tfrac{1}{2} = 420 = R(\alpha, \beta \mid \mu)$$

shown in Table 7.2a.

The F-statistics corresponding to $R(\beta \mid \mu)$ and $R(\alpha \mid \mu, \beta)$ in Table 7.2c are both significant at the 5% level. [The tabulated values are 9.55 and 9.28 for comparing $F(\beta \mid \mu)$ and $F(\alpha \mid \mu, \beta)$ respectively.] We therefore conclude that including both β-effects and α-effects in the model adds significantly to the model's interpretive value.

Table 7.2 shows the analyses of variance for the data of Table 7.1. In contrast, Table 7.3 shows the analysis of variance (excluding mean squares and F-statistics) for the general case, and it also shows the equations from which the expressions for the sums of squares have been dervied.

(*v*) *Ignoring and/or adjusting for effects.* In Tables 7.2b and 7.3b the sums of squares have been described as

$R(\mu)$: due to fitting a mean μ,

$R(\alpha \mid \mu)$: due to fitting α after μ,

and $R(\beta \mid \mu, \alpha)$: due to fitting β after fitting μ and α.

This description carries with it a sequential concept, of first fitting μ, then μ and α and then μ, α and β. An alternative description, similar to that used by some writers, is

$R(\mu)$: due to fitting μ, ignoring α and β,

$R(\alpha \mid \mu)$: due to fitting α, adjusted for μ and ignoring β,

and $R(\beta \mid \mu, \alpha)$: due to fitting β, adjusted for μ and α.

On many occasions, of course, Table 7.2 and 7.3 are shown without the $R(\mu)$ line, and with the SST line reduced by $R(\mu)$ so that it has $N - 1$ degrees of freedom and sum of squares $\text{SST}_m = \mathbf{y}'\mathbf{y} - N\bar{y}_{..}^2$. In that case the mention of μ in the descriptions of $R(\alpha \mid \mu)$ and $R(\beta \mid \mu, \alpha)$ is then often overlooked entirely and they get described as

$R(\alpha \mid \mu)$: due to fitting α, ignoring β,

and $R(\beta \mid \mu, \alpha)$: due to fitting β, adjusted for α.

TABLE 7.3. ANALYSES OF VARIANCE FOR 2-WAY CLASSIFICATION, NO INTERACTION

Source of Variation	Degrees of Freedom[1]	Sum of Squares[2]	Equation
Table 7.3a: For fitting μ, and α and β after μ			
Mean, μ	1	$R(\mu) \quad = n_{..}\bar{y}^2_{..}$	(24)
α and β after μ	$a+b-2$	$R(\alpha, \beta \mid \mu) = \sum_i n_{i.}\bar{y}^2_{i.} + \mathbf{r}'\mathbf{C}^{-1}\mathbf{r} - n_{..}\bar{y}^2_{..}$	(30)
Residual error[3]	N'	$\text{SSE} \quad = \sum_i \sum_j y^2_{ij} - \sum_i n_{i.}\bar{y}^2_{i.} - \mathbf{r}'\mathbf{C}^{-1}\mathbf{r}$	
Total	N	$\text{SST} \quad = \sum_i \sum_j y^2_{ij}$	
Table 7.3b: For fitting μ, α after μ, and β after μ and α			
Mean, μ	1	$R(\mu) \quad = n_{..}\bar{y}^2_{..}$	(24)
α after μ	$a-1$	$R(\alpha \mid \mu) \quad = \sum_i n_{i.}\bar{y}^2_{i.} - n_{..}\bar{y}^2_{..}$	(31)
β after μ and α	$b-1$	$R(\beta \mid \mu, \alpha) = \mathbf{r}'\mathbf{C}^{-1}\mathbf{r}$	(32)
Residual error	N'	$\text{SSE} \quad = \sum_i \sum_j y^2_{ij} - \sum_i n_{i.}\bar{y}^2_{i.} - \mathbf{r}'\mathbf{C}^{-1}\mathbf{r}$	
Total	N	$\text{SST} \quad = \sum_i \sum_j y^2_{ij}$	
Table 7.3c: For fitting μ, β after μ, and α after μ and β			
Mean, μ	1	$R(\mu) \quad = n_{..}\bar{y}^2_{..}$	(24)
β after μ	$b-1$	$R(\beta \mid \mu) \quad = \sum_j n_{.j}\bar{y}^2_{.j} - n_{..}\bar{y}^2_{..}$	(37)
α after β and μ	$a-1$	$R(\alpha \mid \mu, \beta) = \sum_i n_{i.}\bar{y}^2_{i.} + \mathbf{r}'\mathbf{C}^{-1}\mathbf{r} - \sum_j n_{.j}\bar{y}^2_{.j}$	(39)
Residual error	N'	$\text{SSE} \quad = \sum_i \sum_j y^2_{ij} - \sum_i n_{i.}\bar{y}^2_{i.} - \mathbf{r}'\mathbf{C}^{-1}\mathbf{r}$	
Total	N	$\text{SST} \quad = \sum_i \sum_j y^2_{ij}$	

[1] $N \equiv n_{..}$ and $N' = N - a - b + 1$.
[2] $\mathbf{r}'\mathbf{C}^{-1}\mathbf{r}$ is obtained from equations (16)–(18).
[3] Summations are for $i = 1, 2, \ldots, a$ and $j = 1, 2, \ldots, b$.

The omission of μ from descriptions such as these arises from a desire for verbal convenience. The omission is made with the convention that μ is not being ignored, even though it is not being mentioned. However, inclusion of μ in the descriptions is somewhat safer, for then there is no fear of its being overlooked. Furthermore, although in describing $R(\alpha \mid \mu)$ the phrase "ignoring β" is clear and appropriate, the phrase "adjusted for α" in describing $R(\beta \mid \mu, \alpha)$ is not appealing because it may conjure up the idea of adjusting or amending the data in some manner. Since the concept involved is clearly that of fitting β over and above having fitted μ and α, the description "β after μ and α" seems more appropriate. However, the relationship of such descriptions to those involving "ignoring α" and "adjusted for β" should be borne in mind when encountering them in other texts. For example, just as $R(\alpha \mid \mu)$ of Tables 7.2b and 7.3b could be described as the sum of squares for fitting α, adjusted for μ and ignoring β, so also could $R(\beta \mid \mu)$ of Tables 7.2c and 7.3c be called the sum of squares for fitting β, adjusted for μ and ignoring α. But the description of fitting β after μ is preferred.

(*vi*) *Interpretation of results.* From the preceding discussion it should be clear that $F(\alpha \mid \mu)$ and $F(\alpha \mid \mu, \beta)$ are not used for the same purpose [neither are $F(\beta \mid \mu)$ and $F(\beta \mid \mu, \alpha)$]. Distinguishing between these two F's is of paramount importance because it is a distinction that occurs repeatedly in fitting other models. Furthermore, the distinction does not exist with the familiar balanced data situation because, as we shall see subsequently, $F(\alpha \mid \mu)$ and $F(\alpha \mid \mu, \beta)$ are then identical. It occurs only with unbalanced data, and always with such data. That is, the test based on the statistic $F(\alpha \mid \mu)$ is testing the effectiveness (in terms of explaining variation in y) of adding α-effects to the model, over and above μ, whereas $F(\alpha \mid \mu, \beta)$ tests the effectiveness of adding α-effects to the model over and above having μ *and* β-effects in it. These tests are not the same, and neither of them should be described, albeit loosely, as "testing α-effects". The tests must be described more completely, the one as "testing α after μ" and the other as "testing α after μ and β". Similarly $F(\beta \mid \mu)$ and $F(\beta \mid \mu, \alpha)$ are not the same. The former tests "β after μ" and the latter "β after μ and α". Further distinction between F-statistics of this nature will become evident when we consider tests of linear hypotheses to which they relate.

In Table 7.2 all the F-statistics are, at the 5% level, judged significant. From this we conclude that both the α-effects and the β-effects add materially to the explanatory power of the model. However, with other data, conclusions are not always as easily drawn as is this one. For example, suppose in some set of data that, analogous to Table 7.2b, $F(\alpha \mid \mu)$ and $F(\beta \mid \mu, \alpha)$ were both significant but that, analogous to Table 7.2c, neither $F(\beta \mid \mu)$ nor $F(\alpha \mid \mu, \beta)$ were. Admittedly this may happen with only very few sets of data, but since computed F-statistics are just that, namely just functions of data, it is certainly *possible* for such an apparent inconsistency to occur. There then arises the problem of trying to draw conclusions from such a result. To do so is not always easy, and in this regard the ensuing discussion of possible conclusions might by no means meet with universal approval. The problems of interpretation discussed here receive scant mention in most texts for the very reason, one suspects, that they are not definitive and are subject perhaps to personal judgment and certainly to knowledge of the data being analyzed. Also, they are not amenable to exact mathematical treatment. Nevertheless, since they *are* problems of interpretation they arise, in one way or another, on almost any occasion where data are analyzed and for this reason it seems worthwhile to try to reflect on what conclusions might be appropriate in different situations. In attempting to do so, I am all too well aware of leaving myself wide open for criticism. However, at the very worst, exposition of the problems might be of some assistance.

The general problem we consider is what conclusions can be drawn from the various combinations of results that can arise *vis à vis* the significance or non-significance of $F(\alpha \mid \mu)$, $F(\beta \mid \mu, \alpha)$, $F(\beta \mid \mu)$ and $F(\alpha \mid \mu, \beta)$ implicit in Tables 7.3b and 7.3c and illustrated in Tables 7.2b and 7.2c. First, these F-statistics should be considered only if $F(R_m) = F(\alpha, \beta \mid \mu)$ of Table 7.3a is significant. This is so because it is only the significance of $F(R_m)$ which suggests that simultaneous fitting of α and β has explanatory value for the variation in y. However, it does not necessarily mean that *both* α and β are needed in the model. It is the investigation of this aspect of the model that arises from looking at $F(\alpha \mid \mu)$, $F(\beta \mid \mu, \alpha)$, $F(\beta \mid \mu)$ and $F(\beta \mid \mu, \beta)$. There are 16 different situations to consider, as shown in Table 7.4. For $F(\alpha \mid \mu)$ and $F(\beta \mid \alpha, \mu)$ there are 4 possible outcomes: both F's significant, $F(\alpha \mid \mu)$ non-significant and $F(\beta \mid \mu, \alpha)$ significant, the converse of this, and both F's non-significant. These are shown as row headings in Table 7.4. With each of these outcomes a similar 4 outcomes can also occur for $F(\beta \mid \mu)$ and $F(\alpha \mid \mu, \beta)$; they are shown as column headings in Table 7.4. For each of the 16 resulting outcomes, the conclusion to be drawn is shown in the body of the table.

We now indulge in the verbal convenience of omitting μ from our discussion, to use phrases like "α being significant alone" for $F(\alpha \mid \mu)$ being significant, and "α being significant after fitting β" for $F(\alpha \mid \mu, \beta)$ being significant. We do not, however, use phrases like "α being significant" which does not distinguish between $F(\alpha \mid \mu)$ and $F(\alpha \mid \mu, \beta)$ being significant.

The first entry in Table 7.4 corresponds to the case dealt with in Table 7.2: both α and β are significant when fitted either alone or after the other, so the conclusion is to fit both. The second entries in the first row and column are cases of both α and β being significant when fitted after each other, with one of them being significant when fitted alone, and the other not; the conclusion is to fit both. For the second diagonal entry, neither α nor β is significant

TABLE 7.4. SUGGESTED CONCLUSIONS ACCORDING TO SIGNIFICANCE (Sig) AND NON-SIGNIFICANCE (NS) OF F-STATISTICS IN FITTING A MODEL WITH TWO MAIN EFFECTS (α'S AND β'S)—SEE TABLE 7.3

Fitting α and then β after α		Fitting β and then α after β			
	$F(\beta \mid \mu)$:	Sig	NS	Sig	NS
	$F(\alpha \mid \mu, \beta)$:	Sig	Sig	NS	NS
		Effects to be included in model			
$F(\alpha \mid \mu)$: $F(\beta \mid \mu, \alpha)$:	Sig Sig	α and β	α and β	β	Impossible
$F(\alpha \mid \mu)$: $F(\beta \mid \mu, \alpha)$:	NS Sig	α and β	α and β	β	α and β
$F(\alpha \mid \mu)$: $F(\beta \mid \mu, \alpha)$:	Sig NS	α	α	α and β	α
$F(\alpha \mid \mu)$: $F(\beta \mid \mu, \alpha)$:	NS NS	Impossible	α and β	β	neither α nor β

alone but each is significant when fitted after the other; hence we fit both. Similarly, the first entries in the third row and column are cases where one factor (β in the third row and α in the third column) is significant only when fitted alone, but the other is significant when fitted either alone or after the first; hence that other factor—α in the third row (first column) and β in the third column (first row)—is the factor to fit. The second entries in the third row and column are cases where one factor is not significant on its own or after the other, but that other factor is significant on both occasions: it is therefore the factor fitted. Likewise, the last entries in the third row and column are cases where the only significance is of one factor fitted on its own: it is therefore the factor to fit. The diagonal entry in the third row is when both factors (α and β) are significant on their own but neither of them are when fitted after the other. In this case the conclusion is to use both of them. (In some circumstances this choice might be overridden; for example, if determining levels of the α-factor was very costly one might be prepared to use just the β-factor. This, of course, is a consideration that might arise with other entries in Table 7.4, too.) The first two entries in the last row and column of the table are difficult to visualize. Both pairs of entries are situations when fitting the factors in one sequence gives neither F-statistic significant but fitting them in the other sequence gives the F-statistic for fitting the second factor significant. Intuitively one feels that this kind of thing should happen somewhat infrequently. When it does, a reasonable conclusion seems to be to fit both factors, as shown. Finally, the last entry in the table, that in the lower right-hand corner, is when none of the F-statistics are significant, leading to the conclusion to fit neither α nor β.[1]

f. Estimable functions

The basic estimable function for the model (1) is

$$E(y_{ij}) = \mu + \alpha_i + \beta_j \tag{42}$$

and its b.l.u.e. is

$$\widehat{\mu + \alpha_i + \beta_j} = \mu^o + \alpha_i^o + \beta_j^o . \tag{43}$$

Note from this that although individual α's and β's are not estimable, differences between pairs of α's and between pairs of β's are estimable, as are linear functions of these differences. Thus

$$\alpha_i - \alpha_h \text{ is estimable with b.l.u.e. } \widehat{\alpha_i - \alpha_h} = \alpha_i^o - \alpha_h^o ,$$

$$\text{and} \qquad \beta_j - \beta_k \text{ is estimable with b.l.u.e. } \widehat{\beta_j - \beta_k} = \beta_j^o - \beta_k^o . \tag{44}$$

The variances of these b.l.u.e.'s are found from the general result for an estimable function $\mathbf{q}'\mathbf{b}$, that the variance of its b.l.u.e. is $v(\mathbf{q}'\mathbf{b}^o) = \mathbf{q}'\mathbf{G}\mathbf{q}\sigma^2$. Hence if g_{ii} and g_{hh} are the diagonal elements of $\mathbf{G}$ corresponding to α_i and α_h respectively, and g_{ih} is the element at the intersection of the row and column corresponding to α_i and α_h then

$$v(\widehat{\alpha_i - \alpha_h}) = v(\alpha_i^o - \alpha_h^o) = (g_{ii} + g_{hh} - 2g_{ih})\sigma^2. \tag{45}$$

A similar result holds for $v(\beta_j^o - \beta_k^o)$. Furthermore, any linear combination of the estimable functions in (44) is estimable, having for its b.l.u.e. the same linear combination of the b.l.u.e.'s shown in (44). Variances of such b.l.u.e.'s are found in a manner similar to (45).

More generally, if $\mathbf{b} = \{b_s\}$ for $s = 1, 2, \ldots, a + b + 1$ and $\mathbf{G} = \{g_{s,t}\}$ for $s, t = 1, 2, \ldots, a + b + 1$ then, provided $b_s - b_t$ is estimable (i.e., is a difference between two α's or two β's),

$$\widehat{b_s - b_t} = b_s^o - b_t^o, \qquad \text{with } v(\widehat{b_s - b_t}) = (g_{ss} + g_{tt} - 2g_{st})\sigma^2. \tag{46}$$

Example (*continued*). In (11) we have $\alpha_1^o = 26$ and $\alpha_3^o = 14$, so from (44)

$$\widehat{\alpha_1 - \alpha_3} = \alpha_1^o - \alpha_3^o = 26 - 14 = 12.$$

[1] Grateful thanks go to N. S. Urquhart for lengthy discussions on this topic.

And since we earlier derived

$$\mathbf{G} = \frac{1}{12}\begin{bmatrix} 0 & 0 & 0 & 0 & 0 & 0 & 0 & 0 \\ 0 & 7 & 0 & 2 & 3 & -4 & -5 & 0 \\ 0 & 0 & 12 & 0 & 0 & 0 & 0 & 0 \\ 0 & 2 & 0 & 8 & 2 & -4 & -2 & 0 \\ 0 & 3 & 0 & 2 & 7 & -4 & -5 & 0 \\ 0 & -4 & 0 & -4 & -4 & 8 & 4 & 0 \\ 0 & -5 & 0 & -2 & -5 & 4 & 11 & 0 \\ 0 & 0 & 0 & 0 & 0 & 0 & 0 & 0 \end{bmatrix} \tag{47}$$

then $$v(\widehat{\alpha_1 - \alpha_3}) = \tfrac{1}{12}[7 + 8 - 2(2)]\sigma^2 = \tfrac{11}{12}\sigma^2.$$

With σ^2 estimated as $\hat{\sigma}^2 = 4 =$ MSE in Table 7.2, the estimated variance is

$$\hat{v}(\widehat{\alpha_1 - \alpha_3}) = \tfrac{11}{12}(4) = 2\tfrac{2}{3}.$$

g. Tests of hypotheses

As usual, the F-statistic for testing testable hypotheses H: $\mathbf{K}'\mathbf{b} = \mathbf{0}$ is

$$F(H) = Q/s\hat{\sigma}^2 = (\mathbf{K}'\mathbf{b}^\circ)'(\mathbf{K}'\mathbf{G}\mathbf{K})^{-1}\mathbf{K}'\mathbf{b}^\circ/s\hat{\sigma}^2$$

where $$Q = (\mathbf{K}'\mathbf{b}^\circ)'(\mathbf{K}'\mathbf{G}\mathbf{K})^{-1}\mathbf{K}'\mathbf{b}^\circ,$$

using (21) for $\mathbf{G}$, s being the rank and number of rows of $\mathbf{K}'$.

Previous sections have dealt at length with the meaning of the sums of squares in Tables 7.2 and 7.3, interpreting them in terms of reductions in sums of squares due to fitting different models. Their meaning in terms of testing hypotheses is now considered. In this context there is no question of dealing with different models; we are testing hypotheses about the elements of the model (1). First, we show that $F(\beta \mid \mu, \alpha)$ of Table 7.2b is the F-statistic for testing the hypothesis that all β's are equal. Stated as $\beta_j - \beta_b = 0$ for $j = 1, 2, \ldots, b - 1$, the hypothesis can be written as

$$H\colon\ \mathbf{K}'\mathbf{b} = \mathbf{0} \qquad \text{with } \mathbf{K}' = [\mathbf{0}\mathbf{1}\ \ \mathbf{0}\ \ \mathbf{I}_{b-1}\ \ -\mathbf{1}_{b-1}],$$

wherein $\mathbf{K}'$ is partitioned conformably for the product $\mathbf{K}'\mathbf{G}$. Then, with $\mathbf{G}$ of (21)

$$\mathbf{K}'\mathbf{G} = [\mathbf{0}\quad -\mathbf{C}^{-1}\mathbf{M}'\quad \mathbf{C}^{-1}\quad \mathbf{0}]$$

and $$\mathbf{K}'\mathbf{G}\mathbf{K} = \mathbf{C}^{-1}.$$

Also, $$\mathbf{K}'\mathbf{b}^\circ = \mathbf{K}'\mathbf{G}\mathbf{X}\mathbf{y} = (-\mathbf{C}^{-1}\mathbf{M}'\mathbf{y}_\alpha + \mathbf{C}^{-1}\mathbf{y}_\beta)$$

where $\mathbf{y}_\alpha$ is, as in (19), the vector of totals for the a levels of the α-factor; and $\mathbf{y}_\beta = \{y_{.j}\}$ for $j = 1, \ldots, b - 1$ is the vector of totals for the first $b - 1$

evels of the β-factor, as in (22). Then the numerator sum of squares of $F(H)$ is

$$\begin{aligned}
Q &= (\mathbf{K}'\mathbf{b}^o)'(\mathbf{K}'\mathbf{G}\mathbf{K})^{-1}\mathbf{K}'\mathbf{b}^o \\
&= (-\mathbf{C}^{-1}\mathbf{M}'\mathbf{y}_\alpha + \mathbf{C}^{-1}\mathbf{y}_\beta)'(\mathbf{C}^{-1})^{-1}(-\mathbf{C}^{-1}\mathbf{M}'\mathbf{y}_\alpha + \mathbf{C}^{-1}\mathbf{y}_\beta) \\
&= (\mathbf{y}_\beta - \mathbf{N}'\mathbf{D}_a^{-1}\mathbf{y}_a)'\mathbf{C}^{-1}(\mathbf{y}_\beta - \mathbf{N}'\mathbf{D}_a^{-1}\mathbf{y}_a) \\
&= \mathbf{r}'\mathbf{C}^{-1}\mathbf{r}, \text{ by the definition of } \mathbf{r} \text{ in (18)} \\
&= \boldsymbol{\beta}^{o\prime}\mathbf{r}, \text{ by (16)} \\
&= R(\beta \mid \mu, \alpha) \text{ by (32).}
\end{aligned}$$

Example (*continued*). The hypothesis of equality of the β's in the example can be written as

$$H: \quad \beta_1 - \beta_3 = 0$$
$$\beta_2 - \beta_3 = 0\,,$$

i.e., as

$$\mathbf{K}'\mathbf{b} \equiv \begin{bmatrix} 0 & 0 & 0 & 0 & 0 & 1 & 0 & -1 \\ 0 & 0 & 0 & 0 & 0 & 0 & 1 & -1 \end{bmatrix} \mathbf{b} = \mathbf{0}.$$

With $\mathbf{b}^o$ of (11) and (**G**) of (47)

$$\mathbf{K}'\mathbf{b}^o = \begin{bmatrix} -10 \\ -14 \end{bmatrix}$$

and

$$\mathbf{K}'\mathbf{G} = \frac{1}{12}\begin{bmatrix} 0 & -4 & 0 & -4 & -4 & 8 & 4 & 0 \\ 0 & -5 & 0 & -2 & -5 & 4 & 11 & 0 \end{bmatrix}$$

so that

$$\mathbf{K}'\mathbf{G}\mathbf{K} = \frac{1}{12}\begin{bmatrix} 8 & 4 \\ 4 & 11 \end{bmatrix} \quad \text{and} \quad (\mathbf{K}'\mathbf{G}\mathbf{K})^{-1} = \frac{1}{6}\begin{bmatrix} 11 & -4 \\ -4 & 8 \end{bmatrix}.$$

Hence the numerator sum of squares of $F(H)$ is

$$\begin{aligned}
(\mathbf{K}'\mathbf{b}^o)'(\mathbf{K}'\mathbf{G}\mathbf{K})^{-1}\mathbf{K}'\mathbf{b}^o &= [-10 \quad -14]\frac{1}{6}\begin{bmatrix} 11 & -4 \\ -4 & 8 \end{bmatrix}\begin{bmatrix} -10 \\ -14 \end{bmatrix} \\
&= [1100 + 8(196) - 20(14)4]/6 \\
&= 258 = R(\beta \mid \mu, \alpha) \text{ of Table 7.2b.}
\end{aligned}$$

Of course the hypothesis does not have to be stated in exactly the above form to demonstrate this result. For example, stating it as

$$\beta_1 - \beta_2 = 0$$
$$\beta_1 - \beta_3 = 0\,,$$

we have

$$\mathbf{K}' = \begin{bmatrix} 0 & 0 & 0 & 0 & 0 & 1 & -1 & 0 \\ 0 & 0 & 0 & 0 & 0 & 1 & 0 & -1 \end{bmatrix},$$

and hence

$$\mathbf{K}'\mathbf{b}^o = \begin{bmatrix} 4 \\ -10 \end{bmatrix},$$

$$\mathbf{K}'\mathbf{G}\mathbf{K} = \frac{1}{12}\begin{bmatrix} 11 & 4 \\ 4 & 8 \end{bmatrix} \quad \text{and} \quad (\mathbf{K}'\mathbf{G}\mathbf{K})^{-1} = \frac{1}{6}\begin{bmatrix} 8 & -4 \\ -4 & 11 \end{bmatrix}.$$

Then the numerator sum of squares of $F(H)$ is

$$[4 \quad -10]\frac{1}{6}\begin{bmatrix} 8 & -4 \\ -4 & 11 \end{bmatrix}\begin{bmatrix} 4 \\ -10 \end{bmatrix} = [16(8) + 1100 + 8(-10)(-4)]/6 = 258$$

as before.

Thus $R(\beta \mid \mu, \alpha)$ is the numerator sum of squares for the F-statistic for testing H: all β's equal. Similarly $R(\alpha \mid \mu, \beta)$ is the numerator sum of squares for the F-statistic for testing H: all α's equal.

It can similarly be proved (see Exercise 6) that $R(\beta \mid \mu)$ is the numerator sum of squares for testing

$$H\text{: equality of } \beta_j + \frac{1}{n_{.j}}\sum_{i=1}^{a} n_{ij}\alpha_i \qquad \text{for all} \quad j = 1, 2, \ldots, b. \tag{48}$$

For example, with the data of Table 7.1 this hypothesis can be conveniently stated as

$$\begin{aligned} H\text{:}\quad & \beta_1 + \tfrac{1}{3}(\alpha_1 + \alpha_3 + \alpha_4) - [\beta_3 + \tfrac{1}{4}(\alpha_1 + \alpha_2 + \alpha_3 + \alpha_4)] = 0 \\ & \beta_2 + \tfrac{1}{2}(\alpha_1 + \alpha_4) \qquad\quad - [\beta_3 + \tfrac{1}{4}(\alpha_1 + \alpha_2 + \alpha_3 + \alpha_4)] = 0, \end{aligned}$$

i.e., as

$$\mathbf{K}'\mathbf{b} \equiv \begin{bmatrix} 0 & \frac{1}{12} & -\frac{1}{4} & \frac{1}{12} & \frac{1}{12} & 1 & 0 & -1 \\ 0 & \frac{1}{4} & -\frac{1}{4} & -\frac{1}{4} & \frac{1}{4} & 0 & 1 & -1 \end{bmatrix}\mathbf{b} = \mathbf{0}.$$

With $\mathbf{b}^o$ of (11) and $\mathbf{G}$ of (47) we have

$$\mathbf{K}'\mathbf{b}^o = \begin{bmatrix} -7\frac{1}{2} \\ -9 \end{bmatrix}, \qquad \mathbf{K}'\mathbf{G}\mathbf{K} = \frac{1}{12}\begin{bmatrix} 7 & 3 \\ 3 & 9 \end{bmatrix}$$

and

$$(\mathbf{K}'\mathbf{G}\mathbf{K})^{-1} = \frac{2}{9}\begin{bmatrix} 9 & -3 \\ -3 & 7 \end{bmatrix}.$$

Hence

$$(\mathbf{K}'\mathbf{b}^o)'(\mathbf{K}'\mathbf{G}\mathbf{K})^{-1}\mathbf{K}'\mathbf{b}^o = [-7\tfrac{1}{2} \quad -9]\frac{2}{9}\begin{bmatrix} 9 & -3 \\ -3 & 7 \end{bmatrix}\begin{bmatrix} -7\frac{1}{2} \\ -9 \end{bmatrix}$$

$$= \tfrac{2}{9}(\tfrac{1}{4})[9(15^2) + 7(18^2) - 2(15)18(3)]$$

$$= 148\tfrac{1}{2} = R(\beta \mid \mu) \text{ of Table 7.2c.}$$

Hence $F(\beta \mid \mu) = R(\beta \mid \mu)/(b - 1)\hat{\sigma}^2$ is the F-statistic for testing (48). The F-statistic having $R(\alpha \mid \mu)$ of Table 7.2b as its numerator sum of squares tests an analogous hypothesis, namely,

$$F(\alpha \mid \mu) \text{ tests } H\text{: } \quad \alpha_i + \frac{1}{n_{i.}}\sum_{j=1}^{b} n_{ij}\beta_j \text{ equal for all } i.$$

The importance of these results is that $F(\alpha \mid \mu)$ is not a statistic for testing equality of the α's; $F(\alpha \mid \mu, \beta)$ is. The hypothesis that is tested by $F(\alpha \mid \mu)$ is equality of the α's plus averages of the β's, weighted averages using the n_{ij} as weights. Similarly, $F(\beta \mid \mu)$ tests not equality of the β's but equality of the β's plus weighted averages of the α's, as in (48).

h. Models that include restrictions

Since $\mu + \alpha_i + \beta_j$ is estimable, so is $\mu + \frac{1}{a}\sum_{i=1}^{a}\alpha_i + \beta_j$. Therefore, if the model includes the restriction that $\sum_{i=1}^{a}\alpha_i = 0$, then $\mu + \beta_j$ is estimable with b.l.u.e. $\mu^o + \frac{1}{a}\sum_{i=1}^{a}\alpha_i^o + \beta_j^o$, the same as the b.l.u.e. of $\mu + \frac{1}{a}\sum_{i=1}^{a}\alpha_i + \beta_j$ in the unrestricted model. Whether the restriction $\sum_{i=1}^{a}\alpha_i = 0$ is part of the model or not, the estimable functions and their b.l.u.e.'s given in (44) still apply; i.e., $\alpha_i^o - \alpha_h^o$ is still the b.l.u.e. of $\alpha_i - \alpha_h$, and $\beta_j^o - \beta_k^o$ is the b.l.u.e. of $\beta_j - \beta_k$. Similar results hold if the model includes the restriction $\sum_{j=1}^{b}\beta_j = 0$.

The hypothesis of equality of $\beta_j + \frac{1}{n_{.j}}\sum_{i=1}^{a} n_{ij}\alpha_i$ for all $j = 1, 2, \ldots, b$ discussed in the previous section might hint at the possibility of using a model that included the restriction

$$\sum_{i=1}^{a} n_{ij}\alpha_i = 0 \qquad \text{for all } j = 1, 2, \ldots, b. \tag{49}$$

Any value to this suggestion is lost whenever $b \geq a$, for then equations (49) can be solved for the α's (certainly as $\alpha_i = 0$ for all i), regardless of the data. When $b < a$, equations (49) could be used as restrictions in the model, but then only $a - b$ linear functions of the α's would be estimable from the data; furthermore, since equations (49) are data-dependent, in that they are based on the n_{ij}, they suffer from the same deficiencies as do all such restrictions, as explained at the end of Sec. 6.2h.

i. Balanced data

The preceding discussion uses n_{ij} as the number of observations in the ith row (level of the α-factor) and the jth column (level of the β-factor), with all $n_{ij} = 0$ or 1. Although we will show subsequently that much of that discussion applies *in toto* to situations in which the n_{ij} can be any non-negative integers (and hence to $n_{ij} = n$ for all i and j), we here consider just the simplest case of balanced data, $n_{ij} = 1$ for all i and j; i.e., for data like Table 7.1 only without any missing observations.

As is to be expected, there is great simplification of the foregoing results when $n_{ij} = 1$ for all i and j and, of course, the simplifications lead exactly to the familiar calculations in this case [e.g., Kempthorne (1952), p. 72]. A variety of solutions to the normal equations are easily obtainable under these conditions. Those derived from the procedure given above for unbalanced data (see Exercise 7) are

$$\mu^o = 0, \qquad \text{and} \qquad \alpha_i^o = \bar{y}_{i.} - \bar{y}_{..} + \bar{y}_{.b} \qquad \text{for all } i;$$

and

$$\beta_b^o = 0, \qquad \text{and} \qquad \beta_j^o = \bar{y}_{.j} - \bar{y}_{.b} \qquad \text{for} \quad j = 1, 2, \ldots, b-1.$$

Another set of solutions, obtained by use of the "usual constraints"

$$\sum_{i=1}^{a} \alpha_i^o = 0 = \sum_{j=1}^{b} \beta_j^o ,$$

is

$$\mu^o = \bar{y}_{..},$$

$$\alpha_i^o = \bar{y}_{i.} - \bar{y}_{..} \qquad \text{for all } i$$

and

$$\beta_j^o = \bar{y}_{.j} - \bar{y}_{..} \qquad \text{for all } j.$$

In either case the b.l.u.e.'s of differences between α's and β's are

$$\widehat{\alpha_i - \alpha_h} = \bar{y}_{i.} - \bar{y}_{h.}, \qquad \text{with } v(\widehat{\alpha_i - \alpha_h}) = 2\sigma^2/b$$

and

$$\widehat{\beta_j - \beta_k} = \bar{y}_{.j} - \bar{y}_{.k}, \qquad \text{with } v(\widehat{\beta_j - \beta_k}) = 2\sigma^2/a .$$

Differences of this nature are always estimable; if the model includes restrictions $\sum_{i=1}^{a} \alpha_i = 0 = \sum_{j=1}^{b} \beta_j$ paralleling the "usual constraints", then μ, the α_i and β_j are also estimable, with $\hat{\mu} = \bar{y}_{..}$, $\hat{\alpha}_i = \bar{y}_{i.} - \bar{y}_{..}$ and $\hat{\beta}_{.j} = \bar{y}_{.j} - \bar{y}_{..}$.

The most noteworthy consequence of balanced data (all $n_{ij} = 1$) is that Tables 7.3b and 7.3c become identical for *all* data. This is a most important outcome of balanced data. It means that the distinction between $R(\alpha \mid \mu)$ and

$R(\alpha \mid \mu, \beta)$ made in Tables 7.2 and 7.3 no longer occurs, because these two terms both simplify to be the same. So do $R(\beta \mid \mu, \alpha)$ and $R(\beta \mid \mu)$ in those tables: they too simplify to be identically equal. Thus when all $n_{ij} = 1$

$$R(\alpha \mid \mu) = R(\alpha \mid \mu, \beta) = b \sum_{i=1}^{a} \bar{y}_{i.}^2 - ab\bar{y}_{..}^2 = \sum_{i=1}^{a} \sum_{j=1}^{b} (\bar{y}_{i.} - \bar{y}_{..})^2$$

$$\text{and} \quad R(\beta \mid \mu, \alpha) = R(\beta \mid \mu) = a \sum_{j=1}^{b} \bar{y}_{.j}^2 - ab\bar{y}_{..}^2 = \sum_{i=1}^{a} \sum_{j=1}^{b} (\bar{y}_{.j} - \bar{y}_{..})^2, \tag{50}$$

and the analysis of variance becomes as shown in Table 7.5. The sums of squares shown there, namely (50), are familiar expressions. Furthermore, they each have a simple form, are easily calculated and do not involve any matrix manipulations such as those previously described for unbalanced data [e.g., (32) for $R(\beta \mid \mu, \alpha)$]. In addition, because there is no longer any distinction between, for example, $R(\alpha \mid \mu)$ and $R(\alpha \mid \mu, \beta)$, there is no need to distinguish between fitting "α after μ" and "α after μ and β". We are concerned solely with fitting "α after μ" and similarly "β after μ". There is but one analysis of variance table, that shown in Table 7.5, in which $R(\alpha \mid \mu)$ measures the efficacy of having the α-effects in the model and, independently, $R(\beta \mid \mu)$ measures the efficacy of having the β-effects in it.

The convenience of a single analysis of variance (Table 7.5) compared to having two analyses (Tables 7.3b and 7.3c) is obvious: for example, Table 7.4 is no longer pertinent. However, this convenience that occurs with

TABLE 7.5. ANALYSIS OF VARIANCE FOR A 2-WAY CLASSIFICATION WITH NO INTERACTION, WITH BALANCED DATA, ALL $n_{ij} = 1$. (TABLES 7.3b AND 7.3c BOTH SIMPLIFY TO THIS FORM WHEN ALL $n_{ij} = 1$)

Source of Variation	Degrees of Freedom	Sum of Squares		
Mean	1	$R(\mu)$	$= R(\mu)$	$= ab\bar{y}_{..}^2$
α after μ	$a - 1$	$R(\alpha \mid \mu)$	$= R(\alpha \mid \mu, \beta)$	$= \sum_i \sum_j (\bar{y}_{i.} - \bar{y}_{..})^2$
β after μ	$b - 1$	$R(\beta \mid \mu, \alpha)$	$= R(\beta \mid \mu)$	$= \sum_i \sum_j (\bar{y}_{.j} - \bar{y}_{..})^2$
Residual error	$(a - 1)(b - 1)$	SSE	$=$ SSE	$= \sum_i \sum_j (y_{ij} - \bar{y}_{i.} - \bar{y}_{.j} + \bar{y}_{..})^2$
Total	ab	SST	$=$ SST	$= \sum_i \sum_j y_{ij}^2$

balanced data can easily result in a misunderstanding of the analysis of unbalanced data. Students usually encounter balanced data analyses first, such as that in Table 7.5. Explanation in terms of sums of squares of means $\bar{y}_{i.}$ (and $\bar{y}_{.j}$) about the general mean $\bar{y}_{..}$ has much intuitive appeal but, unfortunately, it does not carry over to analyses of unbalanced data. It provides, for example, no explanation as to why there are two analyses of variance for unbalanced data of a 2-way classification, analyses that have different meanings and are calculated differently (*vide* Tables 7.3b and 7.3c). Furthermore, the calculations are quite different from those for balanced data, and the manner of interpreting results is different too; fitting "α after μ and β" in one analysis and "α after μ" in the other. Small wonder that a student may experience disquiet when he views this state of affairs in the light of what has been arduously learned about balanced data. The changes to be made in the analysis and its interpretation appear so large in relation to the cause of it all—having unbalanced instead of balanced data—that the relationship of the analysis for unbalanced data to that for balanced data might, at least initially, not seem at all clear. The relationship is that balanced data are a special case of unbalanced data, and not *vice versa*.

2. THE 2-WAY CLASSIFICATION WITH INTERACTION

Suppose a plant breeder carries out a series of experiments with three different fertilizer treatments on each of four varieties of grain. For each treatment-by-variety combination he plants several $4' \times 4'$ plots. At harvest time he finds that many of the plots have been lost due to being wrongly ploughed up, and all he is left with are the data of Table 7.6. With four of the treatment-variety combinations there are no data at all, and with the others there are varying numbers of plots, ranging from 1 to 4, with a total of 18 plots in all. Table 7.6 shows the yield of each plot, the total yields, the number of plots in each total and the corresponding mean, for each treatment-variety combination having data. Totals, numbers of observations (plots) and means are also shown for the three treatments, the four varieties and for all 18 plots. The symbols for the entries in the table, in terms of the model (see below), are also shown.

a. Model

The equation of a suitable linear model for analyzing data of the nature of Table 7.6 is, as discussed in Chapter 4,

$$y_{ijk} = \mu + \alpha_i + \beta_j + \gamma_{ij} + e_{ijk} \tag{51}$$

TABLE 7.6. WEIGHT[1] OF GRAIN (OUNCES) FROM 4′ × 4′ TRIAL PLOTS

Treatment	Variety 1	2	3	4	Totals
1	8 13 9 30 (3) 10[2] $y_{11.}(n_{11})\bar{y}_{11.}$		12 12 (1) 12 $y_{13.}(n_{13})\bar{y}_{13.}$	7 11 18 (2) 9 $y_{14.}(n_{14})\bar{y}_{14.}$	60 (6) 10 $y_{1..}(n_{1.})\bar{y}_{1..}$
2	6 12 18 (2) 9 $y_{21.}(n_{21})\bar{y}_{21.}$	12 14 26 (2) 13 $y_{22.}(n_{22})\bar{y}_{22.}$			44 (4) 11 $y_{2..}(n_{2.})\bar{y}_{2..}$
3		9 7 16 (2) 8 $y_{32.}(n_{32})\bar{y}_{32.}$	14 16 30 (2) 15 $y_{33.}(n_{33})\bar{y}_{33.}$	10 14 11 13 48 (4) 12 $y_{34.}(n_{34})\bar{y}_{34.}$	94 (8) $11\frac{3}{4}$ $y_{3..}(n_{3.})\bar{y}_{3..}$
Totals	48 (5) 9.6 $y_{.1.}(n_{.1})\bar{y}_{.1.}$	42 (4) $10\frac{1}{2}$ $y_{.2.}(n_{.2})\bar{y}_{.2.}$	42 (3) 14 $y_{.3.}(n_{.3})\bar{y}_{.3.}$	66 (6) 11 $y_{.4.}(n_{.4})\bar{y}_{.4.}$	198 (18) 11 $y_{...}(n_{..})\bar{y}_{...}$

[1] The basic entries in the table are weights from individual plots.
[2] In each triplet of numbers the first is a total weight, the second (in parentheses) is the number of plots in the total and the third is the mean.

for y_{ijk} as the kth observation in the ith treatment and jth variety. In (51) μ is a mean, α_i is the effect of the ith treatment, β_j is the effect of the jth variety, γ_{ij} is the interaction effect for the ith treatment and jth variety and e_{ijk} is the error term. In general we have α_i as the effect due to the ith level of the α-factor, β_j is the effect due to the jth level of the β-factor and γ_{ij} is the interaction effect due to the ith level of the α-factor and the jth level of the β-factor.

TABLE 7.6a. n_{ij}-VALUES OF TABLE 7.6

i	$j = 1$	$j = 2$	$j = 3$	$j = 4$	Totals: $n_{i.}$
1	3	0	1	2	6
2	2	2	0	0	4
3	0	2	2	4	8
Totals: $n_{.j}$	5	4	3	6	$n_{..} = 18$

In the general case there are a levels of the α-factor with $i = 1, \ldots, a$, and b levels of the β-factor, with $j = 1, \ldots, b$; in the example $a = 3$ and $b = 4$.

With balanced data every one of the ab cells of a table such as Table 7.6 would have n observations; furthermore, there would then be ab levels of the γ-factor (the interaction factor) in the data. However, with unbalanced data, when some cells have no observations, as is the case in Table 7.6, there are only as many γ-levels in the data as there are non-empty cells. Let the number of such cells be s; in Table 7.6, $s = 8$. Then, if n_{ij} is the number of observations in the (i, j)th cell (treatment i and variety j), s is the number of cells in which $n_{ij} \neq 0$; i.e., in which $n_{ij} > 0$, in fact, $n_{ij} \geq 1$. For these cells,

$$y_{ij\cdot} = \sum_{k=1}^{n_{ij}} y_{ijk}$$

is the total yield in the (i, j)th cell, and

$$\bar{y}_{ij\cdot} = y_{ij\cdot}/n_{ij}$$

is the corresponding mean. Similarly

$$y_{i\cdot\cdot} = \sum_{j=1}^{b} y_{ij\cdot} \quad \text{and} \quad n_{i\cdot} = \sum_{j=1}^{b} n_{ij}$$

are the total yield and number of observations in the ith treatment. Kindred values for the jth variety are

$$y_{\cdot j\cdot} = \sum_{i=1}^{a} y_{ij\cdot} \quad \text{and} \quad n_{\cdot j} = \sum_{i=1}^{a} n_{ij};$$

and

$$y_{\cdots} = \sum_{i=1}^{a} y_{i\cdot\cdot} = \sum_{j=1}^{b} y_{\cdot j\cdot} = \sum_{i=1}^{a}\sum_{j=1}^{b} y_{ij\cdot} = \sum_{i=1}^{a}\sum_{j=1}^{b}\sum_{k=1}^{n_{ij}} y_{ijk}$$

is the total yield for all plots, the number of observations (plots) therein being

$$n_{\cdot\cdot} = \sum_{i=1}^{a} n_{i\cdot} = \sum_{j=1}^{b} n_{\cdot j} = \sum_{i=1}^{a}\sum_{j=1}^{b} n_{ij}.$$

Examples of these symbols are shown in Table 7.6. The n_{ij} notation used here is entirely parallel with that of the previous section except that there $n_{ij} = 1$ or 0; here $n_{ij} \geq 1$ or $n_{ij} = 0$.

The model equations $\mathbf{y} = \mathbf{Xb} + \mathbf{e}$ for the data of Table 7.6 are given in (52). The headings to the $\mathbf{X}$-matrix, and the dots and partitioning therein, are all in the same style as the model equations in (2) of the preceding section.

b. Normal equations

The normal equations $\mathbf{X}'\mathbf{X}\mathbf{b}^{o} = \mathbf{X}'\mathbf{y}$ corresponding to $\mathbf{y} = \mathbf{Xb} + \mathbf{e}$ of (52) are shown in (53).

$$
\begin{bmatrix} 8 \\ 13 \\ 9 \\ 12 \\ 7 \\ 11 \\ 6 \\ 12 \\ 12 \\ 14 \\ 9 \\ 7 \\ 14 \\ 16 \\ 10 \\ 14 \\ 11 \\ 13 \end{bmatrix}
=
\begin{bmatrix} y_{111} \\ y_{112} \\ y_{113} \\ y_{131} \\ y_{141} \\ y_{142} \\ y_{211} \\ y_{212} \\ y_{221} \\ y_{222} \\ y_{321} \\ y_{322} \\ y_{331} \\ y_{332} \\ y_{341} \\ y_{342} \\ y_{343} \\ y_{344} \end{bmatrix}
=
\begin{array}{c}
\begin{array}{cccccccccccccccc} \mu & \alpha_1 & \alpha_2 & \alpha_3 & \beta_1 & \beta_2 & \beta_3 & \beta_4 & \gamma_{11} & \gamma_{13} & \gamma_{14} & \gamma_{21} & \gamma_{22} & \gamma_{32} & \gamma_{33} & \gamma_{34} \end{array} \\
\left[\begin{array}{c|ccc|cccc|cccccccc}
1 & 1 & . & . & 1 & . & . & . & 1 & . & . & . & . & . & . & . \\
1 & 1 & . & . & 1 & . & . & . & 1 & . & . & . & . & . & . & . \\
1 & 1 & . & . & 1 & . & . & . & 1 & . & . & . & . & . & . & . \\
1 & 1 & . & . & . & . & 1 & . & . & 1 & . & . & . & . & . & . \\
1 & 1 & . & . & . & . & . & 1 & . & . & 1 & . & . & . & . & . \\
1 & 1 & . & . & . & . & . & 1 & . & . & 1 & . & . & . & . & . \\
1 & . & 1 & . & 1 & . & . & . & . & . & . & 1 & . & . & . & . \\
1 & . & 1 & . & 1 & . & . & . & . & . & . & 1 & . & . & . & . \\
1 & . & 1 & . & . & 1 & . & . & . & . & . & . & 1 & . & . & . \\
1 & . & 1 & . & . & 1 & . & . & . & . & . & . & 1 & . & . & . \\
1 & . & . & 1 & . & 1 & . & . & . & . & . & . & . & 1 & . & . \\
1 & . & . & 1 & . & 1 & . & . & . & . & . & . & . & 1 & . & . \\
1 & . & . & 1 & . & . & 1 & . & . & . & . & . & . & . & 1 & . \\
1 & . & . & 1 & . & . & 1 & . & . & . & . & . & . & . & 1 & . \\
1 & . & . & 1 & . & . & . & 1 & . & . & . & . & . & . & . & 1 \\
1 & . & . & 1 & . & . & . & 1 & . & . & . & . & . & . & . & 1 \\
1 & . & . & 1 & . & . & . & 1 & . & . & . & . & . & . & . & 1 \\
1 & . & . & 1 & . & . & . & 1 & . & . & . & . & . & . & . & 1
\end{array}\right]
\end{array}
\begin{bmatrix} \mu \\ \hline \alpha_1 \\ \alpha_2 \\ \alpha_3 \\ \hline \beta_1 \\ \beta_2 \\ \beta_3 \\ \beta_4 \\ \hline \gamma_{11} \\ \gamma_{13} \\ \gamma_{14} \\ \gamma_{21} \\ \gamma_{22} \\ \gamma_{32} \\ \gamma_{33} \\ \gamma_{34} \end{bmatrix}
+
\begin{bmatrix} e_{111} \\ e_{112} \\ e_{113} \\ e_{131} \\ e_{141} \\ e_{142} \\ e_{211} \\ e_{212} \\ e_{221} \\ e_{222} \\ e_{321} \\ e_{322} \\ e_{331} \\ e_{332} \\ e_{341} \\ e_{342} \\ e_{343} \\ e_{344} \end{bmatrix}. \qquad (52)
$$

$$
\begin{array}{c|ccc|cccc|cccccccc}
\mu^o & \alpha_1^o & \alpha_2^o & \alpha_3^o & \beta_1^o & \beta_2^o & \beta_3^o & \beta_4^o & \gamma_{11}^o & \gamma_{13}^o & \gamma_{14}^o & \gamma_{21}^o & \gamma_{22}^o & \gamma_{32}^o & \gamma_{33}^o & \gamma_{34}^o \\
\hline
18 & 6 & 4 & 8 & 5 & 4 & 3 & 6 & 3 & 1 & 2 & 2 & 2 & 2 & 2 & 4 \\
\hline
6 & 6 & \cdot & \cdot & 3 & \cdot & 1 & 2 & 3 & 1 & 2 & \cdot & \cdot & \cdot & \cdot & \cdot \\
4 & \cdot & 4 & \cdot & 2 & 2 & \cdot & \cdot & \cdot & \cdot & \cdot & 2 & 2 & \cdot & \cdot & \cdot \\
8 & \cdot & \cdot & 8 & \cdot & 2 & 2 & 4 & \cdot & \cdot & \cdot & \cdot & \cdot & 2 & 2 & 4 \\
\hline
5 & 3 & 2 & \cdot & 5 & \cdot & \cdot & \cdot & 3 & \cdot & \cdot & 2 & \cdot & \cdot & \cdot & \cdot \\
4 & \cdot & 2 & 2 & \cdot & 4 & \cdot & \cdot & \cdot & \cdot & \cdot & \cdot & 2 & 2 & \cdot & \cdot \\
3 & 1 & \cdot & 2 & \cdot & \cdot & 3 & \cdot & \cdot & 1 & \cdot & \cdot & \cdot & \cdot & 2 & \cdot \\
6 & 2 & \cdot & 4 & \cdot & \cdot & \cdot & 6 & \cdot & \cdot & 2 & \cdot & \cdot & \cdot & \cdot & 4 \\
\hline
3 & 3 & \cdot & \cdot & 3 & \cdot & \cdot & \cdot & 3 & \cdot & \cdot & \cdot & \cdot & \cdot & \cdot & \cdot \\
1 & 1 & \cdot & \cdot & \cdot & \cdot & 1 & \cdot & \cdot & 1 & \cdot & \cdot & \cdot & \cdot & \cdot & \cdot \\
2 & 2 & \cdot & \cdot & \cdot & \cdot & \cdot & 2 & \cdot & \cdot & 2 & \cdot & \cdot & \cdot & \cdot & \cdot \\
2 & \cdot & 2 & \cdot & 2 & \cdot & \cdot & \cdot & \cdot & \cdot & \cdot & 2 & \cdot & \cdot & \cdot & \cdot \\
2 & \cdot & 2 & \cdot & \cdot & 2 & \cdot & \cdot & \cdot & \cdot & \cdot & \cdot & 2 & \cdot & \cdot & \cdot \\
2 & \cdot & \cdot & 2 & \cdot & 2 & \cdot & \cdot & \cdot & \cdot & \cdot & \cdot & \cdot & 2 & \cdot & \cdot \\
2 & \cdot & \cdot & 2 & \cdot & \cdot & 2 & \cdot & \cdot & \cdot & \cdot & \cdot & \cdot & \cdot & 2 & \cdot \\
4 & \cdot & \cdot & 4 & \cdot & \cdot & \cdot & 4 & \cdot & \cdot & \cdot & \cdot & \cdot & \cdot & \cdot & 4
\end{array}
\begin{bmatrix} \mu^o \\ \hline \alpha_1^o \\ \alpha_2^o \\ \alpha_3^o \\ \hline \beta_1^o \\ \beta_2^o \\ \beta_3^o \\ \beta_4^o \\ \hline \gamma_{11}^o \\ \gamma_{13}^o \\ \gamma_{14}^o \\ \gamma_{21}^o \\ \gamma_{22}^o \\ \gamma_{32}^o \\ \gamma_{33}^o \\ \gamma_{34}^o \end{bmatrix}
=
\begin{bmatrix} y_{\cdot\cdot\cdot} \\ \hline y_{1\cdot\cdot} \\ y_{2\cdot\cdot} \\ y_{3\cdot\cdot} \\ \hline y_{\cdot 1\cdot} \\ y_{\cdot 2\cdot} \\ y_{\cdot 3\cdot} \\ y_{\cdot 4\cdot} \\ \hline y_{11\cdot} \\ y_{13\cdot} \\ y_{14\cdot} \\ y_{21\cdot} \\ y_{22\cdot} \\ y_{32\cdot} \\ y_{33\cdot} \\ y_{34\cdot} \end{bmatrix}
=
\begin{bmatrix} 198 \\ \hline 60 \\ 44 \\ 94 \\ \hline 48 \\ 42 \\ 42 \\ 66 \\ \hline 30 \\ 12 \\ 18 \\ 18 \\ 26 \\ 16 \\ 30 \\ 48 \end{bmatrix}
\qquad (53)
$$

Properties of normal equations described in Section 6.4 are evident here. The first row of $\mathbf{X}'\mathbf{X}$, corresponding to the μ-equation, has $n_{..}$ and the $n_{i.}$-, $n_{.j}$- and n_{ij}-values as does the first column and the diagonal. The remaining elements of $\mathbf{X}'\mathbf{X}$, other than zeros, are the n_{ij}-values; and the elements of $\mathbf{X}'\mathbf{y}$ on the righthand side are all the totals $y_{...}$, $y_{i..}$, $y_{.j.}$ and $y_{ij.}$ shown in Table 7.6. As before, the partitioning of $\mathbf{X}'\mathbf{X}$ highlights its form.

c. Solving the normal equations

In contrast to the preceding section, but similar to the examples of Secs. 6.2 and 6.4, the normal equations typified by (53) are easily solved. The number of equations is $p = 1 + a + b + s = 1 + 3 + 4 + 8 = 16$, in (53). But, as is evident in (53), the sum of the α-equations (the three after the first) is identical to the μ-equation; so is the sum of the β-equations. The consequences of this are 2 linear relationships among the rows of $\mathbf{X}'\mathbf{X}$. Also, in the γ-equations the sum of those pertaining to $\gamma_{i'j}$ summed over j equals the $\alpha_{i'}$-equation; e.g., the γ_{11}-, γ_{13}- and γ_{14}-equations sum to the α_1-equation. This is true for all $i' = 1, 2, \ldots, a$, representing further linear relationships, a of them, among rows of $\mathbf{X}'\mathbf{X}$. Similarly, in summing the $\gamma_{ij'}$-equations over i the $\beta_{j'}$-equation is obtained, for all $j' = 1, \ldots, b$. However, of the b relationships represented here, only $b - 1$ of them are linearly independent of those already described, so that the total number of linearly independent relationships is $1 + 1 + a + b - 1 = 1 + a + b$. Hence the rank of $\mathbf{X}'\mathbf{X}$ is $r = 1 + a + b + s - (1 + a + b) = s$. Therefore, in terms of solving the normal equations by the procedure described in (4) of Chapter 6, we set $p - r = 1 + a + b + s - s = 1 + a + b$ elements of $\mathbf{b}^o$ equal to zero. The easiest elements for this purpose are μ^o, all α_i^o (a of them) and all β_j^o (b of them). Setting these equal to zero leaves, from (53), the $s = 8$ equations

$$\begin{aligned} 3\gamma_{11}^o &= 30, & 2\gamma_{22}^o &= 26 \\ \gamma_{13}^o &= 12, & 2\gamma_{32}^o &= 16 \\ 2\gamma_{14}^o &= 18, & 2\gamma_{33}^o &= 30 \\ 2\gamma_{21}^o &= 18, & 4\gamma_{34}^o &= 48. \end{aligned}$$

In general these reduced equations are

$$n_{ij}\gamma_{ij}^o = y_{ij.}$$

with solution

$$\gamma_{ij}^o = \bar{y}_{ij.}\,, \tag{54}$$

for the (i, j)-cells for which $n_{ij} \neq 0$, all s of them. This then, (54), is a solution for $\mathbf{b}^o$: every element is zero except γ_{ij}^o, which takes the value $\bar{y}_{ij.}$, the cell mean. Clearly the solution is simple:

$$\mathbf{b}^{o\prime} = [\mathbf{0}_{1\times(1+a+b)} \quad \bar{\mathbf{y}}'] \tag{55}$$

where $(\bar{\mathbf{y}}')_{1\times s}$ = a vector of all $\bar{y}_{ij\cdot}$'s for which $n_{ij} \neq 0$.

In our example

$$\mathbf{b}^{o\prime} = [0\ 0\ 0\ 0\ 0\ 0\ 0\ 0\ 0\ \bar{y}_{11\cdot}\ \bar{y}_{13\cdot}\ \bar{y}_{14\cdot}\ \bar{y}_{21\cdot}\ \bar{y}_{22\cdot}\ \bar{y}_{32\cdot}\ \bar{y}_{33\cdot}\ \bar{y}_{34\cdot}] \qquad (56)$$

$$= [0\ 0\ 0\ 0\ 0\ 0\ 0\ 0\ 0\ 10\ 12\ 9\ 9\ 13\ 8\ 15\ 12]$$

from Table 7.6.

The simplicity of this solution means that it is virtually unnecessary to derive the generalized inverse of $\mathbf{X}'\mathbf{X}$ that corresponds to $\mathbf{b}^o$. That generalized inverse is, as is evident from (55) and the normal equations (53),

$$\mathbf{G} = \begin{bmatrix} \mathbf{0}_{(1+a+b)\times(1+a+b)} & \mathbf{0}_{(1+a+b)\times s} \\ \mathbf{0}_{s\times(1+a+b)} & \mathbf{D}\{1/n_{ij}\} \end{bmatrix} \qquad (57)$$

where $\mathbf{D}\{1/n_{ij}\}$ is a diagonal matrix of order s of the values $1/n_{ij}$ for the non-zero n_{ij}.

d. Analysis of variance

(*i*) *Basic calculations.* The analysis of variance for the 2-way classification model with interaction is similar to that for the 2-way classification without interaction, discussed in the preceding section. Indeed, the analysis of variance tables are just like those of Tables 7.2 and 7.3, except for the inclusion of an interaction line corresponding to the sum of squares $R(\gamma \mid \mu, \alpha, \beta)$. Calculation of $R(\mu)$, $R(\mu, \alpha)$, $R(\mu, \beta)$ and $R(\mu, \alpha, \beta)$ is the same, except for using $y_{\cdots}$, $y_{i\cdot\cdot}$, $y_{\cdot j\cdot}$ and $y_{ij\cdot}$, respectively, in place of $y_{\cdot\cdot}$, $y_{i\cdot}$, $y_{\cdot j}$ and y_{ij} used in the no-interaction model. Thus, similar to (24), (25) and (36)

$$R(\mu) = n_{\cdot\cdot}\bar{y}^2_{\cdots} = y^2_{\cdots}/n_{\cdot\cdot}, \qquad (58)$$

$$R(\mu, \alpha) = \sum_{i=1}^{a} n_{i\cdot}\bar{y}^2_{i\cdot\cdot} = \sum_{i=1}^{a} y^2_{i\cdot\cdot}/n_{i\cdot}, \qquad (59)$$

and

$$R(\mu, \beta) = \sum_{j=1}^{b} n_{\cdot j}\bar{y}^2_{\cdot j\cdot} = \sum_{j=1}^{b} y^2_{\cdot j\cdot}/n_{\cdot j}. \qquad (60)$$

The model (51) involves the terms μ, α_i, β_j and γ_{ij}. The sum of squares for fitting it is therefore denoted by $R(\mu, \alpha, \beta, \gamma)$ and its value is, as usual, $\mathbf{b}^{o\prime}\mathbf{X}'\mathbf{y}$. With $\mathbf{X}'\mathbf{y}$ and $\mathbf{b}^{o\prime}$ of (53) and (55), respectively, this gives

$$\begin{aligned} R(\mu, \alpha, \beta, \gamma) &= \mathbf{b}^{o\prime}\mathbf{X}'\mathbf{y} \\ &= \bar{\mathbf{y}}'(\text{column vector of } y_{ij\cdot} \text{ totals}) \\ &= \sum_{i=1}^{a}\sum_{j=1}^{b} \bar{y}_{ij\cdot}y_{ij\cdot} \\ &= \sum_{i=1}^{a}\sum_{j=1}^{b} n_{ij}\bar{y}^2_{ij\cdot} = \sum_{i=1}^{a}\sum_{j=1}^{b} y^2_{ij\cdot}/n_{ij}. \end{aligned} \qquad (61)$$

In the second expression of (61) the terms $y_{ij\cdot}^2/n_{ij}$ are defined only for non-zero values of n_{ij} in the data.

The other term needed for the analysis is $R(\mu, \alpha, \beta)$, the sum of squares due to fitting

$$y_{ijk} = \mu + \alpha_i + \beta_j + e_{ijk} \,. \tag{62}$$

This is derived exactly as in equation (26). Thus

$$R(\mu, \alpha, \beta) = \sum_{i=1}^{a} n_{i\cdot}\bar{y}_{i\cdot\cdot}^2 + \mathbf{r}'\mathbf{C}^{-1}\mathbf{r} \tag{63}$$

where $$\mathbf{C} = \{c_{jj'}\} \qquad \text{for } j, j' = 1, 2, \ldots, b-1 \tag{64}$$

with $$c_{jj} = n_{\cdot j} - \sum_{i=1}^{a} n_{ij}^2/n_{i\cdot}\,,$$

$$c_{jj'} = -\sum_{i=1}^{a} n_{ij}n_{ij'}/n_{i\cdot} \qquad \text{for } j \neq j'$$

and $$\mathbf{r} = \{r_j\} = \left\{y_{\cdot j\cdot} - \sum_{i=1}^{a} n_{ij}\bar{y}_{i\cdot\cdot}\right\} \qquad \text{for } j = 1, 2, \ldots, b-1. \tag{65}$$

These are the same calculations as in (16)–(18), using $\bar{y}_{i\cdot\cdot}$ and $y_{\cdot j\cdot}$ in place of $\bar{y}_{i\cdot}$ and $y_{\cdot j}$.

Example. Calculation of (58)–(61) for the data of Table 7.6 is as follows:

$$\begin{aligned} R(\mu) &= 198^2/18 &&= 2{,}178 \\ R(\mu, \alpha) &= 60^2/6 + 44^2/4 + 94^2/8 &&= 2{,}188.5 \\ R(\mu, \beta) &= 48^2/5 + 42^2/4 + 42^2/3 + 66^2/6 &&= 2{,}215.8 \\ R(\mu, \alpha, \beta, \gamma) &= 30^2/3 + 12^2/1 + \cdots + 30^2/2 + 48^2/4 &&= 2{,}260; \end{aligned} \tag{66}$$

and the total sum of squares is, as usual,

$$\sum y^2 = \sum_{i=1}^{a}\sum_{j=1}^{b}\sum_{k=1}^{n_{ij}} y_{ijk}^2 = 8^2 + 13^2 + \cdots + 11^2 + 13^2 = 2{,}316. \tag{67}$$

To facilitate calculation of $R(\mu, \alpha, \beta)$ we use the table of n_{ij}'s shown in Table 7.6a. From this, $\mathbf{C}$ of (64) is

$$\mathbf{C} = \begin{bmatrix} 5 - (3^2/6 + 2^2/4) & -2(2)/4 & -1(3)/6 \\ -2(2)/4 & 4 - (2^2/4 + 2^2/8) & -2(2)/8 \\ -1(3)/6 & -2(2)/8 & 3 - (1^2/6 + 2^2/8) \end{bmatrix}$$

$$= \frac{1}{6}\begin{bmatrix} 15 & -6 & -3 \\ -6 & 15 & -3 \\ -3 & -3 & 14 \end{bmatrix}$$

TABLE 7.7. ANALYSES OF VARIANCE FOR 2-WAY CLASSIFICATION WITH INTERACTION (DATA OF TABLE 7.6)

Source of Variation		Degrees of Freedom		Sum of Squares		Mean Square	F-statistic	
Table 7.7a: For fitting μ, and $\alpha, \beta, \gamma \mid \mu$								
Mean	: μ	1	= 1	$R(\mu)$	= 2,178	2,178	$F(M)$	= 388.9
α, β and γ after μ:	$\alpha, \beta, \gamma \mid \mu$	$s - 1$	= 7	$R(\alpha, \beta, \gamma \mid \mu)$	= 82	11.71	$F(R_m)$	= 2.1
Residual error		$N - s$	= 10	SSE	= 56	5.60		
Total		N	= 18	SST	= 2,316			
Table 7.7b: For fitting μ, then α, then β and then γ: i.e., for fitting μ, $(\alpha \mid \mu)$, $(\beta \mid \mu, \alpha)$ and $(\gamma \mid \mu, \alpha, \beta)$.								
Mean	: μ	1	= 1	$R(\mu)$	= 2,178	2,178	$F(M)$	= 388.9
α after μ	: $\alpha \mid \mu$	$a - 1$	= 2	$R(\alpha \mid \mu)$	= $10\frac{35}{70}$	5.25	$F(\alpha \mid \mu)$	= 0.9
β after μ and α	: $\beta \mid \mu, \alpha$	$b - 1$	= 3	$R(\beta \mid \mu, \alpha)$	= $36\frac{55}{70}$	12.26	$F(\beta \mid \mu, \alpha)$	= 2.2
γ after μ, α and β:	$\gamma \mid \mu, \alpha, \beta$	$s - a - b + 1$	= 2	$R(\gamma \mid \mu, \alpha, \beta)$	= $34\frac{50}{70}$	17.36	$F(\gamma \mid \mu, \alpha, \beta)$	= 3.1
Residual error		$N - s$	= 10	SSE	= 56	5.60		
Total		N	= 18	SST	= 2,316			
Table 7.7c: For fitting μ, then β, then α and then γ: i.e., for fitting μ, $(\beta \mid \mu)$, $(\alpha \mid \mu, \beta)$ and $(\gamma \mid \mu, \alpha, \beta)$.								
Mean	: μ	1	= 1	$R(\mu)$	= 2,178	2,178	$F(M)$	= 388.9
β after μ	: $\beta \mid \mu$	$b - 1$	= 3	$R(\beta \mid \mu)$	= $37\frac{56}{70}$	12.60	$F(\beta \mid \mu)$	= 2.2
α after μ and β	: $\alpha \mid \mu, \beta$	$a - 1$	= 2	$R(\alpha \mid \mu, \beta)$	= $9\frac{34}{70}$	4.74	$F(\alpha \mid \mu, \beta)$	= 0.8
γ after μ, β and α:	$\gamma \mid \mu, \alpha, \beta$	$s - a - b + 1$	= 2	$R(\gamma \mid \mu, \alpha, \beta)$	= $34\frac{50}{70}$	17.36	$F(\gamma \mid \mu, \alpha, \beta)$	= 3.1
Residual error		$N - s$	= 10	SSE	= 56	5.60		
Total		N	= 18	SST	= 2,316			

with
$$\mathbf{C}^{-1} = \frac{1}{126}\begin{bmatrix} 67 & 31 & 21 \\ 31 & 67 & 21 \\ 21 & 21 & 63 \end{bmatrix}.$$

And from Table 7.6 and (65)

$$\mathbf{r} = \begin{bmatrix} 48 - 3(10) - 2(11) \\ 42 - 2(11) - 2(11\tfrac{3}{4}) \\ 42 - 1(10) - 2(11\tfrac{3}{4}) \end{bmatrix} = \begin{bmatrix} -4 \\ -3\tfrac{1}{2} \\ 8\tfrac{1}{2} \end{bmatrix}.$$

Therefore (63), using $R(\mu, \alpha)$ from (66), gives

$$R(\mu, \alpha, \beta) = 2188.5 + [-4 \quad -3\tfrac{1}{2} \quad 8\tfrac{1}{2}]\frac{1}{126}\begin{bmatrix} 67 & 31 & 21 \\ 31 & 67 & 21 \\ 21 & 21 & 63 \end{bmatrix}\begin{bmatrix} -4 \\ -3\tfrac{1}{2} \\ 8\tfrac{1}{2} \end{bmatrix}$$

$$\begin{aligned} &= 2{,}188.5 + \tfrac{1}{4}(1/126)[67(64 + 49) + 63(289) + 62(56) + 42(-15)17] \\ &= 2{,}188\tfrac{1}{2} + 36\tfrac{11}{14} \\ &= 2{,}225\tfrac{2}{7}. \end{aligned} \tag{68}$$

If, quite generally, one wishes to fit the model (62) *ab initio* to data of the nature illustrated in Table 7.6, the procedure just outlined yields the sum of squares for so doing, namely $R(\mu, \alpha, \beta)$. Thus the procedure as described in the preceding section for calculating $R(\mu, \alpha, \beta)$ for the no-interaction model with $n_{ij} = 0$ or 1 is also the basis for calculating $R(\mu, \alpha, \beta)$ whenever data are unbalanced, either when $R(\mu, \alpha, \beta)$ is needed as part of the analysis of variance for the interaction model (51) or when prime interest lies in $R(\mu, \alpha, \beta)$ itself as the reduction in sum of squares due to fitting the no-interaction model (64).

(*ii*) *Fitting different models.* Analyses of variance derived from the sums of squares in (66), (67) and (68) are shown in Table 7.7. Their form is similar to that of Table 7.2. Table 7.7a shows the partitioning of the sum of squares $R(\mu, \alpha, \beta, \gamma)$ into two parts: $R(\mu)$ for fitting only a mean and $R(\alpha, \beta, \gamma \mid \mu)$ for fitting the α-, β- and γ-factors after the mean. For this, $R(\mu)$ is as shown in (66), which yields

$$R(\alpha, \beta, \gamma \mid \mu) = R(\mu, \alpha, \beta, \gamma) - R(\mu) = 2{,}260 - 2{,}178 = 82.$$

Also, the residual error sum of squares is, in the usual manner,

$$\text{SSE} = \Sigma\, y^2 - R(\mu, \alpha, \beta, \gamma) = 2{,}316 - 2{,}260 = 56.$$

These are the terms shown in Table 7.7a. The corresponding F-statistics are also shown. $F(M) = 388.9$ is significant because it exceeds 4.96, the 5% value of the $F_{1,10}$-distribution. Hence we reject the hypothesis H: $E(\bar{y}) = 0$.

On the other hand, $F(R_m) = 2.1$ is less than the 5% value of the $F_{7,10}$-distribution, namely 3.14, and so we conclude that the α-, β- and γ-factors in the model are not effective in explaining variation in the y's over and above that explained by fitting a mean.

The data in Table 7.6 are hypothetical, and in the Table 7.7a analysis of variance $F(R_m) = 2.1$ is not significant. Calculation of the analyses of variance shown in Tables 7.7b and 7.7c is therefore not necessary. Nevertheless, it is instructive to examine the format of these analyses, to see how similar they are to Tables 7.2b and 7.2c. Were $F(R_m)$ of Table 7.7a significant, we would be led to examine whether it was the α-factor, the β-factor, the γ-factor or some combination thereof that was contributing to this significance. After fitting μ, this could be done in one of two ways: either fit μ and α, and then μ, α and β, or fit μ and β, and then μ, α, and β. Either way, γ would be fitted after having fitted μ, α and β. Therefore choice lies in first fitting, after μ, either α or β. This is exactly the situation discussed when describing Table 7.2. Tables 7.7b and 7.7c are therefore similar in format to Tables 7.2b and 7.2c.

Table 7.7b shows the partitioning of $R(\mu, \alpha, \beta, \gamma)$ for fitting μ, then α, then β and then γ, with lines in the analysis of variance for μ, α after μ, β after μ and α, and finally γ after μ, α and β. The sole difference of this from Table 7.2b is the line for the sum of squares due to fitting γ after μ, α and β. This corresponds, of course, to the γ-factor being additional in the interaction model to the α- and β-factors that are in both the interaction and the no-interaction models. The sums of squares for Table 7.7b are, using (66) and (68),

$$R(\mu) = R(\mu) = 2{,}178 = 2{,}178,$$

$$R(\alpha \mid \mu) = R(\mu, \alpha) - R(\mu) = 2{,}188\tfrac{1}{2} - 2{,}178 = 10\tfrac{1}{2},$$

$$R(\beta \mid \mu, \alpha) = R(\mu, \alpha, \beta) - R(\mu, \alpha) = 2{,}225\tfrac{2}{7} - 2{,}188\tfrac{1}{2} = 36\tfrac{11}{14},$$

and

$$R(\gamma \mid \mu, \alpha, \beta) = R(\mu, \alpha, \beta, \gamma) - R(\mu, \alpha, \beta) = 2{,}260 - 2{,}225\tfrac{2}{7} = 34\tfrac{5}{7}.$$

Clearly, these sums of squares add to $R(\mu, \alpha, \beta, \gamma) = 2{,}260$, and are thus a partitioning of this sum of squares. These results are shown in Table 7.7b, with a denominator of 70 for the rational fractions in order to have conformity with Table 7.7c. Naturally, $R(\mu) = 2{,}178$ and $\text{SSE} = \Sigma y^2 - R(\mu, \alpha, \beta, \gamma) = 56$ are the same throughout Table 7.7. Also, the middle three entries of Table 7.7b add to $R(\alpha, \beta, \gamma \mid \mu) = 82$, the middle entry of 7.7a, as do the middle three entries of Table 7.7c. In this way Tables 7.7b and 7.7c are partitionings not only of $R(\mu, \alpha, \beta, \gamma)$ but also of $R(\alpha, \beta, \gamma \mid \mu)$, the sum of squares due to fitting the model over and above the mean.

The analogy between Tables 7.7b and 7.2b is repeated in Tables 7.7c and 7.2c, corresponding to fitting μ, then β, then α and then γ. Thus Table 7.7c has lines in the analysis of variance for μ, $(\beta \mid \mu)$, $(\alpha \mid \mu, \beta)$ and $(\gamma \mid \mu, \alpha, \beta)$. The only difference from Table 7.7b is that $R(\alpha \mid \mu)$ and $R(\beta \mid \mu, \alpha)$ in Table 7.7b are replaced by $R(\beta \mid \mu)$ and $R(\alpha \mid \mu, \beta)$ in Table 7.7c:

$$R(\beta \mid \mu) = R(\mu, \beta) - R(\mu) \qquad = 2{,}215.8 - 2{,}178 \quad = 37\tfrac{56}{70}$$

and $R(\alpha \mid \mu, \beta) = R(\mu, \alpha, \beta) - R(\mu, \beta) = 2{,}225\tfrac{2}{7} - 2{,}215.8 = 9\tfrac{34}{70}$.

The sum of these is, of course, the same as the sum of $R(\alpha \mid \mu)$ and $R(\beta \mid \mu, \alpha)$ in Table 7.7b:

$$R(\beta \mid \mu) + R(\alpha \mid \mu, \beta) = R(\mu, \alpha, \beta) - R(\mu) = R(\alpha \mid \mu) + R(\beta \mid \mu, \alpha);$$

i.e., $$37\tfrac{56}{70} + 9\tfrac{34}{70} = 2{,}225\tfrac{2}{7} - 2{,}178 = 10\tfrac{35}{70} + 36\tfrac{55}{70} = 47\tfrac{2}{7},$$

this sum being

$$R(\mu, \alpha, \beta) - R(\mu) = R(\alpha, \beta \mid \mu).$$

(*iii*) *Computational alternatives.* Equation (63) for $R(\mu, \alpha, \beta)$ is based upon solving the normal equations for the model (62) by "absorbing" the α-equations and solving for $(b - 1)$ β's. This is the procedure described in detail for the no-interaction model in Sec. 1d. As mentioned there, without explicit presentation of details, $R(\mu, \alpha, \beta)$ can also be calculated by solving the normal equations through "absorbing" the β-equations and solving for $(a - 1)$ α's. The calculation of $R(\mu, \alpha, \beta)$ is then as follows.

$$R(\mu, \alpha, \beta) = \sum_{j=1}^{b} n_{.j}\bar{y}^2_{.j.} + \mathbf{u}'\mathbf{T}^{-1}\mathbf{u} \tag{69}$$

where $$\mathbf{T} = \{t_{ii'}\} \qquad \text{for} \quad i, i' = 1, 2, \ldots, a - 1$$

with $$t_{ii} = n_{i.} - \sum_{j=1}^{b} n^2_{ij}/n_{.j}\,,$$

$$t_{ii'} = -\sum_{j=1}^{b} n_{ij}n_{i'j}/n_{.j} \qquad \text{for} \quad i \neq i' \tag{70}$$

and $$\mathbf{u} = \{u_i\} = \left\{y_{i..} - \sum_{j=1}^{b} n_{ij}\bar{y}_{.j.}\right\} \qquad \text{for} \quad i = 1, 2, \ldots, a - 1.$$

Table 7.6 involves 3 α's and 4 β's. For these data it is therefore computationally easier to use (69) instead of (63) for calculating $R(\mu, \alpha, \beta)$ because, in (69), $\mathbf{T}$ has order 2 whereas in (63) $\mathbf{C}$ has order 3. The difference in effort is negligible here, but were there to be many more β's than α's the choice of procedure might be crucial. [This (see Chapter 10) can be the case in variance components analysis, where there may be, say, 2,000 β's and only 12 α's. Then (69), requiring inversion of a matrix of order 11, is clearly preferable to (63), which demands inverting a matrix of order 1,999!]

The two alternative procedures for calculating $R(\mu, \alpha, \beta)$ provide identical numerical results, but different symbolic expressions for certain of the sums of squares in Table 7.7. These expressions are shown in Table 7.8 under the

TABLE 7.8. EQUIVALENT EXPRESSIONS FOR SUMS OF SQUARES IN THE ANALYSIS OF VARIANCE OF THE 2-WAY CLASSIFICATION WITH INTERACTION

Sum of Squares	Degrees of Freedom[1]	Method: Absorbing α's (Use when more α's than β's) See (63) for $\mathbf{r}'\mathbf{C}^{-1}\mathbf{r}$	Method: Absorbing β's (Use when more β's than α's) See (69) for $\mathbf{u}'\mathbf{T}^{-1}\mathbf{u}$
		Fitting α before β (Table 7.7b)	
$R(\mu)$	1	$n_{..}\bar{y}^2_{...}$	$n_{..}\bar{y}^2_{...}$
$R(\alpha \mid \mu)$	$a-1$	$\sum_i n_{i.}\bar{y}^2_{i..} - n_{..}\bar{y}^2_{...}$	$\sum_i n_{i.}\bar{y}^2_{i..} - n_{..}\bar{y}^2_{...}$
$R(\beta \mid \mu, \alpha)$	$b-1$	$\mathbf{r}'\mathbf{C}^{-1}\mathbf{r}$	$\sum_j n_{.j}\bar{y}^2_{.j.} + \mathbf{u}'\mathbf{T}^{-1}\mathbf{u} - \sum_i n_{i.}\bar{y}^2_{i..}$
$R(\gamma \mid \mu, \alpha, \beta)$	$s-a-b+1$	$\sum_i\sum_j n_{ij}\bar{y}^2_{ij.} - \sum_i n_{i.}\bar{y}^2_{i..} - \mathbf{r}'\mathbf{C}^{-1}\mathbf{r}$	$\sum_i\sum_j n_{ij}\bar{y}^2_{ij.} - \sum_j n_{.j}\bar{y}^2_{.j.} - \mathbf{u}'\mathbf{T}^{-1}\mathbf{u}$
SSE	$N-s$	$\sum_i\sum_j\sum_k y^2_{ijk} - \sum_i\sum_j n_{ij}\bar{y}^2_{ij.}$	$\sum_i\sum_j\sum_j y^2_{ijk} - \sum_i\sum_j n_{ij}\bar{y}^2_{ij.}$
SST	N	$\sum_i\sum_j\sum_k y^2_{ijk}$	$\sum_i\sum_j\sum_k y^2_{ijk}$

		Fitting β before α (Table 7.7c)	
$R(\mu)$	1	$n_{..}\bar{y}^2_{...}$	$n_{..}\bar{y}^2_{...}$
$R(\beta \mid \mu)$	$b-1$	$\sum_j n_{.j}\bar{y}^2_{.j.} - n_{..}\bar{y}^2_{...}$	$\sum_j n_{.j}\bar{y}^2_{.j.} - n_{..}\bar{y}^2_{...}$
$R(\alpha \mid \mu, \beta)$	$a-1$	$\sum_i n_{i.}\bar{y}^2_{i..} + \mathbf{r}'\mathbf{C}^{-1}\mathbf{r} - \sum_j n_{.j}\bar{y}^2_{.j.}$	$\mathbf{u}'\mathbf{T}^{-1}\mathbf{u}$
$R(\gamma \mid \mu, \alpha, \beta)$	$s-a-b+1$	$\sum_i\sum_j n_{ij}\bar{y}^2_{ij.} - \sum_i n_{i.}\bar{y}^2_{i..} - \mathbf{r}'\mathbf{C}^{-1}\mathbf{r}$	$\sum_i\sum_j n_{ij}\bar{y}^2_{ij.} - \sum_j n_{.j}\bar{y}^2_{.j.} - \mathbf{u}'\mathbf{T}^{-1}\mathbf{u}$
SSE	$N-s$	$\sum_i\sum_j\sum_k y^2_{ijk} - \sum_i\sum_j n_{ij}\bar{y}^2_{ij.}$	$\sum_i\sum_j\sum_k y^2_{ijk} - \sum_j\sum_j n_{ij}\bar{y}^2_{ij.}$
SST	N	$\sum_i\sum_j\sum_k y^2_{ijk}$	$\sum_i\sum_j\sum_k y^2_{ijk}$

[1] s = number of filled cells.

headings "Absorbing α's" and "Absorbing β's", which describe the method for solving the normal equations implicit in the procedures. Only one of these procedures need be used on any given set of data (although the other always provides a check on the arithmetic involved). The choice of which to use depends on whether there are more or fewer α's than β's. Indeed, even this choice is avoided if we always denote by α the factor which has the larger number of effects. The "Absorbing α's" procedure will then be the one to use. Nevertheless, it is of interest to have the two sets of expressions laid out as they are in Table 7.8.

(*iv*) *Interpretation of results.* The F-statistics in Tables 7.7b and 7.7c, other than $F(M)$, are not significant, as would be expected from the non-significance of $F(R_m)$ in Table 7.7a (see end of Sec. 1e). In general, interpretation of the test statistics $F(\alpha \mid \mu)$, $F(\beta \mid \mu, \alpha)$, $F(\beta \mid \mu)$ and $F(\alpha \mid \mu, \beta)$ in Tables 7.7b and 7.7c is exactly as given in Table 7.4. The possibilities so far as significance and non-significance of the F's is concerned are the same here as there, and interpretation is therefore the same. In addition, Tables 7.7b and 7.7c both have the statistic $F(\gamma \mid \mu, \alpha, \beta)$, which provides a test of the effectiveness (in terms of accounting for variation in y) of fitting the model (51) compared to fitting the model (1). As the difference between the two models is the inclusion of the interaction effects γ_{ij} in (51) the test is often referred to as a test of interaction after fitting main effects. However, as in Table 7.2, so in Table 7.7: interpretation of the F-statistics can be thought of in two ways. The first, already considered, is that of testing the effectiveness of fitting different models, while the second is that of testing linear hypotheses about elements of the model. It is in this latter context that we are better able to consider the meaning of the tests provided by the F's in Table 7.7. First, however, we deal with a limitation on the $R(\quad)$ notation and then, in order to discuss tests of hypotheses, consider estimable functions.

(*v*) *Fitting main effects before interaction.* Notation of the form $R(\alpha \mid \mu) = R(\mu, \alpha) - R(\mu)$ has been defined and freely used in the foregoing. Formally, then, it might seem plausible to define

$$R(\beta \mid \mu, \alpha, \gamma) = R(\mu, \alpha, \beta, \gamma) - R(\mu, \alpha, \gamma).$$

However, before trying to do this a careful look must be taken at the meaning of the interaction γ-factor, for in doing so it will be found that $R(\beta \mid \mu, \alpha, \gamma)$ as formally defined by the notation is identically equal to zero. Evidence of this comes from the models (and corresponding sums of squares) implied in the notation $R(\mu, \alpha, \beta, \gamma) - R(\mu, \alpha, \gamma)$. For $R(\mu, \alpha, \beta, \gamma)$ the model is (51) and

$$R(\mu, \alpha, \beta, \gamma) = \sum_{i=1}^{a} \sum_{j=1}^{b} n_{ij}\bar{y}_{ij\cdot}^2,$$

as in (61). Similarly, in the context of the α's and γ's of (51), the implied model for $R(\mu, \alpha, \gamma)$ is $y_{ijk} = \mu + \alpha_i + \gamma_{ij} + e_{ijk}$. But this is exactly the model of the 2-way nested classification discussed in Sec. 6.4. Hence the corresponding reduction in sum of squares is

$$R(\mu, \alpha, \gamma) = \sum_{i=1}^{a}\sum_{j=1}^{b} n_{ij}\bar{y}_{ij.}^2 , \tag{71}$$

and so $$R(\beta \mid \mu, \alpha, \gamma) = R(\mu, \alpha, \beta, \gamma) - R(\mu, \alpha, \gamma) \equiv 0.$$

Similarly $$R(\mu, \beta, \gamma) = \sum_{i=1}^{a}\sum_{j=1}^{b} n_{ij}\bar{y}_{ij.}^2 = R(\mu, \gamma) \tag{72}$$

and so we also have

$$R(\alpha \mid \mu, \beta, \gamma) \equiv 0 \equiv R(\alpha, \beta \mid \mu, \gamma).$$

From (61), (71) and (72) we see that the reduction in sums of squares due to fitting any model that contains the interaction γ-factor is $\sum_{i=1}^{a}\sum_{j=1}^{b} n_{ij}\bar{y}_{ij.}^2$. More particularly, in (71) and (72) the reduction due to fitting any model which, compared to (51), lacks either α, or β, or both, is equal to $R(\mu, \alpha, \beta, \gamma) = \sum_{i=1}^{a}\sum_{j=1}^{b} n_{ij}\bar{y}_{ij.}^2$. Indeed, as in (72), fitting just (μ and) the γ-factor alone leads to the same reduction in sum of squares. We return to this fact later. Meanwhile, the emphasis here is that in the $R(\ )$ notation there is no such thing as $R(\beta \mid \mu, \alpha, \gamma)$ when γ is the interaction factor between the α- and β-factors. This is the underlying reason for there being only two subsections of Table 7.7 after 7.7a. There, in 7.7b, we have $R(\mu)$, $R(\alpha \mid \mu)$, $R(\beta \mid \mu, \alpha)$ and $R(\gamma \mid \mu, \alpha, \beta)$ based on fitting μ, α, β and γ in that order, and in 7.7c we have $R(\mu)$, $R(\beta \mid \mu)$, $R(\alpha \mid \mu, \beta)$ and $R(\gamma \mid \mu, \alpha, \beta)$ for fitting μ, β, α and γ in that order. Notationally, one might be tempted from this to consider other sequences such as μ, α, γ and β, for example, which would give rise notationally to $R(\mu)$, $R(\alpha \mid \mu)$, $R(\alpha, \gamma \mid \mu)$ and $R(\beta \mid \mu, \alpha, \gamma)$. But since the latter symbol is, as we have seen, identically equal to zero, it is not a sum of squares. As a result, in the fitting of α-, β- and γ-factors, with γ representing α-by-β interactions, we can fit γ only in combination with both α and β. We cannot fit γ unless both α and β are in the model. This is true generally, that in the context of the kind of models being considered here interaction factors can only be fitted when all their corresponding main effects are in the model too. Moreover, only $R(\ )$ symbols adhering to this policy have meaning.

e. Estimable functions

The basic estimable function for the 2-way classification model with interaction is

$$E(y_{ijk}) = \mu + \alpha_i + \beta_j + \gamma_{ij} .$$

Because we frequently refer to this expression we give it a symbol, μ_{ij}:

$$\mu_{ij} = \mu + \alpha_i + \beta_j + \gamma_{ij}\,. \tag{73}$$

Its b.l.u.e. is

$$\hat{\mu}_{ij} = \widehat{\mu + \alpha_i + \beta_j + \gamma_{ij}} = \mu^o + \alpha_i^o + \beta_j^o + \gamma_{ij}^o\,,$$

and since the only non-zero elements in $\mathbf{b}^o$ of (55) are the γ_{ij}^o equal to $\bar{y}_{ij.}$,

$$\hat{\mu}_{ij} = \bar{y}_{ij.}\,. \tag{74}$$

Also $$v(\hat{\mu}_{ij}) = \sigma^2/n_{ij}$$

and $$\text{cov}(\hat{\mu}_{ij}, \hat{\mu}_{i'j'}) = 0 \quad \text{unless} \quad i = i' \text{ and } j = j'. \tag{75}$$

These results are fundamental to the ensuing discussion.

It is clear that μ_{ij}, by its definition in (73), corresponds to the cell in row i and column j of the grid of rows and columns in which the data may be displayed (e.g., Table 7.6). Therefore μ_{ij} is estimable only if the corresponding (i, j)-cell contains observations. This is also clear from (74) wherein $\hat{\mu}_{ij}$, the b.l.u.e. of μ_{ij}, exists only for those cells which have data, i.e., for which there is a $\bar{y}_{ij.}$-value. In saying that μ_{ij} is estimable we therefore implicitly refer only to those μ_{ij}'s corresponding to the cells that have data. The other μ_{ij}'s are not estimable.

Any linear function of the estimable μ_{ij}'s is estimable. However, because of the presence of γ_{ij} in μ_{ij}, differences between rows (α's) or columns (β's) are not estimable. For example, in Table 7.6, $\bar{y}_{11\cdot}$ and $\bar{y}_{21\cdot}$ exist and so μ_{11} and μ_{21} are estimable. Therefore

$$\mu_{11} - \mu_{21} = \alpha_1 - \alpha_2 + \gamma_{11} - \gamma_{21}$$

is estimable. But $\alpha_1 - \alpha_2$ is not. Similarly, $\alpha_1 - \alpha_3 + \gamma_{13} - \gamma_{33}$ and $\beta_1 - \beta_3 + \gamma_{11} - \gamma_{13}$ are estimable but $\alpha_1 - \alpha_3$ and $\beta_1 - \beta_3$ are not. In general,

$$\alpha_i - \alpha_{i'} + \sum_{j=1}^{b} k_{ij}(\beta_j + \gamma_{ij}) - \sum_{j=1}^{b} k_{i'j}(\beta_j + \gamma_{i'j}) \tag{76}$$

for $i \neq i'$ is estimable so long as

$$\sum_{j=1}^{b} k_{ij} = 1 = \sum_{j=1}^{b} k_{i'j} \tag{77}$$

with $\quad k_{ij} = 0$ when $n_{ij} = 0 \quad$ and $\quad k_{i'j} = 0$ when $n_{i'j} = 0$.

Then the b.l.u.e. of (76) is $$\sum_{j=1}^{b} k_{ij}\bar{y}_{ij.} - \sum_{j=1}^{b} k_{i'j}\bar{y}_{i'j.} \tag{78}$$

with variance $$\sum_{j=1}^{b} (k_{ij}^2/n_{ij} + k_{i'j}^2/n_{i'j})\sigma^2.$$

A similar result holds for the β's:

$$\beta_j - \beta_{j'} + \sum_{i=1}^{a} h_{ij}(\alpha_i + \gamma_{ij}) - \sum_{i=1}^{a} h_{ij'}(\alpha_i + \gamma_{ij'}) \tag{79}$$

for $j \neq j'$ is estimable so long as

$$\sum_{i=1}^{a} h_{ij} = 1 = \sum_{i=1}^{a} h_{ij'} \tag{80}$$

where $\quad h_{ij} = 0$ when $n_{ij} = 0 \quad$ and $\quad h_{ij'} = 0$ when $n_{ij'} = 0$,

and the b.l.u.e. of (79) is $\quad \sum_{i=1}^{a} h_{ij}\bar{y}_{ij\cdot} - \sum_{i=1}^{a} h_{ij'}\bar{y}_{ij'\cdot}\,. \quad$ (81)

It is not possible to derive from the μ_{ij}'s an estimable function which is solely a function of either the α's or the β's. The γ's will always be involved. On the other hand, it is possible to derive an estimable function that is a function of only the γ's. Provided the (ij), $(i'j)$ (ij') and $(i'j')$ cells have data in them

$$\begin{aligned} \theta_{ij,i'j'} &\equiv \mu_{ij} - \mu_{i'j} - \mu_{ij'} + \mu_{i'j'} \\ &= \gamma_{ij} - \gamma_{i'j} - \gamma_{ij'} + \gamma_{i'j'} \end{aligned} \tag{82}$$

is estimable, with b.l.u.e.

$$\hat{\theta}_{ij,i'j'} = \bar{y}_{ij\cdot} - \bar{y}_{i'j\cdot} - \bar{y}_{ij'\cdot} + \bar{y}_{i'j'\cdot} \tag{83}$$

and $\quad v(\hat{\theta}_{ij,i'j'}) = (1/n_{ij} + 1/n_{i'j} + 1/n_{ij'} + 1/n_{i'j'})\sigma^2\,.$

Expressions (76), (79) and (82) are the nearest we can come to obtaining estimable functions of intuitively practical value. Differences between row effects cannot be estimated (are not estimable) devoid of interaction effects; they can be estimated only in the presence of average column and interaction effects. For example, with $k_{ij} = 1/m_i$, where m_i is the number of filled cells in the ith row (i.e., $n_{ij} \neq 0$ for m_i values of $j = 1, 2, \ldots, b$), (77) is satisfied and

$$\alpha_i - \alpha_{i'} + \sum_{\substack{j \text{ for} \\ n_{ij} \neq 0}} (\beta_j + \gamma_{ij})/m_i - \sum_{\substack{j \text{ for} \\ n_{i'j} \neq 0}} (\beta_j + \gamma_{i'j})/m_{i'}$$

is estimable, with b.l.u.e.

$$\sum_{\substack{j \text{ for} \\ n_{ij} \neq 0}} \bar{y}_{ij\cdot}/m_i - \sum_{\substack{j \text{ for} \\ n_{i'j} \neq 0}} \bar{y}_{i'j\cdot}/m_{i'}\,.$$

Similarly, because $k_{ij} = n_{ij}/n_{i\cdot}$ also satisfies (77)

$$\alpha_i - \alpha_{i'} + \sum_{j=1}^{b} n_{ij}(\beta_j + \gamma_{ij})/n_{i\cdot} - \sum_{j=1}^{b} n_{i'j}(\beta_j + \gamma_{i'j})/n_{i'\cdot}. \tag{84}$$

is also estimable with b.l.u.e.

$$\sum_j n_{ij}\bar{y}_{ij.}/n_{i.} - \sum_j n_{i'j}\bar{y}_{i'j.}/n_{i'.} = \bar{y}_{i..} - \bar{y}_{i'..}$$

Analogous functions involving $\beta_j - \beta_{j'}$ can be derived from (79).

Examples. Table 7.6 provides the following examples. From (76)–(78)

$$\alpha_1 - \alpha_2 + (\beta_1 + \gamma_{11}) - (\beta_1 + \gamma_{21}) = \alpha_1 - \alpha_2 + \gamma_{11} - \gamma_{21} \tag{85}$$

is estimable with b.l.u.e.

$$\bar{y}_{11.} - \bar{y}_{21.} = 10 - 9 = 1.$$

Similarly $\quad \alpha_1 - \alpha_2 + (\beta_1 + \gamma_{11}) - \frac{1}{2}(\beta_1 + \beta_2 + \gamma_{21} + \gamma_{22})$

is estimable with b.l.u.e.

$$\bar{y}_{11.} - \tfrac{1}{2}(\bar{y}_{21.} + \bar{y}_{22.}) = 10 - \tfrac{1}{2}(9 + 13) = -1.$$

So far as $\alpha_1 - \alpha_2$ is concerned these two estimable functions are the same, but of course they involve different functions of the β's and γ's. In the first, (85), there are no β's due to the fact that for both of rows (treatments) 1 and 2 there are observations in column (variety) 1 (See Table 7.6). An example of (84) is that

$$\alpha_1 - \alpha_3 + [3(\beta_1 + \gamma_{11}) + (\beta_3 + \gamma_{13}) + 2(\beta_4 + \gamma_{14})]/6$$
$$- [2(\beta_2 + \gamma_{32}) + 2(\beta_3 + \gamma_{33}) + 4(\beta_4 + \gamma_{44})]/8$$

is estimable with b.l.u.e.

$$\bar{y}_{1..} - \bar{y}_{3..} = 10 - 11\tfrac{3}{4} = -1\tfrac{3}{4}.$$

Certain other estimable functions deserve mention since they arise in discussing tests of hypotheses corresponding to the F-statistics of Table 7.7. The first is φ_i defined as

$$\varphi_i = \left(n_{i.} - \sum_{j=1}^{b}\frac{n_{ij}^2}{n_{.j}}\right)\alpha_i - \sum_{i'\neq i}^{a}\left(\sum_{j=1}^{b}\frac{n_{ij}n_{i'j}}{n_{.j}}\right)\alpha_{i'}$$
$$+ \sum_{j=1}^{b}\left(n_{ij} - \frac{n_{ij}^2}{n_{.j}}\right)\gamma_{ij} - \sum_{i'\neq i}^{a}\left(\sum_{j=1}^{b}\frac{n_{ij}n_{i'j}}{n_{.j}}\right)\gamma_{i'j}\,. \tag{86}$$

That this *is* estimable is evident from the fact that

$$\varphi_i = \sum_{j=1}^{b}\left[n_{ij}(\mu + \alpha_i + \beta_j + \gamma_{ij}) - \sum_{k=1}^{a}\frac{n_{ij}n_{kj}}{n_{.j}}(\mu + \alpha_k + \beta_j + \gamma_{kj})\right].$$

A similar expression in terms of β's and γ's is also estimable:

$$\psi_j = \left(n_{.j} - \sum_{i=1}^{a} \frac{n_{ij}^2}{n_{i.}}\right)\beta_j - \sum_{j' \neq j}^{b}\left(\sum_{i=1}^{a} \frac{n_{ij}n_{ij'}}{n_{i.}}\right)\beta_{j'}$$
$$+ \sum_{i=1}^{a}\left(n_{ij} - \frac{n_{ij}^2}{n_{i.}}\right)\gamma_{ij} - \sum_{j' \neq j}^{b}\left(\sum_{i=1}^{a} \frac{n_{ij}n_{ij'}}{n_{i.}}\gamma_{ij'}\right). \quad (87)$$

For $\theta_{ij,i'j'}$ defined in (82), functions of estimable θ's are also estimable, naturally. But certain functions of non-estimable θ's are estimable too. For example, with the data of Table 7.6

$$\theta_{11,22} = \mu_{11} - \mu_{21} - \mu_{12} + \mu_{22} \quad \text{and} \quad \theta_{12,33} = \mu_{12} - \mu_{32} - \mu_{13} + \mu_{33}$$

are not estimable, because μ_{12} is not. But the sum of these two θ's does not involve μ_{12} and is estimable; i.e.,

$$\delta = \theta_{11,22} + \theta_{12,33} = \mu_{11} - \mu_{21} + \mu_{22} - \mu_{32} - \mu_{13} + \mu_{33}$$
$$= \mu_{11} - \mu_{13} - \mu_{21} + \mu_{22} - \mu_{32} + \mu_{33} \quad (88)$$

is estimable. In general, if each of two non-estimable θ's involves only a single non-estimable μ_{ij} which is common to both θ's, then the sum (or difference) of those θ's will not involve that μ_{ij} and will be estimable. (88) is an example.

f. Tests of hypotheses

(*i*) *The general hypothesis.* As has already been well established, the F-statistic for testing H: $\mathbf{K}'\mathbf{b} = \mathbf{0}$ is, for $\mathbf{K}'$ of full row rank s^*,

$$F = Q/s^*\hat{\sigma}^2 \quad \text{with} \quad Q = (\mathbf{K}'\mathbf{b}^o)'(\mathbf{K}'\mathbf{G}\mathbf{K})^{-1}\mathbf{K}'\mathbf{b}^o. \quad (89)$$

Furthermore, hypotheses are testable only when they can be expressed in terms of estimable functions—in this case, in terms of the μ_{ij}'s. Thus any testable hypotheses concerning $\mathbf{K}'\mathbf{b}$ will involve linear functions of the μ_{ij}'s; and, by the nature of μ_{ij}, no matter what functions of the α's and β's are involved in $\mathbf{K}'\mathbf{b}$ the functions of the γ_{ij}'s will be the same as those of the μ_{ij}'s. Thus if

$$\boldsymbol{\mu} = \{\mu_{ij}\} \quad \text{and} \quad \boldsymbol{\gamma} = \{\gamma_{ij}\}, \quad \text{for} \quad n_{ij} \neq 0, \quad (90)$$

then when $\mathbf{K}'\mathbf{b} = \mathbf{L}'\boldsymbol{\mu}$ that part of $\mathbf{K}'\mathbf{b}$ which involves $\boldsymbol{\gamma}$ is $\mathbf{L}'\boldsymbol{\gamma}$.

In (55), the only non-zero elements of $\mathbf{b}^o$ are

$$\boldsymbol{\gamma}^o = \{\gamma_{ij}^o\} = \bar{\mathbf{y}} = \{\bar{y}_{ij.}\} \quad \text{for} \quad n_{ij} \neq 0. \quad (91)$$

Similarly, in (57) the only non-null sub-matrix in $\mathbf{G}$ is the diagonal matrix $\mathbf{D}\{1/n_{ij}\}$ corresponding to $\boldsymbol{\gamma}^o$. Therefore, to test the hypothesis

$$H\text{:} \quad \mathbf{K}'\mathbf{b} = \mathbf{0} \text{ equivalent to } \mathbf{L}'\boldsymbol{\mu} = \mathbf{0}, \quad (92)$$

Q of (89) becomes

$$Q = \bar{\mathbf{y}}'\mathbf{L}[\mathbf{L}'\mathbf{D}\{1/n_{ij}\}\mathbf{L}]^{-1}\mathbf{L}'\bar{\mathbf{y}}. \quad (93)$$

Writing $\mathbf{D} \equiv \mathbf{D}\{1/n_{ij}\}$ to simplify notation gives

$$Q = \bar{\mathbf{y}}'\mathbf{L}(\mathbf{L}'\mathbf{D}\mathbf{L})^{-1}\mathbf{L}'\bar{\mathbf{y}}.$$

Example. For Table 7.6

$$\boldsymbol{\gamma}' = [\gamma_{11} \quad \gamma_{13} \quad \gamma_{14} \quad \gamma_{21} \quad \gamma_{22} \quad \gamma_{32} \quad \gamma_{33} \quad \gamma_{34}] \tag{94}$$

and

$$\begin{aligned}\bar{\mathbf{y}}' &= [\bar{y}_{11\cdot} \quad \bar{y}_{13\cdot} \quad \bar{y}_{14\cdot} \quad \bar{y}_{21\cdot} \quad \bar{y}_{22\cdot} \quad \bar{y}_{32\cdot} \quad \bar{y}_{33\cdot} \quad \bar{y}_{34\cdot}] \\ &= [10 \quad 12 \quad 9 \quad 9 \quad 13 \quad 8 \quad 15 \quad 12].\end{aligned} \tag{95}$$

In (85), $\alpha_1 - \alpha_2 + \gamma_{11} - \gamma_{21}$ is estimable, for which $\mathbf{L}'\boldsymbol{\gamma} = \gamma_{11} - \gamma_{21}$ has

$$\mathbf{L}' = [1 \quad 0 \quad 0 \quad -1 \quad 0 \quad 0 \quad 0 \quad 0].$$

Also, from (53)

$$\mathbf{D} = \mathbf{D}\{1/n_{ij}\} = \text{diag}[\tfrac{1}{3} \quad 1 \quad \tfrac{1}{2} \quad \tfrac{1}{2} \quad \tfrac{1}{2} \quad \tfrac{1}{2} \quad \tfrac{1}{2} \quad \tfrac{1}{4}], \tag{96}$$

so that $\mathbf{L}'\mathbf{D} = [\tfrac{1}{3} \quad 0 \quad 0 \quad -\tfrac{1}{2} \quad 0 \quad 0 \quad 0 \quad 0]$

and $\mathbf{L}'\mathbf{D}\mathbf{L} = (\tfrac{1}{3} + \tfrac{1}{2}) = \tfrac{5}{6}$.

Therefore for testing the hypothesis $\alpha_1 - \alpha_2 + \gamma_{11} - \gamma_{21} = 0$ we have

$$Q = (10 - 9)(5/6)^{-1}(10 - 9) = 1.2.$$

In this way we need look at only the γ_{ij} elements of a hypothesis in order to derive $\mathbf{L}'$ and so calculate Q of (89).

(ii) The hypothesis for $F(M)$. In earlier discussing Table 7.7 we interpreted the sums of squares therein as reductions in sums of squares due to fitting different models. Their meaning in terms of testing hypotheses is now considered. In this context we deal not with different models but with hypotheses about just the 2-way classification interaction model, (51). In particular we establish the linear hypotheses corresponding to each of the six different F-statistics in Tables 7.7b and 7.7c, the first of which is $F(M)$.

Results for the general case [e.g., equation (15) of Chapter 6] indicate that $F(M)$ can be used to test the hypothesis H: $E(\bar{y}) = 0$. In the present case this is equivalent to

$$H: \quad \sum_{i=1}^{a}\sum_{j=1}^{b} n_{ij}\mu_{ij} = 0 \qquad \text{for} \quad n_{ij} \neq 0 \tag{97}$$

which, in terms of (92), can be expressed as $\mathbf{L}'\boldsymbol{\mu} = 0$ for $\mathbf{L}'$ being the vector

$$\mathbf{L}' = [n_{11} \quad \cdots \quad n_{ab}] \qquad \text{for those} \quad n_{ij} \neq 0. \tag{98}$$

Hence

$$\mathbf{L}'\mathbf{D} = \mathbf{1}' \qquad \text{with} \qquad \mathbf{L}'\mathbf{D}\mathbf{L} = N. \tag{99}$$

It is then easily shown for (93) that $\mathbf{L}'\bar{\mathbf{y}} = y_{\cdots}$ and so (93) becomes $Q = R(\mu)$, confirming the numerator of $F(M)$.

(iii) Hypotheses for $F(\alpha \mid \mu)$ and $F(\beta \mid \mu)$. We will show that $R(\alpha \mid \mu)$ is the numerator sum of squares for testing

$$H: \quad \frac{1}{n_{i.}} \sum_j n_{ij}\mu_{ij} \text{ equal for all } i,$$

a hypothesis which can also be stated as

$$H: \quad \alpha_i + \frac{1}{n_{i.}} \sum_j n_{ij}(\beta_j + \gamma_{ij}) \text{ equal for all } i. \tag{100}$$

Expressing this as $a - 1$ independent differences

$$H: \quad \frac{1}{n_{1.}} \sum_j n_{1j}\mu_{1j} - \frac{1}{n_{i.}} \sum_j n_{ij}\mu_{ij} = 0 \quad \text{for} \quad i = 2, 3, \ldots, a,$$

it can be seen that for (93) the $(i - 1)$th row of $\mathbf{L}'$ is, for $i = 2, 3, \ldots, a$,

$$\ell'_{i-1} = [\underbrace{n_{11}/n_{1.} \cdots n_{1b}/n_{1.}}_{\substack{\text{corresponding to} \\ n_{1j} \neq 0}} \quad \mathbf{0}' \quad \underbrace{-n_{i1}/n_{i.} \cdots -n_{ib}/n_{i.}}_{\substack{\text{corresponding to} \\ n_{ij} \neq 0}} \quad \mathbf{0}']. \tag{101}$$

From this it can be shown that the $(i - 1)$th element of $\mathbf{L}'\bar{\mathbf{y}}$ is $\bar{y}_{1..} - \bar{y}_{i..}$ and $\mathbf{L}'\mathbf{D}\mathbf{L} = (1/n_{1.})\mathbf{J} + \mathbf{D}\{n_{i.}\}$ for $i = 2, 3, \ldots, a$. Algebraic simplification based on results in Exercise 6 leads to Q of (93) reducing to $R(\alpha \mid \mu)$. Hence (100) is the hypothesis tested by $F(\alpha \mid \mu)$.

Example. For the data of Table 7.6 consider

$$H: \quad \alpha_1 + \tfrac{1}{6}[3(\beta_1 + \gamma_{11}) + (\beta_3 + \gamma_{13}) + 2(\beta_4 + \gamma_{14})] \\ - \{\alpha_2 + \tfrac{1}{4}[2(\beta_1 + \gamma_{21}) + 2(\beta_2 + \gamma_{22})]\} = 0$$

$$\alpha_1 + \tfrac{1}{6}[3(\beta_1 + \gamma_{11}) + (\beta_3 + \gamma_{13}) + 2(\beta_4 + \gamma_{14})] \\ - \{\alpha_3 + \tfrac{1}{8}[2(\beta_2 + \gamma_{32}) + 2(\beta_3 + \gamma_{33}) + 4(\beta_4 + \gamma_{34})]\} = 0,$$

from which

$$\mathbf{L}' = \begin{bmatrix} \frac{3}{6} & \frac{1}{6} & \frac{2}{6} & -\frac{2}{4} & -\frac{2}{4} & 0 & 0 & 0 \\ \frac{3}{6} & \frac{1}{6} & \frac{2}{6} & 0 & 0 & -\frac{2}{8} & -\frac{2}{8} & -\frac{4}{8} \end{bmatrix}, \tag{102}$$

$$\mathbf{L}'\bar{\mathbf{y}} = \begin{bmatrix} \bar{y}_{1..} - \bar{y}_{2..} \\ \bar{y}_{1..} - \bar{y}_{3..} \end{bmatrix} = \begin{bmatrix} 10 - 11 \\ 10 - 11\frac{3}{4} \end{bmatrix} = \begin{bmatrix} -1 \\ -1\frac{3}{4} \end{bmatrix},$$

$$\mathbf{L}'\mathbf{D} = \begin{bmatrix} \frac{1}{6} & \frac{1}{6} & \frac{1}{6} & -\frac{1}{4} & -\frac{1}{4} & 0 & 0 & 0 \\ \frac{1}{6} & \frac{1}{6} & \frac{1}{6} & 0 & 0 & -\frac{1}{8} & -\frac{1}{8} & -\frac{1}{8} \end{bmatrix} \tag{103}$$

$$\text{and} \quad \mathbf{L'DL} = \begin{bmatrix} \frac{1}{6} + \frac{1}{4} & \frac{1}{6} \\ \frac{1}{6} & \frac{1}{6} + \frac{1}{8} \end{bmatrix} \quad \text{with} \quad (\mathbf{L'DL})^{-1} = \frac{4}{9}\begin{bmatrix} 7 & -4 \\ -4 & 10 \end{bmatrix}.$$

$$\text{Hence} \quad Q = [-1 \quad -1\tfrac{3}{4}]\frac{4}{9}\begin{bmatrix} 7 & -4 \\ -4 & 10 \end{bmatrix}\begin{bmatrix} -1 \\ -1\frac{3}{4} \end{bmatrix} = \tfrac{4}{9}(\tfrac{7}{16})(16 + 70 - 32)$$

$$= 10\tfrac{1}{2} = R(\alpha \mid \mu) \text{ of Table 7.7b.}$$

Analogous to the above, $R(\beta \mid \mu)$ is the numerator sum of squares for testing

$$H\colon \quad \beta_j + \frac{1}{n_{.j}}\sum_i n_{ij}(\alpha_i + \gamma_{ij}) \text{ equal for all } j.$$

Exercise 13 provides an example.

(*iv*) *Hypotheses for $F(\alpha \mid \mu, \beta)$ and $F(\beta \mid \mu, \alpha)$.* The hypotheses tested by these F-statistics are, respectively,

$$H\colon \quad \varphi_i = 0 \text{ for all } i \quad \text{and} \quad H\colon \quad \psi_j = 0 \text{ for all } j$$

where φ_i and ψ_j are given by (86) and (87). First note that $\sum_{i=1}^{a} \varphi_i = 0$:

$$\begin{aligned}\sum_{i=1}^{a} \varphi_i &= \sum_{i=1}^{a} \alpha_i\left[n_{i.} - \sum_{j=1}^{b} n_{ij}^2/n_{.j} - \sum_{i' \neq i}^{a} \sum_{j=1}^{b} n_{ij}n_{i'j}/n_{.j}\right] \\ &\quad + \sum_{i=1}^{a}\sum_{j=1}^{b}\gamma_{ij}\left[n_{ij} - n_{ij}^2/n_{.j} - \sum_{i' \neq i} n_{ij}n_{i'j}/n_{.j}\right] \\ &= \sum_{i=1}^{a} \alpha_i\left[n_{i.} - \sum_{j=1}^{b} n_{ij}^2/n_{.j} - \sum_{j=1}^{b} n_{ij}(n_{.j} - n_{ij})/n_{.j}\right] \\ &\quad + \sum_{i=1}^{a}\sum_{j=1}^{b}\gamma_{ij}[n_{ij} - n_{ij}^2/n_{.j} - n_{ij}(n_{.j} - n_{i'j})/n_{.j}] \\ &\equiv 0.\end{aligned}$$

Therefore the hypotheses in $H\colon \varphi_i = 0$ for all i are not independent; stated as a set of independent hypotheses they are

$$H\colon \quad \varphi_i = 0 \quad \text{for} \quad i = 1, 2, \ldots, a-1. \tag{104}$$

Writing this in the form $\mathbf{L}'\boldsymbol{\mu} = \mathbf{0}$ the ith row of $\mathbf{L}'$ is, for $i = 1, \ldots, a-1$, given by

$$\ell_i' = \begin{bmatrix} \{-n_{ij}n_{kj}/n_{.j}\} \text{ for } j = 1, \ldots, b \text{ and } k = 1, \ldots, i-1 \text{ and } n_{kj} \neq 0 \\ \{n_{ij} - n_{ij}^2/n_{.j}\} \text{ for } j = 1, \ldots, b \text{ and } n_{ij} \neq 0 \\ \{-n_{ij}n_{kj}/n_{.j}\} \text{ for } j = 1, \ldots, b \text{ and } k = i+1, \ldots, a \text{ and } n_{kj} \neq 0 \end{bmatrix}'. \tag{105}$$

It is then readily shown that, for (93), the ith element of $\mathbf{L}'\bar{\mathbf{y}}$ is

$$\sum_{k\neq i}\sum_{j}(-n_{ij}n_{kj}/n_{.j})\bar{y}_{kj\cdot} + \sum_{j}(n_{ij} - n_{ij}^2/n_{.j})\bar{y}_{ij\cdot} = y_{i\cdot\cdot} - \sum_{j} n_{ij}\bar{y}_{\cdot j\cdot},$$

and so $$\mathbf{L}'\bar{\mathbf{y}} = \left\{y_{i\cdot\cdot} - \sum_{j=1}^{b} n_{ij}\bar{y}_{\cdot j\cdot}\right\} \quad \text{for} \quad i = 1, \ldots, a-1.$$

Similarly it can be shown that diagonal elements of $\mathbf{L}'\mathbf{DL}$ are

$$n_{i\cdot} - \sum_{j=1}^{b} n_{ij}^2/n_{\cdot j} \quad \text{for} \quad i = 1, 2, \ldots, a-1$$

and off-diagonal elements of $\mathbf{L}'\mathbf{DL}$ are

$$-\sum_{j=1}^{b} n_{ij}n_{i'j}/n_{\cdot j} \quad \text{for} \quad i \neq i' = 1, 2, \ldots, a-1.$$

By analogy, for testing

$$H:\quad \psi_j = 0 \quad \text{for} \quad j = 1, \ldots, b-1$$

we have $$\mathbf{L}'\bar{\mathbf{y}} = \left\{y_{\cdot j\cdot} - \sum_{i=1}^{a} n_{ij}\bar{y}_{i\cdot\cdot}\right\} \quad \text{for} \quad j = 1, 2, \ldots, b-1,$$

with diagonal elements of $\mathbf{L}'\mathbf{DL}$ being

$$n_{\cdot j} - \sum_{i=1}^{a} n_{ij}^2/n_{i\cdot} \quad \text{for} \quad j = 1, 2, \ldots, b-1$$

and off-diagonal elements of $\mathbf{L}'\mathbf{DL}$ being

$$-\sum_{i=1}^{a} n_{ij}n_{ij'}/n_{i\cdot} \quad \text{for} \quad j \neq j' = 1, 2, \ldots, b-1.$$

But in this case we see from (65) that

$$\mathbf{L}'\bar{\mathbf{y}} = \mathbf{r}, \quad \text{and} \quad \mathbf{L}'\mathbf{DL} = \mathbf{C}$$

from (64). Therefore in (93)

$$Q = \mathbf{r}'\mathbf{C}^{-1}\mathbf{r} = R(\beta \mid \mu, \alpha)$$

from (63). Hence $F(\beta \mid \mu, \alpha)$ of Table 7.7b tests $H:\ \psi = 0$, i.e.,

$$H:\quad \left(n_{\cdot j} - \sum_{i=1}^{a} n_{ij}^2/n_{i\cdot}\right)\beta_j - \sum_{j'\neq j}^{b}\left(\sum_{i=1}^{a} n_{ij}n_{ij'}/n_{i\cdot}\right)\beta_{j'}$$
$$+ \sum_{i=1}^{a}(n_{ij} - n_{ij}^2/n_{i\cdot})\gamma_{ij} - \sum_{j'\neq j}^{b}\sum_{i=1}^{a}(n_{ij}n_{ij'}/n_{i\cdot})\gamma_{ij'} = 0 \qquad (106)$$
$$\text{for} \quad j = 1, 2, \ldots, b-1,$$

equivalent to the same hypothesis for $j = 1, 2, \ldots, b$. Correspondingly $F(\alpha \mid \mu, \beta)$ of Table 7.7c tests H: $\varphi_i = 0$, i.e.,

$$H: \quad \left(n_{i\cdot} - \sum_{j=1}^{b} n_{ij}^2/n_{\cdot j}\right)\alpha_i - \sum_{i' \neq i}^{a}\left(\sum_{j=1}^{b} n_{ij}n_{i'j}/n_{\cdot j}\right)\alpha_{i'}$$
$$+ \sum_{j=1}^{b}(n_{ij} - n_{ij}^2/n_{\cdot j})\gamma_{ij} - \sum_{i' \neq i}^{a}\sum_{j=1}^{b}(n_{ij}n_{i'j}/n_{\cdot j})\gamma_{i'j} = 0 \tag{107}$$
$$\text{for} \quad i = 1, 2, \ldots, a - 1,$$

equivalent to the same hypothesis for $i = 1, 2, \ldots, a$. Note that in (106) the coefficients of the β's are the elements $c_{jj'}$ of $\mathbf{C}$ in (64); and the coefficients of the γ's, if summed over i, are also the $c_{jj'}$'s. Analogous properties hold for the coefficients of the α's and γ's in (107).

Example. In accord with (92) the $\mathbf{L}'$ matrix for the hypothesis in (107) is obtained from the coefficients of the γ's, whose terms are

$$\sum_{j=1}^{b}(n_{ij} - n_{ij}^2/n_{\cdot j})\gamma_{ij} - \sum_{i' \neq i}^{a}\sum_{j=1}^{b}(n_{ij}n_{i'j}/n_{\cdot j})\gamma_{i'j} \qquad \text{for} \quad i = 1, 2, \ldots, a - 1.$$

For the data of Tables 7.6 and 7.6a, the value of $\mathbf{L}'$ for the hypothesis (107) is therefore

$$\mathbf{L}' = \begin{bmatrix} 3 - 3^2/5 & 1 - 1^2/3 & 2 - 2^2/6 & -3(2)/5 & 0(2)/4 & 0(2)/4 & -1(2)/3 & -2(4)/6 \\ -2(3)/5 & 0 & 0 & 2 - 2^2/5 & 2 - 2^2/4 & -2^2/4 & 0 & 0 \end{bmatrix}$$

$$= \frac{1}{15}\begin{bmatrix} 18 & 10 & 20 & -18 & 0 & 0 & -10 & -20 \\ -18 & 0 & 0 & 18 & 15 & -15 & 0 & 0 \end{bmatrix}. \tag{108}$$

Thus
$$\mathbf{L}'\mathbf{D} = \frac{1}{15}\begin{bmatrix} 6 & 10 & 10 & -9 & 0 & 0 & -5 & -5 \\ -6 & 0 & 0 & 9 & 7\frac{1}{2} & -7\frac{1}{2} & 0 & 0 \end{bmatrix} \tag{109}$$

so that
$$\mathbf{L}'\mathbf{D}\mathbf{L} = \frac{1}{5}\begin{bmatrix} 16 & -6 \\ -6 & 11 \end{bmatrix} \quad \text{with} \quad (\mathbf{L}'\mathbf{D}\mathbf{L})^{-1} = \frac{1}{28}\begin{bmatrix} 11 & 6 \\ 6 & 16 \end{bmatrix}.$$

Also
$$\mathbf{L}'\bar{\mathbf{y}} = \frac{1}{15}\begin{bmatrix} 180 + 120 + 180 - 162 - 150 - 240 \\ -180 + 162 + 195 - 120 \end{bmatrix}$$
$$= \frac{1}{5}\begin{bmatrix} -24 \\ 19 \end{bmatrix}.$$

Therefore in (93)

$$Q = \tfrac{1}{25}\tfrac{1}{28}[-24 \quad 19]\begin{bmatrix} 11 & 6 \\ 6 & 16 \end{bmatrix}\begin{bmatrix} -24 \\ 19 \end{bmatrix} = \tfrac{16}{700}(396 + 361 - 342)$$
$$= 9\tfrac{34}{70}$$
$$= R(\alpha \mid \mu, \beta) \text{ of Table 7.7c.}$$

(*v*) *Hypotheses for* $F(\gamma \mid \mu, \alpha, \beta)$. The hypothesis tested by $F(\gamma \mid \mu, \alpha, \beta)$ is of the following form, where $s - a - b + 1$ is the degrees of freedom of $R(\gamma \mid \mu, \alpha, \beta)$.

$$H: \left\{\begin{array}{l}\text{Any column vector consisting of } s - a - b + 1 \\ \quad \text{linearly independent functions of the} \\ \theta_{ij,i'j'} = \gamma_{ij} - \gamma_{i'j} - \gamma_{ij'} + \gamma_{i'j'}, \text{ where such} \\ \text{functions are either estimable } \theta\text{'s or estimable} \\ \text{sums or differences of } \theta\text{'s.}\end{array}\right\} = \mathbf{0}. \tag{110}$$

$\theta_{ij,i'j'}$ as used here is as defined in (82), and the estimable sums or differences of θ's are those defined in (88). Writing the hypothesis as

$$\mathbf{L}'\boldsymbol{\gamma} = \mathbf{0}$$

where $\mathbf{L}'$ has order $s - a - b + 1$ by s and rank $s - a - b + 1$, it is clear from the nature of the θ's that $\mathbf{L}'\mathbf{1} = \mathbf{0}$. Furthermore, the equations $\mathbf{L}'\boldsymbol{\gamma} = \mathbf{0}$ have a solution $\gamma_{ij} = \gamma$ for all i and j for which $n_{ij} \neq 0$. The reduced model corresponding to the hypothesis is therefore $E(y_{ijk}) = (\mu + \gamma) + \alpha_i + \beta_j$, for which the reduction in sum of squares is $R(\mu, \alpha, \beta)$. Therefore, in accord with Sec. 3.6d(*ii*),

$$Q = R(\mu, \alpha, \beta, \gamma) - R(\mu, \alpha, \beta) = R(\gamma \mid \mu, \alpha, \beta);$$

and so $F(\gamma \mid \mu, \alpha, \beta)$ tests the hypothesis in (110).

Example. For the data of Table 7.6 we can test the hypothesis

$$H: \begin{cases}\mu_{13} - \mu_{33} - \mu_{14} + \mu_{34} & = 0 \\ \mu_{11} - \mu_{21} - \mu_{12} + \mu_{22} + (\mu_{12} - \mu_{32} - \mu_{13} + \mu_{33}) & = 0.\end{cases} \tag{111}$$

The first of these is $\theta_{13,34} = 0$, in keeping with (82), and the second is $\theta_{11,12} + \theta_{12,33} = 0$ as in (88). In terms of the elements of the model this hypothesis is

$$H: \begin{cases}\gamma_{13} - \gamma_{33} - \gamma_{14} + \gamma_{34} & = 0 \\ \gamma_{11} - \gamma_{21} + \gamma_{22} - \gamma_{13} - \gamma_{32} + \gamma_{33} & = 0,\end{cases} \tag{112}$$

where the second function of γ's is (88). Writing this hypothesis as $\mathbf{L}'\boldsymbol{\gamma} = \mathbf{0}$ gives

$$\mathbf{L}' = \begin{bmatrix} 0 & 1 & -1 & 0 & 0 & 0 & -1 & 1 \\ 1 & -1 & 0 & -1 & 1 & -1 & 1 & 0 \end{bmatrix}, \tag{113}$$

for which $\mathbf{L}'\bar{\mathbf{y}} = \begin{bmatrix} 12 - 9 - 15 + 12 \\ 10 - 12 - 9 + 13 - 8 + 15 \end{bmatrix} = \begin{bmatrix} 0 \\ 9 \end{bmatrix}$

and $$\mathbf{L}'\mathbf{D} = \begin{bmatrix} 0 & 1 & -\frac{1}{2} & 0 & 0 & 0 & -\frac{1}{2} & \frac{1}{4} \\ \frac{1}{3} & -1 & 0 & -\frac{1}{2} & \frac{1}{2} & -\frac{1}{2} & \frac{1}{2} & 0 \end{bmatrix}, \tag{114}$$

so that $\mathbf{L'DL} = \dfrac{1}{12}\begin{bmatrix} 27 & -18 \\ -18 & 40 \end{bmatrix}$ and $(\mathbf{L'DL})^{-1} = \dfrac{1}{63}\begin{bmatrix} 40 & 18 \\ 18 & 27 \end{bmatrix}$.

Hence in (93)

$$\begin{aligned} Q &= [0 \quad 9]\frac{1}{63}\begin{bmatrix} 40 & 18 \\ 18 & 27 \end{bmatrix}\begin{bmatrix} 0 \\ 9 \end{bmatrix} = 81(27)/63 = 243/7 \\ &= 34\tfrac{5}{7} \\ &= R(\gamma \mid \mu, \alpha, \beta) \end{aligned}$$

of Tables 7.7b and 7.7c. Hence (111) is the hypothesis tested by $F(\gamma \mid \mu, \alpha, \beta)$.

Note that hypotheses of this nature involve not only functions of the form $\theta_{ij,i'j'} = \mu_{ij} - \mu_{i'j} - \mu_{ij'} + \mu_{i'j'} = \gamma_{ij} - \gamma_{i'j} - \gamma_{ij'} + \gamma_{i'j'}$, as is the first in (111), but also sums and differences of such functions, as is the second of (111). As has already been explained in the description of δ in (88), these sums and differences are chosen to eliminate a μ_{ij} that is not estimable because its $n_{ij} = 0$. Thus in going from (111) to (112) the non-estimable μ_{12} drops out. Furthermore, note that these expressions can often be derived in more than one way. Thus the second function in (111) is not only

$$\begin{aligned} &\gamma_{11} - \gamma_{21} + \gamma_{22} - \gamma_{13} - \gamma_{32} + \gamma_{33} \\ &\qquad = \mu_{11} - \mu_{21} - \mu_{12} + \mu_{22} + (\mu_{12} - \mu_{32} - \mu_{13} + \mu_{33}), \end{aligned} \tag{115}$$

which eliminates the non-estimable μ_{12}, but it is also

$$\begin{aligned} &\gamma_{11} - \gamma_{21} + \gamma_{22} - \gamma_{13} - \gamma_{32} + \gamma_{33} \\ &\qquad = \mu_{22} - \mu_{32} - \mu_{23} + \mu_{33} - (\mu_{21} - \mu_{11} - \mu_{23} + \mu_{13}) \end{aligned} \tag{116}$$

which eliminates the non-estimable μ_{23}. The exact form of these functions corresponding to any particular $R(\gamma \mid \mu, \alpha, \beta)$ also depends entirely on the available data. The pattern of non-empty cells is the determining factor in establishing which functions of the γ's make up the hypotheses tested by $F(\gamma \mid \mu, \alpha, \beta)$ and which do not. For example, for Table 7.6 one function which is not estimable is

$$\begin{aligned} &\mu_{11} - \mu_{31} - \mu_{13} + \mu_{33} - (\mu_{21} - \mu_{31} - \mu_{24} + \mu_{34}) \\ &\qquad = \mu_{11} - \mu_{13} + \mu_{33} - \mu_{21} + \mu_{24} - \mu_{34} \end{aligned}$$

which eliminates the non-estimable μ_{31} but still retains the non-estimable μ_{24}.

(*vi*) *Reduction to the no-interaction model*. We show here that the hypotheses tested by $F(\alpha \mid \mu)$ and $F(\beta \mid \mu, \alpha)$ in the interaction model reduce to those tested by the same statistics in the no-interaction model—a result which is, of course, to be anticipated.

In the interaction model, $F(\alpha \mid \mu)$ tests the hypothesis (100):

$$H\colon\quad \alpha_i + \frac{1}{n_{i\cdot}}\sum_j n_{ij}(\beta_j + \gamma_{ij}) \text{ equal for all } i.$$

Putting all $\gamma_{ij} = 0$ converts the interaction model into the no-interaction model, and transforms the above hypothesis into

$$H\colon\quad \alpha_i + \sum_j n_{ij}\beta_j/n_{i\cdot} \text{ equal for all } i.$$

This is identical to that tested by $F(\alpha \mid \mu)$ in the no-interaction model, as discussed after the example that follows (48).

Similarly, in the interaction model, the hypothesis tested by $F(\beta \mid \mu, \alpha)$ is that given in (106). Putting all $\gamma_{ij} = 0$ in (106) reduces that hypothesis to

$$\left(n_{\cdot j} - \sum_{i=1}^{a} n_{ij}^2/n_{i\cdot}\right)\beta_j - \sum_{j' \neq j}^{b}\left(\sum_{i=1}^{a} n_{ij}n_{ij'}/n_{i\cdot}\right)\beta_{j'} = 0 \qquad \text{for all } j.$$

This represents $b - 1$ linearly independent equations in b parameters β_1, $\beta_2, \ldots, \beta_b$, the equations being of such a nature that they hold only when the β_j's are all equal, i.e., the hypothesis in (106) reduces to H: equality of all β_j's, the hypothesis tested by $F(\beta \mid \mu, \alpha)$ in the no-interaction model, as indicated in Sec. 7.1g.

(*vii*) *Independence properties*. As indicated in Sec. 5.5g, the sums of squares for testing hypotheses $\mathbf{k}_i'\mathbf{b} = 0$ and $\mathbf{k}_j'\mathbf{b} = 0$ are, on the basis of underlying normality assumptions, independent if $\mathbf{k}_i'\mathbf{G}\mathbf{k}_j = 0$. This property can be used to show that the sums of squares in Table 7.7a are independent, as are those of Table 7.7b and of 7.7c also. To see this, consider $\ell_i'\mathbf{D}\ell_j^*$ where ℓ_i' is a row of $\mathbf{L}'$ for one sum of squares, and $\ell_j^{*\prime}$ is a row of $\mathbf{L}'$for some other sum of squares in the same section of Table 7.7. For example, from (99), $\ell_i'\mathbf{D}$ of $R(\mu)$ is $\mathbf{1}'$ and from (102) an $\ell_j^{*\prime}$ of $R(\alpha \mid \mu)$ is

$$\ell_j^{*\prime} = [\tfrac{3}{6} \quad \tfrac{1}{6} \quad \tfrac{2}{6} \quad -\tfrac{2}{4} \quad -\tfrac{2}{4} \quad 0 \quad 0 \quad 0].$$

Hence
$$\ell_i'\mathbf{D}\ell_j^* = \mathbf{1}'\ell_j^* = 0.$$

The same result will be found true for the other row of $\mathbf{L}'$ in (102), and we conclude that $R(\mu)$ and $R(\alpha \mid \mu)$ are independently distributed. In this way the independence of the $R(\quad)$'s in Table 7.7b is readily established, and similarly for those of Tables 7.7a and 7.7c. Expressions for $\mathbf{L}'\mathbf{D}$ are given in equations (99), (103), (109) and (114) and for $\mathbf{L}'$ in (98), (102), (108) and (113).

g. Models that include restrictions

Since, as in (76),

$$\alpha_i - \alpha_{i'} + \sum_{j=1}^{b} k_{ij}(\beta_j + \gamma_{ij}) - \sum_{j=1}^{b} k_{i'j}(\beta_j + \gamma_{i'j})$$

is estimable, for the k's satisfying (77), it is clear that $\alpha_i - \alpha_{i'}$ is estimable if the model includes restrictions

$$\sum_{j=1}^{b} k_{ij}(\beta_j + \gamma_{ij}) = 0 \quad \text{for all } i, \qquad \text{for} \quad n_{ij} \neq 0. \tag{117}$$

A particular case of this might be when $k_{ij} = n_{ij}/n_{i\cdot}$, as in (84), in which case (117) becomes

$$\sum_{j=1}^{b} n_{ij}(\beta_j + \gamma_{ij}) = 0 \quad \text{for all } i, \qquad \text{for} \quad n_{ij} \neq 0, \tag{118}$$

and the corresponding b.l.u.e. of $\alpha_i - \alpha_{i'}$ is then $\bar{y}_{i\cdot\cdot} - \bar{y}_{i'\cdot\cdot}$. However, there seems to be little merit in having either (117) or (118) as part of a model, for both of them are data-dependent. Both depend on which n_{ij} are non-zero, and (118) is a direct function of the n_{ij}-values. The same thing applies to restrictions that reduce the hypotheses tested by the F-statistics of Table 7.7 to hypotheses that have meaningful interpretation (e.g., a hypothesis of equality of the α's). As inherent parts of a model, these restrictions suffer from the same deficiencies as do all such restrictions, as discussed in Secs. 7.1h and 6.2h.

h. All cells filled

For data having empty cells, such as those of Table 7.6, the nature of which functions are estimable depends entirely on which n_{ij}'s are not zero. For example, with the Table 7.6 data, $\alpha_2 + \gamma_{22} - (\alpha_3 + \gamma_{32})$ is estimable but $\alpha_1 + \gamma_{12} - (\alpha_3 + \gamma_{32})$ is not. In contrast, when all cells are filled, i.e., when there are no empty cells,

$$\alpha_i - \alpha_{i'} + \left(\sum_{j=1}^{b} \gamma_{ij} - \sum_{j=1}^{b} \gamma_{i'j}\right) \Big/ b \tag{119}$$

is estimable for all $i \neq i'$. This is a special case of (76), with $k_{ij} = k_{i'j} = 1/b$ and with b.l.u.e.

$$\widehat{\alpha_i - \alpha_{i'} + \left(\sum_{j=1}^{b} \gamma_{ij} - \sum_{j=1}^{b} \gamma_{i'j}\right) \Big/ b} = \left(\sum_{j=1}^{b} \bar{y}_{ij\cdot} - \sum_{j=1}^{b} \bar{y}_{i'j\cdot}\right) \Big/ b. \tag{120}$$

Hypotheses about (119) can also be tested. Thus

$$H: \ \alpha_i - \alpha_{i'} + \left(\sum_{j=1}^{b} \gamma_{ij} - \sum_{j=1}^{b} \gamma_{i'j}\right) \Big/ b = 0$$

is tested by

$$F = \frac{\left[\sum_{j=1}^{b} (\bar{y}_{ij\cdot} - \bar{y}_{i'j\cdot})\right]^2}{\sum_{j=1}^{b} (1/n_{ij} + 1/n_{i'j})\hat{\sigma}^2}$$

with 1 and $s - a - b + 1$ degrees of freedom. Furthermore, the joint hypothesis

$$H: \quad \alpha_i + \sum_{j=1}^{b} \gamma_{ij}/b \text{ all equal}, \qquad \text{for} \quad i = 1, \ldots, a \tag{121}$$

can also be tested. The F-statistic for doing so is (see Exercise 17)

$$F = \frac{\displaystyle\sum_{i=1}^{a} \frac{\left(\sum_{j=1}^{b} \bar{y}_{ij\cdot}\right)^2}{\sum_{j=1}^{b} 1/n_{ij}} - \left(\sum_{i=1}^{a} \frac{\sum_{j=1}^{b} \bar{y}_{ij\cdot}}{\sum_{j=1}^{b} 1/n_{ij}}\right)^2 \Bigg/ \sum_{i=1}^{a} \frac{1}{\sum_{j=1}^{b} 1/n_{ij}}}{(a-1)\hat{\sigma}^2}. \tag{122}$$

If the model includes the restriction $\sum_{j=1}^{a} \gamma_{ij} = 0$ for all $i = 1, 2, \ldots, a$, then (119) reduces to $\alpha_i - \alpha_{i'}$ and is estimable, with b.l.u.e. given by (120); and the hypothesis (121) becomes H: equality of all α_i's, which is then tested by (122).

Results paralleling (119) through (122) for β's can be obtained in similar fashion.

i. Balanced data

There is great simplification of the preceding results when $n_{ij} = n$ for all i and j, just as in the no-interaction case. The calculations become those of the familiar 2-factor analysis with replication [e.g., Scheffé (1959, p. 110)]. Solutions to the normal equations remain as $\gamma_{ij}^o = \bar{y}_{ij\cdot}$ as the only non-zero elements of $\mathbf{b}^o$. If the model includes restrictions $\sum_{i=1}^{a} \alpha_i = 0, \sum_{j=1}^{b} \beta_j = 0, \sum_{i=1}^{a} \gamma_{ij} = 0$ for all j and $\sum_{j=1}^{b} \gamma_{ij} = 0$ for all i then $\alpha_i - \alpha_{i'}$ and $\beta_j - \beta_{j'}$ are both estimable, with b.l.u.e.'s

$$\widehat{\alpha_i - \alpha_{i'}} = \bar{y}_{i\cdot\cdot} - \bar{y}_{i'\cdot\cdot} \qquad \text{and} \qquad \widehat{\beta_j - \beta_{j'}} = \bar{y}_{\cdot j\cdot} - \bar{y}_{\cdot j'\cdot},$$

with variances $2\sigma^2/bn$ and $2\sigma^2/an$, respectively.

The analysis of variance tables of Tables 7.7 and 7.8 also simplify, just as did Tables 7.2 and 7.3 in the no-interaction case. Thus, $R(\alpha \mid \mu)$ and $R(\alpha \mid \mu, \beta)$ of Table 7.8 become identical:

$$R(\alpha \mid \mu) = R(\alpha \mid \mu, \beta) = bn \sum_{i=1}^{a} (\bar{y}_{i\cdot\cdot} - \bar{y}_{\cdot\cdot\cdot})^2$$

$$\tag{123}$$

and

$$R(\beta \mid \mu, \alpha) = R(\beta \mid \mu) = an \sum_{j=1}^{b} (\bar{y}_{\cdot j\cdot} - \bar{y}_{\cdot\cdot\cdot})^2$$

just as in (50). The resulting, and familiar, analysis of variance is shown in Table 7.9. Its similarity to Table 7.5 is obvious. As there, so now, distinction between fitting "α after μ" and "α after μ and β" is no longer necessary; they are both "α after μ", with reduction in some of squares $R(\alpha \mid \mu)$ as shown in Table 7.9.

In the case of balanced data the numerator of (122) also reduces to $R(\alpha \mid \mu)$ of (123), as is to be expected.

TABLE 7.9. ANALYSIS OF VARIANCE FOR A 2-WAY CLASSIFICATION WITH INTERACTION, WITH BALANCED DATA (ALL $n_{ij} = n$). (BOTH PARTS OF TABLE 7.8 SIMPLIFY TO THIS WHEN $n_{ij} = n$)

Source of Variation	Degrees of Freedom	Sum of Squares
Mean	1	$R(\mu) = abn\bar{y}^2_{...}$
α after μ	$a - 1$	$R(\alpha \mid \mu) = R(\alpha \mid \mu, \beta) = bn\sum_i (\bar{y}_{i..} - \bar{y}_{...})^2$
β after μ	$b - 1$	$R(\beta \mid \mu, \alpha) = R(\beta \mid \mu) = an\sum_j (\bar{y}_{.j.} - \bar{y}_{...})^2$
γ after μ, α and β	$(a - 1)(b - 1)$	$R(\gamma \mid \mu, \alpha, \beta) = n\sum_i\sum_j (\bar{y}_{ij.} - \bar{y}_{i..} - \bar{y}_{.j.} + \bar{y}_{...})^2$
Residual error	$ab(n - 1)$	$\text{SSE} = \sum_i\sum_j\sum_k (y_{ijk} - \bar{y}_{ij.})^2$
Total	abn	$\text{SST} = \sum_i\sum_j\sum_k y^2_{ijk}$

3. INTERPRETATION OF HYPOTHESES

None of the hypotheses (97), (100), (106), (107) or (110) are particularly appealing so far as interpretability is concerned. All of them involve the data themselves—not their magnitudes but the numbers of them, the values of the n_{ij}. For example, (100) is

$$H: \quad \alpha_i + \frac{1}{n_{i\cdot}}\sum_{j=1}^{b} n_{ij}(\beta_j + \gamma_{ij}) \text{ equal for all } i; \tag{124}$$

and the corresponding hypothesis in the no-interaction case is

$$H: \quad \alpha_i + \frac{1}{n_{i.}} \sum_{j=1}^{b} n_{ij}\beta_j \text{ equal for all } i, \tag{125}$$

analogous to (48). In these, the hypotheses involve the n_{ij}'s not only in terms of the weight with which β_j, for example, enters into the hypotheses, but also in relation to whether some of the β_j's enter the hypotheses at all. Thus in (124), if $n_{ip} = 0$, β_p will not occur in the expression containing α_i. In this way, the pattern of the data (the pattern of which n_{ij} are zero and which are not) governs the form of the hypotheses being tested by the F-statistics in Tables 7.2 and 7.7. In Table 7.2, $F(\alpha \mid \mu, \beta)$ and $F(\beta \mid \mu, \alpha)$ test, respectively, differences between α's and differences between β's, but otherwise all hypotheses tested by the F's in Tables 7.2 and 7.7 involve the data through the values of the n_{ij}. In $F(M)$ of both tables, the hypothesis is $H\colon \; E(\bar{y}) = 0$, which is

$$H: \quad \sum_i \sum_j n_{ij}(\mu + \alpha_i + \beta_j + \gamma_{ij})/N = 0,$$

involving weighted means of the elements of the model as they occur in $\bar{y}$. With $F(\alpha \mid \mu)$, the hypothesis involves the α's in the presence of a weighted mean of those β's (and γ's) with which the α's occur in the data; and in $F(\beta \mid \mu)$ the same is true of β's in the presence of weighted means of α's (and γ's). For $F(\alpha \mid \mu, \beta)$ and $F(\beta \mid \mu, \alpha)$ of Table 7.7, the hypotheses involve the n_{ij}'s in the (somewhat complex) manner shown in (106) and (107). For Table 7.7, only in the hypothesis (110) for $F(\gamma \mid \mu, \alpha, \beta)$ are the n_{ij}'s not involved explicitly, but even here their effect is implicit, because their being zero or non-zero determines which functions of the γ's make up the hypothesis.

This dependence of hypotheses on the structure of available data throws doubt on the validity of such hypotheses. Usually an experimenter wishes to test hypotheses that arise from the context of his work and not hypotheses that depend on the pattern of n_{ij}'s in his data. In general, the F-statistics of the analyses in Tables 7.2, 7.3 and 7.7 do, however, rely on the n_{ij}-values of the data. Only if the n_{ij}'s (as they occur in the data) are in direct proportion to the occurrence of the elements of the model in the population might some of the hypotheses corresponding to the analysis of variance F-statistics be valid. This is the case with proportionate subclass numbers, in which case, for example, (125) becomes

$$H: \quad \alpha_i + \sum_{j=1}^{b} p_j\beta_j \text{ equal for all } i,$$

equivalent to $\qquad H: \quad \alpha_i \text{ equal for all } i.$

A feature of the hypotheses (110) tested by $F(\gamma \mid \mu, \alpha, \beta)$ warrants attention. It involves

$$\begin{aligned}\theta_{ij,i'j'} &= \gamma_{ij} - \gamma_{i'j} - \gamma_{ij'} + \gamma_{i'j'}\\ &= \mu_{ij} - \mu_{i'j} - \mu_{ij'} + \mu_{i'j'}\\ &= E(\bar{y}_{ij\cdot}) - E(\bar{y}_{i'j\cdot}) - E(\bar{y}_{ij'\cdot}) + E(\bar{y}_{i'j'\cdot})\\ &= E(\bar{y}_{ij\cdot}) - E(\bar{y}_{i'j\cdot}) - [E(\bar{y}_{ij'\cdot}) - E(\bar{y}_{i'j'\cdot})],\end{aligned}$$

which is a measure of the extent to which the difference between the expected value of the ith and i'th treatments (in terms of Table 7.6) when used on variety j differs from their difference when used on variety j'. This is just the definition of interaction discussed in Sec. 4.3d(ii). Hence we can say that $F(\gamma \mid \mu, \alpha, \beta)$ tests interactions. However, consider what is meant by this; it does not necessarily mean that we are testing the hypothesis that interactions are zero. It would, if the hypothesis were $\theta_{ij,i'j'} = 0$ for sets of various values of i, j, i' and j'. But this is not always so. For example, in (111), part of the hypothesis is $\theta_{11,22} + \theta_{12,33} = 0$ or equivalently, from (116), $\theta_{22,33} - \theta_{21,13} = 0$. As hypotheses, these two statements are clearly not equivalent to hypotheses of θ's being zero. This fact is important. It means, for example, that in testing $\theta_{22,33} - \theta_{21,13} = 0$ each of $\theta_{22,33}$ and $\theta_{21,13}$ could be non-zero (nay, even non-zero and very large but equal) with the hypothesis still being true. The important of this is that $F(\gamma \mid \mu, \alpha, \beta)$ is not, with unbalanced data, testing that interactions are zero. Some interactions can be non-zero, although equal in magnitude (of the same or opposite sign), with the hypothesis tested by $F(\gamma \mid \mu, \alpha, \beta)$ still being true.

4. CONNECTEDNESS

Suppose available data occur as indicated in Table 7.10. If each cell that contains data has only a single observation, the normal equations

TABLE 7.10. PRESENCE OR ABSENCE OF DATA FOR DISCUSSING CONNECTEDNESS (× INDICATES ONE OR MORE OBSERVATIONS; — INDICATES NO OBSERVATIONS)

Level of α-factor	Level of β-factor 1	2	3	4
1	×	×	—	—
2	×	×	—	—
3	—	—	×	×

are as follows.

$$
\begin{array}{cccccccc}
\mu^o & \alpha_1^o & \alpha_2^o & \alpha_3^o & \beta_1^o & \beta_2^o & \beta_3^o & \beta_4^o
\end{array}
$$

$$
\begin{bmatrix}
6 & 2 & 2 & 2 & 2 & 2 & 1 & 1 \\
2 & 2 & \cdot & \cdot & 1 & 1 & \cdot & \cdot \\
2 & \cdot & 2 & \cdot & 1 & 1 & \cdot & \cdot \\
2 & \cdot & \cdot & 2 & \cdot & \cdot & 1 & 1 \\
2 & 1 & 1 & \cdot & 2 & \cdot & \cdot & \cdot \\
2 & 1 & 1 & \cdot & \cdot & 2 & \cdot & \cdot \\
1 & \cdot & \cdot & 1 & \cdot & \cdot & 1 & \cdot \\
1 & \cdot & \cdot & 1 & \cdot & \cdot & \cdot & 1
\end{bmatrix}
\begin{bmatrix}
\mu^o \\ \alpha_1^o \\ \alpha_2^o \\ \alpha_3^o \\ \beta_1^o \\ \beta_2^o \\ \beta_3^o \\ \beta_4^o
\end{bmatrix}
=
\begin{bmatrix}
y_{..} \\ y_{1.} \\ y_{2.} \\ y_{3.} \\ y_{.1} \\ y_{.2} \\ y_{.3} \\ y_{.4}
\end{bmatrix}
\qquad (126)
$$

Subtracting the fourth equation from the first changes (126) to

$$
\begin{array}{cccccccc}
\mu^o & \alpha_1^o & \alpha_2^o & \alpha_3^o & \beta_1^o & \beta_2^o & \beta_3^o & \beta_4^o
\end{array}
$$

$$
\begin{bmatrix}
4 & 2 & 2 & \cdot & 2 & 2 & \cdot & \cdot \\
2 & 2 & \cdot & \cdot & 1 & 1 & \cdot & \cdot \\
2 & \cdot & 2 & \cdot & 1 & 1 & \cdot & \cdot \\
2 & \cdot & \cdot & 2 & \cdot & \cdot & 1 & 1 \\
2 & 1 & 1 & \cdot & 2 & \cdot & \cdot & \cdot \\
2 & 1 & 1 & \cdot & \cdot & 2 & \cdot & \cdot \\
1 & \cdot & \cdot & 1 & \cdot & \cdot & 1 & \cdot \\
1 & \cdot & \cdot & 1 & \cdot & \cdot & \cdot & 1
\end{bmatrix}
\begin{bmatrix}
\mu^o \\ \alpha_1^o \\ \alpha_2^o \\ \alpha_3^o \\ \beta_1^o \\ \beta_2^o \\ \beta_3^o \\ \beta_4^o
\end{bmatrix}
=
\begin{bmatrix}
y_{1.} + y_{2.} \\ y_{1.} \\ y_{2.} \\ y_{3.} \\ y_{.1} \\ y_{.2} \\ y_{.3} \\ y_{.4}
\end{bmatrix}
$$

which can be rewritten as

$$
\begin{bmatrix}
4 & 2 & 2 & 2 & 2 \\
2 & 2 & \cdot & 1 & 1 \\
2 & \cdot & 2 & 1 & 1 \\
2 & 1 & 1 & 2 & \cdot \\
2 & 1 & 1 & \cdot & 2
\end{bmatrix}
\begin{bmatrix}
\mu^o \\ \alpha_1^o \\ \alpha_2^o \\ \beta_1^o \\ \beta_2^o
\end{bmatrix}
=
\begin{bmatrix}
y_{1.} + y_{2.} \\ y_{1.} \\ y_{2.} \\ y_{.1} \\ y_{.2}
\end{bmatrix}
\qquad (127)
$$

and
$$\begin{bmatrix} 2 & 2 & 1 & 1 \\ 1 & 1 & 1 & \cdot \\ 1 & 1 & \cdot & 1 \end{bmatrix} \begin{bmatrix} \mu^o \\ \alpha_3^o \\ \beta_3^o \\ \beta_4^o \end{bmatrix} = \begin{bmatrix} y_{3\cdot} \\ y_{\cdot 3} \\ y_{\cdot 4} \end{bmatrix}. \tag{128}$$

Thus, although the normal equations for the data pattern of Table 7.10 are (126), they can be separated into two sets of equations which, apart from μ, involve quite separate parameters: α_1, α_2, β_1 and β_2 in (127) and α_3, β_3 and β_4 in (128). Furthermore, the data involved in the two sets of equations are also separate: y_{11}, y_{12}, y_{21} and y_{22} in (127) and y_{33} and y_{34} in (128). This separation of the normal equations is brought about by the nature of the data, by the manner in which certain of the cells of the 2-way classification have data and others do not. When it occurs, we say the data are *not connected*, or *disconnected;* otherwise they are *connected*. When data are disconnected the separate sets of data corresponding to the separate sets of normal equations, such as (127) and (128), will be called *disconnected sets of data*. Thus data in the pattern of Table 7.10 are disconnected and there are two disconnected sets of data: one is y_{11}, y_{12}, y_{21} and y_{22} and the other is y_{33} and y_{34}.

The underlying characteristic of disconnected data is that each of its disconnected sets of data can be analyzed separately from the other such sets; each has its own normal equations that can be solved without reference to those of other sets. This is evident in (127) and (128). Certainly each contains μ^o: but since each group of normal equations is less than full rank, they can all be solved with a common μ^o if desired ($\mu^o = 0$ is one possibility).

Disconnectedness of data means not only that each disconnected set of data can be analyzed separately but that all the data cannot be analyzed as a single group of data. For example, as mentioned in Sec. 1d, in the "absorption process" for obtaining $R(\mu, \alpha, \beta)$, the matrix $\mathbf{C}^{-1}$ does not exist for disconnected data. Further evidence of the inability to analyze such data as a single set of data comes from the degrees of freedom of what would otherwise be the analysis. For example, data in the pattern of Table 7.10 would give degrees of freedom for $R(\gamma \mid \mu, \alpha, \beta)$ as $s - a - b + 1 = 6 - 3 - 4 + 1 = 0$. With some patterns of data this value can be negative; thus, were there to be no data in the (1, 1)-cell of Table 7.10, $s - a - b + 1$ would be $5 - 3 - 4 + 1 = -1$, which is clearly meaningless.

Disconnected data have to be analyzed on a within-set basis. This is so whether there is one observation per filled cell or one or more observations. Within each disconnected set of data the appropriate analysis (Table 7.3 or Table 7.8) can be made. From these analyses a pooled analysis can be established. However, in view of the complexity of some of the hypotheses tested by the F-statistics implicit in Tables 7.3 and 7.8, such pooling may be of little practical value. Nevertheless, it is instructive to demonstrate the degrees of freedom for these analyses, as distinct from those that would be given by analyzing the complete data without taking their disconnectedness into account. This pooling is shown in Table 7.11. It is assumed that there are d disconnected sets of data, for the tth of which there are N_t observations, a_t rows, b_t columns and s_t filled cells, with corresponding sums of squares also subscripted by t. The nature of disconnectedness ensures that

$$N = \sum_t N_t, \quad a = \sum_t a_t, \quad b = \sum_t b_t \quad \text{and} \quad s = \sum_t s_t$$

where summation over t is for $t = 1, 2, \ldots, d$. In Table 7.11 we also write

$$p = s - a - b + 1 \quad \text{and} \quad p_t = s_t - a_t - b_t + 1,$$

with

$$p = \sum_t p_t - d + 1. \tag{129}$$

Table 7.11 is based on Table 7.8, for fitting $\alpha \mid \mu$ and $\beta \mid \mu, \alpha$. A similar table, for fitting $\beta \mid \mu$ and $\alpha \mid \mu, \beta$, can also be constructed.

In Table 7.11 the residual sum of squares for the pooled analysis provides an estimator of σ^2 as

$$\hat{\sigma}^2 = \frac{\sum_{t=1}^{d} \text{SSE}_t}{\sum_{t=1}^{d} (N_t - s_t)},$$

which can be used in tests of hypotheses. Also, the first line of the pooled analysis, that for the means, can be partitioned into two terms. Letting

$$m = \text{mean of all data}$$
$$= \sum_t \sqrt{N_t R_t(\mu)} \Big/ \sum_t N_t$$

the partitioning of $\sum_t R_t(\mu)$ with d degrees of freedom is

$$m^2 \sum_t N_t \qquad \text{with 1 degree of freedom}$$

and

$$\sum_t R_t(\mu) - m^2 \sum_t N_t \qquad \text{with } d - 1 \text{ degrees of freedom.}$$

The latter, divided by $(d - 1)\hat{\sigma}^2$, can be used to test the hypothesis of equality of the $E(\bar{y})$'s corresponding to the disconnected sets of data. An example of

TABLE 7.11. POOLING OF ANALYSES OF VARIANCE OF DISCONNECTED SETS OF DATA IN A 2-WAY CLASSIFICATION

tth disconnected set of data[2]			Pooling of d disconnected sets of data[1]		
Source of Variation	d.f.	Sum of Squares	Source of Variation	d.f.	Sum of Squares
μ	1	$R_t(\mu)$	μ: for each set	d	$\sum_t R_t(\mu)$
$\alpha \mid \mu$	$a_t - 1$	$R_t(\alpha \mid \mu)$	$\alpha \mid \mu$, within sets	$\sum_t a_t - d = a - d$	$\sum_t R_t(\alpha \mid \mu)$
$\beta \mid \mu, \alpha$	$b_t - 1$	$R_t(\beta \mid \mu, \alpha)$	$\beta \mid \mu, \alpha$, within sets	$\sum_t b_t - d = b - d$	$\sum_t R_t(\beta \mid \mu, \alpha)$
$\gamma \mid \mu, \alpha, \beta$	p_t	$R_t(\gamma \mid \mu, \alpha, \beta)$	$\gamma \mid \mu, \alpha, \beta$, within sets	$\sum_t p_t = p + d - 1$	$\sum_t R_t(\gamma \mid \mu, \alpha, \beta)$
Residual	$N_t - s_t$	SSE_t	Residual, within sets	$\sum_t (N_t - s_t) = N - s$	$\sum_t \text{SSE}_t$
Total	N_t	$(\sum y^2)_t$	Total	$\sum_t N_t$	$\sum_t (\sum y^2)_t$

[1] $\sum_t$ is for $t = 1, 2, \ldots, d$.

[2] Equation (129) defines p_t and p.

TABLE 7.12. DEGREES OF FREEDOM IN ANALYSIS OF VARIANCE FOR DATA PATTERN OF TABLE 7.10

	Degrees of freedom			
	Analyzed as disconnected data			Analyzed, wrongly, as one set of data, ignoring disconnectedness
	2 disconnected sets		Pooled analysis	
Source of Variation	Set 1 Cells 11, 12, 21, 22	Set 2 Cells 33 and 34		
μ	1	1	2	1
$\alpha \mid \mu$	1	0	1	2
$\beta \mid \mu, \alpha$	1	1	2	3
$\gamma \mid \mu, \alpha, \beta$	1	0	1	0
Residual	$N_1 - 4$	$N_2 - 2$	$N - 6$	$N - 6$
Total	N_1	N_2	N	N

Table 7.11, showing degrees of freedom only, for the data pattern of Table 7.10, is given in Table 7.12.

Estimability of certain functions is greatly affected by disconnectedness. For example, in the case of the no-interaction model of equations (127) and (128) derived from Table 7.10, $\beta_1 - \beta_3$ is not estimable. This is so because β_1 is a parameter in one disconnected set of data and β_3 is a parameter in the other set. In general, functions of parameters that involve parameters relating to different disconnected sets of data are not estimable, whereas functions involving parameters relating to any single set of data can be estimable. For example, in Table 7.10 $\beta_2 - \beta_3$ is not estimable but $\beta_1 - \beta_2$ and $\beta_3 - \beta_4$ are. This is for the no-interaction model. For the interaction model, μ_{ij} of (73) is estimable for all $n_{ij} \neq 0$. But functions of μ_{ij} that involve μ_{ij}'s from different disconnected sets of data are not estimable.

With connected data the rank of $\mathbf{X}$ or, equivalently, of $\mathbf{X'X}$ in the normal equations, is $a + b - 1$ in the no-interaction case. Thus, if data corresponding to Table 7.10 were connected, the rank of $\mathbf{X'X}$ in (126) would be $3 + 4 - 1 = 6$. But because the data are not connected, the rank is $a + b - 1 - (d - 1) = 5$ where d is the number of disconnected sets of data. Equations (127) and (128) illustrate this, their ranks being $2 + 2 - 1 = 3$ and $1 + 2 - 1 = 2$ respectively, summing to 5, the rank of (126). This accounts for the relationship $p = \sum_{t=1}^{d} p_t - (d - 1)$ shown in (129) and in Table 7.11.

It is clear that for a complete set of data to be analyzed by the methods of Table 7.3 or 7.8 (whichever is appropriate) connectedness must be a property of the data. Weeks and Williams (1964) discuss this property for the general k-way classification without interaction and give a procedure for investigating whether or not data are connected. This is given in the next chapter. For data in a 2-way classification it simplifies to the following. Take any cell containing data—the (p, q)th-cell, say. From that cell move along the pth row (in either direction), or up or down the qth column until another filled cell is encountered. Proceed from that cell in the same manner. If, by moving in this fashion, all filled cells can be encountered, then the data are connected; otherwise they are disconnected. If data are disconnected this process will isolate the disconnected set of data containing the original (p, q)th-cell. Restarting the process in some cell not in that set will generate another disconnected set. Continued repetition in this manner yields all the disconnected sets.

Example. In the following array of dots and $\times$'s a dot represents an empty cell and an $\times$ represents a filled cell. The lines joining the $\times$'s isolate the 3 disconnected sets of data.

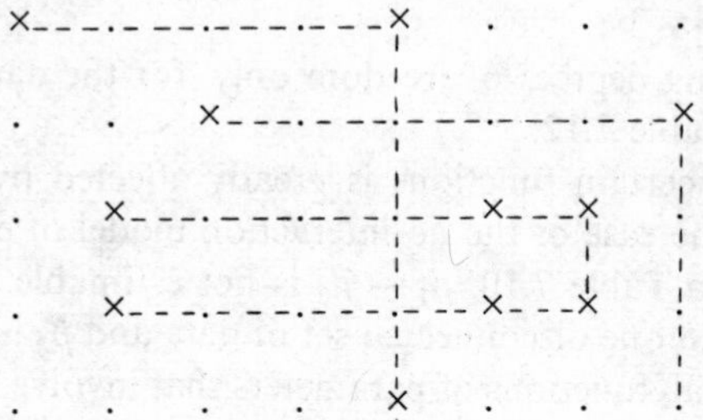

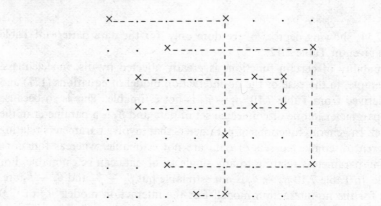

5. μ_{ij}-MODELS

In discussing estimable functions in both the no-interaction and the interaction models of Secs. 1 and 2, great play was made of the fact that $\mu + \alpha_i + \beta_j$ and $\mu_{ij} = \mu + \alpha_i + \beta_j + \gamma_{ij}$ respectively were estimable. In both cases all other estimable functions were linear combinations of these, and in neither case were μ, the α_i, nor the β_j individually estimable—nor the γ_{ij} in the interaction case. In special cases of restricted models, usually with balanced data, these individual elements can become estimable (as discussed in Secs. 1h and 2g) but in general they are not. However, on writing $\mu_{ij} = \mu + \alpha_i + \beta_j$ in the no-interaction model, we can say that in each model the basic underlying estimable function is μ_{ij} (appropriately defined) for $n_{ij} \neq 0$. This fact gives rise to considering what may be called μ_{ij}-models.

A μ_{ij}-model consists of simply writing (in the interaction case)

$$y_{ijk} = \mu_{ij} + e_{ijk}$$

where the e_{ijk}'s have the same distributional properties as before. Then μ_{ij} for $n_{ij} \neq 0$ is estimable, with b.l.u.e. $\bar{y}_{ij.}$ and $v(\hat{\mu}_{ij}) = \sigma^2/n_{ij}$. Any linear function of the estimable μ_{ij}'s is estimable, $\mathbf{k}'\boldsymbol{\mu}$ for example, with b.l.u.e. $\mathbf{k}'\bar{\mathbf{y}}$ and variance $\mathbf{k}'\mathbf{D}\{1/n_{ij}\}\mathbf{k}\sigma^2$; and any hypothesis relating to linear functions of the μ's is testable. Also, the reduction in sum of squares for fitting the model is $R(\mu_{ij}) = \sum_i \sum_j n_{ij}\bar{y}_{ij.}^2$, the same reduction as that in fitting any of the models containing γ_{ij}—see equations (71) and (72).

The simplicity of such a model is readily apparent. There is no confusion over which functions are estimable, what their b.l.u.e.'s are and what hypotheses can be tested. This results from the fact that the μ_{ij}-model is always of full rank with the corresponding value of $\mathbf{X}'\mathbf{X}$ being $\mathbf{D}\{n_{ij}\}$ for $n_{ij} \neq 0$. The normal equations are therefore quite straightforward with simple solutions $\hat{\boldsymbol{\mu}} = \bar{\mathbf{y}}$, where $\boldsymbol{\mu}$ is the vector of μ's and $\bar{\mathbf{y}}$ the corresponding vector of observed cell means.

The μ_{ij}-models have the property that the number of parameters in a model exactly equals the number of filled cells, thus giving rise to the full rank nature of the normal equations. The reason is that the model so specified is not over-specified as it is in using the customary μ, α_i's and β_j's. For example, in the no-interaction model there are, with a rows and b columns, $1 + a + b$ parameters, but only $a + b - 1$ linearly independent means from which to try to estimate them. With the interaction model there are, for s filled cells, $1 + a + b + s$ parameters but only s linearly independent means. In both cases, therefore, there are more parameters in the model than there are linearly independent means in the estimation process. Hence it is impossible to estimate every parameter individually. The μ_{ij}-model is therefore conceptually much easier: there are exactly as many μ_{ij}'s to be estimated as there are observed (cell) means, with a one-to-one correspondence.

From the sampling viewpoint this is appropriate, because to the person whose data are being analyzed the important thing is the s populations corresponding to the s observed sample means $\bar{y}_{ij.}$. Each of these is an estimator of the mean of the population from which the y_{ijk}'s are deemed to be a random sample. These populations are the factor of underlying interest, and therefore the $\bar{y}_{ij.}$'s, the sample means, as the estimators (b.l.u.e.'s) of the population means, are the foundation of the estimation procedure. So far as estimating functions of these population means and testing hypotheses about them is concerned, it is up to the person whose data they are (aided, presumably, by a statistician) to specify in terms of the μ_{ij}'s the functions and hypotheses that are of interest to him. This, of course, is done within the context of the data and what they represent. In short, the situation is no more than that of estimating population means and functions of them and testing hypotheses about them. Just which functions and which hypotheses is determined by the contextual situation of the data. Speed (1969) gives a very complete discussion of this whole topic, and Urquhart *et al.* (1970), in considering certain aspects of it, trace the historical development of linear models as we use them today.

As an example, the experimenter, or person it is whose data are being analyzed, can define row effects as

$$\rho_i = \sum_{j=1}^{b} t_{ij}\mu_{ij} \quad \text{for} \quad n_{ij} \neq 0$$

by giving to t_{ij} any value he pleases. Then the

$$\text{b.l.u.e. of } \rho_i \text{ is } \hat{\rho}_i = \sum_{j=1}^{b} t_{ij}\bar{y}_{ij\cdot}$$

with

$$\hat{v}(\hat{\rho}_i) = \hat{\sigma}^2 \sum_j t_{ij}^2/n_{ij}.$$

The hypothesis H: all ρ_i equal can then be tested using

$$F = \frac{\sum_i^a \hat{\rho}_i^2/v(\hat{\rho}_i) - \left[\sum_i^a \hat{\rho}_i \Big/ v(\hat{\rho}_i)\right]^2 \Big/ \sum_i^a [1/v(\hat{\rho}_i)]}{(a-1)} \tag{130}$$

as given by Henderson (1968). Proof of this result is established in the same manner as is that of equation (122)—see Exercise 17.

Novel as this simplistic approach might seem, it is in essence not at all new, for it long preceded the analysis of variance itself. Urquhart *et al.* (1970) have outlined how Fisher's early development of analysis of variance stemmed from ideas on intra-class correlation. Establishment of models with elements μ, α_i, β_j and so on, such as are currently familiar, followed the analysis of variance and did not precede it. Prior to it there is a plentiful literature on least squares (354 titles in a bibliography dated 1877—*loc. cit.*), based essentially on the estimation of cell means. Any current or future adoption of this handling of linear models would therefore represent no new basic concept. Success in doing this does, however, demand of today's readers a thorough understanding of current procedures.

6. EXERCISES

1. For the data of Table 7.1 obtain $\mathbf{b}^o$ such that $\mu^o = \alpha_1^o = 0$, and check the values of $R(\alpha \mid \mu, \beta)$ and $R(\mu, \alpha, \beta)$ in so doing.

2. Define

$$\mathbf{n}_a' = [n_{1.} \quad \cdots \quad n_{a.}], \qquad \mathbf{m}_a' = [n_{1b} \quad \cdots \quad n_{ab}],$$

$$n_\beta = \begin{bmatrix} n_{.1} \\ \cdot \\ \cdot \\ \cdot \\ n_{.b-1} \end{bmatrix}, \qquad \mathbf{y}_\beta = \begin{bmatrix} y_{.1} \\ \cdot \\ \cdot \\ \cdot \\ y_{.b-1} \end{bmatrix} \quad \text{and} \quad \mathbf{D}_\beta = \begin{bmatrix} n_{.1} & & \mathbf{0} \\ & \ddots & \\ \mathbf{0} & & n_{.b-1} \end{bmatrix}.$$

With these definitions and those given in (19), and using $n_{..}$, $n_{.b}$, $y_{..}$ and $y_{.b}$:
(*a*) Rewrite the normal equations (12).
(*b*) Express $\mathbf{b}^o$ of (20) as $\mathbf{GX'y}$ and so derive $\mathbf{G}$ of (21).
(*c*) Show that $\mathbf{X'XGX'X} = \mathbf{X'X}$.

3. In Table 7.1 change the observation for stove W and pan A from 6 to 12 and repeat the analyses of Table 7.2. What conclusions do you draw?

4. In Table 7.1 change the observation for stove W and pan A from 6 to 15 and repeat the analyses of Table 7.2. What conclusions do you draw?

5. Repeat Exercise 1 for the data of Exercises 3 and 4.

6. (*a*) Derive the inverse of

$$\mathbf{A} = x\mathbf{J} + \mathbf{D}\{y_i\}$$

as

$$\mathbf{A}^{-1} = \{a^{ii}\} \qquad \text{for} \quad i = 1, 2, \ldots, n$$

with

$$a^{ii} = \frac{1}{y_i} - \frac{x}{y_i^2\left(1 + x\sum_{i=1}^{n} 1/y_i\right)}$$

and

$$a^{ii'} = \frac{-x}{y_i y_{i'}\left(1 + x\sum_{i=1}^{n} 1/y_i\right)} \qquad \text{for } i \neq i'.$$

(*b*) With $\bar{x}_{i.} = \sum_{j=1}^{n_i} x_{ij}/n_i$ and $\bar{x}_{..} = \sum_{i=1}^{a}\sum_{j=1}^{n_i} x_{ij}/n_{.}$ show that

$$\sum_{i=2}^{a} (\bar{x}_{1.} - \bar{x}_{i.})^2(n_i - n_i^2/n_.) - \sum_{i \neq i' \neq 1}^{a}\sum^{a} (\bar{x}_{1.} - \bar{x}_{i.})(\bar{x}_{1.} - \bar{x}_{i'.})n_i n_{i'}/n_.$$

$$= \sum_{i=1}^{a} n_i(\bar{x}_{i.} - \bar{x}_{..})^2.$$

(*c*) For the hypothesis

$$H:\quad \text{equality of } \beta_j + \sum_{i=1}^{a} n_{ij}\alpha_i/n_{.j} \qquad \text{for all } j$$

in the no-interaction 2-way classification model, show that the F-statistic reduces, as in Table 7.3c, to

$$F(\beta \mid \mu) = \left(\sum_{j=1}^{b} n_{.j}\bar{y}_{.j}^2 - n_{..}\bar{y}_{..}^2\right) \Big/ (b-1)\hat{\sigma}^2$$

7. When $n_{ij} = 1$ for all i and j, show that the method of solving the normal equations for the no-interaction model by solving equations (14) and (15) leads to solutions

$$\alpha_i^o = \bar{y}_{i.} - \bar{y}_{..} + \bar{y}_{.b} \qquad \text{for all } i$$

and

$$\beta_j^o = \bar{y}_{.j} - \bar{y}_{.b} \qquad \text{for } j = 1, 2, \ldots, b-1,$$

with $\mu^o = 0 = \beta_b^o$.

8. Suppose the lost observations of Table 7.1 are found to be 13 and 5 for pans A and B, respectively, on stove Y and 12 for pan B on stove Z. Solve the normal equations for the complete set of (now balanced) data by the same procedures as used in equations (3)–(11). In doing so, verify the results of Exercise 7.

9. Show that when all $n_{ij} = 1$, the equation $\mathbf{C}\boldsymbol{\beta}^o = \mathbf{r}$ in (16) has

$$\mathbf{C} = a\mathbf{I} - (a/b)\mathbf{J} \qquad \text{and} \qquad \mathbf{r} = a(\bar{\mathbf{y}}_\beta - \bar{y}_{..}\mathbf{1})$$

and hence $\boldsymbol{\beta}^o = \mathbf{C}^{-1}\mathbf{r} = \bar{\mathbf{y}}_\beta - \bar{y}_{.b}\mathbf{1}$,

as obtained in Exercise 7. From this show that (32) becomes

$$R(\beta \mid \alpha, \mu) = \boldsymbol{\beta}^{o\prime}\mathbf{r} = a\sum_{j=1}^{b} \bar{y}_{.j}^2 - ab\bar{y}_{..}^2 = \sum_{i=1}^{a}\sum_{j=1}^{b} (\bar{y}_{.j} - \bar{y}_{..})^2$$

of Table 7.5. Thence show that when $n_{ij} = 1$ for all i and j, Tables 7.3b and 7.3*c* simplify to Table 7.5. (*Note:* All matrices and vectors are of order $b - 1$; $\mathbf{J}$ has all elements unity; and $\bar{\mathbf{y}}_\beta' = [\bar{y}_{.1} \quad \cdots \quad \bar{y}_{.b-1}]$.)

10. Four men and four women play a series of bridge games. At one point in their playing their scores are as shown below.

Bridge Scores (100's)

	Men			
Women	A	B	C	D
P	8	—	9	10
Q	13	—	—	—
R	—	6	14	—
S	12	14	10	24

The blanks are scores which were lost by the scorekeepers. Carry out an analysis of variance procedure to investigate differences between players of the same sex.

11. Use equation (69) and Table 7.8 to confirm Table 7.7.

12. For the data of Table 7.6 establish which of the following functions are estimable and find their b.l.u.e.'s.

    (*a*) $\alpha_2 - \alpha_3 + \beta_1 + \gamma_{21} - \frac{1}{2}(\beta_3 + \beta_4 + \gamma_{33} + \gamma_{34})$

    (*b*) $\beta_2 - \beta_3 + \frac{1}{2}(\alpha_2 - \alpha_1) + \frac{1}{2}(\gamma_{22} + \gamma_{32} - \gamma_{13} - \gamma_{33})$

    (*c*) $\alpha_1 - \alpha_2 + \frac{1}{3}(\beta_1 - \beta_2) + \frac{1}{3}(\gamma_{11} - \gamma_{12})$

    (*d*) $\beta_2 - \beta_3 + \frac{1}{2}(\gamma_{22} + \gamma_{32}) - \frac{1}{3}(\gamma_{13} + 2\gamma_{33})$

    (*e*) $\gamma_{11} - \gamma_{12} - \gamma_{21} + \gamma_{22}$

    (*f*) $\gamma_{11} - \gamma_{14} - \gamma_{21} + \gamma_{22} - \gamma_{32} + \gamma_{34}$

13. Set up a linear hypothesis which is tested by $F(\beta \mid \mu)$ in Table 7.7c and show that its numerator sum of squares is $37\frac{56}{70}$.

14. Set up a linear hypothesis which is tested by $F(\beta \mid \mu, \alpha)$ in Table 7.7b and show that its numerator sum of squares is $36\frac{55}{70}$.

15. For the second function of γ's in (111), find an equivalent function of μ_{ij}'s different from those given in (115) and (116).

16. Formulate a hypothesis, different from that in (111), to be tested by $F(\gamma \mid \mu, \alpha, \beta)$ of Table 7.7.

17. Derive equations (122) and (130), and show that when $n_{ij} = n$ for all i and j they both reduce to $\sum_{i=1}^{a} bn(\bar{y}_{i..} - \bar{y}_{...})^2/(a - 1)\hat{\sigma}^2$.

18. An illustration of unbalanced data used by Elston and Bush (1964) is the following.

Level of α-factor	Level of β-factor: 1	2	3	Total
	Observations			
1	2, 4	3, 5	2, 3	19
2	5, 7	—	3, 1	16
Total	18	8	9	35

Calculate Table 7.8 for these data. An "analysis of variance" given by Elston and Bush shows the following sums of squares.

A	3.125
B	12.208
Interaction	6.125
Error	8.500

Show that the sum of squares designated A is $R(\alpha \mid \mu, \beta)$ and that denoted by B is $R(\beta \mid \mu, \alpha)$. Write down hypotheses tested by the F-statistics available from your calculations and verify their numerator sums of squares.

19. Calculate analyses of variance (Table 7.8) for the following data.

	β-factor		
α-factor	Level 1	Level 2	Level 3
		Observations	
Level 1	13, 9, 8, 14	9, 7	—
Level 2	1, 5, 6	13, 11	6, 12, 7, 11

Establish the hypothesis tested by $F(\gamma \mid \mu, \alpha, \beta)$.

20. Suppose a 2-way classification has only 2 rows and 2 columns. Prove that

$$R(\alpha \mid \mu) = n_{1.}n_{2.}(\bar{y}_{1..} - \bar{y}_{2..})^2/n_{..},$$

$$R(\beta \mid \mu, \alpha) = (y_{.1.} - n_{11}\bar{y}_{1..} - n_{21}\bar{y}_{2..})^2/(n_{11}n_{12}/n_{1.} + n_{21}n_{22}/n_{2.})$$

and

$$R(\gamma \mid \mu, \alpha, \beta) = (\bar{y}_{11.} - \bar{y}_{12.} - \bar{y}_{21.} + \bar{y}_{22})^2/(1/n_{11} + 1/n_{12} + 1/n_{21} + 1/n_{22}).$$

Write down analogous expressions for $R(\beta \mid \mu)$ and $R(\alpha \mid \mu, \beta)$. Illustrate these results by calculating analyses of variance (Table 7.8) for the following data in a 2 × 2 experiment.

	β-factor	
α-factor	Level 1	Level 2
	Observations	
Level 1	9, 10, 14	2, 4, 2, 3, 4
Level 2	63	10, 12, 15, 14, 15, 18

21. In Table 7.1, Table 7.6, Exercises 10, 18 and 19, show that the data are connected.

22. Suppose that data occur in the following cells of a 2-way classification:

(1, 1), (2, 3), (2, 6), (3, 4), (3, 7), (4, 1), (4, 5), (5, 2), (5, 4), (5, 7), (6, 5) and (7, 6).

Establish which sets of data are connected. Write down the degrees of freedom for an analysis of variance of each set of data and for a pooled analysis. What would the degrees of freedom be for wrongly analyzing such data ignoring their disconnectedness? Give examples of estimable and non-estimable functions (assuming a no-interaction model).

23. In the 2-way classification no-interaction model, use the hypotheses tested by $F(\alpha \mid \mu)$ and $F(\alpha \mid \mu, \beta)$ to explain the conclusions suggested in the first two columns of the last row of Table 7.4.

CHAPTER 8

SOME OTHER ANALYSES

Chapters 6 and 7 illustrate applications of the general results of Chapter 5 (models not of full rank) to specific models that often arise in the analysis of unbalanced data. The present chapter briefly discusses three additional topics: the analysis of large-scale survey-type data, the analysis of covariance and some approximate analyses for unbalanced data. In no sense is there an attempt at completeness in the discussion of these topics. They are included for the sake of referring the reader to some of the other analyses available in the literature, with the object of providing him with a connecting link between those expositions and the procedures that have been developed in the earlier chapters of this book.

1. LARGE-SCALE SURVEY-TYPE DATA

Behavioral scientists of many different disciplines often undertake surveys involving the personal interviewing of individuals, heads of households and others. The data collected from such surveys are frequently very extensive. Not only may many people have been interviewed but each of them may have been asked numerous questions, so that the resulting data consist of observations on numerous variables and factors for a large number of people. Some of the problems of analyzing such data by the procedures of Chapter 5 are now discussed. The following example serves as illustration.

a. Example

The Bureau of Labor Statistics Survey of Consumer Expenditures, 1960–61, provides an opportunity for studying patterns of family investment, such as expenditures on equities, durables and human components of the nature

TABLE 8.1. SOME OF THE FACTORS AVAILABLE ON THE DESCRIPTION OF A HOUSEHOLD IN THE BUREAU OF LABOR STATISTICS SURVEY OF CONSUMER EXPENDITURES, 1960–61

Factor	Number of Levels
1. Occupational class of head of household	12
2. Income	11
3. Education of head of household	4
4. Race of head of household	3
5. Number of full-time earners in household	4
6. Family status (just married, 1 child, etc.)	6
7. Family size	6
8. Degree of urbanization	6
9. Geographical region	4
Total number of levels in 9 factors	56

of medical expenses, education and so on. The survey gathered data on many characteristics of each household interviewed. Some of those characteristics, coded as factors with different numbers of levels and used by Brown (1968), are shown in Table 8.1. The basic survey, based on a stratified sampling plan, included some 13,728 family units, of whom 8,577 were deemed by Brown (1968) to be suitable for studying patterns of family investment. (Exclusions were made of those family units not having a male as head of the family, or having only one person in the family, and so on.) Of the many questions of interest in such a study one of particular concern was, "To what extent is expenditure on durables affected by the factors listed in Table 8.1?" One way of attempting to answer this question might be by fitting a linear model to the variable "expenditure on durables".

b. Fitting a linear model

Data of the nature just described should, of course, be subjected to careful preliminary examination before any attempt is made to fit a model involving as many as nine factors, like those of Table 8.1. Various frequency counts and plots of the data, for example, could be included in such examination. Suppose, however, that this examination was made and a linear model, along the lines of Chapters 5–7, was fitted to take account of the factors shown in Table 8.1. Some of the difficulties involved in trying to fit such a model are now discussed.

A model that would have main effects for each of the nine factors of Table

8.1 could also include all possible interactions among those factors. These would include 1,353 first-order interactions, 18,538 second-order interactions, $\cdots$ and 5,474,304 $(= 12 \times 11 \times 4 \times 3 \times 4 \times 6 \times 6 \times 6 \times 4)$ eighth-order interactions [interactions between a level of each of the nine factors—see Sec. 4.3d(iii)]. Two questions are immediately apparent: What is the meaning of a high-order interaction such as one of order 8, and how can we handle large numbers of interactions of this nature?

The answer to the second of these questions allows us to avoid, in large measure, answering the first. Only if the data consist of at least one observation in every sub-most cell of the data, in this case in every one of the 5,474,304 sub-most cells—only then can we handle all the interactions. Since there are only 8,577 observations in the data, the interactions cannot all be considered. This state of affairs is likely to prevail with multi-factor survey data generally, because the number of sub-most cells in the model equals the product of the numbers of levels of all the factors. Furthermore, having data in every sub-most cell requires having data in certain cells that are either empty by definition or, by the nature of the factors, are almost certain to be empty, even in the population. For example, if "just married" of factor 6 in Table 8.1 excludes second marriages and other complications, those households characterized as "just married" cannot have more than 2 full-time workers (factor 5). Similarly, it seems unlikely that, for example, a household of 8 persons (factor 7), with 1 full-time earner (factor 5), where the head of the household did not finish high school (factor 3) and is in the lowest socio-economic occupational class (factor 1)—it seems unlikely that such a household would be in the highest income group (factor 2). Consideration of all possible interactions is therefore rarely feasible.

Even when all cells are filled, and the data could be analyzed using a model that included all interactions, the interpretation of high-order interactions is usually difficult. For example, can we give a reasonable description in terms of the source of our data of what we mean by an eighth-order interaction? I doubt it. Indeed, it is probably fair to say that we would have difficulty in meaningfully describing interactions of order greater than 1, certainly of order greater than 2. First-order interactions can be described and understood reasonably well (see Sec. 4.3d), but interpretation of higher order interactions can present some difficulty. We therefore suffer no great loss if the paucity of data prevents including such interactions in our model. Fortunately, whereas survey data seldom enable all interactions to be included in a model, they are oft-times numerous enough to provide consideration of first-order interactions, which are the interactions we can most readily interpret. This is the case with the data of 8,577 observations in Brown (1968). They are sufficiently numerous to consider the 56 main effects of Table 8.1, together with the corresponding 1,353 first-order interactions—but *not* the 18,538 second-order interactions.

Even when data are sufficient in number to consider first-order interactions we may not want to include them all in the model. For example, with the 9 factors of Table 8.1 there are 36 different kinds of first-order interactions [$\frac{1}{2}n(n-1)$ kinds for n factors]. The choice of which to include in a model is always that of the person whose data are being analyzed. It is he who should know his data well enough to decide which of the interactions should be considered and which should not. Even with just first-order interactions it is a choice which may not always be easy. Moreover, we will see that multi-factor models without any interactions present difficulty enough in interpretation, difficulty that is only further compounded by having interactions.

c. Main-effects-only models

The quandry of which interactions to include in a model can be avoided by omitting them all. The model then involves just the main effects—effects for each of the 56 levels of the 9 factors in Table 8.1. Clearly such a model is a great deal easier, conceptually, than one involving numerous interactions, the choice of which for inclusion in the model may be a matter of question. However, easier though this model appears, it too has some difficulties. The first is an extension of the duality apparent in Tables 7.2b and 7.2c. There, for the 2-way classification, we could consider reductions in sums of squares $R(\alpha \mid \mu)$ and $R(\beta \mid \mu, \alpha)$ or $R(\beta \mid \mu)$ and $R(\alpha \mid \mu, \beta)$: there are 2 sequences in which the main effects can be fitted, either α and then β or β and then α. But with the 9 factors of Table 8.1 there are $9! = 362{,}880$ sequences in which the main effects can be fitted. The choice of which sequence to use in the 2-way classification of Chapter 7 may be immaterial because there are only two sequences and it is relatively easy to look at both, but with 362,880 sequences in the 9-way classification it is essential to decide which few of them are going to be considered. This is a decision for the person whose data are being analyzed—and, again, it is often a decision that is not easy to make. An n-way classification has $n!$ sequences in which the main effects of the n factors can be fitted. Table 8.2 shows the $3! = 6$ sets of reductions in sums of squares that could be calculated for a 3-way classification.

Reductions in sums of squares such as are shown in Table 8.2 are sometimes said to "add up"—they add up to $\text{SST} = \mathbf{y}'\mathbf{y}$, the total uncorrected sums of squares of the observations. Often, of course, the $R(\mu)$ term is not shown in the body of the table but is subtracted from SST to have the other reductions in sums of squares adding up to $\text{SST}_m = \text{SST} - R(\mu) = \Sigma\, y^2 - N\bar{y}^2$.

F-statistics implicit in any of the sets of reductions of sums of squares illustrated in Table 8.2 can be used in either of two ways, as they are in

TABLE 8.2. SETS OF REDUCTIONS IN SUMS OF SQUARES FOR A 3-WAY CLASSIFICATION, MAIN-EFFECTS-ONLY MODEL, WITH MAIN EFFECTS α, β AND γ

$R(\mu)$	$R(\mu)$	$R(\mu)$	$R(\mu)$	$R(\mu)$	$R(\mu)$
$R(\alpha \mid \mu)$	$R(\alpha \mid \mu)$	$R(\beta \mid \mu)$	$R(\beta \mid \mu)$	$R(\gamma \mid \mu)$	$R(\gamma \mid \mu)$
$R(\beta \mid \mu, \alpha)$	$R(\gamma \mid \mu, \alpha)$	$R(\alpha \mid \mu, \beta)$	$R(\gamma \mid \mu, \beta)$	$R(\alpha \mid \mu, \gamma)$	$R(\beta \mid \mu, \gamma)$
$R(\gamma \mid \mu, \alpha, \beta)$	$R(\beta \mid \mu, \alpha, \gamma)$	$R(\gamma \mid \mu, \alpha, \beta)$	$R(\alpha \mid \mu, \beta, \gamma)$	$R(\beta \mid \mu, \alpha, \gamma)$	$R(\alpha \mid \mu, \beta, \gamma)$
SSE[1]	SSE	SSE	SSE	SSE	SSE
SST[1]	SST	SST	SST	SST	SST

[1] SSE $= \mathbf{y}'\mathbf{y} - R(\mu, \alpha, \beta, \gamma)$ and SST $= \mathbf{y}'\mathbf{y} = \Sigma\, y^2$

Chapter 7 for the 2-way classification. There, as discussed in Sec. 7.1e(vi), they are used for testing the effectiveness—in terms of explaining variation in y—of having certain main effect factors in the model. However, just as in Table 7.2 there are 2 possible ways of testing the explanatory power of having α in the model (α before β and α after β), so in Table 8.2 there are, for the 3-way classification, 4 ways of testing the effectiveness of α: based on $R(\alpha \mid \mu)$, $R(\alpha \mid \mu, \beta)$, $R(\alpha \mid \mu, \gamma)$ or $R(\alpha \mid \mu, \beta, \gamma)$. For the n-way classification there are 2^{n-1} ways of testing the effectiveness of a factor in this manner; e.g., $2^8 = 256$ for the 9-way classification of Table 8.1. This is a direct outcome of there being $n!$ sequences in which n main effect factors can be fitted; i.e., $n!$ sets of reductions in sums of squares of the nature illustrated in Table 8.2. The tests of the explanatory power of having any particular main effect in the model therefore depend, very naturally, on the sequence chosen for fitting the main effects.

The F-statistics can also be used, as in Sec. 7.1g, for testing hypotheses about the elements of a main-effects-only model. Here, however, just as in Sec. 7.1g, the only hypotheses that relate to these elements in a clear and simple fashion are those based on fitting one factor after fitting all the others. The hypothesis tested is that the effects of all levels of that factor are equal. For example, in Table 8.2 the hypothesis tested by $F(\alpha \mid \mu, \beta, \gamma)$, based on $R(\alpha \mid \mu, \beta, \gamma)$, is H: α's all equal; similarly $F(\beta \mid \mu, \alpha, \gamma)$ tests H: β's all equal. This is true in general: $F(\alpha \mid \mu, \beta, \gamma, \delta, \ldots, \theta)$ tests H: α's all equal, where $\beta, \gamma, \delta, \ldots, \theta$ represents all the other main effect factors of a model. The other F-statistics that can be calculated provide tests of hypotheses that involve a complex mixture of the effects in the model, just as $(R\beta \mid \mu)$ tests the hypothesis of (48) given in Sec. 7.1g. For example, $F(\alpha \mid \mu, \beta)$ from Table 8.2 will test a hypothesis that involves β's and γ's as well as α's.

Difficulties involved in testing hypotheses by means of reductions in sums of squares that "add up" have just been highlighted: choice of sequence for fitting the factors, and the complex nature of the hypotheses tested by the

F-statistics other than $F(\alpha \mid \mu, \beta, \gamma, \delta, \ldots, \theta)$. However, this in no way affects the use of the general formula

$$F(H) = (\mathbf{K}'\mathbf{b}^o - \mathbf{m})'(\mathbf{K}'\mathbf{G}\mathbf{K})^{-1}(\mathbf{K}'\mathbf{b}^o - \mathbf{m})/s\hat{\sigma}^2$$

for testing any testable hypothesis $\mathbf{K}'\mathbf{b} = \mathbf{m}$ [see equation (70) of Sec. 5.5b]. It is as applicable to situations like that of Table 8.1 as it is to anything discussed in Chapters 5, 6 and 7. As always, of course, one must first ascertain the estimability of $\mathbf{K}'\mathbf{b}$. But within the confines of estimability $F(H)$ can always be used, and its use is not necessarily related to any set of sums of squares that "add up" to SST.

d. Stepwise fitting

When using multiple regression there may, on occasion, be serious doubt about which x-variates from a large available set of x's should be used in the regression model. This difficulty has led to the development of several procedures for letting the data select a "good" set of x-variates, good in the sense of accounting for variation in y in some manner. The various procedures available differ solely in the criterion that each uses for selecting a "good" set. For example, one procedure fits one x-variate, then includes another, and then another and so on. At each step an x-variate is selected, from those not already chosen, which leads to the greatest reduction in the residual sum of squares. A lucid description of this and most of the other procedures is given in Draper and Smith (1966, Chapter 6), with interesting extensions in LaMotte and Hocking (1970) and the references shown there. We give no details of these selection procedures here, but simply point out their application to the fitting of multi-factor models. Instead of applying any one of these selection procedures to single x-variates it can be applied to the sets of dummy (0, 1) variables corresponding to each factor in a model. Then, rather than our having to decide, *a priori*, in which sequence the factors should be fitted, we could use what might be called "stepwise fitting of factors". This would determine, from the data, a sequential fitting of the factors which, in some sense, ranked the factors in decreasing order of importance insofar as accounting for variation in y is concerned. In this way, for example, rather than our selecting one of the sequences implicit in Table 8.2, the data would be used to select one for us. As a result of the stepwise regression technique, the basis of the selection would be using reductions in sums of squares, $R(\ \ )$-terms, as indicators of the extent to which different models account for variation in y.

e. Connectedness

It may sometimes be taken for granted that the difference between the effects of every pair of levels of the same factor is estimable in a main-effects-only model. Indeed this is often so, but it is not universally the case. Sufficient conditions for such differences to be estimable are those set out by

Weeks and Williams (1964) for data to be connected. Suppose there are p factors (and no interactions) in a model, and that the levels of those factors for an observation are denoted by the vector

$$\mathbf{i}' = [i_1 \quad i_2 \quad \cdots \quad i_p].$$

Then two such vectors $\mathbf{i}$ are defined as being *nearly identical* if they are equal in all except one element. Then data sets in which the $\mathbf{i}$-vector of each observation is nearly identical to that of at least one other observation form connected sets of data. A procedure for establishing such sets is given by Weeks and Williams (1964). It is an extension of that given in Sec. 7.4 for the 2-factor model.

As Weeks and Williams (1964) point out in their errata (1965), their conditions for data to be connected are sufficient but not necessary. Data can be connected (in the sense of intra-factor differences between main effects being estimable) without being nearly identical in the manner just described. Fractional factorial experiments are a case in point. For example, suppose for the model

$$y_{ijk} = \mu + \alpha_i + \beta_j + \gamma_k + e_{ijk}$$

with i, j and $k = 1, 2$ we have the data y_{112}, y_{211}, y_{121} and y_{222}. No pair of these 4 observations is nearly identical, and yet

$$E\tfrac{1}{2}(y_{112} - y_{211} + y_{121} - y_{222}) = \alpha_1 - \alpha_2 .$$

Similarly $\beta_1 - \beta_2$ and $\gamma_1 - \gamma_2$ are also estimable and thus all intra-factor differences between main effects are estimable. In that sense the data (which represent a $\frac{1}{2}$-replicate of a 2^3 factorial experiment) are connected, although they have no property of being nearly identical. This exemplifies why the general problem of finding necessary conditions for main effect differences to be estimable remains as yet unsolved.

f. μ_{ij}-models

What has been said about the difficulties of using a main-effects-only model for analyzing large-scale survey-type data applies even more to the analysis of such data using models that include interactions. The sequences in which the factors can be fitted, using reductions in sums of squares that add up to SST, are then more numerous; the hypotheses tested by the resulting F-statistics are more complicated (e.g., see Sec. 7.2f); and the problem of connectedness, in terms of the definition given in Sec. 7.4, is more acute. The example of Table 8.1 illustrates this. There we have 5,474,304 cells in the data, i.e., 5,474,304 different ways in which a household in the survey could be described by the 9 factors of Table 8.1. Yet the total number of households in the survey is only 13,728. Such data will almost assuredly not be connected.

In view of these difficulties with models that include interactions the main-effects-only models appear more feasible, despite their own difficulties, discussed in subsection c. They also have one further problem: that of complete neglect of interactions. In practice this may be a grave omission, because in situations involving many factors, as in Table 8.1, one frequently feels that interactions between the factors do, most assuredly, exist. This being so, it is not very appropriate to ignore them and proceed to make an analysis as if they did not exist. One way out of this predicament is to use the μ_{ij}-model concept discussed in Sec. 7.5. In this we look at the means of the sub-most cells of the data. By "sub-most cells" we mean those cells of the data defined by one level of each of the factors. In the two-way classification of Chapter 7 a sub-most cell is the cell defined by a row and a column; in the 9-way classification of Table 8.1 a sub-most cell is that cell defined by one level of occupational class, one level of income, one level of education of head and so on. The total number of possible sub-most cells is the product of the numbers of levels in the classes—5,474,304 in Table 8.1. The number of sub-most cells in the data is the number of the possible sub-most cells that have data in them. Call this number s. Then, no matter how many factors there are or how many levels each has, the mean of the observations in each sub-most cell is the b.l.u.e. of the population mean for that cell. Thus if $\bar{y}_r$ is the mean of the n_r observations in the rth sub-most cell, for $r = 1, 2, \ldots, s$, then $\bar{y}_r$ is the b.l.u.e. of μ_r, the population mean of that cell. Furthermore, the b.l.u.e. of any linear function $\sum_{r=1}^{s} k_r\mu_r$ is $\sum_{r=1}^{s} k_r\bar{y}_r$ with variance $\sigma^2 \sum_{r=1}^{s} k_r^2/n_r$. Also, any hypothesis concerning a linear function of the μ_r's is testable. Thus

$$H\colon\ \sum_{r=1}^{s} k_r\mu_r = m \tag{1}$$

can be tested by comparing

$$F(H) = \frac{\left(\sum_{r=1}^{s} k_r\bar{y}_r - m\right)^2}{\hat{\sigma}^2 \sum_{r=1}^{s} k_r^2/n_r} \tag{2}$$

against the F-distribution with 1 and $(n_. - s)$ degrees of freedom. The estimator of σ^2 in this expression is the simple within sub-most cell mean square, namely

$$\hat{\sigma}^2 = \sum_{r=1}^{s}\sum_{i=1}^{n_r} (y_{ri} - \bar{y}_r)^2/(n_. - s). \tag{3}$$

The numerator here is, of course, identical to the SSE that would be derived by fitting a model that had in it all possible interactions.

$F(H)$ of (2) provides a means of testing a hypothesis about any linear function of the population sub-most cell means. Just what hypotheses get so tested is the prerogative of the person whose data they are. All he need do is formulate his hypotheses of interest in terms of the sub-most cell means. Whilst this may be no easy task in many cases, it is at least uncomplicated by the confusions of estimability and interactions. Furthermore, hypotheses about sub-most cell population means can be tested simultaneously by a natural extension of the standard results for testing $\mathbf{K}'\mathbf{b} = \mathbf{m}$ in Chapters 3 and 5. Thus if $\boldsymbol{\mu}$ is the vector of sub-most cell population means and $\bar{\mathbf{y}}$ the corresponding vector of observed means, then

$$H: \quad \mathbf{K}'\boldsymbol{\mu} = \mathbf{m}, \tag{4}$$

consisting of s LIN functions $\mathbf{K}'\boldsymbol{\mu}$, is tested by using

$$F(H) = (\mathbf{K}'\bar{\mathbf{y}} - \mathbf{m})'[\mathbf{K}'\mathbf{D}\{1/n_r\}\mathbf{K}]^{-1}(\mathbf{K}'\mathbf{y} - \mathbf{m})/s\hat{\sigma}^2 \tag{5}$$

where $\mathbf{D}\{1/n_r\}$ is the diagonal matrix of the reciprocals of the numbers of observations in the sub-most cells containing data.

Repeated use of (2) and/or (5) does not provide tests whose F-statistics have numerator sums of squares that are independent, as is the case when using sums of squares that "add up", in the manner of Table 8.2. However, as we have seen, hypotheses tested by use of the latter do not involve simple functions of the parameters of the model. In contrast, the hypotheses in (1) and (4), which are tested by means of (2) and (5), are in terms of straightforward linear functions of sub-most cell population means. Further discussion of these procedures can be found in Speed (1969) and Urquhart *et al.* (1970).

2. COVARIANCE

The elements of $\mathbf{X}$ in the equation $\mathbf{y} = \mathbf{X}\mathbf{b} + \mathbf{e}$ used in the regression model (Chapter 3) are observed values of x's corresponding to the vector of observations $\mathbf{y}$. In Chapter 4 we saw how the same equation can be used for linear models involving factors and interactions by using for x's dummy variables that take the values 0 or 1; Chapter 5 gives the general theory and Chapters 6 and 7 contain examples of this. We now consider the case where some of the elements of $\mathbf{X}$ are observed x's and others are dummy (0, 1) variables. Such a situation represents a combining, into one model, of both regression and linear models involving factors and interactions. It is generally referred to as covariance analysis. The basic analysis is that of the factors-and-interactions part of the model suitably amended by the presence of the x-variates—the covariables of the analysis.

General treatment of the model $\mathbf{y} = \mathbf{X}\mathbf{b} + \mathbf{e}$ is given for $\mathbf{X}$ of full column rank in Chapter 3 and for $\mathbf{X}$ not of full column rank in Chapter 5. These two chapters cover regression and what we may call the factors-and-interactions models. With $\mathbf{X}$ having full column rank being just a special case of $\mathbf{X}$ not having full column rank, the procedures of Chapter 5 apply in general for all kinds of $\mathbf{X}$-matrices. In particular they are applicable to the analysis of covariance. Conceptually there is no distinction between the analysis of covariance and what we have already considered. The sole difference is in the form of the elements of $\mathbf{X}$. In regression (Chapter 3) the elements of $\mathbf{X}$ (apart from the column $\mathbf{1}$ corresponding to μ) are observed x's; in factors-and-interactions models (Chapters 5, 6 and 7) the elements of $\mathbf{X}$ are 0 or 1 corresponding to dummy variables. With analysis of covariance, some of the elements of $\mathbf{X}$ are dummy variable 0's and 1's and some are observed values of x-variables. Thus, conceptually, there is nothing new in the analysis of covariance. It involves fitting a model $\mathbf{y} = \mathbf{Xb} + \mathbf{e}$ where some elements of $\mathbf{b}$ are effects corresponding to levels of factors and interactions, in the manner of Chapters 5–7, and some are regression-style coefficients of x-variates, in the manner of Chapter 3. Within this context, the procedures for solving normal equations, establishing estimable functions and their b.l.u.e.'s, testing hypotheses and calculating reductions in sums of squares all follow the same pattern established in Chapter 5 and summarized at the beginning of Chapter 6. No additional concepts are involved. Furthermore, the "recipes" for covariance analysis for balanced data that are to be found in many texts [e.g., Federer (1955, Chapter XVI) and Steel and Torrie (1960, Chapter 15)] are just the consequence of simplifying the general results for unbalanced data.

a. A general formulation

(*i*) *The model.* Distinguishing between the two kinds of parameters that occur in $\mathbf{b}$ when using the model $\mathbf{y} = \mathbf{Xb} + \mathbf{e}$ for covariance analysis will be achieved by partitioning $\mathbf{b}$ into two parts: $\mathbf{a}$ for the general mean μ and the effects corresponding to levels of factors and their interactions, and $\mathbf{b}$ for the regression-style coefficients of the covariates. The corresponding incidence matrices will be $\mathbf{X}$ for the dummy (0, 1) variables and $\mathbf{Z}$ for the values of the covariates. In this way the model is written as

$$\mathbf{y} = \mathbf{Xa} + \mathbf{Zb} + \mathbf{e} \tag{6}$$

where $\mathbf{e} = \mathbf{y} - E(\mathbf{y})$, with $E(\mathbf{e}) = \mathbf{0}$ and $\text{var}(\mathbf{e}) = \sigma^2\mathbf{I}$ in the customary manner. In this formulation $\mathbf{X}$ does not necessarily have full column rank but we will assume, as is usually the case, that $\mathbf{Z}$ does. Thus $\mathbf{X'X}$ has no inverse, where $(\mathbf{Z'Z})^{-1}$ exists. Furthermore, we make the customary and realistic assumption that the columns of $\mathbf{Z}$ are linearly independent of those of $\mathbf{X}$.

(*ii*) *Solving the normal equations.* The normal equations for $\mathbf{a}^\circ$ and $\mathbf{b}^\circ$ are, from (6),

$$\begin{bmatrix} \mathbf{X'X} & \mathbf{X'Z} \\ \mathbf{Z'X} & \mathbf{Z'Z} \end{bmatrix} \begin{bmatrix} \mathbf{a}^\circ \\ \mathbf{b}^\circ \end{bmatrix} = \begin{bmatrix} \mathbf{X'y} \\ \mathbf{Z'y} \end{bmatrix}. \tag{7}$$

Suppose $(\mathbf{X}'\mathbf{X})^-$ is a generalized inverse of $\mathbf{X}'\mathbf{X}$. Then the first equation of (7) gives

$$\begin{aligned}\mathbf{a}^o &= (\mathbf{X}'\mathbf{X})^-(\mathbf{X}'\mathbf{y} - \mathbf{X}'\mathbf{Z}\mathbf{b}^o)\\ &= (\mathbf{X}'\mathbf{X})^-\mathbf{X}'\mathbf{y} - (\mathbf{X}'\mathbf{X})^-\mathbf{X}'\mathbf{Z}\mathbf{b}^o\\ &= \mathbf{a}^* - (\mathbf{X}'\mathbf{X})^-\mathbf{X}'\mathbf{Z}\mathbf{b}^o \qquad (8)\end{aligned}$$

where $\mathbf{a}^* = (\mathbf{X}'\mathbf{X})^-\mathbf{X}'\mathbf{y}$

is the solution of the normal equations for the model without covariate. Substituting for $\mathbf{a}^o$ into (7) gives the solution for $\mathbf{b}^o$:

$$\mathbf{b}^o = \{\mathbf{Z}'[\mathbf{I} - \mathbf{X}(\mathbf{X}'\mathbf{X})^-\mathbf{X}']\mathbf{Z}\}^-\mathbf{Z}'[\mathbf{I} - \mathbf{X}(\mathbf{X}'\mathbf{X})^-\mathbf{X}']\mathbf{y} \qquad (9)$$

where again the superscript minus sign designates a generalized inverse. Substitution of (9) into (8) then gives $\mathbf{a}^o$ explicitly. Solutions (8) and (9) are exactly the same results as would be obtained by using the expression for a generalized inverse of a partitioned matrix given in Sec. 1.7.

Several features of (9) should be noted. First, although $(\mathbf{X}'\mathbf{X})^-$ is not unique, it enters into $\mathbf{b}^o$ only in the form $\mathbf{X}(\mathbf{X}'\mathbf{X})^-\mathbf{X}'$, which is invariant to whatever generalized inverse of $\mathbf{X}'\mathbf{X}$ is used for $(\mathbf{X}'\mathbf{X})^-$. Thus the non-full rank property of $\mathbf{X}$ does not of itself lead to manifold solutions for $\mathbf{b}^o$. Suppose we use $\mathbf{P}$ for

$$\mathbf{P} = \mathbf{I} - \mathbf{X}(\mathbf{X}'\mathbf{X})^-\mathbf{X}' \qquad (10)$$

which, by Theorem 7 of Sec. 1.5a, is both symmetric and idempotent. Then (9) can be written as $\mathbf{b}^o = (\mathbf{Z}'\mathbf{P}\mathbf{Z})^-\mathbf{Z}'\mathbf{P}\mathbf{y}$. Symmetry and idempotency of $\mathbf{P}$ ensure that $\mathbf{Z}'\mathbf{P}\mathbf{Z}$ and $\mathbf{P}\mathbf{Z}$ have the same rank. Furthermore, the properties of $\mathbf{X}$ and $\mathbf{Z}$ given below (6) guarantee that $\mathbf{P}\mathbf{Z}$ has full column rank, and hence $\mathbf{Z}'\mathbf{P}\mathbf{Z}$ is non-singular (see Exercise 4). Therefore $\mathbf{b}^o$ is the sole solution

$$\mathbf{b}^o = \hat{\mathbf{b}} = (\mathbf{Z}'\mathbf{P}\mathbf{Z})^{-1}\mathbf{Z}'\mathbf{P}\mathbf{y}. \qquad (11)$$

(*iii*) *Estimability*. Consideration of the expected value of $\hat{\mathbf{b}}$ of (11) and of $\mathbf{a}^o$ of (8) show that $\mathbf{b}$ is estimable and that $\boldsymbol{\lambda}'\mathbf{a}$ is estimable when $\boldsymbol{\lambda}' = \mathbf{t}'\mathbf{X}$ for some $\mathbf{t}'$; i.e., $\mathbf{b}$ is always estimable and $\boldsymbol{\lambda}'\mathbf{a}$ is estimable whenever it is estimable in the model that has no covariates (See Exercise 4).

(*iv*) *A model for handling the covariates*. The estimator $\hat{\mathbf{b}}$ shown in (11) is the b.l.u.e. of $\mathbf{b}$ in the model (6). By the nature of (11) it is also the b.l.u.e. of $\mathbf{b}$ in the model having equation

$$\mathbf{y} = \mathbf{P}\mathbf{Z}\mathbf{b} + \mathbf{e}. \qquad (12)$$

This, we shall see, provides a convenient method for estimating $\mathbf{b}$.

Recall that in fitting a model of the form $\mathbf{y} = \mathbf{X}\mathbf{a} + \mathbf{e}$ the vector of estimated expected values $\hat{\mathbf{y}}$ corresponding to the vector of observed values $\mathbf{y}$ is $\hat{\mathbf{y}} = \mathbf{X}(\mathbf{X}'\mathbf{X})^-\mathbf{X}'\mathbf{y}$ [equation (10), Sec. 5.2c]. Therefore the vector of residuals, i.e., the vector of deviations of the observed values from their corresponding estimated values, is $\mathbf{y} - \hat{\mathbf{y}} = \mathbf{y} - \mathbf{X}(\mathbf{X}'\mathbf{X})^{-}\mathbf{X}'\mathbf{y}$. This, using $\mathbf{P}$ of (10), gives

$$\mathbf{y} - \hat{\mathbf{y}} = \mathbf{P}\mathbf{y}.$$

Thus $\mathbf{Py}$ is the vector of y-residuals after fitting the model $\mathbf{y} = \mathbf{Xa} + \mathbf{e}$. Similarly, if the jth column of $\mathbf{Z}$ is $\mathbf{z}_j$, the jth column of $\mathbf{PZ}$ in (12) is $\mathbf{Pz}_j$, the vector of $\mathbf{z}_j$-residuals after fitting the model[1] $\mathbf{z}_j = \mathbf{Xa} + \mathbf{e}$. Thus with

$$\mathbf{Z} = \{\mathbf{z}_j\} \qquad \text{for} \quad j = 1, 2, \ldots, q$$

we write $\mathbf{R}_z$ for $\mathbf{PZ}$ and have $\mathbf{R}_z$ as the matrix of residuals:

$$\mathbf{R}_z = \mathbf{PZ} = \{\mathbf{Pz}_j\} = \{\mathbf{z}_j - \hat{\mathbf{z}}_j\} = \{\mathbf{z}_j - \mathbf{X}(\mathbf{X}'\mathbf{X})^{-}\mathbf{X}'\mathbf{z}_j\}. \tag{13}$$

Hence the model (12) is equivalent to the model

$$\mathbf{y} = \mathbf{R}_z\mathbf{b} + \mathbf{e}, \tag{14}$$

and $\hat{\mathbf{b}}$ of (11) is

$$\hat{\mathbf{b}} = (\mathbf{R}_z'\mathbf{R}_z)^{-1}\mathbf{R}_z'\mathbf{y}.$$

$\mathbf{R}_z$ is a matrix of the same order as $\mathbf{Z}$ with its columns being columns of residuals as given in (13); $\mathbf{R}_z'\mathbf{R}_z$ is a matrix of sums of squares and products of z-residuals; and $\mathbf{R}_z'\mathbf{y}$ is a vector of sums of products of z-residuals and the y-observations.

(*v*) *Analyses of variance.* The reduction in sum of squares for fitting a linear model is the inner product of a solution vector and the vector of right-hand sides of the normal equations [e.g., equation (14) of Sec. 5.2f]. Hence the reduction in sum of squares for fitting the model (6) is, from (7), (8) and (11),

$$R(\mathbf{a}, \mathbf{b}) = \mathbf{a}^{o\prime}\mathbf{X}'\mathbf{y} + \hat{\mathbf{b}}'\mathbf{Z}'\mathbf{y}.$$

[$\mathbf{b}$ in the notation $R(\mathbf{a}, \mathbf{b})$ emphasizes the fitting of a vector of coefficients pertaining to the covariates, and $\mathbf{a}$ represents the factors-and-interactions part of the model, including μ.] On substituting for $\mathbf{a}^o$ and $\hat{\mathbf{b}}$ from (8) and (11) and making use of (10), $R(\mathbf{a}, \mathbf{b})$ reduces to

$$\begin{aligned} R(\mathbf{a}, \mathbf{b}) &= \mathbf{y}'\mathbf{X}(\mathbf{X}'\mathbf{X})^{-}\mathbf{X}'\mathbf{y} + \mathbf{y}'\mathbf{PZ}(\mathbf{Z}'\mathbf{PZ})^{-1}\mathbf{Z}'\mathbf{Py} \\ &= \mathbf{y}'\mathbf{X}(\mathbf{X}'\mathbf{X})^{-}\mathbf{X}'\mathbf{y} + \mathbf{y}'\mathbf{R}_z(\mathbf{R}_z'\mathbf{R}_z)^{-1}\mathbf{R}_z'\mathbf{y}. \end{aligned}$$

This is clearly the sum of two reductions:

$$R(\mathbf{a}) = \mathbf{y}'\mathbf{X}(\mathbf{X}'\mathbf{X})^{-}\mathbf{X}'\mathbf{y} = \mathbf{a}^{*\prime}\mathbf{X}'\mathbf{y}, \qquad \text{due to fitting } \mathbf{y} = \mathbf{Xa} + \mathbf{e},$$

and

$$\text{SSRB} = \mathbf{y}'\mathbf{R}_z(\mathbf{R}_z'\mathbf{R}_z)^{-1}\mathbf{R}_z'\mathbf{y} = \hat{\mathbf{b}}'\mathbf{R}_z'\mathbf{y}, \qquad \text{due to fitting } \mathbf{y} = \mathbf{R}_z\mathbf{b} + \mathbf{e};$$

i.e.,

$$R(\mathbf{a}, \mathbf{b}) = R(\mathbf{a}) + \text{SSRB}$$

[1] I am grateful for discussions with N. S. Urquhart.

and so

$$R(\mathbf{b} \mid \mathbf{a}) = R(\mathbf{a}, \mathbf{b}) - R(\mathbf{a}) = \text{SSRB} = \hat{\mathbf{b}}'\mathbf{R}_z'\mathbf{y}.$$

Thus SSRB is the reduction in sum of squares attributable to fitting the covariates, having already fitted the factors-and-interactions part of the model.

Distributional properties of $R(\mathbf{a})$ and $R(\mathbf{b} \mid \mathbf{a})$, based on the usual normality assumptions, come from Theorems 2 and 3 of Sec. 2.5. The idempotency of $\mathbf{X}(\mathbf{X}'\mathbf{X})^{-}\mathbf{X}'$ and of $\mathbf{R}_z(\mathbf{R}_z'\mathbf{R}_z)^{-1}\mathbf{R}_z'$ give

$$R(\mathbf{a})/\sigma^2 \sim \chi^{2\prime}[r(\mathbf{X}), \lambda_a]$$

with $$\lambda_a = \tfrac{1}{2}[\mathbf{a}'\mathbf{X}'\mathbf{X}\mathbf{a} + 2\mathbf{a}'\mathbf{X}'\mathbf{Z}\mathbf{b} + \mathbf{b}'\mathbf{Z}'\mathbf{X}(\mathbf{X}'\mathbf{X})^{-}\mathbf{X}'\mathbf{Z}\mathbf{b}]/\sigma^2;$$

and $$R(\mathbf{b} \mid \mathbf{a})/\sigma^2 \sim \chi^{2\prime}[r(\mathbf{Z}), \tfrac{1}{2}\mathbf{b}'\mathbf{R}_z'\mathbf{R}_z\mathbf{b}/\sigma^2].$$

Also, $R(\mathbf{a})$ and $R(\mathbf{b} \mid \mathbf{a})$ are distributed independently because

$$\mathbf{X}(\mathbf{X}'\mathbf{X})^{-}\mathbf{X}'\mathbf{R}_z(\mathbf{R}_z'\mathbf{R}_z)^{-1}\mathbf{R}_z = \mathbf{0},$$

since $\mathbf{R}_z = \mathbf{PZ}$ and $\mathbf{X}'\mathbf{P} = \mathbf{0}$ by the definition of $\mathbf{P}$ in (10); $R(\mathbf{a})$ and $R(\mathbf{b} \mid \mathbf{a})$ are also independent of

$$\text{SSE} = \mathbf{y}'\mathbf{y} - R(\mathbf{a}, \mathbf{b}) = \mathbf{y}'\mathbf{y} - R(\mathbf{a}) - \text{SSRB}$$

(see Exercise 4), which has a χ^2-distribution:

$$\text{SSE}/\sigma^2 \sim \chi^2_{N-r(\mathbf{X})-r(\mathbf{Z})}.$$

These sums of squares are summarized in Table 8.3a. Mean squares and F-statistics follow in the usual manner.

TABLE 8.3a. ANALYSIS OF VARIANCE FOR FITTING COVARIATES (**b**) AFTER FACTORS AND INTERACTIONS (**a**) IN THE COVARIANCE MODEL $\mathbf{y} = \mathbf{Xa} + \mathbf{Zb} + \mathbf{e}$

Source of Variation	d.f.	Sum of Squares[1]
Factors and interactions	$r(\mathbf{X})$	$R(\mathbf{a}) = \mathbf{y}'\mathbf{X}(\mathbf{X}'\mathbf{X})^{-}\mathbf{X}'\mathbf{y}$
Mean	1	$R(\mu) = N\bar{y}^2$
Factors and interactions (after the mean)	$r(\mathbf{X}) - 1$	$R(\mathbf{a} \mid \mu) = R(\mathbf{a}) - R(\mu)$
Covariates (after factors and interactions)	$r(\mathbf{Z})$	$R(\mathbf{b} \mid \mathbf{a}) = \text{SSRB} = \mathbf{y}'\mathbf{R}_z(\mathbf{R}_z'\mathbf{R}_z)^{-1}\mathbf{R}_z'\mathbf{y}$
Residual error	$N - r(\mathbf{X}) - r(\mathbf{Z})$	$\text{SSE} = \mathbf{y}'\mathbf{y} - R(\mathbf{a}) - \text{SSRB}$
Total	N	$\text{SST} = \mathbf{y}'\mathbf{y}$

[1] $\mathbf{R}_z$ is the matrix of residuals in (13).

The unbiased estimator of σ^2 derived from Table 8.3a is

$$\hat{\sigma}^2 = \text{SSE}/[N - r(\mathbf{X}) - r(\mathbf{Z})].$$

An alternative to the analysis of variance shown in Table 8.3a is to fit the covariates not after the factors and interactions but before them. This necessitates calculating $R(\mathbf{b} \mid \mu) = R(\mu, \mathbf{b}) - R(\mu)$, for which we need $R(\mu, \mathbf{b})$, the reduction in sum of squares due to fitting the model

$$\mathbf{y} = \mu\mathbf{1} + \mathbf{Zb} + \mathbf{e}.$$

This, of course, is simply an intercept regression model, for which

$$\tilde{\mathbf{b}} = (\mathscr{Z}'\mathscr{Z})^{-1}\mathscr{Z}'\mathbf{y} \quad \text{and} \quad \hat{\mu} = \bar{y} - \tilde{\mathbf{b}}'\bar{\mathbf{z}}$$

as in (41) and (42) of Sec. 3.2. In $\tilde{\mathbf{b}}$, $\mathscr{Z}'\mathscr{Z}$ is the matrix of corrected sums of squares and products of the observed z's, and $\mathscr{Z}'\mathbf{y}$ is the vector of corrected sums of products of the z's and the y's. Then $R(\mathbf{b} \mid \mu)$ that we need here is SSR_m of (73) in Sec. 3.4f, so that

$$R(\mathbf{b} \mid \mu) = \mathbf{y}'\mathscr{Z}(\mathscr{Z}'\mathscr{Z})^{-1}\mathscr{Z}'\mathbf{y}.$$

This reduction in sum of squares is for fitting the covariates after the mean.

TABLE 8.3b. ANALYSIS OF VARIANCE FOR FITTING FACTORS AND INTERACTIONS (**a**) AFTER COVARIATES (**b**) IN THE COVARIANCE MODEL $\mathbf{y} = \mathbf{Xa} + \mathbf{Zb} + \mathbf{e}$

Source of Variation	d.f.	Sum of Squares[1]
Mean	1	$R(\mu) = N\bar{y}^2$
Covariates (after mean)	$r(\mathbf{Z})$	$R(\mathbf{b} \mid \mu) = \mathbf{y}'\mathscr{Z}(\mathscr{Z}'\mathscr{Z})^{-1}\mathscr{Z}'\mathbf{y}$
Factors and interactions (after mean and covariates)	$r(\mathbf{X}) - 1$	$R(\mathbf{a} \mid \mu, \mathbf{b}) = R(\mathbf{a} \mid \mu) + \text{SSRB} - R(\mathbf{b} \mid \mu)$
Residual error	$N - r(\mathbf{X}) - r(\mathbf{Z})$	$\text{SSE} = \mathbf{y}'\mathbf{y} - R(\mathbf{a}) - \text{SSRB}$
Total	N	$\text{SST} = \mathbf{y}'\mathbf{y}$

[1] $R(\mathbf{a} \mid \mu)$ and SSRB are given in Table 8.3a.

In addition we need that for fitting the factors and interactions after the mean and covariates:

$$R(\mathbf{a} \mid \mu, \mathbf{b}) = R(\mathbf{a}, \mathbf{b}) - R(\mu, \mathbf{b}),$$

remembering that **a** in this notation *includes* μ. On using $R(\mathbf{a}) + \text{SSRB}$ for $R(\mathbf{a}, \mathbf{b})$ as derived in establishing Table 8.3a, and $R(\mathbf{b} \mid \mu) + R(\mu) = R(\mu, \mathbf{b})$, we have

$$
\begin{aligned}
R(\mathbf{a} \mid \mu, \mathbf{b}) &= R(\mathbf{a}) + \text{SSRB} - R(\mathbf{b} \mid \mu) - R(\mu) \\
&= R(\mathbf{a} \mid \mu) + \text{SSRB} - R(\mathbf{b} \mid \mu).
\end{aligned}
$$

These calculations are summarized in Table 8.3b.

In both Tables 8.3a and 8.3b the terms $R(\mu)$ and $R(\mathbf{a} \mid \mu)$ are those familiarly calculated in the no-covariate model $\mathbf{y} = \mathbf{Xa} + \mathbf{e}$. The additional terms needed are clearly evident.

(*vi*) *Tests of hypotheses.* The distributional properties of $R(\mathbf{b} \mid \mathbf{a})$ and SSE indicate, from (14), that in Table 8.3a

$$
F(\mathbf{b} \mid \mathbf{a}) = \frac{R(\mathbf{b} \mid \mathbf{a})/r(\mathbf{Z})}{\text{SSE}/[N - r(\mathbf{X}) - r(\mathbf{Z})]}
$$

tests the hypothesis H: $\mathbf{b} = \mathbf{0}$.

The hypothesis H: $\mathbf{K}'\mathbf{a} = \mathbf{m}$ is testable provided $\mathbf{K}'\mathbf{a}$ is estimable, in which case the hypothesis can be tested in the usual manner by means of $F(H)$ given in equation (70) of Sec. 5.5b. Using that equation with the solutions $\mathbf{a}^o$ and $\mathbf{b}^o$ given in (8) and (9) necessitates having, for the partitioned matrix shown in (7), the generalized inverse $\mathbf{G}$ that corresponds to those solutions, namely

$$
\mathbf{G} = \begin{bmatrix} \mathbf{X}'\mathbf{X} & \mathbf{X}'\mathbf{Z} \\ \mathbf{Z}'\mathbf{X} & \mathbf{Z}'\mathbf{Z} \end{bmatrix}^{-} = \begin{bmatrix} (\mathbf{X}'\mathbf{X})^{-} & \mathbf{0} \\ \mathbf{0} & \mathbf{0} \end{bmatrix}
+ \begin{bmatrix} -(\mathbf{X}'\mathbf{X})^{-}\mathbf{X}'\mathbf{Z} \\ \mathbf{I} \end{bmatrix} (\mathbf{Z}'\mathbf{P}\mathbf{Z})^{-1}[-\mathbf{Z}'\mathbf{X}(\mathbf{X}'\mathbf{X})^{-} \quad \mathbf{I}]. \tag{15}
$$

This is obtained from (49) of Sec. 1.7. Writing the hypothesis H: $\mathbf{K}'\mathbf{a} = \mathbf{m}$ as

$$
H: \quad [\mathbf{K}' \quad \mathbf{0}] \begin{bmatrix} \mathbf{a} \\ \mathbf{b} \end{bmatrix} = \mathbf{m}
$$

it will be found that the numerator of $F(H)$ then reduces to

$$
\begin{aligned}
Q = (\mathbf{K}'\mathbf{a}^o - \mathbf{m})'[&\mathbf{K}'(\mathbf{X}'\mathbf{X})^{-}\mathbf{K} \\
&+ \mathbf{K}'(\mathbf{X}'\mathbf{X})^{-}\mathbf{X}'\mathbf{Z}(\mathbf{Z}'\mathbf{P}\mathbf{Z})^{-1}\mathbf{Z}'\mathbf{X}(\mathbf{X}'\mathbf{X})^{-}\mathbf{K}]^{-1}(\mathbf{K}'\mathbf{a}^o - \mathbf{m}).
\end{aligned}
$$

We now show that testing H: $\mathbf{K}'\mathbf{a} = \mathbf{0}$ in the no-covariance model has the same numerator sum of squares as does testing H: $\mathbf{K}'[\mathbf{a} + (\mathbf{X}'\mathbf{X})^{-}\mathbf{X}'\mathbf{Z}\mathbf{b}] = \mathbf{0}$ in the covariance model. The solution vector for $\mathbf{a}$ in the no-covariance model is $\mathbf{a}^* = (\mathbf{X}'\mathbf{X})^{-}\mathbf{X}'\mathbf{y}$. From Q of Table 5.9 the numerator sum of squares for testing H: $\mathbf{K}'\mathbf{a} = \mathbf{0}$ in the no-covariance model is therefore

$$
Q = \mathbf{a}^{*\prime}\mathbf{K}[\mathbf{K}'(\mathbf{X}'\mathbf{X})^{-}\mathbf{K}]^{-1}\mathbf{K}'\mathbf{a}^*. \tag{16}
$$

In the covariance model consider the hypothesis

$$H: \quad \mathbf{K}'[\mathbf{a} + (\mathbf{X}'\mathbf{X})^{-}\mathbf{X}'\mathbf{Z}\mathbf{b}] = \mathbf{0} \tag{17}$$

which can be written as

$$H: \quad \mathbf{K}'[\mathbf{I} \quad (\mathbf{X}'\mathbf{X})^{-}\mathbf{X}'\mathbf{Z}]\begin{bmatrix}\mathbf{a}\\ \mathbf{b}\end{bmatrix} = \mathbf{0} \qquad \text{or as} \qquad \mathbf{M}'\begin{bmatrix}\mathbf{a}\\ \mathbf{b}\end{bmatrix} = \mathbf{0}$$

with

$$\mathbf{M}' = \mathbf{K}'[\mathbf{I} \quad (\mathbf{X}'\mathbf{X})^{-}\mathbf{X}'\mathbf{Z}]. \tag{18}$$

This hypothesis can be tested by an F-statistic having numerator sum of squares (see Table 5.9)

$$Q_c = [\mathbf{a}^{o\prime} \quad \hat{\mathbf{b}}']\mathbf{M}(\mathbf{M}'\mathbf{G}\mathbf{M})^{-1}\mathbf{M}'\begin{bmatrix}\mathbf{a}^o\\ \hat{\mathbf{b}}\end{bmatrix}.$$

But from (15) and (18), $\mathbf{M}'\mathbf{G}\mathbf{M} = \mathbf{K}'(\mathbf{X}'\mathbf{X})^{-}\mathbf{K}$, and $[\mathbf{a}^{o\prime} \quad \hat{\mathbf{b}}']\mathbf{M} = \mathbf{a}^{*\prime}\mathbf{K}$ using (8), so that Q_c becomes

$$\begin{aligned} Q_c &= (\mathbf{K}'\mathbf{a}^*)'[\mathbf{K}'(\mathbf{X}'\mathbf{X})^{-}\mathbf{K}]^{-1}\mathbf{K}'\mathbf{a}^* \\ &= Q \text{ of } (16). \end{aligned}$$

Hence the numerator sum of squares for testing $H: \ \mathbf{K}'\mathbf{a} = \mathbf{0}$ in the no-covariance model is also the numerator sum of squares for testing

$$H: \quad \mathbf{K}'[\mathbf{a} + (\mathbf{X}'\mathbf{X})^{-}\mathbf{X}'\mathbf{Z}\mathbf{b}] = \mathbf{0}$$

in the covariance model. This hypothesis appears to be dependent on $(\mathbf{X}'\mathbf{X})^{-}$. It is not, because $\mathbf{K}' = \mathbf{T}'\mathbf{X}$ for some $\mathbf{T}$, since $H: \ \mathbf{K}'\mathbf{a} = \mathbf{0}$ is assumed to be testable.

(*vii*) *Summary*. The preceding development of the analysis of covariance model

$$\mathbf{y} = \mathbf{X}\mathbf{a} + \mathbf{Z}\mathbf{b} + \mathbf{e}$$

can be summarized as follows. First fit

$$\mathbf{y} = \mathbf{X}\mathbf{a} + \mathbf{e}$$

and calculate

$$\mathbf{a}^* = (\mathbf{X}'\mathbf{X})^{-}\mathbf{X}'\mathbf{y} \qquad \text{and} \qquad R(\mathbf{a}) = \mathbf{a}^{*\prime}\mathbf{X}'\mathbf{y}. \tag{19}$$

Then for each column of $\mathbf{Z}$, $\mathbf{z}_j$ say, fit

$$\mathbf{z}_j = \mathbf{X}\mathbf{a} + \mathbf{e}$$

and calculate the $\mathbf{z}_j$-residual vector

$$\mathbf{z}_j - \hat{\mathbf{z}}_j = \mathbf{z}_j - \mathbf{X}(\mathbf{X}'\mathbf{X})^{-}\mathbf{X}'\mathbf{z}_j$$

and the matrix of these residuals

$$\mathbf{R}_z = \{\mathbf{z}_j - \hat{\mathbf{z}}_j\} \qquad \text{for} \quad j = 1, 2, \ldots, q. \tag{20}$$

Fit

$$\mathbf{y} = \mathbf{R}_z\mathbf{b} + \mathbf{e} \tag{21}$$

and calculate

$$\hat{\mathbf{b}} = (\mathbf{R}_z'\mathbf{R}_z)^{-1}\mathbf{R}_z'\mathbf{y}$$

and

$$R(\mathbf{b} \mid \mathbf{a}) = \hat{\mathbf{b}}'\mathbf{R}_z'\mathbf{y}. \tag{22}$$

The solution vector for the covariance model is then

$$\begin{bmatrix} \mathbf{a}^o \\ \hat{\mathbf{b}} \end{bmatrix} = \begin{bmatrix} \mathbf{a}^* - (\mathbf{X}'\mathbf{X})^-\mathbf{X}'\mathbf{Z}\hat{\mathbf{b}} \\ \hat{\mathbf{b}} \end{bmatrix}. \tag{23}$$

From (15) the variance matrices of these solutions are

$$\text{var}(\mathbf{a}^o) = [(\mathbf{X}'\mathbf{X})^- + (\mathbf{X}'\mathbf{X})^-\mathbf{X}'\mathbf{Z}(\mathbf{R}_z'\mathbf{R}_z)^{-1}\mathbf{Z}'\mathbf{X}(\mathbf{X}'\mathbf{X})^-]\sigma^2,$$

$$\text{var}(\hat{\mathbf{b}}) = (\mathbf{R}_z'\mathbf{R}_z)^{-1}\sigma^2 \tag{24}$$

and

$$\text{cov}(\mathbf{a}^o, \hat{\mathbf{b}}) = -(\mathbf{X}'\mathbf{X})^-\mathbf{X}'\mathbf{Z}(\mathbf{R}_z'\mathbf{R}_z)^{-1}\sigma^2.$$

It is clear that, in contrast to fitting an ordinary factors-and-interactions model, the clue to the calculations for a covariance model is the derivation of $\mathbf{R}_z$. Furthermore, calculation of each column of $\mathbf{R}_z$ from the corresponding column of $\mathbf{Z}$ depends solely on the particular factors-and-interactions model being used. No matter what the nature of the covariates, $\mathbf{X}$ is the same for any specific factors-and-interactions model and this determines the derivation of $\mathbf{R}_z$ from $\mathbf{Z}$. When considering the same covariates in different ways for the same factors-and-interactions model the corresponding $\mathbf{Z}$-matrices will be different, but the mode of calculating $\mathbf{R}_z$ on each occasion is always the same. The columns of $\mathbf{R}_z$ are always the vectors of residuals obtained after fitting the no-covariates model to each column of $\mathbf{Z}$. This is illustrated in the examples that follow.

b. The 1-way classification

(*i*) *A single regression.* A simple adaption of equation (23) in Sec. 6.2a gives the equation for a covariance model in the 1-way classification as

$$y_{ij} = \mu + \alpha_i + bz_{ij} + e_{ij} \tag{25}$$

for $i = 1, 2, \ldots, c$ and $j = 1, 2, \ldots, n_i$. In this model μ and the α_i's are the elements of $\mathbf{a}$ of (6), the scalar b is the sole element of $\mathbf{b}$ of (6) and $\mathbf{Z}$ of (6) is a vector $\mathbf{z}$ of the observed values z_{ij} of the covariate, with

$$\mathbf{z}' = [z_{11} \quad z_{12} \quad \cdots \quad z_{1n_1} \quad \cdots \quad z_{i1} \quad z_{i2} \quad \cdots \quad z_{in_i} \quad \cdots \quad z_{c1} \quad z_{c2} \quad \cdots \quad z_{cn_c}], \tag{26}$$

corresponding to the vector of y-observations defined in (26) of Sec. 6.2a.

Fitting the no-covariate form of (25) amounts to fitting the 1-way classification model $y_{ij} = \mu + \alpha_i + e_{ij}$ discussed in Sec. 6.2. There, in equation (31), we see that a solution vector for $\mathbf{a}^*$ of (19) is

$$\mathbf{a}^* = \begin{bmatrix} \mu^* \\ \{\alpha_i^*\} \end{bmatrix} = \begin{bmatrix} 0 \\ \{\bar{y}_{i\cdot}\} \end{bmatrix} \quad \text{for} \quad i = 1, \ldots, c, \tag{27}$$

and from (37) of Sec. 6.2d

$$R(\mathbf{a}) = \sum_{i=1}^{c} y_{i\cdot}^2/n_i\,. \tag{28}$$

Also, the residual corresponding to y_{ij} is

$$y_{ij} - \hat{y}_{ij} = y_{ij} - \mu^* - \alpha_i^* = y_{ij} - \bar{y}_{i\cdot}$$

so that the vector of residuals is

$$\mathbf{y} - \hat{\mathbf{y}} = \{\mathbf{y}_i - \bar{y}_{i\cdot}\mathbf{1}_{n_i}\} = \left\{\begin{bmatrix} y_{i1} - \bar{y}_{i\cdot} \\ y_{i2} - \bar{y}_{i\cdot} \\ \vdots \\ y_{in_i} - \bar{y}_{i\cdot} \end{bmatrix}\right\} \quad \text{for } i = 1, \ldots, c. \tag{29}$$

In fitting (25), $\mathbf{Z}$ of the general model (6) is $\mathbf{z}$ of (26) and so $\mathbf{R}_z$ of (20) is a vector and is, analogous to (29),

$$\mathbf{R}_z = \mathbf{z} - \hat{\mathbf{z}} = \{\mathbf{z}_i - \bar{z}_{i\cdot}\mathbf{1}_{n_i}\} \qquad \text{for} \quad i = 1, 2, \ldots, c.$$

Therefore for $\hat{\mathbf{b}}$ of (22)

$$\mathbf{R}_z'\mathbf{R}_z = \sum_{i=1}^{c}\sum_{j=1}^{n_i}(z_{ij} - \bar{z}_{i\cdot})^2 = \sum_{i=1}^{c}\left(\sum_{j=1}^{n_i} z_{ij}^2 - n_i\bar{z}_{i\cdot}^2\right)$$

and

$$\mathbf{R}_z'\mathbf{y} = \sum_{i=1}^{c}\sum_{j=1}^{n_i}(z_{ij} - \bar{z}_{i\cdot})y_{ij} = \sum_{i=1}^{c}\left(\sum_{j=1}^{n_i} y_{ij}z_{ij} - n_i\bar{y}_{i\cdot}\bar{z}_{i\cdot}\right) \tag{30}$$

so that

$$\hat{b} = \frac{\displaystyle\sum_{i=1}^{c}\left(\sum_{j=1}^{n_i} y_{ij}z_{ij} - n_i\bar{y}_{i\cdot}\bar{z}_{i\cdot}\right)}{\displaystyle\sum_{i=1}^{c}\left(\sum_{j=1}^{n_i} z_{ij}^2 - n_i\bar{z}_{i\cdot}^2\right)}\,. \tag{31}$$

With this value of $\hat{\mathbf{b}}$, $\mathbf{a}^o$ is calculated from (23) as

$$\mathbf{a}^o = \mathbf{a}^* - \hat{b}(\mathbf{X}'\mathbf{X})^-\mathbf{X}'\mathbf{z};$$

i.e.,

$$\begin{bmatrix} \mu^o \\ \{\alpha_i^o\} \end{bmatrix} = \begin{bmatrix} 0 \\ \{\bar{y}_{i\cdot}\} \end{bmatrix} - \hat{b}\begin{bmatrix} 0 \\ \{\bar{z}_{i\cdot}\} \end{bmatrix} = \begin{bmatrix} 0 \\ \{\bar{y}_{i\cdot} - \hat{b}\bar{z}_{i\cdot}\} \end{bmatrix} \quad \text{for} \quad i = 1, \ldots, c. \tag{32}$$

The solution $\alpha_i^o = \bar{y}_{i\cdot} - \hat{b}\bar{z}_{i\cdot}$ is often referred to as an *adjusted mean*—it is the class mean $\bar{y}_{i\cdot}$ adjusted by the class mean of the covariate, using the estimate $\hat{b}$ to make the adjustment.

Examination of (31) and (32) reveals the relationship of these results to ordinary regression analysis. In (31) the numerator of $\hat{b}$ is a sum of terms, each of which is the numerator for estimating the within-class regression of y on z; likewise the denominator of $\hat{b}$ is the sum of the denominators of those within-class regression estimators. Thus $\hat{b}$ is usually referred to as the *pooled within-class regression estimator*. Also, each element in (32)—other than the initial zero—is the within-class intercept estimator using $\hat{b}$ of (31).

The basic calculations for the analysis of variance for fitting the model $E(\mathbf{y}) = \mathbf{Xa}$ in the case of a 1-way classification are, as in Sec. 6.2d,

$$\text{SSR}_{yy} = \sum_{i=1}^{c} n_i \bar{y}_{i\cdot}^2, \qquad \text{SSE}_{yy} = \text{SST}_{yy} - \text{SSR}_{yy}, \quad \text{and} \quad \text{SST}_{yy} = \sum_{i=1}^{c}\sum_{j=1}^{n_i} y_{ij}^2 .$$

We can also calculate

$$\text{SSM}_{yy} = N\bar{y}^2, \qquad \text{SSR}_{m,yy} = \text{SSR}_{yy} - \text{SSM}_{yy},$$

and

$$\text{SST}_{m,yy} = \text{SST}_{yy} - \text{SSM}_{yy} .$$

The subscript yy in these expressions emphasizes that they are functions of squares of the y-observations. Similar functions of the z-observations, and of cross-products of the y's and z's, can also be calculated:

$$\text{SSR}_{yz} = \sum_{i=1}^{c} n_i \bar{y}_{i\cdot}\bar{z}_{i\cdot}, \quad \text{SSE}_{yz} = \text{SST}_{yz} - \text{SSR}_{yz}, \quad \text{and} \quad \text{SST}_{yz} = \sum_{i=1}^{c}\sum_{j=1}^{n_i} y_{ij}z_{ij}$$

and

$$\text{SSM}_{yz} = N\bar{y}\bar{z}, \qquad \text{SSR}_{m,yz} = \text{SSR}_{yz} - \text{SSM}_{yz},$$

and

$$\text{SST}_{m,yz} = \text{SST}_{yz} - \text{SSM}_{yz} .$$

(We do not show explicit expressions for the z's because they are of exactly the same form as those of the y's.) We find these expressions useful in what follows.

First $R(\mathbf{a})$, which for (25) is the reduction due to fitting μ and the α's, is from (28)

$$R(\mu, \alpha) = R(\mathbf{a}) = \text{SSR}_{yy} .$$

Second, from (31),

$$\hat{b} = \text{SSE}_{yz}/\text{SSE}_{zz} \tag{33}$$

so that, from (22) and (30),

$$R(b \mid \mu, \alpha) = R(\mathbf{b} \mid \mathbf{a}) = (\text{SSE}_{yz})^2/\text{SSE}_{zz} . \tag{34}$$

Hence the analysis of variance of Table 8.3a becomes as shown in Table 8.4a.

TABLE 8.4a. ANALYSIS OF VARIANCE FOR FITTING THE COVARIATE AFTER THE CLASS EFFECTS IN THE 1-WAY CLASSIFICATION COVARIANCE MODEL $y_{ij} = \mu + \alpha_i + bz_{ij} + e_{ij}$

Source of Variation	d.f.	Sum of Squares
Mean	1	$R(\mu) = \text{SSM}_{yy}$
α-classes (after mean)	$c - 1$	$R(\alpha \mid \mu) = \text{SSR}_{m,yy}$
Covariate (pooled within-class regression)	1	$R(b \mid \mu, \alpha) = (\text{SSE}_{yz})^2/\text{SSE}_{zz}$
Residual error	$N - c - 1$	$\text{SSE} = \text{SSE}_{yy} - R(b \mid \mu, \alpha)$
Total	N	SST_{yy}

In Table 8.4a the estimated residual variance is

$$\hat{\sigma}^2 = \text{SSE}/(N - c - 1);$$

the hypothesis H: $b = 0$, that the regression slope is zero, is tested using

$$F(b) = R(b \mid \mu, \alpha)/\hat{\sigma}^2, \tag{35}$$

an F-statistic with 1 and $N - c - 1$ degrees of freedom. The F-statistic having $R(\alpha \mid \mu)$ in its numerator does, in the no-covariate model, test the hypothesis H: all α's equal [Sec. 6.2f(iii)]. The corresponding statistic in Table 8.4a tests, from (17), the hypothesis

$$H: \quad \alpha_i + b\bar{z}_{i\cdot} \text{ equal for all } i, \tag{36}$$

the $b\bar{z}_{i\cdot}$ being derived from $(\mathbf{X}'\mathbf{X})^{-}\mathbf{X}'\mathbf{Z}\mathbf{b}$ of (17) in the same way that $\mathbf{a}^{\circ}$ of (32) was derived. This hypothesis represents equality of the α's adjusted for the observed z's.

To derive the equivalent of Table 8.3b for the 1-way classification covariance model, notice first that whenever there is only a single vector as $\mathbf{Z}$, then in Table 8.3b

$$\mathbf{y}'\mathscr{Z} = \text{SST}_{m,yz} \quad \text{and} \quad \mathscr{Z}'\mathscr{Z} = \text{SST}_{m,zz}\,.$$

Hence
$$R(b \mid \mu) = (\text{SST}_{m,yz})^2/\text{SST}_{m,zz}$$

and so Table 8.3b simplifies to Table 8.4b.

The F-statistic for testing

$$H: \quad \alpha_i \text{ equal for all } i$$

TABLE 8.4b. ANALYSIS OF VARIANCE FOR FITTING THE CLASS EFFECTS AFTER THE COVARIATE IN THE 1-WAY CLASSIFICATION COVARIANCE MODEL $y_{ij} = \mu + \alpha_i + bz_{ij} + e_{ij}$

Source of Variation	d.f.	Sum of Squares
Mean	1	$R(\mu) = \text{SSM}_{yy}$
Covariate (after mean)	1	$R(b \mid \mu) = (\text{SST}_{m,yz})^2/\text{SST}_{m,zz}$
α-classes (after mean and covariates)	$c - 1$	$R(\alpha \mid \mu, b) = \text{SSR}_{m,yy} + (\text{SSE}_{yz})^2/\text{SSE}_{zz} - (\text{SST}_{m,yz})^2/\text{SST}_{m,zz}$
Residual error	$N - c - 1$	$\text{SSE} = \text{SSE}_{yy} - R(b \mid \mu, \alpha)$
Total	N	SST_{yy}

could be derived by writing the hypothesis as $\mathbf{K}'\mathbf{a} = \mathbf{0}$ and using the general result for Q given below (15). A possible value for $\mathbf{K}'$ would be $\mathbf{K}' = [\mathbf{0}1 \quad \mathbf{1} \quad -\mathbf{I}]$ of $c - 1$ rows, similar to (59) of Sec. 6.2f(iii). An easier development is to consider the reduced model arising from the hypothesis itself, namely

$$y_{ij} = (\mu + \alpha) + bz_{ij} + e_{ij}\,. \tag{37}$$

This is a model for simple regression, for which the estimator of b is, from equation (14) of Sec. 3.1c,

$$\tilde{b} = \frac{\sum_{i=1}^{c}\sum_{j=1}^{n_i} y_{ij}z_{ij} - N\bar{y}\bar{z}}{\sum_{i=1}^{c}\sum_{j=1}^{n_i} z_{ij}^2 - N\bar{z}^2} = \frac{\text{SST}_{m,yz}}{\text{SST}_{m,zz}}.$$

The reduction in sum of squares for fitting (37) is therefore, using Table 3.3 of Sec. 3.5g,

$$\begin{aligned} R(\mu, b) &= N\bar{y}^2 + \tilde{b}\text{SST}_{m,yz} \\ &= \text{SSM}_{yy} + (\text{SST}_{m,yz})^2/\text{SST}_{m,zz}\,. \end{aligned} \tag{38}$$

The full model is (25), with the reduction in sum of squares being, from Table 8.4a

$$R(\mu, \alpha, b) = \text{SSM}_{yy} + \text{SSR}_{m,yy} + R(b \mid \mu, \alpha); \tag{39}$$

and the F-statistic for testing H: all α's equal in the model (25) has numerator

$$Q = R(\mu, \alpha, b) - R(\mu, b). \tag{40}$$

Using (34), (38) and (39) this becomes $Q = R(\alpha \mid \mu, b)$ of Table 8.4b. Tables similar to 8.4a and 8.4b are to be found in many places; e.g., Federer (1955, p. 485) and Graybill (1961, pp. 385 and 393).

(*ii*) *Example*. Suppose that in the example of Sec. 6.1 the number of children in each family is to be taken into account in studying investment and education. Consider the hypothetical data shown in Table 8.5, the y-values (investment index) being the same as in Table 6.1.

The following basic sums of squares and sums of products can be readily calculated from Table 8.5, those for the y-observations being the same as in Sec. 6.1:

$$\begin{aligned} \text{SSR}_{yy} &= 43{,}997, & \text{SSR}_{zz} &= 95, & \text{SSR}_{yz} &= 2015, \\ \text{SSE}_{yy} &= 82, & \text{SSE}_{zz} &= 6, & \text{SSE}_{yz} &= 3, \\ \text{SST}_{yy} &= 44{,}079, & \text{SST}_{zz} &= 101, & \text{SST}_{yz} &= 2018, \end{aligned} \tag{41}$$

and $\text{SSM}_{yy} = 43{,}687, \quad \text{SSM}_{zz} = 89\tfrac{2}{7}, \quad \text{SSM}_{yz} = 1975.$

These are used in the ensuing calculations.

The pooled regression estimate $\hat{b}$ comes from (33):

$$\hat{b} = \tfrac{3}{6} = \tfrac{1}{2}. \tag{42}$$

Then for $\mathbf{a}^o$ of (32) we need $\mathbf{a}^*$ of (27) which comes from (34) of Sec. 6.2c:

$$\mathbf{a}^{*\prime} = [0 \quad 73 \quad 78 \quad 89]. \tag{43}$$

TABLE 8.5. INVESTMENT INDEX AND NUMBER OF CHILDREN FOR 7 MEN

High School Incomplete		High School Graduate		College Graduate	
Index, y_{1j}	Children, z_{1j}	Index, y_{2j}	Children, z_{2j}	Index, y_{3j}	Children, z_{3j}
74	3	76	2	85	4
68	4	80	4	93	6
77	2				
219	9	156	6	178	10

Hence from (32) and Table 8.5

$$\mathbf{a}^o = \begin{bmatrix} 0 \\ 73 \\ 78 \\ 89 \end{bmatrix} - \tfrac{1}{2}\begin{bmatrix} 0 \\ 3 \\ 3 \\ 5 \end{bmatrix} = \begin{bmatrix} 0 \\ 71\frac{1}{2} \\ 76\frac{1}{2} \\ 86\frac{1}{2} \end{bmatrix}. \tag{44}$$

The analysis of variance in Table 8.4a uses:

$$R(\mu) = \text{SSM}_{yy} = 43{,}687,$$

$$R(\mu, \alpha) = \text{SSR}_{yy} = 43{,}997, \tag{45}$$

and

$$R(b \mid \mu, \alpha) = \text{SSRB} = 3^2/6 = 1\tfrac{1}{2}$$

from (34). Hence Table 8.4a becomes as shown in Table 8.6a. It can be readily checked that $R(\mathbf{a}, \mathbf{b})$ of the general case, which is $R(\mu, \alpha, b)$ here, is

$$\begin{aligned} R(\mu, \alpha, b) &= R(\mathbf{a}, \mathbf{b}) = \mathbf{a}^{o\prime}\mathbf{X}'\mathbf{y} + \hat{b}\mathbf{Z}'\mathbf{y} \\ &= 71\tfrac{1}{2}(219) + 76\tfrac{1}{2}(156) + 86\tfrac{1}{2}(178) + \tfrac{1}{2}(2018) \\ &= 43{,}998\tfrac{1}{2} \\ &= 43{,}687 + 310 + 1\tfrac{1}{2}, \text{ of Table 8.6a} \\ &= \text{SSM}_{yy} + \text{SSR}_{m,yy} + \text{SSRB}, \text{ of Table 8.4a} \end{aligned}$$

as should be the case.

F-statistics available in Table 8.6a can be used for testing hypotheses as follows: from (35)

$$F_{1,3} = \frac{1\frac{1}{2}}{1}\bigg/\frac{80\frac{1}{2}}{3} = .06 \text{ tests } H\colon\quad b = 0;$$

TABLE 8.6a. EXAMPLE OF TABLE 8.4a: DATA OF TABLE 8.5

Source of Variation	d.f.	Sum of Squares
Mean	1	$R(\mu) = 43{,}687$
α-classes (after mean)	2	$R(\alpha \mid \mu) = 310$
Covariate (pooled within-class regression)	1	$R(b \mid \mu, \alpha) = 1\frac{1}{2}$
Residual error	3	$\text{SSE} = 80\frac{1}{2}$
Total	7	$\text{SST}_{yy} = 44{,}079$

TABLE 8.6b. EXAMPLE OF TABLE 8.4b: DATA OF TABLE 8.5

Source of Variation	d.f.	Sum of Squares
Mean	1	$R(\mu) = 43{,}687$
Covariate (after mean)	1	$R(b \mid \mu) = 157.8$
α-classes (after mean and covariate)	2	$R(\alpha \mid \mu, b) = 153.7$
Residual error	3	$\text{SSE} = 80.5$
Total	7	$\text{SST} = 44{,}079$

and from (36)

$$F_{2,3} = \frac{310}{2}\Big/\frac{80\frac{1}{2}}{3} = 5.8 \text{ tests } H: \ \alpha_1 + 3b = \alpha_2 + 3b = \alpha_3 + 5b.$$

Since neither of these F-values exceeds the corresponding 5% critical values of 10.13 and 9.55 respectively, both hypotheses are not rejected.

To calculate Table 8.4b we get, using (41),

$$R(b \mid \mu) = (2018 - 1975)^2/(101 - 89\tfrac{2}{7}) = 43^2/11\tfrac{5}{7} = 157.8.$$

Hence, by subtraction from the sum of two terms of Table 8.6a,

$$R(\alpha \mid \mu, b) = 310 + 1\tfrac{1}{2} - 157.8 = 153.7,$$

and Table 8.4b becomes as shown in Table 8.6b. Since

$$F_{2,3} = \frac{R(\alpha \mid \mu, b)}{2(80.5/3)} = \frac{153.7(3)}{161} = 2.86$$

is less than the corresponding 5% critical value of 9.55 the hypothesis $H: \ \alpha_1 = \alpha_2 = \alpha_3$ in the covariate model is not rejected.

(iii) The intra-class regression model

In (25) we applied the general procedure for covariance analysis to the 1-way classification with a solitary covariate and a single regression coefficient b. We now show how the general procedure applies when the covariate occurs in the model in some fashion other than the simple case of (25). One alternative (an easy one) is considered here and two others are contemplated in Exercise 8. In all three cases $\mathbf{a}^*$ and $R(\mathbf{a})$ are the same as for the model (25).

The model based on (25) assumes the same regression slope for all classes. This need not necessarily be the case. An obvious alternative is the model

$$y_{ij} = \mu + \alpha_i + b_i z_{ij} + e_{ij} \tag{46}$$

in which there is a different regression for each class. It can be called an *intra-class regression model.*

The general procedure proceeds quite straightforwardly for this model. Compared to (25), $\mathbf{a}^*$ and $R(\mathbf{a})$ remain the same, but $\mathbf{b}$ and $\mathbf{Z}$ are changed. $\mathbf{b}$ is now a vector of the regression slopes and $\mathbf{Z}$ is an $N \times c$ matrix:

$$\mathbf{Z} = \begin{bmatrix} \mathbf{z}_1 & \mathbf{0} & \cdots & \mathbf{0} \\ \mathbf{0} & \mathbf{z}_2 & \cdots & \mathbf{0} \\ \cdot & \cdot & \cdot & \cdot \\ \cdot & \cdot & \cdot & \cdot \\ \cdot & \cdot & \cdot & \cdot \\ \mathbf{0} & \mathbf{0} & \cdots & \mathbf{z}_c \end{bmatrix} = \mathbf{D}\{\mathbf{z}_i\} = \sum_{i=1}^{c}{}^{+} \mathbf{z}_i, \tag{47}$$

for $\mathbf{z}_i$ being the vector of n_i observed z's in the ith class.

Applying to each column of $\mathbf{Z}$ in (47) the derivation of the corresponding vector of residuals shown in (29) for $\mathbf{y}$, it is clear that $\mathbf{R}_z$ of (20) is

$$\mathbf{R}_z = \begin{bmatrix} \mathbf{z}_1 - \bar{z}_{1\cdot}\mathbf{1}_{n_1} & \cdots & \mathbf{0} \\ \cdot \quad \cdot & & \cdot \\ \cdot & \cdot & \cdot \\ \cdot & \cdot & \cdot \\ \mathbf{0} & \cdots & \mathbf{z}_c - \bar{z}_{c\cdot}\mathbf{1}_{n_c} \end{bmatrix} = \sum_{i=1}^{c}{}^{+} \{\mathbf{z}_i - \bar{z}_{i\cdot}\mathbf{1}_{n_i}\}. \tag{48}$$

Hence for $\hat{\mathbf{b}}$ of (22), $\mathbf{R}_z'\mathbf{R}_z$ is the diagonal matrix

$$\begin{aligned} \mathbf{R}_z'\mathbf{R}_z &= \mathbf{D}\{(\mathbf{z}_i - \bar{z}_{i\cdot}\mathbf{1}_{n_i})'(\mathbf{z}_i - \bar{z}_{i\cdot}\mathbf{1}_{n_i})\} \\ &= \mathbf{D}\left\{\sum_{j=1}^{n_i} z_{ij}^2 - n_i\bar{z}_{i\cdot}^2\right\} \qquad \text{for} \quad i = 1, 2, \ldots, c. \end{aligned}$$

Similarly,

$$\begin{aligned} \mathbf{R}_z'\mathbf{y} &= \{(\mathbf{z}_i - \bar{z}_{i\cdot}\mathbf{1}_{n_i})'\mathbf{y}_i\} \\ &= \left\{\sum_{j=1}^{n_i} y_{ij}z_{ij} - n_i\bar{y}_{i\cdot}\bar{z}_{i\cdot}\right\} \qquad \text{for} \quad i = 1, 2, \ldots, c. \end{aligned}$$

On defining

$$(\mathrm{SSE}_{zz})_i = \sum_{j=1}^{n_i} z_{ij}^2 - n_i\bar{z}_{i\cdot}^2 \quad \text{and} \quad (\mathrm{SSE}_{yz})_i = \sum_{j=1}^{n_i} y_{ij}z_{ij} - n_i\bar{y}_{i\cdot}\bar{z}_{i\cdot} \tag{49}$$

we then have

$$\mathbf{R}_z'\mathbf{R}_z = \mathbf{D}\{(\mathrm{SSE}_{zz})_i\} \quad \text{and} \quad \mathbf{R}_z'\mathbf{y} = \{(\mathrm{SSE}_{yz})_i\} \tag{50}$$

so that

$$\hat{\mathbf{b}} = (\mathbf{R}_z'\mathbf{R}_z)^{-1}\mathbf{R}_z'\mathbf{y} = \left\{\frac{(\mathrm{SSE}_{yz})_i}{(\mathrm{SSE}_{zz})_i}\right\},$$

i.e.,

$$\hat{b}_i = \frac{(\mathrm{SSE}_{yz})_i}{(\mathrm{SSE}_{zz})_i}, \qquad \text{for} \quad i = 1, 2, \ldots, c. \tag{51}$$

Then with $\mathbf{a}^*$ of (27) we get $\mathbf{a}^o$ from (23) as

$$\mathbf{a}^o = \begin{bmatrix} \mu^o \\ \{\alpha_i^o\} \end{bmatrix} = \begin{bmatrix} 0 \\ \{\bar{y}_{i\cdot} - \hat{b}_i \bar{z}_{i\cdot}\} \end{bmatrix} \quad \text{for} \quad i = 1, 2, \ldots, c. \tag{52}$$

Thus, from (51), we see that $\hat{b}_i$ is the within-class regression estimator of y on z within the ith class, and α_i^o in (52) is the corresponding intercept estimator for that class. Notice, too, from the definitions in (49) and the result in (51), that the sums of the numerators and denominators of the $\hat{b}_i$ are, respectively, the numerator and denominator of the pooled within-class estimator of (33).

For the model (46) we have

$$R(\mu, \alpha) = R(\mathbf{a}) = \sum y_{i\cdot}^2/n_i = \text{SSR}_{yy}$$

as before, in (28); and from (22)

$$R(\mathbf{b} \mid \mu, \alpha) = \hat{\mathbf{b}}'\mathbf{R}_z'\mathbf{y}, = \sum_{i=1}^{c} \frac{(\text{SSE}_{yz})_i^2}{(\text{SSE}_{zz})_i} \tag{53}$$

from (50) and (51). These are the reductions to be used in the analysis of variance for fitting (46), along the lines of Table 8.3a. However, it is more instructive to also incorporate Table 8.4a and establish a test of the hypothesis H: all b_i's equal, for the model (46). This is readily achieved by subtracting $R(b \mid \mu, \alpha)$ of Table 8.4a from $R(\mathbf{b} \mid \mu, \alpha)$ of (53); i.e.,

$$R(\mathbf{b} \mid \mu, \alpha) - R(b \mid \mu, \alpha)$$

is the numerator for testing the hypothesis H: all b_i's equal, in the model (46). The complete analysis is shown in Table 8.7.

TABLE 8.7. ANALYSIS OF VARIANCE FOR FITTING THE MODEL $y_{ij} = \mu + \alpha_i + b_i z_{ij} + e_{ij}$ FOR THE 1-WAY CLASSIFICATION

Source of Variation	d.f.	Sum of Squares
Mean	1	$R(\mu) = \text{SSM}_{yy}$
α-classes (after mean)	$c - 1$	$R(\alpha \mid \mu) = \text{SSR}_{m,yy}$
Covariate (within-class)	c	$R(\mathbf{b} \mid \mu, \alpha) = \sum_i \frac{(\text{SSE}_{yz})_i^2}{(\text{SSE}_{zz})_i}$
Pooled	1	$R(b \mid \mu, \alpha) = \frac{(\text{SSE}_{yz})^2}{\text{SSE}_{zz}}$
Difference (H: b_i's equal)	$c - 1$	$R(\mathbf{b} \mid \mu, \alpha) - R(b \mid \mu, \alpha)$
Residual error	$N - 2c$	$\text{SSE} = \text{SSE}_{yy} - R(\mathbf{b} \mid \mu, \alpha)$
Total	N	SST_{yy}

With
$$\hat{\sigma}^2 = \frac{\text{SSE}}{N - 2c},$$
$$F = \frac{R(\mathbf{b} \mid \mu, \alpha) - R(b \mid \mu, \alpha)}{(c - 1)\hat{\sigma}^2} \tag{54}$$

can be used to test H: all b_i's equal. Non-rejection of this hypothesis can lead to estimating the pooled b as in (33). The F-statistic based on (40) then provides a test, under the assumption of equal b_i's, of the hypothesis that the α_i's are equal. The statistic
$$F = R(b \mid \mu, \alpha)/\hat{\sigma}^2 \tag{55}$$
is also available for testing the hypothesis that this pooled b is zero. Of course, using it conditionally in this manner, conditional on (54) being non-significant, changes the nominal probability level of any critical value used for (55) from that customarily associated with it.

When the hypothesis H: b_i's all equal is rejected, a test of the hypothesis H: α_i's all equal can be developed, although interpretation of equal α's and unequal b's, i.e., of equal intercept and unequal slopes, is often not easy. It implies a model in the form of a pencil of regression lines through the common intercept. Development of the test is left to the reader (see Exercise 9). In this case (17) takes the form (see Exercise 10)
$$H: \quad \alpha_i + b_i\bar{z}_{i.} \text{ equal for all } i$$
which can be tested by $R(\alpha \mid \mu)/(c - 1)\hat{\sigma}^2$.

(*iv*) *Example* (*continued*). The Table 8.5 data readily yield estimates of the within-class regression slopes from (51) as
$$\hat{b}_1 = -\tfrac{9}{2} = -4\tfrac{1}{2}, \qquad \hat{b}_2 = \tfrac{4}{2} = 2 \qquad \text{and} \qquad \hat{b}_3 = \tfrac{8}{2} = 4,$$
so that from (53)
$$R(\mathbf{b} \mid \mu, \alpha) = \frac{(-9)^2}{2} + \frac{4^2}{2} + \frac{8^2}{2} = 80\tfrac{1}{2}.$$
Hence
$$\text{SSE} = \text{SSE}_{yy} - R(\mathbf{b} \mid \mu, \alpha) = 82 - 80\tfrac{1}{2} = 1\tfrac{1}{2}.$$

Table 8.7 therefore becomes as shown in Table 8.8 (based on Table 8.6a). The residual error sum of squares is very small in this example because two of the classes for which within-class regressions have been estimated have only two sets of observations (see Table 8.5) and so the estimation for those classes is a perfect fit. The only contribution to the residual error is from the one class having three observations. Table 8.5 is, of course, a trivial example but is intended solely for illustrating derivation of the analysis and not for any intrinsic value; this is true of all the examples.

TABLE 8.8. EXAMPLE OF TABLE 8.7: DATA OF TABLE 8.5 (SEE TABLE 8.6a ALSO)

Source of Variation	d.f.		Sum of Squares	
Mean	1		$R(\mu) = 43{,}687$	
α-classes (after mean)	2		$R(\alpha \mid \mu) = 310$	
Covariate (within-class)	3		$R(\mathbf{b} \mid \mu, \alpha) = 80\frac{1}{2}$	
Pooled		1		$R(b \mid \mu, \alpha) = 1\frac{1}{2}$
Difference		2		Difference $= 79$
Residual error	1		SSE $= 1\frac{1}{2}$	
Total	7		SST $= 44{,}079$	

(*v*) *Another example.*[1] Consider the case of just 2 classes in a 1-way classification. Then $R(\alpha \mid \mu)$ reduces to $n_1 n_2(\bar{y}_{1\cdot} - \bar{y}_{2\cdot})^2/n_{\cdot}$, and the hypothesis tested by $R(\alpha \mid \mu)$ in Table 8.7 is $H\colon\ \alpha_1 + b_1\bar{z}_1 = \alpha_2 + b_2\bar{z}_2$. Suppose that the observed means in the two classes are the same, $\bar{y}_{1\cdot} = \bar{y}_{2\cdot}$, or nearly so. Then $R(\alpha \mid \mu) = 0$ and the hypothesis is not rejected. The conclusion must not be drawn from this, however, that there is no significant difference between the classes at other values of z. Differences between $\alpha_1 + b_1 z$ and $\alpha_2 + b_2 z$ may be very real for certain values of z. Suppose, for example, that the estimated regression lines have the appearance of Figure 8.1. For certain values of z greater than z_0 the adjusted value of y for class 2 might be significantly greater than that for class 1; and similarly for certain values of z

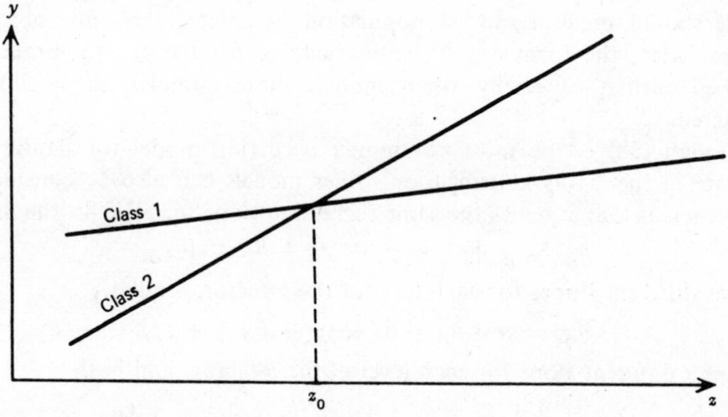

Figure 8.1. Estimated regression lines of y on z for two classes.

[1] I am grateful to E. C. Townsend for bringing this to my notice.

less than z_0 the mean adjusted y-response for class 2 may be significantly less than class 1. A numerical illustration of this is provided in Exercise 11.

c. The 2-way classification (with interaction)

The purpose of this section is to indicate very briefly how the general results of the preceding subsections a and b can be applied to the 2-way classification (with interaction) in the same way that they have been applied to the 1-way classification in subsection b.

We take as the starting point $\mathbf{a}^*$ and $R(\mathbf{a})$ for the no-covariate 2-way classification (with interaction) model, the model discussed in Sec. 7.2. From equations (55) and (61) of Secs. 7.2c and 7.2d(i) respectively

$$\mathbf{a}^* = \begin{bmatrix} \mathbf{0} \\ \bar{\mathbf{y}} \end{bmatrix} \qquad \text{and} \qquad R(\mathbf{a}) = \sum_{i=1}^{c} \sum_{j=1}^{b} y_{ij\cdot}^2 / n_{ij}\,, \tag{56}$$

where $\bar{\mathbf{y}}$ is the vector of cell means, $\bar{y}_{ij\cdot}$. We also have

$$y_{ijk} - \hat{y}_{ijk} = y_{ijk} - \bar{y}_{ij\cdot} \tag{57}$$

as a typical element in the vector of residuals for fitting the no-covariate model. It defines the basis for deriving $\mathbf{R}_z$, whose columns are vectors of residuals obtained from the columns of $\mathbf{Z}$.

A frequently-seen model for covariance in the 2-way classification is

$$y_{ijk} = \mu + \alpha_i + \beta_j + \gamma_{ij} + bz_{ijk} + e_{ijk}\,. \tag{58}$$

Often just the no-interaction case is considered, with γ_{ij} omitted, and sometimes the term in the covariate is in the form $b(z_{ij} - \bar{z}_{..})$ rather than bz_{ij} [see, for example, Federer (1955, p. 487) and Steel and Torrie (1960, p. 309)]. The form bz_{ij} seems preferable because then the equation of the model does not involve a sample (i.e., observed) mean, this being appropriate since models should be in terms of population parameters and not observed samples. Also, the form bz_{ij} is more tractable for the general procedure described earlier, especially when models more complex than (58) are considered.

Although (58) is the most commonly occurring model for handling a covariate in the 2-way classification, other models can also be considered. Thus, whereas (58) assumes the same regression slope for all cells, the model

$$y_{ijk} = \mu + \alpha_i + \beta_j + \gamma_{ij} + b_i z_{ijk} + e_{ijk} \tag{59}$$

assumes different slopes for each level of the α-factor. Similarly,

$$y_{ijk} = \mu + \alpha_i + \beta_j + \gamma_{ij} + b_j z_{ijk} + e_{ijk} \tag{60}$$

assumes a different slope for each level of the β-factor; and both

$$y_{ijk} = \mu + \alpha_i + \beta_j + \gamma_{ij} + (b_i + b_j)z_{ijk} + e_{ijk} \tag{61}$$

and

$$y_{ijk} = \mu + \alpha_i + \beta_j + \gamma_{ij} + b_{ij}z_{ijk} + e_{ijk} \tag{62}$$

assume different slopes for each (i, j)-cell.

Each of these five models, (58)–(62), can be handled by the general method based on $\mathbf{a}^*$ and $R(\mathbf{a})$ of (56), and on deriving each column of $\mathbf{R}_z$ from the procedure indicated in (57). The exact form of $\mathbf{Z}$ in the general model (6) is determined by the form of the b-coefficients in (58)–(62). For example, in (58), $\mathbf{Z}$ is an $N \times 1$ vector, of all the observed z's; in (59), for c levels of the α-factor, it is an $N \times c$ matrix, of the same form as (48); and so on. Whichever of the models (58)–(62) is fitted, the analyses of variance of Tables 8.3a and 8.3b can be used.

Furthermore, fitting successive ones of the models (58)–(62) can also be accommodated, in the same way that Table 8.4a was utilized in developing Table 8.7 when fitting $y_{ij} = \mu + \alpha_i + b_i z_{ij} + e_{ij}$ after having fitted $y_{ij} = \mu + \alpha_i + bz_{ij} + e_{ij}$. For each of (58)–(62), $R(\mathbf{a})$ of Table 8.3a is calculated as in (56) and represents $R(\mu, \alpha, \beta, \gamma)$, which can be partitioned in either of the two ways indicated in Table 7.8. The hypotheses corresponding to those partitionings are derived by means of (17), from the hypotheses tested in the no-covariate model, discussed in Secs. 7.2f(ii)–(v). [In no-interaction analogues of (58)–(62), $R(\mathbf{a})$ of Table 8.3a is $R(\mu, \alpha, \beta)$ of (26) in Sec. 7.1e(i), and can be partitioned as indicated in Table 7.3.] Details, although lengthy, are quite straightforward. A numerical example is provided in Exercise 12.

Covariance procedures for multiple covariates are simple extensions of the methods for one covariate and follow the general procedures discussed above.

3. DATA HAVING ALL CELLS FILLED

Analysis of unbalanced data is more difficult than that of balanced data, for the very reason that they are unbalanced. Interpretation of the analysis is often more difficult also. These difficulties can sometimes be avoided for data which, although unbalanced, are not too far removed from being balanced. In such cases it is sometimes possible to make minor modifications to the data so as to be able to use a balanced data analysis. The decision as to whether to do this or not is, of course, a matter very much open to question, namely, When are unbalanced data "not too far removed" from being balanced? It seems unlikely that this can ever be resolved satisfactorily. Nevertheless, the advantages of using balanced data analyses are so great that one would like to use them whenever feasible: they are easily carried out and usually easy to interpret—especially in comparison to analogous unbalanced data analyses.

The disadvantage of modifying unbalanced data so as to be able to use a balanced data analysis is that doing so introduces a measure of approximation into the analyses—its degree depending on the extent to which the unbalanced data have been modified in order to permit the balanced analysis. However, with the advantages of balanced data analyses being so attractive they may, on occasion, outweigh the disadvantage of some degree of approximation, particularly when the latter might well be deemed small. Instances in which this might be so are outlined below. To simplify presentation they are given in terms of examples of the 2-way crossed classification.

a. Estimating missing observations

If all n_{ij}'s except a few are the same, it is often reasonable to estimate missing observations. For example, suppose with 2 rows and 3 columns that the numbers of observations are as shown in Table 8.9. Data of this nature often arise from what set out to be a planned experiment (in Table 8.9, of 6 observations per cell) and ended up with a few observations missing. Such data are unbalanced, but so slightly as to render the temptation of modifying them to make them balanced irresistible. This can be done by estimating the missing observations, in this case one observation for the cell in the first row and third column. One procedure for doing this is to suppose that observation is u, say, and choose u so as to minimize the residual sum of squares. Had n_{13} been 6 and not 5 this residual, on the basis of an interactions model [see equation (61) of Sec. 7.2d(i)], would have been

$$\sum_{i=1}^{2}\sum_{j=1}^{3}\sum_{k=1}^{6} y_{ijk}^2 - (1/6)\sum_{i=1}^{2}\sum_{j=1}^{3} y_{ij\cdot}^2 \,.$$

As it is, we now take the residual as

$$\begin{aligned}\text{SSE} = \sum_{i=1}^{2}\sum_{j=1}^{2}\sum_{k=1}^{6} y_{ijk}^2 + \sum_{k=1}^{6} y_{23k}^2 + \sum_{k=1}^{5} y_{13k}^2 + u^2 \\ - (1/6)\left[\sum_{i=1}^{2}\sum_{j=1}^{2} y_{ij\cdot}^2 + y_{23\cdot}^2 + (y_{13\cdot} + u)^2\right].\end{aligned}$$

Solution of $\partial(\text{SSE})/\partial u = 0$ leads to

$$u = \sum_{k=1}^{5} y_{13k}/5 = \bar{y}_{13\cdot}$$

and so the missing observation in the (1, 3)-cell is estimated by the mean of the observations that are there.

TABLE 8.9. n_{ij}-VALUES

6	6	5
6	6	6

Of course the form of the results arising from such a process depends on the model used, since this determines the residual sum of squares. Had the model for the data of Table 8.9 been that of no interaction the error sum of squares would have been (see Sec. 7.1)

$$\sum_{i=1}^{2}\sum_{j=1}^{2}\sum_{k=1}^{6} y_{ijk}^2 + \sum_{k=1}^{6} y_{23k}^2 + \sum_{k=1}^{5} y_{13k}^2 + u^2 - (1/18)[(y_{1..} + u)^2 + y_{2..}^2]$$
$$- (1/12)[y_{.1.}^2 + y_{.2.}^2 + (y_{.3.} + u)^2] + (1/36)(y_{...} + u)^2.$$

Minimization of this with respect to u leads, in the general case of a rows, b columns and n observations per cell in all cells except one, the (i, j)th-cell, to

$$u_{ij} = \frac{ax_{i..} + bx_{.j.} - x_{...}}{ab(n-1) + (a-1)(b-1)}. \tag{63}$$

This is equivalent to the result given by Federer (1955, p. 134, equation V-52) for $n = 1$, who also gives results for more than one missing observation when $n = 1$. (These are the procedures referred to at the beginning of Sec. 7.1.)

Reference can be made to Bartlett (1937) for a generalization of the above procedure which depends on a covariance technique. In the model $\mathbf{y} = \mathbf{Xa} + \mathbf{Zb} + \mathbf{e}$ this involves the following:

(i) in $\mathbf{y}$ include each missing observation as an observation of zero,
(ii) in $\mathbf{b}$ include, negatively, a parameter for each missing observation,
(iii) in $\mathbf{Z}$ have one column for each parameter mentioned in (ii), all entries being zero except for a single unity corresponding to the y-value of zero specified in (i).

It will be found that the normal equations of this covariance model are satisfied by the estimated missing observations derived by minimizing residual error sums of squares as described earlier. For example, for the data of Table 8.9 this covariance model (without row-by-column interactions) has the following normal equations:

$$\begin{bmatrix} 36 & 18 & 18 & 12 & 12 & 12 & 1 \\ 18 & 18 & \cdot & 6 & 6 & 6 & 1 \\ 18 & \cdot & 18 & 6 & 6 & 6 & \cdot \\ 12 & 6 & 6 & 12 & \cdot & \cdot & \cdot \\ 12 & 6 & 6 & \cdot & 12 & \cdot & \cdot \\ 12 & 6 & 6 & \cdot & \cdot & 12 & 1 \\ 1 & 1 & \cdot & \cdot & \cdot & 1 & 1 \end{bmatrix} \begin{bmatrix} \mu^o \\ \alpha_1^o \\ \alpha_2^o \\ \beta_1^o \\ \beta_2^o \\ \beta_3^o \\ -u^o \end{bmatrix} = \begin{bmatrix} y_{...} \\ y_{1..} \\ y_{2..} \\ y_{.1.} \\ y_{.2.} \\ y_{.3.} \\ 0 \end{bmatrix} \tag{64}$$

It can be shown (see Exercise 13) that the appropriate form of (63) is a solution to (64). Although this procedure leads to the same results as minimizing residual sums of squares it is often computationally much easier, because it can be applied directly by means of the analysis of covariance procedures (see Sec. 2).

Estimates of missing observations are used just as if they were data. The only change to be made in the balanced data analysis of the combined data (observed and missing) is in the degrees of freedom for the residual error sum of squares. They are calculated as for balanced data and then reduced by the number of missing observations that have been estimated. Thus in an interaction analysis of data like those of Table 8.9 the residual error sum of squares for 6 observations in every cell would be $6(5) = 30$, but with one estimated missing observation it is reduced to 29.

b. Setting data aside

If the numbers of observations in the sub-most cells differ from each other by only a few, it might not be unreasonable to randomly set aside data from appropriate cells in order to reduce all cells to having the same number of observations in each. A balanced data analysis is then readily available on the data so reduced. For example, in data having the n_{ij}-values of Table 8.10 it might be reasonable to randomly set aside observations in order to reduce each cell to 11 observations. Disadvantages in this method are all too clear. The first is the inevitable indecisiveness implicit in the suggestion of doing this only when the n_{ij} differ "by only a few". It begs the question "What is a few?", to which there is no clear-cut answer. All one can say is that the method might be tolerable for n_{ij}-values like those of Table 8.10 but not for some like those of Table 8.11. Too much data would have to be set aside. Of course it can be strongly argued that no data should ever be ignored. That is so, except that all good rules do have their exceptions. Accepting the fact that balanced data analyses are preferred over those for unbalanced data, it appears to this writer that randomly setting aside data in cases having n_{ij}-values like those of Table 8.10 is probably not unreasonable—especially if the within-cell variation is small. Although a clear definition of when to do this and when not to cannot be given, there will surely be occasions when it seems reasonably safe to do so, and at least on these occasions it would seem to be an acceptable

TABLE 8.10. n_{ij}-VALUES

14	11	13
11	13	15

TABLE 8.11. n_{ij}-VALUES

10	17	21
19	22	9

procedure. After all, for the person whose data they are, the ease of interpretation of a balanced data analysis is surely worthwhile.

The method does not involve discarding of data—nor is it described as such—only setting it aside. The implication is that after setting data aside and making a balanced data analysis those data can be returned and the process repeated. Random selection of data for setting aside can be made again and another analysis calculated. It will, of course, not be statistically independent of the first analysis and if the conclusions stemming from it are not in agreement with those of the first analysis, then the second analysis has brought confusion and not enlightenment. If further analyses in this manner bring additionally different conclusions then confusion is compounded. However, in cases where only "a few" observations are being set aside, and especially where within-cell variance is small, this confusion would seem unlikely to arise very often. Indeed, if such confusion does occur, one might be suspicious that some of the observations that had been set aside are outliers and perhaps should be treated as such. Indeed, outliers should probably be set aside, permanently, in the first place. Nevertheless, the method must be used with caution. At worst one can always retreat to the unbalanced data analysis.

c. Analyses of means

(*i*) *Unweighted means analysis*. An easily calculated analysis when all sub-most cells are filled is to treat the means of those cells as observations and subject them to a balanced data analysis, as suggested by Yates (1934). This is, of course, only an approximate analysis with, as usual, the degree of approximation depending on the extent to which the unbalanced data are not balanced. The calculations for the analysis are straightforward. It is known as the unweighted means analysis and proceeds as follows.

Suppose the model for y_{ijk} is, as in equation (51) of Sec. 7.2a,

$$y_{ijk} = \mu + \alpha_i + \beta_j + \gamma_{ij} + e_{ijk} .$$

For each cell calculate the mean

$$x_{ij} = \bar{y}_{ij\cdot} = \sum_{k=1}^{n_{ij}} y_{ijk}/n_{ij} .$$

Then the unweighted means analysis is as shown in Table 8.12.

Several facets of Table 8.12 are worth noting. First, the means of the x_{ij}'s are calculated in the usual manner: e.g., $\bar{x}_{i\cdot} = \sum_{j=1}^{b} x_{ij}/b$. Second, the residual error sum of squares, SSE, is exactly as calculated in the model for y_{ijk} of

TABLE 8.12. UNWEIGHTED MEANS ANALYSIS FOR A 2-WAY CROSSED CLASSIFICATION

Source of Variation	d.f.	Sum of Squares	Mean Square
Rows	$a-1$	$SSA_u = b\sum_{i=1}^{a}(\bar{x}_{i.} - \bar{x}_{..})^2$	MSA_u
Columns	$b-1$	$SSB_u = a\sum_{j=1}^{b}(\bar{x}_{.j} - \bar{x}_{..})^2$	MSB_u
Interaction	$(a-1)(b-1)$	$SSAB_u = \sum_{i=1}^{a}\sum_{j=1}^{b}(\bar{x}_{ij} - \bar{x}_{i.} - \bar{x}_{.j} + \bar{x}_{..})^2$	$MSAB_u$
Residual error	$N-ab$	$SSE = \sum_i\sum_j\sum_k(y_{ijk} - \bar{y}_{ij.})^2$	MSE

Sec. 7.2. Third, the sums of squares do not add up to SST $= \Sigma y^2$; the first three, SSA_u, SSB_u and $SSAB_u$ add to $\Sigma\Sigma\, x_{ij}^2 - x_{..}^2/ab$, but all four do not add to SST. Fourth, the sums of squares SSA_u and SSB_u do not have χ^2-distributions, nor are they independent of SSE. Expected values of the mean squares are as follows.

$$
\begin{aligned}
E(MSA_u) &= \frac{b}{a-1}\sum_{i=1}^{a}[\alpha_i + \bar{\gamma}_{i\cdot} - (\bar{\alpha}. + \bar{\gamma}..)]^2 + n_h\sigma_e^2 \\
E(MSB_u) &= \frac{a}{b-1}\sum_{j=1}^{b}[\beta_j + \bar{\gamma}._j - (\bar{\beta}. + \bar{\gamma}..)]^2 + n_h\sigma_e^2 \\
E(MSAB_u) &= \frac{1}{(a-1)(b-1)}\sum_{i=1}^{a}\sum_{j=1}^{b}(\gamma_{ij} - \bar{\gamma}_{i\cdot} - \bar{\gamma}._j + \bar{\gamma}..)^2 + n_h\sigma_e^2 \\
E(MSE) &= \sigma_e^2,
\end{aligned} \tag{65}
$$

with
$$n_h = \frac{1}{ab}\sum_{i=1}^{a}\sum_{j=1}^{b}\frac{1}{n_{ij}},$$

$1/n_h$ being the harmonic mean of all ab n_{ij}'s.

Since the mean squares of (65) do not have χ^2-distributions their ratios do not provide F-statistics for testing hypotheses. However, Gosslee and Lucas (1965) suggest that they provide reasonably satisfactory F-statistics using amended degrees of freedom for the numerator mean squares. For example, the numerator degrees of freedom suggested for MSA_u/MSE is

$$f_a' = \frac{(a-1)^2\left(\sum_{i=1}^{a} 1/h_{i\cdot}\right)^2}{(\sum 1/h_{i\cdot})^2 + (a-2)a\sum 1/h_{i\cdot}^2} \tag{66}$$

where
$$h_{i\cdot} = \frac{1}{b}\sum_{j=1}^{b}\frac{1}{n_{ij}}$$

with $1/h_{i\cdot}$ being the harmonic mean of the n_{ij}'s of the cells of the ith row. The origin of (66) in Gosslee and Lucas (1965) is that of equating the first two moments of MSA_u to the first two moments of a χ^2-distribution, in the manner of Sec. 2.4i. Although these amended degrees of freedom modify MSA_u/MSE to be an approximate F-statistic, we see from (65) that the hypothesis it tests is equality of $\alpha_i + \bar{\gamma}_{i\cdot}$ for all i. This is not equivalent to H: equality of all α_i, unless we assume as a restriction on the model that $\bar{\gamma}_{i\cdot} = 0$ for all i. Alternatively, and indeed very reasonably, we can interpret the test as testing equality of the row effects in the presence of their average interaction effects, a hypothesis that may often be of interest.

The question attaching to any approximate analysis suggested as a substitute for the exact unbalanced data analysis remains: When can the unweighted means analysis be used? As usual, there is no decisive answer (apart from requiring, trivially, that all $n_{ij} > 0$). Since the unweighted means analysis uses cell means as if they were observations with uniform sampling error, a criterion for using the analysis is to require that these sampling errors be approximately the same. This demands, since the sampling error of a cell mean is proportional to $1/\sqrt{n_{ij}}$, that the values of $1/\sqrt{n_{ij}}$ be approximately equal. What is meant by "equal" in this context is necessarily vague. For example, the values of $1/\sqrt{n_{ij}}$ are approximately equal for the cells of Table 8.11 and for those of Table 8.13, but not for Table 8.14. Unweighted means analyses would therefore seem appropriate for data having the n_{ij}-values of Table 8.11 or 8.13 but not for those of Table 8.14, wherein $1/\sqrt{9}$ is more than four times as large as $1/\sqrt{200}$. Maybe a ratio of 2:1 could be tolerated in the values of $1/\sqrt{n_{ij}}$, for using the unweighted means analysis, but probably not a ratio as large as 4:1. The appropriate analysis for Table 8.14 is the unbalanced data analysis.

(*ii*) *Example.* Suppose data for 2 rows and 3 columns are as shown in Table 8.15. The layout of data follows the same style as Table 7.6: each triplet of numbers represents a total of observations, the number of observations in that total (in parentheses) and the corresponding mean.

TABLE 8.13. n_{ij}-VALUES

192	250	175
320	168	270

TABLE 8.14. n_{ij}-VALUES

10	17	200
130	22	9

TABLE 8.15. AN EXAMPLE OF 2 ROWS AND 3 COLUMNS

Row	Column 1	Column 2	Column 3	Total
1	7	2	3	
	11	4	9	
		6		
	18 (2) 9	12 (3) 4	12 (2) 6	42 (7) 6
2	11	15	38	
	14	16	46	
	17	19		
		22		
	42 (3) 14	72 (4) 18	84 (2) 42	198 (9) 22
Total	60 (5) 12	84 (7) 12	96 (4) 24	240 (16) 15

The unweighted analysis of means of these data is based on the cell means, summarized in Table 8.16. Fitting the model

$$x_{ij} = \mu + \alpha_i + \beta_j + e_{ij}$$

to the values of Table 8.16 gives

$$R(\mu) = 93^2/6 = 1{,}441\tfrac{1}{2}$$

$$R(\mu, \alpha) = (19^2 + 74^2)/3 = 1{,}945\tfrac{2}{3}$$

$$R(\mu, \beta) = (23^2 + 22^2 + 48^2)/2 = 1{,}658\tfrac{1}{2}$$

and

$$R(\mu, \alpha, \beta) = 9^2 + 4^2 + \cdots + 42^2 = 2{,}417\,.$$

TABLE 8.16. CELL MEANS OF TABLE 8.15

Row	Column 1	Column 2	Column 3	Total
1	9	4	6	19
2	14	18	42	74
Total	23	22	48	93

TABLE 8.17. EXAMPLE OF TABLE 8.12 UNWEIGHTED MEANS ANALYSIS OF DATA OF TABLE 8.15

Source of Variation	d.f.	Sum of Squares	
Rows	1	$\text{SSA}_u = 1{,}945\frac{2}{3} - 1{,}441\frac{1}{2}$	$= 504\frac{1}{6}$
Columns	2	$\text{SSB}_u = 1{,}658\frac{1}{2} - 1{,}441\frac{1}{2}$	$= 217$
Interaction	2	$\text{SSAB}_u = 2{,}417 - 1{,}945\frac{2}{3} - 1{,}658\frac{1}{2} + 1{,}441\frac{1}{2}$	$= 254\frac{1}{3}$
Residual error	10	SSE	$= 114$

From these the first three terms of Table 8.12 are calculated as shown in Table 8.17. The last term, SSE, comes directly from the data of Table 8.15 as

$$\text{SSE} = (7^2 + 11^2 - 18^2/2) + \cdots + (38^2 + 46^2 - 84^2/2) = 114,$$

the sum of the within-cell sums of squares.

F-statistics can be calculated in the usual fashion. By (66) the amended degrees of freedom for MSA_u/MSE are $f'_a = 1$, because (66) simplifies to unity when $a = 2$. To illustrate the calculation of (66) we derive the comparable expression for f'_b as follows.

$$\frac{1}{h_{.1}} = \tfrac{1}{2}(\tfrac{1}{2} + \tfrac{1}{3}) = \tfrac{5}{12}; \qquad \frac{1}{h_{.2}} = \tfrac{1}{2}(\tfrac{1}{3} + \tfrac{1}{4}) = \tfrac{7}{24}; \qquad \frac{1}{h_{.3}} = \tfrac{1}{2}(\tfrac{1}{2} + \tfrac{1}{2}) = \tfrac{1}{2};$$

$$\sum_{j=1}^{3} 1/h_{.j}^2 = (100 + 49 + 144)/24^2 = 293/24^2;$$

$$\sum_{j=1}^{3} 1/h_{.j} = (10 + 7 + 12)/24 = 29/24;$$

$$f'_b = \frac{(3-1)^2(29/24)^2}{(29/24)^2 + (3-2)3(293/24^2)} = \frac{3{,}364}{1{,}720} = 1.96.$$

(*iii*) *Weighted squares of means.* An alternative analysis of means is that known as the weighted squares of means, devised by Yates (1934), who also suggested the unweighted analysis just discussed. The advantage of the weighted analysis is that it provides mean squares that *do* have χ^2-distributions. Hence F-statistics are available for hypothesis testing.

The analysis is based on sums of squares of the means $x_{ij} = \bar{y}_{ij.}$ defined earlier, weighting the terms in those sums of squares in inverse proportion

TABLE 8.18. WEIGHTED MEANS ANALYSIS FOR A 2-WAY CROSSED CLASSIFICATION

Source of Variation	d.f.	Sum of Squares[1]	Mean Square
Rows	$a-1$	$\mathrm{SSA}_w = \sum_{i=1}^{a} w_i(\bar{x}_{i.} - \bar{x}_{[1]})^2$	MSA_w
Columns	$b-1$	$\mathrm{SSB}_w = \sum_{j=1}^{b} v_j(\bar{x}_{.j} - \bar{x}_{[2]})^2$	MSB_w
Interaction	$(a-1)(b-1)$	$\mathrm{SSAB}_w = \mathrm{SSAB}_u$ of Table 8.12 $= \sum_{i=1}^{a}\sum_{j=1}^{b}(x_{ij} - \bar{x}_{i.} - \bar{x}_{.j} + \bar{x}_{..})^2$	$\mathrm{MSAB}_w = \mathrm{MSAB}_u$
Residual error	$N - ab$	$\mathrm{SSE} = \sum_i\sum_j\sum_k (y_{ijk} - \bar{y}_{ij.})^2$	MSE

[1] $w_i = \left(\frac{1}{b^2}\sum_{j=1}^{b}\frac{1}{n_{ij}}\right)^{-1}$ and $\bar{x}_{[1]} = \Sigma\, w_i\bar{x}_{i.}/\Sigma\, w_i$. $\quad v_j = \left(\frac{1}{a^2}\sum_{i=1}^{a}\frac{1}{n_{ij}}\right)^{-1}$ and $\bar{x}_{[2]} = \Sigma\, v_j\bar{x}_{.j}/\Sigma\, v_j$.

to the variance of the term concerned. Thus instead of

$$\text{SSA}_u = b\sum_{i=1}^{a}(\bar{x}_{i\cdot} - \bar{x}_{\cdot\cdot})^2$$

of Table 8.12, we use

$$\text{SSA}_w = \sum_{i=1}^{a} w_i(\bar{x}_{i\cdot} - \bar{x}_{[1]})^2$$

where w_i is $\sigma^2/v(\bar{x}_{i\cdot})$ and $\bar{x}_{[1]}$ is the weighted mean of the $\bar{x}_{i\cdot}$'s weighted by the w_i. Details are shown in Table 8.18. The sums of squares in this table do not add up to SST $= \Sigma\, y^2$, just as the ones in Table 8.12 do not. When divided by σ^2 they do have χ^2-distributions and so the F-statistics MSA_w/MSE, MSB_w/MSE and MSAB_w/MSE provide exact tests of hypotheses concerning the α's, β's and γ's. The exact form of the hypotheses is ascertained by considering expected values of the mean squares. They are

$$E(\text{MSA}_w) = \frac{1}{a-1}\sum_{i=1}^{a} w_i\left[\alpha_i + \bar{\gamma}_{i\cdot} - \frac{\Sigma\, w_i(\alpha_i + \bar{\gamma}_{i\cdot})}{\Sigma\, w_i}\right]^2 + \sigma_e^2$$

$$(67)$$

and $$E(\text{MSB}_w) = \frac{1}{b-1}\sum_{j=1}^{b} v_j\left[\beta_j + \bar{\gamma}_{\cdot j} - \frac{\Sigma\, v_j(\beta_j + \bar{\gamma}_{\cdot j})}{\Sigma\, v_j}\right]^2 + \sigma_e^2.$$

Hence $$F = \text{MSA}_w/\text{MSE}$$

tests the hypothesis

$$H: \quad (\alpha_i + \bar{\gamma}_{i\cdot}) \text{ all equal.} \qquad (68)$$

As with the unweighted analysis of means [Table 8.12 and expected values (65)] so here, the hypothesis (68) involves the $\bar{\gamma}_{i\cdot}$'s. If, as a restriction on the model we assume $\bar{\gamma}_{i\cdot} = 0$ for all i, the hypothesis is then one of weighted equality of the α_i's, where the weights are the w_i's. Alternatively, without any restriction, it is a hypothesis of weighted equality of the row effects in the presence of their average interaction effects. The important difference from the unweighted analysis is, though, that the F-statistics of Table 8.18 have exact F-distributions whereas those of Table 8.12 have only approximate F-distributions.

We return to Tables 8.12 and 8.18 when discussing variance components in Chapter 10.

(*iv*) *Example* (*continued*). Calculation of Table 8.18 for the data of Tables 8.15 and 8.16 is as follows.

$$w_1 = [\tfrac{1}{9}(\tfrac{1}{2} + \tfrac{1}{3} + \tfrac{1}{2})]^{-1} = \tfrac{27}{4} \quad \text{and} \quad w_2 = [\tfrac{1}{9}(\tfrac{1}{3} + \tfrac{1}{4} + \tfrac{1}{2})]^{-1} = \tfrac{108}{13};$$

$$v_1 = [\tfrac{1}{4}(\tfrac{1}{2} + \tfrac{1}{3})]^{-1} = \tfrac{24}{5}, \qquad v_2 = [\tfrac{1}{4}(\tfrac{1}{3} + \tfrac{1}{4})]^{-1} = \tfrac{48}{7}$$

TABLE 8.19. EXAMPLE OF TABLE 8.18 WEIGHTED SQUARES OF MEANS ANALYSIS OF DATA OF TABLE 8.15

Source of Variation	d.f.	Sum of Squares
Rows	1	$SSA_w = 1{,}251\frac{21}{29}$
Columns	2	$SSB_w = 488\frac{26}{137}$
Interaction	2	$SSAB_w = 254\frac{1}{3} = SSAB_u$ of Table 8.17
Residual error	10	$SSE = 114$

and

$$v_3 = [\tfrac{1}{4}(\tfrac{1}{2} + \tfrac{1}{2})]^{-1} = 4;$$

$$\bar{x}_{[1]} = \frac{\frac{27}{4}(\frac{19}{3}) + \frac{108}{13}(\frac{74}{3})}{\frac{27}{4} + \frac{108}{13}} = \frac{477}{29};$$

$$SSA_w = \tfrac{27}{4}(\tfrac{19}{3} - \tfrac{477}{29})^2 + \tfrac{108}{13}(\tfrac{74}{3} - \tfrac{477}{29})^2 = 1{,}251\tfrac{21}{29};$$

$$\bar{x}_{[2]} = \frac{\frac{24}{5}(\frac{23}{2}) + \frac{48}{7}(\frac{22}{2}) + 4(\frac{48}{2})}{\frac{24}{5} + \frac{48}{7} + 4} = \frac{1{,}983}{137};$$

$$SSB_w = \tfrac{24}{5}(\tfrac{23}{2} - \tfrac{1{,}983}{137})^2 + \tfrac{48}{7}(\tfrac{22}{2} - \tfrac{1{,}983}{137})^2 + 4(\tfrac{48}{2} - \tfrac{1{,}983}{137})^2 = \tfrac{66{,}882}{137} = 488\tfrac{26}{137}.$$

Table 8.18 therefore becomes as shown in Table 8.19.

d. Separate analyses

Suppose data had the n_{ij}-values shown in Table 8.20. For purposes of discussion dashed lines divide the cells into four sets, A, B, C and D. The only appropriate way of analyzing the complete set of data represented by the n_{ij}-values of Table 8.20 would be to use an unbalanced data analysis. This is clear from the empty cells and widely disparate values of the non-zero n_{ij}'s. Such an analysis, using the interaction model of Sec. 7.2, would provide no testable hypothesis concerning row (or column) effects unencumbered by

TABLE 8.20. n_{ij}-VALUES

(A)	27	32	0	3	1	(B)
	11	12	2	0	2	
(C)	1	0	27	16	24	(D)
	0	8	15	21	22	

interactions. Bearing this in mind, notice that in the four cells labeled A and in the six labeled D, all cells are filled. Also, in B and C there are few data and several empty cells. This prompts the suggestion of making two separate analyses, one of the cells A and the other of cells D, using an analysis of means in both cases. In analyzing A, comparison between rows 1 and 2 and between columns 1 and 2 could be made, and from analyzing D comparisons among rows 3 and 4, and among columns 3, 4 and 5 could be made. Of course, comparisons that cut across these groups of rows and columns are precluded by such an analysis, but then the only alternative, an unbalanced data analysis, provides no satisfactory information on such comparisons anyway, in the interaction model. Therefore little would seem to be lost by analysing just A and D.

When data of the nature alluded to in Table 8.20 occur, one might immediately question the process by which n_{ij}-values of such disparate sizes and groupings have arisen. Be that as it may, in analyzing large-scale survey data such as are discussed in Sec. 1, the suggestion has sometimes been made of analyzing just the all-cells-filled subsets of cells that occur throughout the data. Whilst such a suggestion may be open to criticism, it might not be unreasonable in a small situation like that envisaged in Table 8.20—should it ever arise. It amounts to analyzing sets of data that are what might be called "*weakly connected*". In Table 8.20 cells labeled B and C do have data in them, but very small amounts compared to the A and D. Were B and C to contain no data at all then the sets A and D would be disconnected sets of data and they would *have* to be analyzed separately. As it is, analyzing A and D separately and ignoring B and C would be easy both to compute and to interpret, and for these reasons it may be preferable to analyzing the complete data as one analysis.

4. EXERCISES

1. Fit different models to the data given in the exercises of Chapter 4, using them to illustrate aspects of Sec. 1 of this chapter.

2. (*a*) Use equation (22) to confirm (42).
   (*b*) Write down the normal equations for the data of Table 8.5 using (25) as the model. Derive the solution given by (42) and (44).

3. Derive the distributions of $R(\mathbf{a})$ and SSRB shown in Table 8.3a, and show that $R(\mathbf{a})$, SSRB and $\mathbf{y}'\mathbf{y} - R(\mathbf{a}) -$ SSRB are pair-wise independent.

4. For the general covariance model of Sec. 2a(i), prove that $\mathbf{Z}'\mathbf{PZ}$ is non-singular, for $\mathbf{P}$ of equation (10), and hence show that $\mathbf{b}$ is estimable. Show also that $\boldsymbol{\lambda}'\mathbf{a}$ is estimable under the same conditions that it is estimable in the model without covariates.

5. Use equation (70) of Sec. 5.5b to confirm the F-statistic, based on (40), for testing H: all α_i's equal.

6. Find the means of $\mathbf{a}^o$ and $\hat{\mathbf{b}}$ of (8) and (11) and also their variances and covariance, and show the relationship of these to (15).

7. Graybill (1961, p. 392) gives the F-statistic for testing H: all α's equal in the 1-way classification, with one covariate as (in our notation)

$$\frac{1}{(c-1)\hat{\sigma}^2}\left\{\text{SSR}_{m,yy} + \text{SSE}_{yy} - \frac{(\text{SSR}_{m,yz} + \text{SSE}_{yz})^2}{\text{SSR}_{m,zz} + \text{SSE}_{zz}} - \left[\text{SSE}_{yy} - \frac{(\text{SSE}_{yz})^2}{\text{SSE}_{zz}}\right]\right\}.$$

Show the equivalence of this to $R(\alpha \mid \mu, b)$ of Table 8.4b.

8. For the data of Table 8.5 fit each of the following models and calculate the analyses of variance of Tables 8.4a and 8.4b. Suggest appropriate hypotheses and test them.
   (*a*) The covariate affects y linearly, in the same manner for all high school graduates as it does college graduates, but differently for those who did not complete high school.
   (*b*) The covariate affects y in both a linear and a quadratic manner, the same for everyone.

9. For the model (46) develop a test of the hypothesis H: all α_i's equal and illustrate it with the data of Table 8.5 [Urquhart (1969b)].

10. Show that in Table 8.7 the hypothesis H: $\alpha_i + b_i\bar{z}_{i.}$ equal for all i is tested by $R(\alpha \mid \mu)/(c-1)\hat{\sigma}^2$.

11. Townsend (1969) gives data concerning an experiment designed to determine if the usual lecture-type classroom presentation could be replaced by a programmed text. A class of 62 sophomore students was divided randomly into two groups, with one group receiving the usual lectures while the other was given a programmed textbook for independent study. At the end of the semester both groups were given the same examination. In addition to final examination score (x_1), a measurement of I.Q. (x_2) was recorded for each student. (Other educational studies indicate that performance may be linearly related to I.Q.) Using the basic calculations shown in Table 8.21, carry out a covariance analysis testing any hypotheses you think suitable.

TABLE 8.21. TWO GROUPS OF STUDENTS

Totals	Received Lecture	Received Programmed Text
n	31	31
Σx_1	2,139	2,149
Σx_2	3,100	3,100
Σx_1^2	148,601	157,655
Σx_2^2	318,990	319,920
$\Sigma x_1 x_2$	216,910	224,070

12. The following table shows milligrams of seed planted, corresponding to the yield data in Table 7.6.

	Variety			
Treatment	1	2	3	4
1	2	—	7	3
	4			5
	3			
2	5	6	—	—
	3	4		
3	—	6	6	4
		2	8	6
				5
				7

Use these data to fit covariance models (58)–(62) to the data of Table 7.6.

13. Solve equations (64) for solutions that satisfy $\alpha_1^o + \alpha_2^o = 0$ and $\beta_1^o + \beta_2^o + \beta_3^o = 0$, showing that the resulting solution for μ^o is the form given in (63).

14. Show that SSA_w of Table 8.18 has a χ^2-distribution independent of that of SSE.

15. Calculate the exact unbalanced data analyses for the data of Table 8.15 and compare them with Tables 8.17 and 8.19.

16. Derive $R(\alpha \mid \mu, b)$ of Table 8.4b as the numerator sum of squares for testing H: $\mathbf{K}'\mathbf{a} = \mathbf{0}$ using $\mathbf{K}' = [\mathbf{0}\mathbf{1} \quad \mathbf{1} \quad -\mathbf{I}]$ of $c - 1$ columns.

17. Derive an expression for SSE of Tables 8.3a and 8.3b which suggests that it is the residual error sum of squares for fitting a linear model to $\mathbf{Py}$. Describe the model.

18. Show that the error sum of squares in Tables 8.4a and 8.4b is the same as that for fitting the model $\mathbf{y} - \hat{\mathbf{b}}\mathbf{z} = \mathbf{Xa} + \mathbf{e}$ for $\hat{\mathbf{b}}$ of (33) and that the solution $\mathbf{a}^o$ is that given before equation (32).

CHAPTER 9

INTRODUCTION TO VARIANCE COMPONENTS

Interest in the models of Chapters 5–8 lies mainly in estimating (and testing hypotheses about) linear functions of the effects in the models. These effects are what we call fixed effects, and the models are correspondingly called fixed effects models. There are, however, situations where we have no interest in linear functions of effects but where, by the nature of the data and their derivation, the things of prime interest concerning the effects are variances. Effects of this nature are called random effects, and certain of the models involving them are called random effects models. Other models, involving a mixture of fixed effects and random effects, are called mixed models.

Distinguishing between fixed effects and random effects is the first major topic of this chapter. It is undertaken by means of examples designed to illustrate differences between the two kinds of effects. Discussion of these examples emphasizes the meaning and use of different models in different situations, rather than mathematical details.

The variances associated with random effects are called variance components. The estimation of variance components from balanced data is the chapter's second major topic. Succeeding chapters deal with the more difficult topic of estimating variance components from unbalanced data.

1. FIXED AND RANDOM MODELS

Although the models of Chapters 5–8 are fixed effects models they have not been called that until now. Discussion of fixed effects and random effects therefore begins with a fixed effects model to confirm the use of this name.

a. A fixed effects model

A classic experiment in agricultural research concerns testing the efficacy of nitrogen (N), potash (P) and potassium (K) on crop yield. Suppose an experiment of this kind involved 24 plants, with 6 plants receiving nitrogen, 6 plants getting potash, 6 plants potassium and 6 plants getting no fertilizer at all, these being considered as control (C). A suitable model for analyzing this experiment would be the 1-way classification model (see Sec. 6.2)

$$y_{ij} = \mu + \alpha_i + e_{ij}\,, \tag{1}$$

where y_{ij} is the jth observation on the ith treatment, with μ being a mean, α_i being the effect of treatment i and e_{ij} being an error term in the usual way.

Analysis of this experiment can lead to estimating $\alpha_1 - \alpha_4$, for example, and to testing the hypothesis H: $\alpha_1 - \alpha_4 = 0$. In studying differences of this nature consider the treatments that are being dealt with. They are 4, very specific treatments of interest: in using them we have no thought for any other fertilizers and our interest lies solely in studying N, P and K in relation to each other and to no fertilizer. This, for example, would be particularly so if our experiment was a field trial laid out on a farmer's land with a view to demonstrating to him the value of those three fertilizers. In doing this there would be no thought for other fertilizers. This is the concept of fixed effects. Our attention is fixed upon just the treatments in the experiment, upon these and no others, and so the effects are called *fixed effects*. Furthermore, because all the effects in the model are fixed effects (apart from the error terms, which are always random) the model is called the *fixed effects model*. It is often referred to as *Model I*, so named by Eisenhart (1947).

The manner in which data are obtained always affects the inferences that can be drawn from them. We therefore consider a sampling process pertinent to this fixed effects model, in which the α's are the fixed effects of the four specific treatments, N, P, K and C (control). The data are envisaged as being one possible set of data involving these same treatments that could be derived in repetitions of the experiment, repetitions in which the e's on each occasion would be a random sample from a population of error terms distributed as $(\mathbf{0}, \sigma_e^2\mathbf{I})$.[1] It is the randomness associated with obtaining the e's that provides the means for making inferences about functions of the α_i's and about σ_e^2.

b. A random effects model

Suppose a laboratory experiment designed to study the maternal ability of mice uses litter weights of ten-day-old litters as a measure of maternal ability, after the manner of Young *et al.* (1965). Six litters from each of four dams, all of one breed, constitute the data. A suitable model for analyzing the data is the 1-way classification model

$$y_{ij} = \mu + \delta_i + e_{ij}\,, \tag{2}$$

[1] From this point on we use σ_e^2 in place of σ^2 for the residual error variance.

where y_{ij} is the weight of the jth litter from the ith dam, δ_i being the effect due to the ith dam and e_{ij} the customary error term.

Consider the δ_i's and the dams they represent. The data relate to maternal ability, a variable that is assuredly subject to biological variation from animal to animal. The prime concern of the experiment is therefore unlikely to center on specifically the 4 female mice used in the experiment. After all, they are only a sample from a large population of mice, the females of the breed, each of which has some ability in a maternal capacity. The animals that are in the experiment are therefore envisaged as a random sample of 4 from a population of females.

In the fertilizer experiment previously described, each fertilizer is of specific importance and interest, with no thought for it being a sample from a population of fertilizers. But in the mouse experiment each mouse is merely a sample (of one) from a population of female mice. Nothing important has conditioned our choosing any one mouse over another, and we have no specific interest in the difference between any one of our 4 mice and any other of them. Interest does lie, however, in the extent to which maternal ability varies throughout the population of mice, and to this end our model is directed.

The sampling process involved in obtaining such data is taken as being such that any one of many possible sets of data could be derived from repetitions of the data-gathering process. But now, in concentrating attention on repetitions, we do not confine ourselves to always having the same 4 mice—we imagine getting a random sample of 4 on each occasion from the population of mice. And furthermore, for whatever 4 mice we get on any occasion we envisage getting a random sample of e's from a population of errors, just as with the fixed model. Thus our concept of the error terms is the same in both models. But whereas in the fixed model we conceive of always having the same α's, the same treatments, now, in the case of the mice data, we think of taking a random sample of mice on each occasion. Thus the δ_i's of our data are a random sample from a population of δ's. Hence, so far as the data are concerned, the δ_i's therein are random variables, which in this context we call *random effects*. And the model is correspondingly called the *random effects model* or, sometimes, the *random model*. Eisenhart (1947) called it *Model II*, a name that continues to receive widespread use.

In each model the error terms are a random sample from a population distributed as $(\mathbf{0}, \sigma_e^2\mathbf{I})$. But whereas in the fixed effects model the α's represent effects of specific treatments, in the random model the δ's are also a random sample, from a population distributed as $(\mathbf{0}, \sigma_\delta^2\mathbf{I})$. Furthermore, sampling of the δ's is assumed to be independent of that of the e's and so covariances between δ's and e's are zero. Also, if the distribution of the δ's was to have a non-zero mean μ_δ, we could rewrite the model (2) as

$$y_{ij} = (\mu + \mu_\delta) + (\delta_i - \mu_\delta) + e_{ij} . \qquad (3)$$

Then, on defining $\mu + \mu_\delta$ as the mean and $\delta_i - \mu_\delta$ as the dam effect, the latter would have a zero mean. There is therefore no loss in generality in taking the mean of the δ's in (2) to be zero.

With the δ's and e's of (2) being random variables with variances σ_δ^2 and σ_e^2, respectively, the variance of an observation is, from (2), $\sigma_y^2 = \sigma_\delta^2 + \sigma_e^2$. The variances σ_δ^2 and σ_e^2 are accordingly called *variance components;* each is a variance in its own right and is a component of σ_y^2. The model is sometimes referred to as a *variance component model.* Estimation of the variance components and inferences about them are the objectives of using such a model.

c. Other examples

(*i*) *Of treatments and varieties.* The fixed effects model of equation (1) relates to 4 fertilizer treatments. Suppose this experiment is expanded, to using each of the 4 treatments on 6 different plants of each of 3 varieties of the plant. The 2-way classification (with interaction) model of Sec. 7.2 would then be suitable for y_{ijk}, the yield of the kth plant of the jth variety receiving the ith treatment. Thus we write

$$y_{ijk} = \mu + \alpha_i + \beta_j + \gamma_{ij} + e_{ijk} , \qquad (4)$$

where μ is a general mean, α_i is the effect on yield of the ith treatment, β_j is the effect of the jth variety, γ_{ij} is the interaction and e_{ijk} is the usual error term. Just as treatment effects α_i were earlier described as fixed effects, so they are now. Similarly, the variety effects β_j are also fixed effects because, in this experiment, interest in varieties centers solely on the 3 varieties being used. There is no thought that they are a random sample from some population of varieties. Thus both the α_i and the β_j and their interactions are considered as fixed effects and we have a fixed effects model.

(*ii*) *Of mice and men.* Suppose the mouse experiment had been supervised by three laboratory technicians, one for each successive pair of litters that the mice had. One possible model for the resulting data would be

$$y_{ijk} = \mu + \delta_i + \tau_j + \theta_{ij} + e_{ijk} , \qquad (5)$$

where y_{ijk} is the weight of the kth litter from the ith dam when being cared for by the jth technician: δ_i is the effect on litter weight of the ith dam and τ_j is the effect of the jth technician, and θ_{ij} is the interaction. We earlier explained how δ_i is a random effect, representing the maternal capacity of the ith dam chosen randomly from a population of (female) mice. It is not difficult to imagine τ_j being a random effect of similar nature. A laboratory experiment has to be cared for, and usually there is little interest so far as the experiment itself is concerned in who the technician is attending to it. He can reasonably be thought of as a random sample (of one) from some population of laboratory technicians, so that in the whole experiment we have a random sample of 3 technicians. The τ_j are correspondingly random effects, with zero mean and variance σ_τ^2. Similarly, the interaction effects are also random, with zero mean and variance σ_θ^2; and all covariances are taken as zero. Thus all elements in the model (5)—save μ—are random effects and we have a random model. The parameters of interest, apart from μ, are σ_δ^2, σ_τ^2 and σ_θ^2, representing the influence of dam, technician and dam-by-technician interaction, respectively, on the variance of y. That part of the variance not accounted for by these effects is σ_e^2, the residual error variance, in the usual manner.

(*iii*) *Of cows and bulls.* Another example of the random model arises in dairy cow breeding. With the advent of artificial insemination, a bull can sire offspring in many different places simultaneously and have progeny in numerous different herds. When the females among these progeny themselves calve and start to give milk, analyses of their milk yields can be made. A suitable model for y_{ijk}, the milk yield of the kth daughter in herd i sired by bull j, is

$$y_{ijk} = \mu + \alpha_i + \beta_j + \gamma_{ij} + e_{ijk}\,. \tag{6}$$

α_i is the effect on yield of the cow's being in herd i, β_j is the effect of bull j, γ_{ij} is the interaction effect and e_{ijk} is the customary random error term. In this case all effects are considered random: the herds involved in the data are assumed to be a random sample from a population of herds, the bulls are taken as being random sample of bulls, and the interaction effects are assumed to be random, too. These effects are also considered to be mutually independent, with variances σ_α^2, σ_β^2, σ_γ^2 and σ_e^2 respectively. The animal breeder is interested in estimating these variances so that he can estimate the ratio $4\sigma_\beta^2/(\sigma_\alpha^2 + \sigma_\beta^2 + \sigma_\gamma^2 + \sigma_e^2)$, a ratio which is important in bringing about increased milk production through selective breeding.

2. MIXED MODELS

A general mean μ (a fixed effect) and error terms e (random) occur in all the preceding examples, as they do in most models. Apart from these, all effects in each of the preceding models are either fixed or random. We now consider models where some of the effects (other than μ and e) are fixed and some are random. Such models are called *mixed models.* Of course, any model containing a fixed effect μ and random error terms is truly a mixed model, but the description is usually reserved for models whose effects other than μ and e's are a mixture of fixed and random effects. In some situations, as we shall see (Sec. 10.8), it is convenient to treat all models as though they were mixed models, but generally the distinction is made between fixed, random and mixed models as described here. We give some examples of mixed models.

(*i*) *Of mice and diets.* Suppose in the mouse experiment that instead of the mice being cared for by three different technicians one man supervised the whole experiment. Suppose, further, that three specially prepared diets were used, with the purpose of the experiment being to compare the three diets. Then, if y_{ijk} is the kth litter weight of the ith dam when receiving diet j, the model will be

$$y_{ijk} = \mu + \delta_i + \varphi_j + \gamma_{ij} + e_{ijk} \,. \tag{7}$$

Now, though, because the diets are three specific diets of interest, the φ_j effects representing those diets are fixed effects. As before, the δ_i—the dam effects—are random. Thus (7) is a model containing fixed effects φ_j and random effects δ_i. This is a *mixed model*, a mixture of fixed and random effects.

Notice that (7) includes interaction effects γ_{ij} for interactions between dams and diets. Since dams are being taken as a random effect it is logical that these interactions are random also. Thus the model has φ_j as fixed effects, and the δ_i and γ_{ij} as random, having zero means and variances σ_δ^2 and σ_γ^2 respectively.

(*ii*) *Of treatments and crosses.* In an experiment concerning fertilizers, suppose that 6 plants of each of 20 replicate crosses of 2 varieties of the crop (early- and late-ripening tomatoes, say) are used. Each cross would be a random sample from the infinite number of times that the two varieties could be crossed. Equation (4) could still be the equation of the model, but β_j would now be a random effect for the jth replicate cross, and γ_{ij} would be the (random) interaction effect between the ith fertilizer treatment and the jth cross. Thus equation (4), formerly appropriate to a fixed effects model, is now suited to a mixed model. The equation of the model is unchanged but the meanings of some of its terms have changed.

(*iii*) *On measuring shell velocities.* Thompson (1963), following Grubbs (1948), discusses the problem of using several instruments to simultaneously measure the muzzle velocity of firing a random sample of shells from a manufacturer's stock. A suitable model for y_{ij}, the velocity of the ith shell as recorded by the jth measuring instrument, is

$$y_{ij} = \mu + \alpha_i + \beta_j + e_{ij} \,.$$

In this, α_i is the effect of the ith shell and β_j is the bias in instrument j. Since the shells fired are a random sample of shells the α_i are random effects; and because the instruments used are the only instruments of interest, the β_j are fixed effects. So again we have a mixed model.

3. FIXED OR RANDOM?

Equation (4) for the treatments and varieties example is indistinguishable from (6) for the bulls and herds example. But the models involved are different in the two cases because of the interpretation attributed to the effects: in one case fixed and in the other random. In these and the other examples discussed most of the effects are categorically fixed or random. Thus fertilizer treatments are clearly fixed effects, as are diets and measuring instruments. Similarly, mice, bulls and artillery shells are random effects. But how about the laboratory technicians, where three of them cared for the mice; or the herds wherein the bulls' progeny were being milked? In each case these effects have been assumed random. But this might not always be so. With the technicians, for example, the situation might have been not that each one came and went as a random sample of employees, so to speak, but that all were available and we wanted to assess differences between those three specific technicians. In that case the technician effects in equation (5) would be fixed effects, not random. Similarly with the herd effects in equation (6). Analyses of data of such situations usually involve hundreds of herds that are considered a random sample from some larger population of herds. But were the situation to be one of analyzing just a few herds, five or six say, wherein the sole interest lay in just those herds, then herd effects in (6) would more appropriately be fixed and not random. Thus we see that the situation to which a model applies is the deciding factor in determining whether the effects of a factor are fixed or random.

In some situations the decision as to whether certain effects are fixed or random is not immediately obvious. Take the case of year effects, for example, in studying wheat yields: are the effects of years on yield to be considered fixed or random? The years themselves are unlikely to be random, for they will probably be a group of consecutive years over which the data have been gathered or the experiments run. But the effects on yield may reasonably be considered random—unless, perhaps, one is interested in comparing specific years for some purpose.

In endeavoring to decide whether a set of effects is fixed or random, the context of the data, the manner in which they were gathered and the environment from which they came are the determining factors. In considering these points the important question is that of inference: are inferences going to be drawn from these data about *just* these levels of the factor? "Yes"—then the effects are to be considered as fixed effects. "No"—then, presumably, inferences will be made not just about the levels occurring in the data but about some population of levels of the factor from which those in the data are presumed to have come; and so the effects are considered as being random. Thus when inferences are going to be confined to the effects in the model the effects are considered fixed; and when inferences will be made about a population of effects from which those in the data are considered to be a random sample then the effects are considered as random.

It is to be emphasized that the assumption of randomness does not carry with it the assumption of normality. Often this assumption *is* made for random effects, but it is a separate assumption made subsequent to that of assuming effects are random. Although most estimation procedures for variance components do not require normality, if distributional properties of the resulting estimators are to be investigated then normality of the random effects is often assumed.

4. FINITE POPULATIONS

Random effects occurring in data are assumed to be from a population of effects. The populations are usually considered to have infinite size as is, for example, the population of all possible crosses between two varieties of tomato. They could be crossed an infinite number of times. However, the definition of random effects does not demand infinite populations of such effects. They can be finite. In addition, finite populations may be very large, indeed so large as to be considered infinite for most purposes; an example would be all the mice in New York State on July 4, 1970! Hence random effects factors can have conceptual populations of three kinds insofar as their size is concerned: infinite, finite but so large as to be deemed infinite, and finite.

We shall be concerned with random effects coming solely from populations assumed to be of infinite size, either because this is the case or because, although finite, the population is large enough to be taken as infinite. These are the most oft-occurring situations found in practical problems. Finite populations, a propos variance components, are discussed in several places, e.g., Bennett and Franklin (1954, p. 404) and Gaylor and Hartwell (1969). Rules for converting the estimation procedure of any infinite-population situation into one of finite populations are given in Searle and Fawcett (1970).

5. INTRODUCTION TO ESTIMATION

The estimation of variance components from balanced data relies almost exclusively on one method. For unbalanced data there are several methods each of which simplifies to the method used for balanced data. We therefore consider balanced data first. For any random (or mixed) model the method of estimating variance components relies on the mean squares of the analysis of variance for the corresponding fixed effects model. The general procedure is to calculate the analysis of variance as if the model were a fixed effects model and then derive the expected values of the mean squares under the random (or mixed) model. Certain of the expected values will be linear functions of the variance components. Equating these expected mean squares to their calculated (observed) values leads to linear equations in the variance components, the solutions to which are taken as the estimators of those components. This method of estimating variance components is known as the *analysis of variance method.*

Mean squares in analyses of variance are quadratic forms in the observations. Their expected values can therefore be derived from Theorem 1 of Chapter 2, wherein **V** is the variance-covariance matrix of the observations. Although for balanced data this is not the easiest method of calculating expected values of mean squares, it is instructive to demonstrate the form of the **V**-matrix for a simple random model. It is the basis of such matrices for unbalanced data for which Theorem 1 of Chapter 2 is of utmost importance. We illustrate by means of the mouse example of Sec. 1b.

a. Variance matrix structures

In all the fixed effects models of Chapters 5–8 the covariance matrix of the observations, var(**y**), has been of the form $\sigma_e^2\mathbf{I}_N$. However, this is not so for random (and mixed) models, because the covariance structure of the random effects determines the variance-covariance matrix of the vector of observations.

Suppose we rewrite the model for the mouse example, equation (2), as

$$y_{ij} = \mu + \alpha_i + e_{ij}, \tag{8}$$

where μ and e_{ij} are the same as in (2) and α_i is now used in place of δ_i. Thus α_i is a random effect, with zero mean and variance σ_α^2, and is independent of the e's and the other α's; i.e., $E(\alpha_i\alpha_k) = 0$ for $i \neq k$, and $E(\alpha_i e_{i'j'}) = 0$ for all i, i' and j'. From this we have

$$\operatorname{cov}(y_{ij}, y_{i'j'}) = \begin{cases} \sigma_\alpha^2 + \sigma_e^2 & \text{for } i = i', \quad j = j', \\ \sigma_\alpha^2 & \text{for } i = i', \quad j \neq j', \\ 0 & \text{for } i \neq i'. \end{cases}$$

Hence, for example, the variance-covariance matrix of the 6 observations on the first dam is

$$\text{var}\begin{bmatrix} y_{11} \\ y_{12} \\ y_{13} \\ y_{14} \\ y_{15} \\ y_{16} \end{bmatrix} = \begin{bmatrix} \sigma_\alpha^2+\sigma_e^2 & \sigma_\alpha^2 & \sigma_\alpha^2 & \sigma_\alpha^2 & \sigma_\alpha^2 & \sigma_\alpha^2 \\ \sigma_\alpha^2 & \sigma_\alpha^2+\sigma_e^2 & \sigma_\alpha^2 & \sigma_\alpha^2 & \sigma_\alpha^2 & \sigma_\alpha^2 \\ \sigma_\alpha^2 & \sigma_\alpha^2 & \sigma_\alpha^2+\sigma_e^2 & \sigma_\alpha^2 & \sigma_\alpha^2 & \sigma_\alpha^2 \\ \sigma_\alpha^2 & \sigma_\alpha^2 & \sigma_\alpha^2 & \sigma_\alpha^2+\sigma_e^2 & \sigma_\alpha^2 & \sigma_\alpha^2 \\ \sigma_\alpha^2 & \sigma_\alpha^2 & \sigma_\alpha^2 & \sigma_\alpha^2 & \sigma_\alpha^2+\sigma_e^2 & \sigma_\alpha^2 \\ \sigma_\alpha^2 & \sigma_\alpha^2 & \sigma_\alpha^2 & \sigma_\alpha^2 & \sigma_\alpha^2 & \sigma_\alpha^2+\sigma_e^2 \end{bmatrix}$$

$$= \sigma_e^2\mathbf{I} + \sigma_\alpha^2\mathbf{J}. \tag{9}$$

We meet this form of matrix repeatedly: $\lambda_1\mathbf{I} + \lambda_2\mathbf{J}$ where λ_1 and λ_2 are scalars (usually variances) and $\mathbf{J}$ is a square matrix with every element unity. In the present case it is the covariance matrix of the set of 6 litter weights from each dam. Since the weights are independent, as between one dam and another, the covariance matrix of all 24 weights can be partitioned as

$$\text{var}(\mathbf{y}) = \begin{bmatrix} \sigma_e^2\mathbf{I} + \sigma_\alpha^2\mathbf{J} & \mathbf{0} & \mathbf{0} & \mathbf{0} \\ \mathbf{0} & \sigma_e^2\mathbf{I} + \sigma_\alpha^2\mathbf{J} & \mathbf{0} & \mathbf{0} \\ \mathbf{0} & \mathbf{0} & \sigma_e^2\mathbf{I} + \sigma_\alpha^2\mathbf{J} & \mathbf{0} \\ \mathbf{0} & \mathbf{0} & \mathbf{0} & \sigma_e^2\mathbf{I} + \sigma_\alpha^2\mathbf{J} \end{bmatrix},$$

where $\mathbf{I}$ and $\mathbf{J}$ have order equal to the number of observations in the classes, in this case 6. This matrix is the direct sum of four matrices of the form shown in (9). Using $\sum^+$ to denote the operation of direct sum, as in Sec. 6.2a, we write

$$\text{var}(\mathbf{y}) = \sum_{i=1}^{4}{}^{+} (\sigma_e^2\mathbf{I} + \sigma_\alpha^2\mathbf{J}), \tag{10}$$

a notation we have frequent occasion to use, especially with unbalanced data in the form $\sum_{i=1}^{a}{}^{+} (\sigma_e^2\mathbf{I}_i + \sigma_\alpha^2\mathbf{J}_i)$ where $\mathbf{I}_i$ and $\mathbf{J}_i$ are then of order n_i.

b. Analyses of variance

The 1-way classification model of Sec. 6.2d is suitable for the fertilizer experiment discussed in Sec. 1a. Its analysis of variance is shown in Table 9.1, based on Table 6.4.

The basic use of Table 9.1 is to summarize calculation of the F-statistic MSR_m/MSE for testing H: all α's equal. The lower section of the table contains the expected values of the mean squares and is usually not shown for fixed effects models. Nevertheless, its presence emphasizes the hypothesis

TABLE 9.1. ANALYSIS OF VARIANCE FOR 4 FERTILIZER TREATMENTS EACH USED ON 6 PLANTS

Source of Variation	d.f.	Sum of Squares	Mean Square
Mean	1	$\text{SSM} = R(\mu) = 24\bar{y}_{..}^2$	$\text{MSM} = \text{SSM}/1$
Treatments	3	$\text{SSR}_m = R(\alpha \mid \mu) = \sum_{i=1}^{4} 6(\bar{y}_{i.} - \bar{y}_{..})^2$	$\text{MSR}_m = \text{SSR}_m/3$
Residual error	20	$\text{SSE} = \text{SST} - R(\mu, \alpha) = \sum_{i=1}^{4}\sum_{j=1}^{6} (y_{ij} - \bar{y}_{i.})^2$	$\text{MSE} = \text{SSE}/20$
Total	24	$\text{SST} = \sum_{i=1}^{4}\sum_{j=1}^{6} y_{ij}^2$	

Expected mean squares

$$E(\text{MSM}) = 24\left(\mu + \tfrac{1}{4}\sum_{i=1}^{4} \alpha_i\right)^2 + \sigma_e^2$$

$$E(\text{MSR}_m) = \tfrac{6}{3}\sum_{i=1}^{4}\left(\alpha_i - \tfrac{1}{4}\sum_{i=1}^{4} \alpha_i\right)^2 + \sigma_e^2$$

$$E(\text{MSE}) = \sigma_e^2$$

that can be tested by the F-statistic. This is so because, for $F = Q/s\hat{\sigma}^2$ used so much in earlier chapters,

$$F \sim F'\{s, N - r, [E(Q) - s\sigma^2]/2\sigma^2\},$$

as can be shown from applying Theorems 1 and 2 of Chapter 2 to Q (see Exercise 4). Therefore the hypothesis concerning s LIN estimable functions which makes $[E(Q) - s\sigma^2]$ zero is tested by comparing $F = Q/s\hat{\sigma}^2$ against the central $F(s, N - r)$-distribution.

Example. In Table 9.1

$$E(\text{SSR}_m) = 6\sum_{i=1}^{4} (\alpha_i - \bar{\alpha}_.)^2 + 3\sigma_e^2 .$$

Hence, $F = \text{SSR}_m/3\hat{\sigma}_e^2$ tests the hypothesis that makes $6\sum (\alpha_i - \bar{\alpha}_.)^2$ zero, namely $H\colon\ \alpha_1 = \alpha_2 = \alpha_3 = \alpha_4$.

Although expected mean squares are helpful in indicating the hypotheses tested by the corresponding F-statistics, they are shown in Table 9.1 for comparison with the random model case of the mouse experiment of Sec. 1a. Since the fixed effects analogue of the model for this experiment is the same as the model for the fertilizer experiment, the variance components of the mouse data are estimated from the analysis of variance shown in Table 9.2. It is identical to Table 9.1 except for the section of expected values of mean squares. In both cases these expected values can be obtained from Theorem 1 of Chapter 2. For Table 9.1 the covariance matrix $\mathbf{V} = \text{var}(\mathbf{y})$ is $\sigma^2\mathbf{I}$, and for Table 9.2 it is $\mathbf{V}$ of (10). An alternative (and often easier) derivation is the "brute force" one of substituting the equation of the model into the mean squares and then taking expectations, using the assumptions of the appropriate model in each case (see Exercise 1). In practice, neither of these methods need be used for balanced data, because simple rules of thumb then apply, as given in Sec. 6. Neither method is therefore illustrated here. Illustration for the 2-way classification, balanced data, is given in Sec. 7, and for unbalanced data in Chapter 10.

TABLE 9.2. ANALYSIS OF VARIANCE OF 4 DAMS EACH HAVING 6 LITTERS

Source of Variation	d.f.	Sum of Squares	Mean Square
Mean	1	$\text{SSM} = R(\mu) = 24\bar{y}_{..}^2$	$\text{MSM} = \text{SSM}/1$
Dams	3	$\text{SSR}_m = R(\alpha \mid \mu) = \sum_{i=1}^{4} 6(\bar{y}_{i.} - \bar{y}_{..})^2$	$\text{MSR}_m = \text{SSR}_m/3$
Residual error	20	$\text{SSE} = \text{SST} - R(\mu, \alpha) = \sum_{i=1}^{4}\sum_{j=1}^{6}(y_{ij} - \bar{y}_{i.})^2$	$\text{MSE} = \text{SSE}/20$
Total	24	$\text{SST} = \sum_{i=1}^{4}\sum_{j=1}^{6} y_{ij}^2$	

Expected mean squares

$$E(\text{MSM}) = 24\mu^2 + 6\sigma_\alpha^2 + \sigma_e^2$$

$$E(\text{MSR}_m) = 6\sigma_\alpha^2 + \sigma_e^2$$

$$E(\text{MSE}) = \sigma_e^2$$

The feature that distinguishes Table 9.2 from Table 9.1 is that the expected mean squares differ. $E(\text{MSR}_m)$ in Table 9.2 contains a term in σ_α^2, the variance component pertaining to the dams, whereas in Table 9.1 it contains a quadratic function of the fixed effects. This difference in the two tables (also evident in the expected value of MSM) arises solely because of the differing models: in Table 9.2 the α_i's are random effects and in Table 9.1 they are fixed effects.

It is possible to make the tables look even more alike: if in Table 9.1 we define

$$s_\alpha^2 = \frac{\sum_{i=1}^{4}\left(\alpha_i - \frac{1}{4}\sum_{i=1}^{4}\alpha_i\right)^2}{3}$$

then $E(\text{MSR}_m) = 6s_\alpha^2 + \sigma_e^2$ comparable to $E(\text{MSR}_m) = 6\sigma_\alpha^2 + \sigma_e^2$ of Table 9.2. Defining and using s_α^2 in this fashion has no particular merit other than emphasizing that the quadratic in the α_i's in $E(\text{MSR}_m)$ in Table 9.1, the fixed effects model, is tantamount to a sample variance of the treatment effects α_1, α_2, α_3 and α_4. While this kind of relationship is true for balanced data it does not hold for unbalanced data. More importantly, the one-to-one correspondence between s^2 and σ^2 so illustrated does not necessarily exist, even with balanced data, in more complex experimental designs than the one considered here. It is therefore misleading to use s_α^2 in the belief that this apparent correspondence is universal. It is not.

c. Estimation

The residual error variance in the fixed effects model of Table 9.1 is estimated in the usual way by $\hat{\sigma}_e^2 = \text{MSE}$. This is tantamount to the analysis of variance method of estimating variance components by equating mean squares to their expected values. It is continued in Table 9.2 to give not only

$$\hat{\sigma}_e^2 = \text{MSE} \quad \text{but also} \quad 6\hat{\sigma}_\alpha^2 + \hat{\sigma}_e^2 = \text{MSR}_m .$$

The solutions to these equations are

$$\hat{\sigma}_e^2 = \text{MSE} \quad \text{and} \quad \hat{\sigma}_\alpha^2 = (\text{MSR}_m - \text{MSE})/6,$$

and they are the estimators of σ_e^2 and σ_α^2.

The preceding example is the simplest illustration of estimating variance components from balanced data of a random (or mixed) model. It extends easily to other balanced data situations. In the analysis of variance there will be as many mean squares whose expectations do not involve fixed effects as there are variance components to be estimated. Equating each of these mean squares to their expected values gives a set of linear equations in the variance components, the solutions to which are the estimators of the variance components. In Table 9.2, for example, $E(\text{MSM})$ involves μ, but the other expected mean squares do not and so they yield the estimators of the variance components of the model. With random models the only expected mean square involving fixed effects is $E(\text{MSM})$, that for the mean. In mixed models there will also be others, but there will also be sufficient expected mean squares that do not involve fixed effects to provide equations that yield estimators of the variance components. This is the analysis of variance method of estimating variance components.

The procedure of "equating mean squares to their expected values" is a special case of the more general procedure of equating quadratic forms to their expected values, as used in a variety of ways with unbalanced data. These are discussed in Chapter 10. For balanced data the "obvious" quadratic forms to use are the analysis of variance mean squares and, indeed, it turns out that the resulting estimators have several optimal properties. Since derivation of the estimators depends upon availability of expected mean squares we turn first to these and the rules which enable them to be written down on sight. Subsequently we consider the properties of the estimators.

6. RULES FOR BALANCED DATA

Discussion is confined to factorial designs, consisting of crossed and nested classifications and combinations thereof, where the number of observations in all of the sub-most subclasses is the same. Situations of partially balanced data, such as in Latin squares, balanced incomplete blocks and their extensions are thus excluded. Otherwise, the rules of thumb for setting up analysis of variance tables apply to any combination of any number of crossed and/or nested classifications. These rules lay out procedures for determining (i) the lines in the analysis of variance, (ii) their degrees of freedom, (iii) formulae for calculating sums of squares and (iv) expected values of mean squares. Most of the rules are based on Henderson (1959, 1969) except that Rule 9 comes from Millman and Glass (1967), who rely heavily on the Henderson paper for a similar set of rules.

The description of the rules is purposefully brief, with no attempt at substantiation. For this the reader is referred to Lum (1954) and Schultz (1955).

a. Establishing analysis of variance tables

(*i*) *Factors and levels.* The analysis of variance table is described in terms of factors A, B, C, ..., with the number of levels in them being n_a, n_b, n_c, ... respectively. When one factor is nested within another the notation will be $C{:}B$ for factor C within factor B, $C{:}BA$ for C within AB subclasses and so on. A letter on the left of the colon represents the nested factor and those on the right of the colon represent the factors within which the nested factor is found. With a nested factor, C for example, n_c is the number of levels of factor C within each of the factors in which it is nested. Factors that are not nested, namely those forming cross classifications, will be called crossed factors.

Within every sub-most subclass of the data there are assumed to be the same number of observations, n_w, either one or more than one. In either case these observations can, as Millman and Glass (1967) point out, be referred to as replications within all other subclasses. Following Henderson (1959) we refer to these as the "within" factor, using the notation $W{:}ABC\ldots$, the number of levels of the "within" factor (i.e., number of replicates) being n_w. The total number of observations is then the product of the n's, namely $N = n_a n_b n_c \ldots n_w$.

(*ii*) *Lines in the analysis of variance table*

Rule 1. There is one line for each factor (crossed or nested), for each interaction, and for "within".

(*iii*) *Interactions.* Interactions are obtained symbolically as products of factors, both factorial and nested. All products of 2, 3, 4, . . . factors are considered. For the sake of generality all crossed factors are assumed to have a colon to the right of the symbol; e.g., $A{:}$, $B{:}$ and so on.

Rule 2. Every interaction is of the form $ABC\ldots{:}\,XYZ\ldots$, where $ABC\ldots$ is the product on the left of the colon of the factors being combined and $XYZ\ldots$ is the product on the right of the colon of the factors so associated with A, B and $C\ldots$.

Rule 3. Repeated letters on the right of the colon are replaced by one of their kind.

Rule 4. If any letter occurs on both sides of a colon that interaction does not exist.

Examples.

Factors	*Interaction*	
A and B	AB	(Rule 2)
A and $C{:}B$	$AC{:}B$	(Rule 2)
$A{:}B$ and $C{:}B$	$AC{:}BB \equiv AC{:}B$	(Rule 3)
$A{:}B$ and $B{:}DE$	$AB{:}BDE$, nonexistent	(Rule 4)

The symbolic form $W{:}ABC\ldots$ for replicates does, by Rule 4, result in no interactions involving W. Furthermore, the line in the analysis of variance labeled $W{:}ABC\ldots$, being the "within" line, is the residual error line.

(*iv*) *Degrees of freedom.* Each line in an analysis of variance table refers either to a crossed factor (such as A:), to a nested factor (such as $C{:}B$) or to an interaction (e.g., $AC{:}B$). Any line can therefore be typified by the general expression given for an interaction in Rule 2, namely $ABC\ldots{:}XYZ\ldots$.

Rule 5. Degrees of freedom for the line denoted by

$$AB{:}XY \text{ are } (n_a - 1)(n_b - 1)n_x n_y .$$

The rule is simple. Degrees of freedom are the product of terms like $(n_a - 1)$ for every letter A on the left of the colon and of terms like n_x for every letter X on the right of the colon.

Rule 6. The sum of all degrees of freedom is $N - 1$, with $N = n_w n_a n_b n_c \ldots$.

(*v*) *Sums of squares.* The symbols that specify a line in the analysis of variance are used to establish the corresponding sum of squares. The basic elements are taken to be the uncorrected sums of squares with notation:

$$1 \equiv CF = N\bar{y}^2$$

and $\quad a, ab, abc \equiv$ uncorrected sums of squares for the A-factor and the AB and ABC subclasses, respectively.

Rule 7. The sum of squares for the line denoted by

$$AB{:}XY \text{ is } (a - 1)(b - 1)xy = abxy - axy - bxy + xy .$$

Again the rule is simple: symbolically, a sum of squares is the product of terms like $(a - 1)$ for every letter A on the left of the colon and of terms like x for every letter X on the right of the colon. This rule is identical to Rule 5 for degrees of freedom: if in the expression for degrees of freedom every n_f is replaced by f, the resulting expansion is, symbolically, the sum of squares: e.g.,

$$(n_a - 1)(n_b - 1)n_x n_y \text{ becomes } (a - 1)(b - 1)xy = abxy - axy - bxy + xy.$$

After expansion, interpretation of these products of lower case letters is as uncorrected sums of squares.

Note that all sums of squares are expressed essentially in terms of crossed factors. Even when a factor is nested, sums of squares are expressed in terms of uncorrected sums of squares calculated as if the nested factor were a crossed factor. For example, the sum of squares for $A{:}B$ (A within B) is $(a - 1)b = ab - b$, where ab is the uncorrected sum of squares of the AB subclasses.

Rule 8. The total of all sums of squares is $\sum y^2 - CF$, where $\sum y^2$ represents the sum of squares of the individual observations, $wabc\ldots$ in the above notation, and where CF is the correction factor.

TABLE 9.3. EXAMPLE OF RULES 1–8: ANALYSIS OF VARIANCE FOR FACTORS A, B, $C:B$ AND $W:ABC$

Line (Rules 1–4)	Degrees of Freedom (Rule 5)	Sum of Squares (Rule 7)
A	$n_a - 1$	$(a - 1) = a - 1$
B	$n_b - 1$	$(b - 1) = b - 1$
$C:B$	$(n_c - 1)n_b$	$(c - 1)b = bc - b$
AB	$(n_a - 1)(n_b - 1)$	$(a - 1)(b - 1) = ab - a - b + 1$
$AC:B$	$(n_a - 1)(n_c - 1)n_b$	$(a - 1)(c - 1)b = abc - ab - bc + b$
$W:ABC$	$(n_w - 1)n_a n_b n_c$	$(w - 1)abc = wabc - abc$
Total	$N - 1$ (Rule 6)	$\sum y^2 - CF \equiv wabc$ (Rule 8)

Example. Table 9.3 shows the analysis of variance derived from these rules for the case of two crossed classifications A and B, a classification C nested within B, namely $C:B$, and the within factor $W:ABC$. Application of these rules is indicated at appropriate points in the table.

b. Calculating sums of squares

The uncorrected sums of squares denoted by lower case letters such as a and ab have so far been defined solely in words; for example, ab is the uncorrected sum of squares for AB subclasses. Henderson (1959, 1969) has no formal, algebraic definition of these terms—and in some sense it is unnecessary so to do, since "everyone knows" what is meant by this: the uncorrected sum of squares for the AB subclasses is the sum over all such subclasses of the square of each subclass total, the sum being divided by the number of observations in such a subclass (the same number in each). However, Millman and Glass (1967) give a neat procedure for formalizing this. It starts from an expression for the total of all the observations. We state the rule using as an example the uncorrected sum of squares bc in a situation where x_{hijk} is the observation in levels h, i, j and k of factors A, B, C and W respectively.

Rule 9. (*a*) Write down the total of all observations:

$$\sum_{h=1}^{n_a} \sum_{i=1}^{n_b} \sum_{j=1}^{n_c} \sum_{k=1}^{n_w} x_{hijk} .$$

(*b*) Re-order the summation signs so that those pertaining to the letters in the symbolic form of the uncorrected sum of squares of interest (bc, in this case) come first, and enclose the remainder of the sum in parentheses:

$$\sum_{i=1}^{n_b} \sum_{j=1}^{n_c} \left(\sum_{h=1}^{n_a} \sum_{k=1}^{n_w} x_{hijk} \right).$$

(*c*) Square the parenthesis and divide by the product of the n's therein. The result is the required sum of squares:

e.g.,
$$bc = \frac{\sum_{i=1}^{n_b} \sum_{j=1}^{n_c} \left(\sum_{h=1}^{n_a} \sum_{k=1}^{n_w} x_{hijk} \right)^2}{n_a n_w}.$$

As a workable rule this is patently simple.

c. Expected values of mean squares, *E*(MS)

Mean squares are sums of squares divided by degrees of freedom. Expected values of mean squares, to be denoted generally by E(MS), are obtained by the following rules.

(*i*) *Completely random models.*

Rule 10. Denote variances by σ^2 with appropriate subscripts. There will be as many σ^2's, with corresponding subscripts, as there are lines in the analysis table. The variance corresponding to the W-factor is the error variance: $\sigma^2_{w:abc\ldots} = \sigma^2_e$.

Example. Where there is an $AC{:}B$ interaction, there is a variance $\sigma^2_{ac:b}$.

When $n_w = 1$, there is no W-line in the analysis of variance, although it may be appropriate to envisage σ^2_w as existing.

Rule 11. Whenever a σ^2 appears in any E(MS) its coefficient is the product of all n's whose subscripts do not occur in the subscript of that σ^2.

Example. The coefficient of $\sigma^2_{ac:b}$ is n_w when the factors are A, B, $C{:}B$ and $W{:}ABC$.

This rule implies that the coefficient of $\sigma^2_{w:abc\ldots}$ is always unity.

Rule 12. Each E(MS) contains only those σ^2's (with coefficients) whose subscripts include all letters pertaining to the MS.

Example. For the $AC{:}B$ line $E[\text{MS}(AC{:}B)] = n_w\sigma^2_{ac:b} + \sigma^2_{w:abc}$.

According to this rule $\sigma^2_e = \sigma^2_{w:abc\ldots}$ occurs in every E(MS) expression.

The above examples of Rules 10–12 are part of the expected values shown in Table 9.4. These are the expected values, under the random model, of the mean squares of the analysis of variance of Table 9.3.

(*ii*) *Fixed effects and mixed models.*

Rule 13. Treat the model as completely random, except that σ^2-terms corresponding to fixed effects and interactions of fixed effects get changed into quadratic functions of these fixed effects. All other σ^2-terms remain, including those pertaining to interactions of fixed and random effects.

TABLE 9.4. EXAMPLE OF RULES 10–12: EXPECTED VALUES, UNDER THE RANDOM MODEL, OF MEAN SQUARES OF TABLE 9.3.

Mean Square	Variances (Rule 10) and Coefficients (Rule 11)					
	$n_b n_c n_w \sigma_a^2$	$n_a n_c n_w \sigma_b^2$	$n_a n_w \sigma_{c:b}^2$	$n_c n_w \sigma_{ab}^2$	$n_w \sigma_{ac:b}^2$	$\sigma_{w:abc}^2 = \sigma_e^2$
	Terms included (Rule 12)					
MS(A)	*			*	*	*
MS(B)		*	*	*	*	*
MS($C:B$)			*		*	*
MS(AB)				*	*	*
MS($AC:B$)					*	*
MS($W:ABC$)						*

* denotes a σ^2-term that is included; e.g., $n_b n_c n_w \sigma_a^2$ is part of $E[\text{MS}(A)]$.

This rule is equivalent to that given by Henderson (1969) but differs from Henderson (1959), where it is stated that some σ^2-terms "disappear" from some of the expectations of mean squares. Explanation of this difference is included in the discussion of the 2-way classification that now follows.

7. THE 2-WAY CLASSIFICATION

Chapter 7 deals fully with the analysis of unbalanced data from the fixed effects model of the 2-way classification. The analysis of variance for balanced data shown there is repeated here as Table 9.5, using new symbols for the sums of squares and mean squares. SSA, for example, is the sum of squares for the A-factor (after μ), with $\text{SSA} = R(\alpha \mid \mu) = R(\alpha \mid \mu, \beta) = bn \sum_{i=1}^{a} (\bar{y}_{i..} - \bar{y}_{...})^2$, as in Table 7.9. Expected values of these sums of squares are now developed for the fixed, random and mixed models, both as illustration of the "brute force" method of deriving such expectations and for discussing certain aspects of the mixed model.

The equation of the model is, as in Chapter 7,

$$y_{ijk} = \mu + \alpha_i + \beta_j + \gamma_{ij} + e_{ijk} \tag{11}$$

with $i = 1, 2, \ldots, a$, $j = 1, 2, \ldots, b$ and, since we are considering balanced data, $k = 1, 2, \ldots, n$. To establish expected values of the sums of squares in

TABLE 9.5. ANALYSIS OF VARIANCE FOR A 2-WAY CLASSIFICATION INTERACTION MODEL, WITH BALANCED DATA (see Table 7.9)

Source of Variation	d.f.	Sum of Squares
Mean	1	$\text{SSM} = N\bar{y}^2_{...}$
A-factor	$a - 1$	$\text{SSA} = bn\sum_{i=1}^{a}(\bar{y}_{i..} - \bar{y}_{...})^2$
B-factor	$b - 1$	$\text{SSB} = an\sum_{j=1}^{b}(\bar{y}_{.j.} - \bar{y}_{...})^2$
AB interaction	$(a - 1)(b - 1)$	$\text{SSAB} = n\sum_{i=1}^{a}\sum_{j=1}^{b}(\bar{y}_{ij.} - \bar{y}_{i..} - \bar{y}_{.j.} + \bar{y}_{...})^2$
Residual error	$ab(n - 1)$	$\text{SSE} = \sum_{i=1}^{a}\sum_{j=1}^{b}\sum_{k=1}^{n}(y_{ijk} - \bar{y}_{ij.})^2$
Total	$N = abn$	$\text{SST} = \sum_{i=1}^{a}\sum_{j=1}^{b}\sum_{k=1}^{n}y^2_{ijk}$

Mean Squares
$\text{MSM} = \text{SSM}$
$\text{MSA} = \text{SSA}/(a - 1)$
$\text{MSB} = \text{SSB}/(b - 1)$
$\text{MSAB} = \text{SSAB}/(a - 1)(b - 1)$
$\text{MSE} = \text{SSE}/ab(n - 1)$

Table 9.5 first write down the various means. They involve using

$$\bar{\alpha}. = \sum_{i=1}^{a}\alpha_i\Big/a, \qquad \bar{\beta}. = \sum_{j=1}^{b}\beta_j\Big/b$$

and $\bar{\gamma}_{.i}$, $\bar{\gamma}_{.j}$ and $\bar{\gamma}_{..}$ defined in analogous manner. Hence from (11)

$$\begin{aligned}
\bar{y}_{i..} &= \mu + \alpha_i + \bar{\beta}. + \bar{\gamma}_{i.} + \bar{e}_{i..}\,,\\
\bar{y}_{.j.} &= \mu + \bar{\alpha}. + \beta_j + \bar{\gamma}_{.j} + \bar{e}_{.j.}\,,\\
\bar{y}_{ij.} &= \mu + \alpha_i + \beta_j + \gamma_{ij} + \bar{e}_{ij.}
\end{aligned} \tag{12}$$

and

$$\bar{y}_{...} = \mu + \bar{\alpha}. + \bar{\beta}. + \bar{\gamma}_{..} + \bar{e}_{...}\,.$$

Substituting (11) and (12) into Table 9.5 gives

$$\begin{aligned}
\text{SSM} &= N(\mu + \bar{\alpha}. + \bar{\beta}. + \bar{\gamma}.. + \bar{e}...)^2,\\
\text{SSA} &= bn\sum_{i=1}^{a}(\alpha_i - \bar{\alpha}. + \bar{\gamma}_{i\cdot} - \bar{\gamma}.. + \bar{e}_{i\cdot\cdot} - \bar{e}...)^2,\\
\text{SSB} &= an\sum_{j=1}^{b}(\beta_j - \bar{\beta}. + \bar{\gamma}_{\cdot j} - \bar{\gamma}.. + \bar{e}_{\cdot j\cdot} - \bar{e}...)^2, \qquad (13)\\
\text{SSAB} &= n\sum_{i=1}^{a}\sum_{j=1}^{b}(\gamma_{ij} - \bar{\gamma}_{i\cdot} - \bar{\gamma}_{\cdot j} + \bar{\gamma}.. + \bar{e}_{ij\cdot} - \bar{e}_{i\cdot\cdot} - \bar{e}_{\cdot j\cdot} + \bar{e}...)^2
\end{aligned}$$

and $$\text{SSE} = \sum_{i=1}^{a}\sum_{j=1}^{b}\sum_{k=1}^{n}(e_{ijk} - \bar{e}_{ij\cdot})^2.$$

Now, no matter what model we use, fixed, random or mixed, we take the error terms as having zero mean and variance σ_e^2 and being independent of one another. Furthermore, the expected value of the product of an error term with μ, an α, a β or a γ is zero. [If the effects are fixed the products have zero expectation because $E(e_{ijk}) = 0$ and, when any of the effects are random, products with e-terms have zero expectation because of assuming independence.] Finally, expected values of squares and products of means of the e's are such that, for example,

$$E(\bar{e}_{i\cdot\cdot}^2) = \sigma_e^2/bn,$$
$$E(\bar{e}_{i\cdot\cdot}\bar{e}...) = E(\bar{e}_{\cdot j\cdot}\bar{e}...) = E(\bar{e}_{ij\cdot}\bar{e}...) = \sigma_e^2/abn$$

and $$E(\bar{e}_{i\cdot\cdot}\bar{e}_{\cdot j\cdot}) = \sigma_e^2/abn.$$

Hence for the terms in (13)

$$\begin{aligned}
E(\bar{e}_{i\cdot\cdot} - \bar{e}...)^2 &= (a-1)\sigma_e^2/abn,\\
E(\bar{e}_{\cdot j\cdot} - \bar{e}...)^2 &= (b-1)\sigma_e^2/abn,\\
E(\bar{e}_{ij\cdot} - \bar{e}_{i\cdot\cdot} - \bar{e}_{\cdot j\cdot} + \bar{e}...)^2 &= (a-1)(b-1)\sigma_e^2/abn \qquad (14)
\end{aligned}$$

and $$E(e_{ijk} - \bar{e}_{ij\cdot})^2 = (n-1)\sigma_e^2/n\,.$$

Consequently, on taking expected values of (13), and dividing by degrees of freedom to convert them to mean squares, we get

$$\begin{aligned}
E(\text{MSM}) &= EN(\mu + \bar{\alpha}. + \bar{\beta}. + \bar{\gamma}..)^2 + \sigma_e^2\,,\\
E(\text{MSA}) &= \frac{bn}{a-1}\sum_{i=1}^{a}E(\alpha_i - \bar{\alpha}. + \bar{\gamma}_{i\cdot} - \bar{\gamma}..)^2 + \sigma_e^2\,,\\
E(\text{MSB}) &= \frac{an}{b-1}\sum_{j=1}^{b}E(\beta_j - \bar{\beta}. + \bar{\gamma}_{\cdot j} - \bar{\gamma}..) + \sigma_e^2\,, \qquad (15)\\
E(\text{MSAB}) &= \frac{n}{(a-1)(b-1)}\sum_{i=1}^{a}\sum_{j=1}^{b}E(\gamma_{ij} - \bar{\gamma}_{i\cdot} - \bar{\gamma}_{\cdot j} + \bar{\gamma}..)^2 + \sigma_e^2
\end{aligned}$$

and $$E(\text{MSE}) = \sigma_e^2\,.$$

These results hold whether the model is fixed, random or mixed. Each model determines the consequence of the expectation operations shown on the right-hand sides of (15).

a. The fixed effects model

In the fixed effects model all the α's, β's and γ's are fixed effects. Therefore the expectation operations on the right-hand sides of (15) just involve dropping the E symbol. The results are shown in Table 9.6. They have been derived, be it noted, without making any use of the "usual restrictions" on elements of the model.

Suppose we consider a model

$$y_{ijk} = \mu' + \alpha'_i + \beta'_j + \gamma'_{ij} + e_{ijk} \tag{16}$$

in which the "usual restrictions" are part of the model, namely

$$\sum_{i=1}^{a} \alpha'_i = 0, \qquad \sum_{j=1}^{b} \beta'_j = 0,$$

and

$$\sum_{i=1}^{a} \gamma'_{ij} = 0, \text{ for all } j, \qquad \sum_{j=1}^{b} \gamma'_{ij} = 0, \text{ for all } i. \tag{17}$$

Before using these restrictions the expected mean squares will be those of Table 9.6, with primes on the α's, β's and γ's. After using the restrictions (17) the expectations reduce to those of Table 9.7, because (17) implies $\bar{\alpha}'_. = 0$, $\bar{\beta}'_. = 0$, $\bar{\gamma}'_{.j} = 0$ for all j and $\bar{\gamma}'_{i.} = 0$ for all i.

The apparent difference between Tables 9.6 and 9.7 can be shown to be

TABLE 9.6. EXPECTED MEAN SQUARES OF A 2-WAY CLASSIFICATION INTERACTION MODEL, WITH BALANCED DATA. (see Table 9.5)

Fixed effects model
$E(\text{MSM}) = N(\mu + \bar{\alpha}_. + \bar{\beta}_. + \bar{\gamma}_{..})^2 + \sigma_e^2$
$E(\text{MSA}) = \dfrac{bn}{a-1}\sum_{i=1}^{a}(\alpha_i - \bar{\alpha}_. + \bar{\gamma}_{i.} - \bar{\gamma}_{..})^2 + \sigma_e^2$
$E(\text{MSB}) = \dfrac{an}{b-1}\sum_{j=1}^{b}(\beta_j - \bar{\beta}_. + \bar{\gamma}_{.j} - \bar{\gamma}_{..})^2 + \sigma_e^2$
$E(\text{MSAB}) = \dfrac{n}{(a-1)(b-1)}\sum_{i=1}^{a}\sum_{j=1}^{b}(\gamma_{ij} - \bar{\gamma}_{i.} - \bar{\gamma}_{.j} + \bar{\gamma}_{..})^2 + \sigma_e^2$
$E(\text{MSE}) = \sigma_e^2$

TABLE 9.7. EXPECTED MEAN SQUARES OF A 2-WAY CLASSIFICATION INTERACTION MODEL, WITH BALANCED DATA. (see Table 9.5)

Fixed effects model, that includes the restrictions

$$\sum_{i=1}^{a} \alpha'_i = 0 = \sum_{j=1}^{b} \beta'_j = \sum_{i=1}^{a} \gamma'_{ij} \text{ for all } j, \text{ and } \sum_{j=1}^{b} \gamma'_{ij} = 0 \text{ for all } i.$$

$$E(\text{MSM}) = N\mu'^2 + \sigma_e^2$$

$$E(\text{MSA}) = \frac{bn}{a-1}\sum_{i=1}^{a} \alpha_i'^2 + \sigma_e^2$$

$$E(\text{MSB}) = \frac{an}{b-1}\sum_{j=1}^{b} \beta_j'^2 + \sigma_e^2$$

$$E(\text{MSAB}) = \frac{n}{(a-1)(b-1)}\sum_{i=1}^{a}\sum_{j=1}^{b} \gamma_{ij}'^2 + \sigma_e^2$$

$$E(\text{MSE}) = \sigma_e^2$$

just that, apparent and not real. Suppose we rewrite the model as

$$y_{ijk} = \mu_{ij} + e_{ijk} \tag{18}$$

$$= \bar{\mu}_{..} + (\bar{\mu}_{i.} - \bar{\mu}_{..}) + (\bar{\mu}_{.j} - \bar{\mu}_{..}) + (\mu_{ij} - \bar{\mu}_{i.} - \bar{\mu}_{.j} + \bar{\mu}_{..}) + e_{ijk}. \tag{19}$$

Then, on defining

$$\mu' = \bar{\mu}_{..}, \quad \alpha'_i = \bar{\mu}_{i.} - \bar{\mu}_{..}, \quad \beta'_j = \bar{\mu}_{.j} - \bar{\mu}_{..} \quad \text{and} \quad \gamma'_{ij} = \mu_{ij} - \bar{\mu}_{i.} - \bar{\mu}_{.j} + \bar{\mu}_{..} \tag{20}$$

equation (19) is identical to (16), and α'_i, β'_j and γ'_{ij}, by their definition in (20), satisfy (17); for example, $\sum_{i=1}^{a} \alpha'_i = \sum_{i=1}^{a} (\bar{\mu}_{i.} - \bar{\mu}_{..}) = 0$. Therefore the definitions in (20) are consistent with the expected mean squares of Table 9.7 and so, for example,

$$E(\text{MSA}) = \frac{bn}{a-1}\sum_{i=1}^{a} \alpha_i'^2 + \sigma_e^2 .$$

But in comparing (18) and (11) note that

$$\mu_{ij} = \mu + \alpha_i + \beta_j + \gamma_{ij} \tag{21}$$

and so with

$$\alpha'_i = \bar{\mu}_{i.} - \bar{\mu}_{..}$$

from (20) we have

$$\alpha'_i = \mu + \alpha_i + \bar{\beta}_. + \bar{\gamma}_{.i} - (\mu + \bar{\alpha}_. + \bar{\beta}_. + \bar{\gamma}_{..}),$$

i.e.,

$$\alpha'_i = \alpha_i - \bar{\alpha}_. + \bar{\gamma}_{i.} - \bar{\gamma}_{..} . \tag{22}$$

Thus $\sum_{i=1}^{a} \alpha_i'^2$ of E(MSA) in Table 9.7 has the same meaning as does

$$\sum_{i=1}^{a} (\alpha_i - \bar{\alpha}. + \bar{\gamma}_{i\cdot} - \bar{\gamma}_{\cdot\cdot})^2$$

of E(MSA) in Table 9.6. Hence interpretation of the F-statistic MSA/MSE is the same whether one uses Table 9.6 or 9.7: MSA/MSE tests the significance of α-effects in the presence of (or, plus the average of) interaction effects. In Table 9.7 the symbols are defined, as in (17), so that these averages are zero whereas in Table 9.6 they are not so defined. The equivalence of the expressions for E(MSA) in Tables 9.6 and 9.7, resting as it does on (22), can also be demonstrated for the other entries in the two tables, based upon

$$\beta_j' = \beta_j - \bar{\beta}. + \bar{\gamma}_{\cdot j} - \bar{\gamma}_{\cdot\cdot}\,,$$

$$\gamma_{ij}' = \gamma_{ij} - \bar{\gamma}_{i\cdot} - \bar{\gamma}_{\cdot j} + \bar{\gamma}_{\cdot\cdot} \tag{23}$$

and

$$\mu' = \mu + \bar{\alpha}. + \bar{\beta}. + \bar{\gamma}_{\cdot\cdot}\,.$$

Defining effects that satisfy "the usual restrictions" in the manner of (20) has the effect of simplifying Table 9.6 to the form of Table 9.7. But this simplification occurs only for balanced data. It does not occur for unbalanced data because the sums of squares used with such data (e.g., Table 7.8) have expected values that do not involve the means of the effects in such a simple manner as with balanced data (see Table 9.6). Restrictions that are in terms of weighted sums of the effects are sometimes suggested for unbalanced data, although these have no simplifying effect when there are empty cells, as is often the case with unbalanced data.

A special case of the simplifying effect of the "usual restrictions" (20) that is of some interest is E(MSM). In Table 9.6

$$E(\text{MSM}) = N(\mu + \bar{\alpha}. + \bar{\beta}. + \bar{\gamma}_{\cdot\cdot})^2 + \sigma_e^2 = N[E(\bar{y})]^2 + \sigma_e^2\,, \tag{24}$$

consistent with the hypothesis H: $E(\bar{y}) = 0$ that can be tested by the F-statistic. In Table 9.7 the expected value is

$$E(\text{MSM}) = N\mu'^2 + \sigma_e^2$$

consistent with testing, in *that* model, H: $E(\bar{y}) \equiv \mu' = 0$. This is the origin of the concept of "testing the mean" by the F-statistic $F(M)$ = MSM/MSE, referred to in earlier chapters. There, with unbalanced data, we saw how the meaning of this phrase was best described in terms of testing H: $E(\bar{y}) = 0$. With balanced data, that description is still appropriate when the model has no "usual restrictions", as is evident in (24), but when the model does include such restrictions the hypothesis H: $E(\bar{y}) = 0$ reduces to H: $\mu' = 0$ and thus gets described as "testing the mean".

TABLE 9.8. EXPECTED MEAN SQUARES OF A 2-WAY CLASSIFICATION INTERACTION MODEL, WITH BALANCED DATA. (see Table 9.5)

Random effects model
$E(\text{MSM}) = abn\mu^2 + bn\sigma_\alpha^2 + an\sigma_\beta^2 + n\sigma_\gamma^2 + \sigma_e^2$
$E(\text{MSA}) = bn\sigma_\alpha^2 + n\sigma_\gamma^2 + \sigma_e^2$
$E(\text{MSB}) = an\sigma_\beta^2 + n\sigma_\gamma^2 + \sigma_e^2$
$E(\text{MSAB}) = n\sigma_\gamma^2 + \sigma_e^2$
$E(\text{MSE}) = \sigma_e^2$

b. The random effects model

All the α-, β- and γ-effects in the model are random in the random effects model, with zero means and variances σ_α^2, σ_β^2 and σ_γ^2 respectively so that, for example

$$E(\alpha_i) = 0 \quad \text{and} \quad E(\alpha_i^2) = \sigma_\alpha^2 . \tag{25}$$

The effects are also assumed to be uncorrelated with each other; e.g.,

$$E(\alpha_i\beta_j) = 0 = E(\alpha_i\gamma_{ij}) \quad \text{and} \quad E(\alpha_i\alpha_{i'}) = 0 \quad \text{for} \quad i \neq i'. \tag{26}$$

Furthermore, similar to (14)

$$E(\alpha_i - \bar{\alpha}.)^2 = (a-1)\sigma_\alpha^2/a . \tag{27}$$

Similar results hold for the β's and γ's. Using them in (15) gives the expectations shown in Table 9.8.

Estimation of the variance components from Table 9.8 is achieved by equating mean squares to their expected values, the resulting solutions for the components being the estimators. This gives

$$\begin{aligned} \hat{\sigma}_e^2 &= \text{MSE}, & \hat{\sigma}_\beta^2 &= (\text{MSB} - \text{MSAB})/an, \\ \hat{\sigma}_\gamma^2 &= (\text{MSAB} - \text{MSE})/n, & \hat{\sigma}_\alpha^2 &= (\text{MSA} - \text{MSAB})/bn. \end{aligned} \tag{28}$$

c. The mixed model

Suppose the α-effects are fixed effects and the β's and γ's are random. Then the expectation operations on the right-hand sides of (15) involve dropping the E symbol insofar as it pertains to α's and using properties like those of (25), (26) and (27) for the β's and γ's. This leads to the results shown in Table 9.9.

TABLE 9.9. EXPECTED MEAN SQUARES OF A 2-WAY CLASSIFICATION INTERACTION MODEL, WITH BALANCED DATA

Mixed model: α's fixed, β's and γ's random	
$E(\text{MSM}) = abn(\mu + \bar{\alpha}_.)^2$	$+\, an\sigma_\beta^2 + n\sigma_\gamma^2 + \sigma_e^2$
$E(\text{MSA}) = \dfrac{bn}{a-1}\sum_{i=1}^{a}(\alpha_i - \bar{\alpha}_.)^2$	$+\, n\sigma_\gamma^2 + \sigma_e^2$
$E(\text{MSB}) =$	$an\sigma_\beta^2 + n\sigma_\gamma^2 + \sigma_e^2$
$E(\text{MSAB}) =$	$n\sigma_\gamma^2 + \sigma_e^2$
$E(\text{MSE}) =$	σ_e^2

The difference between the random and mixed models is that the α's are random effects in the random model and are fixed effects in the mixed model. Since only the first two equations in (15) involve α's, only the first two entries in Table 9.9 differ from the corresponding entries in Table 9.8, and then only through having quadratic terms in the α's instead of terms in σ_α^2.

The expectations in Table 9.9 are arrived at without making any use of the "usual restrictions" on elements of the model, just as are the expectations in Table 9.6 for the fixed effects model. However, if the restriction $\sum_{i=1}^{a} \alpha_i = 0$ is taken as part of the mixed model then E(MSA) of Table 9.9 reduces to

$$E(\text{MSA}) = \frac{bn}{a-1}\sum_{i=1}^{a}\alpha_i^2 + n\sigma_\gamma^2 + \sigma_e^2\,,$$

the quadratic in the α's being similar to that of Table 9.8.

An alternative mixed model that is often used is

$$y_{ijk} = \mu'' + \alpha_i'' + \beta_j'' + \gamma_{ij}'' + e_{ijk} \tag{29}$$

with the restriction

$$\sum_{i=1}^{a} \gamma_{ij}'' = \gamma_{\cdot j}'' = 0 \qquad \text{for all } j. \tag{30}$$

In (29) the α″'s are fixed effects and the β″'s and γ″'s are random effects with zero means and variances $\sigma_{\beta''}^2$ and $\gamma_{\gamma''}^2$, respectively, and with the β″'s and γ″'s being uncorrelated with each other and of the e's. All this is exactly the same as in the mixed model described earlier, except for (30). This restriction implies a covariance between certain of the γ″'s, namely between γ_{ij}'' and $\gamma_{i'j}''$ for $i \neq i'$. Suppose this covariance is the same,

$$\text{cov}(\gamma_{ij}'', \gamma_{i'j}'') = c, \text{for all} \quad i \neq i' \text{ and } j. \tag{31}$$

Then, from (30)

$$v\left(\sum_{i=1}^{a} \gamma''_{ij}\right) = 0$$

and so $$a\sigma^2_{\gamma''} + a(a-1)c = 0,$$

giving $$c = -\sigma^2_{\gamma''}/(a-1). \tag{32}$$

Note that this covariance pertains only to γ'''s within the same level of the β-factor, arising as it does from (30). The covariance between γ'''s in the same level of the α-factor is zero as usual:

$$\text{cov}(\gamma''_{ij}, \gamma''_{ij'}) = 0 \qquad \text{for all} \quad i \text{ and } j \neq j'. \tag{33}$$

Prior to utilizing (30), the expected mean squares for the model (29) can be derived from equations (15) with double prime superscripts on μ, the α's, β's and γ's. Upon invoking $\bar{\gamma}''_{.j} = 0$ from (30), and hence $\bar{\gamma}''_{..} = 0$, equations (15) become

$$E(\text{MSM}) = N(\mu'' + \bar{\alpha}''_{.})^2 + NE(\bar{\beta}''^2_{.}) + \sigma^2_e\,,$$

$$E(\text{MSA}) = \frac{bn}{a-1}\left[\sum_{i=1}^{a}(\alpha''_i - \bar{\alpha}''_{.})^2 + \sum_{i=1}^{a} E(\bar{\gamma}''_{i.})^2\right] + \sigma^2_e\,,$$

$$E(\text{MSB}) = \frac{an}{b-1}\sum_{j=1}^{b} E(\beta''_j - \bar{\beta}''_{.})^2 + \sigma^2_e\,, \tag{34}$$

$$E(\text{MSAB}) = \frac{n}{(a-1)(b-1)}\sum_{i=1}^{a}\sum_{j=1}^{b} E(\gamma''_{ij} - \bar{\gamma}''_{i.})^2 + \sigma^2_e$$

and $$E(\text{MSE}) = \sigma^2_e\,.$$

In carrying out the expectation operations in E(MSA) and E(MSAB), use is made of (31), (32) and (33) to give

$$E(\bar{\gamma}''_{i.})^2 = \sigma^2_{\gamma''}[1/b + b(b-1)0/b^2] = \sigma^2_{\gamma''}/b$$

and $$E(\gamma''_{ij} - \bar{\gamma}''_{i.})^2 = \sigma^2_{\gamma''}(1 + 1/b - 2/b) = (b-1)\sigma^2_{\gamma''}/b\,.$$

As a result, expressions (34) reduce to those shown in Table 9.10.

The results in Table 9.10 differ from those in Table 9.9 in two important ways: E(MSB) and E(MSM) do not contain $\sigma^2_{\gamma''}$, and the term in $\sigma^2_{\gamma''}$ that does occur in E(MSA) and E(MSAB) includes the fraction $a/(a-1)$. The first of these differences, the absence of $\sigma^2_{\gamma''}$ from, particularly, E(MSB), is the reason for Rule 13 at the end of Sec. 6 differing from the first edition of Henderson (1959, 1969) but being the same as the second. The first edition specifies a general rule which leads to the absence of $\sigma^2_{\gamma''}$ from E(MSB) on the basis of $\gamma''_{.j} = 0$, as in (30), whereas the second specifies a general rule which retains σ^2_{γ} in E(MSB) as in Table 9.9, using a model that has no restriction like (30).

TABLE 9.10. EXPECTED MEAN SQUARES OF A 2-WAY CLASSIFICATION INTERACTION MODEL, WITH BALANCED DATA

Mixed model, with restrictions on interaction effects: $\gamma''_{.j} = 0$ for all j.

$$E(\text{MSM}) = N(\mu'' + \bar{\alpha}''_{.})^2 + an\sigma^2_{\beta''} + \quad + \sigma^2_e$$

$$E(\text{MSA}) = \frac{bn}{a-1}\sum_{i=1}^{a}(\alpha''_i - \bar{\alpha}''_{.})^2 + n\left(\frac{a}{a-1}\right)\sigma^2_{\gamma''} + \sigma^2_e$$

$$E(\text{MSB}) = an\sigma^2_{\beta''} + \sigma^2_e$$

$$E(\text{MSAB}) = n\left(\frac{a}{a-1}\right)\sigma^2_{\gamma''} + \sigma^2_e$$

$$E(\text{MSE}) = \sigma^2_e$$

This dual approach to the mixed model is evident in many places. For example, Mood (1950, p. 344) and Kirk (1968, p. 137) use the Table 9.9 expectations whereas Anderson and Bancroft (1952, p. 339), Scheffé (1959, p. 269), Graybill (1961, p. 398) and Snedecor and Cochran (1967, p. 367) use those akin to Table 9.10. Mood and Graybill (1963) do not discuss the topic. Although results like Table 9.10 predominate in the literature, those of Table 9.9 are consistent with the results for unbalanced data and this fact, as Hartley and Searle (1969) point out, is strong argument for using Table 9.9.

The second difference between Tables 9.9 and 9.10 is the occurrence of $a/(a-1)$ in the terms in the interaction variance component in Table 9.10. This is a consequence of the restriction $\gamma''_{.j} = 0$ of (30) as shown also, for example, in Steel and Torrie (1960, pp. 214, 246).

A relationship between Tables 9.9 and 9.10 can be established as follows. The model for Table 9.9 is

$$y_{ijk} = \mu + \alpha_i + \beta_j + \gamma_{ij} + e_{ijk}$$

with the α's as fixed effects and the β's and γ's random. Suppose it is rewritten as

$$y_{ijk} = \mu + \alpha_i + \beta_j + \bar{\gamma}_{.j} + \gamma_{ij} - \bar{\gamma}_{.j} + e_{ijk}\,.$$

On defining $\mu'' = \mu$, $\alpha''_i = \alpha_i$,

$$\beta''_j = \beta_j + \bar{\gamma}_{.j} \quad \text{and} \quad \gamma''_{ij} = \gamma_{ij} - \bar{\gamma}_{.j} \tag{35}$$

we have the model (29), corresponding to Table 9.10. This is so because, from (35),

$$\gamma''_{.j} = \gamma_{.j} - \gamma_{.j} = 0$$

as in (30). Other properties of the γ'''s are also evident. First, from (35),

$$\sigma^2_{\beta''} = \sigma^2_\beta + \sigma^2_\gamma/a \tag{36}$$

and

$$\sigma^2_{\gamma''} = \sigma^2_\gamma(1 + 1/a - 2/a) = (a - 1)\sigma^2_\gamma/a, \tag{37}$$

giving

$$\sigma^2_\gamma = \frac{a}{a-1}\,\sigma^2_{\gamma''}. \tag{38}$$

Also,

$$\begin{aligned}
\text{cov}(\beta''_j, \gamma''_{ij}) &= \sigma^2_\gamma(1/a - 1/a) = 0, \\
\text{cov}(\beta''_j, \gamma''_{ij'}) &= 0 \text{ for } j \neq j', \\
\text{cov}(\gamma''_{ij}, \gamma''_{i'j}) &= \sigma^2_\gamma(-1/a - 1/a + 1/a) \\
&= -\sigma^2_\gamma/a \\
&= -\sigma^2_{\gamma''}/(a-1) \text{ from (38)},
\end{aligned}$$

this being the same as in (31) and (32). Also

$$\text{cov}(\gamma''_{ij}, \gamma''_{i'j'}) = 0$$

as in (33). Hence properties of the β'''s and γ'''s defined in (35) are exactly those attributed to (29) and (30) in deriving Table 9.10. And substituting (36) and (38) into Table 9.10 yields Table 9.9.[1]

The question of which model to use, that leading to Table 9.9 or the one leading to Table 9.10, remains open and has not been considered. Lengthy discussion of the model (29) and (30) that leads to Table 9.10 is to be found in such papers as Wilk and Kempthorne (1955, 1956) and Cornfield and Tukey (1956) as well as in Scheffé (1959). The model that leads to Table 9.9 is the one customarily used for unbalanced data. More than this will not be said. The object of this section has been to show a relationship between the two different models. In either model the variance components are estimated from the last three mean squares of the appropriate table, either 9.9 or 9.10.

8. ESTIMATING VARIANCE COMPONENTS FROM BALANCED DATA

The method of estimating variance components from balanced data has been discussed and illustrated in terms of the 1-way and 2-way classifications. Extension to multi-way classifications is straightforward. The rules of Sec.6

[1] Conversations with C. R. Henderson, R. R. Hocking and N. S. Urquhart on this topic are gratefully acknowledged.

TABLE 9.11. ANALYSIS OF VARIANCE FOR 1-WAY CLASSIFICATION RANDOM MODEL, n OBSERVATIONS IN EACH OF a CLASSES, $N = an$

Source of Variation	d.f.	Sum of Squares	Mean Square	Expected value of mean square
Mean	1	$SSM = T_\mu$	$MSM = SSM$	$N\mu^2 + n\sigma_\alpha^2 + \sigma_e^2$
Classes	$a - 1$	$SSA = T_A - T_\mu$	$MSA = SSA/(a-1)$	$n\sigma_\alpha^2 + \sigma_e^2$
Residual error	$a(n-1)$	$SSE = T_o - T_A$	$MSE = SSE/a(n-1)$	σ_e^2
Total	an	$SST = T_o$		

determine both the appropriate analyses of variance and their expected mean squares. The latter are equated to observed mean squares for obtaining estimators. Properties of estimators derived in this fashion are now discussed. For illustration, we use the 1-way classification random model, the analysis of variance for which is shown in Table 9.11, a generalization of Table 9.2. In this we envisage data consisting of a classes with n observations in each.

The notation

$$T_o = \sum_{i=1}^{a}\sum_{j=1}^{n} y_{ij}^2, \qquad T_A = n\sum_{i=1}^{a} \bar{y}_{i.}^2 \quad \text{and} \quad T_\mu = N\bar{y}_{..}^2, \tag{39}$$

with $N = an$, is introduced and used in Table 9.11 because it refers to the basic calculations required, it simplifies writing of the analysis of variance table and it extends conveniently to unbalanced data. Each T-term is a total uncorrected sum of squares, with subscript indicating the factor it refers to: o for the observations, A for the A-factor and μ for $T_\mu = R(\mu)$.

Estimation of σ_α^2 and σ_e^2 follows from Table 9.11 in the same way that it does from Table 9.2:

$$\hat{\sigma}_e^2 = \text{MSE} \quad \text{and} \quad \hat{\sigma}_\alpha^2 = (\text{MSA} - \text{MSE})/n\,. \tag{40}$$

Notation From hereon the use of ^ over a symbol to denote best linear unbiased estimation is abandoned. Henceforth it simply means "an estimator of".

a. Unbiasedness and minimum variance

Variance components estimators obtained from balanced data by the analysis of variance method are unbiased, be the model mixed or random. Suppose that $\mathbf{m} = \{M_i\}$, for $i = 1, 2, \ldots, k$, is the vector of mean squares in the analysis such that $E(\mathbf{m})$ does not involve fixed effects and that $\boldsymbol{\sigma}^2$ is the vector of variance components to be estimated, with $E(\mathbf{m}) = \mathbf{P}\boldsymbol{\sigma}^2$ for $\mathbf{P}$ non-singular. Then $\mathbf{m} = \mathbf{P}\boldsymbol{\sigma}^2$ are the equations to be solved as

$$\hat{\boldsymbol{\sigma}}^2 = \mathbf{P}^{-1}\mathbf{m} \tag{41}$$

for the variance components estimators. They are unbiased because

$$E(\hat{\boldsymbol{\sigma}}^2) = \mathbf{P}^{-1}E(\mathbf{m}) = \mathbf{P}^{-1}\mathbf{P}\boldsymbol{\sigma}^2 = \boldsymbol{\sigma}^2.$$

It is to be emphasized that although the property of unbiasedness applies here to both random and mixed models we are concerned in this section just with balanced data. Unbiasedness is not a property of the analogous estimation procedure for unbalanced data from mixed models. We return to this point in the next chapter, noting now only that not even this simplest of properties, unbiasedness, is universally true for analysis of variance estimators of variance components.

The estimators in $\hat{\boldsymbol{\sigma}}^2$ of (41) have the smallest variance of all estimators which are both quadratic functions of the observations and unbiased. This is the property of minimum variance quadratic unbiasedness presented in Graybill and Hultquist (1961). Under normality assumptions, the estimators in (41) have the smallest variance from among all unbiased estimators, both those that are quadratic functions of the observations and those that are not. This result is discussed by Graybill (1954) and Graybill and Wortham (1956). These papers, and the minimum variance properties they establish, apply only to estimators from balanced data. Discussion of similar properties for estimators from unbalanced data appears to be limited to the 1-way classification, as in Townsend (1968) and Harville (1969a).

b. Negative estimates

A variance component is, by definition, positive. Nevertheless, estimates derived from (41) can be negative. A simple example illustrates this. Suppose three observations in each of two classes are those of Table 9.12. Then, as in (39),

$$\begin{aligned} T_A &= 51^2/3 + 45^2/3 &&= 1{,}542 \\ T_\mu &= 96^2/6 &&= 1{,}536 \\ T_0 &= 19^2 + 17^2 + 15^2 + 25^2 + 5^2 + 15^2 &&= 1{,}750. \end{aligned}$$

TABLE 9.12. HYPOTHETICAL DATA OF A 1-WAY CLASSIFICATION, 3 OBSERVATIONS IN 2 CLASSES

Class	Observations			Total
1	19	17	15	$51 = y_{1.}$
2	25	5	15	$45 = y_{2.}$
				$96 = y_{..}$

TABLE 9.13. ANALYSIS OF VARIANCE IN DATA IN TABLE 9.12

Source	d.f.	Sum of Squares	Mean Square	Expected Mean Square
Mean	1	1,536 = 1,536	1,536	
Classes	1	1,542 − 1,536 = 6	6	$3\sigma_\alpha^2 + \sigma_e^2$
Residual error	4	1,750 − 1,542 = 208	52	σ_e^2
Total	6	1,750 = 1,750		

The analysis of variance for the data of Table 9.12 is shown in Table 9.13. Hence, as in (40),

$$\hat{\sigma}_e^2 = 52 \quad \text{and} \quad \hat{\sigma}_\alpha^2 = (6 - 52)/3 = -15\tfrac{1}{3}. \tag{42}$$

This demonstrates how negative estimates can arise from the analysis of variance method. There is nothing intrinsic in the method to prevent it. This is so not only with a simple case such as (42) but also in many-factored models, both with balanced data and with unbalanced data.

It is clearly embarrassing to estimate a variance component as negative, since interpretation of a negative estimate of a non-negative parameter is obviously a problem. Several courses of action exist, few of them satisfactory.

(i) Accept the estimate, despite its distastefulness, and use it as evidence that the true value of the component is zero. Although this interpretation may be appealing, the unsatisfying nature of the negative estimate still remains. This is particularly so if the negative estimate is used in estimating a sum of components. The estimated sum can be less than the estimate of an individual component. For example, from (42) we have the estimated sum of the components as $\hat{\sigma}_\alpha^2 + \hat{\sigma}_e^2 = 52 - 15\frac{1}{3} = 36\frac{1}{3} < \hat{\sigma}_e^2$.

(ii) Accept the negative estimate as evidence that the true value of the corresponding component is zero and hence, as the estimate, use zero in place of the negative value. Although this seems a logical replacement such a truncation procedure disturbs the properties of the estimates as otherwise obtained. For example, they are no longer unbiased.

(iii) Use the negative estimate as indication of a zero component to ignore that component in the model, but retain the factor so far as the lines in the analysis of variance table are concerned. This leads to ignoring the component estimated as negative and re-estimating the others. Thompson (1961, 1962) gives rules for doing this, known as "pooling minimal mean squares with predecessors", and gives an application in Thompson and Moore (1963).

(iv) Interpret the negative estimate as indication of a wrong model and re-examine the source of one's data to look for a new model. In this connection, Searle and Fawcett (1970) suggest that finite population models may be viable alternatives because they sometimes give positive estimates when infinite population models have yielded negative estimates. Their use is likely to be of limited extent, however. In contrast, Nelder (1954) suggests that at least for split plot and randomized block designs, randomization theory indicates that negative variance components can occur in some situations. Such an apparent inconsistency can arise from the intra-block correlation of plots being less than the inter-block correlation.

(v) Interpret the negative estimate as throwing question on the method which yielded it, and use some other method that yields non-negative estimators. Two possibilities exist. One is to use Bayes procedures, for which the reader is referred to Hill (1965, 1967), Tiao and Tan (1965, 1966), and Tiao and Box (1967) and to Harville (1969b) for commentary thereon. A second possibility is to use maximum likelihood estimators, as suggested by Herbach (1959) and Thompson (1962); these are discussed at the end of this chapter.

(vi) Take the negative estimate as indication of insufficient data, and follow the statistician's last hope: collect more data and analyze them, either on their own or pooled with those that yielded the negative estimate. If the estimate from the pooled data is negative that would be additional evidence that the corresponding component is indeed zero.

Obtaining a negative estimate from the analysis of variance method is solely a consequence of the data and the method. It in no way depends on any implied distributional assumption, normality or otherwise. However, when normality *is* assumed, it is possible in certain cases to derive the probability of obtaining a negative estimate. This is discussed in Sec. 9e below.

9. NORMALITY ASSUMPTIONS

No particular form for the distribution of the error terms or of the random effects in the model has been assumed up to now. All the preceding results in this chapter are true for any distribution. We now make the normality assumptions, namely that the e's and each set of random effects in the model are normally distributed, with zero means and variance-covariance structure discussed earlier. That is, the effects of each random factor have a variance-covariance matrix that is their variance (component) multiplied by an identity matrix; and effects of each random factor are independent of those of every other factor and of the error terms. Under these conditions we assume normality.

a. Distribution of mean squares

Let f, SS, and M be the degrees of freedom, sum of squares and mean square

$$M = \mathrm{SS}/f \tag{43}$$

in a line of an analysis of variance of *balanced data*. Under the normality assumptions just described it can be shown that

$$\begin{aligned} &\mathrm{SS}/E(M) \sim \chi^2(f), \text{ and the SS-terms are pairwise independent.} \\ \text{Hence} \quad &fM/E(M) \sim \chi^2(f), \text{ and the } M\text{'s are pairwise independent.} \end{aligned} \tag{44}$$

Result (44) can be derived by writing $\mathrm{SS}/E(M)$ as a quadratic form $\mathbf{y}'\mathbf{A}\mathbf{y}$ in the observations $\mathbf{y}$, and applying Theorems 2 and 4 of Chapter 2. In applying these theorems to random or mixed models, $\mathbf{V}$ is not $\sigma_e^2\mathbf{I}$, as it is in the fixed model, but is a matrix whose elements are functions of the σ^2's of the model, as illustrated in (9) and (10). Nevertheless, for the $\mathbf{A}$-matrices involved in expressing each $\mathrm{SS}/E(M)$ as a quadratic $\mathbf{y}'\mathbf{A}\mathbf{y}$ it will be found that $\mathbf{AV}$ is always idempotent. Furthermore, for the random model, $\boldsymbol{\mu}$ has the form $\mu\mathbf{1}$ and $\boldsymbol{\mu}'\mathbf{A}\boldsymbol{\mu} = \mu\mathbf{1}'\mathbf{A}\mathbf{1}\mu$ will, by the nature of $\mathbf{A}$, always be zero. Hence the χ^2's are central, as indicated in (44). For the mixed model, (44) will also apply for all sums of squares whose expected values do not involve fixed effects; those that do involve fixed effects will be non-central χ^2's.

Example. The variance-covariance matrix for the 1-way classification model of Table 9.11 is

$$\mathbf{V} = \sigma_e^2\mathbf{I} + \sigma_\alpha^2 \sum_{i=1}^{a}{}^{+} \mathbf{J} \tag{45}$$

where $\mathbf{I}$ has order $N = an$ and $\mathbf{J}$ has order n, and (45) is a generalization of (10). Now for Table 9.11, with $\mathbf{J}_N$ being a $\mathbf{J}$-matrix of order N, the terms of (44) are

$$\mathrm{SS} = \mathrm{SSA} = \mathbf{y}'\left(n^{-1}\sum_{i=1}^{a}{}^{+}\mathbf{J} - N^{-1}\mathbf{J}_N\right)\mathbf{y} \tag{46}$$

and $\quad E(M) = E(\mathrm{MSA}) = n\sigma_\alpha^2 + \sigma_e^2,$

so that

$$\mathrm{SS}/E(M) = \mathbf{y}'\mathbf{A}\mathbf{y} \quad \text{with} \quad \mathbf{A} = \frac{n^{-1}\sum_{i=1}^{a}{}^{+}\mathbf{J} - N^{-1}\mathbf{J}_N}{n\sigma_\alpha^2 + \sigma_e^2}. \tag{47}$$

Hence, using properties of $\mathbf{J}$-matrices such as $\mathbf{1}'\mathbf{J} = n\mathbf{1}'$ and $\mathbf{J}^2 = n\mathbf{J}$ [e.g., Searle (1966, p. 197)],

$$\begin{aligned} \mathbf{AV} &= \left[\sigma_e^2\left(n^{-1}\sum_{i=1}^{a}{}^{+}\mathbf{J} - N^{-1}\mathbf{J}_N\right) + \sigma_\alpha^2\left(\sum_{i=1}^{a}{}^{+}\mathbf{J} - nN^{-1}\mathbf{J}_N\right)\right]\Big/\left(n\sigma_\alpha^2 + \sigma_e^2\right) \\ &= \sum_{i=1}^{a}{}^{+} n^{-1}\mathbf{J} - N^{-1}\mathbf{J}_N. \end{aligned} \tag{48}$$

It is easily shown that $(\mathbf{AV})^2 = \mathbf{AV}$, i.e., that $\mathbf{AV}$ is idempotent; and from (47) $\mathbf{1'A} = \mathbf{0}$. Hence

$$\text{SSA}/E(\text{MSA}) \sim \chi^{2'}[r(\mathbf{AV}), 0] = \chi^2(a-1), \tag{49}$$

the rank of $\mathbf{AV}$ being its trace, namely $a - 1$, as is evident from (48).

There are, of course, easier ways of deriving (49), but the intermediary steps (45)–(48) have useful generalizations in the case of unbalanced data.

b. Distribution of estimators

Equating mean squares to their expected values as a method of deriving variance component estimators gives estimators that are linear functions of the mean squares. These mean squares have the properties given in (44). The resulting variance components estimators are therefore linear functions of multiplies of χ^2-variables, some of them with negative coefficients. No closed form exists for the distribution of such functions and, furthermore, the coefficients are themselves functions of the population variance components.

Example. In Table 9.11

$$\frac{(a-1)\text{MSA}}{n\sigma_\alpha^2 + \sigma_e^2} \sim \chi^2(a-1)$$

and, independently,

$$\frac{a(n-1)\text{MSE}}{\sigma_e^2} \sim \chi^2(an-a).$$

Therefore

$$\hat{\sigma}_\alpha^2 = \frac{\text{MSA} - \text{MSE}}{n}$$

$$\sim \frac{n\sigma_\alpha^2 + \sigma_e^2}{n(a-1)}\chi^2(a-1) - \frac{\sigma_e^2}{an(n-1)}\chi^2(an-a). \tag{50}$$

The exact form of the distribution of (50) cannot be derived, both because its second term is negative and because σ_α^2 and σ_e^2 occur in the coefficients and are unknown. This state of affairs is true for these kinds of variance components estimators generally. Were the coefficients of the χ^2's known, the methods of Robinson (1965) or of Wang (1967) could be employed to obtain the distributions as infinite series expansions.

In contrast to other components the distribution of $\hat{\sigma}_e^2$ is always known exactly, under normality assumptions:

$$\hat{\sigma}_e^2 = \text{MSE} \sim \frac{\sigma_e^2}{f_{\text{MSE}}}\chi^2(f_{\text{MSE}}) \tag{51}$$

where f_{MSE} is the degrees of freedom associated with MSE.

Generalization of (50) arises from (41), which is $\hat{\boldsymbol{\sigma}}^2 = \mathbf{P}^{-1}\mathbf{m}$. The elements of $\mathbf{m}$ follow (44) and so, for example, $M_i \sim E(M_i)f_i^{-1}\chi^2(f_i)$. Now write

$$\mathbf{C} = \text{diag}\{f_i^{-1}\chi^2(f_i)\} \qquad \text{for} \quad i = 1, 2, \ldots, k$$

where there are k lines in the analysis of variance being used. Then from (41)

$$\hat{\boldsymbol{\sigma}}^2 \sim \mathbf{P}^{-1}\mathbf{C}E(\mathbf{m}) \sim \mathbf{P}^{-1}\mathbf{C}\mathbf{P}\boldsymbol{\sigma}^2. \tag{52}$$

In this way the vector of estimators is expressed as a vector of multiples of central χ^2-variables.

c. Tests of hypotheses

Expected values of mean squares (derived by the rules of Sec. 6) will suggest which mean squares are the appropriate denominators for testing hypotheses that certain variance components are zero. Thus in Table 9.9, MSAB/MSE is appropriate for testing the hypothesis H: $\sigma_\gamma^2 = 0$; and MSB/MSAB is the F-statistic for testing H: $\sigma_\beta^2 = 0$. In the random model all ratios of mean squares have central F-distributions, because all mean squares follow (44). In the mixed model the same is true of ratios of mean squares whose expected values contain no fixed effects.

The table of expected values will not always suggest the "obvious" denominator for testing a hypothesis. For example, suppose in Table 9.4 we wished to test the hypothesis $\sigma_b^2 = 0$. From that table we have, using M_1, M_2, M_3 and M_4 respectively for MS(B), MS(C:B), MS(AB) and MS(AC:B),

$$\begin{aligned}
E(M_1) &= k_1\sigma_b^2 + k_2\sigma_{c:b}^2 + k_3\sigma_{ab}^2 + k_4\sigma_{ac:b}^2 + \sigma_e^2 \\
E(M_2) &= \phantom{k_1\sigma_b^2 + {}} k_2\sigma_{c:b}^2 \phantom{{}+ k_3\sigma_{ab}^2} + k_4\sigma_{ac:b}^2 + \sigma_e^2 \\
E(M_3) &= \phantom{k_1\sigma_b^2 + k_2\sigma_{c:b}^2 + {}} k_3\sigma_{ab}^2 + k_4\sigma_{ac:b}^2 + \sigma_e^2 \\
E(M_4) &= \phantom{k_1\sigma_b^2 + k_2\sigma_{c:b}^2 + k_3\sigma_{ab}^2 + {}} k_4\sigma_{ac:b}^2 + \sigma_e^2
\end{aligned}$$

where we have here written the coefficients of the σ^2's, the products of n's shown in the column heading of Table 9.4, as k's: e.g., $k_1 = n_a n_c n_w$. It is clear from these expected values that no mean square in the table is suitable as a denominator to M_1 for an F-statistic to test H: $\sigma_b^2 = 0$, because there is no mean square whose expected value is $E(M_1)$ with the σ_b^2 term omitted, namely

$$E(M_1) - k_1\sigma_b^2 = k_2\sigma_{c:b}^2 + k_3\sigma_{ab}^2 + k_4\sigma_{ac:b}^2 + \sigma_e^2. \tag{53}$$

However, there is a linear function of the other mean squares whose expected value equals $E(M_1) - k_1\sigma_b^2$, *viz.*

$$E(M_2) + E(M_3) - E(M_4) = k_2\sigma_{c:b}^2 + k_3\sigma_{ab}^2 + k_4\sigma_{ac:b}^2 + \sigma_e^2. \tag{54}$$

From this we show how to use the mean squares in (53) and (54) to calculate a ratio that is approximately distributed as a central F-distribution.

In (54) some of the mean squares are involved negatively. But using (53) it is clear that

$$E(M_1) + E(M_4) = k_1\sigma_b^2 + E(M_2) + E(M_3).$$

From this let us generalize to

$$E(M_r + \cdots + M_s) = k\sigma_\alpha^2 + E(M_m + \cdots + M_n) \tag{55}$$

and consider testing the hypothesis H: $\sigma_\alpha^2 = 0$ where σ_α^2 is any component of a model. The statistic suggested by Satterthwaite (1946), for testing this hypothesis is

$$F = \frac{M'}{M''} = \frac{M_r + \cdots + M_s}{M_m + \cdots + M_n}, \quad \text{which is approximately} \sim F(p, q) \tag{56}$$

$$\text{where} \quad p = \frac{(M_r + \cdots + M_s)^2}{M_r^2/f_r + \cdots + M_s^2/f_s} \quad \text{and} \quad q = \frac{(M_m + \cdots + M_n)^2}{M_m^2/f_m + \cdots + M_n^2/f_n}. \tag{57}$$

In p and q, the term f_i is the degrees of freedom associated with the mean square M_i. Furthermore, of course, p and q are not necessarily integers and so, in comparing F against tabulated values of the F-distribution, interpolation will be necessary.

The basis of this test is that both numerator and denominator of (56) are distributed approximately as multiples of central χ^2-variables (each mean square in the analysis is distributed as a multiple of a central χ^2). Furthermore, in (56) there is no mean square that occurs in both numerator and denominator, which are therefore independent, and so F of (56) is distributed approximately as $F(p, q)$ as shown.

Both M' and M'' in (56) are sums of mean squares and, as Satterthwaite (1946) showed, $pM'/E(M')$ is distributed approximately as a central χ^2 with p degrees of freedom for p of (57). (A similar result holds for M'' with q degrees of freedom.) More generally, consider the case where some mean squares are included negatively. Suppose

$$M_0 = M_1 - M_2$$

where M_1 and M_2 are now *sums* of mean squares having f_1 and f_2 degrees of freedom respectively. Let

$$\rho = E(M_1)/E(M_2) \qquad \text{and} \qquad \hat{\rho} = M_1/M_2 \geq 1,$$

and

$$\hat{f}_0 = (\hat{\rho} - 1)^2/(\hat{\rho}/f_1 + 1/f_2)^2.$$

Then, simulation studies by Gaylor and Hopper (1969) suggest that

$$\frac{\hat{f}_0 M_0}{E(M_0)} \text{ is approximately} \sim \chi^2(f_0)$$

provided
$$\rho > F_{f_2,f_1,0.975}, \qquad f_1 \leq 100 \qquad \text{and} \qquad f_1 \leq 2f_2 .$$
They further suggest that $\rho > F_{f_2,f_1,0.975}$ "appears to be fulfilled reasonably well" when
$$\hat{\rho} > F_{f_2,f_1,0.975} \times F_{f_1,f_2,0.50} .$$
Under these conditions, Satterthwaite's procedure in (56) and (57) can be used on functions of mean squares that involve differences as well as sums.

d. Confidence intervals

The inability to derive exact distributions does not preclude the use of approximate confidence intervals and, in some cases, of exact intervals. A method for obtaining approximate confidence intervals for a linear function of expected mean squares is that given by Graybill (1961, p. 369). Define $\chi^2_{n,L}$ and $\chi^2_{n,U}$ as lower and upper points of a $(1 - \alpha)\%$ region of the $\chi^2(n)$-distribution such that
$$\Pr\{\chi^2_{n,L} \leq \chi^2(n) \leq \chi^2_{n,U}\} = 1 - \alpha. \tag{58}$$
Then for any constants k_i, such that $\sum k_i M_i > 0$, the approximate confidence interval on $\sum k_i E(M_i)$ is given by
$$\Pr\left\{\frac{n \sum k_i M_i}{\chi^2_{r,U}} \leq \sum k_i E(M_i) \leq \frac{n \sum k_i M_i}{\chi^2_{r,L}}\right\} = 1 - \alpha$$
where
$$r = \frac{(\sum k_i M_i)^2}{\sum k_i^2 M_i^2/f_i},$$
analogous to (57). Since r will seldom be an integer, $\chi^2_{r,L}$ and $\chi^2_{r,U}$ are obtained from tables of the central χ^2-distribution, using either interpolation or the nearest (or next largest) integer to r. A correction to the tabulated χ^2-values when $r < 30$ is given by Welch (1956) and recommended by Graybill (1961, p. 370), where details may be found. Other methods for finding simultaneous confidence intervals on ratios of variance components are to be found in Broemeling (1969).

Suppose M_1 and M_2 are two mean squares having the properties of (44) and such that
$$E(M_1) = \theta + \sigma_e^2 \qquad \text{and} \qquad E(M_2) = \sigma_e^2 .$$
Suppose f_1 and f_2 are the respective degrees of freedom of M_1 and M_2 and let
$$F = M_1/M_2 .$$
Then, with $F_{f_1,f_2,\alpha}$ being the upper $\alpha\%$ point on the $F(f_1, f_2)$ distribution, i.e., a fraction $\alpha\%$ of the distribution lying beyond $F_{f_1,f_2,\alpha}$, write
$$\alpha_1 + \alpha_2 = \alpha$$
and
$$F_1 = F_{f_2,f_1,\alpha_1}, \qquad F_2 = F_{f_1,f_2,\alpha_2},$$
$$F_1' = F_{\infty,f_1,\alpha_1}, \qquad F_2' = F_{f_1,\infty,\alpha_2}.$$
An approximate $(1 - \alpha)\%$ confidence interval on θ given by Scheffé (1959,

p. 235), similar to that of Bulmer (1957), is

$$\frac{M_2(F - F_2)(F + F_2 - F_2')}{FF_2'} < \theta < \frac{M_2(F - 1/F_1)(F + 1/F_1 - 1/F_1')}{F/F_1'}.$$

When $F < F_2$ the lower limit is taken as zero and when $F < 1/F_1$ the interval is taken as zero. Scheffé indicates that this interval can be "seriously invalidated by non-normality, especially of the random effects" for which M_1 is the mean square.

Although only approximate confidence intervals can be placed on variance components generally, there are some instances where exact intervals can be derived. The most notable is for σ_e^2, based on the χ^2-distribution of (51). It yields the interval contained in the statement

$$\Pr\left\{\frac{\text{SSE}}{\chi^2_{f_{\text{SSE}},U}} \leq \sigma_e^2 \leq \frac{\text{SSE}}{\chi^2_{f_{\text{SSE}},L}}\right\} = 1 - \alpha \tag{59}$$

where $f_{\text{SSE}} = f_{\text{MSE}}$ and the χ^2-values are derived from tables as in (58).

Other exact confidence intervals readily available are those for the 1-way classification shown in Table 9.14. The first entry there is the appropriate

TABLE 9.14. CONFIDENCE INTERVALS ON VARIANCE COMPONENTS AND FUNCTIONS THEREOF, IN THE 1-WAY CLASSIFICATION, RANDOM MODEL, BALANCED DATA (see Table 9.11)

Parameter	Exact Confidence Interval[1]		Confidence Coefficient
	Lower Limit	Upper Limit	
σ_e^2	$\dfrac{\text{SSE}}{\chi^2_{a(n-1),U}}$	$\dfrac{\text{SSE}}{\chi^2_{a(n-1),L}}$	$1 - \alpha$
σ_α^2	$\dfrac{\text{SSA}(1 - F_U/F)}{n\chi^2_{a-1,U}}$	$\dfrac{\text{SSA}(1 - F_L/F)}{n\chi^2_{a-1,L}}$	$1 - 2\alpha$
$\dfrac{\sigma_\alpha^2}{\sigma_\alpha^2 + \sigma_e^2}$	$\dfrac{F/F_U - 1}{n + F/F_U - 1}$	$\dfrac{F/F_L - 1}{n + F/F_L - 1}$	$1 - \alpha$
$\dfrac{\sigma_e^2}{\sigma_\alpha^2 + \sigma_e^2}$	$\dfrac{n}{n + F/F_L - 1}$	$\dfrac{n}{n + F/F_U - 1}$	$1 - \alpha$
$\dfrac{\sigma_\alpha^2}{\sigma_e^2}$	$\dfrac{F/F_U - 1}{n}$	$\dfrac{F/F_L - 1}{n}$	$1 - \alpha$

[1] Notation: $F = \text{MSA}/\text{MSE}$

$$\Pr\{\chi^2_{n,L} \leq \chi^2(n) \leq \chi^2_{n,U}\} = 1 - \alpha$$

$$\Pr\{F_L \leq F[a - 1, a(n - 1)] \leq F_U\} = 1 - \alpha$$

form of (59). The last three entries are equivalent intervals for different ratio functions, all based on the fact that for $F = \text{MSA/MSE}$

$$\sigma_e^2 F/(n\sigma_\alpha^2 + \sigma_e^2) \sim F[a-1, a(n-1)]. \tag{60}$$

The interval for $\sigma_e^2/(\sigma_\alpha^2 + \sigma_e^2)$ is given by Graybill (1961, p. 379) and that for $\sigma_\alpha^2/\sigma_e^2$ by Scheffé (1959, p. 229). The second entry in the table, the interval for σ_α^2, is given by Williams (1962) and stems from combining (60) and the distribution of SSE/σ_e^2.

e. Probability of negative estimates

Consider two mean squares M_1 and M_2 of the nature described in (44). Suppose $E(M_1 - M_2) = k\sigma^2$ so that

$$\hat{\sigma}^2 = (M_1 - M_2)/k.$$

Then the probability of $\hat{\sigma}^2$ being negative is

$$\begin{aligned}\Pr\{\hat{\sigma}^2 \text{ is negative}\} &= \Pr\{M_1/M_2 < 1\}\\ &= \Pr\left\{\frac{M_1/E(M_1)}{M_2/E(M_2)} < \frac{E(M_2)}{E(M_1)}\right\}\\ &= \Pr\left\{F(f_1, f_2) < \frac{E(M_2)}{E(M_1)}\right\}.\end{aligned} \tag{61}$$

This provides a means of calculating the probability that an estimator of the form $\hat{\sigma}^2 = (M_1 - M_2)/k$ will be negative. It requires giving values to the variance components being estimated because $E(M_2)$ and $E(M_1)$ are functions of the components. However, in using a series of arbitrary values for these components, calculation of (61) provides some general indication of the probability of obtaining a negative estimate. The development of this procedure is given by Leone *et al.* (1968). Clearly, it could also be extended to use the approximate F-statistic of (56) for finding the probability that the estimate of σ_α^2 of (55) would be negative.

Example. For the 1-way classification of Table 9.11 equation (61) is

$$\begin{aligned}\Pr\{\hat{\sigma}_\alpha^2 < 0\} &= \Pr\{F_{a-1,a(n-1)} < \sigma_e^2/(\sigma_e^2 + n\sigma_\alpha^2)\}\\ &= \Pr\{F_{a-1,a(n-1)} < 1/(1 + n\rho)\}\end{aligned}$$

where $\rho = \sigma_\alpha^2/\sigma_e^2$.

f. Sampling variances of estimators

Sampling variances of variance component estimators that are linear functions of χ^2-variables can be derived even though the distribution functions of the estimators, generally speaking, cannot be. The variances are, of course, functions of the unknown components.

(*i*) *Derivation.* With the estimators being linear functions of mean squares they are linear functions of quadratic forms of the observations and hence are themselves quadratic forms of the observations. Theorem 1 of Chapter 2 could therefore be used to derive their variances. This is the procedure used with unbalanced data, in the next chapter. However, with balanced data the mean squares are independent with known distributions, as in (44), and variances of linear functions of them are therefore easily derived. By writing an estimator as

$$\hat{\sigma}^2 = \sum k_i M_i$$

we have, from (44), $\text{cov}(M_i M_{i'}) = 0$ for $i \neq i'$ and

$$v(M_i) = 2f_i[E(M_i)/f_i]^2 = 2[E(M_i)]^2/f_i\,.$$

Hence
$$v(\hat{\sigma}^2) = 2\sum \frac{k_i^2[E(M_i)]^2}{f_i}\,. \tag{62}$$

Example. In the 1-way classification of Table 9.11

$$\hat{\sigma}_\alpha^2 = (\text{MSA} - \text{MSE})/n$$

and so from (62)

$$v(\hat{\sigma}_\alpha^2) = \frac{2}{n^2}\left[\frac{(n\sigma_\alpha^2 + \sigma_e^2)^2}{a-1} + \frac{\sigma_e^4}{a(n-1)}\right]. \tag{63}$$

Similarly, from (51)

$$v(\hat{\sigma}_e^2) = \frac{2\sigma_e^4}{f_{\text{MSE}}}\,, \tag{64}$$

which, for Table 9.11, is

$$v(\hat{\sigma}_e^2) = \frac{2\sigma_e^4}{a(n-1)}\,. \tag{65}$$

(*ii*) *Covariance matrix.* Mean squares in the analysis of variance of balanced data are distributed independently of one another, as noted in (44). They therefore have zero covariances. But such is not necessarily the case with variance component estimators that are linear functions of these mean squares. Such estimators usually have non-zero covariances. For example, in the 1-way classification we have, from (40),

$$\text{cov}(\hat{\sigma}_\alpha^2, \hat{\sigma}_e^2) = -v(\text{MSE})/n \tag{66}$$

$$= -2\sigma_e^4/an(n-1). \tag{67}$$

In general, the variance-covariance matrix of the vector of estimators is, from (41),

$$\text{var}(\hat{\boldsymbol{\sigma}}^2) = \mathbf{P}^{-1}\,\text{var}(\mathbf{m})\mathbf{P}^{-1\prime}. \tag{68}$$

Because the mean squares are independent var($\mathbf{m}$) is diagonal, which we write as

$$\text{var}(\mathbf{m}) = \mathbf{D} = \text{diag}\{2[E(M_i)]^2/f_i\} \quad \text{for} \quad i = 1, 2, \ldots, k. \tag{69}$$

Then

$$\text{var}(\hat{\boldsymbol{\sigma}}^2) = \mathbf{P}^{-1}\mathbf{D}\mathbf{P}^{-1\prime}, \tag{70}$$

each element being a quadratic function of the variance components, as is $[E(M_i)]^2$ in (69).

(*iii*) *Unbiased estimation.* The quadratic nature of the elements of var($\hat{\boldsymbol{\sigma}}^2$) just noted makes estimation of them in any optimal manner not easy. The simplest and most oft-used procedure is that of replacing $E(M_i)$ in $\mathbf{D}$ by M_i. Thus from (69) we write

$$\mathbf{D}_1 = \text{diag}\{2M_i^2/f_i\} \quad \text{for} \quad i = 1, 2, \ldots, k \tag{71}$$

and then have

$$\widetilde{\text{var}}(\hat{\boldsymbol{\sigma}}^2) = \mathbf{P}^{-1}\mathbf{D}_1\mathbf{P}^{-1\prime}. \tag{72}$$

These estimators have no known desirable properties. They are not even unbiased.

Unbiased estimation of var($\hat{\boldsymbol{\sigma}}^2$) can, however, be readily obtained from (71) through replacing f_i therein by $f_i + 2$. Thus with

$$\mathbf{D}_2 = \text{diag}\{2M_i^2/(f_i + 2)\} \quad \text{for} \quad i = 1, \ldots, k, \tag{73}$$

we have

$$\widehat{\text{var}}(\hat{\boldsymbol{\sigma}}^2) = \mathbf{P}^{-1}\mathbf{D}_2\mathbf{P}^{-1\prime} \tag{74}$$

as an unbiased estimator of var($\hat{\boldsymbol{\sigma}}^2$). For example, from (63) and (65)

$$\hat{v}(\hat{\sigma}_e^2) = \frac{2\hat{\sigma}_e^4}{a(n-1)+2}$$

and

$$\hat{v}(\hat{\sigma}_\alpha^2) = \frac{2}{n^2}\left[\frac{(n\hat{\sigma}_\alpha^2 + \hat{\sigma}_e^2)^2}{a+1} + \frac{\hat{\sigma}_e^4}{a(n-1)+2}\right]$$

are unbiased estimators of the variances of $\hat{\sigma}_e^2$ and $\hat{\sigma}_\alpha^2$.

The reason that (74) gives an unbiased estimator of var($\hat{\boldsymbol{\sigma}}^2$) is as follows. For any mean square M, with degrees of freedom f,

$$v(M) = 2[E(M)]^2/f$$

and, by definition,

$$v(M) = E(M^2) - [E(M)]^2.$$

Hence

$$E(M^2) = (1 + 2/f)[E(M)]^2$$

and so

$$\frac{M^2}{f+2} \text{ is an unbiased estimator of } \frac{[E(M)]^2}{f}.$$

Therefore, using $M_i^2/(f_i + 2)$ in place of $[E(M_i)]^2/f_i$ in (69), as is done in (73), makes $\mathbf{D}_2$ an unbiased estimator of $\mathbf{D}$ and hence $\mathbf{P}^{-1}\mathbf{D}_2\mathbf{P}^{-1\prime}$ of (74) is an unbiased estimator of var($\hat{\boldsymbol{\sigma}}^2$) $= \mathbf{P}^{-1}\mathbf{D}\mathbf{P}^{-1\prime}$.

g. Maximum likelihood estimation

Estimating parameters of a fixed effects model by the method of maximum likelihood leads in many cases (under normality assumptions) to the same estimators as do the methods of least squares and best linear unbiased estimation. One would hope that with variance component estimation it would lead to the analysis of variance estimators. But such is not the case. The ability of analysis of variance estimators to be negative shows that they cannot be maximum likelihood estimators, because the latter would be derived by maximizing the likelihood over the parameter space, which is non-negative so far as variance components are concerned. Maximum likelihood estimators must therefore be non-negative. The problem of deriving maximum likelihood estimation is therefore not as straightforward for variance components as it is for the parameters of a fixed effects model; and indeed, with unbalanced data, explicit estimators cannot be obtained. Some of the results available for balanced data, notably in the 1-way classification, are now discussed.

The likelihood of the sample of observations in the 1-way classification of Table 9.11 is

$$L = (2\pi)^{-\frac{1}{2}an}\,|\mathbf{V}|^{-\frac{1}{2}}\exp\{-\tfrac{1}{2}(\mathbf{y}-\mu\mathbf{1})'\mathbf{V}^{-1}(\mathbf{y}-\mu\mathbf{1})\}, \tag{75}$$

where $\mathbf{V}$ of (45) can be rewritten as

$$\mathbf{V} = \sum_{i=1}^{a}{}^{+}\,(\sigma_e^2\mathbf{I} + \sigma_\alpha^2\mathbf{J})$$

with $\mathbf{I}$ and $\mathbf{J}$ being of order n. Then

$$|\mathbf{V}| = \prod_{i=1}^{a}|(\sigma_e^2\mathbf{I} + \sigma_\alpha^2\mathbf{J})| = [\sigma_e^{2(n-1)}(\sigma_e^2 + n\sigma_\alpha^2)]^a$$

and

$$\mathbf{V}^{-1} = \sum_{i=1}^{a}{}^{+}\left[\frac{1}{\sigma_e^2}\mathbf{I} - \frac{\sigma_\alpha^2}{\sigma_e^2(\sigma_e^2 + n\sigma_\alpha^2)}\mathbf{J}\right].$$

Substituting for $|\mathbf{V}|$ and $\mathbf{V}^{-1}$ into (75) leads, after a little simplification, to

$$L = \frac{\exp -\frac{1}{2}\left[\dfrac{\text{SSE}}{\sigma_e^2} + \dfrac{\text{SSA}}{\sigma_e^2 + n\sigma_\alpha^2} + \dfrac{an(\bar{y}_{..} - \mu)^2}{\sigma_e^2 + n\sigma_\alpha^2}\right]}{(2\pi)^{\frac{1}{2}an}(\sigma_e^2)^{\frac{1}{2}a(n-1)}(\sigma_e^2 + n\sigma_\alpha^2)^{\frac{1}{2}a}}. \tag{76}$$

Equating to zero the differentials of $\log L$ with respect to μ, σ_α^2 and σ_e^2 and denoting the solutions by $\tilde{\mu}$, $\tilde{\sigma}_\alpha^2$ and $\tilde{\sigma}_e^2$ gives $\tilde{\mu} = \bar{y}_{..}$, and

$$a(\tilde{\sigma}_e^2 + n\tilde{\sigma}_\alpha^2) = \text{SSA} \qquad \text{and} \qquad a(n-1)\tilde{\sigma}_e^2 = \text{SSE}. \tag{77}$$

In doing this the maximization (of L) has not been restricted to positive values of $\tilde{\sigma}_\alpha^2$ and $\tilde{\sigma}_e^2$; hence the solutions to (77)

$$\tilde{\sigma}_e^2 = \text{SSE}/a(n-1) = \text{MSE}$$

TABLE 9.15. ESTIMATORS OF VARIANCE COMPONENTS IN THE 1-WAY CLASSIFICATION, RANDOM MODEL, WITH BALANCED DATA

Methods of Estimation	Conditions	Estimators of σ_α^2	Estimators of σ_e^2
Analysis of variance	None	$(\text{MSA} - \text{MSE})/n$	MSE
Maximum likelihood (Herbach, 1959)	$\frac{a-1}{a}\text{MSA} \geq \text{MSE}$	$\left(\frac{a-1}{a}\text{MSA} - \text{MSE}\right)\Big/ n$	MSE
	$\frac{a-1}{a}\text{MSA} < \text{MSE}$	0	$\frac{\text{SST}}{an}$
Restricted maximum likelihood (Thompson, 1962)	$\text{MSA} \geq \text{MSE}$	$(\text{MSA} - \text{MSE})/n$	MSE
	$\text{MSA} < \text{MSE}$	0	$\frac{\text{SST}}{an - 1}$

and
$$\tilde{\sigma}_\alpha^2 = \frac{(\text{SSA}/a - \tilde{\sigma}_e^2)}{n} = \frac{(1 - 1/a)\text{MSA} - \text{MSE}}{n}$$

are not maximum likelihood estimators. Herbach (1959) shows that when $\tilde{\sigma}_\alpha^2$ is negative, i.e., $(1 - 1/a)\text{MSA} < \text{MSE}$, the maximum likelihood estimator of σ_α^2 is 0 and that of σ_e^2 is SST/an. This result is shown in Table 9.15, along with that of Thompson (1962), who uses a restricted maximum likelihood procedure, confined to just that portion of the set of sufficient statistics which is location invariant. This is the basis for Thompson's procedure mentioned in Sec. 8b of pooling minimal mean squares with predecessors when the analysis of variance method yields negative estimates.

10. EXERCISES

1. Suppose you have balanced data from a model having factors A, B, C within AB-subclasses, and D within C. Set up the analysis of variance table, and give expected values of mean squares for (i) the random model, (ii) the mixed model when A is a fixed effects factor and (iii) the mixed model when both A and B are fixed effects factors.

2. Repeat Exercise 1 for a model having factors A, B, D, and C within AB.
3. A split-plot experiment, whose main plots form a randomized complete blocks design, can be analyzed with the model

$$y_{ijk} = \mu + \alpha_i + \rho_j + \delta_{ij} + \beta_k + \theta_{ik} + e_{ijk}\,.$$

Set up the analysis of variance table, and give expected values of mean squares for the following cases:
(*a*) Random model.
(*b*) Mixed model, ρ's and δ's random.
(*c*) Mixed model, only the β's fixed.
(*d*) Mixed model, only the α's fixed.

4. Show that $F = Q/s\hat{\sigma}^2$ as used in earlier chapters [e.g., equation (21) of Chapter 6] is distributed as $F'\{s, N - r, [E(Q) - s\sigma^2]/2\sigma^2\}$.
5. Use Theorems 2 and 4 of Chapter 2 to show that $\text{SSA}/(\sigma_e^2 + n\sigma_\alpha^2)$ and SSE/σ_e^2 of Table 9.11 are distributed independently as central χ^2-variables.
6. For the random model

$$y_{ijk} = \mu + \alpha_i + \beta_{ij} + e_{ijk}\,,$$

with balanced data, derive explicit expressions for (i) the analysis of variance estimators of the variance components and, under normality assumptions, (ii) the variances of those estimators and (iii) unbiased estimators of those variances.

7. Repeat Exercise 6 for the models

$$(a)\ y_{ijk} = \mu + \alpha_i + \beta_j + e_{ijk}$$

and

$$(b)\ y_{ijk} = \mu + \alpha_i + \beta_j + (\alpha\beta)_{ij} + e_{ijk}\,.$$

8. When the α_i in Exercise 6 are fixed effects show that the generalized least squares normal equations for $\mu + \alpha_i$ lead to $\widehat{\mu + \alpha_i} = \bar{y}_{i..}\,.$

CHAPTER 10

METHODS OF ESTIMATING VARIANCE COMPONENTS FROM UNBALANCED DATA

Estimation of variance components from balanced data rests almost entirely on one method, the analysis of variance method described at length in the preceding chapter. In contrast, there are several methods available for use with unbalanced data, a number of which are now described. They are presented largely in general terms and are illustrated by means of the 1-way and the 2-way (crossed) classifications. Most of the illustrations are of individual aspects of the methods and not of complete analyses. The purpose of the chapter is to describe methodology without the clutter of lengthy details of specific cases. This should enable the reader to direct his attention to basic procedures rather than being diverted to their numerous details in individual applications. These are given in the next chapter, where, with little or no discussion of methodology, we have gathered together specific results available in the literature and shown them in full detail. The present chapter is therefore a chronicle of the various methods and the following one is a catalogue of the available consequences of applying those methods to specific cases.

1. EXPECTATIONS OF QUADRATIC FORMS

The analysis of variance method of estimating variance components from balanced data is based on equating mean squares of analyses of variance to their expected values. This is a well-defined method with balanced data because there is only one analysis of variance for any particular model. For example, with the 2-way classification interaction model the only analysis of variance is that of Table 7.9 (or equivalently Table 9.5). However, with unbalanced data for that same model there are two analyses of variance, namely the two parts of Table 7.8, one for fitting α before β and the other for fitting β before α. This is so in general; there can be several, maybe many, ways of partitioning a total sum of squares. (Table 8.2 is an example.) On the face of it there are no criteria for choosing any one of these partitionings over the others when it comes to using one of them for purposes of estimating variance components. We return to this matter subsequently, for the moment noticing only that there is, with unbalanced data, no uniquely "obvious" set of sums of squares or quadratic forms in the observations that can be optimumly used for estimating variance components. There is instead a variety of quadratic forms that can be used, each of them in the method of equating observed quadratic forms to their expected values. We therefore begin by considering the expected value of the general quadratic form $\mathbf{y}'\mathbf{Q}\mathbf{y}$.[1]

The general linear model is taken, as usual, to be

$$\mathbf{y} = \mathbf{X}\mathbf{b} + \mathbf{e} \tag{1}$$

where $\mathbf{y}$ is $N \times 1$ (N observations) and, for the sake of generality,

$$\text{var}(\mathbf{y}) = \mathbf{V}.$$

Then, from Theorem 1 of Sec. 2.5a, the expected value of the quadratic form $\mathbf{y}'\mathbf{Q}\mathbf{y}$ is

$$E(\mathbf{y}'\mathbf{Q}\mathbf{y}) = \text{tr}(\mathbf{Q}\mathbf{V}) + E(\mathbf{y}')\mathbf{Q}E(\mathbf{y}). \tag{2}$$

We look at this in terms of the model (1) being successively a fixed effects model, a mixed model and a random model.

In every case $\mathbf{b}$ represents all the effects in the model, be they fixed, random or mixed. Also, in each model we take $E(\mathbf{e}) = \mathbf{0}$, so that var($\mathbf{e}$) is $E(\mathbf{e}\mathbf{e}') = \sigma_e^2\mathbf{I}$. Furthermore, when $\mathbf{b}$ is a vector of fixed effects, $E(\mathbf{b}\mathbf{e}') = \mathbf{b}E(\mathbf{e}') = \mathbf{0}$; and when $\mathbf{b}$ includes elements that are random effects we assume they have zero means, and zero covariance with the elements in $\mathbf{e}$; thus at all times $E(\mathbf{b}\mathbf{e}') = E(\mathbf{e}\mathbf{b}') = \mathbf{0}$.

a. Fixed effects models

In the usual fixed effects model $\mathbf{b}$ is a vector of fixed effects with $E(\mathbf{y}) = \mathbf{X}\mathbf{b}$ and $\mathbf{V} = \sigma_e^2\mathbf{I}_N$. Then (2) becomes

$$E(\mathbf{y}'\mathbf{Q}\mathbf{y}) = \mathbf{b}'\mathbf{X}'\mathbf{Q}\mathbf{X}\mathbf{b} + \sigma_e^2\,\text{tr}(\mathbf{Q}) \tag{3}$$

Examples. Two well-known applications of (3) are when $\mathbf{Q} = \mathbf{I}_N$, giving

$$E(\mathbf{y}'\mathbf{y}) = \mathbf{b}'\mathbf{X}'\mathbf{X}\mathbf{b} + N\sigma_e^2;$$

[1] The matrix $\mathbf{Q}$ used here is not to be confused with the scalar Q used earlier for the numerator sum of squares in hypothesis testing.

and when $\mathbf{Q}$ is $\mathbf{X}(\mathbf{X}'\mathbf{X})^{-}\mathbf{X}'$, for which $\mathbf{y}'\mathbf{Q}\mathbf{y}$ is the reduction in sum of squares $R(\mathbf{b})$, giving

$$\begin{aligned} E[R(\mathbf{b})] &= \mathbf{b}'\mathbf{X}'\mathbf{X}\mathbf{b} + \sigma_e^2 \operatorname{tr}[\mathbf{X}(\mathbf{X}'\mathbf{X})^{-}\mathbf{X}'] \\ &= \mathbf{b}'\mathbf{X}'\mathbf{X}\mathbf{b} + \sigma_e^2 r(\mathbf{X}), \end{aligned}$$

because $\mathbf{X}(\mathbf{X}'\mathbf{X})^{-}\mathbf{X}'$ is idempotent and has the same rank as $\mathbf{X}$ (Theorem 7 of Sec. 1.5). Hence $E[\mathbf{y}'\mathbf{y} - R(\mathbf{b})] = [N - r(\mathbf{X})]\sigma_e^2$, the familiar result for a residual sum of squares (see Sec. 5.2e).

b. Mixed models

In a mixed model we partition $\mathbf{b}'$ as

$$\mathbf{b}' = [\mathbf{b}_1' \quad \mathbf{b}_A' \quad \mathbf{b}_B' \quad \cdots \quad \mathbf{b}_K'] \tag{4}$$

where $\mathbf{b}_1$ contains all the fixed effects of the model (including the mean μ) and where the other $\mathbf{b}$'s each represent a set of random effects for the factors $A, B, C, \ldots, K$ respectively. Although only single subscripts are used, interaction effects and/or nested-factor effects are not excluded by this notation. They are considered merely as factors, each identified by a single subscript rather than the letters of the corresponding main effects. For example, the AB-interaction effects might be in the vector $\mathbf{b}_G$.

The model (1) is written in terms of (4) as

$$\mathbf{y} = \mathbf{X}_1\mathbf{b}_1 + \mathbf{X}_A\mathbf{b}_A + \mathbf{X}_B\mathbf{b}_B + \cdots + \mathbf{X}_K\mathbf{b}_K + \mathbf{e};$$

i.e., as

$$\mathbf{y} = \mathbf{X}_1\mathbf{b}_1 + \sum_{\theta=A}^{K} \mathbf{X}_\theta\mathbf{b}_\theta + \mathbf{e} \tag{5}$$

where $\mathbf{X}$ has been partitioned conformably for the product $\mathbf{X}\mathbf{b}$ and where θ in the summation takes the values $A, B, \ldots, K$. For the random effects we make the two initial assumptions: that they have zero means and that the effects of each random factor have zero covariance with those of every other factor. Thus we write $E(\mathbf{b}_\theta) = \mathbf{0}$ and from (5) obtain

$$E(\mathbf{y}) = \mathbf{X}_1\mathbf{b}_1 \tag{6}$$

and

$$\mathbf{V} = \operatorname{var}(\mathbf{y}) = \sum_{\theta=A}^{K} \mathbf{X}_\theta \operatorname{var}(\mathbf{b}_\theta)\mathbf{X}_\theta' + \sigma_e^2\mathbf{I}_N \tag{7}$$

where $\mathbf{I}_N$ is an identity matrix of order N, and $\operatorname{var}(\mathbf{b}_\theta)$ is the covariance matrix of the random effects of the θ-factor. These effects are usually assumed to be uncorrelated, with uniform variance σ_θ^2, so that

$$\operatorname{var}(\mathbf{b}_\theta) = \sigma_\theta^2\mathbf{I}_{N_\theta} \qquad \text{for} \quad \theta = A, B, \ldots, K, \tag{8}$$

there being N_θ different effects of the θ-factor in the data, i.e., N_θ levels of that factor. Thus in (7)

$$\mathbf{V} = \sum_{\theta=A}^{K} \mathbf{X}_\theta\mathbf{X}_\theta'\sigma_\theta^2 + \sigma_e^2\mathbf{I}_N. \tag{9}$$

Hence from (6) and (9) the expected value of the quadratic form in (2) is

$$E(\mathbf{y}'\mathbf{Q}\mathbf{y}) = (\mathbf{X}_1\mathbf{b}_1)'\mathbf{Q}\mathbf{X}_1\mathbf{b}_1 + \sum_{\theta=A}^{K} \sigma_\theta^2 \operatorname{tr}(\mathbf{Q}\mathbf{X}_\theta\mathbf{X}_\theta') + \sigma_e^2 \operatorname{tr}(\mathbf{Q}). \qquad (10)$$

c. Random effects models

In a random model all effects are taken to be random—all, that is, save μ, the general mean. The expression (10) just developed for $E(\mathbf{y}'\mathbf{Q}\mathbf{y})$ for the mixed model can therefore be used for the random model, by letting $\mathbf{b}_1$ be the scalar μ and $\mathbf{X}_1$ be a vector of 1's denoted by $\mathbf{1}$. Thus for the random model

$$E(\mathbf{y}'\mathbf{Q}\mathbf{y}) = \mu^2\mathbf{1}'\mathbf{Q}\mathbf{1} + \sum_{\theta=A}^{K} \sigma_\theta^2 \operatorname{tr}(\mathbf{Q}\mathbf{X}_\theta\mathbf{X}_\theta') + \sigma_e^2 \operatorname{tr}(\mathbf{Q}). \qquad (11)$$

d. Applications

Applying these general results to particular models involves partitioning $\mathbf{b}$ into sub-vectors each of which contains effects pertaining to all levels of one complete classification (or interaction of classifications) involved in the linear model. In this way expressions (3), (10) and (11) represent the general results for the fixed, mixed and random models respectively. With their aid, expectations of quadratic forms can be readily obtained for any of the three models. For example, suppose we had

$$y_{ijkh} = \mu + \alpha_i + \beta_j + \gamma_k + \delta_{jk} + e_{ijkh}\,.$$

In vector form this could be written as

$$\mathbf{y} = \mu\mathbf{1} + \mathbf{X}_A\mathbf{b}_A + \mathbf{X}_B\mathbf{b}_B + \mathbf{X}_C\mathbf{b}_C + \mathbf{X}_D\mathbf{b}_D + \mathbf{e}$$

where $\mathbf{b}_A$ is the vector of α-effects, $\mathbf{b}_B$ is the vector of β's, and $\mathbf{b}_C$ and $\mathbf{b}_D$ are vectors of the γ- and δ-terms respectively. In this way the results in (3), (10) and (11) can be applied to finding expectations of any quadratic form $\mathbf{y}'\mathbf{Q}\mathbf{y}$ of the observations $\mathbf{y}$.

2. ANALYSIS OF VARIANCE METHOD (HENDERSON'S METHOD 1)

The analysis of variance method with balanced data consists of equating mean squares to their expected values. Essentially the same procedure is used with unbalanced data.

We begin by discussing the method in terms of an example, the 2-way classification interaction model. Although not the simplest example that could be used it illustrates facets of the method that cannot be demonstrated with a simpler one. Many details of deriving estimators for the 2-way classification interaction model are given here but the complete results are not. They are shown in Chapter 11. In this chapter we give just those details necessary for illustrating the method and its various aspects.

a. Model and notation

The model for the 2-way classification with interaction is

$$y_{ijk} = \mu + \alpha_i + \beta_j + \gamma_{ij} + e_{ijk} \tag{12}$$

where y_{ijk} is the kth observation in the ith level of the A-factor and the jth level of the B-factor; $i = 1, 2, \ldots, a, j = 1, 2, \ldots, b$ and $k = 1, 2, \ldots, n_{ij}$, with s of the n_{ij}-values being non-zero. A complete description of the fixed effects case of the model is given in Sec. 7.2a. In the random model, which we now consider, the α_i's, β_j's and γ_{ij}'s are all assumed to be random with zero means and variances $\sigma_\alpha^2\mathbf{I}_a$, $\sigma_\beta^2\mathbf{I}_b$ and $\sigma_\gamma^2\mathbf{I}_s$ respectively. This means, for example, that

$$E(\alpha_i) = 0, \qquad E(\alpha_i^2) = \sigma_\alpha^2 \qquad \text{and} \qquad E(\alpha_i\alpha_{i'}) = 0 \qquad \text{for} \quad i \neq i', \tag{13}$$

with similar results for the β's and γ's. Also, all covariances between pairs of non-identical random variables are assumed zero. The e-terms follow the usual prescription: $E(\mathbf{e}) = \mathbf{0}$, $\text{var}(\mathbf{e}) = \sigma_e^2\mathbf{I}_N$ and the covariance of every e with every random effect is zero.

b. Analogous sums of squares

The analysis of variance for balanced data in the model (12) is shown as Table 9.5. It contains a term

$$\text{SSA} = bn\sum_{i=1}^{a}(\bar{y}_{i\cdot\cdot} - \bar{y}_{\cdot\cdot\cdot})^2 = \sum_{i=1}^{a}\frac{y_{i\cdot\cdot}^2}{bn} - \frac{y_{\cdot\cdot\cdot}^2}{abn}, \tag{14}$$

the bar and dot notation of totals and means being the same as defined in Sec. 7.2a. The term analogous to (14) for unbalanced data is

$$\text{SSA} = \sum_{i=1}^{a}\frac{y_{i\cdot\cdot}^2}{n_{i\cdot}} - \frac{y_{\cdot\cdot\cdot}^2}{n_{\cdot\cdot}}. \tag{15}$$

This is one of the terms used for estimating variance components by the analysis of variance method from unbalanced data. In similar manner the other terms are

$$\text{SSB} = \sum_{j=1}^{b}\frac{y_{\cdot j\cdot}^2}{n_{\cdot j}} - \frac{y_{\cdot\cdot\cdot}^2}{n_{\cdot\cdot}} \tag{16}$$

$$\text{SSAB} = \sum_{i=1}^{a}\sum_{j=1}^{b}\frac{y_{ij\cdot}^2}{n_{ij}} - \sum_{i=1}^{a}\frac{y_{i\cdot\cdot}^2}{n_{i\cdot}} - \sum_{j=1}^{b}\frac{y_{\cdot j\cdot}^2}{n_{\cdot j}} + \frac{y_{\cdot\cdot\cdot}^2}{n_{\cdot\cdot}} \tag{17}$$

and

$$\text{SSE} = \sum_{i=1}^{a}\sum_{j=1}^{b}\sum_{k=1}^{n_{ij}} y_{ijk}^2 - \sum_{i=1}^{a}\sum_{j=1}^{b}\frac{y_{ij\cdot}^2}{n_{ij}}. \tag{18}$$

The analysis of variance method of variance component estimation for unbalanced data then involves equating (15)–(18) to their expected values. Before considering derivation of these expected values comments about these SS-terms are in order.

(*i*) *Empty cells.* Since n_{ij} is the number of observations in a cell it can, as we have seen, be zero. The summations in SSAB and SSE that involve n_{ij} in the denominator are therefore defined only for the (i, j) combinations for which n_{ij} is non-zero; i.e., the summing is over only those s cells that have observations in them. The possibility of zero denominators is thus removed.

(*ii*) *Balanced data.* It is clear that when the data are balanced, i.e., $n_{ij} = n$ for all i and j, then (15) reduces to (14). In similar fashion (16), (17) and (18) reduce to the corresponding analysis of variance sums of squares for balanced data shown in Table 9.5.

(*iii*) *A negative "sum of squares".* Expressions (15)–(18) have been established solely by analogy with the analysis of variance of balanced data. In general not all such analogous expressions are sums of squares. For example, SSAB of (17) is not always positive (see Exercise 1 of Chapter 2) and so it is not a sum of squares. We might therefore refer to (15)–(18) and their counterparts in more complicated models as *analogous sums of squares* and the method as the *analogous analysis of variance method.* It is, however, conventionally called the analysis of variance method, or Henderson's Method 1, after Henderson (1953).

(*iv*) *Uncorrected sums of squares.* Because, in general, the SS-terms are not sums of squares we deal with them in terms of uncorrected sums of squares, to be denoted by T's, as introduced for balanced data in equation (39) of Sec. 9.8. Thus for the SS-terms of (15)–(18) we define

$$T_A = \sum_{i=1}^{a} \frac{y_{i\cdot\cdot}^2}{n_{i\cdot}} \quad \text{and} \quad T_B = \sum_{j=1}^{b} \frac{y_{\cdot j\cdot}^2}{n_{\cdot j}},$$

$$T_{AB} = \sum_{i=1}^{a} \sum_{j=1}^{b} \frac{y_{ij\cdot}^2}{n_{ij}} \quad \text{and} \quad T_\mu = \frac{y_{\cdot\cdot\cdot}^2}{n_{\cdot\cdot}}, \tag{19}$$

with

$$T_0 = \sum_{i=1}^{a} \sum_{j=1}^{b} \sum_{k=1}^{n_{ij}} y_{ijk}^2 .$$

Apart from T_μ for the correction factor for the mean and T_0 for the total sum of squares of all observations, the subscript to a T denotes the factor it applies to and provides easy recognition of the calculating required. For example,

$$T_A = \sum_{\substack{\text{levels of} \\ A\text{-factor}}} \frac{(\text{total } y \text{ for a level of the } A\text{-factor})^2}{\text{no. of observations in that total}} . \tag{20}$$

Similarly, T_{AB} is calculated by an expression equivalent to (20) only with "A-factor" replaced by "AB-factor". With the T's of (19) the SS-terms in

(15)–(18) are

$$\begin{aligned} \text{SSA} &= T_A - T_\mu \quad &\text{and}\quad \text{SSB} &= T_B - T_\mu\,, \\ \text{SSAB} &= T_{AB} - T_A - T_B + T_\mu \quad &\text{and}\quad \text{SSE} &= T_o - T_{AB}\,. \end{aligned} \tag{21}$$

In this form the SS-terms are handled with relative ease, since the T's are positive definite quadratic forms with manageable matrices.

c. Expectations

Variance components are estimated by equating observed values of terms like (15)–(18) to their expected values. The observed values are calculated from the T's and the expected values of (15)–(18), which are quadratic forms in the observations, could be derived by using Theorem 1 of Sec. 2.5; so could the expected values of the T's. However, the "brute force" method illustrated for balanced data in Sec. 9.7 is probably no more lengthy than using the theorem, especially when simplifications arising from the model are fully utilized. We therefore illustrate by deriving $E(\text{SSA}) = E(T_A) - E(T_\mu)$, and then give a generalization. The derivation of $E(T_A)$ *in extenso* serves as a guide to deriving expected values of T's generally.

(*i*) *An example.* We obtain

$$E(\text{SSA}) = E(T_A) - E(T_\mu)$$

by substituting the model (12) into T_A and T_μ of (19) and then taking expectations. First, for T_A, we have

$$y_{i\cdot\cdot} = \sum_{j=1}^{b}\sum_{k=1}^{n_{ij}} y_{ijk} = n_{i\cdot}\mu + n_{i\cdot}\alpha_i + \sum_{j=1}^{b} n_{ij}\beta_j + \sum_{j=1}^{b} n_{ij}\gamma_{ij} + e_{i\cdot\cdot}\,. \tag{22}$$

Hence on squaring and expanding the right-hand side of (22) and dividing by $n_{i\cdot}$, we get

$$\begin{aligned} \frac{y_{i\cdot\cdot}^2}{n_{i\cdot}} &= n_{i\cdot}\mu^2 + n_{i\cdot}\alpha_i^2 + \frac{\sum_{j=1}^{b} n_{ij}^2\beta_j^2}{n_{i\cdot}} + \frac{\sum_{j=1}^{b} n_{ij}^2\gamma_{ij}^2}{n_{i\cdot}} + \frac{e_{i\cdot\cdot}^2}{n_{i\cdot}} \\ &\quad + \frac{\sum_{j=1}^{b}\sum_{j'\neq j}^{b} n_{ij}n_{ij'}\beta_j\beta_{j'}}{n_{i\cdot}} + \frac{\sum_{j=1}^{b}\sum_{j'\neq j}^{b} n_{ij}n_{ij'}\gamma_{ij}\gamma_{ij'}}{n_{i\cdot}} \\ &\quad + 2\Bigg[\mu n_{i\cdot}\alpha_i + \mu\sum_{j=1}^{b} n_{ij}\beta_j + \mu\sum_{j=1}^{b} n_{ij}\gamma_{ij} + \mu e_{i\cdot\cdot} + \alpha_i\sum_{j=1}^{b} n_{ij}\beta_j \\ &\quad + \alpha_i\sum_{j=1}^{b} n_{ij}\gamma_{ij} + \alpha_i e_{i\cdot\cdot} + \frac{\left(\sum_{j=1}^{b} n_{ij}\beta_j\right)\left(\sum_{j=1}^{b} n_{ij}\gamma_{ij}\right)}{n_{i\cdot}} \\ &\quad + \frac{\left(\sum_{j=1}^{b} n_{ij}\beta_j\right)e_{i\cdot\cdot}}{n_{i\cdot}} + \frac{\left(\sum_{j=1}^{b} n_{ij}\gamma_{ij}\right)e_{i\cdot\cdot}}{n_{i\cdot}}\Bigg]. \end{aligned} \tag{23}$$

Expression (23) holds true no matter which effects in the model are fixed and which are random.

Consider taking the expected value of (23) under a random model. Products involving μ go to zero because the other term in such products is a random variable having zero expectation; e.g., $E(\mu n_{i\cdot}\alpha_i) = \mu n_{i\cdot}E(\alpha_i) = 0$. Products of random variables also have zero expectation, because all covariances and expected values are zero; e.g., $E\left(\alpha_i \sum_{j=1}^{b} n_{ij}\beta_j\right) = \sum_{j=1}^{b} n_{ij}E(\alpha_i\beta_j)$ and $E(\alpha_i\beta_j) = \text{cov}(\alpha_j\beta_j) + E(\alpha_i)E(\beta_j) = 0$; similarly $\sum_{j=1}^{b}\sum_{j'\neq j}^{b} n_{ij}n_{ij'}E(\beta_j\beta_{j'}) = 0$. The only non-zero terms are the expected values of all squared terms which, apart from μ^2, become variances. These are the only non-zero terms remaining in $E(y_{i\cdot\cdot}^2/n_{i\cdot})$ and so

$$E\left(\frac{y_{i\cdot\cdot}^2}{n_{i\cdot}}\right) = n_{i\cdot}\mu^2 + n_{i\cdot}\sigma_\alpha^2 + \frac{\sum_{j=1}^{b} n_{ij}^2}{n_{i\cdot}}\sigma_\beta^2 + \frac{\sum_{j=1}^{b} n_{ij}^2}{n_{i\cdot}}\sigma_\gamma^2 + \sigma_e^2, \tag{24}$$

the last term being σ_e^2 because

$$E\left(\frac{e_{i\cdot\cdot}^2}{n_{i\cdot}}\right) = \sum_{j=1}^{b}\frac{\sum_{k=1}^{n_{ij}} E(e_{ijk}^2)}{n_{i\cdot}} = \frac{n_{i\cdot}\sigma_e^2}{n_{i\cdot}} = \sigma_e^2,$$

with the cross-products in the e's having zero expectation. Hence, summing (24) gives

$$E(T_A) = \sum_{i=1}^{a} E\left(\frac{y_{i\cdot\cdot}^2}{n_{i\cdot}}\right)$$

$$= N\mu^2 + N\sigma_\alpha^2 + \sum_{i=1}^{a}\frac{\sum_{j=1}^{b} n_{ij}^2}{n_{i\cdot}}\sigma_\beta^2 + \sum_{i=1}^{a}\frac{\sum_{j=1}^{b} n_{ij}^2}{n_{i\cdot}}\sigma_\gamma^2 + a\sigma_e^2. \tag{25}$$

The extended form (23) shows clearly how (24) and (25) are derived; and it is particularly useful when we come to the case of mixed models where not all cross-product terms have an expected value of zero; e.g., see equation (30). However, the consequences of the expected values of the model [e.g., (13)] make it easy to go directly from (22) to (25). Thus for T_μ we write

$$y_{\cdots} = N\mu + \sum_{i=1}^{a} n_{i\cdot}\alpha_i + \sum_{j=1}^{b} n_{\cdot j}\beta_j + \sum_{i=1}^{a}\sum_{j=1}^{b} n_{ij}\gamma_{ij} + e_{\cdots}$$

and so

$$E(T_\mu) = E\left(\frac{y_{\cdots}^2}{N}\right) = N\mu^2 + \frac{\sum_{i=1}^{a} n_{i\cdot}^2}{N}\sigma_\alpha^2 + \frac{\sum_{j=1}^{b} n_{\cdot j}^2}{N}\sigma_\beta^2 + \frac{\sum_{i=1}^{a}\sum_{j=1}^{b} n_{ij}^2}{N}\sigma_\gamma^2 + \sigma_e^2 \tag{26}$$

and hence

$$E(\text{SSA}) = E(T_A) - E(T_\mu)$$

$$= \left(N - \frac{\sum_{i=1}^{a} n_{i\cdot}^2}{N}\right)\sigma_\alpha^2 + \left(\sum_{i=1}^{a}\frac{\sum_{j=1}^{b} n_{ij}^2}{n_{i\cdot}} - \frac{\sum_{j=1}^{b} n_{\cdot j}^2}{N}\right)\sigma_\beta^2$$

$$+ \left(\sum_{i=1}^{a}\frac{\sum_{j=1}^{b} n_{ij}^2}{n_{i\cdot}} - \frac{\sum_{i=1}^{a}\sum_{j=1}^{b} n_{ij}^2}{N}\right)\sigma_\gamma^2 + (a-1)\sigma_e^2 . \qquad (27)$$

Expected values of SSB and SSAB can be obtained in like fashion. Together with $E(\text{SSE}) = (N - s)\sigma_e^2$ the four expected values, when equated to their corresponding observed values, provide four equations in the four required variance components.

A noticeable aspect of (27) is that it has a non-zero coefficient for every variance component in the model, whereas with balanced data the comparable expected value contains no term in σ_β^2 [see $E(\text{MSA})$ in Table 9.8]. The term in σ_β^2 in (27) does, of course, reduce to zero for balanced data; i.e., when

$$n_{ij} = n, \qquad n_{i\cdot} = bn, \qquad n_{\cdot j} = an \qquad \text{and} \qquad n_{\cdot\cdot} = N = abn, \qquad (28)$$

the coefficient of σ_β^2 in (27) is

$$\sum_{i=1}^{a}\frac{\sum_{j=1}^{b} n_{ij}^2}{n_{i\cdot}} - \frac{\sum_{j=1}^{b} n_{\cdot j}^2}{N} = a\left(\frac{bn^2}{bn}\right) - \frac{ba^2n^2}{abn} = an - an = 0.$$

Similarly the coefficient of σ_α^2 in (27) becomes

$$N - \frac{\sum_{i=1}^{a} n_{i\cdot}^2}{N} = abn - \frac{ab^2n^2}{abn} = bn(a-1),$$

and that of σ_γ^2 reduces to $n(a-1)$. Hence (27) for balanced data becomes

$$E(\text{SSA}) = (a-1)(bn\sigma_\alpha^2 + n\sigma_\gamma^2 + \sigma_e^2)$$

as is implicit in Table 9.8.

(*ii*) *Mixed models*. Suppose that in the 2-way classification the A-factor is a fixed effects factor. Then the α_i's of the model are fixed effects, and the expected values of the SS-terms of (21) differ from their values under the random model. For example, in taking the expected value of (23) to obtain $E(T_A)$ we have, with the α's as fixed effects,

$$\begin{aligned} E(n_{i\cdot}\alpha_i^2) &= n_{i\cdot}\alpha_i^2 , \text{ and not } n_{i\cdot}\sigma_\alpha^2 \text{ as in (24)};\\ E(2\mu n_{i\cdot}\alpha_i) &= 2\mu n_{i\cdot}\alpha_i , \text{ and not 0 as in (24).} \end{aligned} \qquad (29)$$

Other terms in (23) involving α_i will have zero expectation, just as they did in the derivation of (24) but now for a different reason: for example, $E(\alpha_i\beta_j) = 0$ in (24) because the α's and β's were random variables with zero means and covariances. In the mixed model $E(\alpha_i\beta_j)$ is still equal to 0, but because $E(\alpha_i\beta_j) = \alpha_i E(\beta_j) = \alpha_i(0)$.

Equations (29) mean that in the mixed model, instead of the terms $N\mu^2 + N\sigma_\alpha^2$,

$$E(T_A) \text{ contains } N\mu^2 + \sum_{i=1}^{a} n_{i\cdot}\alpha_i^2 + 2\mu\sum_{i=1}^{a} n_{i\cdot}\alpha_i\,. \tag{30}$$

Similarly it can be shown that

$$E(T_\mu) \text{ contains } N\mu^2 + \frac{\left(\sum_{i=1}^{a} n_{i\cdot}\alpha_i\right)^2}{N} + 2\mu\sum_{i=1}^{a} n_{i\cdot}\alpha_i\,. \tag{31}$$

Therefore

$$E(\text{SSA}) = E(T_A) - E(T_\mu) \text{ contains } \sum_{i=1}^{a} n_{i\cdot}\alpha_i^2 - \frac{\left(\sum_{i=1}^{a} n_{i\cdot}\alpha_i\right)^2}{N} = \theta_1\,, \text{ say.} \tag{32}$$

Carrying through the same process for SSB shows that

$$E(\text{SSB}) = E(T_B) - E(T_\mu) \text{ contains} \sum_{j=1}^{b} \frac{\left(\sum_{i=1}^{a} n_{ij}\alpha_i\right)^2}{n_{\cdot j}} - \frac{\left(\sum_{i=1}^{a} n_{i\cdot}\alpha_i\right)^2}{N} = \theta_2, \text{ say.} \tag{33}$$

The important thing to notice here is that $\theta_1 \neq \theta_2$, so that $E(\text{SSA} - \text{SSB})$ is not free of the fixed effects in the way that $E(T_A - T_\mu)$ is of $N\mu^2$. This is true generally: in mixed models, expected values of the SS-terms contain functions of the fixed effects that cannot be eliminated by considering linear combinations of the terms. Thus the analysis of variance method cannot be used for mixed models.

There are two obvious ways of overcoming the above difficulty, but both are deviants from the true mixed model and must therefore be considered as unsatisfactory. The first is to ignore the fixed effects altogether and eliminate them from the model: what remains is a model that is completely random, for which estimation of the variance components can be made. The second possibility is to assume the fixed effects are in fact random, and then treat the model as if it were completely random. In the resulting estimation process, components for the fixed effect factors will be estimated and can be ignored. In using either of these possibilities we deal with random models, for which the estimation process is suitable. But the variance component estimators will, in both cases, be biased because their expectations under the true, mixed model will not equal the variance components of that model—they will include quadratic functions of the fixed effects. Despite this, if the models which these approximations invoke are in any way acceptable alternatives to the mixed model then the approximations may be of some use. Furthermore, they utilize the relatively easy arithmetic of the analysis of variance method, which is sometimes advantageous in face of the greater complexity of other analyses of mixed models (see Sec. 3).

(*iii*) *General results.* General rules for obtaining expected values of T-terms in random models are now developed. To do so we write the model as

$$\mathbf{y} = \mu\mathbf{1} + \sum_{\theta=A}^{K} \mathbf{X}_\theta \mathbf{b}_\theta + \mathbf{e} \tag{34}$$

which is (5) with $\mathbf{X}_1 = \mathbf{1}$ and $\mathbf{b}_1$ taken as the scalar μ. To derive $E(T_A)$ from (20) we define

$$\left.\begin{aligned} y.(A_i) &= \text{total of} \\ n(A_i) &= \text{number of} \end{aligned}\right\} \text{observations in the } i\text{th level of the } A\text{-factor}$$

and have, from (20),

$$T_A = \sum_{i=1}^{N_A} \frac{[y.(A_i)]^2}{n(A_i)}. \tag{35}$$

Now, just as in Sec. 6.5, define $n(A_i, \theta_j)$ as the number of observations in the ith level of the A-factor and the jth level of the θ-factor. Also define b_{θ_j} as the jth element of $\mathbf{b}_\theta$, and $e.(A_i)$ as the total of the error terms corresponding to $y.(A_i)$. Then using (34) in (35) gives

$$T_A = \sum_{i=1}^{N_A} \frac{\left[n(A_i)\mu + \sum_{\theta=A}^{K}\sum_{j=1}^{N_\theta} n(A_i, \theta_j) b_{\theta_j} + e.(A_i)\right]^2}{n(A_i)}. \tag{36}$$

Taking expected values of this gives (i) a term in μ^2:

$$\sum_{i=1}^{N_A} \frac{[n(A_i)]^2 \mu^2}{n(A_i)} = \mu^2 \sum_{i=1}^{N_A} n(A_i) = N\mu^2; \tag{37}$$

(ii) a term in σ_θ^2, for $\theta = A, B, \ldots, K$:

$$k(\sigma_\theta^2, T_A)\sigma_\theta^2 = \sum_{i=1}^{N_A} \frac{\sum_{j=1}^{N_\theta} [n(A_i, \theta_j)]^2}{n(A_i)} \sigma_\theta^2, \tag{38}$$

so defining $k(\sigma_\theta^2, T_A)$ as the coefficient of σ_θ^2 in $E(T_A)$; and (iii) a term in σ_e^2:

$$\sum_{=1}^{N_A} \frac{n(A_i)\sigma_e^2}{n(A_i)} = \sigma_e^2 \sum_{i=1}^{N_A} 1 = N_A \sigma_e^2. \tag{39}$$

Example. We derive $E(T_A)$ of (25) from (39). The terms $N\mu^2$ and $N_A\sigma_e^2$ need no demonstration. The others are

$$k(\sigma_\alpha^2, T_A)\sigma_\alpha^2 = \sum_{i=1}^{a} \frac{[n(\alpha_i, \alpha_i)]^2}{n(\alpha_i)} \sigma_\alpha^2 = \sum_{i=1}^{a} n_{i\cdot}\sigma_\alpha^2 = N\sigma_\alpha^2;$$

$$k(\sigma_\beta^2, T_A)\sigma_\beta^2 = \sum_{i=1}^{a} \frac{\sum_{j=1}^{b}[n(\alpha_i, \beta_j)]^2}{n(\alpha_i)} \sigma_\beta^2 = \sum_{i=1}^{a} \frac{\sum_{j=1}^{b} n_{ij}^2}{n_{i\cdot}} \sigma_\beta^2;$$

and
$$k(\sigma_\gamma^2, T_A)\sigma_\gamma^2 = \sum_{i=1}^{a} \frac{\sum_{j=1}^{b}[n(\alpha_i, \gamma_{ij})]^2}{n(\alpha_i)} \sigma_\gamma^2 = \sum_{i=1}^{a} \frac{\sum_{j=1}^{b} n_{ij}^2}{n_{i\cdot}} \sigma_\gamma^2 .$$

Similarly the terms in $E(T_\mu)$ are, for example, of the form:

$$k(\sigma_\alpha^2, T_\mu)\sigma_\alpha^2 = \frac{\sum_{i=1}^{a}[n(\alpha_i)]^2}{N} \sigma_\alpha^2 = \frac{\sum_{i=1}^{a} n_{i\cdot}^2}{N} \sigma_\alpha^2$$

as in (26).

(*iv*) *Calculation by "synthesis"*. Calculating coefficients of σ^2's in terms like E(SSA) and $E(T_A)$ without first requiring the algebraic form of these coefficients can be achieved by a method developed by Hartley (1967). The method applies to calculating coefficients of the σ^2's in expected values of any quadratic form that is homogeneous in the observations **y**, and it requires no distributional properties of the model. He has called it the method of "synthesis". We describe it in terms of calculating T_A of the preceding example.

Write T_A of (35) as

$$T_A = \sum_{i=1}^{N_A} \frac{[y.(A_i)]^2}{n(A_i)} = \mathbf{y}'\mathbf{Q}_A\mathbf{y} = T_A(\mathbf{y}) \tag{40}$$

and define
$$\mathbf{x}(\theta, j) = j\text{th column of } \mathbf{X}_\theta . \tag{41}$$

Then the method of synthesis derives $k(\sigma_\theta^2, T_A)$, the coefficient of σ_θ^2 in $E(T_A)$, as

$$k(\sigma_\theta^2, T_A) = \sum_{j=1}^{N_\theta} T_A[\mathbf{x}(\theta, j)]; \tag{42}$$

i.e., using each column of $\mathbf{X}_\theta$ as a column of data (all 0's and 1's) calculate T_A, and sum the results over all columns of $\mathbf{X}_\theta$. The sum is the coefficient of σ_θ^2 in $E(T_A)$, namely $k(\sigma_\theta^2, T_A)$ of (38).

This procedure can be used numerically without recourse to explicit algebraic forms of the coefficients $k(\sigma_\theta^2, T_A)$, and since it applies to any quadratic form in the place of T_A it can also be used directly on the SS-terms.

Thus, paraphrasing Hartley's words: we can apply the analysis of variance method in turn to each of the N_θ columns of $\mathbf{X}_\theta$ used as data. Single out a particular quadratic $f(\mathbf{y})$ and form the sum of the $f(\mathbf{y})$ over the N_θ analyses of variance, to obtain $k[\sigma_\theta^2, f(\mathbf{y})]$, the coefficient of σ_θ^2 in $E[f(\mathbf{y})]$. Carrying out $\sum_{\theta=A}^{K} N_\theta$ analyses of variance and summing them appropriately therefore gives all the coefficients of the σ^2's in the expected quadratics. Since many of the "observations" in these analyses will be zero, any computer procedure designed for this task should take account of this many-zeroed feature of the "data".

Equivalence of (42) to (38) is readily shown. The jth column of $\mathbf{X}_\theta$, namely $\mathbf{x}(\theta, j)$, has $n(\theta_j)$ ones in it and $N - n(\theta_j)$ zeros. Therefore, using $\mathbf{x}(\theta, j)$ as the vector $\mathbf{y}$ in $y_{\cdot}(A_i)$ of (40) we require the total of the "observations" in $\mathbf{x}(\theta, j)$ that are in the ith level of A. These "observations" will consist of $n(A_i, \theta_j)$ ones and $n(A_i) - n(A_i, \theta_j)$ zeros; their total is thus $n(A_i, \theta_j)$. Therefore from (40)

$$T_A[\mathbf{x}(\theta, j)] = \sum_{i=1}^{N_A} \frac{[n(A_i, \theta_j)]^2}{n(A_i)}$$

and summing this over j, as in (42), yields (38).

This method of "synthesis" can also be applied to calculating variances of variance component estimators [see Sec. 2d(iii) following], and it has been extended by Rao (1968) to general incidence matrices and to mixed models.

d. Sampling variances of estimators

The analogous sums of squares [in the manner of (14)–(17)] used in the analysis of variance method for unbalanced data are the SS-terms and they do not, under normality assumptions, have χ^2-distributions. Nor are they distributed independently of one another. The only sum of squares with a known distribution is SSE, which follows a χ^2-distribution in the usual manner and has zero covariance with the other SS-terms. $\hat{\sigma}_e^2 = \text{SSE}/(N - s)$ therefore has a similar distribution. The other estimators, which are linear functions of the SS-terms, have distributions that are unknown. Despite this, variances of these estimators, under normality assumptions, can be derived. Suppose we define

$\mathbf{c}$ = vector of SS-terms, but not SSE,

$\boldsymbol{\sigma}^2$ = vector of σ^2's, but not σ_e^2

and $\mathbf{f}$ = vector of "degrees of freedom", the coefficients of σ_e^2 in $E(\mathbf{c})$.

The vector of SS-terms is therefore $[\mathbf{c}' \quad \text{SSE}]$, and equating this to its expected value yields the variance components estimators. Suppose $\mathbf{P}$ is the matrix of coefficients of variance components (other than σ_e^2) in $E(\mathbf{c})$. Then we can write

$$E\begin{bmatrix}\mathbf{c}\\ \text{SSE}\end{bmatrix} = \begin{bmatrix}\mathbf{P} & \mathbf{f}\\ \mathbf{0} & N-s\end{bmatrix}\begin{bmatrix}\boldsymbol{\sigma}^2\\ \sigma_e^2\end{bmatrix} \tag{43}$$

and equating $\begin{bmatrix}\mathbf{c}\\ \text{SSE}\end{bmatrix}$ to its expected values gives the estimators

$$\hat{\sigma}_e^2 = \text{SSE}/(N-s)$$

and

$$\hat{\boldsymbol{\sigma}}^2 = \mathbf{P}^{-1}(\mathbf{c} - \hat{\sigma}_e^2\mathbf{f}). \tag{44}$$

These expressions provide a means for deriving variances of the estimators.

(*i*) *Derivation.* The distribution of SSE/σ_e^2 is χ^2_{N-s} with variance $2(N-s)$ and so, from (44),

$$v(\hat{\sigma}_e^2) = 2\sigma_e^4/(N-s). \tag{45}$$

Now SSE (and hence $\hat{\sigma}_e^2$) has zero covariance with every element of $\mathbf{c}$, i.e., with every other SS-term. Therefore, from (44)

$$\text{cov}(\hat{\boldsymbol{\sigma}}^2, \hat{\sigma}_e^2) = -\mathbf{P}^{-1}\mathbf{f}v(\hat{\sigma}_e^2) \tag{46}$$

and

$$\text{var}(\hat{\boldsymbol{\sigma}}^2) = \mathbf{P}^{-1}[\text{var}(\mathbf{c}) + v(\hat{\sigma}_e^2)\mathbf{f}\mathbf{f}']\mathbf{P}^{-1\prime}.$$

In addition, since the SS-terms are linear functions of the T's, we can, with

$$\mathbf{t} = \text{vector of } T\text{'s}, \qquad \text{write } \mathbf{c} = \mathbf{Ht} \tag{47}$$

for some matrix $\mathbf{H}$ (that is quite unrelated to $\mathbf{H} = \mathbf{GX'X}$ of previous chapters). In the case of the 2-way classification, for example, $\mathbf{H}$ is the matrix of the transformation of the T's to SSA, SSB and SSAB shown in (21). Hence

$$\text{var}(\mathbf{c}) = \mathbf{H}\,\text{var}(\mathbf{t})\mathbf{H}'$$

and

$$\text{var}(\hat{\boldsymbol{\sigma}}^2) = \mathbf{P}^{-1}[\mathbf{H}\,\text{var}(\mathbf{t})\mathbf{H}' + v(\hat{\sigma}_e^2)\mathbf{f}\mathbf{f}']\mathbf{P}^{-1\prime}. \tag{48}$$

This is the result derived in Searle (1958) and utilized in the general case by Blischke (1968). Its application in any particular situation requires obtaining only var($\mathbf{t}$), the variance-covariance matrix of the T's. $\mathbf{P}$ is the matrix of coefficients of the σ^2's in the expected values of the SS-terms, $\mathbf{H}$ is the matrix expressing the relationship between the SS-terms and the T's, and $\mathbf{f}$ is the vector of the "degrees of freedom" in the SS-terms, the coefficients of σ_e^2 in the expected values of the SS-terms.

Deriving elements of var($\mathbf{t}$) involves cumbersome algebra, although the basis of two different methods for doing so is quite straightforward. For both methods we assume normality, i.e., that

$$\mathbf{y} \sim N(\mu\mathbf{1}, \mathbf{V}), \tag{49}$$

and first show the manner in which μ^2 occurs in the variances and covariances of the T's. From (40) it can be shown that

$$\mathbf{Q}_A = \sum_{i=1}^{N_A}{}^{+} \frac{1}{n(A_i)}\mathbf{J}_{n(A_i)}; \tag{50}$$

i.e., that $\mathbf{Q}_A$ is a diagonal matrix of square matrices of order $n(A_i)$ with every element being $1/n(A_i)$. This kind of result applies not just to the A-factor but to every factor θ of the model (34). For two factors A and B we then have, from Chapter 2,

$$v(T_A) = 2\,\mathrm{tr}(\mathbf{VQ}_A)^2 + 4\mu^2\mathbf{1}'\mathbf{Q}_A\mathbf{VQ}_A\mathbf{1}$$

and a similar expression for $v(T_B)$; and

$$\mathrm{cov}(T_A, T_B) = 2\,\mathrm{tr}(\mathbf{VQ}_A\mathbf{VQ}_B) + 4\mu^2\mathbf{1}'\mathbf{Q}_A\mathbf{VQ}_B\mathbf{1}\,.$$

But from (50)

$$\mathbf{1}'\mathbf{Q}_A = \mathbf{1}'$$

with the same true of $\mathbf{Q}_B$ also, and so

$$v(T_A) = 2\,\mathrm{tr}(\mathbf{VQ}_A)^2 + 4\mu^2\mathbf{1}'\mathbf{V1}$$

and
$$\mathrm{cov}(T_A, T_B) = 2\,\mathrm{tr}(\mathbf{VQ}_A\mathbf{VQ}_B) + 4\mu^2\mathbf{1}'\mathbf{V1}.$$

Hence $4\mu^2\mathbf{1}'\mathbf{V1}$ is part of all the variances and covariances of the T's. However, because in $\mathbf{c} = \mathbf{Ht}$ the T's are used only in terms of differences between them, the $4\mu^2\mathbf{1}'\mathbf{V1}$ term in the above expressions can be ignored. This is equivalent to assuming $\mu = 0$ and it gives

$$v(T_A) = 2\,\mathrm{tr}(\mathbf{VQ}_A)^2 \tag{51}$$

and
$$\mathrm{cov}(T_A, T_B) = 2\,\mathrm{tr}(\mathbf{VQ}_A\mathbf{VQ}_B). \tag{52}$$

From these the elements of var($\mathbf{t}$) can be obtained, as has been done for several specific cases whose details are given in the next chapter.

Blischke (1966, 1968) obtains the same elements of var($\mathbf{t}$) by using the fact that for normal variables u and v

$$\mathrm{cov}(u^2, v^2) = 2[\mathrm{cov}(u, v)]^2$$

(see Exercise 23 of Chapter 2). Therefore, since T_A and T_B are weighted sums of squares of normally distributed random variables their covariance,

$$\mathrm{cov}(T_A, T_B) = \mathrm{cov}\left\{\sum_{i=1}^{N_A} \frac{[y_{.}(A_i)]^2}{n(A_i)},\quad \sum_{j=1}^{N_B} \frac{[y_{.}(B_j)]^2}{n(B_j)}\right\},$$

is, assuming $\mu = 0$,

$$\mathrm{cov}(T_A, T_B) = \sum_{i=1}^{N_A}\sum_{j=1}^{N_B} \frac{2\{\mathrm{cov}[y_{.}(A_i), y_{.}(B_j)]\}^2}{n(A_i)n(B_j)}\,.$$

A special case of this is

$$\mathrm{var}(T_A) = \sum_{i=1}^{N_A} \frac{2\{\mathrm{var}[y_{.}(A_i)]\}^2}{[n(A_i)]^2} + \sum_{i\neq i'}^{N_A} \frac{2\{\mathrm{cov}[y_{.}(A_i), y_{.}(A_{i'})]\}^2}{n(A_i)n(A_{i'})}\,.$$

Whether these expressions or their equivalent matrix forms (51) and (52) are used, the ensuing algebra for specific cases is cumbersome and tedious, as is evident from the results listed in the next chapter.

The extent of the elements in var(**t**) is one of the difficulties in deriving that matrix. An r-way classification random model, with all interactions (see Exercise 5), involves $2^{r-1}(2^r + 1)$ different elements in var(**t**), each element being a linear function of the same number of squares and products of variance components. Thus a square matrix of order $2^{r-1}(2^r + 1)$ of coefficients is involved. For $r = 2$, 3, 4 and 5 this matrix has order 10, 36, 136 and 528 respectively. Its elements for $r = 2$ and $r = 3$ are shown in Chapter 11.

(*ii*) *Estimation.* When the elements of var(**t**) have been derived from (51) and (52) they can be used in (48) to give var($\hat{\boldsymbol{\sigma}}^2$). However, the elements of var(**t**) are quadratic functions of the unknown variance components. The problem of estimating var($\hat{\boldsymbol{\sigma}}^2$) therefore remains. A common procedure is to replace the variance components in var($\hat{\boldsymbol{\sigma}}^2$) by their estimates and use the resulting value of var($\hat{\boldsymbol{\sigma}}^2$) as the estimator of var($\hat{\boldsymbol{\sigma}}^2$). As an estimator, this has no known desirable properties—other than being relatively easy to compute. A small numerical example is discussed in Searle (1961b).

Unbiased estimators of the variances and covariances of the variance components estimators, i.e., of (45), (46) and (48), can be derived as follows. First array (45), (46) and the elements of the upper triangular half of (48) in a vector **v**:

$$\mathbf{v} = \text{vector of variances and covariances of all } \hat{\sigma}^2\text{'s.}$$

Similarly array the squares of all the σ^2's and the products of every pair of them in another vector $\boldsymbol{\gamma}$:

$$\boldsymbol{\gamma} = \text{vector of all squares and products of } \sigma^2\text{'s;}$$

for example, in the 1-way classification with components σ_α^2 and σ_e^2,

$$\mathbf{v}' = [v(\hat{\sigma}_\alpha^2) \quad v(\hat{\sigma}_e^2) \quad \text{cov}(\hat{\sigma}_\alpha^2, \hat{\sigma}_e^2)]$$

and

$$\boldsymbol{\gamma}' = [\sigma_\alpha^4 \quad \sigma_e^4 \quad \sigma_\alpha^2\sigma_e^2].$$

Then, because of (45), (46) and (48), every element in **v** is a linear combination of the elements in $\boldsymbol{\gamma}$, and so, for some matrix **A** say,

$$\mathbf{v} = \mathbf{A}\boldsymbol{\gamma}. \tag{53}$$

With an r-way classification random model that has all possible interactions **A** of (53) has order $2^{r-1}(2^r + 1)$. However, **A** is not the matrix referred to at the end of subsection (*i*) where the different elements of var(**t**) were envisaged as a vector, $\mathbf{B}\boldsymbol{\gamma}$ say. In (53) it is $v(\hat{\sigma}_e^2)$ and the elements of cov($\hat{\boldsymbol{\sigma}}^2$, $\hat{\sigma}_e^2$), and var($\hat{\boldsymbol{\sigma}}^2$) being written as $\mathbf{A}\boldsymbol{\gamma}$. The matrices **A** and **B** have the same order but are not equal.

Unbiased estimation of $\mathbf{v}$ is derived from (53). First note that every variance component estimator in $\hat{\boldsymbol{\sigma}}^2$ of (44) is unbiased and so, for example, on writing $\hat{\sigma}_A^4$ for $(\hat{\sigma}_A^2)^2$ we have

$$E(\hat{\sigma}_A^4) = v(\hat{\sigma}_A^2) + \sigma_A^4 \,. \tag{54}$$

Similarly

$$E(\hat{\sigma}_A^2\hat{\sigma}_B^2) = \text{cov}(\hat{\sigma}_A^2, \hat{\sigma}_B^2) + \sigma_A^2\sigma_B^2 \,. \tag{55}$$

Writing $\hat{\boldsymbol{\gamma}}$ as the vector of squares and products of the $\hat{\sigma}^2$'s corresponding to $\boldsymbol{\gamma}$ we have, from (54) and (55), that

$$E(\hat{\boldsymbol{\gamma}}) = \mathbf{v} + \boldsymbol{\gamma}. \tag{56}$$

It will then be found that replacing $\boldsymbol{\gamma}$ in (53) by $\hat{\boldsymbol{\gamma}} - \hat{\mathbf{v}}$ and calling the resulting expression $\hat{\mathbf{v}}$ yields $\hat{\mathbf{v}}$ as an unbiased estimator of $\mathbf{v}$; i.e.,

$$\hat{\mathbf{v}} = \mathbf{A}(\hat{\boldsymbol{\gamma}} - \hat{\mathbf{v}}) \tag{57}$$

gives

$$\hat{\mathbf{v}} = (\mathbf{I} + \mathbf{A})^{-1}\mathbf{A}\hat{\boldsymbol{\gamma}} \tag{58}$$

as an unbiased estimator of $\mathbf{v}$. Utilizing (53) and (56) in taking the expected value of (58) shows that $E(\hat{\mathbf{v}}) = \mathbf{v}$. The elements of $\hat{\mathbf{v}}$ in (58) are therefore unbiased estimators of the variances and covariances of the analysis of variance estimators of the variance components.

The derivation of $\hat{\mathbf{v}}$ is described by Mahamunulu (1963) in terms of (57), namely of replacing every σ_A^4-term in (53) by $\hat{\sigma}_A^4 - \hat{v}(\hat{\sigma}_A^2)$ and every $\sigma_A^2\sigma_B^2$-term in (53) by $\hat{\sigma}_A^2\hat{\sigma}_B^2 - \widehat{\text{cov}}(\hat{\sigma}_A^2, \hat{\sigma}_B^2)$, and calling the resulting expression $\hat{\mathbf{v}}$. The result given in (58) is the form derived by Ahrens (1965).

The nature of (45) ensures that (58) yields

$$\hat{v}(\hat{\sigma}_e^2) = 2\hat{\sigma}_e^4/(N - s + 2). \tag{59}$$

This can, of course, be derived directly from (45) using the counterpart of (54) for σ_e^2. In the same way, (58) also yields

$$\widehat{\text{cov}}(\hat{\boldsymbol{\sigma}}^2, \hat{\sigma}_e^2) = -\mathbf{P}^{-1}\mathbf{f}\hat{v}(\hat{\sigma}_e^2) \tag{60}$$

as an unbiased estimator of (46). The remaining terms in $\hat{\mathbf{v}}$ are unbiased estimators of elements of $\text{var}(\hat{\boldsymbol{\sigma}}^2)$ of (48).

Example. The analysis of variance for the 1-way classification model is derived in Sec. 6.2d. Denoting SSR_m given there as SSA, we have

$$\text{SSA} = \sum_{i=1}^{a} n_i\bar{y}_{i\cdot}^2 - N\bar{y}_{\cdot\cdot}^2 \,, \qquad \text{with} \quad E(\text{SSA}) = (N - \sum n_i^2/N)\sigma_\alpha^2 + (a-1)\sigma_e^2$$

and

$$\text{SSE} = \sum_{i=1}^{a}\sum_{j=1}^{n} y_{ij}^2 - \sum_{i=1}^{a} n_i\bar{y}_{i\cdot}^2 \,, \quad \text{with} \quad E(\text{SSE}) = (N-a)\sigma_e^2 \,.$$

Therefore, from (43), **P** and **f** are scalars,

$$\mathbf{P} = N - \sum n_i^2/N \quad \text{and} \quad \mathbf{f} = a - 1. \tag{61}$$

The estimators are

$$\hat{\sigma}_e^2 = \frac{\text{SSE}}{N - a} \quad \text{and} \quad \hat{\sigma}_\alpha^2 = \frac{\text{SSA} - (a-1)\hat{\sigma}_e^2}{N - \sum n_i^2/N}. \tag{62}$$

The variances and covariance of these estimators are [see Crump (1947) and Searle (1956)]

$$\begin{aligned} v(\hat{\sigma}_e^2) &= k_1\sigma_e^4, \quad \text{for} \quad k_1 = 2/(N-a) \\ \text{cov}(\hat{\sigma}_\alpha^2, \hat{\sigma}_e^2) &= k_2\sigma_e^4, \quad \text{for} \quad k_2 = -2(a-1)/[(N-a)(N-S_2/N)] \\ \text{and} \qquad v(\hat{\sigma}_\alpha^2) &= k_3\sigma_e^4 + k_4\sigma_e^2\sigma_\alpha^2 + k_5\sigma_\alpha^4 \end{aligned} \tag{63}$$

with

$$k_3 = \frac{2N^2(N-1)(a-1)}{(N^2-S_2)^2(N-a)}, \quad k_4 = \frac{4N}{N^2 - S_2} \quad \text{and} \quad k_5 = \frac{2(N^2S_2 + S_2^2 - 2NS_3)}{(N^2 - S_2)^2}.$$

where $S_2 = \sum_{i=1}^{a} n_i^2$ and $S_3 = \sum_{i=1}^{a} n_i^3$. Therefore (53) is

$$\begin{bmatrix} v(\hat{\sigma}_e^2) \\ \text{cov}(\hat{\sigma}_\alpha^2, \hat{\sigma}_e^2) \\ v(\hat{\sigma}_\alpha^2) \end{bmatrix} = \begin{bmatrix} k_1 & 0 & 0 \\ k_2 & 0 & 0 \\ k_3 & k_4 & k_5 \end{bmatrix} \begin{bmatrix} \sigma_e^4 \\ \sigma_\alpha^2\sigma_e^2 \\ \sigma_\alpha^4 \end{bmatrix}$$

and so (58) is

$$\begin{aligned} &\begin{bmatrix} \hat{v}(\hat{\sigma}_e^2) \\ \widehat{\text{cov}}(\hat{\sigma}_\alpha^2, \hat{\sigma}_e^2) \\ \hat{v}(\hat{\sigma}_\alpha^2) \end{bmatrix} \\ &= \begin{bmatrix} 1 + k_1 & 0 & 0 \\ k_2 & 1 & 0 \\ k_3 & k_4 & 1 + k_5 \end{bmatrix}^{-1} \begin{bmatrix} k_1 & 0 & 0 \\ k_2 & 0 & 0 \\ k_3 & k_4 & k_5 \end{bmatrix} \begin{bmatrix} \hat{\sigma}_e^4 \\ \hat{\sigma}_e^2\hat{\sigma}_\alpha^2 \\ \hat{\sigma}_\alpha^4 \end{bmatrix} \\ &= \frac{1}{(1+k_1)(1+k_5)} \begin{bmatrix} k_1(1+k_5) & 0 & 0 \\ k_2(1+k_5) & 0 & 0 \\ k_3 - k_2k_4 & k_4(1+k_1) & k_5(1+k_1) \end{bmatrix} \begin{bmatrix} \hat{\sigma}_e^4 \\ \hat{\sigma}_e^2\hat{\sigma}_\alpha^2 \\ \hat{\sigma}_\alpha^4 \end{bmatrix}. \end{aligned} \tag{64}$$

From (64)

$$\hat{v}(\hat{\sigma}_e^2) = \frac{k_1(1+k_5)\hat{\sigma}_e^4}{(1+k_1)(1+k_5)} = \frac{k_1\hat{\sigma}_e^4}{1+k_1} = \frac{2\sigma_e^4}{N - a + 2}$$

on substituting for k_1, a result that is in keeping with (59). Similarly, from (64)

$$\widehat{\text{cov}}(\hat{\sigma}_\alpha^2, \hat{\sigma}_e^2) = \frac{k_2}{1 + k_1}\,\hat{\sigma}_e^4 = \frac{k_2}{k_1}\left(\frac{k_1\hat{\sigma}_e^4}{1 + k_1}\right) = \frac{k_2}{k_1}\,\hat{v}(\hat{\sigma}_e^2),$$

which agrees with (60) because, from (63) and (61), $k_2/k_1 = -\mathbf{P}^{-1}\mathbf{f}$ of (60).

(*iii*) *Calculation by synthesis.* The "synthesis" method of calculating numerical coefficients of σ^2's in expected values of quadratic forms has been described in Sec. 2c(iv). It can also be applied to calculating coefficients of squares and products of σ^2's in variances and covariances. We give the procedure for obtaining $E(T_AT_B)$ from which $\text{cov}(T_A, T_B)$ can then be obtained, using $E(T_A)$ and $E(T_B)$ based upon (42).

We first write $\mathbf{e} = \mathbf{X}_0\mathbf{b}_0$ with $\mathbf{X}_0 = \mathbf{I}$ and $\mathbf{b}_0 = \mathbf{e}$ so that the model (34) becomes $\mathbf{y} = \mu\mathbf{1} + \sum_{\theta=0}^{K} \mathbf{X}_\theta\mathbf{b}_\theta$. Then Hartley (1967) derives $E(T_AT_B)$ in the form

$$E(T_AT_B) = \sum_{\theta,\varphi=0}^{K} k(\sigma_\theta^2\sigma_\varphi^2, T_AT_B)\sigma_\theta^2\sigma_\varphi^2 + \sum_{\theta=0}^{K} h(\mu_{4,\theta}, T_AT_B)\mu_{4,\theta}$$

where, by definition

$$\mu_{4,0} = E(e_i^4) \qquad \text{for} \quad i = 1, 2, \ldots, N$$

and, for $\theta = A, B, \ldots, K$

$$\mu_{4,\theta} = E(b_{\theta_j}^4) \qquad \text{for} \quad j = 1, 2, \ldots, N_\theta\,.$$

With these definitions the coefficients in $E(T_AT_B)$ given by Hartley (1967) are

$$\begin{aligned} h(\mu_{4,\theta}, T_AT_B) &= \text{coefficient of } \mu_{4,\theta} \text{ in } E(T_AT_B) \\ &= \sum_{j=1}^{N\theta} T_A[\mathbf{x}(\theta, j)]T_B[\mathbf{x}(\theta, j)], \end{aligned} \tag{65}$$

$$\begin{aligned} k(\sigma_\theta^4, T_AT_B) &= \text{coefficient of } \sigma_\theta^4 \text{ in } E(T_AT_B) \\ &= \sum_{j=1}^{N\theta}\sum_{j'<j}^{N\theta} T_A[\mathbf{x}(\theta, j) + \mathbf{x}(\theta, j')]T_B[\mathbf{x}(\theta, j) + \mathbf{x}(\theta, j')] \\ &\qquad - (N_\theta - 5)h(\mu_{4,\theta}, T_AT_B), \end{aligned} \tag{66}$$

$$\begin{aligned} k(\sigma_\theta^2\sigma_\varphi^2, T_AT_B) &= \text{coefficient of } \sigma_\theta^2\sigma_\varphi^2 \text{ in } E(T_AT_B) \\ &= \frac{1}{2}\sum_{j=1}^{N\theta}\sum_{j'=1}^{N\varphi} T_A[\mathbf{x}(\theta, j) + \mathbf{x}(\varphi, j')]T_B[\mathbf{x}(\theta, j) + \mathbf{x}(\varphi, j')] \\ &\qquad - N_\theta h(\mu_{4,\varphi}, T_AT_B) - N_\varphi h(\mu_{4,\theta}, T_AT_B). \end{aligned} \tag{67}$$

Thus for $h(\mu_{4,\theta}, T_AT_B)$ we use columns of $\mathbf{X}_\theta$ as "data" vectors in T_A and T_B. In $k(\sigma_\theta^4, T_AT_B)$ we add pairs of different columns of $\mathbf{X}_\theta$ and use the sums as "data" vectors in T_A and T_B. And for $k(\sigma_\theta^2\sigma_\varphi^2, T_AT_B)$ we add, in all possible combinations, a column of $\mathbf{X}_\theta$ and a column of $\mathbf{X}_\varphi$ and use these sums as "data" vectors in T_A and T_B. These results are quite general and apply to any quadratic forms of the observations, including the use of T_A in place of T_B to obtain $E(T_A^2)$ and hence $v(T_A)$. Furthermore, the results are all in terms of variances and fourth moments, and no particular form of distribution has been assumed for the random variables. The formulae are well suited computationally for obtaining coefficients, numerically, in specific situations, although with large amounts of data the calculations would be extensive. They could also be used to find coefficients algebraically, although in most cases the details involved would be quite tedious. A simple example follows.

Example. Hartley (1967) illustrates his results by finding the variance of

$$s^2 = s^2(x) = \left(\sum_{i=1}^{n} x_i^2 - n\bar{x}^2\right) \Big/ (n-1)$$

as

$$v(s^2) = E(s^2s^2) - \sigma^4$$

where

$$E(s^2s^2) = k_{00}\sigma^4 + h_0\mu_{4,0}\,.$$

By (65)

$$h_0 = \sum_{j=1}^{N_0} \{s^2[\mathbf{x}(0,j)]\}^2$$

$$= \sum_{j=1}^{n} [s^2(\text{column of } \mathbf{I}_n)]^2$$

$$= n\left[\frac{1-n(1/n)^2}{n-1}\right]^2 = \frac{1}{n}.$$

And by (66)

$$k_{00} = \sum_{j=1}^{N_0} \sum_{j'<j}^{N_0} \{s^2[\mathbf{x}(0,j) + \mathbf{x}(0,j')]\}^2 - (N_0 - 5)h_0$$

$$= \sum_{j=1}^{n} \sum_{j'<j}^{n} [s^2(\text{sum of 2 columns of } \mathbf{I}_n)]^2 - (n-5)/n$$

$$= \frac{(n-1)n}{2}\left[\frac{2-n(2/n)^2}{(n-1)}\right]^2 - \frac{n-5}{n} = \frac{n^2-2n+3}{n(n-1)}.$$

Hence

$$v(s^2) = k_{00}\sigma^4 + h_0\mu_{4,0} - \sigma^4$$

$$= \left[\frac{n^2-2n+3}{n(n-1)} - 1\right]\sigma^4 + \frac{\mu_{4,0}}{n} = \frac{3-n}{n(n-1)}\,\sigma^4 + \frac{\mu_{4,0}}{n}$$

as can be obtained directly. With normality assumptions $\mu_{4,0} = 3\sigma^4$ and the result reduces to the familiar $v(s^2) = 2\sigma^4/(n-1)$.

3. ADJUSTING FOR BIAS IN MIXED MODELS

We indicated in Sec. 2c(ii) that with unbalanced data the analysis of variance method for mixed models leads to biased estimators of variance components. There is, of course, a dual problem with mixed models—estimation of both the fixed effects and the variance components of the random effects. We here confine attention to estimating just the variance components. In some situations this is exactly what might be done in practice; with genetic data, for example, effects that are often considered fixed, such as year effects, might be of little interest compared to the genetic variance components. On the other hand, if trends in the year effects were of interest, their estimation together with that of the variance components would be considered simultaneously. This dual estimation problem is considered subsequently.

The method known as Method 2 in Henderson (1953) first uses the data to estimate the fixed effects of the model. The data are then adjusted by these estimators and the variance components are estimated from the data so adjusted. The whole procedure is designed so that the resulting variance components estimators are not biased by the presence of the fixed effects in the model, as are the analysis of variance estimators. The method certainly provides unbiased estimators but, as has been shown by Searle (1968), the method is not uniquely defined. Furthermore, certain simplified forms of it, of which Henderson's Method 2 is one special case, cannot be used whenever the model includes interactions between the fixed effects and the random effects. These points we now consider, closely following Searle (1968), the details of which are not repeated here.

a. General method

We consider the general model (34) in the form

$$\mathbf{y} = \mu\mathbf{1} + \mathbf{X}_f\mathbf{b}_f + \mathbf{X}_r\mathbf{b}_r + \mathbf{e} \tag{68}$$

where all fixed effects other than μ are represented by $\mathbf{b}_f$ and all random effects by $\mathbf{b}_r$. We take $E(\mathbf{b}_r) = \mathbf{0}$ and so $E(\mathbf{b}_r\mathbf{b}_r') = \text{var}(\mathbf{b}_r)$, the variance-covariance matrix of the random effects. Suppose an estimator of the fixed effects $\mathbf{b}_f$ is $\tilde{\mathbf{b}}_f = \mathbf{Ly}$. Then $\mathbf{z} = \mathbf{y} - \mathbf{X}_f\tilde{\mathbf{b}}_f$ is a vector of the data adjusted by the estimator $\tilde{\mathbf{b}}_f$. Substitution from (68) shows that the model for $\mathbf{z}$ contains no terms in $\mathbf{b}_f$ provided $\mathbf{L}$ is a generalized inverse of $\mathbf{X}_f$. Under this condition the analysis of variance method applied to $\mathbf{z}$ will yield unbiased estimators of the variance components. However, the fact that $\mathbf{L}$ has only to be a generalized inverse of $\mathbf{X}_f$ indicates the lack of uniqueness in the method.

b. A simplification

The calculations involved in applying the analysis of variance method to $\mathbf{y}$, particularly those involving the random effects $\mathbf{X}_r\mathbf{b}_r$, have been documented in the preceding section. In $\mathbf{z} = \mathbf{y} - \mathbf{X}_f\tilde{\mathbf{b}}_f$ the term in the random effects is $\mathbf{X}_r - \mathbf{X}_f\mathbf{L}\mathbf{X}_r$. Were $\mathbf{X}_f\mathbf{L}\mathbf{X}_r$ to be null, applying the analysis of variance method to $\mathbf{z}$ would, so far as random effects are concerned, be the same as applying the method to $\mathbf{y}$. More specifically, suppose we choose $\mathbf{L}$ such that the model for $\mathbf{z}$ is

$$\mathbf{z} = \mu^*\mathbf{1} + \mathbf{X}_r\mathbf{b}_r + \mathbf{Z}\mathbf{e}, \tag{69}$$

for μ^* being a scalar (not necessarily equal to μ) and for $\mathbf{Z}$ being some matrix. The analysis of variance method applied to (69) would then involve no fixed effects and, although treatment of the error terms in (69) would differ from that of the error terms in (68), treatment of the random effects would be the same as when using (68). Apart from calculations relating to σ_e^2, therefore, using the analysis of variance method on (69) would be the same as using it on (68) with the fixed effects ignored. To achieve this it has been shown (Searle, 1968) that $\mathbf{L}$ need not be a generalized inverse of $\mathbf{X}_f$ but has to satisfy three conditions:

$$\mathbf{X}_f\mathbf{L}\mathbf{X}_r = \mathbf{0}, \tag{70}$$

$$\mathbf{X}_f\mathbf{L} \text{ having its row sums equal} \tag{71}$$

and

$$\mathbf{X}_f - \mathbf{X}_f\mathbf{L}\mathbf{X}_f \text{ having all its rows the same.} \tag{72}$$

Although the non-unique condition on $\mathbf{L}$, that $\mathbf{X}_f\mathbf{L}\mathbf{X}_f = \mathbf{X}_f$, has been replaced by these three conditions, they do not necessarily determine $\mathbf{L}$ uniquely. Furthermore, an implication of these conditions is that the model for $\mathbf{y}$ must not contain interactions between fixed and random effects. This is a severe limitation on the method.

c. A special case: Henderson's Method 2

The procedure described by Henderson (1953) as Method 2 is simply one specific way in which the simpler form of the generalized method can be carried out. That is, Henderson's Method 2 estimates $\mathbf{b}_f$ as $\tilde{\mathbf{b}}_f = \mathbf{L}\mathbf{y}$ using an $\mathbf{L}$ that satisfies (70), (71) and (72), and then uses the analysis of variance method on $\mathbf{y} - \mathbf{X}_f\tilde{\mathbf{b}}_f$. Through being just one way of executing the simpler form of the generalized method, Henderson's Method 2 suffers from the limitation already alluded to, that it cannot be used whenever the model contains interactions between fixed and random effects. Although this is not stated explicitly by Henderson (1953) it is true of his example ,wherein the fixed effects in a study of dairy production records were years and the random effects were herds, sires and herd-by-sire interactions. There were no interactions of years with herds and/or sires.

In using Henderson's Method 2 we first estimate $\mathbf{b}_f$ by least squares assuming, temporarily and for this purpose only, that $\mu = 0$ and that the random effects are fixed. This leads to the equations

$$\mathbf{X}'\mathbf{X}\begin{bmatrix}\tilde{\mathbf{b}}_f \\ \tilde{\mathbf{b}}_r\end{bmatrix} = \begin{bmatrix}\mathbf{X}_f'\mathbf{X}_f & \mathbf{X}_f'\mathbf{X}_r \\ \mathbf{X}_r'\mathbf{X}_f & \mathbf{X}_r'\mathbf{X}_r\end{bmatrix}\begin{bmatrix}\tilde{\mathbf{b}}_f \\ \tilde{\mathbf{b}}_r\end{bmatrix} = \begin{bmatrix}\mathbf{X}_f'\mathbf{y} \\ \mathbf{X}_r'\mathbf{y}\end{bmatrix}. \tag{73}$$

It is the manner in which (73) is solved that leads to the solution $\tilde{\mathbf{b}}_f$ being $\tilde{\mathbf{b}}_f = \mathbf{L}\mathbf{y}$ with $\mathbf{L}$ satisfying (70), (71) and (72). The essential part of the solution is picking a generalized inverse of $\mathbf{X}'\mathbf{X}$ in the manner described at the end of Sec. 1.1, doing it in such a way that in striking out rows and columns of $\mathbf{X}'\mathbf{X}$ to reduce it to full rank as many as possible must be rows and columns through $\mathbf{X}_f'\mathbf{X}_f$. Details of this process and the reasons for its satisfying (70)–(72) are given in Searle (1968). Despite being able to specify the method in this manner it nevertheless suffers from the deficiencies already alluded to: it is not uniquely specified and it cannot be used in the presence of interactions between fixed and random effects. Hence its use is not recommended.

4. FITTING CONSTANTS METHOD (HENDERSON'S METHOD 3)

Fitting the linear models of Chapters 5–8 is often referred to as the technique of fitting constants, as mentioned in Chapter 4, because the effects of fixed effects models are sometimes called constants. A third method of estimating variance components that we now describe is based on the fitting of these models and is accordingly called the *fitting constants method*, or Henderson's Method 3, after Henderson (1953). For whatever model is being used the method uses reductions in sums of squares due to fitting both this model and different sub-models thereof, in the manner of Chapters 6, 7 and 8. These reductions, the $R(\ )$-terms of those chapters, are used in exactly the same manner as are the SS-terms, the analogous sums of squares of the analysis of variance method, namely estimating the variance components by equating each computed reduction to its expected value—its expected value under the full model. We describe the general properties of the method and then illustrate its application in the 2-way classification. The presentation follows closely that of Searle (1968).

a. General properties

We rewrite the general model $\mathbf{y} = \mathbf{X}\mathbf{b} + \mathbf{e}$ as

$$\mathbf{y} = \mathbf{X}_1\mathbf{b}_1 + \mathbf{X}_2\mathbf{b}_2 + \mathbf{e}, \tag{74}$$

where the partitioning simply divides $\mathbf{b}$ into two groups of effects, $\mathbf{b}_1$ and $\mathbf{b}_2$, with no thought for whether the groups represent fixed or random effects. This is considered subsequently. The reduction in sum of squares due to fitting this model will be denoted by $R(\mathbf{b}_1, \mathbf{b}_2)$. For the moment we are concerned with finding the expected values of $R(\mathbf{b}_1, \mathbf{b}_2)$ and of the reduction in sum of squares due to fitting the sub-model

$$\mathbf{y} = \mathbf{X}_1\mathbf{b}_1 + \mathbf{e}. \tag{75}$$

Both expectations will be taken under the full model, (74).

Denoting the reduction in sum of squares due to fitting (75) by $R(\mathbf{b}_1)$, we write

$$R(\mathbf{b}_2 \mid \mathbf{b}_1) = R(\mathbf{b}_1, \mathbf{b}_2) - R(\mathbf{b}_1) \tag{76}$$

in the manner of Sec. 6.3a. We will show that the expected value of (76) under the model (74) involves only σ_e^2 and

$$E(\mathbf{b}_2\mathbf{b}_2') = \text{var}(\mathbf{b}_2) + E(\mathbf{b}_2)E(\mathbf{b}_2'), \tag{77}$$

and it does not involve $\mathbf{b}_1$. Consequently the fitting constants method, by judicious choice of sub-models represented by $\mathbf{b}_1$ in (76), yields unbiased estimators of the variance components of the full model, estimators that are uncomplicated by any fixed effects that may be in the model.

First we slightly modify equation (2) for $E(\mathbf{y}'\mathbf{Q}\mathbf{y})$. In the general model $\mathbf{y} = \mathbf{X}\mathbf{b} + \mathbf{e}$ the vector $\mathbf{b}$ can be fixed, random or mixed. Adopting the convention that for a fixed effect $E(b_i) = b_i$ enables $E(\mathbf{b})$ to be defined whatever the nature of $\mathbf{b}$, and so from (2)

$$\begin{aligned} E(\mathbf{y}'\mathbf{Q}\mathbf{y}) &= \text{tr}[\mathbf{Q}\{\mathbf{X}\,\text{var}(\mathbf{b})\mathbf{X}' + \sigma_e^2\mathbf{I}\}] + E(\mathbf{b}')\mathbf{X}'\mathbf{Q}\mathbf{X}E(\mathbf{b}) \\ &= \text{tr}[\mathbf{X}'\mathbf{Q}\mathbf{X}E(\mathbf{b}\mathbf{b}')] + \sigma_e^2\,\text{tr}(\mathbf{Q}). \end{aligned}$$

In this form $E(\mathbf{y}'\mathbf{Q}\mathbf{y})$ is suitable for considering the models (74) and (75).

In fitting (74) the reduction in sum of squares is, as in equation (14) of Sec. 5.2f,

$$R(\mathbf{b}_1, \mathbf{b}_2) = \mathbf{y}'\mathbf{X}(\mathbf{X}'\mathbf{X})^-\mathbf{X}'\mathbf{y} \tag{78}$$

where $(\mathbf{X}'\mathbf{X})^-$ is a generalized inverse of $\mathbf{X}'\mathbf{X}$. Taking the expectation of (78) gives

$$\begin{aligned} ER(\mathbf{b}_1, \mathbf{b}_2) &= \text{tr}\{(\mathbf{X}'\mathbf{X})E(\mathbf{b}\mathbf{b}')\} + \sigma_e^2 r(\mathbf{X}) \\ &= \text{tr}\left\{\begin{bmatrix} \mathbf{X}_1'\mathbf{X}_1 & \mathbf{X}_1'\mathbf{X}_2 \\ \mathbf{X}_2'\mathbf{X}_1 & \mathbf{X}_2'\mathbf{X}_2 \end{bmatrix} E(\mathbf{b}\mathbf{b}')\right\} + \sigma_e^2 r(\mathbf{X}). \end{aligned}$$

Similarly, when fitting (75) the reduction in sum of squares is

$$R(\mathbf{b}_1) = \mathbf{y}'\mathbf{X}_1(\mathbf{X}_1'\mathbf{X}_1)^-\mathbf{X}_1'\mathbf{y},$$

with $$ER(\mathbf{b}_1) = \text{tr}\{\mathbf{X}'\mathbf{X}_1(\mathbf{X}_1'\mathbf{X}_1)^-\mathbf{X}_1'\mathbf{X}E(\mathbf{bb}')\} + \sigma_e^2 r(\mathbf{X}_1)$$

$$= \text{tr}\left\{\begin{bmatrix}\mathbf{X}_1'\mathbf{X}_1 \\ \mathbf{X}_2'\mathbf{X}_1\end{bmatrix}(\mathbf{X}_1'\mathbf{X}_1)^-[\mathbf{X}_1'\mathbf{X}_1 \quad \mathbf{X}_1'\mathbf{X}_2]E(\mathbf{bb}')\right\} + \sigma_e^2 r(\mathbf{X}_1)$$

$$= \text{tr}\left\{\begin{bmatrix}\mathbf{X}_1'\mathbf{X}_1 & \mathbf{X}_1'\mathbf{X}_2 \\ \mathbf{X}_2'\mathbf{X}_1 & \mathbf{X}_2'\mathbf{X}_1(\mathbf{X}_1'\mathbf{X}_1)^-\mathbf{X}_1'\mathbf{X}_2\end{bmatrix}E(\mathbf{bb}')\right\} + \sigma_e^2 r(\mathbf{X}_1).$$

Hence the expected value of $R(\mathbf{b}_2 \mid \mathbf{b}_1)$ is

$$E[R(\mathbf{b}_2 \mid \mathbf{b}_1)] = E[R(\mathbf{b}_1, \mathbf{b}_2) - R(\mathbf{b}_1)]$$
$$= \text{tr}\{\mathbf{X}_2'[\mathbf{I} - \mathbf{X}_1(\mathbf{X}_1'\mathbf{X}_1)^-\mathbf{X}_1']\mathbf{X}_2 E(\mathbf{b}_2\mathbf{b}_2')\} + \sigma_e^2[r(\mathbf{X}) - r(\mathbf{X}_1)]. \quad (79)$$

As forecast, the only **b**-term involved here is $\mathbf{b}_2$; i.e., the expectation of $R(\mathbf{b}_2 \mid \mathbf{b}_1)$ is a function simply of $E(\mathbf{b}_2\mathbf{b}_2')$ and σ_e^2. It involves neither $E(\mathbf{b}_1\mathbf{b}_1')$ nor $E(\mathbf{b}_1\mathbf{b}_2')$. Note, too, that this result has been derived without any assumptions on the form of $E(\mathbf{bb}')$.

The consequences of (79) are important. It means that if the **b**-vector of one's model can be partitioned into two parts $\mathbf{b}_1$ and $\mathbf{b}_2$ where $\mathbf{b}_2$ contains just random effects, then $ER(\mathbf{b}_2 \mid \mathbf{b}_1)$ as given in (79) contains only σ_e^2 and the variance components relating to those random effects. Thus, when $\mathbf{b}_1$ represents all the fixed effects, $ER(\mathbf{b}_2 \mid \mathbf{b}_1)$ contains no terms due to those fixed effects. This is the value of the method of fitting constants to the mixed model: it yields estimates of the variance components unaffected by the fixed effects. Furthermore, in the random model, where $\mathbf{b}_1$ contains random effects, $ER(\mathbf{b}_2 \mid \mathbf{b}_1)$ contains no terms arising from $\text{var}(\mathbf{b}_1)$ nor, more importantly, any terms arising from any covariance between the elements of $\mathbf{b}_1$ and $\mathbf{b}_2$. Hence, even if the model is such that terms in $\mathbf{b}_1$ are correlated with terms in $\mathbf{b}_2$ the expectation in (79) does not involve this correlation—it depends solely on the second moments of the elements in $\mathbf{b}_2$ (and on σ_e^2).

Compared with the analysis of variance method the immediate importance of the fitting constants method lies in its appropriateness for the mixed model, for which it yields variance component estimators that are unbiased by the fixed effects. It is therefore the preferred method for mixed models. Its disadvantage is that it involves calculating generalized inverses of matrices that will be very large in models having large numbers of effects in them, a difficulty that can arise in calculating not only reductions in sums of squares but also coefficients of the σ^2's in their expectations. Hartley's (1967) method of synthesis, described in Sec. 2d(iii), can be used as one means of calculation, and other available short cuts are described by Gaylor *et al.* (1970).

Application of the method to the 2-way classification is now considered.

b. The 2-way classification

The equation of the 2-way classification interaction model is shown in (12). Reductions in the sum of squares for fitting the fixed effects version of this model, and sub-models of it, are arrayed in Table 7.8. Included there are

$$\begin{aligned} R(\alpha \mid \mu) &= R(\mu, \alpha) &&- R(\mu) \\ R(\beta \mid \mu, \alpha) &= R(\mu, \alpha, \beta) &&- R(\mu, \alpha) \\ R(\gamma \mid \mu, \alpha, \beta) &= R(\mu, \alpha, \beta, \gamma) &&- R(\mu, \alpha, \beta) \\ \text{and} \quad \text{SSE} &= \textstyle\sum y^2 &&- R(\mu, \alpha, \beta, \gamma). \end{aligned} \tag{80}$$

These terms can be used in the fitting constants method of estimating variance components in a mixed or random effects version of the 2-way classification model. To do so requires their expected values.

(*i*) *Expected values.* The expected value of SSE in (80) is, as usual, $(N - s)\sigma_e^2$. Taking $R(\gamma \mid \mu, \alpha, \beta)$ next, its expected value can be derived from (79). But expected values of $R(\alpha \mid \mu)$ and $R(\gamma \mid \mu, \alpha, \beta)$ cannot be obtained directly from (79). This is because (79) is the expected value of $R(\mathbf{b}_2 \mid \mathbf{b}_1) = R(\mathbf{b}_1, \mathbf{b}_2) - R(\mathbf{b}_1)$, which is the difference between two $R(\cdot)$-terms one of which is for the full model and the other of which is for a sub-model. This is the only kind of $R(\cdot \mid \cdot)$-term to which (79) applies; $R(\gamma \mid \mu, \alpha, \beta)$ of (80) is an example. In contrast, (79) does not apply to $R(\cdot \mid \cdot)$-terms that are differences between two $R(\cdot)$-terms that are both for sub-models. For this reason, with the full model involving μ, α, β and γ, (79) does not apply to $R(\alpha \mid \mu)$ and $R(\beta \mid \mu, \alpha)$ of (80).

Although (79) cannot be used directly on $R(\alpha \mid \mu)$ and $R(\beta \mid \mu, \alpha)$ it can be utilized by considering certain sums of the terms in (80) that involve $R(\alpha \mid \mu)$ and $R(\beta \mid \mu, \alpha)$. For example, (79) applies to

$$R(\alpha, \beta, \gamma \mid \mu) = R(\mu, \alpha, \beta, \gamma) - R(\mu) \tag{81}$$

which is the sum of the first three terms in (80):

$$R(\alpha, \beta, \gamma \mid \mu) = R(\alpha \mid \mu) + R(\beta \mid \mu, \alpha) + R(\gamma \mid \mu, \alpha, \beta).$$

Similarly, (79) applies to

$$R(\beta, \gamma \mid \mu, \alpha) = R(\mu, \alpha, \beta, \gamma) - R(\mu, \alpha) \tag{82}$$

which is

$$R(\beta, \gamma \mid \mu, \alpha) = R(\beta \mid \mu, \alpha) + R(\gamma \mid \mu, \alpha, \beta).$$

Equating observed values of $R(\cdot \mid \cdot)$-terms to their expected values to obtain variance component estimators, using (81) and (82) in place of $R(\alpha \mid \mu)$ and $R(\beta \mid \mu, \alpha)$ of (80), yields equations that are linear combinations of those that would arise from using (80). The estimators will therefore be the same. The form taken by the expected values of these reductions, (81), (82) and the last two terms of (80), is shown in Table 10.1. So also are computing formulae for the reductions. These, and the h-coefficients of the σ^2's [which would be derived from (79)], are discussed subsequently. The coefficients of the σ_e^2's have already been obtained from (79).

TABLE 10.1. REDUCTIONS IN SUM OF SQUARES FOR ESTIMATING VARIANCE COMPONENTS IN A 2-WAY CLASSIFICATION INTERACTION, RANDOM MODEL, UNBALANCED DATA

Reduction in sum of squares	Computing formula[1]	Expected values[2]
$R(\alpha,\beta,\gamma \mid \mu) = R(\mu,\alpha,\beta,\gamma) - R(\mu)$	$= T_{AB} - T_\mu$	$h_1\sigma_\alpha^2 + h_2\sigma_\beta^2 + h_3\sigma_\gamma^2 + (s-1)\sigma_e^2$
$R(\beta,\gamma \mid \mu,\alpha) = R(\mu,\alpha,\beta,\gamma) - R(\mu,\alpha)$	$= T_{AB} - T_A$	$h_4\sigma_\beta^2 + h_5\sigma_\gamma^2 + (s-a)\sigma_e^2$
$R(\gamma \mid \mu,\alpha,\beta) = R(\mu,\alpha,\beta,\gamma) - R(\mu,\alpha,\beta)$	$= T_{AB} - R(\mu,\alpha,\beta)$	$h_6\sigma_\gamma^2 + \quad s^*\sigma_e^2$
SSE $= \sum y^2 - R(\mu,\alpha,\beta,\gamma)$	$= T_o - T_{AB}$	$(N-s)\sigma_e^2$

The T's are defined in (19), and $R(\mu,\alpha,\beta)$ is defined in (63) of Sec. 7.2d(i).
The h's come from (79) and are given in Sec. 11.4e.
$s^* = s - a - b + 1$.

(*ii*) *Estimation.* The nature of (79) and of the reductions shown in Table 10.1 ensures that the expectations of those reductions involve successively more variance components, one at a time, reading from the bottom up. Estimation of the components from Table 10.1 is therefore quite straightforward:

$$\hat{\sigma}_e^2 = \text{SSE}/(N-s)$$

$$\hat{\sigma}_\gamma^2 = [R(\gamma \mid \mu, \alpha, \beta) - (s-a-b+1)\hat{\sigma}_e^2]/h_6$$

$$\hat{\sigma}_\beta^2 = [R(\beta, \gamma \mid \mu, \alpha) - h_5\hat{\sigma}_\gamma^2 - (s-a)\hat{\sigma}_e^2]/h_4 \tag{83}$$

and $$\hat{\sigma}_\alpha^2 = [R(\alpha, \beta, \gamma \mid \mu) - h_2\hat{\sigma}_\beta^2 - h_3\hat{\sigma}_\gamma^2 - (s-1)\hat{\sigma}_e^2]/h_1 .$$

These estimators are easily calculated once the R's and h's have been obtained. To this we now turn.

(*iii*) *Calculation.* Expressions for calculating the $R(\cdot)$-terms of Table 10.1 are given at equations (58)–(63) of Sec. 7.2d(i). Most of them are the same as the T's given in (19) of this chapter; i.e.,

$$R(\mu) = T_\mu, \qquad R(\mu, \alpha) = T_A,$$

$$R(\mu, \alpha, \beta, \gamma) = T_{AB} \quad \text{and} \quad \sum y^2 = T_o . \tag{84}$$

These are easily calculated, as in (19), and lead to the computing formulae shown in Table 10.1. It is noticeable that the only term which is not part of the analysis of variance method is $R(\mu, \alpha, \beta)$. Calculation of this is given at equations (63)–(65) in Section 7.2d(i) and is repeated again in Chapter 11. For the moment we are concerned with general methodology rather than its specific applications. Details of calculating $R(\mu, \alpha, \beta)$ and the h's of Table 10.1 are therefore left until Chapter 11. In passing we may note that, because the reductions in Table 10.1 are largely functions of T's, most of the h's are correspondingly functions of coefficients of σ^2's in expected values of T's. A general expression for these coefficients is given in (38). Full details are shown in Chapter 11.

c. Too many equations

Table 10.1 contains no term $R(\mu, \beta) = T_B$. This is because the table is based on the reductions in sum of squares shown in (80). These in turn come from the first part of Table 7.8, which deals, in the fixed effects model, with the fitting of α before β, a context in which $R(\mu, \beta)$ does not arise. On the other hand, $R(\mu, \beta) = R(\mu) + R(\beta \mid \mu)$, comes from the second part of Table 7.8, concerned with fitting β before α. Observe, however, that there is nothing sacrosanct about either part of that table so far as estimation of variance components in the random model is concerned. In (80) we have used the first part, but we could have just as well used the second. Rearrangement of the reductions in sums of squares therein, in the manner of Table 10.1, yields Table 10.2. It is exactly the same as Table 10.1 except for the second entry which involves $R(\mu, \beta)$ instead of $R(\mu, \alpha)$. Equating the reductions to their

TABLE 10.2. AN ALTERNATIVE SET OF REDUCTIONS IN SUM OF SQUARES FO ESTIMATING VARIANCE COMPONENTS IN A 2-WAY CLASSIFICATION INTERAC TION, RANDOM MODEL, UNBALANCED DATA

Reduction in sum of squares	Computing formula[1]	Expected value[2]
$R(\alpha, \beta, \gamma \mid \mu) = R(\mu, \alpha, \beta, \gamma) - R(\mu)$	$= T_{AB} - T_\mu$	$h_1\sigma_\alpha^2 + h_2\sigma_\beta^2 + h_3\sigma_\gamma^2 + (s-1)\sigma$
$R(\alpha, \gamma \mid \mu, \beta) = R(\mu, \alpha, \beta, \gamma) - R(\mu, \beta)$	$= T_{AB} - T_B$	$h_7\sigma_\alpha^2 \quad + h_8\sigma_\gamma^2 + (s-b)\sigma$
$R(\gamma \mid \mu, \alpha, \beta) = R(\mu, \alpha, \beta, \gamma) - R(\mu, \alpha, \beta)$	$= T_{AB} - R(\mu, \alpha, \beta)$	$h_6\sigma_\gamma^2 + \quad s^*\sigma$
SSE $= \sum y^2 - R(\mu, \alpha, \beta, \gamma)$	$= T_o - T_{AB}$	$(N-s)\sigma_e^2$

[1] The T's are defined in (19), and $R(\mu, \alpha, \beta)$ is defined in (63) of Sec. 7.2d(i).
[2] The h's come from (79) and are given in Sec. 11.4e.
$s^* = s - a - b + 1$.

expected values yields the following estimators of the variance components:

$$\begin{aligned}
\hat{\sigma}_e^2 &= \mathrm{SSE}/(N - s) \\
\hat{\sigma}_\gamma^2 &= [R(\gamma \mid \mu, \alpha, \beta) - (s - a - b + 1)\hat{\sigma}_e^2]/h_6 \\
\hat{\sigma}_\alpha^2 &= [R(\alpha, \gamma \mid \mu, \beta) - h_8\hat{\sigma}_\gamma^2 - (s - b)\hat{\sigma}_e^2]/h_7
\end{aligned} \tag{85}$$

and $$\hat{\sigma}_\beta^2 = [R(\alpha, \beta, \gamma \mid \mu) - h_1\hat{\sigma}_\alpha^2 - h_3\hat{\sigma}_\gamma^2 - (s - 1)\hat{\sigma}_e^2]/h_2 .$$

The estimators $\hat{\sigma}_e^2$ and $\hat{\sigma}_\gamma^2$ in (85) are the same as those in (83), but $\hat{\sigma}_\alpha^2$ and $\hat{\sigma}_\beta^2$ are not. The question immediately arises as to which estimators should be used, (83) or (85)? Unfortunately there is no satisfactory answer to this question; indeed, there is almost no answer at all. Whereas in the fixed effects model there is often good reason for choosing between fitting β after α and fitting α after β, there appears to be no criteria for making this choice when using the reductions in sums of squares to estimate variance components in the random model. It means, in effect, that we can have, in the fitting constants method, more equations than variance components; for example, Tables 10.1 and 10.2 provide between them five equations in four variance components.

This is an unsolved difficulty with the fitting constants method of estimation: it can yield more equations than there are components to be estimated, and it provides no guidance as to which equations should be used. Furthermore, it is a difficulty that applies quite generally to the method and can assume some magnitude in multi-classification models, where many different sets of reductions in sums of squares can be available. For example, there are six sets in a 3-way classification model (see Table 8.2). Not only can each of these sets be used on its own, but combinations of terms from them can also be used. In Tables 10.1 and 10.2, for example, the last two lines are the same; these, and the second line from each table, could therefore be used to provide estimators. This is the principle of the procedures considered by Harville (1967) and Low (1964).

A criterion that could have some appeal for deciding on which reductions to use is that they should add up to the total sum of squares corrected for the mean, $\mathrm{SST}_m = \sum y^2 - T_\mu$. Although the reductions listed in Tables 10.1 and 10.2 do not meet this requirement explicitly, they are linear combinations of reductions that do so and therefore provide the same estimators; e.g., the terms in Table 10.1 are linear combinations of those in (80) which do add to SST_m. One feature of this criterion is that the resulting estimators come from reductions that account for the total observed variability in the y's, and they are reductions with known properties in fixed effects models. This criterion would confine us to using sets of reductions like those of Tables 10.1 and 10.2 and would preclude using combinations of terms from these tables. On the other hand, using combinations is attractive, because, for example, Table 10.1 excludes $R(\mu, \beta)$ and Table 10.2 excludes $R(\mu, \alpha)$, terms which one feels, intuitively, should not be omitted.

Knowing, as we do, certain properties of the analysis of variance estimators with balanced data suggests that whatever reductions are used for estimating variance components from unbalanced data, they should be such as to reduce the resulting estimators to the analysis of variance estimators when the data *are* balanced, i.e., when the n_{ij}'s are all equal. However, this criterion is of little help in selecting which set of reductions to use with unbalanced data because all sets reduce to the analysis of variance of balanced data when the n's are equal. For example, (80) reduces to Table 7.9 when $n_{ij} = n$ for all i and j.

One possible way of overcoming the situation of having more equations than variance components is to apply "least squares" as suggested by Robson (1957). Arraying all calculated reductions as a vector $\mathbf{r}$ let us suppose that $E(\mathbf{r}) = \mathbf{A}\boldsymbol{\sigma}^2$. Then $\mathbf{r} = \mathbf{A}\hat{\boldsymbol{\sigma}}^2$ are the equations we would like to solve for $\hat{\boldsymbol{\sigma}}^2$. However, when there are more equations than variance components these equations will usually not be consistent.[1] Nevertheless, provided the reductions in $\mathbf{r}$ are linearly independent and $\mathbf{A}$ thus has full column rank, we could estimate $\hat{\boldsymbol{\sigma}}^2$ by "least squares" as $\hat{\boldsymbol{\sigma}}^2 = (\mathbf{A}'\mathbf{A})^{-1}\mathbf{A}'\mathbf{r}$.

d. Mixed models

The fitting constants method of estimation applies equally as well to mixed models as to random models. Indeed, for mixed models, it provides unbiased estimators which the analysis of variance method does not. This, as has already been explained, arises from (79). Based on that result we use only those reductions that have no fixed effects in their expected values. For example, in the 2-way classification model with α's as fixed effects we would use the last three lines of Table 10.1. They will, by (79), have no fixed effects in their expectations, and they provide unbiased estimators of σ_e^2, σ_γ^2 and σ_β^2. The one entry in Table 10.2 that differs from Table 10.1 is $R(\alpha, \gamma \mid \mu, \beta)$ and it has, by (79), an expected value that is not free of the fixed α-effects and so cannot be used. The last three lines of Table 10.1 are therefore the basis of estimation in the 2-way classification mixed model having α's as fixed effects.

The principles illustrated here are quite straightforward and extend readily to multi-classification mixed models.

Variations on Henderson's Method 2, of adjusting for bias in mixed models, can be mentioned here. That method, as shown in equation (73), temporarily assumes the random effects are fixed, for purposes of solving normal equations for the fixed effects. An alternative is to temporarily ignore the random effects, and solve normal equations for the fixed effects as $\tilde{\mathbf{b}}_f = (\mathbf{X}_f'\mathbf{X}_f)^-\mathbf{X}_f'\mathbf{y}$.

[1] Thanks go to D. A. Harville for bringing this to my attention.

The data are then adjusted to be

$$\mathbf{z} = \mathbf{y} - \mathbf{X}_f\tilde{\mathbf{b}}_f = [\mathbf{I} - \mathbf{X}_f(\mathbf{X}_f'\mathbf{X}_f)^-\mathbf{X}_f']\mathbf{y} = [\mathbf{I} - \mathbf{X}_f(\mathbf{X}_f'\mathbf{X}_f)^-\mathbf{X}_f'](\mathbf{X}_r\mathbf{b}_r + \mathbf{e}),$$

using (68) as the model for **y**. With **z**, two possibilities are available: the analysis of variance method and the fitting constants method. The latter was suggested by Zelen (1968) as being equivalent to using the fitting constants method directly on **y**. Details of this are demonstrated in Searle (1969); see also Exercise 4.

e. Sampling variances of estimators

Each $R(\cdot)$ reduction used in the fitting constants method can be expressed in the form $\mathbf{y}'\mathbf{X}(\mathbf{X}'\mathbf{X})^-\mathbf{X}'\mathbf{y}$ for some matrix **X**. On the basis of normality assumptions, both for the error terms and the random effects, the sampling variance of each reduction can therefore be obtained from Theorem 2 of Chapter 2. Covariances between reductions can be derived in similar manner:

$$\text{cov}(\mathbf{y}'\mathbf{P}\mathbf{y}, \mathbf{y}'\mathbf{Q}\mathbf{y}) = 2\,\text{tr}(\mathbf{PVQV}) + 4\boldsymbol{\mu}'\mathbf{PVQ}\boldsymbol{\mu}$$

when $\mathbf{y} \sim N(\boldsymbol{\mu}, \mathbf{V})$. In this way, sampling variances of variance components estimators can be developed, since the estimators are linear combinations of these reductions. The details are somewhat lengthy, involving extensive matrix manipulations. Rohde and Tallis (1969) give general results applicable to components of both variance and covariance. Specific cases have been discussed by Low (1964) and Harville (1969c).

5. ANALYSIS OF MEANS METHODS

Data in which every subclass of the model contains observations can, in fixed effects models, be analyzed in terms of the means of the sub-most subclasses. Two such analyses are discussed in Sec. 8.3c. The mean squares of those analyses can also be used for estimating variance components in random and mixed models. Expected values of these mean squares for the random model are shown in Table 10.3. Estimators of the variance components are obtained in the usual manner of equating the mean squares to their expected values. The estimators are unbiased. Through being quadratic forms in the observations, their variances could, under normality assumptions, be obtained from Theorem 1 of Chapter 2. The variances could also be derived by the method of "synthesis" described in Sec. 2d(iii). For mixed models, only the mean squares whose expectations contain no fixed effects will be used for estimating the variance components. For example, if the α's are fixed effects in the 2-way classification, MSA_u or MSA_w of Table 10.3 will not be used.

TABLE 10.3. EXPECTED VALUES OF MEAN SQUARES IN TWO ANALYSES OF MEANS OF THE 2-WAY CLASSIFICATION INTERACTION RANDOM MODEL HAVING ALL $n_{ij} > 0$

a. Unweighted means analysis **(Table 8.12)**[1],

$$\begin{aligned}
E(\text{MSA}_u) &= b\sigma_\alpha^2 + \sigma_\gamma^2 + n_h\sigma_e^2\\
E(\text{MSB}_u) &= a\sigma_\beta^2 + \sigma_\gamma^2 + n_h\sigma_e^2\\
E(\text{MSAB}_u) &= \sigma_\gamma^2 + n_h\sigma_e^2\\
E(\text{MSE}) &= \sigma_e^2
\end{aligned}$$

b. Weighted means analysis **(Table 8.18)**[2],

$$\begin{aligned}
E(\text{MSA}_w) &= \frac{1}{(a-1)b}\left(\sum_{i=1}^a w_i - \sum_{i=1}^a w_i^2 \Big/ \sum_{i=1}^a w_i\right)(b\sigma_\alpha^2 + \sigma_\gamma^2) + \sigma_e^2\\
E(\text{MSB}_w) &= \frac{1}{a(b-1)}\left(\sum_{j=1}^b v_j - \sum_{j=1}^b v_j^2 \Big/ \sum_{j=1}^b v_j\right)(a\sigma_\beta^2 + \sigma_\gamma^2) + \sigma_e^2\\
E(\text{MSAB}_w) &= \sigma_\gamma^2 + n_h\sigma_e^2\\
E(\text{MSE}) &= \sigma_e^2
\end{aligned}$$

[1] $n_h = \sum_{i=1}^a \sum_{j=1}^b n_{ij}^{-1} \Big/ ab.$

[2] $w_i = b^2 \Big/ \sum_{j=1}^b n_{ij}^{-1}; \quad v_j = a^2 \Big/ \sum_{i=1}^a n_{ij}^{-1}.$

Extension of Table 10.3 to multi-way classifications depends upon extension of Tables 8.12 and 8.18. This is particularly straightforward for the unweighted means analysis of Table 8.12. However, the need for having data in every subclass of the model still remains. Analyses of means cannot be made otherwise.

6. SYMMETRIC SUMS METHODS

A method of estimating variance components based on symmetric sums of products of the observations, rather than sums of squares, has been suggested by Koch (1967a, 1968). The method uses the fact that expected values of products of observations are linear functions of the variance components. Sums of these products (and hence means of them) therefore provide unbiased estimators of the components. We illustrate in terms of the 1-way classification.

Consider the random model for the 1-way classification $y_{ij} = \mu + \alpha_i + e_{ij}$, where $E(\alpha_i) = E(e_{ij}) = 0$, $E(\alpha_i^2) = \sigma_\alpha^2$ and $E(e_{ij}^2) = \sigma_e^2$ for all i and j, and all covariances are zero. Then expected values of products of observations are as follows:

$$\begin{aligned} E(y_{ij}y_{i'j'}) &= \mu^2 + \sigma_\alpha^2 + \sigma_e^2 && \text{when} \quad i = i' \text{ and } j = j'; \\ &= \mu^2 + \sigma_\alpha^2 && \text{when} \quad i = i' \text{ and } j \neq j'; \\ &= \mu^2 && \text{when} \quad i \neq i'. \end{aligned} \tag{86}$$

Estimators are derived from means of the different products in (86):

$$\widehat{\mu^2} + \hat{\sigma}_\alpha^2 + \hat{\sigma}_e^2 = \sum_{i=1}^{a}\sum_{j=1}^{n_i} y_{ij}^2 \Big/ N \tag{87}$$

$$\begin{aligned} \widehat{\mu^2} + \hat{\sigma}_\alpha^2 &= \sum_{i=1}^{a}\sum_{j=1}^{n_i}\sum_{j'\neq j}^{n_{i'}} y_{ij}y_{ij'} \Big/ \sum_{i=1}^{a} n_i(n_i - 1) \\ &= \left(\sum_{i=1}^{a} y_{i\cdot}^2 - \sum_{i=1}^{a}\sum_{j=1}^{n_i} y_{ij}^2\right) \Big/ (S_2 - N) \end{aligned} \tag{88}$$

where $S_2 = \sum_{i=1}^{a} n_i^2$, and

$$\begin{aligned} \widehat{\mu^2} &= \sum_{i=1}^{a}\sum_{i'\neq i}^{a}\sum_{j=1}^{n_i}\sum_{j'=1}^{n_{i'}} y_{ij}y_{i'j'} \Big/ \sum_{i=1}^{a}\sum_{i'\neq i}^{a} n_i n_{i'} \\ &= \left(y_{\cdot\cdot}^2 - \sum_{i=1}^{a} y_{i\cdot}^2\right) \Big/ (N^2 - S_2). \end{aligned} \tag{89}$$

Estimators $\hat{\sigma}_e^2$ and $\hat{\sigma}_\alpha^2$ are easily obtained from these expressions.

These estimators are unbiased and consistent, and they are identical to the analysis of variance estimators in the case of balanced data. However, their variances are functions of μ, a deficiency noted by Koch (1968). Evidence of this is seen in $\hat{\sigma}_e^2$ by using (87) and (88) to write

$$\hat{\sigma}_e^2 = \mathbf{y}'\left(k_1\mathbf{I}_N - k_2 \sum_{i=1}^{a}{}^{+} \mathbf{J}_{ni}\right)\mathbf{y} \tag{90}$$

where $\quad k_1 = S_2/N(S_2 - N) \quad$ and $\quad k_2 = 1/(S_2 - N). \qquad (91)$

In deriving the variance of (90) from Theorem 1 of Chapter 2, it will be found that the term in μ^2 is

$$\begin{aligned} 4\mu^2\mathbf{1}'\left(k_1\mathbf{I}_N - k_2\sum_{i=1}^{a}{}^{+} \mathbf{J}_{n_i}\right)&\left(\sigma_e^2\mathbf{I}_N + \sigma_\alpha^2\sum_{i=1}^{a}{}^{+} \mathbf{J}_{n_i}\right)\left(k_1\mathbf{I}_N - k_2\sum_{i=1}^{a}{}^{+} \mathbf{J}_{n_i}\right)\mathbf{1} \\ &= 4\mu^2\sum_{i=1}^{a}(\sigma_e^2 + n_i\sigma_\alpha^2)n_i(k_1 - n_ik_2)^2 . \end{aligned}$$

This is non-zero for unequal n_i although zero when the n_i are equal. Hence for unbalanced data the variance of $\hat{\sigma}_e^2$ derived from (87) and (88) is a function of μ, as may also be shown for $\hat{\sigma}_\alpha^2$. This is clearly unsatisfactory.

This difficulty is overcome by Koch (1968), who suggests that, instead of using symmetric sums of products, symmetric sums of squares of differences should be used. Thus in the 1-way classification

$$E(y_{ij} - y_{i'j'})^2 = 2\sigma_e^2 \qquad \text{when} \quad i = i' \quad \text{and} \quad j \neq j';$$
$$= 2(\sigma_e^2 + \sigma_\alpha^2) \qquad \text{when} \quad i \neq i'.$$

Estimators are therefore derived from

$$2\hat{\sigma}_e^2 = \sum_{i=1}^{a} \sum_{j=1}^{n_i} \sum_{j' \neq j}^{n_i} (y_{ij} - y_{ij'})^2 \Big/ \sum_{i=1}^{a} n_i(n_i - 1)$$

and (92)

$$2(\hat{\sigma}_e^2 + \hat{\sigma}_\alpha^2) = \sum_{i=1}^{a} \sum_{i' \neq i}^{a} \sum_{j=1}^{n_i} \sum_{j'=1}^{n_{i'}} (y_{ij} - y_{i'j'})^2 \Big/ \sum_{i=1}^{a} \sum_{i' \neq i}^{a} n_i n_{i'}\,.$$

The resulting estimators have variances that are free of μ, because (92) contains no terms in μ. The estimators are unbiased and for balanced data reduce to the analysis of variance estimators.

A by-product of (89) is a procedure given by Koch (1967b) for obtaining an unbiased estimator of μ from an unbiased estimator of μ^2. Suppose the latter is $\widehat{\mu^2} = q(\mathbf{y})$, a quadratic function of the observations as is, for example, (89). Then

$$E(\widehat{\mu^2}) = E[q(\mathbf{y})] = \mu^2.$$

From this it can be shown that for scalars θ and g

$$E[q(\mathbf{y} + \theta\mathbf{1})] = q(\mathbf{y}) + 2g\theta + \theta^2 = \widehat{\mu^2} + 2g\theta + \theta^2.$$

Minimizing this with respect to θ gives $\theta = -g$ with the minimum value being $\widehat{\mu^2} - g^2$. This suggests taking $\hat{\mu} = g$, i.e., taking the estimator of μ as half the coefficient of θ in $q(\mathbf{y} + \theta\mathbf{1})$ where $\widehat{\mu^2} = q(\mathbf{y})$ derived when estimating variance components. That this gives an unbiased estimator is easily seen:

$$\hat{\mu} = g = [q(\mathbf{y} + \theta\mathbf{1}) - q(y) - \theta^2]/2\theta$$

and so $$E(\hat{\mu}) = [(\mu + \theta)^2 - \mu^2 - \theta^2]/2\theta = \mu.$$

As an example, we have in (89) an estimator of μ^2 which is

$$q(\mathbf{y}) = \left(y_{..}^2 - \sum_{i=1}^{a} y_{i.}^2\right) \Big/ (N^2 - S_2).$$

Thus

$$q(\mathbf{y} + \theta\mathbf{1}) = \left[(y_{..} + N\theta)^2 - \sum_{i=1}^{a} (y_{i\cdot} + n_i\theta)^2\right] \Big/ (N^2 - S_2),$$

from which the estimator of μ, taken as half the coefficient of θ, is

$$\hat{\mu} = \left(Ny_{..} - \sum_{i=1}^{a} n_i y_{i\cdot}\right) \Big/ (N^2 - S_2).$$

It does, of course, reduce to $\bar{y}_{..}$ with balanced data.

7. INFINITELY MANY QUADRATICS

If the reader has gained an impression from the preceding sections that there are many quadratic forms of the observations that can be used for estimating variance components from unbalanced data, then he has judged the situation correctly. There are infinitely many quadratic forms that can be used in the manner of the analysis of variance method, namely equating observed values of quadratic forms to their expected values and solving the resulting equations to get estimators of the variance components. This procedure is widely used, as we have seen, but it has a serious deficiency: it gives no criteria for selecting the quadratic forms to be used. The only known property that the method gives to the resulting estimators is that they are universally unbiased for random models and, with the fitting constants method, unbiased for mixed models.

Even the property of unbiasedness is of questionable value. As a property of estimators it has been borrowed from fixed effects estimation, but in the context of variance component estimation it may not be appropriate. In estimating fixed effects, the basis of desiring unbiasedness of our estimators is the concept of repetition of data and associated estimates. This basis is often not valid with unbalanced data from random models—repeated data, perhaps, but not necessarily with the same pattern of unbalancedness or with the same set of (random) effects in the data. Replications of data are not, therefore, just replications of any existing data structure. Mean unbiasedness may therefore no longer be pertinent, and replacing it with some other criterion might be considered. Modal unbiasedness is one possibility, suggested by Searle (1968, discussion), although Harville (1969b) doubts if modally unbiased estimators exist and questions the justification of such a criterion on decision-theoretic grounds. Nevertheless, as Kempthorne (1968) points out, mean unbiasedness in estimating fixed effects "... leads to residuals which do not contain systematic effects and is therefore valuable ... and is fertile mathematically in that it reduces the class of candidate statistics (or estimates)". However, "... in the variance component problem it does not lead to a fertile smaller class of statistics".

All the estimation methods that have been discussed reduce to the analysis of variance method when the data are balanced. This and unbiasedness of the resulting estimators are the only known properties of the methods. Otherwise, the quadratic forms involved in each method have been selected solely because they seemed "reasonable" in one way or another. However, "reasonableness" of the quadratic forms in each case provides little or no comparison of any properties of the estimators that result from the different methods. Probably the simplest idea would be to compare sampling variances. Unfortunately this comparison soon becomes bogged down in algebraic complexity. Not only are the variances in any way tractable only if normality is assumed but also, just as with balanced data, the variances themselves are functions of the variance components. The complexity of the variances is evident in (63) which, aside from $v(\hat{\sigma}_e^2) = 2\sigma_e^4/(N - s)$, is the simplest example of a sampling variance of a variance component estimator obtained from unbalanced data. Suppose we rewrite (63) as

$$v(\hat{\sigma}_\alpha^2) = \frac{2N}{N^2 - \sum n_i^2}\left\{\frac{N(N-1)(a-1)\sigma_e^4}{N^2 - \sum n_i^2} + 2\sigma_e^2\sigma_\alpha^2 + \frac{[N^2 \sum n_i^2 + (\sum n_i^2)^2 - 2N \sum n_i^3]\sigma_\alpha^4}{N(N^2 - \sum n_i^2)}\right\}. \quad (93)$$

It is clear that studying the behaviour of this variance as a function of N, the total number of observations, of a, the number of classes, of n_i, the number of observations in the ith class for $i = 1, 2, \ldots, a$, and of σ_α^2 and σ_e^2—doing this is no small task, let alone comparing it with some equally as complex a function that is the variance of some other estimator. And this is the simplest example of unbalanced data. It is easy to understand, therefore, how it is that analytic comparison of the variances of different estimators presents great difficulties.

As a result of the analytical difficulties just described comparisons of estimators available in the literature have largely been in terms of numerical studies. These, though, are not without their difficulties also, and results can be costly to attain. Kussmaul and Anderson (1967) have studied a special case of the 2-way nested classification which makes it a particular form of the 1-way classification. A study of the latter by Anderson and Crump (1967) suggests that the unweighted means estimator of σ_α^2 appears, for very unbalanced data, to have larger variance than does the analysis of variance estimator for small values of $\rho = \sigma_\alpha^2/\sigma_e^2$, but that it has smaller variance for large ρ. The 2-way classification interaction model has been studied by Bush and Anderson (1963) in terms of several cases of what can well be called

TABLE 10.4. VALUES OF n_{ij} IN SOME 6×6 DESIGNS USED BY BUSH AND ANDERSON (1963)

Design Number																	
S22						C18						I24					
2	1	0	0	0	0	1	1	1	0	0	0	1	1	0	0	0	0
1	2	1	0	0	0	1	1	1	0	0	0	1	1	0	0	0	0
0	1	2	1	0	0	0	1	1	1	0	0	2	1	0	0	0	0
0	0	1	2	1	0	0	0	1	1	1	0	1	2	0	0	0	0
0	0	0	1	2	1	0	0	0	1	1	1	1	1	2	1	1	1
0	0	0	0	1	2	0	0	0	1	1	1	1	1	2	1	1	1

planned unbalancedness. For example, in the case of 6 rows and 6 columns, three of the designs used are those shown in Table 10.4. Designs such as these were used to compare the analysis of variance, the fitting constants and the weighted means methods of estimation. Comparisons were made, by way of variances of the estimators, both of different designs as well as of different estimation procedures, over a range of values of the underlying variance components. For the designs used, the general trend of the results is that, for values of the error variance much larger than the other components, the analysis of variance method estimators have smallest variance, but otherwise the fitting constants method estimators have.

Even with present-day computing facilities, making comparisons such as those made by Bush and Anderson is no small task. Nevertheless, as samples of unbalanced data generally, the examples they used (their designs) are of somewhat limited extent. This, of course, is the difficulty with numerical comparisons: planning sets of n_{ij}-values that will provide comparisons that are informative about unbalanced data in general. Even in the 1-way classification there are infinitely many sets of n_i-values available for (93) for studying the behavior of $v(\hat{\sigma}_\alpha^2)$—along with also varying the values of a and of σ_α^2 and σ_e^2. There is difficulty enough in planning a series of these values that in any sense "covers the field", a difficulty that is simply multiplied when one comes to consider higher-order classifications such as those handled by Bush and Anderson (1963). Neither analytic nor numeric comparisons of estimators are therefore easily resolved.

The one thing that *can* be done is to go back to the grounds on which "reasonableness" was judged appropriate in establishing the methods. The situation is summarized by Searle (1971). "The analysis of variance method commends itself because it is the obvious analogue of the analysis of variance of balanced data, and it is easy to use; some of its terms are not sums of squares, and it gives biased estimators in mixed models. The generalized form of Henderson's Method 2 makes up for this deficiency, but is not uniquely defined and his specific definition of it cannot be used when there are interactions between fixed and random effects. The fitting constants method uses sums of squares that have non-central χ^2-distributions in the fixed effects model, and it gives unbiased estimators in mixed models; but it can involve more quadratics than there are components to be estimated; and it can also involve extensive computing" (inverting matrices of order equal to the number of random effects in the model). For data in which all subclasses are filled the analysis of means methods have the advantage of being easier to compute than the fitting constants method; the unweighted means analysis is especially easy. All of the methods reduce, for balanced data, to the analysis of variance method, and all of them can yield negative estimates. Little more than this can be said by way of comparing the methods. The problem awaits thorough investigation.

8. MAXIMUM LIKELIHOOD FOR MIXED MODELS

In Sec. 9.2 we mentioned that all models could, in fact, be called mixed models. This is so because every model usually has both a general mean μ, which is a fixed effect, and error terms $\mathbf{e}$, which are random. Thus although by its title this section might appear to be devoted to only one class of models it does in fact apply to all linear models.

The fitting constants method of estimating variance components gives unbiased estimators of the components even for mixed models. However, it is only a method for estimating the variance components of the model and gives no guidance on the problem of estimating the fixed effects. Were the variance components of the model known there would, of course, be no problem in estimating estimable functions of the fixed effects from a solution of the normal equations $\mathbf{X'V^{-1}Xb^\circ} = \mathbf{X'V^{-1}y}$ of the generalized least squares procedure. In these equations $\mathbf{V}$ is the variance-covariance matrix of $\mathbf{y}$, the elements of $\mathbf{V}$ being functions of the (assumed known) variance components. However, when these components are unknown, as is usually the case, we have the problem of wanting to estimate, simultaneously, both the fixed effects and the variance components of the model.

At least two courses of action are available. (i) Use the fitting constants method to estimate the variance components, and then use the resulting estimates in place of the true components in $\mathbf{V}$ in the generalized least squares equations for the fixed effects. (ii) Estimate the fixed effects and the variance components simultaneously, with a unified procedure such as maximum likelihood. In both cases recourse has usually to be made to an iterative procedure with its attendant computing requirements, which can be extensive, although some progress has been made analytically, the results of which we now indicate.

a. Estimating fixed effects

Let us write the model

$$\mathbf{y} = \mathbf{Xb} + \mathbf{Zu} + \mathbf{e} \tag{94}$$

where $\mathbf{b}$ is the vector of fixed effects, $\mathbf{u}$ is the vector of random effects, $\mathbf{X}$ and $\mathbf{Z}$ are the corresponding incidence matrices and $\mathbf{e}$ is the vector of random error terms. The random effects and the error terms are assumed, in the usual way, to have zero means, to be uncorrelated and, in this case, to have variance-covariance matrices

$$\text{var}(\mathbf{u}) = E(\mathbf{uu}') = \mathbf{D} \quad \text{and} \quad \text{var}(\mathbf{e}) = E(\mathbf{ee}') = \mathbf{R} \tag{95}$$

that are assumed known. Then from (94)

$$\mathbf{V} = \text{var}(\mathbf{y}) = \mathbf{ZDZ}' + \mathbf{R}. \tag{96}$$

We also assume that $\mathbf{V}$ is non-singular. The normal equations stemming from generalized least squares are then

$$\mathbf{X}'\mathbf{V}^{-1}\mathbf{Xb}^o = \mathbf{X}'\mathbf{V}^{-1}\mathbf{y} \tag{97}$$

with solution

$$\mathbf{b}^o = (\mathbf{X}'\mathbf{V}^{-1}\mathbf{X})^-\mathbf{X}'\mathbf{V}^{-1}\mathbf{y}. \tag{98}$$

If $\mathbf{V}$ is singular, $\mathbf{V}^{-1}$ in (97) and (98) is replaced by $\mathbf{V}^-$, as in (138) of Sec. 5.8b. Under normality assumptions for the u's and e's (98) also represents the maximum likelihood solution.

Calculating (98) involves $\mathbf{V}^{-1}$, a matrix of order equal to the number of observations, which can be very large, perhaps many thousands. Having obtained $\mathbf{V}^{-1}$, then a generalized inverse $(\mathbf{X}'\mathbf{V}^{-1}\mathbf{X})^-$ is needed also, although this will be a lesser task because its order is the number of levels of the fixed effects. The difficulty with (97) and (98) is therefore that of calculating $\mathbf{V}^{-1}$. In the fixed effects case $\mathbf{V}$ usually has the form $\sigma_e^2\mathbf{I}_N$ or, with a little more generality, it may be diagonal. In either case inversion of $\mathbf{V}$ is simple. But in general, $\mathbf{V} = \mathbf{ZDZ}' + \mathbf{R}$ of (96) is not diagonal, even if $\mathbf{D}$ and $\mathbf{R}$ are, and so $\mathbf{V}^{-1}$ is not always easy to calculate. However, as indicated by Henderson *et al.* (1959), a set of equations not involving $\mathbf{V}^{-1}$ can be established, alternative to (97), for deriving $\mathbf{b}^o$. This we now show.

Suppose that in (94) the effects represented by $\mathbf{u}$ were in fact fixed and not random. Then, because $\text{var}(\mathbf{e}) = \mathbf{R}$, the normal equations for the now completely fixed effects model would be

$$\begin{bmatrix}\mathbf{X}'\\ \mathbf{Z}'\end{bmatrix}\mathbf{R}^{-1}[\mathbf{X} \quad \mathbf{Z}]\begin{bmatrix}\tilde{\mathbf{b}}\\ \tilde{\mathbf{u}}\end{bmatrix} = \begin{bmatrix}\mathbf{X}'\\ \mathbf{Z}'\end{bmatrix}\mathbf{R}^{-1}\mathbf{y}.$$

i.e.,

$$\begin{bmatrix}\mathbf{X}'\mathbf{R}^{-1}\mathbf{X} & \mathbf{X}'\mathbf{R}^{-1}\mathbf{Z}\\ \mathbf{Z}'\mathbf{R}^{-1}\mathbf{X} & \mathbf{Z}'\mathbf{R}^{-1}\mathbf{Z}\end{bmatrix}\begin{bmatrix}\tilde{\mathbf{b}}\\ \tilde{\mathbf{u}}\end{bmatrix} = \begin{bmatrix}\mathbf{X}'\mathbf{R}^{-1}\mathbf{y}\\ \mathbf{Z}'\mathbf{R}^{-1}\mathbf{y}\end{bmatrix} \tag{99}$$

where we use the notation $\tilde{\mathbf{b}}$ in contrast to $\mathbf{b}^o$ to distinguish a solution of (99) from one of (97).

Suppose that we amend equations (99) by adding $\mathbf{D}^{-1}$ to the lower right-hand sub-matrix $\mathbf{Z}'\mathbf{R}^{-1}\mathbf{Z}$ of the matrix on the left. This gives

$$\begin{bmatrix} \mathbf{X}'\mathbf{R}^{-1}\mathbf{X} & \mathbf{X}'\mathbf{R}^{-1}\mathbf{Z} \\ \mathbf{Z}'\mathbf{R}^{-1}\mathbf{X} & \mathbf{Z}'\mathbf{R}^{-1}\mathbf{Z} + \mathbf{D}^{-1} \end{bmatrix} \begin{bmatrix} \mathbf{b}^* \\ \mathbf{u}^* \end{bmatrix} = \begin{bmatrix} \mathbf{X}'\mathbf{R}^{-1}\mathbf{y} \\ \mathbf{Z}'\mathbf{R}^{-1}\mathbf{y} \end{bmatrix} \tag{100}$$

where solutions to these equations are distinguished by the asterisk notation. Then it can be shown that the solutions $\mathbf{b}^*$ to (100) are identical to the solutions $\mathbf{b}^o$ of (97). In this way, (100) provides a means of deriving $\mathbf{b}^o$ without having to invert $\mathbf{V}$. We have only to invert $\mathbf{D}$ and $\mathbf{R}$, which are usually diagonal, and then to solve (100) which has as many equations as there are both fixed and random effects in the model. This is usually considerably fewer than the number of observations, and so (100) is easier to solve than (97).

The equivalence of $\mathbf{b}^*$ of (100) to $\mathbf{b}^o$ of (98) is readily demonstrated. From (100)

$$\mathbf{u}^* = (\mathbf{Z}'\mathbf{R}^{-1}\mathbf{Z} + \mathbf{D}^{-1})^{-1}(\mathbf{Z}'\mathbf{R}^{-1}\mathbf{y} - \mathbf{Z}'\mathbf{R}^{-1}\mathbf{X}\mathbf{b}^*)$$

and so

$$\mathbf{X}'[\mathbf{R}^{-1} - \mathbf{R}^{-1}\mathbf{Z}(\mathbf{Z}'\mathbf{R}^{-1}\mathbf{Z} + \mathbf{D}^{-1})^{-1}\mathbf{Z}'\mathbf{R}^{-1}]\mathbf{X}\mathbf{b}^* \\ = \mathbf{X}'[\mathbf{R}^{-1} - \mathbf{R}^{-1}\mathbf{Z}(\mathbf{Z}'\mathbf{R}^{-1}\mathbf{Z} + \mathbf{D}^{-1})^{-1}\mathbf{Z}'\mathbf{R}^{-1}]\mathbf{y},$$

which in writing

$$\mathbf{W} = \mathbf{R}^{-1} - \mathbf{R}^{-1}\mathbf{Z}(\mathbf{Z}'\mathbf{R}^{-1}\mathbf{Z} + \mathbf{D}^{-1})^{-1}\mathbf{Z}'\mathbf{R}^{-1}$$

becomes

$$\mathbf{X}'\mathbf{W}\mathbf{X}\mathbf{b}^* = \mathbf{X}'\mathbf{W}\mathbf{y}. \tag{101}$$

But

$$\begin{aligned} \mathbf{W}\mathbf{V} &= [\mathbf{R}^{-1} - \mathbf{R}^{-1}\mathbf{Z}(\mathbf{Z}'\mathbf{R}^{-1}\mathbf{Z} + \mathbf{D}^{-1})^{-1}\mathbf{Z}'\mathbf{R}^{-1}](\mathbf{Z}\mathbf{D}\mathbf{Z}' + \mathbf{R}) \\ &= \mathbf{R}^{-1}\mathbf{Z}\mathbf{D}\mathbf{Z}' + \mathbf{I} - \mathbf{R}^{-1}\mathbf{Z}(\mathbf{Z}'\mathbf{R}^{-1}\mathbf{Z} + \mathbf{D}^{-1})^{-1}(\mathbf{Z}'\mathbf{R}^{-1}\mathbf{Z}\mathbf{D}\mathbf{Z}' + \mathbf{Z}') \\ &= \mathbf{R}^{-1}\mathbf{Z}\mathbf{D}\mathbf{Z}' + \mathbf{I} - \mathbf{R}^{-1}\mathbf{Z}(\mathbf{Z}'\mathbf{R}^{-1}\mathbf{Z} + \mathbf{D}^{-1})^{-1}(\mathbf{Z}'\mathbf{R}^{-1}\mathbf{Z} + \mathbf{D}^{-1})\mathbf{D}\mathbf{Z}' \\ &= \mathbf{I}, \end{aligned}$$

and so $\mathbf{W} = \mathbf{V}^{-1}$. Therefore equations (101) and (97) are the same and so the solution $\mathbf{b}^*$ to (101), which is part of the solution to (100), is a solution to (97) given in (98). Equation (100), with its computational advantages over (97), can therefore be used to derive a solution to (97).

Equations (100) are easily described. They are simply the normal equations of the model assuming all effects fixed, namely equations (99), modified by adding the inverse of the variance-covariance matrix of the random effects $\mathbf{u}$ to the sub-matrix that is the coefficient of $\tilde{\mathbf{u}}$ in the "$\tilde{\mathbf{u}}$-equations"—i.e., by adding $\mathbf{D}^{-1}$ to $\mathbf{Z}'\mathbf{R}^{-1}\mathbf{Z}$, as in (100). This is particularly simple in certain special cases. For example, when $\mathbf{R} = \text{var}(\mathbf{e}) = \sigma_e^2\mathbf{I}_N$, as is so often assumed, equations (99) are

$$\begin{bmatrix} \mathbf{X'X} & \mathbf{X'Z} \\ \mathbf{Z'X} & \mathbf{Z'Z} \end{bmatrix} \begin{bmatrix} \tilde{\mathbf{b}} \\ \tilde{\mathbf{u}} \end{bmatrix} = \begin{bmatrix} \mathbf{X'y} \\ \mathbf{Z'y} \end{bmatrix} \tag{102}$$

and equations (100) are

$$\begin{bmatrix} \mathbf{X'X} & \mathbf{X'Z} \\ \mathbf{Z'X} & \mathbf{Z'Z} + \sigma_e^2\mathbf{D}^{-1} \end{bmatrix} \begin{bmatrix} \mathbf{b}^* \\ \mathbf{u}^* \end{bmatrix} = \begin{bmatrix} \mathbf{X'y} \\ \mathbf{Z'y} \end{bmatrix}. \tag{103}$$

Furthermore, $\mathbf{D}$ is often diagonal of the form

$$\mathbf{D} = \text{diag}\{\sigma_\theta^2\mathbf{I}_{N\theta}\} \qquad \text{for} \quad \theta = A, B, \ldots, K$$

where $A, B, \ldots, K$ are the random factors, the factor θ having N_θ levels and variance σ_θ^2. In this case $\sigma_e^2\mathbf{D}^{-1}$ of (103) requires just adding $\sigma_e^2/\sigma_\theta^2$ to appropriate diagonal elements of $\mathbf{Z'Z}$. In particular, if there is only one random factor (103) becomes

$$\begin{bmatrix} \mathbf{X'X} & \mathbf{X'Z} \\ \mathbf{Z'X} & \mathbf{Z'Z} + (\sigma_e^2/\sigma_\theta^2)\mathbf{I} \end{bmatrix} \begin{bmatrix} \mathbf{b}^* \\ \mathbf{u}^* \end{bmatrix} = \begin{bmatrix} \mathbf{X'y} \\ \mathbf{Z'y} \end{bmatrix}. \tag{104}$$

This formulation of the maximum likelihood solution $\mathbf{b}^o = \mathbf{b}^*$ applies, of course, only when the variance components are known, although just their values relative to σ_e^2 need be known in most applications as, for example, in (104). However, together with the fitting constants method of estimating variance components free of the fixed effects, (100) and its simplified forms provide a framework for estimating both the fixed effects and the variance components of a mixed model.

Equations (100) arise from the joint density of $\mathbf{y}$ and $\mathbf{u}$ which, on assuming $\mathbf{e} \sim N(\mathbf{0}, \mathbf{R})$ and $\mathbf{u} \sim N(\mathbf{0}, \mathbf{D})$, is

$$\begin{aligned} f(\mathbf{y}, \mathbf{u}) &= g(\mathbf{y} \mid \mathbf{u})h(\mathbf{u}) \\ &= C \exp[-\tfrac{1}{2}(\mathbf{y} - \mathbf{Xb} - \mathbf{Zu})'\mathbf{R}^{-1}(\mathbf{y} - \mathbf{Xb} - \mathbf{Zu})] \exp[-\tfrac{1}{2}\mathbf{u}'\mathbf{D}^{-1}\mathbf{u}] \end{aligned}$$

where C is a constant. Maximizing with respect to $\mathbf{b}$ and $\mathbf{u}$ leads at once to (100).

The solution for $\mathbf{b}^*$ in (100) is of interest because $\mathbf{b}$ is a vector of fixed effects in the model (94). However, even though $\mathbf{u}$ is a vector of random variables in (94), the solution for $\mathbf{u}^*$ in (100) is, in many situations, of interest also. It is an estimator of the conditional mean of $\mathbf{u}$ given $\mathbf{y}$. This we now show. First, from (94) and (95) we have $\text{cov}(\mathbf{u}, \mathbf{y}') = \mathbf{DZ}'$. Then, on assuming normality,

$$E(\mathbf{u} \mid \mathbf{y}) = E(\mathbf{u}) + \text{cov}(\mathbf{u}, \mathbf{y}')[\text{var}(\mathbf{y})]^{-1}[\mathbf{y} - E(\mathbf{y})] = \mathbf{DZ'V}^{-1}(\mathbf{y} - \mathbf{Xb}).$$

Hence from (100),

$$\begin{aligned}\mathbf{u}^* &= (\mathbf{Z}'\mathbf{R}^{-1}\mathbf{Z} + \mathbf{D}^{-1})^{-1}\mathbf{Z}'\mathbf{R}^{-1}(\mathbf{y} - \mathbf{X}\mathbf{b}^*)\\ &= (\mathbf{Z}'\mathbf{R}^{-1}\mathbf{Z} + \mathbf{D}^{-1})^{-1}\mathbf{Z}'\mathbf{R}^{-1}\mathbf{V}\mathbf{V}^{-1}(\mathbf{y} - \mathbf{X}\mathbf{b}^*)\\ &= (\mathbf{Z}'\mathbf{R}^{-1}\mathbf{Z} + \mathbf{D}^{-1})^{-1}\mathbf{Z}'\mathbf{R}^{-1}(\mathbf{Z}\mathbf{D}\mathbf{Z}' + \mathbf{R})\mathbf{V}^{-1}(\mathbf{y} - \mathbf{X}\mathbf{b}^*)\\ &= (\mathbf{Z}'\mathbf{R}^{-1}\mathbf{Z} + \mathbf{D}^{-1})^{-1}(\mathbf{Z}'\mathbf{R}^{-1}\mathbf{Z} + \mathbf{D}^{-1})\mathbf{D}\mathbf{Z}'\mathbf{V}^{-1}(\mathbf{y} - \mathbf{X}\mathbf{b}^*)\\ &= \mathbf{D}\mathbf{Z}'\mathbf{V}^{-1}(\mathbf{y} - \mathbf{X}\mathbf{b}^*),\end{aligned}$$

which is exactly $E(\mathbf{u} \mid \mathbf{y})$ with $\mathbf{b}$ replaced by $\mathbf{b}^*$, which we know is the maximum likelihood estimator of $\mathbf{b}$. Hence $\mathbf{u}^* = \widehat{E(\mathbf{u} \mid \mathbf{y})}$ is the maximum likelihood estimator of the mean of $\mathbf{u}$, for a given set of observations $\mathbf{y}$. It is, as mentioned by Henderson *et al.* (1959) and further discussed in Henderson (1963), the "estimated genetic merit" used by animal breeders. In their case $\mathbf{u}$ is a vector of genetic merit values of a series of animals from whom $\mathbf{y}$ is the vector of production records, and the problem is to use $\mathbf{y}$ to get estimated values of $\mathbf{u}$ in order to decide which animals are best in some sense.

b. Fixed effects and variance components

Maximum likelihood equations for estimating variance components from unbalanced data cannot be solved explicitly. The equations for the simplest case possible illustrate this. Consider the 1-way classification as described in Sec. 6. With

$$\mathbf{V} = \text{var}(\mathbf{y}) = \sigma_e\mathbf{I}_N + \sigma_\alpha^2 \sum_{i=1}^{a}{}^{+}\mathbf{J}_{n_i}\,.$$

as used there,

$$|\mathbf{V}| = \sigma_e^{2(N-a)} \prod_{i=1}^{a} (\sigma_e^2 + n_i\sigma_\alpha^2)$$

and
$$\mathbf{V}^{-1} = (1/\sigma_e^2)\mathbf{I}_N + \sum_{i=1}^{a}{}^{+}\frac{1}{n_i}\left(\frac{1}{\sigma_e^2 + n_i\sigma_\alpha^2} - \frac{1}{\sigma_e^2}\right)\mathbf{J}_{n_i}\,.$$

The likelihood function, on the basis of normality, is

$$(2\pi)^{-\frac{1}{2}N}|\mathbf{V}|^{-\frac{1}{2}} \exp\{-\tfrac{1}{2}(\mathbf{y} - \mu\mathbf{1})'\mathbf{V}^{-1}(\mathbf{y} - \mu\mathbf{1})\},$$

and, after substituting for $|\mathbf{V}|$ and $\mathbf{V}^{-1}$, the logarithm of this reduces to

$$\begin{aligned}L = \tfrac{1}{2}N\log(2\pi) - \tfrac{1}{2}(N - a)\log\sigma_e^2 - \tfrac{1}{2}\sum_{i=1}^{a}\log(\sigma_e^2 + n_i\sigma_\alpha^2)\\ - \tfrac{1}{2}(1/\sigma_e^2)\sum_{i=1}^{a}\sum_{j=1}^{n_i}(y_{ij} - \bar{y}_{i.})^2 - \tfrac{1}{2}\sum_{i=1}^{a}\frac{n_i(\bar{y}_{i.} - \mu)^2}{\sigma_e^2 + n_i\sigma_\alpha^2}\,.\end{aligned}$$

Equating to zero the differentials of L with respect to μ, σ_e^2 and σ_α^2 gives, formally, the equations whose solutions (to be denoted by $\tilde{\mu}$, $\tilde{\sigma}_e^2$ and $\tilde{\sigma}_\alpha^2$) are the maximum likelihood estimators. These equations are as follows:

$$\tilde{\mu} = \frac{\sum_{i=1}^{a} \frac{n_i \bar{y}_{i\cdot}}{\tilde{\sigma}_e^2 + n_i \tilde{\sigma}_\alpha^2}}{\sum_{i=1}^{a} \frac{n_i}{\tilde{\sigma}_e^2 + n_i \tilde{\sigma}_\alpha^2}},$$

$$\frac{N-a}{\tilde{\sigma}_e^2} + \sum_{i=1}^{a} \frac{1}{\tilde{\sigma}_e^2 + n_i \tilde{\sigma}_\alpha^2} - \frac{\sum_{i=1}^{a}\sum_{j=1}^{n_i} (y_{ij} - \bar{y}_{i\cdot})^2}{\tilde{\sigma}_e^4} - \sum_{i=1}^{a} \frac{n_i(\bar{y}_{i\cdot} - \tilde{\mu})^2}{(\tilde{\sigma}_e^2 + n_i \tilde{\sigma}_\alpha^2)^2} = 0$$

and
$$\sum_{i=1}^{a} \frac{n_i}{\tilde{\sigma}_e^2 + n_i \tilde{\sigma}_\alpha^2} - \sum_{i=1}^{a} \frac{n_i^2(\bar{y}_{i\cdot} - \tilde{\mu})^2}{(\tilde{\sigma}_e^2 + n_i \tilde{\sigma}_\alpha^2)^2} = 0.$$

Clearly these equations have no explicit solution for $\tilde{\mu}$, $\tilde{\sigma}_e^2$ and $\tilde{\sigma}_\alpha^2$. They do, of course, reduce to the simpler equations of balanced data given in equation (77) of Sec. 9.9g, when $n_i = n$ for all i. Even if solutions could be found in the unbalanced data case, the problem of using them to derive a non-negative estimator of σ_α^2 must also be considered, just as it is at the end of Sec. 9.9g for balanced data.

Explicit maximum likelihood estimators must therefore be despaired of. However, Hartley and Rao (1967) have developed a general set of equations from which specific estimates are obtained by iteration, involving extensive computations. We give their equations and mention how they indicate a solution may be obtained. To do so we rewrite the model (94) using

$$\mathbf{Zu} = \sum_{\theta=A}^{K} \mathbf{Z}_\theta \mathbf{u}_\theta,$$

where $\mathbf{u}_\theta$ is the vector of random effects of the θ-factor. Then defining γ_θ as $\gamma_\theta = \sigma_\theta^2/\sigma_e^2$ for $\theta = A, B, \ldots, K$, and $\mathbf{H}$ as

$$\mathbf{H} = \mathbf{I}_N + \sum_{\theta=A}^{K} \gamma_\theta \mathbf{Z}_\theta \mathbf{Z}_\theta', \tag{105}$$

$\mathbf{V}$ of (96) is $\mathbf{V} = \sigma_e^2\mathbf{H}$. On assuming normality the logarithm of the likelihood is

$$-\tfrac{1}{2}N \log(2\pi) - \tfrac{1}{2}N \log \sigma_e^2 - \tfrac{1}{2} \log |\mathbf{H}| - (\mathbf{y} - \mathbf{Xb})'\mathbf{H}^{-1}(\mathbf{y} - \mathbf{Xb})/2\sigma_e^2,$$

and equating to zero the differentials of this with respect to σ_e^2, the γ_θ and the elements of $\mathbf{b}$, gives the following equations:

$$\mathbf{X}'\tilde{\mathbf{H}}^{-1}\mathbf{X}\tilde{\mathbf{b}} = \mathbf{X}'\tilde{\mathbf{H}}^{-1}\mathbf{y}, \tag{106}$$

$$\tilde{\sigma}_e^2 = (\mathbf{y} - \mathbf{X}\tilde{\mathbf{b}})'\tilde{\mathbf{H}}^{-1}(\mathbf{y} - \mathbf{X}\tilde{\mathbf{b}})/N \tag{107}$$

and
$$\operatorname{tr}(\tilde{\mathbf{H}}^{-1}\mathbf{Z}_\theta\mathbf{Z}_\theta') = (\mathbf{y} - \mathbf{X}\tilde{\mathbf{b}})'\tilde{\mathbf{H}}^{-1}\mathbf{Z}_\theta\mathbf{Z}_\theta'\tilde{\mathbf{H}}^{-1}(\mathbf{y} - \mathbf{X}\tilde{\mathbf{b}})/\tilde{\sigma}_e^2 \quad \text{for} \quad \theta = A, B, \ldots, K. \tag{108}$$

These equations have to be solved for the elements of $\tilde{\mathbf{b}}$, the error variance $\tilde{\sigma}_e^2$ and the variance components inherent in $\tilde{\mathbf{H}}$. Hartley and Rao (1967) indicate how this can be achieved, either by the method of steepest ascent or by obtaining an alternative form for (108) which are the difficult equations to handle. Equations (106) and (107) are, of course, recognizable as the maximum likelihood equations for the fixed effects and the error variance; and they are easily solved if values of the $\tilde{\gamma}_\theta$'s are available for $\tilde{\mathbf{H}}$. Thus is iteration established via equations (106), (107) and (108).

c. Large sample variances

General expressions for large sample variances of maximum likelihood estimators of variance components have been obtained, under normality assumptions, by Searle (1970). They can be derived despite the fact that the estimators themselves cannot be obtained explicitly. On using the model (94), with $\text{var}(\mathbf{Zu} + \mathbf{e}) = \mathbf{V}$ as in (96) and with $\mathbf{y} \sim N(\mathbf{Xb}, \mathbf{V})$, the likelihood of the sample is

$$(2\pi)^{-\frac{1}{2}N}|\mathbf{V}|^{-\frac{1}{2}} \exp\{-\tfrac{1}{2}(\mathbf{y} - \mathbf{Xb})'\mathbf{V}^{-1}(\mathbf{y} - \mathbf{Xb})\}.$$

Apart from a constant the logarithm of this is

$$L = -\tfrac{1}{2}\log|\mathbf{V}| - \tfrac{1}{2}(\mathbf{y} - \mathbf{Xb})'\mathbf{V}^{-1}(\mathbf{y} - \mathbf{Xb}). \tag{109}$$

Suppose the model has p fixed effects and q variance components represented by $\boldsymbol{\sigma}^2 = \{\sigma_i^2\}$ for $i = 1, 2, \ldots, q$, one element of $\boldsymbol{\sigma}^2$ being σ_e^2. Then [see Wald (1943)] the variance-covariance matrix of the large sample maximum likelihood estimators of the p elements of $\mathbf{b}$ and the q variance components is

$$\begin{bmatrix} \text{var}(\tilde{\mathbf{b}}) & \text{cov}(\tilde{\mathbf{b}},\tilde{\boldsymbol{\sigma}}^2) \\ \text{cov}(\boldsymbol{\sigma}^2,\tilde{\mathbf{b}}) & \text{var}(\boldsymbol{\sigma}^2) \end{bmatrix} = \begin{bmatrix} -E(\mathbf{L}_{bb}) & -E(\mathbf{L}_{b\sigma^2}) \\ -E(\mathbf{L}_{b\sigma^2})' & -E(\mathbf{L}_{\sigma^2\sigma^2}) \end{bmatrix}^{-1}. \tag{110}$$

In (110) $\tilde{\mathbf{b}}$ and $\tilde{\boldsymbol{\sigma}}^2$ are maximum likelihood estimators of $\mathbf{b}$ and $\boldsymbol{\sigma}^2$ respectively, and the left-hand side is a statement of their covariance matrix. The right-hand side of (110) shows how to derive this covariance matrix. In its submatrices $\mathbf{L}_{bb}$, for example, is the $p \times p$ matrix of second differentials of $\mathbf{L}$ of (109) with respect to elements of $\mathbf{b}$. Definition of $\mathbf{L}_{b\sigma^2}$ and $\mathbf{L}_{\sigma^2\sigma^2}$ follows in similar manner.

The nature of (109) is such that, after some algebraic manipulations, (110) yields the following results:

$$\text{var}(\tilde{\mathbf{b}}) = (\mathbf{X}'\mathbf{V}^{-1}\mathbf{X})^{-1}, \tag{111}$$

$$\text{cov}(\tilde{\mathbf{b}}, \tilde{\boldsymbol{\sigma}}^2) = \mathbf{0} \tag{112}$$

and

$$\text{var}(\tilde{\boldsymbol{\sigma}}^2) = 2\left\{\text{tr}\left(\mathbf{V}^{-1}\frac{\partial\mathbf{V}}{\partial\sigma_i^2}\mathbf{V}^{-1}\frac{\partial\mathbf{V}}{\partial\sigma_j^2}\right) \quad \text{for} \quad i, j = 1 \cdots q\right\}^{-1}. \tag{113}$$

Searle (1970) gives details of deriving these results.

The three results (111)–(113) merit attention. First, (111) corresponds to the variance of $\mathbf{b}^{o}$ in (98) and therefore comes as no surprise. Nevertheless, it indicates that for unbalanced data from any mixed model the variance-covariance matrix of the maximum likelihood (under normality) estimators of the fixed effects is what it would be if the variance components were known and were not having to be estimated. Second, (112) shows that covariances between large sample maximum likelihood estimators of fixed effects and variance components are zero. The simplest case of this relates to the mean of a sample and the sample variance; under normality they are distributed independently. The generalization of this result is (112), which is therefore no surprise either. However, the generality of the context of its derivation is to be observed. Finally, (113) gives the variance-covariance matrix of the large sample maximum likelihood estimators of the variance components. It is, we notice, quite free of $\mathbf{X}$, the incidence matrix of the fixed effects. Its form, as evident in (113), is the inverse of a matrix whose typical element is the trace of a product of matrices $\mathbf{V}^{-1}$ and derivatives of $\mathbf{V}$ with respect to the variance components.

Example. Consider N observations from the model $y_i = \mu + e_i$ with $\mathbf{e} \sim N(\mathbf{0}, \sigma^2\mathbf{I}_N)$. Then $\mathbf{V} = \sigma^2\mathbf{I}_N$, $\mathbf{V}^{-1} = (1/\sigma^2)\mathbf{I}_N$ and $\mathbf{V}_{\sigma^2} = \mathbf{I}_N$. Hence from (113)

$$\text{var}(\tilde{\sigma}^2) = 2\{\text{tr}[(1/\sigma^2)\mathbf{I}_N\mathbf{I}_N]^2\}^{-1} = 2(N/\sigma^4)^{-1} = 2\sigma^4/N,$$

as is well known. Additional results stemming from (113) are shown in the next chapter.

9. MIXED MODELS HAVING ONE RANDOM FACTOR

The mixed model (94) has several simplifying features when it has only one factor that is random. We assume that in

$$\mathbf{y} = \mathbf{X}\mathbf{b} + \mathbf{Z}\mathbf{u} + \mathbf{e} \tag{114}$$

$r(\mathbf{X}) = r$, with $\mathbf{b}$ representing $q \geq r$ fixed effects and $\mathbf{u}$, in representing the random effects, contains t effects for just one random factor, having variance σ_u^2. As a result, $\mathbf{Z}$ has full column rank, t, with its columns summing to $\mathbf{1}$, the same as do certain columns of $\mathbf{X}$. This is assumed to be the only linear relationship of the columns of $\mathbf{Z}$ to those of $\mathbf{X}$. Hence

$$r[\mathbf{X} \quad \mathbf{Z}] = r(\mathbf{X}) + t - 1 = r + t - 1.$$

Also, by the nature of $\mathbf{Z}$, the matrix $\mathbf{Z}'\mathbf{Z}$ is diagonal, of order t, with $(\mathbf{Z}'\mathbf{Z})^{-1}$ existing.

Since the model is a mixed model, estimation is by the fitting constants method, using

$$\text{SSE} = \mathbf{y}'\mathbf{y} - R(\mathbf{b}, \mathbf{u}) \qquad \text{and} \qquad R(\mathbf{u} \mid \mathbf{b}) = R(\mathbf{b}, \mathbf{u}) - R(\mathbf{b}),$$

with
$$E(\text{SSE}) = [N - (r + t - 1)]\sigma_e^2 \tag{115}$$

in the usual manner and, from (79),

$$E[R(\mathbf{u} \mid \mathbf{b})] = \sigma_u^2 \operatorname{tr}[\mathbf{Z}'\mathbf{Z} - \mathbf{Z}'\mathbf{X}(\mathbf{X}'\mathbf{X})^{-}\mathbf{X}'\mathbf{Z}] + \sigma_e^2[r(\mathbf{X}) + t - 1 - r(\mathbf{X})]. \tag{116}$$

Hence estimators are

$$\hat{\sigma}_e^2 = \frac{\mathbf{y}'\mathbf{y} - R(\mathbf{b}, \mathbf{u})}{N - r(\mathbf{X}) - t + 1} \tag{117}$$

and
$$\hat{\sigma}_u^2 = \frac{R(\mathbf{u} \mid \mathbf{b}) - \hat{\sigma}_e^2(t - 1)}{\operatorname{tr}[\mathbf{Z}'\mathbf{Z} - \mathbf{Z}'\mathbf{X}(\mathbf{X}'\mathbf{X})^{-}\mathbf{X}'\mathbf{Z}]}, \tag{118}$$

as given by Cunningham and Henderson (1968) for the case of $\mathbf{X}$ having full column rank. For a particular case of ensuring the non-singularity of $\mathbf{X}'\mathbf{X}$ through appropriate "constraints" (see Sec. 5.7), Cunningham (1969) gives a simple expression for the denominator of (118).

A computational difficulty in the preceding formulation is

$$R(\mathbf{b}, \mathbf{u}) = \mathbf{y}'[\mathbf{X} \quad \mathbf{Z}]\begin{bmatrix} \mathbf{X}'\mathbf{X} & \mathbf{X}'\mathbf{Z} \\ \mathbf{Z}'\mathbf{X} & \mathbf{Z}'\mathbf{Z} \end{bmatrix}^{-}\begin{bmatrix} \mathbf{X}' \\ \mathbf{Z}' \end{bmatrix}\mathbf{y}. \tag{119}$$

Because $\mathbf{Z}$ has as many columns as there are random effects in the data, and the random effects can be very numerous, calculation of (119) may often be onerous. However, a generalization of the "absorption process" described in Chapter 7 for the 2-way classification permits of easier calculation as follows. With

$$R(\mathbf{u}) = \mathbf{y}'\mathbf{Z}(\mathbf{Z}'\mathbf{Z})^{-1}\mathbf{Z}'\mathbf{y}, \tag{120}$$

which is easy to calculate, we find that

$$R(\mathbf{b} \mid \mathbf{u}) = R(\mathbf{b}, \mathbf{u}) - R(\mathbf{u}) \tag{121}$$

simplifies, after substitution from (119) and (120), to

$$R(\mathbf{b} \mid \mathbf{u}) = \mathbf{b}^{o\prime}\mathbf{X}'[\mathbf{I} - \mathbf{Z}(\mathbf{Z}'\mathbf{Z})^{-1}\mathbf{Z}']\mathbf{y} \tag{122}$$

where
$$\mathbf{b}^o = \mathbf{Q}^{-}\mathbf{X}'[\mathbf{I} - \mathbf{Z}(\mathbf{Z}'\mathbf{Z})^{-1}\mathbf{Z}']\mathbf{y} \tag{123}$$

with
$$\mathbf{Q} = \mathbf{X}'\mathbf{X} - \mathbf{X}'\mathbf{Z}(\mathbf{Z}'\mathbf{Z})^{-1}\mathbf{Z}'\mathbf{X}. \tag{124}$$

Because $\mathbf{Q}$ and $\mathbf{b}^o$ have q rows, (122) is easier to compute than (119). Using (120) and (122) we then calculate $R(\mathbf{b}, \mathbf{u})$ as

$$R(\mathbf{b}, \mathbf{u}) = R(\mathbf{b} \mid \mathbf{u}) + R(\mathbf{u})$$

and hence, for (118), calculate $R(\mathbf{u} \mid \mathbf{b})$ as

$$R(\mathbf{u} \mid \mathbf{b}) = R(\mathbf{b} \mid \mathbf{u}) + R(\mathbf{u}) - R(\mathbf{b}), \tag{125}$$

where

$$R(\mathbf{b}) = \mathbf{y}'\mathbf{X}(\mathbf{X}'\mathbf{X})^{-}\mathbf{X}'\mathbf{y} \tag{126}$$

is also easily computed.

Results (122)–(126) are similar to those summarized by Cunningham and Henderson (1968) for a model in which $\mathbf{X}$ is assumed to have full column rank. This restriction is, as we see, not necessary. The crucial result is (122), derived from (121) by substituting from (119) and (120) using

$$\begin{bmatrix} \mathbf{X}'\mathbf{X} & \mathbf{X}'\mathbf{Z} \\ \mathbf{Z}'\mathbf{X} & \mathbf{Z}'\mathbf{Z} \end{bmatrix}^{-} = \begin{bmatrix} \mathbf{0} & \mathbf{0} \\ \mathbf{0} & (\mathbf{Z}'\mathbf{Z})^{-1} \end{bmatrix} + \begin{bmatrix} \mathbf{I} \\ -(\mathbf{Z}'\mathbf{Z})^{-1}\mathbf{Z}'\mathbf{X} \end{bmatrix} \mathbf{Q}^{-}[\mathbf{I} - \mathbf{X}'\mathbf{Z}(\mathbf{Z}'\mathbf{Z})^{-1}] \tag{127}$$

with $\mathbf{Q}$ of (124). In carrying out this derivation it will be found that $\mathbf{b}^o$ of (123) is a solution to

$$\begin{bmatrix} \mathbf{X}'\mathbf{X} & \mathbf{X}'\mathbf{Z} \\ \mathbf{Z}'\mathbf{X} & \mathbf{Z}'\mathbf{Z} \end{bmatrix} \begin{bmatrix} \mathbf{b}^o \\ \mathbf{u}^o \end{bmatrix} = \begin{bmatrix} \mathbf{X}'\mathbf{y} \\ \mathbf{Z}'\mathbf{y} \end{bmatrix}. \tag{128}$$

These are least squares normal equations for $\mathbf{b}^o$ and $\mathbf{u}^o$ assuming that $\mathbf{u}$ is a vector of fixed rather than random effects. Recall, however, that comparable equations for getting maximum likelihood solutions for the fixed effects are, from (104),

$$\begin{bmatrix} \mathbf{X}'\mathbf{X} & \mathbf{X}'\mathbf{Z} \\ \mathbf{Z}'\mathbf{X} & \mathbf{Z}'\mathbf{Z} + \lambda\mathbf{I} \end{bmatrix} \begin{bmatrix} \mathbf{b}^* \\ \mathbf{u}^* \end{bmatrix} = \begin{bmatrix} \mathbf{X}'\mathbf{y} \\ \mathbf{Z}'\mathbf{y} \end{bmatrix} \tag{129}$$

where

$$\lambda = \sigma_e^2/\sigma_u^2 . \tag{130}$$

Since (129) is formally the same as (128) except for $\mathbf{Z}'\mathbf{Z} + \lambda\mathbf{I}$ replacing $\mathbf{Z}'\mathbf{Z}$, Cunningham and Henderson (1968) suggested making this replacement throughout the whole variance component estimation process described in (117) through (126). The result is an iterative procedure based on the maximum likelihood equations implicit in (129). Thus (117) and (118) would become

$$\sigma_e^{*2} = \frac{\mathbf{y}'\mathbf{y} - R^*(\mathbf{b}, \mathbf{u})}{N - r(\mathbf{X}) - t + 1} \tag{131}$$

and

$$\sigma_u^{*2} = \frac{R^*(\mathbf{u} \mid \mathbf{b}) - \sigma_e^{*2}(t-1)}{\mathrm{tr}[\mathbf{Z}'\mathbf{Z} + \lambda\mathbf{I} - \mathbf{Z}'\mathbf{X}(\mathbf{X}'\mathbf{X})^{-}\mathbf{X}'\mathbf{Z}]} . \tag{132}$$

The comparable definitions of the R^*-terms are

$$R^*(\mathbf{b}, \mathbf{u}) = R^*(\mathbf{b} \mid \mathbf{u}) + R^*(\mathbf{u}) \tag{133}$$

derived from using

$$\mathbf{P} = \mathbf{Z}'\mathbf{Z} + \lambda\mathbf{I} \tag{134}$$

in place of $\mathbf{Z}'\mathbf{Z}$ in (120)–(126). Thus from (120)

$$R^*(\mathbf{u}) = \mathbf{y}'\mathbf{Z}\mathbf{P}^{-1}\mathbf{Z}'\mathbf{y}, \tag{135}$$

and from (122) through (124)

$$R^*(\mathbf{b} \mid \mathbf{u}) = \mathbf{y}'(\mathbf{I} - \mathbf{Z}\mathbf{P}^{-1}\mathbf{Z}')\mathbf{X}[\mathbf{X}'(\mathbf{I} - \mathbf{Z}\mathbf{P}^{-1}\mathbf{Z}')\mathbf{X}]^{-}\mathbf{X}'(\mathbf{I} - \mathbf{Z}\mathbf{P}^{-1}\mathbf{Z}')\mathbf{y}; \tag{136}$$

and (126) remains the same,

$$R^*(\mathbf{b}) = \mathbf{y}'\mathbf{X}(\mathbf{X}'\mathbf{X})^{-}\mathbf{X}'\mathbf{y}. \tag{137}$$

Then for (132), just as in (125),

$$R^*(\mathbf{u} \mid \mathbf{b}) = R^*(\mathbf{b} \mid \mathbf{u}) + R^*(\mathbf{u}) - R^*(\mathbf{b}). \tag{138}$$

The replacement of $\mathbf{Z}'\mathbf{Z}$ by $\mathbf{P} = \mathbf{Z}'\mathbf{Z} + \lambda\mathbf{I}$ as just described is based on the premise that the expected values of SSE* $= \mathbf{y}'\mathbf{y} - R^*(\mathbf{b}, \mathbf{u})$ and $R^*(\mathbf{u} \mid \mathbf{b})$ are those of SSE and $R(\mathbf{u} \mid \mathbf{b})$ shown in (115) and (116) with $\mathbf{Z}'\mathbf{Z}$ replaced by $\mathbf{P}$. Unfortunately, as Thompson (1969) has pointed out, this is not so, and consequently (131) and (132) are not unbiased estimators. Derivation of unbiased estimators, as indicated by Thompson (1969), proceeds as follows. Notice, first, that from (134)

$$\begin{aligned}\mathbf{P}^{-1}\mathbf{Z}'(\mathbf{Z}\mathbf{Z}'\sigma_u^2 + \sigma_e^2\mathbf{I}) &= \mathbf{P}^{-1}(\mathbf{Z}'\mathbf{Z}\sigma_u^2 + \sigma_e^2\mathbf{I})\mathbf{Z}' \\ &= \mathbf{P}^{-1}(\mathbf{Z}'\mathbf{Z} + \lambda\mathbf{I})\sigma_u^2\mathbf{Z}', \quad \text{from (130)} \\ &= \mathbf{P}^{-1}\mathbf{P}\mathbf{Z}'\sigma_u^2, \quad \text{from (134)} \\ &= \mathbf{Z}'\sigma_u^2;\end{aligned} \tag{139}$$

second, from (114)

$$E(\mathbf{y}\mathbf{y}') = \mathbf{X}\mathbf{b}\mathbf{b}'\mathbf{X}' + \mathbf{Z}\mathbf{Z}'\sigma_u^2 + \sigma_e^2\mathbf{I}. \tag{140}$$

Hence, using $E(\mathbf{y}'\mathbf{A}\mathbf{y}) = \text{tr}[\mathbf{A}E(\mathbf{y}\mathbf{y}')]$, the expected value of (135) is

$$\begin{aligned}E[R^*(\mathbf{u})] &= \text{tr}[\mathbf{Z}\mathbf{P}^{-1}\mathbf{Z}'E(\mathbf{y}\mathbf{y}')] \\ &= \text{tr}[\mathbf{Z}\mathbf{P}^{-1}\mathbf{Z}'\mathbf{X}\mathbf{b}\mathbf{b}'\mathbf{X}' + \mathbf{Z}\mathbf{Z}'\sigma_u^2].\end{aligned} \tag{141}$$

Similarly, with

$$\mathbf{T} = \mathbf{I} - \mathbf{Z}\mathbf{P}^{-1}\mathbf{Z}', \tag{142}$$

(139) gives $\mathbf{T}(\mathbf{Z}\mathbf{Z}'\sigma_u^2 + \sigma_e^2\mathbf{I}) = \sigma_e^2\mathbf{I}$ so that from (136) and (140) we have

$$\begin{aligned}E[R^*(\mathbf{b} \mid \mathbf{u})] &= \text{tr}[\mathbf{T}\mathbf{X}(\mathbf{X}'\mathbf{T}\mathbf{X})^{-}\mathbf{X}'\mathbf{T}E(\mathbf{y}\mathbf{y}')] \\ &= \text{tr}[\mathbf{T}\mathbf{X}(\mathbf{X}'\mathbf{T}\mathbf{X})^{-}\mathbf{X}'\mathbf{T}\mathbf{X}\mathbf{b}\mathbf{b}'\mathbf{X}' + \mathbf{T}\mathbf{X}(\mathbf{X}'\mathbf{T}\mathbf{X})^{-}\mathbf{X}'\sigma_e^2] \\ &= \text{tr}[\mathbf{T}\mathbf{X}\mathbf{b}\mathbf{b}'\mathbf{X}' + \mathbf{T}\mathbf{X}(\mathbf{X}'\mathbf{T}\mathbf{X})^{-}\mathbf{X}'\sigma_e^2],\end{aligned} \tag{143}$$

and from (137) and (140)

$$E[R^*(\mathbf{b})] = \text{tr}[\mathbf{X}(\mathbf{X}'\mathbf{X})^-\mathbf{X}'E(\mathbf{yy}')]$$
$$= \text{tr}[\mathbf{Xbb}'\mathbf{X}' + \mathbf{X}(\mathbf{X}'\mathbf{X})^-\mathbf{X}'(\mathbf{ZZ}'\sigma_u^2 + \sigma_e^2\mathbf{I})]. \tag{144}$$

Therefore, from (138), using (141), (143) and (144),

$$E[R^*(\mathbf{u} \mid \mathbf{b})] = \text{tr}\{(\mathbf{ZP}^{-1}\mathbf{Z}' + \mathbf{T} - \mathbf{I})\mathbf{Xbb}'\mathbf{X}' + [\mathbf{I} - \mathbf{X}(\mathbf{X}'\mathbf{X})^-\mathbf{X}']\mathbf{ZZ}'\sigma_u^2$$
$$+ [\mathbf{TX}(\mathbf{X}'\mathbf{TX})^-\mathbf{X}' - \mathbf{X}(\mathbf{X}'\mathbf{X})^-\mathbf{X}']\sigma_e^2\}$$
$$= \sigma_u^2\,\text{tr}[\mathbf{Z}'\mathbf{Z} - \mathbf{Z}'\mathbf{X}(\mathbf{X}'\mathbf{X})^-\mathbf{X}'\mathbf{Z}]$$
$$+ \sigma_e^2\,\text{tr}[\mathbf{X}'\mathbf{TX}(\mathbf{X}'\mathbf{TX})^- - \mathbf{X}'\mathbf{X}(\mathbf{X}'\mathbf{X})^-]. \tag{145}$$

Now by Lemma 1 in Sec. 1.2c, $\text{tr}[\mathbf{XTX}'(\mathbf{XTX}')^-] = r(\mathbf{XTX}')$. Furthermore, $\mathbf{T}$ has full rank (its inverse being $\mathbf{ZZ}'/\lambda + \mathbf{I}$), and so $r(\mathbf{XTX}') = r(\mathbf{X})$; and $\text{tr}[\mathbf{X}'\mathbf{X}(\mathbf{X}'\mathbf{X})^-] = r(\mathbf{X})$. Hence the last term of (145) is zero and so

$$E[R^*(\mathbf{u} \mid \mathbf{b})] = \sigma_u^2\,\text{tr}[\mathbf{Z}'\mathbf{Z} - \mathbf{Z}'\mathbf{X}(\mathbf{X}'\mathbf{X})^-\mathbf{X}'\mathbf{Z}].$$

Also, from (140)–(143)

$$E[\mathbf{y}'\mathbf{y} - R^*(\mathbf{b}, \mathbf{u})] = E[\mathbf{y}'\mathbf{y} - R^*(\mathbf{u}) - R^*(\mathbf{b} \mid \mathbf{u})] = [N - r(\mathbf{X})]\sigma_e^2.$$

Therefore, in place of (131) and (132) estimators for σ_e^2 and σ_u^2 are

$$\tilde{\sigma}_e^2 = \frac{\mathbf{y}'\mathbf{y} - [R^*(\mathbf{u}) + R^*(\mathbf{b} \mid \mathbf{u})]}{N - r(\mathbf{X})} \tag{146}$$

and

$$\tilde{\sigma}_u^2 = \frac{R^*(\mathbf{u}) + R^*(\mathbf{b} \mid \mathbf{u}) - R^*(\mathbf{b})}{\text{tr}[\mathbf{Z}'\mathbf{Z} - \mathbf{Z}'\mathbf{X}(\mathbf{X}'\mathbf{X})^-\mathbf{X}'\mathbf{Z}]}. \tag{147}$$

These results, given by Thompson (1969) for $\mathbf{X}$ of full column rank, provide an iterative procedure because, through $\mathbf{P}$ of (134), the reductions $R^*(\mathbf{u})$ and $R^*(\mathbf{b} \mid \mathbf{u})$ of (135) and (136) involve $\lambda = \sigma_u^2/\sigma_e^2$. Estimation is therefore achieved by taking an initial value of λ, calculating (146) and (147), using the results to get a next value of λ and repeating the process.

The replacement of $\mathbf{Z}'\mathbf{Z}$ by $\mathbf{P} = \mathbf{Z}'\mathbf{Z} + \lambda\mathbf{I}$ in the fitting constants method of estimation does not lead from (117) and (118) to (131) and (132) because in the method so modified $R^*(\mathbf{b}, \mathbf{u})$ is not, as Thompson (1969) points out, a reduction in sum of squares due to solving (129). It is true that

$$R^*(\mathbf{b}, \mathbf{u}) = \mathbf{b}^{*\prime}\mathbf{X}'\mathbf{y} + \mathbf{u}^{*\prime}\mathbf{Z}'\mathbf{y}.$$

However, the right-hand side of this equation is the reduction in sum of squares only when the equation from which it stems, (129) in this case, is, for some matrix $\mathbf{W}$, of the form

$$\mathbf{W}'\mathbf{W}\begin{bmatrix}\mathbf{b}^*\\ \mathbf{u}^*\end{bmatrix} = \mathbf{W}'\mathbf{y}.$$

By observation, (129) is not of this form. Furthermore, as shown by Thompson (1969), the reduction in sum of squares after solving (129) is

$$\mathbf{y}'\mathbf{y} - (\mathbf{y} - \mathbf{X}\mathbf{b}^* - \mathbf{Z}\mathbf{u}^*)'(\mathbf{y} - \mathbf{X}\mathbf{b}^* - \mathbf{Z}\mathbf{u}^*) = R^*(\mathbf{u}, \mathbf{b}) + \lambda\mathbf{u}^{*\prime}\mathbf{u}^*.$$

The calculations involved in the estimators (117), and (118), are summarized in Sec. 11.7b and those for the estimators (146) and (147) are in Sec. 11.7c.

10. BEST QUADRATIC UNBIASED ESTIMATION

The variance component analogue of the best linear unbiased estimator of a function of fixed effects is a best quadratic unbiased estimator (BQUE) of a variance component. By this we mean a quadratic function of the observations that is an unbiased estimator of the component, and of all such estimators it is the one having minimum variance. BQUE's of variance components from balanced data are those derived by the analysis of variance method, as has been discussed in Sec. 9.8a. Derivation of such estimators from unbalanced data is, however, more difficult—a situation that is not unexpected. Ideally we would like estimators that are uniformly "best" for all values of the variance components. In general, no such uniformly BQUE's exist. However, Townsend and Searle (1971) have obtained locally BQUE's for the variance components in a 1-way classification with $\mu = 0$, and from these they have suggested approximate BQUE's for the $\mu \neq 0$ model. We here outline the development for the $\mu = 0$ case.

The model $y_{ij} = \alpha_i + e_{ij}$ is written, similar to (94), as $\mathbf{y} = \mathbf{Z}\boldsymbol{\alpha} + \mathbf{e}$ with $\mathbf{V}$ of (96) being $\mathbf{V} = \sigma_\alpha^2\mathbf{Z}\mathbf{Z}' + \sigma_e^2\mathbf{I}$. Suppose we let the desired estimators of σ_e^2 and σ_α^2 be

$$\hat{\sigma}_e^2 = \mathbf{y}'\mathbf{A}\mathbf{y} \quad \text{and} \quad \hat{\sigma}_\alpha^2 = \mathbf{y}'\mathbf{B}\mathbf{y}$$

such that

$$E(\hat{\sigma}_e^2) = \text{tr}(\mathbf{A}\mathbf{V}) = \sigma_e^2 \quad \text{and} \quad E(\hat{\sigma}_\alpha^2) = \text{tr}(\mathbf{B}\mathbf{V}) = \sigma_\alpha^2 \tag{148}$$

and that

$$v(\hat{\sigma}_e^2) = 2\text{tr}(\mathbf{A}\mathbf{V})^2 \quad \text{and} \quad v(\hat{\sigma}_\alpha^2) = 2\text{tr}(\mathbf{B}\mathbf{V})^2 \quad \text{be minimized.} \tag{149}$$

The problem is then to find matrices $\mathbf{A}$ and $\mathbf{B}$ such that (149) is satisfied subject to (148). Upon obtaining the canonical form of $\mathbf{V}$ under orthogonal similarity as $\mathbf{P}'\mathbf{V}\mathbf{P} = \mathbf{D}$ where $\mathbf{P}$ is orthogonal and $\mathbf{D}$ is the diagonal matrix of latent roots of $\mathbf{V}$, we find that satisfying (148) and (149) demands minimizing $2\text{tr}(\mathbf{D}\mathbf{Q})^2$ subject to $\sigma_e^2 = \text{tr}(\mathbf{D}\mathbf{Q})$ and minimizing $2\text{tr}(\mathbf{D}\mathbf{R})^2$ subject to $\sigma_\alpha^2 = \text{tr}(\mathbf{D}\mathbf{R})$ where $\mathbf{Q} = \mathbf{P}'\mathbf{A}\mathbf{P}$ and $\mathbf{R} = \mathbf{P}'\mathbf{B}\mathbf{P}$. The latent roots of $\mathbf{V}$ are

σ_e^2, with multiplicity $N - a$, and $\sigma_e^2 + n_i\sigma_\alpha^2$ for $i = 1, 2, \ldots, a$; the corresponding latent vectors are the columns of the matrix $\Sigma^+ \mathbf{G}_i$, where $\mathbf{G}_i'$ is the last $(n_i - 1)$ rows of a Helmert matrix of order n_i (see Sec. 2.1), and the columns of $\mathbf{Z}$. The minimization procedure leads, after some algebraic simplification, to the following results. Define

$$\rho = \sigma_\alpha^2/\sigma_e^2, \qquad r = \sum_{i=1}^{a} \frac{1}{(1 + n_i\rho)^2} + N - a,$$

$$s = \sum_{i=1}^{a} \frac{n_i^2}{(1 + n_i\rho)^2} \quad \text{and} \quad t = \sum_{i=1}^{a} \frac{n_i}{(1 + n_i\rho)^2}.$$

Then the BQUE's are

$$\hat{\sigma}_e^2 = \frac{1}{rs - t^2}\left[\sum_{i=1}^{a} \frac{s - tn_i}{(1 + n_i\rho)^2}\frac{y_{i\cdot}^2}{n_i} + s(\text{SSE})\right]$$

and

$$\hat{\sigma}_\alpha^2 = \frac{1}{rs - t^2}\left[\sum_{i=1}^{a} \frac{rn_i - t}{(1 + n_i\rho)^2}\frac{y_{i\cdot}^2}{n_i} - t(\text{SSE})\right], \tag{150}$$

where SSE is the usual error sum of squares, $\Sigma\Sigma\, y_{ij}^2 - \Sigma\, n_i\bar{y}_{i\cdot}^2$.

These estimators are functions of the variance components through being functions of the ratio $\rho = \sigma_\alpha^2/\sigma_e^2$. The variances of the estimators are identical to those of the large sample maximum likelihood estimators, and the limits of the estimators as $\rho \to 0$ are the Koch (1968) estimators given in (92). The limit of $\hat{\sigma}_e^2$ as $\rho \to \infty$ is the analysis of variance method estimator of σ_e^2.

Comparison of the BQUE's with other estimators is difficult not only because their variances are functions of the unknown variance components but also because the BQUE's themselves, as in (150), are functions of those components. Townsend (1968) therefore compared the BQUE's with the analysis of variance method (ANOVA) estimators numerically. In doing so he used a range of values of ρ, both for the actual BQUE's (assuming ρ known) and for approximate BQUE's using a prior estimate, or guess, ρ_0 of ρ, in the estimation procedure. He found that considerable reduction in the variance of estimates of σ_α^2 can be achieved if the approximate BQUE is used rather than the ANOVA estimator. Furthermore, this advantage can be gained even when rather inaccurate prior estimates (guesses) of ρ are used as ρ_0. The reduction in variance appears to be greatest when the data are severely unbalanced and ρ is either small or large, and it appears smallest for values of ρ that are moderately small. In some cases there is actually no reduction in variance, when the ANOVA is a BQUE for some specific ρ. Details of these comparisons are to be found in Townsend (1968). The estimators, their variances and suggested expressions for the $\mu \neq 0$ model, taken from Townsend (1968), are shown in Sec. 11.1f.

11. EXERCISES

1. In equation (32) show that for balanced data $\theta_1 = bn(\Sigma\alpha_i^2 - a\bar{\alpha}_{\cdot}^2)$, and that for the α's random θ_1 is the term in σ_α^2 in (27).

2. Establish result (33) and show that for balanced data $\theta_2 = 0$ and for the α's random

$$\theta_2 = \left(\sum_{j=1}^{b} \frac{\sum_{i=1}^{a} n_{ij}^2}{n_{.j}} - \frac{\sum_{i=1}^{a} n_{i.}^2}{N} \right) \sigma_\alpha^2$$

3. Explain why, for nested classification models, there is only one way of carrying out the fitting constants method of estimating variance components and that it is equivalent to the analysis of variance method.

4. In fitting $\mathbf{y} = \mu\mathbf{1} + \mathbf{X}_f\mathbf{b}_f + \mathbf{X}_1\mathbf{b}_1 + \mathbf{X}_2\mathbf{b}_2 + \mathbf{e}$ show that $R(\mathbf{b}_1 \mid \mathbf{b}_f)$ equals $R(\mathbf{b}_1)_z$ when fitting $\mathbf{z} = \mathbf{W}\mathbf{y} = \mathbf{W}\mathbf{X}_1\mathbf{b}_1 + \mathbf{W}\mathbf{X}_2\mathbf{b}_2 + \mathbf{W}\mathbf{e}$, where $\mathbf{W} = \mathbf{I} - \mathbf{X}_f(\mathbf{X}_f'\mathbf{X}_f)^-\mathbf{X}_f'$. Show also that the reduction in sum of squares due to fitting $\mathbf{z} = \mathbf{W}\mathbf{X}_1\mathbf{b}_1 + \mathbf{W}\mathbf{e}$ by generalized least squares is $R(\mathbf{b}_1)_z$.

5. In the r-way classification random model, having all possible interactions, show that var($\mathbf{t}$) has $2^{r-1}(2^r + 1)$ different elements.

6. Derive, from first principles, $v(s^2)$ given at the end of Sec. 10.2.

7. Show that the estimators $\hat{\sigma}_e^2$ and $\hat{\sigma}_\alpha^2$ given by (87)–(89) are the analysis of variance method estimators for balanced data.

8. Find the variance of $\hat{\sigma}_e^2$ given in (90), checking the term in $4\mu^2$ given below (91). Check your results for balanced data.

9. Find the variance of the estimator $\hat{\sigma}_\alpha^2$ that can be derived from (88) and (89).

10. Show that for balanced data the estimators in (90) and (92) simplify to be the analysis of variance method estimators.

11. Use (38) to derive

$$E(T_o) = N\left(\mu^2 + \sum_{\theta=A}^{K} \sigma_\theta^2 + \sigma_e^2\right)$$

and

$$E(T_\mu) = N\mu^2 + \sum_{\theta=A}^{K} \left[\sum_j n^2(\theta_j) \right] \sigma_\theta^2 + \sigma_e^2.$$

12. Derive equation (122) and the last equation in Sec. 9.

CHAPTER 11

VARIANCE COMPONENT ESTIMATION FROM UNBALANCED DATA: FORMULAE

This chapter catalogues detailed formulae resulting from the application of methods discussed in Chapter 10 to specific models. Just the results available in the literature are included, with reference to their source. The amount of information available therefore varies from model to model, depending on what results have been found in the literature. Bayesian results are not included here, for the methodology behind them has not been discussed in Chapter 10. The reader interested in this is referred to Hill (1965, 1967), Tiao and Tan (1965, 1966) and Tiao and Box (1967). Other than this exclusion, every attempt has been made to make the following list complete but there can, of course, be no guarantee of this. The results are given without comment, in a notation that has been made as uniform and as consistent with Chapters 9 and 10 as possible. The results for each model have been gathered together, starting with the 1-way classification and ending with the 3-way crossed classification.

The equation of the model is shown for each situation discussed. Limits of the subscripts are shown also and thereafter, for typographical convenience, many of the summations are written using just these subscripts. For example, with $i = 1, 2, \ldots, a$, we write

$$N = \sum_{i=1}^{a} n_i \quad \text{as} \quad N = \textstyle\sum_i n_i \,.$$

1. THE 1-WAY CLASSIFICATION

a. Model

$$y_{ij} = \mu + \alpha_i + e_{ij} \,,$$

$$i = 1, 2, \ldots, a \quad \text{and} \quad j = 1, 2, \ldots, n_i \,, \quad \text{with} \quad N = \textstyle\sum_i n_i \,.$$

b. Analysis of variance estimators

Calculate $$T_o = \sum_{i=1}^{a}\sum_{j=1}^{n_i} y_{ij}^2, \quad T_A = \sum_{i=1}^{a} y_{i\cdot}^2/n_i, \quad T_\mu = \frac{y_{\cdot\cdot}^2}{N},$$

$$S_2 = \sum_i n_i^2 \quad \text{and} \quad S_3 = \sum_i n_i^3.$$

Then $$\hat{\sigma}_e^2 = (T_o - T_A)/(N - a)$$

and $$\hat{\sigma}_\alpha^2 = [T_A - T_\mu - (a-1)\hat{\sigma}_e^2]/(N - S_2/N).$$

c. Variances of analysis of variance estimators (under normality)

$$v(\hat{\sigma}_e^2) = 2\sigma_e^4/(N - a)$$

$$v(\hat{\sigma}_\alpha^2) = \frac{2\sigma_e^4 N^2(N-1)(a-1)}{(N-a)(N^2 - S_2)^2} + \frac{4\sigma_e^2\sigma_\alpha^2 N}{N^2 - S_2} + \frac{2\sigma_\alpha^4(N^2 S_2 + S_2^2 - 2NS_3)}{(N^2 - S_2)^2}$$

$$\text{cov}(\hat{\sigma}_\alpha^2, \hat{\sigma}_e^2) = -N(a-1)v(\hat{\sigma}_e^2)/(N^2 - S_2).$$

(Searle, 1956)

d. Variances of large sample maximum likelihood estimators (with normality)

Calculate

$$\rho = \sigma_\alpha^2/\sigma_e^2, \quad w_i = n_i/(1 + n_i\rho)$$

and $$D = N\sum_i w_i^2 - (\sum_i w_i)^2.$$

Then $$v(\tilde{\sigma}_e^2) = 2\sigma_e^4(\sum_i w_i^2)/D,$$

$$v(\tilde{\sigma}_\alpha^2) = 2\sigma_e^4[N - a + \sum_i w_i^2/n_i^2]/D$$

and $$\text{cov}(\tilde{\sigma}_\alpha^2, \tilde{\sigma}_e^2) = -2\sigma_e^4(\sum_i w_i^2/n_i)/D.$$

(Crump 1951; Searle, 1956)

e. Symmetric sums estimators

$$\widehat{\mu^2} = (y_{\cdot\cdot}^2 - \sum_i y_{i\cdot}^2)/(N^2 - S_2)$$

$$\mathring{\sigma}_e^2 = (\sum_i n_i \sum_j y_{ij}^2 - \sum_i y_{i\cdot}^2)/(S_2 - N)$$

$$\mathring{\sigma}_\alpha^2 = (NT_o - \sum_i n_i \sum_j y_{ij}^2)/(N^2 - S_2) - \widehat{\mu^2} - \mathring{\sigma}_e^2.$$

(Koch, 1968)

f. BQUE's for the model with $\mu = 0$ (under normality)

With $\rho = \sigma_\alpha^2/\sigma_e^2$ calculate

$$q_i = 1/(1 + n_i\rho),$$

$$r = \sum_i q_i^2 + (N - a), \quad s = \sum_i n_i^2 q_i^2 \quad \text{and} \quad t = \sum_i n_i q_i^2.$$

Then $${}_B\hat{\sigma}_e^2 = [\sum_i (s - tn_i)q_i^2 n_i \bar{y}_{i\cdot}^2 + s(T_o - T_A)]/(rs - t^2)$$

and $${}_B\hat{\sigma}_\alpha^2 = [\sum_i (rn_i - t)q_i^2 n_i \bar{y}_{i\cdot}^2 - t(T_o - T_A)]/(rs - t^2)$$

with variances

$$v({}_B\hat{\sigma}_e^2) = 2s\sigma_e^4/(rs - t^2) \quad \text{and} \quad v({}_B\hat{\sigma}_\alpha^2) = 2r\sigma_e^4/(rs - t^2).$$

For an approximate BQUE of σ_α^2, assign some value ρ_0 to ρ and using ρ_0 in place of ρ denote q_i by $q_{i,0}$.

Then $${}_{B,0}\hat{\sigma}_\alpha^2 = [\textstyle\sum_i(r_0 n_i - t_0)q_{i,0}^2 n_i \bar{y}_{i\cdot}^2 - t_0(T_o - T_A)]/(r_0 s_0 - t_0^2)$$

with variance

$$v({}_{B,0}\hat{\sigma}_\alpha^2) = 2\sigma_e^4[\textstyle\sum_i(r_0 n_i - t_0)^2 q_{i,0}^4/q_i^2 + (N - a)t_0^2]/(r_0 s_0 - t_0^2)^2.$$

(Townsend, 1968)

g. BQUE's for the model with $\mu \neq 0$

They are unknown. A suggested estimator for σ_α^2, based on using results from the $\mu = 0$ model in the general estimator given by Tukey (1957), is as follows.

Calculate $$w_i = \frac{rn_i^2 - tn_i}{(rs - t^2)(1 + n_i\rho)^2} \quad \text{and} \quad u_i = \frac{n_i q_i}{\sum_i n_i q_i},$$

$$\theta_i = w_i - 2u_i w_i + u_i^2 \quad \text{and} \quad \bar{y}_u = \textstyle\sum_i u_i \bar{y}_{i\cdot},$$

and $$\varphi_{ii'} = u_i u_{i'} - u_i w_{i'} - u_{i'} w_i, \quad \text{for } i \neq i'$$

Then $${}_B\hat{\hat{\sigma}}_\alpha^2 = [\textstyle\sum_i w_i(\bar{y}_{i\cdot} - \bar{y}_u)^2 - (\sum_i \theta_i/n_i)(T_o - T_A)/(N - a)]/\sum_i \theta_i$$

with variance

$$\begin{aligned} v({}_B\hat{\hat{\sigma}}_\alpha^2) = {} & 2\sigma_e^4[\textstyle\sum_i \theta_i^2 + (\sum_i \theta_i/n_i)^2/(N - a) + \sum_i \sum_{i' \neq i} n_i n_{i'} \varphi_{ii'}] \\ & + 2\sigma_e^2\sigma_\alpha^2[2\textstyle\sum_i n_i \theta_i^2 + \sum_i \sum_{i' \neq i}(n_i + n_{i'})\varphi_{ii'}] \\ & + 2\sigma_\alpha^4[\textstyle\sum_i n_i^2 \theta_i^2 + \sum_i \sum_{i' \neq i} n_i n_{i'} \varphi_{ii'}]. \end{aligned}$$

(Townsend, 1968)

2. THE 2-WAY NESTED CLASSIFICATION

a. Model

$$y_{ijk} = \mu + \alpha_i + \beta_{ij} + e_{ijk},$$

$$i = 1, 2, \ldots, a, \quad j = 1, 2, \ldots, b_i \quad \text{and} \quad k = 1, 2, \ldots, n_{ij},$$

with $$b. = \textstyle\sum_i b_i \quad \text{and} \quad N = \sum_i \sum_j n_{ij}.$$

b. Analysis of variance estimators

Calculate

$$k_1 = \textstyle\sum_i n_{i\cdot}^2/N, \quad k_3 = \sum_i \sum_j n_{ij}^2/N, \quad k_{12} = \sum_i(\sum_j n_{ij}^2/n_{i\cdot}),$$

$$T_A = \textstyle\sum_i y_{i\cdot\cdot}^2/n_{i\cdot}, \quad T_{AB} = \sum_i \sum_j y_{ij\cdot}^2/n_{ij},$$

$$T_o = \textstyle\sum_i \sum_j \sum_k y_{ijk}^2 \quad \text{and} \quad T_\mu = y_{\cdots}^2/N.$$

Then $\hat{\sigma}_e^2 = (T_o - T_{AB})/(N - b.)$

$$\hat{\sigma}_\beta^2 = [T_{AB} - T_A - (b. - a)\hat{\sigma}_e^2]/(N - k_{12})$$

$$\hat{\sigma}_\alpha^2 = [T_A - T_\mu - (k_{12} - k_3)\hat{\sigma}_\beta^2 - (a - 1)\hat{\sigma}_e^2]/(N - k_1).$$

(Searle, 1961)

c. Variances of analysis of variance estimators (under normality)

$$v(\hat{\sigma}_e^2) = 2\sigma_e^4/(N - b.).$$

Calculate

$$k_4 = \textstyle\sum_i\sum_j n_{ij}^3 \qquad k_5 = \sum_i(\sum_j n_{ij}^3/n_{i\cdot})$$

$$k_6 = \textstyle\sum_i(\sum_j n_{ij}^2)^2/n_{i\cdot} \qquad k_7 = \sum_i(\sum_j n_{ij}^2)^2/n_{i\cdot}^2$$

$$k_8 = \textstyle\sum_i n_{i\cdot}(\sum_j n_{ij}^2) \qquad k_9 = \sum_i n_{i\cdot}^3$$

and

$$\lambda_1 = (N - k_{12})^2[k_1(N + k_1) - 2k_9/N],$$

$$\lambda_2 = k_3[N(k_{12} - k_3)^2 + k_3(N - k_{12})^2] + (N - k_3)^2 k_7 - 2(N - k_3)[(k_{12} - k_3)k_5 + (N - k_{12})k_6/N] + 2(N - k_{12})(k_{12} - k_3)k_4/N,$$

$$\lambda_3 = [(N - k_{12})^2(N - 1)(a - 1) - (N - k_3)^2(a - 1)(b. - a) + (k_{12} - k_3)^2(N - 1)(b. - a)]/(N - b.),$$

$$\lambda_4 = (N - k_{12})^2[k_3(N + k_1) - 2k_8/N],$$

$$\lambda_5 = (N - k_{12})^2(N - k_1)$$

and

$$\lambda_6 = (N - k_{12})(N - k_3)(k_{12} - k_3).$$

Then

$$v(\hat{\sigma}_\alpha^2) = \frac{2(\lambda_1\sigma_\alpha^4 + \lambda_2\sigma_\beta^4 + \lambda_3\sigma_e^4 + 2\lambda_4\sigma_\alpha^2\sigma_\beta^2 + 2\lambda_5\sigma_\alpha^2\sigma_e^2 + 2\lambda_6\sigma_\beta^2\sigma_e^2)}{(N - k_1)^2(N - k_{12})^2}$$

$$v(\hat{\sigma}_\beta^2) = \frac{2(k_7 + Nk_3 - 2k_5)\sigma_\beta^4 + 4(N - k_{12})\sigma_\beta^2\sigma_e^2 + 2(b. - a)(N - a)\sigma_e^4/(N - b.)}{(N - k_{12})^2}$$

$$\text{cov}(\hat{\sigma}_\alpha^2, \hat{\sigma}_e^2) = [(k_{12} - k_3)(b. - a)/(N - k_{12}) - (a - 1)]v(\hat{\sigma}_e^2)/(N - k_1)$$

$$\text{cov}(\hat{\sigma}_\beta^2, \hat{\sigma}_e^2) = -(b. - a)v(\hat{\sigma}_e^2)/(N - k_{12})$$

$$\text{cov}(\hat{\sigma}_\alpha^2, \hat{\sigma}_\beta^2) = \{2[k_5 - k_7 + (k_6 - k_4)/N]\sigma_\beta^4 + 2(a - 1)(b. - a)\sigma_e^4/(N - b.) - (N - k_{12})(k_{12} - k_3)v(\hat{\sigma}_\beta^2)\}/(N - k_1)(N - k_{12}).$$

(Searle, 1961)

d. Variances of large sample maximum likelihood estimators (with normality)

$$\begin{bmatrix} v(\tilde{\sigma}_\alpha^2) & \text{cov}(\tilde{\sigma}_\alpha^2, \tilde{\sigma}_\beta^2) & \text{cov}(\tilde{\sigma}_\alpha^2, \tilde{\sigma}_e^2) \\ \text{cov}(\tilde{\sigma}_\alpha^2, \tilde{\sigma}_\beta^2) & v(\tilde{\sigma}_\beta^2) & \text{cov}(\tilde{\sigma}_\beta^2, \tilde{\sigma}_e^2) \\ \text{cov}(\tilde{\sigma}_\alpha^2, \tilde{\sigma}_e^2) & \text{cov}(\tilde{\sigma}_\beta^2, \tilde{\sigma}_e^2) & v(\tilde{\sigma}_e^2) \end{bmatrix} = 2 \begin{bmatrix} t_{\alpha\alpha} & t_{\alpha\beta} & t_{\alpha e} \\ t_{\alpha\beta} & t_{\beta\beta} & t_{\beta e} \\ t_{\alpha e} & t_{\beta e} & t_{ee} \end{bmatrix}^{-1}$$

with
$$m_{ij} = n_{ij}\sigma_\beta^2 + \sigma_e^2 ,$$
$$A_{ipq} = \textstyle\sum_j (n_{ij}^p / m_{ij}^q), \text{ for integers } p \text{ and } q,$$
$$q_i = 1 + \sigma_\alpha^2 A_{i11} ,$$

and
$$t_{\alpha\alpha} = \textstyle\sum_i A_{i11}^2 / q_i^2 ,$$
$$t_{\alpha\beta} = \textstyle\sum_i A_{i22} / q_i^2 ,$$
$$t_{\alpha e} = \textstyle\sum_i A_{i12} / q_i^2 ,$$
$$t_{\beta\beta} = \textstyle\sum_i (A_{i22} - 2\sigma_\alpha^2 A_{i33}/q_i + \sigma_\alpha^4 A_{i22}^2 / q_i^2),$$
$$t_{\beta e} = \textstyle\sum_i (A_{i12} - 2\sigma_\alpha^2 A_{i23}/q_i + \sigma_\alpha^4 A_{i12} A_{i22} / q_i^2)$$

and
$$t_{ee} = \textstyle\sum_i (A_{i02} - 2\sigma_\alpha^2 A_{i13}/q_i + \sigma_\alpha^4 A_{i12}^2 / q_i^2) + (N - b.)/\sigma_e^4 .$$

(Searle, 1970)

e. Symmetric sums estimators

$$\widehat{\mu^2} = (y_{\cdots}^2 - \textstyle\sum_i y_{i\cdot\cdot}^2)/(N^2 - Nk_1)$$
$$g_\alpha = \textstyle\sum_i (y_{i\cdot\cdot}^2 - \sum_j y_{ij\cdot}^2)/N(k_1 - k_3)$$
$$\hat{\sigma}_e^2 = (\textstyle\sum_i \sum_j n_{ij} \sum_k y_{ijk}^2 - \sum_i \sum_j y_{ij\cdot}^2)/N(k_3 - 1)$$
$$\hat{\sigma}_\beta^2 = \textstyle\sum_i \sum_j (n_{i\cdot} - n_{ij}) \sum_k y_{ijk}^2 / N(k_1 - k_3) - g_\alpha - \hat{\sigma}_e^2$$
$$\hat{\sigma}_\alpha^2 = \textstyle\sum_i (N - n_{i\cdot}) \sum_j \sum_k y_{ijk}^2 / N(N - k_1) - \widehat{\mu^2} - \hat{\sigma}_\beta^2 - \hat{\sigma}_e^2 .$$

(Koch, 1968)

3. THE 3-WAY NESTED CLASSIFICATION

a. Model

$$y_{ijkm} = \mu + \alpha_i + \beta_{ij} + \gamma_{ijk} + e_{ijkm} ,$$

$i = 1, 2, \ldots, a$, $\quad j = 1, 2, \ldots, b_i$, $\quad k = 1, 2, \ldots, c_{ij}$,

and
$$m = 1, 2, \ldots, n_{ijk} ,$$
with

$$b. = \textstyle\sum_i b_i , \quad c_{i\cdot} = \sum_j c_{ij} , \quad c_{\cdot j} = \sum_i c_{ij} \quad \text{and} \quad N = \sum_i \sum_j \sum_k n_{ijk} .$$

b. Analysis of variance estimators

Calculate $k_1 = \sum_i n_{i..}^2/N \qquad k_2 = \sum_i\sum_j n_{ij.}^2/N$

$$k_3 = \sum_i\sum_j\sum_k n_{ijk}^2/N \qquad k_4 = \sum_i\sum_j n_{ij.}^2/n_{i..}$$

$$k_5 = \sum_i\sum_j\sum_k n_{ijk}^2/n_{i..} \qquad k_6 = \sum_i\sum_j\sum_k n_{ijk}^2/n_{ij.}$$

and $v_1 = N - k_1 \qquad v_2 = k_4 - k_2 \qquad v_3 = k_5 - k_3 \qquad v_4 = a - 1$

$$v_5 = N - k_4 \qquad v_6 = k_6 - k_5 \qquad v_7 = b. - a \qquad v_8 = N - k_6$$

$$v_9 = c.. - b. \qquad v_{10} = N - c.. \,.$$

Then with

$$T_o = \sum_i\sum_j\sum_k\sum_m y_{ijkm}^2\,, \qquad T_A = \sum_i y_{i...}^2/n_{i..}\,,$$

$$T_{AB} = \sum_i\sum_j y_{ij..}^2/n_{ij.}, \quad T_{ABC} = \sum_i\sum_j\sum_k y_{ijk.}^2/n_{ijk} \quad \text{and} \quad T_\mu = y_{....}^2/N,$$

$$\hat\sigma_e^2 = (T_o - T_{ABC})/v_{10}$$

$$\hat\sigma_\gamma^2 = (T_{ABC} - T_{AB} - v_9\hat\sigma_e^2)/v_8$$

$$\hat\sigma_\beta^2 = (T_{AB} - T_A - v_7\hat\sigma_e^2 - v_6\hat\sigma_\gamma^2)/v_5$$

$$\hat\sigma_\alpha^2 = (T_A - T_\mu - v_4\hat\sigma_e^2 - v_3\hat\sigma_\gamma^2 - v_2\hat\sigma_\beta^2)/v_1\,.$$

(Mahamunulu, 1963)

c. Variances of analysis of variance estimators (under normality)

$$v(\hat\sigma_e^2) = 2\sigma_e^4/v_{10}$$

$$\operatorname{cov}(\hat\sigma_\alpha^2, \hat\sigma_e^2) = [v_2(v_7v_8 - v_6v_9) + v_5(v_3v_9 - v_4v_8)]v(\hat\sigma_e^2)/v_1v_5v_8$$

$$\operatorname{cov}(\hat\sigma_\beta^2, \hat\sigma_e^2) = -(v_7v_8 - v_6v_9)v(\hat\sigma_e^2)/v_5v_8$$

$$\operatorname{cov}(\hat\sigma_\gamma^2, \hat\sigma_e^2) = -v_9v(\hat\sigma_e^2)/v_8\,.$$

Calculate

$$k_7 = \sum_i n_{i..}^3 \qquad k_8 = \sum_i\sum_j n_{ij.}^3$$

$$k_9 = \sum_i\sum_j\sum_k n_{ijk}^3 \qquad k_{10} = \sum_i(\sum_j\sum_k n_{ijk}^3)/n_{i..}$$

$$k_{11} = \sum_i\sum_j(\sum_k n_{ijk}^3)/n_{ij.} \qquad k_{12} = \sum_i(\sum_j n_{ij.}^3)/n_{i..}$$

$$k_{13} = \sum_i(\sum_j n_{ij.}^2)^2/n_{i..} \qquad k_{14} = \sum_i(\sum_j\sum_k n_{ijk}^2)^2/n_{i..}$$

$$k_{15} = \sum_i\sum_j(\sum_k n_{ijk}^2)^2/n_{ij.} \qquad k_{16} = \sum_i\{\sum_j n_{ij.}(\sum_k n_{ijk}^2)\}/n_{i..}$$

$$k_{17} = \sum_i(\sum_j n_{ij.}^2)(\sum_j\sum_k n_{ijk}^2)/n_{i..} \qquad k_{18} = \sum_i\{\sum_j(\sum_k n_{ijk}^2)^2/n_{ij.}\}/n_{i..}$$

$$k_{19} = \sum_i\sum_j(\sum_k n_{ijk}^2)^2/n_{ij.}^2 \qquad k_{20} = \sum_i(\sum_j n_{ij.}^2)(\sum_j\sum_k n_{ijk}^2)/n_{i..}^2$$

$$k_{21} = \sum_i(\sum_j\sum_k n_{ijk}^2)^2/n_{i..}^2 \qquad k_{22} = \sum_i(\sum_j n_{ij.}^2)^2/n_{i..}^2$$

$$k_{23} = \sum_i n_{i..}(\sum_j n_{ij.}^2) \qquad k_{24} = \sum_i n_{i..}(\sum_j\sum_k n_{ijk}^2)$$

$$k_{25} = \sum_i\sum_j n_{ij.}(\sum_k n_{ijk}^2)$$

and $\qquad \Delta_1 = k_{19} + k_{21} - 2k_{18}\,, \qquad \Delta_2 = Nk_3 + k_{19} - 2k_{11}\,,$

$$\Delta_3 = k_{10} - k_{18}\,, \qquad \Delta_4 = k_{11} - k_{19} \qquad \text{and} \qquad \Delta_5 = (k_9 - k_{15})/N\,.$$

Also calculate

$$d_1 = v_8^2(Nk_2 + k_{22} - 2k_{12})$$
$$d_2 = v_8^2\Delta_1 + v_6^2\Delta_2 + 2v_6v_8(\Delta_3 - \Delta_4)$$
$$d_3 = (v_7v_8 - v_6v_9)^2/v_{10} + v_7v_8^2 + v_6^2v_9$$
$$d_4 = v_8^2(Nk_3 + k_{20} - 2k_{16}), \qquad d_5 = v_5v_8^2 \qquad \text{and} \qquad d_6 = v_6v_8(v_6 + v_8)$$

and

$$g_1 = v_5d_5(Nk_1 + k_1^2 - 2k_7/N)$$
$$g_2 = v_5d_5(k_{22} + k_2^2 - 2k_{13}/N) + v_2^2d_1 - 2v_2d_5[k_{12} - k_{22} - (k_8 - k_{13})/N]$$
$$g_3 = v_5d_5(k_{21} + k_3^2 - 2k_{14}/N) + v_2^2v_8^2\Delta_1 + (v_2v_6 - v_3v_5)^2\Delta_2 - 2v_2d_5[k_{18} - k_{21} - (k_{15} - k_{14})/N] + 2v_8(v_2v_6 - v_3v_5)[v_5(\Delta_3 - \Delta_5) - v_2(\Delta_4 - \Delta_3)]$$
$$g_4 = v_5d_5(a - 1) + v_2^2v_7v_8^2 + v_9(v_2v_6 - v_3v_5)^2 + [v_4v_5v_8 - v_2v_7v_8 + v_9(v_2v_6 - v_3v_5)]^2/v_{10}$$
$$g_5 = v_5d_5(Nk_2 + k_1k_2 - 2k_{23}/N)$$
$$g_6 = v_5d_5(Nk_3 + k_1k_3 - 2k_{24}/N)$$
$$g_7 = v_1v_5d_5$$
$$g_8 = v_5d_5(k_{20} + k_2k_3 - 2k_{17}/N) + v_2^2v_8^2(Nk_3 + k_{20} - 2k_{16}) - 2v_2d_5[k_{16} - k_{20} - (k_{25} - k_{17})/N]$$
$$g_9 = v_2d_5(v_2 + v_5) \qquad \text{and} \qquad g_{10} = v_8[v_8(v_3v_5^2 + v_2^2v_6) + (v_2v_6 - v_3v_5)^2].$$

Then

$$v(\hat\sigma_\alpha^2) = 2(g_1\sigma_\alpha^4 + g_2\sigma_\beta^4 + g_3\sigma_\gamma^4 + g_4\sigma_e^4 + 2g_5\sigma_\alpha^2\sigma_\beta^2 + 2g_6\sigma_\alpha^2\sigma_\gamma^2 + 2g_7\sigma_\alpha^2\sigma_e^2 + 2g_8\sigma_\beta^2\sigma_\gamma^2 + 2g_9\sigma_\beta^2\sigma_e^2 + 2g_{10}\sigma_\gamma^2\sigma_e^2)/v_1^2v_5^2v_8^2$$
$$v(\hat\sigma_\beta^2) = 2(d_1\sigma_\beta^4 + d_2\sigma_\gamma^4 + d_3\sigma_e^4 + 2d_4\sigma_\beta^2\sigma_\gamma^2 + 2d_5\sigma_\beta^2\sigma_e^2 + 2d_6\sigma_\gamma^2\sigma_e^2)/v_5^2v_8^2$$
$$v(\hat\sigma_\gamma^2) = 2[\Delta_2\sigma_\gamma^4 + v_9(v_9 + v_{10})\sigma_e^4/v_{10} + 2v_8\sigma_\gamma^2\sigma_e^2]/v_8^2$$
$$\text{cov}(\hat\sigma_\beta^2, \hat\sigma_\gamma^2) = [2(\Delta_4 - \Delta_3)\sigma_\gamma^4 + 2v_7v_9\sigma_e^4/v_{10} - v_6v_8v(\hat\sigma_\gamma^2)]/v_5v_8$$
$$\text{cov}(\hat\sigma_\alpha^2, \hat\sigma_\gamma^2) = \{2[v_5(\Delta_3 - \Delta_5) - v_2(\Delta_4 - \Delta_3)]\sigma_\gamma^4 + 2v_9(v_4v_5 - v_2v_7)\sigma_e^4/v_{10} - v_8(v_3v_5 - v_2v_6)v(\hat\sigma_\gamma^2)\}/v_1v_5v_8$$

and

$$v_1v_5v_8[\text{cov}(\hat\sigma_\alpha^2, \hat\sigma_\beta^2)] = 2(k_{12} - k_{22} - (k_8 - k_{13})/N]\sigma_\beta^4 + 2[k_{18} - k_{21} - (k_{15} - k_{14})/N - v_6(\Delta_3 - \Delta_5) - v_3(\Delta_4 - \Delta_3)]\sigma_\gamma^4 + 2[k_{16} - k_{20} - (k_{25} - k_{17})/N]\sigma_\beta^2\sigma_\gamma^2 + 2[v_4v_7v_8 - v_9(v_4v_6 + v_3v_7)]\sigma_e^4/v_{10} - v_2v_5v_8v(\hat\sigma_\beta^2) + v_3v_6v_8v(\hat\sigma_\gamma^2).$$

(Mahamunulu, 1963)

4. THE 2-WAY CLASSIFICATION WITH INTERACTION, RANDOM MODEL

a. Model

$$y_{ijk} = \mu + \alpha_i + \beta_j + \gamma_{ij} + e_{ijk}\,,$$

$$i = 1, 2, \ldots, a, \qquad j = 1, 2, \ldots, b \qquad \text{and} \qquad k = 1, 2, \ldots, n_{ij}\,,$$

with $\qquad n_{ij} > 0$ for s (i, j)-cells $\qquad$ and $\qquad \sum_i\sum_j n_{ij} = N\,.$

b. Analysis of variance estimators

Calculate Table 11.1 and

$$T_o = \sum_i\sum_j\sum_k y_{ijk}^2\,, \qquad T_A = \sum_i y_{i\cdot\cdot}^2/n_{i\cdot}\,, \qquad T_B = \sum_j y_{\cdot j\cdot}^2/n_{\cdot j}\,,$$

$$T_{AB} = \sum_i\sum_j y_{ij\cdot}^2/n_{ij} \qquad \text{and} \qquad T_\mu = y_{\cdot\cdot\cdot}^2/N\,.$$

TABLE 11.1. ANALYSIS OF VARIANCE ESTIMATION OF VARIANCE COMPONENTS IN THE 2-WAY CROSSED CLASSIFICATION INTERACTION RANDOM MODEL

Terms needed for calculating estimators and their variances. For estimators only, calculate just k_1, k_2, k_3, k_4 and k_{23}.

$k_1 = \sum_i n_{i\cdot}^2$	$k_2 = \sum_j n_{\cdot j}^2$
$k_3 = \sum_i(\sum_j n_{ij}^2)/n_{i\cdot}$	$k_4 = \sum_j(\sum_i n_{ij}^2)/n_{\cdot j}$
$k_5 = \sum_i n_{i\cdot}^3$	$k_6 = \sum_j n_{\cdot j}^3$
$k_7 = \sum_i(\sum_j n_{ij}^2)^2/n_{i\cdot}$	$k_8 = \sum_j(\sum_i n_{ij}^2)^2/n_{\cdot j}$
$k_9 = \sum_i(\sum_j n_{ij}^2)^2/n_{i\cdot}^2$	$k_{10} = \sum_j(\sum_i n_{ij}^2)^2/n_{\cdot j}^2$
$k_{11} = \sum_i(\sum_j n_{ij}^3)/n_{i\cdot}$	$k_{12} = \sum_j(\sum_i n_{ij}^3)/n_{\cdot j}$
$k_{13} = \sum_i(\sum_j n_{ij}^2)(\sum_j n_{ij}n_{\cdot j})/n_{i\cdot}$	$k_{14} = \sum_j(\sum_i n_{ij}^2)(\sum_i n_{ij}n_{i\cdot})/n_{\cdot j}$
$k_{15} = \sum_i(\sum_j n_{ij}n_{\cdot j})^2/n_{i\cdot}$	$k_{16} = \sum_j(\sum_i n_{ij}n_{i\cdot})^2/n_{\cdot j}$
$k_{17} = \sum_i(\sum_j n_{ij}^2 n_{\cdot j})/n_{i\cdot}$	$k_{18} = \sum_j(\sum_i n_{ij}^2 n_{i\cdot})/n_{\cdot j}$
$k_{19} = \sum_i(\sum_j n_{ij}^2)n_{i\cdot}$	$k_{20} = \sum_j(\sum_i n_{ij}^2)n_{\cdot j}$
$k_{21} = \sum_i\sum_{i'\neq i}(\sum_j n_{ij}n_{i'j})^2/n_{i\cdot}n_{i'\cdot}$	$k_{22} = \sum_j\sum_{j'\neq j}(\sum_i n_{ij}n_{ij'})^2/n_{\cdot j}n_{\cdot j'}$
$k_{23} = \sum_i\sum_j n_{ij}^2$	$k_{24} = \sum_i\sum_j n_{ij}^3$
$k_{25} = \sum_i\sum_j n_{ij}n_{i\cdot}n_{\cdot j}$	$k_{26} = \sum_i\sum_j n_{ij}^2/n_{i\cdot}n_{\cdot j}$
$k_{27} = \sum_i\sum_j n_{ij}^3/n_{i\cdot}n_{\cdot j}$	$k_{28} = \sum_i\sum_j n_{ij}^4/n_{i\cdot}n_{\cdot j}$

$k_r' = k_r/N$ for all r.

Then
$$\hat{\sigma}_e^2 = (T_o - T_{AB})/(N - s)$$

and with
$$\mathbf{P} = \begin{bmatrix} N - k_1' & k_3 - k_2' & k_3 - k_{23}' \\ k_4 - k_1' & N - k_2' & k_4 - k_{23}' \\ k_1' - k_4 & k_2' - k_3 & N - k_3 - k_4 + k_{23}' \end{bmatrix}$$

$$\hat{\boldsymbol{\sigma}}^2 = \begin{bmatrix} \hat{\sigma}_\alpha^2 \\ \hat{\sigma}_\beta^2 \\ \hat{\sigma}_\gamma^2 \end{bmatrix} = \mathbf{P}^{-1} \begin{bmatrix} T_A - T_\mu - (a-1)\hat{\sigma}_e^2 \\ T_B - T_\mu - (b-1)\hat{\sigma}_e^2 \\ T_{AB} - T_A - T_B + T_\mu - (s - a - b + 1)\hat{\sigma}_e^2 \end{bmatrix}$$

as in (44) of Sec. 10.2d. This is equivalent to calculating

$$\delta_A = [T_{AB} - T_A - (s - a)\hat{\sigma}_e^2]/(N - k_3)$$

and
$$\delta_B = [T_{AB} - T_B - (s - b)\hat{\sigma}_e^2]/(N - k_4)$$

with which

$$\hat{\sigma}_\gamma^2 = [(N - k_1')\delta_B + (k_3 - k_2')\delta_A - \{T_A - T_\mu - (a-1)\hat{\sigma}_e^2\}]/(N - k_1' - k_2' + k_{23}'),$$

$$\hat{\sigma}_\beta^2 = \delta_A - \hat{\sigma}_\gamma^2 \quad \text{and} \quad \hat{\sigma}_\alpha^2 = \delta_B - \hat{\sigma}_\gamma^2 .$$

(Searle, 1958)

c. Variances of analysis of variance estimators (under normality)

$$v(\hat{\sigma}_e^2) = 2\sigma_e^4/(N - s).$$

For **P** given above and for **H** and **f** being

$$\mathbf{H} = \begin{bmatrix} 1 & 0 & 0 & -1 \\ 0 & 1 & 0 & -1 \\ -1 & -1 & 1 & 1 \end{bmatrix} \quad \text{and} \quad \mathbf{f} = \begin{bmatrix} a - 1 \\ b - 1 \\ s - a - b + 1 \end{bmatrix}$$

$$\text{var}(\hat{\boldsymbol{\sigma}}^2) = \mathbf{P}^{-1}[\mathbf{H}\,\text{var}(\mathbf{t})\mathbf{H}' + v(\hat{\sigma}_e^2)\mathbf{f}\mathbf{f}']\mathbf{P}^{-1\prime},$$

and $\text{cov}(\hat{\boldsymbol{\sigma}}^2, \hat{\sigma}_e^2) = -\mathbf{P}^{-1}\mathbf{f}v(\hat{\sigma}_e^2)$, from (46) on p. 434

where
$$\text{var}(\mathbf{t}) = \text{var}\begin{bmatrix} T_A \\ T_B \\ T_{AB} \\ T_\mu \end{bmatrix}.$$

Var(**t**) has 10 different elements; each element is a function of the 10 squares and products of σ_α^2, σ_β^2, σ_γ^2 and σ_e^2. The 10×10 matrix of these coefficients is shown in Table 11.2. Apart from N, a, b, s and unity Table 11.2 involves only

TABLE 11.2. ANALYSIS OF VARIANCE ESTIMATION OF VARIANCE COMPONENTS IN THE 2-WAY CROSSED CLASSIFICATION INTERACTION RANDOM MODEL

Coefficients of squares and products of variance components in var(**t**), the variance–covariance matrix of the T's, the uncorrected sums of squares.
k_r for $r = 1, 2, \ldots, 28$ given in Table 11.1
$k'_r = k_r/N$ for all r

	$v(T_A)$	$v(T_B)$	$v(T_{AB})$	$v(T_\mu)$	$\text{cov}(T_A, T_B)$	$\text{cov}(T_A, T_{AB})$	$\text{cov}(T_A, T_\mu)$	$\text{cov}(T_B, T_{AB})$	$\text{cov}(T_B, T_\mu)$	$\text{cov}(T_{AB}, T_\mu)$
$2\sigma_\alpha^4$	k_1	$k_{22} + k_{10}$	k_1	$(k'_1)^2$	k_{18}	k_1	k'_5	k_{18}	k'_{16}	k'_5
$2\sigma_\beta^4$	$k_{21} + k_9$	k_2	k_2	$(k'_2)^2$	k_{17}	k_{17}	k'_{15}	k_2	k'_6	k'_6
$2\sigma_\gamma^4$	k_9	k_{10}	k_{23}	$(k'_{23})^2$	k_{28}	k_{11}	k'_7	k_{12}	k'_8	k'_{24}
$2\sigma_e^4$	a	b	s	1	k_{26}	a	1	b	1	1
$4\sigma_\alpha^2\sigma_\beta^2$	k_{23}	k_{23}	k_{23}	$k'_1k'_2$	k_{23}	k_{23}	k'_{25}	k_{23}	k'_{25}	k'_{25}
$4\sigma_\alpha^2\sigma_\gamma^2$	k_{23}	k_{10}	k_{23}	$k'_1k'_{23}$	k_{12}	k_{23}	k'_{19}	k_{12}	k'_{14}	k'_{19}
$4\sigma_\alpha^2\sigma_e^2$	N	k_4	N	k'_1	k_4	N	k'_1	k_4	k'_1	k'_1
$4\sigma_\beta^2\sigma_\gamma^2$	k_9	k_{23}	k_{23}	$k'_2k'_{23}$	k_{11}	k_{11}	k'_{13}	k_{23}	k'_{20}	k'_{20}
$4\sigma_\beta^2\sigma_e^2$	k_3	N	N	k'_2	k_3	k_3	k'_2	N	k'_2	k'_2
$4\sigma_\gamma^2\sigma_e^2$	k_3	k_4	N	k'_{23}	k_{27}	k_3	k'_{23}	k_4	k'_{23}	k'_{23}

28 different terms. These are shown in Table 11.1. An example of using Table 11.2 is

$$v(T_{AB}) = 2[k_1\sigma_\alpha^4 + k_2\sigma_\beta^4 + k_{23}\sigma_\gamma^4 + s\sigma_e^4 + 2(k_{23}\sigma_\alpha^2\sigma_\beta^2 + k_{23}\sigma_\alpha^2\sigma_\gamma^2 + N\sigma_\alpha^2\sigma_e^2 + k_{23}\sigma_\beta^2\sigma_\gamma^2 + N\sigma_\beta^2\sigma_e^2 + N\sigma_\gamma^2\sigma_e^2)].$$

d. Symmetric sums estimators

Calculate

$$h_A = [\sum_i\sum_j(n_{\cdot j} - n_{ij})\sum_k y_{ijk}^2 - (\sum_j y_{\cdot j\cdot}^2 - \sum_i\sum_j y_{ij\cdot}^2)]/(k_2 - k_{23})$$
$$h_B = [\sum_i\sum_j(n_{i\cdot} - n_{ij})\sum_k y_{ijk}^2 - (\sum_i y_{i\cdot\cdot}^2 - \sum_i\sum_j y_{ij\cdot}^2)]/(k_1 - k_{23})$$

and

$$h_{AB} = [\sum_i\sum_j(N - n_{i\cdot} - n_{\cdot j} + n_{ij})\sum_k y_{ijk}^2 - (y_{\cdot\cdot\cdot}^2 - \sum_i y_{i\cdot\cdot}^2 - \sum_j y_{\cdot j\cdot}^2 + \sum_i\sum_j y_{ij\cdot}^2)]/(N^2 - k_1 - k_2 + k_{23}).$$

Then $\hat\sigma_e^2 = \sum_i\sum_j n_{ij}(\sum_k y_{ijk}^2 - n_{ij}\bar y_{ij\cdot}^2)/(k_{23} - N),$

$$\hat\sigma_\alpha^2 = h_{AB} - h_B, \qquad \hat\sigma_\beta^2 = h_{AB} - h_A$$

and $\hat\sigma_\gamma^2 = h_A + h_B - h_{AB} - \hat\sigma_e^2.$

(Koch, 1968)

e. Fitting constants method estimators

Label the factor having the smaller number of levels in the data as the β-factor, with b levels.

Calculate $R(\mu, \alpha, \beta)$ and h_6 as in Table 11.3. Also, using Table 11.1,

calculate $h_1 = N - k_1'$, $\quad h_2 = N - k_2'$, $\quad h_3 = N - k_{23}'$,

$$h_4 = N - k_3 = h_5 \quad \text{and} \quad h_7 = N - k_4 = h_8\,.$$

Then $\hat\sigma_e^2 = (T_o - T_{AB})/(N - s)$

and $\hat\sigma_\gamma^2 = [T_{AB} - R(\mu, \alpha, \beta) - (s - a - b + 1)\hat\sigma_e^2]/h_6\,.$

Estimators of σ_α^2 and σ_β^2 come from using *any two* of the following based on Tables 10.1 and 10.2

$$\hat\sigma_\alpha^2 = [T_{AB} - T_B - (s - b)\hat\sigma_e^2]/h_7 - \hat\sigma_\gamma^2, \qquad [10.2]$$

$$\hat\sigma_\beta^2 = [T_{AB} - T_A - (s - a)\hat\sigma_e^2]/h_4 - \hat\sigma_\gamma^2 \qquad [10.1]$$

and $h_1\hat\sigma_\alpha^2 + h_2\hat\sigma_\beta^2 = T_{AB} - T_\mu - h_3\hat\sigma_\gamma^2 - (s - 1)\hat\sigma_e^2\,.$ [10.1 and 10.2]

This procedure is derived in Sec. 10.4. In Table 11.3 the calculations for $R(\mu, \alpha, \beta)$ are the same as those in Sec. 7.2, and the calculations for h_6

TABLE 11.3. COMPUTING FORMULAE FOR THE TERMS NEEDED IN THE FITTING CONSTANTS METHOD OF ESTIMATING VARIANCE COMPONENTS ADDITIONAL TO THOSE NEEDED IN THE ANALYSIS OF VARIANCE METHOD; FOR THE 2-WAY CLASSIFICATION, MIXED OR RANDOM MODELS

To calculate $R(\mu, \alpha, \beta)$ compute	To calculate h_6 compute
For $j = 1, \ldots, b$ $c_{jj} = n_{.j} - \sum_{i=1}^{a} \frac{n_{ij}^2}{n_{i.}}$ $c_{jj'} = -\sum_{i=1}^{a} \frac{n_{ij}n_{ij'}}{n_{i.}}, j \neq j'$ $\left(\text{Check: } \sum_{j'=1}^{b} c_{jj'} = 0\right)$ $r_j = y_{.j.} - \sum_{i=1}^{a} n_{ij}\bar{y}_{i..}$ $\left(\text{Check: } \sum_{j=1}^{b} r_j = 0\right)$	For $i = 1, \ldots, a$ $\lambda_i = \sum_{j=1}^{b} n_{ij}^2/n_{i.}$ For $i = 1, \ldots, a$ and $j, j' = 1, \ldots, b$ $f_{i,jj} = (n_{ij}^2/n_{i.})(\lambda_i + n_{i.} - 2n_{ij})$ $f_{i,jj'} = (n_{ij}n_{ij'}/n_{i.})(\lambda_i - n_{ij} - n_{ij'})$ for $j \neq j'$ $\left(\text{Check: } \sum_{j=1}^{b} f_{i,jj'} = 0\right)$
For $j, j' = 1, 2, \ldots, (b-1)$ $\mathbf{C} = \{c_{jj'}\}$ and $\mathbf{C}^{-1} = \{c^{jj'}\}$ $\mathbf{r} = \{r_j\}$ Then $t_B = \mathbf{r}'\mathbf{C}^{-1}\mathbf{r} = R(\beta \mid \mu, \alpha)$ and $R(\mu, \alpha, \beta) = T_A + t_B.$	For $i = 1, \ldots, a$ and $j, j' = 1, \ldots, (b-1)$ $\mathbf{F}_i = \{f_{i,jj'}\}$ Then $k^* = \sum_{i=1}^{a} \lambda_i + \text{tr}(\mathbf{C}^{-1} \sum_{i=1}^{a} \mathbf{F}_i)$ and $h_6 = N - k^*.$

come from Searle and Henderson (1961). Simplification of general results in Rohde and Tallis (1969) would yield variances of these estimators.

f. Analysis of means estimators

These methods can be used only if all cells have data in them; i.e., $s = ab$, and $n_{ij} \geq 1$ for all $i = 1, 2, \ldots, a$ and $j = 1, 2, \ldots, b$.

Two possible analyses are the unweighted and the weighted means analyses. Details of the mean squares in these analyses, and of their expected values, are found in Chapters 8 and 10, as indicated in Table 11.4. Equating expected values to calculated values in either method yields variance component estimators. Similar handling of the other analyses considered by Gosslee and Lucas (1965) provides additional methods of estimation.

TABLE 11.4. ANALYSIS OF MEANS METHODS FOR ESTIMATING VARIANCE COMPONENTS WHEN ALL CELLS CONTAIN DATA

	Mean Squares	
Method	Calculation	Expected Values
Unweighted means	Table 8.12, Sec. 8.3c(i)	Table 10.3a, Sec. 10.5
Weighted means	Table 8.18, Sec. 8.3c(iii)	Table 10.3b, Sec. 10.5

Estimators from the unweighted means method are, from Table 10.3a of Sec. 10.5,

$$\hat{\sigma}_e^2 = \text{MSE}, \qquad \hat{\sigma}_\beta^2 = (\text{MSB}_u - \text{MSAB}_u)/a\,,$$
$$\hat{\sigma}_\gamma^2 = \text{MSAB}_u - n_h\hat{\sigma}_e^2\,, \qquad \hat{\sigma}_\alpha^2 = (\text{MSA}_u - \text{MSAB}_u)/b\,.$$

The mean squares used here are defined in Table 8.12 of Sec. 8.3.

Variances of these estimators utilize

$$N_1 = ab\textstyle\sum_i\sum_j(1/n_{ij}) = a^2b^2n_h\,, \qquad N_2 = ab\sum_i\sum_j(1/n_{ij}^2),$$
$$N_3 = a\textstyle\sum_i\sum_j\sum_s(1/n_{ij}n_{is}), \qquad N_4 = b\sum_i\sum_j\sum_r(1/n_{ij}n_{rj})$$

and

$$N_5 = \textstyle\sum_i\sum_j\sum_r\sum_s(1/n_{ij}n_{rs}).$$

Then

$$v(\hat{\sigma}_e^2) = 2\sigma_e^4/(N - ab),$$
$$\begin{aligned}v(\hat{\sigma}_\gamma^2) = n_h^2v(\hat{\sigma}_e^2) &+ 2(\sigma_\gamma^4 + 2n_h\sigma_\gamma^2\sigma_e^2)/[(a-1)(b-1)]\\ &+ 2[(a-2)(b-2)N_2 + (a-2)N_3\\ &+ (b-2)N_4 + N_5]\sigma_e^4/[ab(a-1)(b-1)]^2,\end{aligned}$$
$$\begin{aligned}v(\hat{\sigma}_\beta^2) = 2\{(1 - 1/a)\sigma_\beta^4 &+ [\sigma_\beta^2 + (\sigma_\gamma^2 - \sigma_\beta^2)/a]^2\\ &+ 2n_h[\sigma_\beta^2 + (\sigma_\gamma^2 - \sigma_\beta^2)/a]\sigma_e^2\}/[(a-1)(b-1)]\\ &+ 2[N_5 - N_3/a + (b-2)(N_4 - N_2/a)]\sigma_e^4/[ab(a-1)(b-1)]^2,\end{aligned}$$

and

$$\begin{aligned}v(\hat{\sigma}_\alpha^2) = 2\{(1 - 1/b)\sigma_\alpha^4 &+ [\sigma_\alpha^2 + (\sigma_\gamma^2 - \sigma_\alpha^2)/b]^2\\ &+ 2n_h[\sigma_\alpha^2 + (\sigma_\gamma^2 - \sigma_\alpha^2)/b]\sigma_e^2\}/[(a-1)(b-1)]\\ &+ 2[N_5 - N_4/b + (a-2)(N_3 - N_2/b)]\sigma_e^4/[ab(a-1)(b-1)]^2.\end{aligned}$$

(Hirotsu, 1966)

5. THE 2-WAY CLASSIFICATION WITH INTERACTION, MIXED MODEL

a. Model

$$y_{ijk} = \mu + \alpha_i + \beta_j + \gamma_{ij} + e_{ijk}\,, \qquad \beta_j\text{'s taken as fixed effects.}$$

$$i = 1, 2, \ldots, a, \qquad j = 1, 2, \ldots, b \qquad \text{and} \qquad k = 1, 2, \ldots, n_{ij}\,,$$

with $n_{ij} > 0$ for s (i, j)-cells and $\sum_i \sum_j n_{ij} = N$.

The model is exactly the same as the random model case of the preceding section, except that the β's are taken as fixed effects. They are assumed to be fewer in number than the random effects in the data.

b. Fitting constants method estimators

Calculate T_o, T_{AB} and T_B of the analysis of variance method and $R(\mu, \alpha, \beta)$ and h_6 of Table 11.3. Then

$$\hat{\sigma}_e^2 = (T_o - T_{AB})/(N - s),$$

$$\hat{\sigma}_\gamma^2 = [T_{AB} - R(\mu, \alpha, \beta) - (s - a - b + 1)\hat{\sigma}_e^2]/h_6$$

and

$$\hat{\sigma}_\alpha^2 = [T_{AB} - T_B - (s - b)\hat{\sigma}_e^2]/(N - k_4) - \hat{\sigma}_\gamma^2\,.$$

These estimators arise from the last three lines of Table 10.2 of Sec. 10.4, as discussed in that section.

c. Fixed effects estimators

Writing the model as

$$\mathbf{y} = \mu\mathbf{1} + \mathbf{X}\boldsymbol{\alpha} + \mathbf{Z}\boldsymbol{\beta} + \mathbf{W}\boldsymbol{\gamma} + \mathbf{e}$$

where $\boldsymbol{\alpha}$, $\boldsymbol{\beta}$ and $\boldsymbol{\gamma}$ are the vectors of α-, β-, and γ-effects, we can use (103) of Sec. 10.8a to write down equations

$$\begin{bmatrix} \mathbf{X}'\mathbf{X} + (\sigma_e^2/\sigma_\alpha^2)\mathbf{I} & \mathbf{X}'\mathbf{Z} & \mathbf{X}'\mathbf{W} \\ \mathbf{Z}'\mathbf{X} & \mathbf{Z}'\mathbf{Z} & \mathbf{Z}'\mathbf{W} \\ \mathbf{W}'\mathbf{X} & \mathbf{W}'\mathbf{Z} & \mathbf{W}'\mathbf{W} + (\sigma_e^2/\sigma_\gamma^2)\mathbf{I} \end{bmatrix} \begin{bmatrix} \boldsymbol{\alpha}^o \\ \boldsymbol{\beta}^o \\ \boldsymbol{\gamma}^o \end{bmatrix} = \begin{bmatrix} \mathbf{X}'\mathbf{y} \\ \mathbf{Z}'\mathbf{y} \\ \mathbf{W}'\mathbf{y} \end{bmatrix}.$$

When the variance components are known, solutions to these for $\boldsymbol{\beta}^o$ are maximum likelihood. When $\sigma_e^2/\sigma_\alpha^2$ and $\sigma_e^2/\sigma_\gamma^2$ are unknown, using $\hat{\sigma}_e^2/\hat{\sigma}_\alpha^2$ and $\hat{\sigma}_e^2/\hat{\sigma}_\gamma^2$ in their place provides equations that can be solved, although their solution is not maximum likelihood.

6. THE 2-WAY CLASSIFICATION WITHOUT INTERACTION, RANDOM MODEL

a. Model

$$y_{ijk} = \mu + \alpha_i + \beta_j + e_{ijk},$$

$$i = 1, 2, \ldots, a, \quad j = 1, 2, \ldots, b \quad \text{and} \quad k = 1, 2, \ldots, n_{ij},$$

with $\quad n_{ij} > 0$ for s (i, j)-cells $\quad$ and $\quad N = \sum_i \sum_j n_{ij}$.

b. Analysis of variance estimators

Calculate

$$T_o = \sum_i \sum_j \sum_k y_{ijk}^2, \qquad T_\mu = y_{...}^2/N,$$

$$T_A = \sum_i y_{i..}^2/n_{i.} \quad \text{and} \quad T_B = \sum_j y_{.j.}^2/n_{.j}.$$

Using Table 11.1 calculate

$$\lambda_1 = (N - k_1')/(N - k_4) \quad \text{and} \quad \lambda_2 = (N - k_2')/(N - k_3).$$

Then $$\hat{\sigma}_e^2 = \frac{\lambda_2(T_o - T_A) + \lambda_1(T_o - T_B) - (T_o - T_\mu)}{\lambda_2(N - a) + \lambda_1(N - b) - (N - 1)},$$

$$\hat{\sigma}_\alpha^2 = [T_o - T_B - (N - b)\hat{\sigma}_e^2]/(N - k_4)$$

and $$\hat{\sigma}_\beta^2 = [T_o - T_A - (N - a)\hat{\sigma}_e^2]/(N - k_3).$$

c. Variances of analysis of variance estimators (under normality)

Writing

$$\mathbf{Q} = \begin{bmatrix} N - k_1' & k_3 - k_2' & a - 1 \\ k_4 - k_1' & N - k_2' & b - 1 \\ k_1' - k_4 & k_2' - k_3 & N - a - b + 1 \end{bmatrix} \quad \text{and} \quad \hat{\boldsymbol{\sigma}}^2 = \begin{bmatrix} \hat{\sigma}_\alpha^2 \\ \hat{\sigma}_\beta^2 \\ \hat{\sigma}_e^2 \end{bmatrix}$$

it can be shown that the estimators are solutions to

$$\mathbf{Q}\hat{\boldsymbol{\sigma}}^2 = \begin{bmatrix} T_A - T_\mu \\ T_B - T_\mu \\ T_o - T_A - T_B + T_\mu \end{bmatrix} = \mathbf{Ht} + \begin{bmatrix} 0 \\ 0 \\ T_o - T_{AB} \end{bmatrix}$$

for **Ht** of Sec. 4c.

When $n_{ij} = 0$ or 1, $T_{AB} = T_o$ and $\mathbf{Q}\hat{\boldsymbol{\sigma}}^2 = \mathbf{Ht}$ so that

$$\text{var}(\hat{\boldsymbol{\sigma}}^2) = \mathbf{Q}^{-1}\mathbf{H}\,\text{var}(\mathbf{t})\mathbf{H}'\mathbf{Q}^{-1\prime}.$$

Var(**t**) will be calculated exactly as in Tables 11.1 and 11.2 except with $\sigma_\gamma^2 = 0$.

When $n_{ij} \geq 0$, T_{AB} exists even though not used in the estimation procedure. Nevertheless

$$\hat{\boldsymbol{\sigma}}^2 = \mathbf{Q}^{-1}\left[\mathbf{Ht} + \begin{pmatrix} 0 \\ 0 \\ T_o - T_{AB} \end{pmatrix}\right].$$

Furthermore, $T_o - T_{AB}$ has variance $2\sigma_e^4(N - s)$ and is independent of every element in $\mathbf{Ht}$, whether $\sigma_\gamma^2 = 0$ or not. Therefore

$$\text{var}(\hat{\boldsymbol{\sigma}}^2) = \mathbf{Q}^{-1}\mathbf{H}\,\text{var}(\mathbf{t})\mathbf{H}'\mathbf{Q}^{-1\prime} + 2\,\mathbf{q}_3\mathbf{q}_3'\sigma_e^4(N - s)$$

where $\mathbf{q}_3$ is column 3 of $\mathbf{Q}^{-1}$. As with the $n_{ij} = 0$ or 1 case, var(**t**) is calculated from Tables 11.1 and 11.2 using $\sigma_\gamma^2 = 0$.

d. Symmetric sums estimators

Calculate

$$h_A = \textstyle\sum_i(n_{i\cdot}\sum_j y_{ij}^2 - y_{i\cdot}^2)/(k_1 - N)$$
$$h_B = \textstyle\sum_j(n_{\cdot j}\sum_i y_{ij}^2 - y_{\cdot j}^2)/(k_2 - N)$$

and
$$h_{AB} = [\textstyle\sum_i\sum_j(N - n_{i\cdot} - n_{\cdot j} + n_{ij})y_{ij}^2 - (y_{\cdot\cdot}^2 - \sum_i y_{i\cdot}^2 - \sum_j y_{\cdot j}^2 + \sum_i\sum_j y_{ij}^2)]/(N^2 - k_1 - k_2 + N).$$

Then $\hat{\sigma}_\alpha^2 = h_{AB} - h_B$, $\hat{\sigma}_\beta^2 = h_{AB} - h_A$ and $\hat{\sigma}_e^2 = h_A + h_B - h_{AB}$.

Koch (1968) gives these results for just the case of $n_{ij} = 0$ or 1, but presumably they would extend quite directly to the $n_{ij} \geq 0$ case. Denominators of the h's, for example, would have their N's (not N^2) replaced by k_{23}.

e. Fitting constants method estimators

Calculate $R(\mu, \alpha, \beta)$ of Table 11.3, and from Table 11.1

$$h_1 = N - k_1' \qquad h_2 = N - k_2'$$
$$h_4 = N - k_3 \qquad h_7 = N - k_4 .$$

Then
$$\hat{\sigma}_e^2 = [T_o - R(\mu, \alpha, \beta)]/(N - a - b + 1)$$

and σ_α^2 and σ_β^2 are estimated by using *any two* of the following

$$\hat{\sigma}_\alpha^2 = [R(\mu, \alpha, \beta) - T_B - (a - 1)\hat{\sigma}_e^2]/h_7 ,$$
$$\hat{\sigma}_\beta^2 = [R(\mu, \alpha, \beta) - T_A - (b - 1)\hat{\sigma}_e^2]/h_4$$

and
$$h_1\hat{\sigma}_\alpha^2 + h_2\hat{\sigma}_\beta^2 = R(\mu, \alpha, \beta) - T_\mu - (a + b - 2)\hat{\sigma}_e^2 .$$

These expressions come from Tables 10.1 and 10.2 by not using $R(\mu, \alpha, \beta, \gamma)$ and having $\sigma_\gamma^2 = 0$.

f. Variances of fitting constants method estimators (under normality)

For estimators obtained without using the last equation of the preceding section Low (1964) derives the following variances and covariances. Calculate

$$N' = N - a - b + 1$$

and, with the aid of Table 11.1,

$$f_1 = k_1 - 2k_{18} + \sum_i \sum_{i'} (\sum_j n_{ij} n_{i'j}/n_{.j})^2$$

and

$$f_2 = k_2 - 2k_{17} + \sum_j \sum_{j'} (\sum_i n_{ij} n_{ij'}/n_{i.})^2.$$

Then

$$v(\hat{\sigma}_e^2) = 2\sigma_e^4/N',$$

$$\text{cov}(\hat{\sigma}_\alpha^2, \hat{\sigma}_e^2) = -(a-1)v(\hat{\sigma}_e^2)/h_7,$$

$$\text{cov}(\hat{\sigma}_\beta^2, \hat{\sigma}_e^2) = -(b-1)v(\hat{\sigma}_e^2)/h_4,$$

$$v(\hat{\sigma}_\alpha^2) = 2[\sigma_e^4(N-b)(a-1)/N' + 2h_7\sigma_e^2\sigma_\alpha^2 + f_1\sigma_\alpha^4]/h_7^2,$$

$$v(\hat{\sigma}_\beta^2) = 2[\sigma_e^4(N-a)(b-1)/N' + 2h_4\sigma_e^2\sigma_\beta^2 + f_2\sigma_\beta^4]/h_4^2$$

and

$$\text{cov}(\hat{\sigma}_\alpha^2, \hat{\sigma}_\beta^2) = 2\sigma_e^4[k_{26} - 1 + (a-1)(b-1)/N']/h_4 h_7.$$

(Low, 1964)

7. MIXED MODELS WITH ONE RANDOM FACTOR

a. Model

$$\mathbf{y} = \mathbf{Xb} + \mathbf{Zu} + \mathbf{e},$$

where $r(\mathbf{X}) = r$, $\mathbf{Z}$ has full column rank t, and $\mathbf{u}$ represents the effects of a single random factor having variance σ_u^2. The estimation procedures summarized below come from Sec. 10.9, based upon Cunningham and Henderson (1968) and Thompson (1969).

b. Fitting constants method estimators

Calculate

$$T_o = \mathbf{y}'\mathbf{y},$$

$$\mathbf{M} = \mathbf{X}'[\mathbf{I} - \mathbf{Z}(\mathbf{Z}'\mathbf{Z})^{-1}\mathbf{Z}'],$$

$$\mathbf{Q} = \mathbf{MX}, \quad \mathbf{Q}^- \quad \text{and} \quad \mathbf{b}^o = \mathbf{Q}^-\mathbf{My}.$$

$$R(\mathbf{b} \mid \mathbf{u}) = \mathbf{b}^{o\prime}\mathbf{My},$$

$$R(\mathbf{u}) = \mathbf{y}'\mathbf{Z}(\mathbf{Z}'\mathbf{Z})^{-1}\mathbf{Z}'\mathbf{y} \quad \text{and} \quad R(\mathbf{b}) = \mathbf{y}'\mathbf{X}(\mathbf{X}'\mathbf{X})^-\mathbf{X}'\mathbf{y},$$

$$R(\mathbf{b}, \mathbf{u}) = R(\mathbf{b} \mid \mathbf{u}) + R(\mathbf{u}) \quad \text{and} \quad R(\mathbf{u} \mid \mathbf{b}) = R(\mathbf{b}, \mathbf{u}) - R(\mathbf{b}).$$

Also calculate

$$c = \text{tr}[\mathbf{Z}'\mathbf{Z} - \mathbf{Z}'\mathbf{X}(\mathbf{X}'\mathbf{X})^-\mathbf{X}'\mathbf{Z}].$$

Then

$$\hat{\sigma}_e^2 = [T_o - R(\mathbf{b}, \mathbf{u})]/(N - r - t + 1) \quad \text{and} \quad \hat{\sigma}_u^2 = [R(\mathbf{u} \mid \mathbf{b}) - (t - 1)\hat{\sigma}_e^2]/c.$$

c. An iterative procedure

Assign an initial value to

$$\lambda = \sigma_e^2/\sigma_u^2,$$

and calculate

$$\mathbf{P} = \mathbf{Z}'\mathbf{Z} + \lambda\mathbf{I} \quad \text{and} \quad \mathbf{T} = \mathbf{I} - \mathbf{Z}\mathbf{P}^{-1}\mathbf{Z}',$$

$$R^*(\mathbf{b} \mid \mathbf{u}) = \mathbf{y}'\mathbf{T}\mathbf{X}(\mathbf{X}'\mathbf{T}\mathbf{X})^-\mathbf{X}'\mathbf{T}\mathbf{y},$$

$$R^*(\mathbf{u}) = \mathbf{y}'\mathbf{Z}'\mathbf{P}^{-1}\mathbf{Z}\mathbf{y} \quad \text{and} \quad R^*(\mathbf{b}) = R(\mathbf{b}) = \mathbf{y}'\mathbf{X}(\mathbf{X}'\mathbf{X})^-\mathbf{X}'\mathbf{y},$$

$$R^*(\mathbf{b}, \mathbf{u}) = R^*(\mathbf{b} \mid \mathbf{u}) + R^*(\mathbf{u}) \quad \text{and} \quad R^*(\mathbf{u} \mid \mathbf{b}) = R^*(\mathbf{b}, \mathbf{u}) - R^*(\mathbf{b}).$$

Also calculate c as above, and then

$$\tilde{\sigma}_e^2 = [T_0 - R^*(\mathbf{b}, \mathbf{u})]/(N - r) \quad \text{and} \quad \tilde{\sigma}_u^2 = R^*(\mathbf{u} \mid \mathbf{b})/c.$$

Use these to evaluate $\tilde{\lambda} = \tilde{\sigma}_e^2/\tilde{\sigma}_u^2$ and with this value of λ repeat the calculation of $\tilde{\sigma}_e^2$ and $\tilde{\sigma}_u^2$.

8. THE 2-WAY CLASSIFICATION WITHOUT INTERACTION, MIXED MODEL

a. Model

$$y_{ijk} = \mu + \alpha_i + \beta_j + e_{ijk}, \quad \beta_j\text{'s taken as fixed effects.}$$

$$i = 1, 2, \ldots, a, \quad j = 1, 2, \ldots, b \quad \text{and} \quad k = 1, 2, \ldots, n_{ij},$$

with $\quad n_{ij} > 0$ for s (i, j)-cells and $N = \sum_i\sum_j n_{ij}$.

b. Fitting constants method estimators

$$\hat{\sigma}_e^2 = [T_o - R(\mu, \alpha, \beta)]/(N - a - b + 1)$$

$$\hat{\sigma}_\alpha^2 = [R(\mu, \alpha, \beta) - T_B - (a - 1)\hat{\sigma}_e^2]/h_7.$$

The variances and covariance of these estimators are as in Sec. 6f.

c. An iterative procedure

Write the model as

$$\mathbf{y} = \mathbf{X}\mathbf{b} + \mathbf{Z}\mathbf{u} + \mathbf{e},$$

where $\mathbf{b}$ is the vector of μ and the fixed β_j-effects and $\mathbf{u}$ is the vector of random α_i-effects. Then the prescriptions of Sec. 7c apply exactly.

d. Fixed effects estimators

In terms of the preceding model the maximum likelihood solution for $\mathbf{b}' = [\mu \quad \beta_1 \quad \cdots \quad \beta_b]$ will be

$$\mathbf{b}^* = (\mathbf{XTX}')^{-}\mathbf{X}'\mathbf{Ty}$$

with

$$\mathbf{T} = \mathbf{I} - \mathbf{Z}[\mathbf{Z}'\mathbf{Z} + (\sigma_e^2/\sigma_\alpha^2)\mathbf{I}]^{-1}\mathbf{Z}'.$$

9. THE 3-WAY CLASSIFICATION, RANDOM MODEL

a. Model

Using notation such as $(\alpha\beta)_{ij}$ for the interaction effect peculiar to the ith α-level and the jth β-level write the model as

$$y_{ijkh} = \mu + \alpha_i + \beta_j + \gamma_k + (\alpha\beta)_{ij} + (\alpha\gamma)_{ik} + (\beta\gamma)_{jk} + (\alpha\beta\gamma)_{ijk} + e_{ijkh}\,,$$

with $i = 1, 2, \ldots, a, \quad j = 1, 2, \ldots, b,$

$$k = 1, 2, \ldots, c \quad \text{and} \quad h = 1, 2, \ldots, n_{ijk}\,.$$

b. Analysis of variance estimators

$$\hat{\sigma}_e^2 = (T_o - T_{ABC})/(N - s)$$

where s is the number of filled cells. The other seven components are estimated from calculating the elements of

$$\mathbf{t}' = [T_A \quad T_B \quad T_C \quad T_{AB} \quad T_{AC} \quad T_{BC} \quad T_{ABC} \quad T_\mu]$$

and of $E(\mathbf{t}')$. The T's are calculated as in (35) of Sec. 10.2c(iii), and the coefficients of the σ^2's in the $E(T)$-terms are obtained from (38) of the same section. The T's and their expected values are shown in Table 11.5. Equating the calculated values of

$$\begin{array}{ll} T_A \quad - T_\mu & T_C \; - T_\mu \\ T_B \quad - T_\mu & T_{AC} - T_A - T_C + T_\mu \\ T_{AB} \; - T_A - T_B + T_\mu & T_{BC} - T_B - T_C + T_\mu \end{array}$$

and

$$T_{ABC} - T_{AB} - T_{AC} - T_{BC} + T_A + T_B + T_C - T_\mu$$

to their expected values yields, together with $\hat{\sigma}_e^2$, estimators of the components

$$\boldsymbol{\sigma}^{2\prime} = [\sigma_A^2 \quad \sigma_B^2 \quad \sigma_C^2 \quad \sigma_{AB}^2 \quad \sigma_{AC}^2 \quad \sigma_{BC}^2 \quad \sigma_{ABC}^2 \quad \sigma_e^2].$$

TABLE 11.5. 3-WAY CLASSIFICATION RANDOM MODEL. T'S AND THEIR EXPECTED VALUES

T-term[1]	(Uncorrected sum of squares)	[2]Coefficients of μ^2 and σ^2's in $E(T)$								
		μ^2	σ^2_A	σ^2_B	σ^2_C	σ^2_{AB}	σ^2_{AC}	σ^2_{BC}	σ^2_{ABC}	σ^2_e
T_A	$\Sigma_i(y^2_{i\ldots}/n_{i\cdot\cdot})$	N	N	$c_{i(j)}$	$c_{i(k)}$	$c_{i(j)}$	$c_{i(k)}$	$c_{i(jk)}$	$c_{i(jk)}$	a
T_B	$\Sigma_j(y^2_{\cdot j\cdot\cdot}/n_{\cdot j\cdot})$	N	$c_{j(i)}$	N	$c_{j(k)}$	$c_{j(i)}$	$c_{j(ik)}$	$c_{j(k)}$	$c_{j(ik)}$	b
T_C	$\Sigma_k(y^2_{\cdot\cdot k\cdot}/n_{\cdot\cdot k})$	N	$c_{k(i)}$	$c_{k(j)}$	N	$c_{k(ij)}$	$c_{k(i)}$	$c_{k(j)}$	$c_{k(ij)}$	c
T_{AB}	$\Sigma_i\Sigma_j(y^2_{ij\cdot\cdot}/n_{ij\cdot})$	N	N	N	$c_{ij(k)}$	N	$c_{ij(k)}$	$c_{ij(k)}$	$c_{ij(k)}$	s_{ab}
T_{AC}	$\Sigma_i\Sigma_k(y^2_{i\cdot k\cdot}/n_{i\cdot k})$	N	N	$c_{ik(j)}$	N	$c_{ik(j)}$	N	$c_{ik(j)}$	$c_{ik(j)}$	s_{ac}
T_{BC}	$\Sigma_j\Sigma_k(y^2_{\cdot jk\cdot}/n_{\cdot jk})$	N	$c_{jk(i)}$	N	N	$c_{jk(i)}$	$c_{jk(i)}$	N	$c_{jk(i)}$	s_{bc}
T_{ABC}	$\Sigma_i\Sigma_j\Sigma_k(y^2_{ijk\cdot}/n_{ijk})$	N	N	N	N	N	N	N	N	s
T_μ	$y^2_{\cdots\cdot}/N$	N	d_i	d_j	d_k	d_{ij}	d_{ik}	d_{jk}	d_{ijk}	1
T_o	$\Sigma_i\Sigma_j\Sigma_k\Sigma_h y^2_{ijkh}$	N	N	N	N	N	N	N	N	N

[1] Formulae for the T's come from equation (35) of Chapter 10.

[2] Coefficients come from equation (38) of Chapter 10. For example,

$$c_{i(j)} = \Sigma_i(\Sigma_j n^2_{ij\cdot})/n_{i\cdot\cdot} \quad \text{and} \quad c_{k(ij)} = \Sigma_k(\Sigma_i\Sigma_j n^2_{ij\cdot})/n_{\cdot\cdot k}\,.$$

s_{ab}, s_{ac}, s_{bc} are, respectively, the number of AB-, AC- and BC-subclasses containing data.

s = number of ABC-subclasses containing data.

$d_i = \Sigma_i n^2_{i\cdot\cdot}/N$, $d_{jk} = \Sigma_j\Sigma_k n^2_{\cdot jk}/N$, etc.

c. Variances of analysis of variance estimators (under normality)

The variance-covariance matrix of **t**, var(**t**), is 8×8 with 36 different elements, as discussed in Sec. 10.2d(i). Each element is a linear function of the 36 squares and products of the σ^2's in $\boldsymbol{\sigma}^2$. The elements of the resulting 36×36 matrix are shown in Table 11.6, prepared by W. R. Blischke as an appendix to Blischke (1968). It is published here with his kind permission.

The three factors are denoted by numbers 1, 2, 3 instead of letters A, B, C. For example, T_1 and T_{23} correspond to T_A and T_{BC} respectively and σ_3^2 is σ_C^2. The entries of Table 11.6 are defined in terms of the n_{ijh}, using the usual "dot" notation, and the additional notation

$$w_{ijhstu} = n_{ijh}n_{stu} ,$$

with the convention that an asterisk in the fourth, fifth or sixth subscript is used to indicate that that subscript is equated to the first, second, or third subscript, respectively, prior to summation. Thus, for example,

$$w_{ij\cdot st*} = \sum_h n_{ijh}n_{sth} , \qquad w_{i\cdot\cdot i\cdot *} = \sum_{j,h,t} n_{ijh}n_{ith} = \sum_h n_{i\cdot h}^2 ,$$

and

$$\sum_i \frac{w_{i\cdot\cdot i\cdot *}^2}{w_{i\cdot\cdot i\cdot\cdot}} = \sum_i \frac{\left(\sum_h w_{i\cdot hi\cdot h}\right)^2}{w_{i\cdot\cdot i\cdot\cdot}} = \sum_i \frac{1}{n_{i\cdot\cdot}^2}\left(\sum_h n_{i\cdot h}^2\right)^2 .$$

The symbol $A_{i,j}$ is used to denote the table entry in the ith row and jth column. Throughout the table it is understood that summations extend over all subscripts.

TABLE 11.6. COEFFICIENTS OF SQUARES AND PRODUCTS OF VARIANCE COMPONENTS IN VARIANCES AND COVARIANCES OF UNCORRECTED SUMS OF SQUARES (T'S) IN THE 3-WAY CROSSED CLASSIFICATION

Coefficients are denoted by $A_{i,j}$ for $i, j = 1, 2, \ldots, 36$. $A_{i,j}$ is the coefficient of the title of row i in $\frac{1}{2}$cov(title of column j); e.g., the $A_{1,2}$ entry is the coefficient of σ_1^4 in $\frac{1}{2}\text{cov}(T_2) \equiv \frac{1}{2}\text{var}(T_2)$, and the entry $A_{9,11}$ is the coefficient of $2\sigma_1^2\sigma_2^2$ in $\frac{1}{2}\text{cov}(T_1, T_{12})$. $\sigma_0^2 \equiv \sigma_e^2$ throughout.

	Column	1	2	3	4	5
	(1/2)Cov	T_1	T_2	T_3	T_{12}	T_{13}
Row	Product					
1	σ_1^4	$w_{\mathrm{...*..}}$	$\sum \frac{w^2_{\mathrm{ij.st.}}}{w_{\mathrm{.j..t.}}}$	$\sum \frac{w^2_{\mathrm{i.hs.u}}}{w_{\mathrm{..h..u}}}$	$A_{1,1}$	$A_{1,1}$
2	σ_2^4	$\sum \frac{w^2_{\mathrm{ij.st.}}}{w_{\mathrm{i..s..}}}$	$w_{\mathrm{....*.}}$	$\sum \frac{w^2_{\mathrm{.jh.tu}}}{w_{\mathrm{..h..u}}}$	$A_{2,2}$	$\sum \frac{w^2_{\mathrm{i.hs*u}}}{w_{\mathrm{i.hs.u}}}$
3	σ_3^4	$\sum \frac{w^2_{\mathrm{i.hs.u}}}{w_{\mathrm{i..s..}}}$	$\sum \frac{w^2_{\mathrm{.jh.tu}}}{w_{\mathrm{.j..t.}}}$	$w_{\mathrm{.....*}}$	$\sum \frac{w^2_{\mathrm{ij.st*}}}{w_{\mathrm{ij.st.}}}$	$A_{3,3}$
4	σ_{12}^4	$\sum \frac{w^2_{\mathrm{i..i*.}}}{w_{\mathrm{i..i..}}}$	$\sum \frac{w^2_{\mathrm{.j.*j.}}}{w_{\mathrm{.j..j.}}}$	$\sum \frac{w^2_{\mathrm{ijhstu}}}{w_{\mathrm{..h..u}}}$	$A_{9,1}$	$\sum \frac{w^2_{\mathrm{i.hi*u}}}{w_{\mathrm{i.hi.u}}}$
5	σ_{13}^4	$\sum \frac{w^2_{\mathrm{i..i.*}}}{w_{\mathrm{i..i..}}}$	$\sum \frac{w^2_{\mathrm{ijhstu}}}{w_{\mathrm{.j..t.}}}$	$\sum \frac{w^2_{\mathrm{i.hs.u}}}{w_{\mathrm{..h..h}}}$	$\sum \frac{w^2_{\mathrm{ij.it*}}}{w_{\mathrm{ij.it.}}}$	$A_{10,1}$
6	σ_{23}^4	$\left(\sum \frac{w_{\mathrm{i..i**}}}{n_{\mathrm{i..}}}\right)^2$	$\sum \frac{w^2_{\mathrm{.j..j*}}}{w_{\mathrm{.j..j.}}}$	$\sum \frac{w^2_{\mathrm{.jh.tu}}}{w_{\mathrm{..h..h}}}$	$\sum \frac{w^2_{\mathrm{ij.sj*}}}{w_{\mathrm{ij.sj.}}}$	$\sum \frac{w^2_{\mathrm{i.hs*h}}}{w_{\mathrm{i.hs.h}}}$
7	σ_{123}^4	$\sum \frac{w^2_{\mathrm{i..i**}}}{w_{\mathrm{i..i..}}}$	$\sum \frac{w^2_{\mathrm{.j.*j*}}}{w_{\mathrm{.j..j.}}}$	$\sum \frac{w^2_{\mathrm{..h**h}}}{w_{\mathrm{..h..h}}}$	$\sum \frac{w^2_{\mathrm{ij.ij*}}}{w_{\mathrm{ij.ij.}}}$	$\sum \frac{w^2_{\mathrm{i.hi*h}}}{w_{\mathrm{i.hi.h}}}$
8	σ_0^4	I	J	H	m_{12}	m_{13}
9	$2\sigma_1^2\sigma_2^2$	$w_{\mathrm{...**.}}$	$A_{1,1}$	$\sum_{i,j}\left(\sum_h \frac{w_{\mathrm{i.h.jh}}}{n_{\mathrm{..h}}}\right)^2$	$A_{9,1}$	$A_{9,1}$
10	$2\sigma_1^2\sigma_3^2$	$w_{\mathrm{...*.*}}$	$\sum_{i,h}\left(\sum_j \frac{w_{\mathrm{ij..jh}}}{n_{\mathrm{.j.}}}\right)^2$	$A_{10,1}$	$A_{10,1}$	$A_{10,1}$
11	$2\sigma_1^2\sigma_{12}^2$	$A_{9,1}$	$A_{9,1}$	$\sum_{i,s,j}\left(\sum_h \frac{w_{\mathrm{i.hsjh}}}{n_{\mathrm{..h}}}\right)^2$	$A_{9,1}$	$A_{9,1}$
12	$2\sigma_1^2\sigma_{13}^2$	$A_{10,1}$	$\sum_{i,s,h}\left(\sum_j \frac{w_{\mathrm{ij.sjh}}}{n_{\mathrm{.j.}}}\right)^2$	$A_{5,3}$	$A_{10,1}$	$A_{10,1}$

TABLE 11.6 (*Continued*)

Row	Column (1/2)Cov Product	6 T_{23}	7 T_{123}	8 T_f	9 T_1,T_2	10 T_1,T_3	11 T_1,T_{12}
1	σ_1^4	$\sum \frac{w^2_{.jh*tu}}{w_{.jh.tu}}$	$A_{1,1}$	$\left(\frac{A_{1,1}}{n}\right)^2$	$\sum \frac{w^2_{ij.i..}}{w_{.j.i..}}$	$\sum \frac{w^2_{i.hi..}}{w_{..hi..}}$	$A_{1,1}$
2	σ_2^4	$A_{2,2}$	$A_{2,2}$	$\left(\frac{A_{2,2}}{n}\right)^2$	$\sum \frac{w^2_{ij..j.}}{w_{i...j.}}$	$\sum \frac{w^2_{i.h.*.}}{w_{i....h}}$	$A_{2,9}$
3	σ_3^4	$A_{3,3}$	$A_{3,3}$	$\left(\frac{A_{3,3}}{n}\right)^2$	$\sum \frac{w^2_{ij...*}}{w_{i...j.}}$	$\sum \frac{w^2_{i.h..h}}{w_{i....h}}$	$\sum \frac{w^2_{ij.s.*}}{w_{ij.s..}}$
4	σ_{12}^4	$\sum \frac{w^2_{.jh*ju}}{w_{.jh.ju}}$	$A_{9,1}$	$\left(\frac{A_{9,1}}{n}\right)^2$	$\sum \frac{w^2_{ij.ij.}}{w_{i...j.}}$	$\sum \frac{w^2_{i.hi*.}}{w_{i....h}}$	$A_{17,9}$
5	σ_{13}^4	$\sum \frac{w^2_{.jh*th}}{w_{.jh.th}}$	$A_{10,1}$	$\left(\frac{A_{10,1}}{n}\right)^2$	$\sum \frac{w^2_{ij.i.*}}{w_{i...j.}}$	$\sum \frac{w^2_{i.hi.h}}{w_{i....h}}$	$\sum \frac{w^2_{ij.i.*}}{w_{ij.i..}}$
6	σ_{23}^4	$A_{16,2}$	$A_{16,2}$	$\left(\frac{A_{16,2}}{n}\right)^2$	$\sum \frac{w^2_{ij..j*}}{w_{i...j.}}$	$\sum \frac{w^2_{i.h.*h}}{w_{i....h}}$	$\sum \frac{w^2_{ij.sj*}}{w_{ij.s..}}$
7	σ_{123}^4	$\sum \frac{w^2_{.jh*jh}}{w_{.jh.jh}}$	$A_{13,1}$	$\left(\frac{A_{13,1}}{n}\right)^2$	$\sum \frac{w^2_{ij.ij*}}{w_{i...j.}}$	$\sum \frac{w^2_{i.hi*h}}{w_{i....h}}$	$\sum \frac{w^2_{ij.ij*}}{w_{ij.i..}}$
8	σ_0^4	m_{23}	m_{123}	1	$\sum \frac{w_{ij.ij.}}{w_{i...j.}}$	$\sum \frac{w_{i.hi.h}}{w_{i....h}}$	I
9	$2\sigma_1^2\sigma_2^2$	$A_{9,1}$	$A_{9,1}$	$\frac{A_{1,1}A_{2,2}}{n^2}$	$A_{9,1}$	$\sum \frac{w_{ijhi.h}n_{.j.}}{n_{..h}}$	$A_{9,1}$
10	$2\sigma_1^2\sigma_3^2$	$A_{10,1}$	$A_{10,1}$	$\frac{A_{1,1}A_{3,3}}{n^2}$	$\sum \frac{w_{ijhij.}n_{..h}}{n_{.j.}}$	$A_{10,1}$	$A_{10,1}$
11	$2\sigma_1^2\sigma_{12}^2$	$A_{4,6}$	$A_{9,1}$	$\frac{A_{1,1}A_{9,1}}{n^2}$	$\sum \frac{w^2_{ij.ij.}}{w_{.j.ij.}}$	$\sum \frac{w_{i.hijh}n_{ij.}}{n_{i.h}}$	$A_{9,1}$
12	$2\sigma_1^2\sigma_{13}^2$	$A_{5,6}$	$A_{10,1}$	$\frac{A_{1,1}A_{10,1}}{n^2}$	$\sum \frac{w_{ij.ijh}n_{i.h}}{n_{ij.}}$	$\sum \frac{w^2_{i.hi.h}}{w_{i.h..h}}$	$A_{10,1}$

TABLE 11.6 (*Continued*)

Row	Column (1/2)Cov Product	12 T_1,T_{13}	13 T_1,T_{23}	14 T_1,T_{123}	15 T_1,T_f	16 T_2,T_3
1	σ_1^4	$A_{1,1}$	$\sum \frac{w^2_{ijhi..}}{w_{.jhi..}}$	$A_{1,1}$	$\sum \frac{w^2_{i..i..}}{w_{i.....}}$	$\sum \frac{w^2_{.jh*..}}{w_{.j...h}}$
2	σ_2^4	$\sum \frac{w^2_{i.hs*.}}{w_{i.hs..}}$	$\sum \frac{w^2_{ijh.j.}}{w_{.jhi..}}$	$A_{2,9}$	$\sum \frac{w^2_{i...*.}}{w_{i.....}}$	$\sum \frac{w^2_{ijh.j.}}{w_{i.h.j.}}$
3	σ_3^4	$A_{3,10}$	$\sum \frac{w^2_{ijh..h}}{w_{.jhi..}}$	$A_{3,10}$	$\sum \frac{w^2_{i....*}}{w_{i.....}}$	$\sum \frac{w^2_{.jh..h}}{w_{.j...h}}$
4	σ_{12}^4	$\sum \frac{w^2_{i.hi*.}}{w_{i.hs..}}$	$\sum \frac{w^2_{ijhij.}}{w_{i...jh}}$	$A_{17,9}$	$\sum \frac{w^2_{i..i*.}}{w_{i.....}}$	$\sum \frac{w^2_{.jh*j.}}{w_{.j...h}}$
5	σ_{13}^4	$A_{23,10}$	$\sum \frac{w^2_{ijhi.h}}{w_{i...jh}}$	$A_{23,10}$	$\sum \frac{w^2_{i..i.*}}{w_{i.....}}$	$\sum \frac{w^2_{.jh*.h}}{w_{.j...h}}$
6	σ_{23}^4	$\sum \frac{w^2_{i.hs*h}}{w_{i.hs..}}$	$\sum \frac{w^2_{ijh.jh}}{w_{i...jh}}$	$A_{6,13}$	$\sum \frac{w^2_{i...**}}{w_{i.....}}$	$\sum \frac{w^2_{.jh.jh}}{w_{.j...h}}$
7	σ_{123}^4	$\sum \frac{w^2_{i.hi*h}}{w_{i..i.h}}$	$\sum \frac{w^2_{ijhijh}}{w_{i...jh}}$	$A_{34,13}$	$\sum \frac{w^2_{i..i**}}{w_{i.....}}$	$\sum \frac{w^2_{.jh*jh}}{w_{.j...h}}$
8	σ_0^4	I	$\sum \frac{w_{ijhijh}}{w_{i...jh}}$	I	1	$\sum \frac{w_{.jh.jh}}{w_{.j...h}}$
9	$2\sigma_1^2\sigma_2^2$	$A_{9,1}$	$\sum \frac{w^2_{ijh.j.}}{w_{.jh.j.}}$	$A_{9,1}$	$\sum \frac{w_{.j.*..}\, n_{.j.}}{n}$	$\sum \frac{w_{i.h.*h}\, n_{i..}}{n_{..h}}$
10	$2\sigma_1^2\sigma_3^2$	$A_{10,1}$	$\sum \frac{w^2_{ijh..h}}{w_{.jh..h}}$	$A_{10,1}$	$\sum \frac{w_{..h*..}\, n_{..h}}{n}$	$\sum \frac{w_{ij..j*}\, n_{i..}}{n_{.j.}}$
11	$2\sigma_1^2\sigma_{12}^2$	$A_{9,1}$	$\sum \frac{w^2_{ijhij.}}{w_{.jhij.}}$	$A_{9,1}$	$\sum \frac{w^2_{ij.i..}}{w_{i.....}}$	$\sum \frac{w_{.jh*..}\, w_{.jh*j.}}{w_{.j...h}}$
12	$2\sigma_1^2\sigma_{13}^2$	$A_{10,1}$	$\sum \frac{w^2_{ijhi.h}}{w_{.jhi.h}}$	$A_{10,1}$	$\sum \frac{w^2_{i.hi..}}{w_{i.....}}$	$\sum \frac{w_{.jh*..}\, w_{.jh*.h}}{w_{.j...h}}$

TABLE 11.6 (*Continued*)

Row	Column (1/2)Cov Product	17 T_2, T_{12}	18 T_2, T_{13}	19 T_2, T_{23}	20 T_2, T_{123}
1	σ_1^4	$A_{1,9}$	$\sum \frac{w_{ijhi..}^2}{w_{i.h.j.}}$	$\sum \frac{w_{.jh*t.}^2}{w_{.jh..u}}$	$A_{1,9}$
2	σ_2^4	$A_{2,2}$	$\sum \frac{w_{ijh.j.}^2}{w_{i.h.j.}}$	$A_{2,2}$	$A_{2,2}$
3	σ_3^4	$\sum \frac{w_{ij..t*}^2}{w_{ij..t.}}$	$\sum \frac{w_{ijh..h}^2}{w_{i.h.j.}}$	$A_{3,16}$	$A_{3,16}$
4	σ_{12}^4	$A_{11,9}$	$\sum \frac{w_{ijhij.}^2}{w_{.j.i.h}}$	$\sum \frac{w_{.jh*j.}^2}{w_{.jh.j.}}$	$A_{11,9}$
5	σ_{13}^4	$\sum \frac{w_{ij.it*}^2}{w_{ij..t.}}$	$\sum \frac{w_{ijhi.h}^2}{w_{.j.i.h}}$	$\sum \frac{w_{.jh*th}^2}{w_{.jh.t.}}$	$A_{5,18}$
6	σ_{23}^4	$\sum \frac{w_{ij..j*}^2}{w_{ij..j.}}$	$\sum \frac{n_{ijh}^2 n_{.jh}}{w_{.j.i.h}}$	$A_{24,16}$	$A_{24,16}$
7	σ_{123}^4	$\sum \frac{w_{ij.ij*}^2}{w_{ij..j.}}$	$\sum \frac{w_{ijhijh}^2}{w_{.j.i.h}}$	$\sum \frac{w_{.jh*jh}^2}{w_{.j..jh}}$	$A_{32,18}$
8	σ_0^4	J	$\sum \frac{w_{ijhijh}}{w_{.j.i.h}}$	J	J
9	$2\sigma_1^2\sigma_2^2$	$A_{9,1}$	$\sum \frac{w_{ijhi..}^2}{w_{i.hi..}}$	$A_{9,1}$	$A_{9,1}$
10	$2\sigma_1^2\sigma_3^2$	$\sum \frac{w_{.j.*.h} n_{.jh}}{n_{.j.}}$	$\sum \frac{w_{i....h} n_{ijh}^2}{w_{i.h.j.}}$	$A_{10,17}$	$A_{10,17}$
11	$2\sigma_1^2\sigma_{12}^2$	$A_{11,9}$	$\sum \frac{w_{i..ij.} n_{ijh}^2}{w_{i.h.j.}}$	$A_{4,19}$	$A_{11,9}$
12	$2\sigma_1^2\sigma_{13}^2$	$A_{12,9}$	$\sum \frac{w_{ijhi..}^2}{w_{.j.i..}}$	$\sum \frac{w_{.jh*t.} w_{.jh*th}}{w_{.jh.t.}}$	$A_{12,9}$

TABLE 11.6 (*Continued*)

Row	Column (1/2)Cov Product	21 T_2,T_f	22 T_3,T_{12}	23 T_3,T_{13}	24 T_3,T_{23}
1	σ_1^4	$\sum\frac{w^2_{.j.*..}}{w_{.j....}}$	$\sum\frac{w^2_{ijhi..}}{w_{ij...h}}$	$A_{1,10}$	$\sum\frac{w^2_{.jh*.u}}{w_{.jh..u}}$
2	σ_2^4	$\sum\frac{w^2_{.j..j.}}{w_{.j....}}$	$\sum\frac{w^2_{ijh.j.}}{w_{ij...h}}$	$\sum\frac{w^2_{i.h.*u}}{w_{i.h..u}}$	$A_{2,16}$
3	σ_3^4	$\sum\frac{w^2_{.j...*}}{w_{.j....}}$	$\sum\frac{w^2_{ijh..h}}{w_{ij...h}}$	$A_{3,3}$	$A_{3,3}$
4	σ_{12}^4	$\sum\frac{w^2_{.j.*j.}}{w_{.j....}}$	$\sum\frac{w^2_{ijhij.}}{w_{ij...h}}$	$\sum\frac{w^2_{.jh*ju}}{w_{.jh..u}}$	$\sum\frac{w^2_{i.hi*u}}{w_{i.h..u}}$
5	σ_{13}^4	$\sum\frac{w^2_{.j.*.*}}{w_{.j....}}$	$\sum\frac{w^2_{ijhi.h}}{w_{ij...h}}$	$A_{12,10}$	$\sum\frac{w^2_{.jh*.h}}{w_{.jh..h}}$
6	σ_{23}^4	$\sum\frac{w^2_{.j..j*}}{w_{.j....}}$	$\sum\frac{w^2_{ijh.jh}}{w_{ij...h}}$	$\sum\frac{w^2_{i.h.*h}}{w_{i.h..h}}$	$A_{19,16}$
7	σ_{123}^4	$\sum\frac{w^2_{.j.*j*}}{w_{.j....}}$	$\sum\frac{w^2_{ijhijh}}{w_{ij...h}}$	$\sum\frac{w^2_{i.hi*h}}{w_{i.h..h}}$	$\sum\frac{w^2_{.jh*jh}}{w_{.jh..h}}$
8	σ_0^4	1	$\sum\frac{w_{ijhijh}}{w_{ij...h}}$	H	H
9	$2\sigma_1^2\sigma_2^2$	$A_{9,15}$	$\sum\frac{w_{i...j.}n^2_{ijh}}{w_{ij...h}}$	$\sum\frac{w_{ij.i.h}n_{.jh}}{n_{..h}}$	$A_{9,23}$
10	$2\sigma_1^2\sigma_3^2$	$\sum\frac{w_{.j.*..}w_{j...*}}{w_{.j....}}$	$\sum\frac{w^2_{ijhi..}}{w_{ij.i..}}$	$A_{10,1}$	$A_{10,1}$
11	$2\sigma_1^2\sigma_{12}^2$	$\sum\frac{w_{.j.*..}w_{.j.*j.}}{w_{.j....}}$	$\sum\frac{w^2_{ijhi..}}{w_{..hi..}}$	$A_{11,10}$	$\sum\frac{w_{.jh*.u}w_{.jh*ju}}{w_{.jh..h}}$
12	$2\sigma_1^2\sigma_{13}^2$	$\sum\frac{w_{.j.*..}w_{.j.*.*}}{w_{.j....}}$	$\sum\frac{w_{ijhi..}n^2_{ijh}}{w_{ij...h}}$	$A_{12,10}$	$A_{5,24}$

TABLE 11.6 (*Continued*)

	Column (1/2)Cov	25 T_3, T_{123}	26 T_3, T_f	27 T_{12}, T_{13}	28 T_{12}, T_{23}	29 T_{12}, T_{123}	30 T_{12}, T_f
ow	Product						
1	σ_1^4	$A_{1,10}$	$\sum \frac{w^2_{..h*..}}{w_{..h...}}$	$A_{1,1}$	$A_{3,22}$	$A_{1,1}$	$A_{1,15}$
2	σ_2^4	$A_{2,16}$	$\sum \frac{w^2_{..h.*.}}{w_{..h...}}$	$A_{2,18}$	$A_{2,2}$	$A_{2,2}$	$A_{2,21}$
3	σ_3^4	$A_{3,3}$	$\sum \frac{w^2_{..h..h}}{w_{..h...}}$	$A_{1,13}$	$A_{1,13}$	$A_{1,13}$	$\sum \frac{w^2_{ij...*}}{w_{ij....}}$
4	σ_{12}^4	$A_{4,22}$	$\sum \frac{w^2_{..h**.}}{w_{..h...}}$	$A_{17,18}$	$A_{11,13}$	$A_{9,1}$	$\sum \frac{w^2_{ij.ij.}}{w_{ij....}}$
5	σ_{13}^4	$A_{12,10}$	$\sum \frac{w^2_{..h*.h}}{w_{..h...}}$	$A_{23,22}$	$\sum \frac{w^2_{ijhith}}{w_{ij..th}}$	$A_{23,22}$	$\sum \frac{w^2_{ij.i.*}}{w_{ij....}}$
6	σ_{23}^4	$A_{19,16}$	$\sum \frac{w^2_{..h.*h}}{w_{..h...}}$	$\sum \frac{w^2_{ijhsjh}}{w_{ij.s.h}}$	$A_{24,22}$	$A_{24,22}$	$\sum \frac{w^2_{ij..j*}}{w_{ij....}}$
7	σ_{123}^4	$A_{29,22}$	$\sum \frac{w^2_{..h**h}}{w_{..h...}}$	$\sum \frac{w^2_{ijhijh}}{w_{ij.i.h}}$	$\sum \frac{w^2_{ijhijh}}{w_{ij..jh}}$	$A_{25,22}$	$\sum \frac{w^2_{ij.ij*}}{w_{ij....}}$
8	σ_0^4	H	1	$\sum \frac{w_{ijhijh}}{w_{ij.i.h}}$	$\sum \frac{w_{ijhijh}}{w_{ij..jh}}$	m_{12}	1
9	$2\sigma_1^2\sigma_1^2$	$A_{9,23}$	$\sum \frac{w_{..h*..}w_{..h.*.}}{w_{..h...}}$	$A_{9,1}$	$A_{9,1}$	$A_{9,1}$	$A_{9,15}$
.0	$2\sigma_1^2\sigma_2^2$	$A_{10,1}$	$A_{10,15}$	$A_{10,1}$	$A_{10,1}$	$A_{10,1}$	$A_{10,15}$
.1	$2\sigma_1^2\sigma_{12}^2$	$A_{11,10}$	$\sum \frac{w_{..h*..}w_{..h**.}}{w_{..h...}}$	$A_{9,1}$	$A_{11,13}$	$A_{9,1}$	$A_{11,15}$
.2	$2\sigma_1^2\sigma_{13}^2$	$A_{12,10}$	$\sum \frac{w_{..h*..}w_{..h*.h}}{w_{..h...}}$	$A_{10,1}$	$A_{11,14}$	$A_{10,1}$	$A_{12,15}$

TABLE 11.6 (*Continued*)

Row	Column (1/2)Cov Product	31 T_{13},T_{23}	32 T_{13},T_{123}	33 T_{13},T_f	34 T_{23},T_{123}	35 T_{23},T_f	36 T_{123},T_f
1	σ_1^4	$A_{3,22}$	$A_{1,1}$	$A_{1,15}$	$A_{3,22}$	$\sum \frac{w_{.jh*..}^2}{w_{.jh...}}$	$A_{1,15}$
2	σ_2^4	$A_{2,18}$	$A_{2,18}$	$\sum \frac{w_{i.h.*.}^2}{w_{i.h...}}$	$A_{2,2}$	$A_{2,21}$	$A_{2,21}$
3	σ_3^4	$A_{3,3}$	$A_{3,3}$	$A_{3,26}$	$A_{3,3}$	$A_{3,26}$	$A_{3,26}$
4	σ_{12}^4	$\sum \frac{w_{ijhiju}^2}{w_{i.h.ju}}$	$A_{17,18}$	$\sum \frac{w_{i.hi*.}^2}{w_{i.h...}}$	$A_{11,13}$	$\sum \frac{w_{.jh*j.}^2}{w_{.jh...}}$	$A_{4,30}$
5	σ_{13}^4	$A_{11,14}$	$A_{10,1}$	$\sum \frac{w_{i.hi.h}^2}{w_{i.h...}}$	$A_{11,14}$	$\sum \frac{w_{.jh*.h}^2}{w_{.jh...}}$	$A_{5,33}$
6	σ_{23}^4	$A_{19,18}$	$A_{19,18}$	$\sum \frac{w_{i.h.*h}^2}{w_{i.h...}}$	$A_{16,2}$	$\sum \frac{w_{.jh.jh}^2}{w_{.jh...}}$	$A_{6,35}$
7	σ_{123}^4	$\sum \frac{w_{ijhijh}^2}{w_{i.h.jh}}$	$A_{21,5}$	$\sum \frac{w_{i.hi*h}^2}{w_{i.h...}}$	$A_{15,6}$	$\sum \frac{w_{.jh*jh}^2}{w_{.jh...}}$	$\sum \frac{w_{ijhijh}^2}{w_{ijh...}}$
8	σ_0^4	$\sum \frac{w_{ijhijh}}{w_{i.h.jh}}$	m_{13}	1	m_{23}	1	1
9	$2\sigma_1^2\sigma_2^2$	$A_{9,1}$	$A_{9,1}$	$A_{9,15}$	$A_{9,1}$	$A_{9,15}$	$A_{9,15}$
10	$2\sigma_1^2\sigma_3^2$	$A_{10,1}$	$A_{10,1}$	$A_{10,15}$	$A_{10,1}$	$A_{10,15}$	$A_{10,15}$
11	$2\sigma_1^2\sigma_{12}^2$	$A_{11,13}$	$A_{9,1}$	$A_{11,15}$	$A_{11,13}$	$\sum \frac{w_{.jh*..}w_{.jh*j.}}{w_{.jh...}}$	$A_{11,15}$
12	$2\sigma_1^2\sigma_{13}^2$	$A_{11,14}$	$A_{10,1}$	$A_{12,15}$	$A_{11,14}$	$\sum \frac{w_{.jh*..}w_{.jh*j.}}{w_{.jh...}}$	$A_{12,15}$

TABLE 11.6 (*Continued*)

Row	Column (1/2)Cov Product	1 T_1	2 T_2	3 T_3	4 T_{12}	5 T_{13}
13	$2\sigma_1^2\sigma_{23}^2$	$w_{...***}$	$\sum \frac{w^2_{ij..jh}}{w_{.j..j.}}$	$\sum \frac{w^2_{i.h.jh}}{w_{..h..h}}$	$A_{13,1}$	$A_{13,1}$
14	$2\sigma_1^2\sigma_{123}^2$	$A_{13,1}$	$\sum \frac{w^2_{ij.sjh}}{w_{.j..j.}}$	$\sum \frac{w^2_{i.hsjh}}{w_{..h..h}}$	$A_{13,1}$	$A_{13,1}$
15	$2\sigma_1^2\sigma_0^2$	n	$\sum \frac{w_{.j.*j.}}{n_{.j.}}$	$\sum \frac{w_{..h*.h}}{n_{..h}}$	n	n
16	$2\sigma_2^2\sigma_3^2$	$\sum_{j,h}\left(\sum_i \frac{w_{ij.i.h}}{n_{i..}}\right)^2$	$w_{....**}$	$A_{16,2}$	$A_{16,2}$	$A_{16,2}$
17	$2\sigma_2^2\sigma_{12}^2$	$A_{2,1}$	$A_{9,1}$	$\sum_{i,j,t}\left(\sum_h \frac{w_{.jhith}}{n_{..h}}\right)^2$	$A_{9,1}$	$A_{4,5}$
18	$2\sigma_2^2\sigma_{13}^2$	$\sum \frac{w^2_{ij.i.h}}{w_{i..i..}}$	$A_{13,1}$	$A_{18,1}$	$A_{13,1}$	$A_{13,1}$
19	$2\sigma_2^2\sigma_{23}^2$	$\sum_{j,t,h}\left(\sum_i \frac{w_{ij.ith}}{n_{i..}}\right)^2$	$A_{16,2}$	$A_{16,2}$	$A_{16,2}$	$A_{6,5}$
20	$2\sigma_2^2\sigma_{123}^2$	$\sum \frac{w^2_{ij.ith}}{w_{i..i..}}$	$A_{13,1}$	$A_{17,3}$	$A_{13,1}$	$A_{7,5}$
21	$2\sigma_2^2\sigma_0^2$	$\sum \frac{w_{i..i*.}}{n_{i..}}$	n	$\sum \frac{w_{..h.*h}}{n_{..h}}$	n	$\sum \frac{w_{i.hi*h}}{n_{i.h}}$
22	$2\sigma_3^2\sigma_{12}^2$	$A_{18,1}$	$A_{13,2}$	$A_{13,1}$	$A_{13,1}$	$A_{13,1}$
23	$2\sigma_3^2\sigma_{13}^2$	$A_{5,1}$	$\sum_{i,h,u}\left(\sum_j \frac{w_{.jhiju}}{n_{.j.}}\right)^2$	$A_{10,1}$	$A_{5,4}$	$A_{10,1}$
24	$2\sigma_3^2\sigma_{23}^2$	$\sum_{j,h,u}\left(\sum_i \frac{w_{i.hiju}}{n_{i..}}\right)^2$	$\sum \frac{w^2_{.j..j*}}{w_{.j..j.}}$	$A_{16,2}$	$A_{6,4}$	$A_{16,2}$

TABLE 11.6 (*Continued*)

Row	Column (1/2)Cov Product	6 T_{23}	7 T_{123}	8 T_f	9 T_1, T_2	10 T_1, T_{13}
13	$2\sigma_1^2\sigma_{23}^2$	$A_{13,1}$	$A_{13,1}$	$\frac{A_{1,1}A_{16,2}}{n^2}$	$\sum \frac{w_{ij..j*}n_{ij.}}{n_{.j.}}$	$\sum \frac{w_{..h*jh}n_{.jh}}{n_{..h}}$
14	$2\sigma_1^2\sigma_{123}^2$	$A_{7,6}$	$A_{13,1}$	$\frac{A_{1,1}A_{13,1}}{n^2}$	$\sum \frac{w_{ij.ijh}^2}{w_{ij..j.}}$	$\sum \frac{w_{ijhi.h}^2}{w_{i.h..h}}$
15	$2\sigma_1^2\sigma_0^2$	$\sum \frac{w_{.jh*jh}}{n_{.jh}}$	n	$\frac{A_{1,1}}{n}$	$A_{15,2}$	$A_{15,3}$
16	$2\sigma_2^2\sigma_3^2$	$A_{16,2}$	$A_{16,2}$	$\frac{A_{2,2}A_{3,3}}{n^2}$	$\sum \frac{w_{ij...*}n_{ij.}}{n_{i..}}$	$\sum \frac{w_{i..ij*}n_{.j.}}{n_{i..}}$
17	$2\sigma_2^2\sigma_{12}^2$	$A_{9,1}$	$A_{9,1}$	$\frac{A_{2,2}A_{9,1}}{n^2}$	$\sum \frac{w_{ij.ij.}^2}{w_{i..ij.}}$	$\sum \frac{w_{i.h.*.}w_{i.hi*.}}{w_{i....h}}$
18	$2\sigma_2^2\sigma_{13}^2$	$A_{13,1}$	$A_{13,1}$	$\frac{A_{2,2}A_{10,1}}{n^2}$	$\sum \frac{w_{ij.i.*}n_{ij.}}{n_{i..}}$	$\sum \frac{w_{i.hi.h}w_{i.h.*.}}{w_{i....h}}$
19	$2\sigma_2^2\sigma_{23}^2$	$A_{16,2}$	$A_{16,2}$	$\frac{A_{2,2}A_{16,2}}{n^2}$	$\sum \frac{w_{ij..j*}n_{ij.}}{n_{i..}}$	$\sum \frac{w_{i.h.*.}w_{i.h.*h}}{w_{i....h}}$
20	$2\sigma_2^2\sigma_{123}^2$	$A_{13,1}$	$A_{13,1}$	$\frac{A_{2,2}A_{13,1}}{n^2}$	$\sum \frac{w_{ijhij.}^2}{w_{ij.i..}}$	$\sum \frac{w_{i.hi*h}w_{i.h.*.}}{w_{i....h}}$
21	$2\sigma_2^2\sigma_0^2$	n	n	$\frac{A_{2,2}}{n}$	$A_{21,1}$	$\sum \frac{w_{i.h.*.}n_{i.h}}{w_{i....h}}$
22	$2\sigma_3^2\sigma_{12}^2$	$A_{13,1}$	$A_{13,1}$	$\frac{A_{3,3}A_{4,1}}{n^2}$	$\sum \frac{w_{ij.ij.}w_{ij...*}}{w_{i...j.}}$	$A_{18,9}$
23	$2\sigma_3^2\sigma_{13}^2$	$A_{10,1}$	$A_{10,1}$	$\frac{A_{3,3}A_{10,1}}{n^2}$	$\sum \frac{w_{ij...*}w_{ij.i.*}}{w_{i...j.}}$	$\sum \frac{w_{i.hi.h}^2}{w_{i..i.h}}$
24	$2\sigma_3^2\sigma_{23}^2$	$A_{16,2}$	$A_{16,2}$	$\frac{A_{3,3}A_{16,2}}{n^2}$	$\sum \frac{w_{ij..j*}w_{ij...*}}{w_{i...j.}}$	$\sum \frac{w_{i.h.*h}n_{i.h}}{n_{i..}}$

TABLE 11.6 (*Continued*)

Row	Column (1/2)Cov Product	11 T_1,T_{12}	12 T_1,T_{13}	13 T_1,T_{23}	14 T_1,T_{123}	15 T_1,T_f
13	$2\sigma_1^2\sigma_{23}^2$	$A_{13,1}$	$A_{13,1}$	$A_{13,1}$	$A_{13,1}$	$\sum \frac{w_{i...**}n_{i..}}{n}$
14	$2\sigma_1^2\sigma_{123}^2$	$A_{13,1}$	$A_{13,1}$	$\sum \frac{w^2_{ijhijh}}{w_{.jhijh}}$	$A_{13,1}$	$\sum \frac{w_{ijhi..}}{w_{...i..}}$
15	$2\sigma_1^2\sigma_0^2$	n	n	$A_{15,6}$	n	$\frac{A_{1,1}}{n}$
16	$2\sigma_2^2\sigma_3^2$	$\sum \frac{w_{i...j*}n_{ij.}}{n_{i..}}$	$A_{16,11}$	$\sum \frac{w_{.j...h}n^2_{ijh}}{w_{i...jh}}$	$A_{16,11}$	$\sum \frac{w_{i...*.}w_{i....*}}{w_{...i..}}$
17	$2\sigma_2^2\sigma_{12}^2$	$A_{17,9}$	$A_{4,12}$	$\sum \frac{w_{ij..j.}n^2_{ijh}}{w_{i...jh}}$	$A_{17,9}$	$\sum \frac{w_{i...*.}w_{i..i*.}}{w_{...i..}}$
18	$2\sigma_2^2\sigma_{13}^2$	$A_{18,9}$	$A_{18,9}$	$\sum \frac{w_{i.h.j.}n^2_{ijh}}{w_{i...jh}}$	$A_{18,9}$	$\sum \frac{w_{i...*.}w_{i..i.*}}{w_{...i..}}$
19	$2\sigma_2^2\sigma_{23}^2$	$A_{19,9}$	$\sum \frac{w_{i.hs*.}w_{i.hs*h}}{w_{i.hs..}}$	$\sum \frac{w^2_{ijh.j.}}{w_{i...j.}}$	$A_{19,9}$	$\sum \frac{w_{i...*.}w_{i...**}}{w_{...i..}}$
20	$2\sigma_2^2\sigma_{123}^2$	$A_{20,9}$	$\sum \frac{w_{i.hi*h}w_{i.hi*.}}{w_{i.hi..}}$	$\sum \frac{n^3_{ijh}n_{.j.}}{w_{i...jh}}$	$A_{20,9}$	$\sum \frac{w_{i...*.}w_{i..i**}}{w_{...i..}}$
21	$2\sigma_2^2\sigma_0^2$	$A_{21,1}$	$A_{21,1}$	$\sum \frac{n^2_{ijh}n_{.j.}}{w_{i...jh}}$	$A_{21,1}$	$\frac{A_{2,2}}{n}$
22	$2\sigma_3^2\sigma_{12}^2$	$A_{18,9}$	$A_{18,9}$	$\sum \frac{w_{ij...h}n^2_{ijh}}{w_{i...jh}}$	$A_{18,9}$	$\sum \frac{w_{i....*}w_{i..i*.}}{w_{...i..}}$
23	$2\sigma_3^2\sigma_{13}^2$	$A_{5,11}$	$A_{23,10}$	$\sum \frac{w_{i.h..h}n^2_{ijh}}{w_{i...jh}}$	$A_{23,10}$	$\sum \frac{w_{i....*}w_{i..i.*}}{w_{...i..}}$
24	$2\sigma_3^2\sigma_{23}^2$	$\sum \frac{w_{ij.s.*}w_{ij.sj*}}{w_{ij.s..}}$	$A_{24,10}$	$\sum \frac{w^2_{ijh..h}}{w_{i....h}}$	$A_{24,10}$	$\sum \frac{w_{i....*}w_{i..i**}}{w_{...i..}}$

TABLE 11.6 (*Continued*)

Row	Column (1/2)Cov Product	16 T_2,T_3	17 T_2,T_{12}	18 T_2,T_{13}	19 T_2,T_{23}	20 T_2,T_1
13	$2\sigma_1^2\sigma_{23}^2$	$\sum \frac{w_{.jh.jh}w_{.jh*..}}{w_{.j...h}}$	$A_{13,9}$	$\sum \frac{w_{i...jh}n_{ijh}^2}{w_{.j.i.h}}$	$A_{13,9}$	$A_{13,}$
14	$2\sigma_1^2\sigma_{123}^2$	$\sum \frac{w_{.jh*..}w_{.jh*jh}}{w_{.j...h}}$	$A_{14,9}$	$\sum \frac{n_{ijh}^3 n_{i..}}{w_{.j.i.h}}$	$\sum \frac{w_{.jh*j.}w_{.jh*jh}}{w_{.jh.j.}}$	$A_{14,}$
15	$2\sigma_1^2\sigma_0^2$	$\sum \frac{w_{.jh*..}n_{.jh}}{w_{.j...h}}$	$A_{15,2}$	$\sum \frac{n_{ijh}^2 n_{i..}}{w_{.j.i.h}}$	$A_{15,2}$	$A_{15,2}$
16	$2\sigma_2^2\sigma_3^2$	$A_{16,2}$	$A_{16,2}$	$\sum \frac{w_{ijh..h}^2}{w_{i.h..h}}$	$A_{16,2}$	$A_{16,2}$
17	$2\sigma_2^2\sigma_{12}^2$	$\sum \frac{w_{.jh*j.}n_{.jh}}{n_{..h}}$	$A_{9,1}$	$\sum \frac{w_{ijhij.}^2}{w_{i.hij.}}$	$A_{9,1}$	$A_{9,1}$
18	$2\sigma_2^2\sigma_{13}^2$	$A_{13,9}$	$A_{13,1}$	$A_{13,1}$	$A_{13,1}$	$A_{13,1}$
19	$2\sigma_2^2\sigma_{23}^2$	$\sum \frac{w_{.jh.jh}^2}{w_{..h.jh}}$	$A_{16,2}$	$\sum \frac{w_{ijh.jh}^2}{w_{i.h.jh}}$	$A_{16,2}$	$A_{16,2}$
20	$2\sigma_2^2\sigma_{123}^2$	$\sum \frac{w_{ijh.jh}^2}{w_{..h.jh}}$	$A_{13,1}$	$\sum \frac{w_{ijhijh}^2}{w_{i.hijh}}$	$A_{13,1}$	$A_{13,1}$
21	$2\sigma_2^2\sigma_0^2$	$A_{21,3}$	n	$A_{21,5}$	n	n
22	$2\sigma_3^2\sigma_{12}^2$	$A_{13,9}$	$A_{13,9}$	$\sum \frac{w_{ij...h}n_{ijh}^2}{w_{.j.i.h}}$	$A_{17,16}$	$A_{13,9}$
23	$2\sigma_3^2\sigma_{13}^2$	$\sum \frac{w_{.jhijh}n_{i.h}}{n_{.j.}}$	$\sum \frac{w_{ij..t*}w_{ij.it*}}{w_{ij..t.}}$	$\sum \frac{w_{ijh..h}^2}{w_{.j...h}}$	$A_{23,16}$	$A_{23,16}$
24	$2\sigma_3^2\sigma_{23}^2$	$\sum \frac{w_{.jh.jh}^2}{w_{.j..jh}}$	$A_{6,17}$	$\sum \frac{n_{ijh}^2 w_{.jh..h}}{w_{.j.i.h}}$	$A_{24,16}$	$A_{24,16}$

TABLE 11.6 (*Continued*)

Column (1/2)Cov Product	21 T_2, T_f	22 T_3, T_{12}	23 T_3, T_{13}	24 T_3, T_{23}	25 T_3, T_{123}
$2\sigma_1^2\sigma_{23}^2$	$\sum \frac{w_{.j.*..}w_{.j..j*}}{w_{....j.}}$	$\sum \frac{w_{i...jh}n_{ijh}^2}{w_{ij...h}}$	$A_{13,10}$	$A_{13,10}$	$A_{13,10}$
$2\sigma_1^2\sigma_{123}^2$	$\sum \frac{w_{.j.*..}w_{.j.*j*}}{w_{....j.}}$	$\sum \frac{n_{ijh}^3 n_{i..}}{w_{ij...h}}$	$A_{14,10}$	$\sum \frac{w_{.jh*.h}w_{.jh*jh}}{w_{.jh..h}}$	$A_{14,10}$
$2\sigma_1^2\sigma_0^2$	$\frac{A_{1,1}}{n}$	$\sum \frac{n_{ijh}^2 n_{i..}}{w_{ij...h}}$	$A_{15,3}$	$A_{15,3}$	$A_{15,3}$
$2\sigma_2^2\sigma_3^2$	$\sum \frac{w_{..hi*.}n_{..h}}{n}$	$\sum \frac{w_{ijh.j.}^2}{w_{ij..j.}}$	$A_{16,2}$	$A_{16,2}$	$A_{16,2}$
$2\sigma_2^2\sigma_{12}^2$	$\sum \frac{w_{ij..j.}^2}{w_{....j.}}$	$\sum \frac{w_{ijh.j.}^2}{w_{.j...h}}$	$\sum \frac{w_{i.h.*u}w_{i.hi*u}}{w_{i.h..u}}$	$A_{17,16}$	$A_{17,16}$
$2\sigma_2^2\sigma_{13}^2$	$\sum \frac{w_{.j.*.*}n_{.j.}}{n}$	$\sum \frac{w_{i.h.j.}n_{ijh}^2}{w_{ij...h}}$	$A_{13,10}$	$A_{13,10}$	$A_{13,10}$
$2\sigma_2^2\sigma_{23}^2$	$\sum \frac{w_{.jh.j.}^2}{w_{....j.}}$	$\sum \frac{w_{.j..jh}n_{ijh}^2}{w_{ij...h}}$	$A_{6,23}$	$A_{19,16}$	$A_{19,16}$
$2\sigma_2^2\sigma_{123}^2$	$\sum \frac{w_{ijh.j.}^2}{w_{....j.}}$	$\sum \frac{n_{ijh}^3 n_{.j.}}{w_{ij...h}}$	$\sum \frac{w_{i.h.*h}w_{i.hi*h}}{w_{i.h..h}}$	$A_{20,16}$	$A_{20,16}$
$2\sigma_2^2\sigma_0^2$	$\frac{A_{2,2}}{n}$	$\sum \frac{n_{ijh}^2 n_{.j.}}{w_{ij...h}}$	$A_{21,3}$	$A_{21,3}$	$A_{21,3}$
$2\sigma_3^2\sigma_{12}^2$	$\sum \frac{w_{.j...*}w_{.j.*j.}}{w_{....j.}}$	$A_{13,1}$	$A_{13,1}$	$A_{13,1}$	$A_{13,1}$
$2\sigma_3^2\sigma_{13}^2$	$\sum \frac{w_{.j...*}w_{.j.*.*}}{w_{....j.}}$	$\sum \frac{w_{ijhi.h}^2}{w_{i.hij.}}$	$A_{10,2}$	$A_{10,2}$	$A_{10,1}$
$2\sigma_3^2\sigma_{23}^2$	$\sum \frac{w_{.j...*}w_{.j..j*}}{w_{....j.}}$	$\sum \frac{w_{ijh.jh}^2}{w_{.jhij.}}$	$A_{16,2}$	$A_{16,2}$	$A_{16,2}$

TABLE 11.6 (*Continued*)

Row	Column (1/2)Cov Product	26 T_3,T_f	27 T_{12},T_{13}	28 T_{12},T_{23}	29 T_{12},T_{123}	30 T_{12},T_f	31 T_{13},T_{23}
13	$2\sigma_1^2\sigma_{23}^2$	$\sum \frac{w_{..h*..}w_{..h.*h}}{w_{..h...}}$	$A_{13,1}$	$A_{13,1}$	$A_{13,1}$	$A_{13,15}$	$A_{13,1}$
14	$2\sigma_1^2\sigma_{123}^2$	$\sum \frac{w_{..h*..}w_{..h**h}}{w_{..h...}}$	$A_{13,1}$	$A_{14,13}$	$A_{13,1}$	$A_{14,15}$	$A_{14,13}$
15	$2\sigma_1^2\sigma_0^2$	$\frac{A_{1,1}}{n}$	n	$A_{15,6}$	n	$\frac{A_{1,1}}{n}$	$A_{15,6}$
16	$2\sigma_2^2\sigma_3^2$	$A_{16,21}$	$A_{16,2}$	$A_{16,2}$	$A_{16,2}$	$A_{16,21}$	$A_{16,2}$
17	$2\sigma_2^2\sigma_{12}^2$	$\sum \frac{w_{..h.*.}w_{..h**.}}{w_{..h...}}$	$A_{17,18}$	$A_{9,1}$	$A_{9,1}$	$A_{17,21}$	$A_{17,18}$
18	$2\sigma_2^2\sigma_{13}^2$	$\sum \frac{w_{..h.*.}w_{..h*.h}}{w_{..h...}}$	$A_{13,1}$	$A_{13,1}$	$A_{13,1}$	$A_{18,21}$	$A_{13,1}$
19	$2\sigma_2^2\sigma_{23}^2$	$\sum \frac{w_{..h.*.}w_{..h.*h}}{w_{..h...}}$	$A_{19,18}$	$A_{16,2}$	$A_{16,2}$	$A_{19,21}$	$A_{19,18}$
20	$2\sigma_2^2\sigma_{123}^2$	$\sum \frac{w_{..h.*.}w_{..h**h}}{w_{..h...}}$	$A_{20,18}$	$A_{13,1}$	$A_{13,1}$	$A_{20,21}$	$A_{20,18}$
21	$2\sigma_2^2\sigma_0^2$	$\frac{A_{2,2}}{n}$	$A_{21,5}$	n	n	$\frac{A_{2,2}}{n}$	$A_{21,5}$
22	$2\sigma_3^2\sigma_{12}^2$	$\sum \frac{w_{..h**.}n_{..h}}{n}$	$A_{13,1}$	$A_{13,1}$	$A_{13,1}$	$A_{22,26}$	$A_{13,1}$
23	$2\sigma_3^2\sigma_{13}^2$	$\sum \frac{w_{..h*.h}n_{..h}}{n}$	$A_{23,22}$	$A_{23,22}$	$A_{23,22}$	$\sum \frac{w_{ij...*}w_{ij.i.*}}{w_{...ij.}}$	$A_{10,1}$
24	$2\sigma_3^2\sigma_{23}^2$	$\sum \frac{w_{..h.*h}n_{..h}}{n}$	$A_{24,22}$	$A_{24,22}$	$A_{24,22}$	$\sum \frac{w_{ij...*}w_{ij..j*}}{w_{...ij.}}$	$A_{16,2}$

TABLE 11.6 (*Continued*)

Row	Column (1/2)Cov Product	32 T_{13},T_{123}	33 T_{13},T_f	34 T_{23},T_{123}	35 T_{23},T_f	36 T_{123},T_f
13	$2\sigma_1^2\sigma_{23}^2$	$A_{13,1}$	$A_{13,15}$	$A_{13,1}$	$A_{13,15}$	$A_{13,15}$
14	$2\sigma_1^2\sigma_{123}^2$	$A_{13,1}$	$A_{14,15}$	$A_{14,13}$	$\sum \frac{w_{.jh*..}w_{.jh*jh}}{w_{....jh}}$	$A_{14,15}$
15	$2\sigma_1^2\sigma_0^2$	n	$\frac{A_{1,1}}{n}$	$A_{15,6}$	$\frac{A_{1,1}}{n}$	$\frac{A_{1,1}}{n}$
16	$2\sigma_2^2\sigma_3^2$	$A_{16,2}$	$A_{16,21}$	$A_{16,2}$	$A_{16,21}$	$A_{16,21}$
17	$2\sigma_2^2\sigma_{12}^2$	$A_{17,18}$	$\sum \frac{w_{i.h.*.}w_{i.hi*.}}{w_{...i.h}}$	$A_{9,1}$	$A_{17,21}$	$A_{17,21}$
18	$2\sigma_2^2\sigma_{13}^2$	$A_{13,1}$	$A_{18,21}$	$A_{13,1}$	$A_{18,21}$	$A_{15,21}$
19	$2\sigma_2^2\sigma_{23}^2$	$A_{19,18}$	$\sum \frac{w_{i.h.*.}w_{i.h.*h}}{w_{...i.h}}$	$A_{16,2}$	$A_{19,21}$	$A_{19,21}$
20	$2\sigma_2^2\sigma_{123}^2$	$A_{20,18}$	$\sum \frac{w_{i.h.*.}w_{i.hi*h}}{w_{...i.h}}$	$A_{13,1}$	$A_{20,21}$	$A_{20,21}$
21	$2\sigma_2^2\sigma_0^2$	$A_{21,5}$	$\frac{A_{2,2}}{n}$	n	$\frac{A_{2,2}}{n}$	$\frac{A_{2,2}}{n}$
22	$2\sigma_3^2\sigma_{12}^2$	$A_{13,1}$	$A_{22,26}$	$A_{13,1}$	$A_{22,26}$	$A_{22,26}$
23	$2\sigma_3^2\sigma_{13}^2$	$A_{10,1}$	$A_{23,26}$	$A_{10,1}$	$A_{23,26}$	$A_{23,26}$
24	$2\sigma_3^2\sigma_{23}^2$	$A_{16,2}$	$A_{24,26}$	$A_{16,2}$	$A_{24,26}$	$A_{24,26}$

TABLE 11.6 (*Continued*)

Row	Column (1/2)Cov Product	1 T_1	2 T_2	3 T_3	4 T_{12}	5 T_{13}	6 T_{23}	7 T_{123}
25	$2\sigma_3^2\sigma_{123}^2$	$\sum \frac{w_{i.hiju}^2}{w_{i..i..}}$	$\sum \frac{w_{.jhiju}^2}{w_{.j..j.}}$	$A_{13,1}$	$A_{7,4}$	$A_{13,1}$	$A_{13,1}$	$A_{13,1}$
26	$2\sigma_3^2\sigma_0^2$	$\sum \frac{w_{i..i.*}}{n_{i..}}$	$\sum \frac{w_{.j..j*}}{n_{.j.}}$	n	$\sum \frac{w_{ij.ij*}}{n_{ij.}}$	n	n	n
27	$2\sigma_{12}^2\sigma_{13}^2$	$A_{18,1}$	$A_{14,2}$	$A_{14,3}$	$A_{13,1}$	$A_{13,1}$	$A_{7,6}$	$A_{13,1}$
28	$2\sigma_{12}^2\sigma_{23}^2$	$A_{20,1}$	$A_{13,2}$	$\sum \frac{w_{.jhith}^2}{w_{..h..h}}$	$A_{13,1}$	$A_{7,5}$	$A_{13,1}$	$A_{13,1}$
29	$2\sigma_{12}^2\sigma_{123}^2$	$A_{20,1}$	$A_{14,2}$	$A_{4,3}$	$A_{13,1}$	$A_{7,5}$	$A_{7,6}$	$A_{13,1}$
30	$2\sigma_{12}^2\sigma_0^2$	$A_{21,1}$	$A_{15,2}$	$[A_{4,3}]^{1/2}$	n	$A_{21,5}$	$A_{15,6}$	n
31	$2\sigma_{13}^2\sigma_{23}^2$	$A_{25,1}$	$A_{25,2}$	$A_{13,3}$	$A_{7,4}$	$A_{13,1}$	$A_{13,1}$	$A_{13,1}$
32	$2\sigma_{13}^2\sigma_{123}^2$	$A_{25,1}$	$A_{5,2}$	$\sum \frac{w_{i.hsjh}^2}{w_{..h..h}}$	$A_{7,4}$	$A_{13,1}$	$A_{7,6}$	$A_{13,1}$
33	$2\sigma_{13}^2\sigma_0^2$	$A_{26,1}$	$[A_{5,2}]^{1/2}$	$A_{15,3}$	$A_{26,4}$	n	n	n
34	$2\sigma_{23}^2\sigma_{123}^2$	$A_{6,1}$	$A_{5,2}$	$A_{28,3}$	$A_{7,4}$	$A_{7,5}$	$A_{13,1}$	$A_{13,1}$
35	$2\sigma_{23}^2\sigma_0^2$	$[A_{6,1}]^{1/2}$	$A_{26,2}$	$A_{15,3}$	$A_{26,4}$	$A_{21,5}$	$A_{15,6}$	n
36	$2\sigma_{123}^2\sigma_0^2$	$[A_{6,1}]^{1/2}$	$[A_{5,2}]^{1/2}$	$[A_{4,3}]^{1/2}$	$A_{26,4}$	$A_{21,5}$	$A_{15,6}$	n

TABLE 11.6 (*continued*)

Row	Column (1/2)Cov Product	8 T_f	9 T_1, T_2	10 T_1, T_3	11 T_1, T_{12}	12 T_1, T_{13}
25	$2\sigma_3^2\sigma_{123}^2$	$\frac{A_{3,3}A_{13,1}}{n^2}$	$\sum \frac{n_{iju}^2 w_{ij...*}}{w_{i...j.}}$	$\sum \frac{w_{i.hijh}^2}{w_{i..i.h}}$	$\sum \frac{n_{iju}^2 w_{ij.i.*}}{w_{ij.i..}}$	$A_{25,10}$
26	$2\sigma_3^2\sigma_0^2$	$\frac{A_{3,3}}{n}$	$\sum \frac{w_{ij...*}n_{ij.}}{w_{i...j.}}$	$A_{26,1}$	$A_{26,1}$	$A_{26,1}$
27	$2\sigma_{12}^2\sigma_{13}^2$	$\frac{A_{9,1}A_{10,1}}{n^2}$	$\sum \frac{n_{ij.}^2 w_{ij.i.*}}{w_{i...j.}}$	$\sum \frac{n_{i.h}^2 w_{i.hi*.}}{w_{i....h}}$	$A_{18,9}$	$A_{18,9}$
28	$2\sigma_{12}^2\sigma_{23}^2$	$\frac{A_{9,1}A_{16,2}}{n^2}$	$\sum \frac{n_{ij.}^2 w_{ij..j*}}{w_{i...j.}}$	$\sum \frac{w_{i.hi*.}w_{i.h.*h}}{w_{i....h}}$	$A_{20,9}$	$A_{20,13}$
29	$2\sigma_{12}^2\sigma_{123}^2$	$\frac{A_{9,1}A_{13,1}}{n^2}$	$\sum \frac{w_{ij.ijh}^2}{w_{i...j.}}$	$\sum \frac{n_{ijh}^2 w_{i.hi*.}}{w_{i....h}}$	$A_{20,9}$	$A_{20,13}$
30	$2\sigma_{12}^2\sigma_0^2$	$\frac{A_{9,1}}{n}$	$\sum \frac{n_{ij.}^3}{w_{i...j.}}$	$\sum \frac{w_{i.hi*.}n_{i.h}}{w_{i....h}}$	$A_{21,1}$	$A_{21,1}$
31	$2\sigma_{13}^2\sigma_{23}^2$	$\frac{A_{10,1}A_{16,2}}{n^2}$	$\sum \frac{w_{ij..j*}w_{ij.i.*}}{w_{i...j.}}$	$\sum \frac{n_{i.h}^2 w_{i.h.*h}}{w_{i....h}}$	$A_{25,11}$	$A_{25,10}$
32	$2\sigma_{13}^2\sigma_{123}^2$	$\frac{A_{10,1}A_{13,1}}{n^2}$	$\sum \frac{n_{ijh}^2 w_{ij.i.*}}{w_{i...j.}}$	$\sum \frac{w_{ijhi.h}^2}{w_{i....h}}$	$A_{25,11}$	$A_{25,10}$
33	$2\sigma_{13}^2\sigma_0^2$	$\frac{A_{10,1}}{n}$	$\sum \frac{w_{ij.i.*}n_{ij.}}{w_{i...j.}}$	$\sum \frac{n_{i.h}^3}{w_{i....h}}$	$A_{26,1}$	$A_{26,1}$
34	$2\sigma_{23}^2\sigma_{123}^2$	$\frac{A_{16,2}A_{13,1}}{n^2}$	$\sum \frac{n_{ijh}^2 w_{ij..j*}}{w_{i...j.}}$	$\sum \frac{n_{ijh}^2 w_{i.h.*h}}{w_{i....h}}$	$A_{7,11}$	$A_{7,12}$
35	$2\sigma_{23}^2\sigma_0^2$	$\frac{A_{16,2}}{n}$	$\sum \frac{w_{ij..j*}n_{ij.}}{w_{i...j.}}$	$\sum \frac{w_{i.h.*h}n_{i.h}}{w_{i....h}}$	$A_{35,1}$	$A_{35,1}$
36	$2\sigma_{123}^2\sigma_0^2$	$\frac{A_{13,1}}{n}$	$\sum \frac{n_{ijh}^2 n_{ij.}}{w_{i...j.}}$	$\sum \frac{n_{ijh}^2 n_{i.h}}{w_{i....h}}$	$A_{35,1}$	$A_{35,1}$

TABLE 11.6 (*Continued*)

Row	Column (1/2)Cov Product	13 T_1, T_{23}	14 T_1, T_{123}	15 T_1, T_f	16 T_2, T_3	17 T_2, T_{12}
25	$2\sigma_3^2\sigma_{123}^2$	$\sum \frac{n_{ijh}^3 n_{..h}}{w_{i...jh}}$	$A_{25,10}$	$\sum \frac{w_{i....*} w_{i..i**}}{w_{...i..}}$	$\sum \frac{w_{.jhijh}^2}{w_{.jh.j.}}$	$\sum \frac{w_{ij..j*} w_{ij.i}}{w_{ij..j.}}$
26	$2\sigma_3^2\sigma_0^2$	$\sum \frac{n_{ijh}^2 n_{..h}}{w_{i...jh}}$	$A_{26,1}$	$\frac{A_{3,3}}{n}$	$A_{26,2}$	$A_{26,2}$
27	$2\sigma_{12}^2\sigma_{13}^2$	$\sum \frac{n_{ijh}^2 w_{ij.i.h}}{w_{i...jh}}$	$A_{10,9}$	$\sum \frac{w_{ij.i.h}^2}{w_{...i..}}$	$\sum \frac{w_{.jh*j.} w_{.jh*.h}}{w_{.j...h}}$	$A_{14,9}$
28	$2\sigma_{12}^2\sigma_{23}^2$	$A_{20,9}$	$A_{20,9}$	$\sum \frac{w_{i..i*.} w_{i...**}}{w_{...i..}}$	$\sum \frac{w_{.jh.jh} w_{.jh*j.}}{w_{.j...h}}$	$A_{13,9}$
29	$2\sigma_{12}^2\sigma_{123}^2$	$\sum \frac{n_{ijh}^3 n_{ij.}}{w_{i...jh}}$	$A_{20,9}$	$\sum \frac{w_{ij.ith}^2}{w_{...i..}}$	$\sum \frac{w_{.jh*j.} w_{.jh*jh}}{w_{.j...h}}$	$A_{14,9}$
30	$2\sigma_{12}^2\sigma_0^2$	$\sum \frac{n_{ijh}^2 n_{ij.}}{w_{i...jh}}$	$A_{21,1}$	$\frac{A_{9,1}}{n}$	$\sum \frac{w_{.jh*j.} n_{.jh}}{w_{.j...h}}$	$A_{15,2}$
31	$2\sigma_{13}^2\sigma_{23}^2$	$A_{25,10}$	$A_{25,10}$	$\sum \frac{w_{i..i.*} w_{i...**}}{w_{...i..}}$	$\sum \frac{n_{.jh}^2 w_{.jh*.h}}{w_{.j...h}}$	$A_{25,17}$
32	$2\sigma_{13}^2\sigma_{123}^2$	$\sum \frac{n_{ijh}^3 n_{i.h}}{w_{i...jh}}$	$A_{25,10}$	$\sum \frac{w_{i.hiju}^2}{w_{...i..}}$	$\sum \frac{w_{.jh*jh} w_{.jh*.h}}{w_{.j...h}}$	$A_{17,17}$
33	$2\sigma_{13}^2\sigma_0^2$	$\sum \frac{n_{ijh}^2 n_{i.h}}{w_{i...jh}}$	$A_{26,1}$	$\frac{A_{10,1}}{n}$	$\sum \frac{w_{.jh*.h} n_{.jh}}{w_{.j...h}}$	$A_{33,2}$
34	$2\sigma_{23}^2\sigma_{123}^2$	$\sum \frac{w_{ijhijh}^2}{w_{ijhi..}}$	$A_{34,13}$	$\sum \frac{w_{i...**} w_{i..i**}}{w_{...i..}}$	$\sum \frac{w_{ijh.jh}^2}{w_{.j...h}}$	$A_{25,17}$
35	$2\sigma_{23}^2\sigma_0^2$	$A_{35,1}$	$A_{35,1}$	$\frac{A_{16,2}}{n}$	$\sum \frac{n_{.jh}^3}{w_{.j...h}}$	$A_{26,2}$
36	$2\sigma_{123}^2\sigma_0^2$	$\sum \frac{n_{ijh}^3}{w_{i...jh}}$	$A_{35,1}$	$\frac{A_{13,1}}{n}$	$\sum \frac{w_{.jh*jh} n_{.jh}}{w_{.j...h}}$	$A_{33,2}$

TABLE 11.6 (*Continued*)

Row	Column (1/2)Cov Product	18 T_2,T_{13}	19 T_2,T_{23}	20 T_2,T_{123}	21 T_2,T_f	22 T_3,T_{12}	23 T_3,T_{13}
25	$2\sigma_3^2\sigma_{123}^2$	$\sum \frac{n_{ijh}^3 n_{..h}}{w_{.j.i.h}}$	$A_{25,16}$	$A_{25,16}$	$\sum \frac{w_{.j...*}w_{.j.*j*}}{w_{....j.}}$	$\sum \frac{w_{ijhijh}^2}{w_{ijhij.}}$	$A_{13,1}$
26	$2\sigma_3^2\sigma_0^2$	$\sum \frac{n_{ijh}^2 n_{..h}}{w_{.j.i.h}}$	$A_{26,2}$	$A_{26,2}$	$\frac{A_{3,3}}{n}$	$A_{26,4}$	n
27	$2\sigma_{12}^2\sigma_{13}^2$	$A_{14,9}$	$A_{14,19}$	$A_{14,9}$	$\sum \frac{w_{.j.*j.}w_{.j.*.*}}{w_{....j.}}$	$A_{14,10}$	$A_{14,10}$
28	$2\sigma_{12}^2\sigma_{23}^2$	$\sum \frac{n_{ijh}^2 w_{ij..jh}}{w_{.j.i.h}}$	$A_{13,9}$	$A_{13,9}$	$\sum \frac{w_{ij..jh}^2}{w_{....j.}}$	$A_{20,16}$	$A_{20,23}$
29	$2\sigma_{12}^2\sigma_{123}^2$	$\sum \frac{n_{ijh}^3 n_{ij.}}{w_{.j.i.h}}$	$A_{14,19}$	$A_{14,9}$	$\sum \frac{w_{ij.sjh}^2}{w_{....j.}}$	$\sum \frac{w_{ijhijh}^2}{w_{..hijh}}$	$A_{17,23}$
30	$2\sigma_{12}^2\sigma_0^2$	$\sum \frac{n_{ijh}^2 n_{ij.}}{w_{.j.i.h}}$	$A_{15,2}$	$A_{15,2}$	$\frac{A_{9,1}}{n}$	$A_{30,3}$	$A_{30,3}$
31	$2\sigma_{13}^2\sigma_{23}^2$	$A_{25,16}$	$A_{25,16}$	$A_{25,16}$	$\sum \frac{w_{.j..j*}w_{.j.*.*}}{w_{....j.}}$	$\sum \frac{n_{ijh}^2 w_{i.h.jh}}{w_{ij...h}}$	$A_{13,10}$
32	$2\sigma_{13}^2\sigma_{123}^2$	$\sum \frac{w_{ijhijh}^2}{w_{ijh.j.}}$	$A_{7,19}$	$A_{32,18}$	$\sum \frac{w_{.j.*.*}w_{.j.*j*}}{w_{....j.}}$	$\sum \frac{n_{ijh}^3 n_{i.h}}{w_{ij...h}}$	$A_{14,10}$
33	$2\sigma_{13}^2\sigma_0^2$	$A_{33,2}$	$A_{33,2}$	$A_{33,2}$	$\frac{A_{10,1}}{n}$	$\sum \frac{n_{ijh}^2 n_{i.h}}{w_{ij...h}}$	$A_{15,3}$
34	$2\sigma_{23}^2\sigma_{123}^2$	$\sum \frac{n_{ijh}^3 n_{.jh}}{w_{.j.i.h}}$	$A_{25,16}$	$A_{25,16}$	$\sum \frac{w_{.jhijh}^2}{w_{....j.}}$	$\sum \frac{n_{ijh}^3 n_{.jh}}{w_{ij...h}}$	$A_{20,23}$
35	$2\sigma_{23}^2\sigma_0^2$	$\sum \frac{n_{ijh}^2 n_{.jh}}{w_{.j.i.h}}$	$A_{26,2}$	$A_{26,2}$	$\frac{A_{16,2}}{n}$	$\sum \frac{n_{ijh}^2 n_{.jh}}{w_{ij...h}}$	$A_{21,3}$
36	$2\sigma_{123}^2\sigma_0^2$	$\sum \frac{n_{ijh}^3}{w_{.j.i.h}}$	$A_{33,2}$	$A_{33,2}$	$\frac{A_{13,1}}{n}$	$\sum \frac{w_{ijhijh}^2}{w_{ij...h}}$	$A_{30,3}$

TABLE 11.6 (*Continued*)

Row	Column (1/2)Cov Product	24 T_3, T_{23}	25 T_3, T_{123}	26 T_3, T_f	27 T_{12}, T_{13}	28 T_{12}, T_{23}	29 T_{12}, T_{123}
25	$2\sigma_3^2\sigma_{123}^2$	$A_{13,1}$	$A_{13,1}$	$\sum \frac{w_{ijh..h}^2}{w_{.....h}}$	$A_{25,22}$	$A_{25,22}$	$A_{25,22}$
26	$2\sigma_3^2\sigma_0^2$	n	n	$\frac{A_{3,3}}{n}$	$A_{26,4}$	$A_{26,4}$	$A_{26,4}$
27	$2\sigma_{12}^2\sigma_{13}^2$	$A_{14,24}$	$A_{14,10}$	$\sum \frac{w_{..h*.h} w_{..h**.}}{w_{.....h}}$	$A_{13,1}$	$A_{14,13}$	$A_{13,1}$
28	$2\sigma_{12}^2\sigma_{23}^2$	$A_{20,16}$	$A_{20,16}$	$\sum \frac{w_{..h.*h} w_{..h**.}}{w_{.....h}}$	$A_{20,18}$	$A_{13,1}$	$A_{13,1}$
29	$2\sigma_{12}^2\sigma_{123}^2$	$A_{7,24}$	$A_{29,22}$	$\sum \frac{w_{..h**.} w_{..h**h}}{w_{.....h}}$	$A_{20,18}$	$A_{14,13}$	$A_{13,1}$
30	$2\sigma_{12}^2\sigma_0^2$	$A_{30,3}$	$A_{30,3}$	$\frac{A_{9,1}}{n}$	$A_{21,5}$	$A_{15,6}$	n
31	$2\sigma_{13}^2\sigma_{23}^2$	$A_{13,10}$	$A_{13,10}$	$\sum \frac{w_{i.h.jh}^2}{w_{.....h}}$	$A_{25,22}$	$A_{25,22}$	$A_{25,22}$
32	$2\sigma_{13}^2\sigma_{123}^2$	$A_{14,24}$	$A_{14,10}$	$\sum \frac{w_{i.hsjh}^2}{w_{.....h}}$	$A_{25,22}$	$A_{7,28}$	$A_{25,22}$
33	$2\sigma_{13}^2\sigma_0^2$	$A_{15,3}$	$A_{15,3}$	$\frac{A_{10,1}}{n}$	$A_{26,4}$	$\sum \frac{n_{ijh}^3}{w_{ij..jh}}$	$A_{26,4}$
34	$2\sigma_{23}^2\sigma_{123}^2$	$A_{20,16}$	$A_{20,16}$	$\sum \frac{w_{.jhith}^2}{w_{.....h}}$	$A_{7,27}$	$A_{25,22}$	$A_{25,22}$
35	$2\sigma_{23}^2\sigma_0^2$	$A_{21,3}$	$A_{21,3}$	$\frac{A_{16,2}}{n}$	$\sum \frac{n_{ijh}^3}{w_{ij.i.h}}$	$A_{26,4}$	$A_{26,4}$
36	$2\sigma_{123}^2\sigma_0^2$	$A_{30,3}$	$A_{30,3}$	$\frac{A_{13,1}}{n}$	$A_{35,27}$	$A_{33,28}$	$A_{26,4}$

TABLE 11.6 (*Continued*)

Row	Column (1/2)Cov Product	30 T_{12}, T_f	31 T_{13}, T_{23}	32 T_{13}, T_{123}	33 T_{13}, T_f	34 T_{23}, T_{123}
25	$2\sigma_3^2\sigma_{123}^2$	$\sum \frac{w_{ij...*}w_{ij.ij*}}{w_{...ij.}}$	$A_{13,1}$	$A_{13,1}$	$A_{25,26}$	$A_{13,1}$
26	$2\sigma_3^2\sigma_0^2$	$\frac{A_{3,3}}{n}$	n	n	$\frac{A_{3,3}}{n}$	n
27	$2\sigma_{12}^2\sigma_{13}^2$	$\sum \frac{w_{ij.ijh}n_{i.h}}{n}$	$A_{14,13}$	$A_{13,1}$	$A_{27,3}$	$A_{14,3}$
28	$2\sigma_{12}^2\sigma_{23}^2$	$\sum \frac{w_{ij.ijh}n_{.jh}}{n}$	$A_{20,18}$	$A_{20,18}$	$\sum \frac{w_{i.hi*.}w_{i.h.*h}}{w_{...i.h}}$	$A_{13,1}$
29	$2\sigma_{12}^2\sigma_{123}^2$	$\sum \frac{w_{ijhij.}^2}{w_{...ij.}}$	$A_{7,31}$	$A_{20,18}$	$\sum \frac{w_{i.hi*.}w_{i.hi*h}}{w_{...i.h}}$	$A_{14,13}$
30	$2\sigma_{12}^2\sigma_0^2$	$\frac{A_{9,1}}{n}$	$\sum \frac{n_{ijh}^3}{w_{i.h.jh}}$	$A_{21,5}$	$\frac{A_{9,1}}{n}$	$A_{15,6}$
31	$2\sigma_{13}^2\sigma_{23}^2$	$\sum \frac{w_{ij.i.*}w_{ij..j*}}{w_{...ij.}}$	$A_{13,1}$	$A_{13,1}$	$\sum \frac{w_{i.hijh}n_{.jh}}{n}$	$A_{13,1}$
32	$2\sigma_{13}^2\sigma_{123}^2$	$\sum \frac{w_{ij.i.*}w_{ij.ij*}}{w_{...ij.}}$	$A_{14,13}$	$A_{13,1}$	$\sum \frac{w_{ijhi.h}^2}{w_{...i.h}}$	$A_{14,13}$
33	$2\sigma_{13}^2\sigma_0^2$	$\frac{A_{10,1}}{n}$	$A_{15,6}$	n	$\frac{A_{10,1}}{n}$	$A_{15,6}$
34	$2\sigma_{23}^2\sigma_{123}^2$	$\sum \frac{w_{ij..j*}w_{ij.ij*}}{w_{...ij.}}$	$A_{20,18}$	$A_{20,18}$	$\sum \frac{w_{i.h.*h}w_{i.hi*h}}{w_{...i.h}}$	$A_{13,1}$
35	$2\sigma_{23}^2\sigma_0^2$	$\frac{A_{16,2}}{n}$	$A_{21,5}$	$A_{21,5}$	$\frac{A_{16,2}}{n}$	n
36	$2\sigma_{123}^2\sigma_0^2$	$\frac{A_{13,1}}{n}$	$A_{30,31}$	$A_{21,5}$	$\frac{A_{13,1}}{n}$	$A_{15,6}$

TABLE 11.6 (*Continued*)

Row	Column (1/2)Cov Product	35 T_{23}, T_f	36 T_{123}, T_f
25	$2\sigma_3^2\sigma_{123}^2$	$A_{25,26}$	$A_{25,26}$
26	$2\sigma_3^2\sigma_0^2$	$\frac{A_{3,3}}{n}$	$\frac{A_{3,3}}{n}$
27	$2\sigma_{12}^2\sigma_{13}^2$	$\sum \frac{w_{.jh*j.}\, w_{.jh*.h}}{w_{....jh}}$	$A_{27,30}$
28	$2\sigma_{12}^2\sigma_{23}^2$	$A_{28,30}$	$A_{28,3}$
29	$2\sigma_{12}^2\sigma_{123}^2$	$\sum \frac{w_{.jh*j.}\, w_{.jh*jh}}{w_{....jh}}$	$A_{29,30}$
30	$2\sigma_{12}^2\sigma_0^2$	$\frac{A_{9,1}}{n}$	$\frac{A_{9,1}}{n}$
31	$2\sigma_{13}^2\sigma_{23}^2$	$A_{31,33}$	$A_{31,33}$
32	$2\sigma_{13}^2\sigma_{123}^2$	$\sum \frac{w_{.jh*.h}\, w_{.jh*jh}}{w_{....jh}}$	$A_{32,33}$
33	$2\sigma_{13}^2\sigma_0^2$	$\frac{A_{10,1}}{n}$	$\frac{A_{10,1}}{n}$
34	$2\sigma_{23}^2\sigma_{123}^2$	$\sum \frac{w_{ijh.jh}^2}{w_{....jh}}$	$A_{34,33}$
35	$2\sigma_{23}^2\sigma_0^2$	$\frac{A_{16,2}}{n}$	$\frac{A_{16,2}}{n}$
36	$2\sigma_{123}^2\sigma_0^2$	$\frac{A_{13,1}}{n}$	$\frac{A_{13,1}}{n}$

LITERATURE CITED

Ahrens, H. (1965). Standardfehler geschatzter Varianzkomponenten eines unbalanzie Versuchplanes in r-stufiger hierarchischer Klassifikation. *Monatsb. Deut. Akad. Wiss. Berlin*, **7**, 89–94.

Anderson, R. L., and T. A. Bancroft. (1952). *Statistical Theory in Research*. McGraw-Hill, New York.

Anderson, R. L., and P. P. Crump. (1967). Comparisons of designs and estimation procedures for estimating parameters in a two-stage nested process. *Technometrics*, **9**, 499–516.

Anderson, T. W. (1958). *An Introduction to Multivariate Statistical Analysis*. Wiley, New York.

Anscombe, F. J. and J. W. Tukey. (1963). The examination and analysis of residuals. *Technometrics*, **5**, 141–160.

Banerjee, K. S. (1964). A note on idempotent matrices. *Ann. Math. Stat.*, **35,** 880–882.

Bartlett, M. S. (1937). Some examples of statistical methods in research in agriculture and applied biology. *J. Roy. Stat. Soc. Supp.*, **4,** 137–183.

Bennett, C. A., and N. L. Franklin. (1954). *Statistical Analysis in Chemistry and the Chemical Industry*. Wiley, New York.

Blischke, W. R. (1966). Variances of estimates of variance components in a three-way classification. *Biometrics*, **22,** 553–565.

Blischke, W. R. (1968). Variances of moment estimators of variance components in the unbalanced r-way classification. *Biometrics*, **24,** 527–540.

Boullion, T. L., and P. L. Odell, Editors. (1968). *Theory and Application of Generalized Inverse Matrices*. (Proceedings of a symposium.) Mathematics Series No. 4, Texas Technical College, Lubbock, Texas.

Broemeling, L. D. (1969). Confidence regions for variance ratios of random models. *J. Am. Stat. Assoc.*, **64,** 660–664.

Brown, K. H. (1968). Social class and family investment. Ph.D. Thesis, Cornell University, Ithaca, N.Y.

Bulmer, M. G. (1957). Approximate confidence limits for components of variance. *Biometrika*, **44,** 159–167.

Bush, N., and R. L. Anderson. (1963). A comparison of three different procedures for estimating variance components. *Technometrics*, **5,** 421–440.

Chipman, J. S. (1964). On least squares with insufficient observations. *J. Am. Stat. Assoc.*, **59,** 1078–1111.

Cochran, W. G. (1934). The distribution of quadratic forms in a normal system with applications to the analysis of variance. *Proc. Cambridge Philos. Soc.*, **30,** 178–191.

Cornfield, J., and J. W. Tukey. (1956). Average values of mean squares in factorials. *Ann. Math. Stat.*, **27,** 907–949.

Cox, D. R., and E. J. Snell. (1968). A general definition of residuals. *J. Roy. Stat. Soc.* (*B*), **30**, 248–275.

Cramer, H. (1951). *Mathematical Methods of Statistics.* Princeton University Press.

Crump, S. L. (1947). The estimation of components of variance in multiple classifications. Ph.D. Thesis, Iowa State College Library, Ames, Iowa.

Crump, S. L. (1951). Present status of variance component analysis. *Biometrics*, **7**, 1–16.

Cunningham, E. P. (1969). A note on the estimation of variance components by the method of fitting constants. *Biometrika*, **56**, 683–684.

Cunningham, E. P., and C. R. Henderson. (1968). An iterative procedure for estimating fixed effects and variance components in mixed model situations. *Biometrics*, **24**, 13–25. Corrigenda, *Biometrics*, **25**, 777–778, 1969.

Draper, N. R., and H. Smith. (1966). *Applied Regression Analysis.* Wiley, New York.

Eisenhart, C. (1947). The assumptions underlying the analysis of variance. *Biometrics*, **3**, 1–21.

Elston, R. C., and N. Bush. (1964). The hypotheses that can be tested when there are interactions in an analysis of variance model. *Biometrics*, **20**, 681–698.

Evans, D. A. (1969). Personal communication.

Federer, W. T. (1955). *Experimental Design.* Macmillan, New York.

Gaylor, D. W., and T. D. Hartwell. (1969). Expected mean squares for nested classifications. *Biometrics*, **25**, 427–430.

Gaylor, D. W., and F. N. Hopper. (1969). Estimating the degrees of freedom for linear combinations of mean squares by Satterthwaite's formula. *Technometrics*, **11**, 691–706.

Gaylor, D. W., H. L. Lucas, and R. L. Anderson. (1970). Calculation of expected mean squares by the abbreviated Doolittle and square root methods. *Biometrics*, **26**, 641–656.

Goldman, A. J., and M. Zelen. (1964). Weak generalized inverses and minimum variance linear unbiased estimators. *J. Res. Nat. Bur. of Standards*, **68B**, 151–172.

Good, I. J. (1963). On the independence of quadratic expressions (with an appendix by L. R. Welch). *J. Roy. Stat. Soc.* (*B*), **25**, 377–382.

Good, I. J. (1966). On the independence of quadratic expressions: corrigenda. *J. Roy. Stat. Soc.* (*B*), **28**, 584.

Good, I. J. (1969). Conditions for a quadratic form to have a chi-squared distribution. *Biometrika*, **56**, 215–216.

Gosslee, D. G., and H. L. Lucas. (1965). Analysis of variance of disproportionate data when interaction is present. *Biometrics*, **21**, 115–133.

Graybill, F. A. (1954). On quadratic estimates of variance components. *Ann. Math. Stat.*, **25**, 367–372.

Graybill, F. A. (1961). *An Introduction to Linear Statistical Models.* Vol. I. McGraw-Hill, New York.

Graybill, F. A., and R. A. Hultquist. (1961). Theorems concerning Eisenhart's Model II. *Ann. Math. Stat.*, **32,** 261–269.

Graybill, F. A., and G. Marsaglia. (1957). Idempotent matrices and quadratic forms in the general linear hypothesis. *Ann. Math. Stat.*, **28,** 678–686.

Graybill, F. A., C. D. Meyer and R. J. Painter. (1966). Note on the computation of the generalized inverse of a matrix. *SIAM Review*, **8,** 522–524.

Graybill, F. A., and A. W. Wortham. (1956). A note on uniformly best unbiased estimators for variance components. *J. Am. Stat. Assoc.*, **51,** 266–268.

Greville, T. N. E. (1966). Note on the generalized inverse of a matrix product. *SIAM Review*, **8,** 518–521. Erratum, *SIAM Review*, **9,** 249, 1967.

Grubbs, F. E. (1948). On estimating precision of measuring instruments and product variability. *J. Am. Stat. Assoc.*, **43,** 243–264.

Hartley, H. O. (1967). Expectation, variances and covariances of ANOVA mean squares by 'synthesis'. *Biometrics*, **23,** 105–114, and Corrigenda, 853.

Hartley, H. O., and J. N. K. Rao. (1967). Maximum likelihood estimation for the mixed analysis of variance model. *Biometrika*, **54,** 93–108.

Hartley, H. O., and S. R. Searle. (1969). On interaction variance components in mixed models. *Biometrics*, **25,** 573–576.

Harville, D. A. (1967). Estimability of variance components for the 2-way classification with interaction. *Ann. Math. Stat.*, **38,** 1508–1519.

Harville, D. A. (1969a). Quadratic unbiased estimation of variance components for the one-way classification. *Biometrika*, **56,** 313–326.

Harville, D. A. (1969b). Variance component estimation for the unbalanced one-way random classification—a critique. *Aerospace Research Laboratories, ARL-69-0180*, Wright-Patterson Air Force Base, Ohio.

Harville, D. A. (1969c). Variances of variance-component estimators for the unbalanced 2-way cross-classification with application to balanced incomplete block designs. *Ann. Math. Stat.*, **40,** 408–416.

Henderson, C. R. (1953). Estimation of variance and covariance components. *Biometrics*, **9,** 226–252.

Henderson, C. R. (1963). Selection index and expected genetic advance. *Statistical Genetics in Plant Breeding*, National Academy of Sciences, National Research Council publication No. 982.

Henderson, C. R. (1968). Personal communication.

Henderson, C. R. (1959, 1969). Design and analysis of animal husbandry experiments. Chapter 1 of *Techniques and Procedures in Animal Science Research* (1st Ed., 1959; 2nd Ed., 1969), American Society of Animal Science.

Henderson, C. R., O. Kempthorne, S. R. Searle, and C. N. Von Krosigk. (1959). Estimation of environmental and genetic trends from records subject to culling. *Biometrics*, **15,** 192–218.

Herbach, L. H. (1959). Properties of type II analysis of variance tests. *Ann. Math. Stat.*, **30,** 939–959.

Hill, B. M. (1965). Inference about variance components in the one-way model. *J. Am. Stat. Assoc.*, **60,** 806–825.

Hill, B. M. (1967). Correlated errors in the random model. *J. Am. Stat. Assoc.*, **62,** 1387–1400.

Hirotsu, C. (1966). Estimating variance components in a two-way layout with unequal numbers of observations. *Rept. Stat. Appl. Res., Union of Japanese Scientists and Engineers* (*JUSE*), **13,** No. 2, 29–34.

Hogg, R. V., and A. T. Craig. (1965). *Introduction to Mathematical Statistics.* Macmillan, New York.

John, P. W. M. (1964). Pseudo-inverses in analysis of variance. *Ann. Math. Stat.*, **35,** 895–896.

Kempthorne, O. (1952). *The Design and Analysis of Experiments.* Wiley, New York.

Kempthorne, O. (1968). Discussion of Searle [1968]. *Biometrics*, **24,** 782–784.

Khatri, C. G. (1963). Further contributions to Wishartness and independence of second degree polynomials in normal vectors. *J. Indian Stat. Assoc.*, **1,** 61–70.

Khatri, C. G. (1968). Some results for the singular normal multivariate regression models. *Sankhyā, Ser. A*, **30,** 267–280.

Kirk, R. E. (1968). *Experimental Design: Procedures for the Behavioral Science.* Brooks/Cole, Belmont, California.

Koch, G. G. (1967a). A general approach to the estimation of variance components. *Technometrics*, **9,** 93–118.

Koch, G. G. (1967b). A procedure to estimate the population mean in random effects models. *Technometrics*, **9,** 577–586.

Koch, G. G. (1968). Some further remarks on "A general approach to the estimation of variance components". *Technometrics*, **10,** 551–558.

Kussmaul, K., and R. L. Anderson. (1967). Estimation of variance components in two-stage nested designs with complete samples. *Technometrics*, **9,** 373–390.

LaMotte, L. R., and R. R. Hocking. (1970). Computational efficiency in the selection of regression variables. *Technometrics*, **12,** 83–94.

Lancaster, H. O. (1954). Traces and cumulants of quadratic forms in normal variables. *J. Roy. Stat. Soc.* (*B*), **16,** 247–254.

Lancaster, H. O. (1965). The Helmert matrices. *Am. Math. Monthly*, **72,** 4–12.

Leone, F. C., L. S. Nelson, N. L. Johnson, and S. Eisenstat. (1968). Sampling distributions of variance components. II. Empirical studies of unbalanced nested designs. *Technometrics*, **10,** 719–738.

Li, C. C. (1964). *Introduction to Experimental Statistics.* McGraw-Hill, New York.

Low, L. Y. (1964). Sampling variances of estimates of components of variance from a non-orthogonal two-way classification. *Biometrika*, **51,** 491–494.

Loynes, R. M. (1966). On idempotent matrices. *Ann. Math. Stat.*, **37,** 295–296.

Loynes, R. M. (1969). On Cox and Snell's general definition of residuals. *J. Roy. Stat. Soc.* (*B*), **31,** 103–106.

Lum, Mary D. (1954). Rules for determining error terms in hierarchal and partially hierarchal models. Wright Air Development Center, Dayton, Ohio.

Mahamunulu, D. M. (1963). Sampling variances of the estimates of variance components in the unbalanced 3-way nested classification. *Ann. Math. Stat.*, **34,** 521–527.

McElroy, F. W. (1967). A necessary and sufficient condition that ordinary least squares estimators be best linear unbiased. *J. Am. Stat. Assoc.*, **62,** 1302–1304.

Miller, R. G. (1966). *Simultaneous Statistical Inference*. McGraw-Hill, New York.

Millman, J., and G. V. Glass. (1967). Rules of thumb for writing the anova table. *J. of Educational Measurement*, **4,** 41–51.

Mood, A. M. (1950). *Introduction to the Theory of Statistics*. McGraw-Hill, New York.

Mood, A. M., and F. A. Graybill. (1963). *Introduction to the Theory of Statistics* (2nd Ed.). McGraw-Hill, New York.

Moore, E. H. (1920). On the reciprocal of the general algebraic matrix. *Bull. Am. Math. Soc.*, **26,** 394–395.

Nelder, J. A. (1954). The interpretation of negative components of variance. *Biometrika*, **41,** 544–548.

Patnaik, P. B. (1949). The noncentral χ^2 and F distributions and their approximations. *Biometrika*, **36,** 202–232.

Penrose, R. A. (1955). A generalized inverse for matrices. *Proc. Cambridge Philos. Soc.*, **51,** 406–413.

Plackett. R. L. (1960). *Principles of Regression Analysis*. Oxford University Press.

Rao, C. R. (1962). A note on a generalized inverse of a matrix with applications to problems in mathematical statistics. *J. Roy. Stat. Soc.* (*B*), **24,** 152–158.

Rao, C. R. (1965). *Linear Statistical Inference and its Applications*. Wiley, New York.

Rao, C. R. (1966). Generalized inverse for matrices and its applications in mathematical statistics. *Research Papers in Statistics*, Festschrift for J. Neyman, Ed., F. N. David. Wiley, New York.

Rao, J. N. K. (1968). On expectations, variances and covariances of anova mean squares by 'synthesis'. *Biometrics*, **24,** 963–978.

Rayner, A. A., and D. Livingstone. (1965). On the distribution of quadratic forms in singular normal variables. *S. African J. Agric. Sci.*, **8,** 357–370.

Rayner, A. A., and R. M. Pringle. (1967). A note on generalized inverses in the linear hypothesis not of full rank. *Ann. Math. Stat.*, **38,** 271–273.

Robinson, J. (1965). The distribution of a general quadratic form in normal variables. *Australian J. of Stat.*, **7,** 110–114.

Robson, D. S. (1957). Personal communication.

Rohde, C. A. (1964). Contributions to the theory, computation and applications of generalized inverses. *Institute of Statistics, Mimeo. Series No. 392*, University of North Carolina at Raleigh.

Rohde, C. A. (1965). Generalized inverses of partitioned matrices. *J. Soc. Ind. App. Math.*, **13,** 1033–1035.

Rohde, C. A. (1966). Some results on generalized inverses. *SIAM Review*, **8,** 201–205.

Rohde, C. A., and G. M. Tallis. (1969). Exact first- and second-order moments of estimates of components of covariance. *Biometrika*, **56,** 517–526.

Satterthwaite, F. E. (1946). An approximate distribution of estimates of variance components. *Biometrics Bulletin*, **2,** 110–114.

Scheffé, H. (1959). *The Analysis of Variance*. Wiley, New York.

Schultz, E. F., Jr. (1955). Rules of thumb for determining expectations of mean squares in analysis of variance. *Biometrics*, **11**, 123–135.

Searle, S. R. (1956). Matrix methods in variance and covariance components analysis. *Ann. Math. Stat.*, **27**, 737–748.

Searle, S. R. (1958). Sampling variances of estimates of components of variance. *Ann. Math. Stat.*, **29**, 167–178.

Searle, S. R. (1961a). Variance components in the unbalanced 2-way nested classification. *Ann. Math. Stat.*, **32**, 1161–1166.

Searle, S. R. (1961b). Estimating the heritability of butterfat production. *J. Agr. Sci.*, **57**, 289–294.

Searle, S. R. (1966). *Matrix Algebra for the Biological Sciences*. Wiley, New York.

Searle, S. R. (1968). Another look at Henderson's methods of estimating variance components. *Biometrics*, **24**, 749–788.

Searle, S. R. (1969). Variance component estimation: proof of equivalence of alternative method 4 to Henderson's method 3. *Paper No. BU-260-M in the Mimeo. Series, Biometrics Unit*, Cornell University, Ithaca, N.Y.

Searle, S. R. (1970). Large sample variances of maximum likelihood estimators of variance components. *Biometrics*, **26**, 505–524.

Searle, S. R. (1971). Topics in variance component estimation. *Biometrics*, **27**, 1–76.

Searle, S. R., and R. F. Fawcett. (1970). Expected mean squares in variance components models having finite populations. *Biometrics*, **26**, 243–254.

Searle, S. R., and W. H. Hausman. (1970). *Matrix Algebra for Business and Economics*. Wiley, New York.

Searle, S. R., and C. R. Henderson. (1961). Computing procedures for estimating components of variance in the two-way classification, mixed model. *Biometrics*, **17**, 607–616. Corrigenda, *Biometrics*, **23**, 852, 1967.

Searle, S. R., and Jon G. Udell. (1970). The use of regression on dummy variables in management research. *Management Science: Application*, **16**, B, 397–409.

Seelye, C. J. (1958). Conditions for a positive-definite quadratic form established by induction. *Am. Math. Monthly*, **65**, 355-356.

Shanbhag, D. N. (1966). On the independence of quadratic forms. *J. Roy. Stat. Soc. (B)*, **28**, 582–583.

Shanbhag, D. N. (1968). Some remarks concerning Khatri's result on quadratic forms. *Biometrika*, **55**, 593–595.

Snedecor, G. W., and W. G. Cochran. (1967). *Statistical Methods*. (6th Ed.). Iowa State University Press, Ames, Iowa.

Speed, F. M. (1969). A new approach to the analysis of linear models. Ph.D. Thesis, Texas A & M University, College Station, Texas.

Steel, R. G. D., and J. H. Torrie. (1960). *Principles and Procedures of Statistics*. McGraw-Hill, New York.

Styan, G. P. H. (1969). Notes on the distribution of quadratic forms in singular normal variables. *Technical Report 122*, University of Minnesota, Minneapolis, Minn.

Tang, P. C. (1938). The power function of the analysis of variance tests with tables and illustrations of their use. *Stat. Research Memoirs*, **2,** 126–149.

Thiel, H. (1968). A simplification of the BLUS procedure for analyzing regression disturbances. *J. Am. Stat. Assoc.*, **63,** 242–251.

Thompson, R. (1969). Iterative estimation of variance components for non-orthogonal data. *Biometrics*, **25,** 767–773.

Thompson, W. A., Jr. (1961). Negative estimates of variance components: an introduction. *Bulletin, International Institute of Statistics*, **34,** 1–4.

Thompson, W. A., Jr. (1962). The problem of negative estimates of variance components. *Ann. Math. Stat.*, **33,** 273–289.

Thompson, W. A., Jr. (1963). Precision of simultaneous measurement procedures. *J. Am. Stat. Assoc.*, **58,** 474–479.

Thompson, W. A., Jr., and James R. Moore. (1963). Non-negative estimates of variance components. *Technometrics*, **5,** 441–450.

Tiao, G. C., and G. E. P. Box. (1967). Bayesian analysis of a three-component hierarchical design model. *Biometrika*, **54,** 109–125.

Tiao, G. C., and W. Y. Tan. (1965). Bayesian analysis of random-effects models in the analysis of variance. I. Posterior distribution of variance components. *Biometrika*, **52,** 37–53.

Tiao, G. C., and W. Y. Tan. (1966). Bayesian analysis of random-effects models in the analysis of variance. II. Effect of autocorrelated errors. *Biometrika*, **53,** 477–495.

Townsend, E. C. (1968). Unbiased estimators of variance components in simple unbalanced designs. Ph.D. Thesis, Cornell University, Ithaca, N.Y.

Townsend, E. C. (1969). Lecture notes, Statistics 362, University of West Virginia, Morgantown, W.Va.

Townsend, E. C., and S. R. Searle. (1971). Best quadratic unbiased estimation of variance components from unbalanced data in the 1-way classification. *Biometrics*, (in press).

Tukey, J. W. (1957). Variances of variance components: II. The unbalanced single classification. *Ann. Math. Stat.*, **28,** 43–56.

Urquhart, N. S. (1968). Computations of generalized inverse matrices which satisfy special conditions. *SIAM Review*, **10,** 216–218.

Urquhart, N. S. (1969a). The nature of the lack of uniqueness of generalized inverse matrices. *SIAM Review*, **11,** 268–271.

Urquhart, N. S. (1969b). Lecture notes, Cornell University, Ithaca, N.Y.

Urquhart, N. S., D. L. Weeks, and C. R. Henderson. (1970). Estimation associated with linear models: a revisitation. *Paper BU-195*, *Biometrics Unit*, Cornell University, Ithaca, N.Y.

Vogler, L. E., and K. A. Norton. (1957). Graphs and tables of the significance levels $F(\gamma_1, \gamma_2, p)$ for the Fisher-Snedecor variance ratio. *Report No. 5069*, *Nat. Bur. of Standards*, Boulder, Colo.

Wald, A. (1943). Tests of statistical hypotheses concerning several parameters when the number of variables is large. *Trans. Am. Math. Soc.*, **54,** 426–482.

Wang, Y. Y. (1967). A comparison of several variance component estimators. *Biometrika*, **54**, 301–305.

Weeks, D. L., and D. R. Williams. (1964). A note on the determination of connectedness in an *N*-way cross classification. *Technometrics*, **6**, 319–324. Errata, *Technometrics*, **7**, 281, 1965.

Welch, B. L. (1956). On linear combinations of several variances. *J. Am. Stat. Assoc.*, **51**, 132–148.

Wilk, M. B., and O. Kempthorne. (1955). Fixed, mixed and random models. *J. Am. Stat. Assoc.*, **50**, 1144–1167.

Wilk, M. B., and O. Kempthorne. (1956). Some aspects of the analysis of factorial experiments in a completely randomized design. *Ann. Math. Stat.*, **27**, 950–985.

Wilks, S. S. (1962). *Mathematical Statistics*. Wiley, New York.

Williams, E. J. (1959). *Regression Analysis*. Wiley, New York.

Williams, J. S. (1962). A confidence interval for variance components. *Biometrika*, **49**, 278–281.

Yates, F. (1934). The analysis of multiple classifications with unequal numbers in the different classes. *J. Am. Stat. Assoc.*, **29**, 51–66.

Young, C. W., J. E. Legates, and B. R. Farthing. (1965). Pre- and post-natal influences on growth, prolificacy and maternal performance in mice. *Genetics*, **52**, 553–561.

Zelen, M. (1968). Discussion of Searle [1968]. *Biometrics*, **24**, 779–780.

Zyskind, G., and F. B. Martin. (1969). A general Gauss-Markoff theorem for linear models with arbitrary non-negative covariance structure. *SIAM J. of Applied Mathematics*, **17**, 1190–1202.

STATISTICAL TABLES

The following abridged tables of the normal, and central t-, χ^2- and F-distributions are given solely for convenience. They in no way represent detailed coverage of these distributions.

TABLE 1. VALUES P_x AND x ON THE NORMAL DISTRIBUTION

$N(0, 1)$-distribution

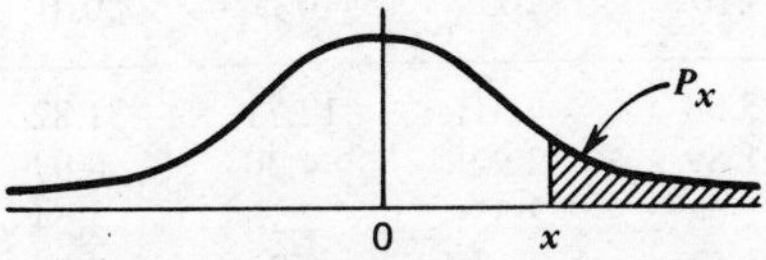

$P_x = \Pr\{N(0, 1)\text{-variable} \geq x\}$

P_x	x	P_x	x	P_x	x	P_x	x	P_x	x
.50	0.00	**.050**	**1.64**	.030	1.88	.020	2.05	**.010**	**2.33**
.45	0.13	.048	1.66	.029	1.90	.019	2.07	.009	2.37
.40	0.25	.046	1.68	.028	1.91	.018	2.10	.008	2.41
.35	0.39	.044	1.71	.027	1.93	.017	2.12	.007	2.46
.30	0.52	.042	1.73	.026	1.94	.016	2.14	.006	2.51
.25	0.67	.040	1.75	**.025**	**1.96**	.015	2.17	**.005**	**2.58**
.20	0.84	.038	1.77	.024	1.98	.014	2.20	.004	2.65
.15	1.04	.036	1.80	.023	2.00	.013	2.23	.003	2.75
.10	**1.28**	.034	1.83	.022	2.01	.012	2.26	.002	2.88
.05	**1.64**	.032	1.85	.021	2.03	.011	2.29	**.001**	**3.09**
								0.000	∞

Bold-face values are those often (but not exclusively) used when testing hypotheses and/or establishing confidence intervals.

Source: Table 2 of Lindley and Miller (1958), *Cambridge Elementary Statistical Tables*, published by Cambridge University Press, with kind permission of the authors and publishers.

TABLE 2. VALUES OF $t_{n,\alpha}$ ON THE $t(n)$-DISTRIBUTION

$t(n)$-distribution

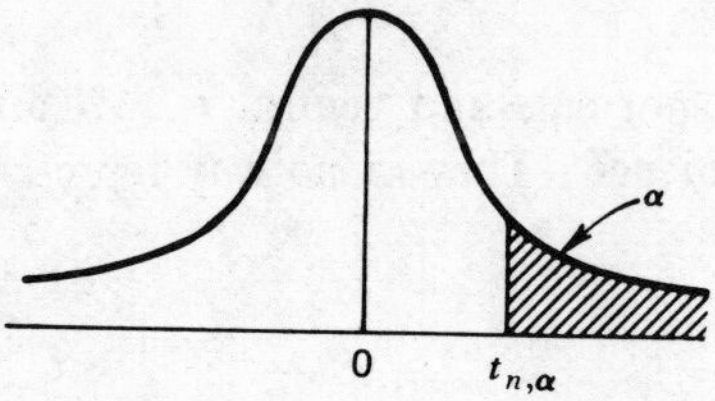

Note: $\Pr\{t(n)\text{-variable} \geq t_{n,\alpha}\} = \alpha$

$\Pr\{t(n)\text{-variable} \leq -t_{n,\alpha}\} = \alpha$

$\Pr\{|t(n)\text{-variable}| \geq t_{n,\alpha}\} = 2\alpha$

n	α				
(d.f.)	.10	.05	.025	.010	.005
1	3.08	6.31	12.71	31.82	63.66
2	1.89	2.92	4.30	6.97	9.92
3	1.64	2.35	3.18	4.54	5.84
4	1.53	2.13	2.78	3.75	4.60
5	1.48	2.02	2.57	3.36	4.03
6	1.44	1.94	2.45	3.14	3.71
7	1.42	1.89	2.36	3.00	3.50
8	1.40	1.86	2.31	2.90	3.36
9	1.38	1.83	2.26	2.82	3.25
10	1.37	1.81	2.23	2.76	3.17
12	1.36	1.78	2.18	2.68	3.06
14	1.34	1.76	2.14	2.62	2.98
16	1.34	1.75	2.12	2.58	2.92
18	1.33	1.73	2.10	2.55	2.88
20	1.32	1.72	2.09	2.53	2.84
30	1.31	1.70	2.04	2.46	2.75
40	1.30	1.68	2.02	2.42	2.70
60	1.30	1.67	2.00	2.39	2.66
120	1.29	1.66	1.98	2.36	2.62
∞ [$N(0, 1)$]	1.28	1.64	1.96	2.33	2.58

Source: Table 2 is adapted from Table III of Fisher and Yates (1963), *Statistical Tables for Biological, Agricultural and Medical Research*, 6th Ed., published by Oliver and Boyd, Edinburgh, with kind permission of the authors and publishers.

TABLE 3. VALUES OF $\chi^2_{n,\alpha}$ ON THE $\chi^2(n)$-DISTRIBUTION

$$\Pr\{\chi^2(n)\text{-variable} \geq \chi^2_{n,\alpha}\} = \alpha$$

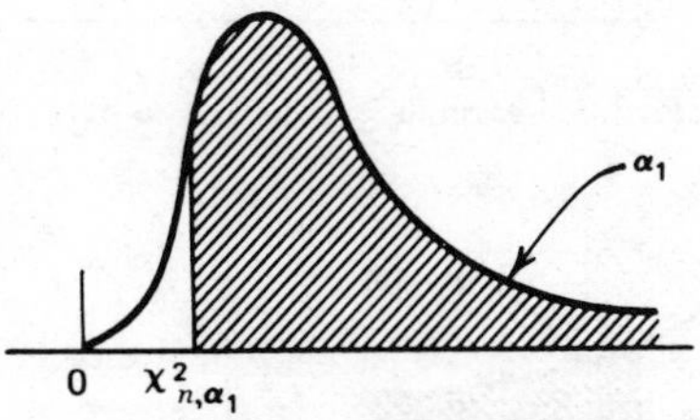

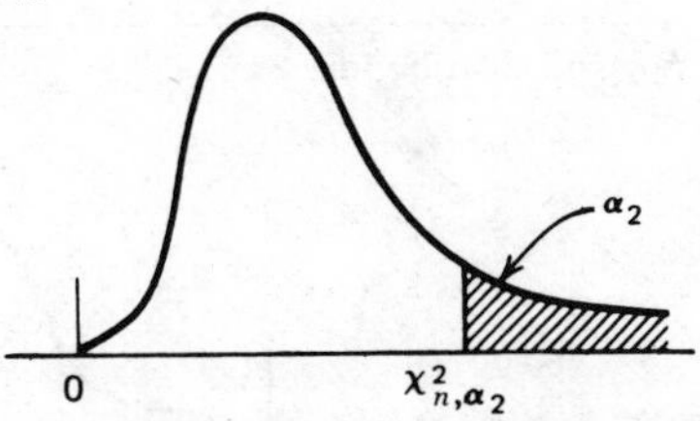

n	α_1			α_2		
(d.f.)	.99	.975	.95	.05	.025	.01
			Values of $\chi^2_{n,\alpha}$			
1	.000157	.000982	.00393	3.84	5.02	6.63
2	.020	.051	.103	5.99	7.38	9.21
3	.115	.216	.352	7.81	9.35	11.34
4	.297	.484	.711	9.49	11.14	13.28
5	.554	.831	1.145	11.07	12.83	15.09
6	.872	1.24	1.64	12.59	14.45	16.81
7	1.24	1.69	2.17	14.07	16.01	18.48
8	1.65	2.18	2.73	15.51	17.53	20.09
9	2.09	2.70	3.33	16.92	19.02	21.67
10	2.56	3.25	3.94	18.31	20.48	23.21
11	3.05	3.82	4.57	19.68	21.92	24.73
12	3.57	4.40	5.23	21.03	23.34	26.22
13	4.11	5.01	5.89	22.36	24.74	27.69
14	4.66	5.63	6.57	23.68	26.12	29.14
15	5.23	6.26	7.26	25.00	27.49	30.58
20	8.26	9.59	10.85	31.41	34.17	37.57
25	11.52	13.12	14.61	37.65	40.65	44.31
30	14.95	16.79	18.49	43.77	46.98	50.89
40	22.16	24.43	26.51	55.76	59.34	63.69
50	29.71	32.36	34.76	67.50	71.42	76.15
60	37.48	40.48	43.19	79.08	83.30	88.38
70	45.44	48.76	51.74	90.53	95.02	100.4
80	53.54	57.15	60.39	101.9	106.6	112.3
90	61.75	65.65	69.13	113.1	118.1	124.1
100	70.06	74.22	77.93	124.3	129.6	135.8

Source: Abridged from Table 8 of Pearson and Hartley (1954), *Biometrika Tables for Statisticians, Volume I*, published at the Cambridge University Press for *Biometrika* Trustees, with kind permission of the authors and publishers.

TABLE 4a. VALUES OF $F_{n_1, n_2, \alpha}$ ON THE $F(n_1, n_2)$-DISTRIBUTION

$F(n_1, n_2)$-distribution

$\Pr\{F(n_1, n_2)\text{-variable} \geq F_{n_1, n_2, \alpha}\} = \alpha = .05$

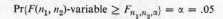

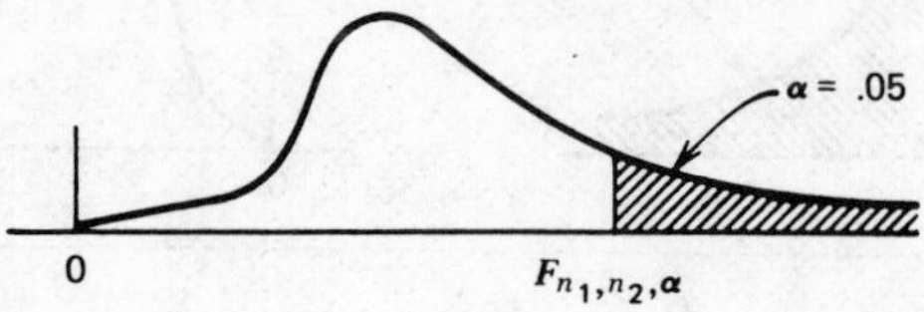

$\alpha = .05$

n_2 (denom. d.f.)	n_1 (numerator d.f.) 1	2	4	6	8	10	12	24	∞
	$[t_{n_2, .025}]^2$	Values of $F_{n_1, n_2, \alpha}$							
1	161.4	199.5	224.6	234.0	238.9	241.9	243.9	249.1	254.3
2	18.51	19.00	19.25	19.33	19.37	19.40	19.41	19.45	19.50
3	10.13	9.55	9.12	8.94	8.85	8.79	8.74	8.64	8.53
4	7.71	6.94	6.39	6.16	6.04	5.96	5.91	5.77	5.63
5	6.61	5.79	5.19	4.95	4.82	4.74	4.68	4.53	4.36
6	5.99	5.14	4.53	4.28	4.15	4.06	4.00	3.84	3.67
7	5.59	4.74	4.12	3.87	3.73	3.64	3.57	3.41	3.23
8	5.32	4.46	3.84	3.58	3.44	3.35	3.28	3.12	2.93
9	5.12	4.26	3.63	3.37	3.23	3.14	3.07	2.90	2.71
10	4.96	4.10	3.48	3.22	3.07	2.98	2.91	2.74	2.54
11	4.84	3.98	3.36	3.09	2.95	2.85	2.79	2.61	2.40
12	4.75	3.89	3.26	3.00	2.85	2.75	2.69	2.51	2.30
13	4.67	3.81	3.18	2.92	2.77	2.67	2.60	2.42	2.21
14	4.60	3.74	3.11	2.85	2.70	2.60	2.53	2.35	2.13
15	4.54	3.68	3.06	2.79	2.64	2.54	2.48	2.29	2.07
20	4.35	3.49	2.87	2.60	2.45	2.35	2.28	2.08	1.84
25	4.24	3.39	2.76	2.49	2.34	2.24	2.16	1.96	1.71
30	4.17	3.32	2.69	2.42	2.27	2.16	2.09	1.89	1.62
40	4.08	3.23	2.61	2.34	2.18	2.08	2.00	1.79	1.51
60	4.00	3.15	2.53	2.25	2.10	1.99	1.92	1.70	1.39
120	3.92	3.07	2.45	2.17	2.02	1.91	1.83	1.61	1.25
∞	3.84	3.00	2.37	2.10	1.94	1.83	1.75	1.52	1.00

Source: Abridged from Table 18 of Pearson and Hartley (1954), *Biometrika Tables for Statisticians, Volume I*, published at the Cambridge University Press for *Biometrika* Trustees, with kind permission of the authors and publishers.

TABLE 4b. VALUES OF $F_{n_1,n_2,\alpha}$ ON THE $F(n_1, n_2)$-DISTRIBUTION

$\Pr\{F(n_1, n_2)\text{-variable} \geq F_{n_1,n_2,\alpha}\} = \alpha = .01$

$F(n_1, n_2)$-distribution

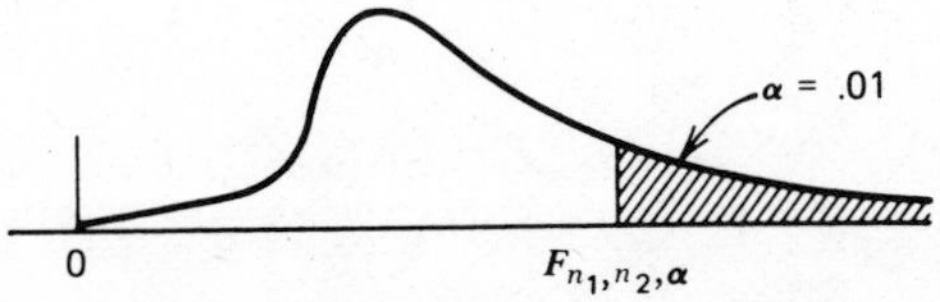

$\alpha = .01$

n_2 (denom. d.f.)	1	2	4	6	8	10	12	24	∞
	$[t_{n_2,.005}]^2$	Values of $F_{n_1,n_2,\alpha}$							
1	4052	5000	5625	5859	5982	6056	6106	6235	6366
2	98.50	99.00	99.25	99.33	99.37	99.40	99.42	99.46	99.50
3	34.12	30.82	28.71	27.91	27.49	27.23	27.05	26.60	26.13
4	21.20	18.00	15.98	15.21	14.80	14.55	14.37	13.93	13.46
5	16.26	13.27	11.39	10.67	10.29	10.05	9.89	9.47	9.02
6	13.75	10.92	9.15	8.47	8.10	7.87	7.72	7.31	6.88
7	12.25	9.55	7.85	7.19	6.84	6.62	6.47	6.07	5.65
8	11.26	8.65	7.01	6.37	6.03	5.81	5.67	5.28	4.86
9	10.56	8.02	6.42	5.80	5.47	5.26	5.11	4.73	4.31
10	10.04	7.56	5.99	5.39	5.06	4.85	4.71	4.33	3.91
11	9.65	7.21	5.67	5.07	4.74	4.54	4.40	4.02	3.60
12	9.33	6.93	5.41	4.82	4.50	4.30	4.16	3.78	3.36
13	9.07	6.70	5.21	4.62	4.30	4.10	3.96	3.59	3.17
14	8.86	6.51	5.04	4.46	4.14	3.94	3.80	3.43	3.00
15	8.68	6.36	4.89	4.32	4.00	3.80	3.67	3.29	2.87
20	8.10	5.85	4.43	3.87	3.56	3.37	3.23	2.86	2.42
25	7.77	5.57	4.18	3.63	3.32	3.13	2.99	2.62	2.17
30	7.56	5.39	4.02	3.47	3.17	2.98	2.84	2.47	2.01
40	7.31	5.18	3.83	3.29	2.99	2.80	2.66	2.29	1.80
60	7.08	4.98	3.65	3.12	2.82	2.63	2.50	2.12	1.60
120	6.85	4.79	3.48	2.96	2.66	2.47	2.34	1.95	1.38
∞	6.63	4.61	3.32	2.80	2.51	2.32	2.18	1.79	1.00

Source: Abridged from Table 18 of Pearson and Hartley (1954), *Biometrika Tables for Statisticians, Volume I*, published at the Cambridge University Press for the *Biometrika* Trustees, with kind permission of the authors and publishers.

Index

The index is to be used in conjunction with the Table of Contents. Entries in the index that refer to chapters are itemized in the Table of Contents.

Applied Probability and Statistics (Continued)

LANCASTER · An Introduction to Medical Statistics
LEWIS · Stochastic Point Processes
MANN, SCHAFER and SINGPURWALLA · Methods for Statistical Analysis of Reliability and Life Data
MILTON · Rank Order Probabilities: Two-Sample Normal Shift Alternatives
OTNES and ENOCHSON · Digital Time Series Analysis
PRENTER · Splines and Variational Methods
RAO and MITRA · Generalized Inverse of Matrices and Its Applications
SARD and WEINTRAUB · A Book of Splines
SEAL · Stochastic Theory of a Risk Business
SEARLE · Linear Models
THOMAS · An Introduction to Applied Probability and Random Processes
WHITTLE · Optimization under Constraints
WONNACOTT and WONNACOTT · Econometrics
YOUDEN · Statistical Methods for Chemists
ZELLNER · An Introduction to Bayesian Inference in Econometrics

Tracts on Probability and Statistics

BILLINGSLEY · Ergodic Theory and Information
BILLINGSLEY · Convergence of Probability Measures
CRAMÉR and LEADBETTER · Stationary and Related Stochastic Processes
JARDINE and SIBSON · Mathematical Taxonomy
KINGMAN · Regenerative Phenomena
RIORDAN · Combinatorial Identities
TAKACS · Combinatorial Methods in the Theory of Stochastic Processes